Nuclear Chemistry

Theory and Applications

Nuclear Chemistry
Theory and Applications

by

GREGORY R. CHOPPIN

Department of Chemistry,
Florida State University,
Tallahassee, Florida 32306, USA

and

JAN RYDBERG

Department of Nuclear Chemistry,
Chalmers University of Technology,
S-41296 Goteborg, Sweden

PERGAMON PRESS

Oxford · New York · Toronto · Sydney · Paris · Frankfurt

U.K. Pergamon Press Ltd., Headington Hill Hall, Oxford OX3 0BW, England

U.S.A. Pergamon Press Inc., Maxwell House, Fairview Park, Elmsford, New York 10523, U.S.A.

CANADA Pergamon of Canada, Suite 104, 150 Consumers Road, Willowdale, Ontario M2J 1P9, Canada

AUSTRALIA Pergamon Press (Aust.) Pty. Ltd., P.O. Box 544, Potts Point, N.S.W. 2011, Australia

FRANCE Pergamon Press SARL, 24 rue des Ecoles, 75240 Paris, Cedex 05, France

FEDERAL REPUBLIC OF GERMANY Pergamon Press GmbH, 6242 Kronberg–Taunus Hammerweg 6, Federal Republic of Germany

---

First edition 1980

British Library Cataloguing in Publication Data

Choppin, Gregory Robert
Nuclear chemistry.
1. Radiochemistry
I. Title II. Rydberg, J
541′.38 QD601.2 79–40371

ISBN 0-08-023826-2 hardcover
ISBN 0-08-023823-8 flexicover

Printed in Great Britain by A. Wheaton & Co. Ltd., Exeter

Contents

Foreword

There is no universally accepted definition for the term "nuclear chemistry". For purposes of our text we regard nuclear chemistry in its broadest context as an interdisciplinary subject with roots in physics, biology, and chemistry. The basic aspects include among others (i) nuclear reactions and energy levels, (ii) the types and energetics of radioactive decay, (iii) the formation and properties of radioactive elements, (iv) the effect of individual isotopes on chemical and physical properties, and (v) the effects of nuclear radiation on matter. Research in (i) and (ii) is often indistinguishable in purpose and practice from that in nuclear physics, although for nuclear chemists chemical techniques may play a significant role. (iii) and (iv) can be classified as *radiochemistry* and *isotope chemistry*, while (v) falls in the classification of *radiation chemistry.*

Applied aspects of nuclear chemistry involve production of radioactive isotopes, radiation processing, radiation conservation of foods, etc., as well as all parts of the nuclear fuel cycle such as uranium recovery, isotope separation, reactions in the fuel elements, processing of spent fuel elements, waste handling, and effects of radiation on reactor materials. Radiation health aspects and techniques for remote control are other important fields.

Knowledge in nuclear chemistry is an essential tool for research, development, and control in many areas of chemistry and technology (tracer methods, activation analysis, control gauges in industry, etc.), medicine (radiopharmaceuticals, nuclear medicine, radioimmuno assay, etc.), geology, and archeology (radioactive dating).

Nuclear chemistry taught in Europe and the United States has mainly been based on the excellent comprehensive textbooks by Haissinsky (1964), Friedlander, Kennedy, and Miller (1964), Lieser (1969), Harvey (1969), and McKay (1971), as well as smaller books, e.g. by Carswell (1967), the authors (Choppin 1961, Rydberg 1966), and others. Since these books were written, nuclear science and technology have expanded significantly and are of importance to most countries. This has created a demand for a more modern text. We hope our book will satisfy some of this demand. It differs from the earlier books in that a larger emphasis is put on the role of the chemist in the nuclear energy field. Also, we treat radiation risks and protection aspects more thoroughly than has been the practice in earlier texts. It seems that courses in nuclear chemistry in Europe place more emphasis on radiochemistry while those in the United States tend to emphasize nuclear physics. We have attempted to provide coverage of both areas in our text as well as of the technological aspects.

Hopefully, this text can serve both for an upper level undergraduate as well as for a graduate course. To this dual purpose we have attempted to write simply, covering a broad array of topics, yet providing adequate depth and thoroughness of basic concepts and of detail of applied topics. The mathematics is rather uncomplicated. Emphasis has been placed on principles with a few illustrative examples rather than to present a large number of cases in various fields; such collections are easily available (e.g. in IAEA conference proceedings). Though it is up to the teacher to select those sections that he feels most important for his particular course, we offer suggestions for a reduced text suitable for a shorter course; the chapters and sections that we feel can be omitted in such a course are marked with an asterisk in the contents lists of each chapter. Both this shorter course and the full text have been taught by the authors. Depending on the major emphases of the course and its length, deletions beyond those indicated would be quite possible. For example, in a course directed primarily to radiochemistry, Chapters 6 and 9 could be omitted. A short course concerned primarily with the nuclear physics aspects would include these chapters but might omit Chapters 16 and 20.

Each chapter ends with exercises and a literature list. In the exercises we have tried to select representative problems for the preceding text. Instead of making a few composite problems, which many students would consider more difficult to solve, we prefer to offer several problems where each treats only a smaller part of the text.

The literature list of each chapter covers recent publications, textbooks, and articles, which may guide the student deeper into the field. They do not contain documentary information necessary for the text, although some figures and tables in the literature are recommended explicitly. In addition, a more general literature list of book series, journals, and reference tables is presented in Appendix A.

In the appendixes we have added some material which we think is outside the scope of a normal textbook in nuclear chemistry but nevertheless is useful to many readers of our book. By courtesy of Prof. W. Seelmann-Eggebert we also include an isotope chart in Appendix N.

We adhere to the SI system, though derived units (or cgs units) sometimes are used to simplify the reading of older or technical literature. In two particular cases we deviate from this rule: we use grams per mole for molecular weight, and, commonly, curies for radioactivity.

Many friends and co-workers have helped us in discussions, manuscript readings, checking exercises, editorial efforts, etc. For this we express our gratitude to Lehna Andersson, Karin Brodén, Hubert Eschrich, Ingela Hagström, Kurt Lidén, Jan-Olov Liljenzin, Siv Johnson, Ulla Olofsson, Gunnar Skarnemark, Ingvor Svantesson, Stig Wingefors, Göran Persson, Heino Kipatsi, Nils-Göran Sjöstrand, Erik Hellstrand, Sue Eaker, Becca James, Patricia Baisden, Eva Jomar, and our students in the classes in nuclear chemistry at CTH and FSU. This text has been many years in preparation during which time our wives, Britta and Ann, were very understanding, for which we are most grateful.

Göteborg, Sweden
Tallahassee, Florida, USA

CHAPTER 1

Beginnings of Nuclear Science

Contents

1.1. Origins of the atomic theory

In the sixth century BC the Greek philosopher Thales proposed a theory that all matter in the universe was composed of water in varying forms, concentrations, and combinations. About 100 years later, Empedocles developed Thales' idea further by suggesting that there are actually four elements: earth, air, fire, and water. Other Greek philosophers such as Leucippos and Democritos in the fifth century BC proposed that matter consisted of solid, small, and indivisible spheres which were called *atoms* after the Greek word for unbreakable, but other theories concerning the nature of matter also developed down the centuries. These ideas of the nature of matter were not advanced much further until the seventeenth and eighteenth centuries when inductive experimentation provided a sounder basis for speculation.

In 1661 R. Boyle (1627–1691) wrote in his book the *Sceptical Chemist* that an *element* could be defined as a substance which cannot be broken down into simpler substances and hence is of ultimate simplicity. In his view a few elements made up all the complex substances of nature by means of various combinations. In the latter part of the eighteenth century A. Lavoisier (1743–1794) initiated the age of modern chemistry when he made clear the value of quantitative experimentation. He also compiled a list of elements which included 31 substances, of which it has been shown subsequently that 26 are true elements (the other five were metal oxides).

With the recognition of the elemental nature of some substances and with the statement of the *law of conservation of mass* based on quantitative experimentation, chemistry moved rapidly to a sounder formulation of atomic theory beginning with J. Dalton (1766–1844) in the first years of the nineteenth century. Through the work of many chemists and physicists such as A. M. Ampère (1775–1836), A. Avogadro (1776–1856), J. J. Berzelius (1779–1840), and J. L. Gay-Lussac (1778–1850) new elements were identified, and—more importantly—the distinction between elements and compounds became clear. The development of the *kinetic gas theory* based on the assumption that moving atoms were the cause of pressure, temperature, heat conductivity, viscosity, etc., of gases (R. Clausius 1822–1888, L. Boltzmann 1844–1906) made it possible to estimate the size of gaseous atoms (10^{-9}–10^{-10} m, J. Loschmidt 1865) and their approximate weight (10^{-25}–10^{-27} kg), from which the number of atoms (or molecules) per atom weight unit (the Avogadro number)† could be calculated. Through the work of other scientists such as M. Faraday (1791–1867) and D. Mendeleev (1834–1907), the nineteenth century became an extremely fruitful period for progress in the understanding of the nature of matter.

In the last quarter of the nineteenth century experiments by J. J. Thomson (1856–1940) and others demonstrated that the idea of solid indivisible atoms was not acceptable for explaining many phenomena which suggested the presence of electrical charges or of electrons in atoms. Thomson proposed a "plum pudding" model in which the negative electrons were embedded in a sphere of positive charge. However, this model had many inadequacies and never received wide scientific acceptance. The confusion about the nature of the atom was increased by the discoveries of H. Becquerel and E. Rutherford in the first years of this century.

1.2. Radioactive elements

In 1895 W. Roentgen discovered that when cathode rays (i.e. electrons) struck the wall of an evacuated glass tube, it caused the wall material to emit visible light (fluoresce), while at the same time a very penetrating radiation was produced. The name *X-ray* was given to this radiation. Learning about this, H. Becquerel, who had been interested in the fluorescent spectra of materials, immediately decided to investigate the possibility that the fluorescence observed in some salts when exposed to sunlight also caused emission of X-rays. Crystals of potassium uranyl sulfate were placed on top of photographic plates, wrapped in black paper, and the assembly was exposed to the sunlight. After development of some of the photographic plates, Becquerel concluded (erroneously) from the presence of black spots under the crystals that fluorescence in the crystals led to the emission of X-rays, which penetrated the wrapping paper and resulted in the blackening of the photographic plate. However, Becquerel soon found that the radiation causing the blackening was not "a transformation of solar energy" because it was found to occur even with assemblies that had not been exposed to light; the uranyl salt obviously produced radiation spontaneously. This radiation, which was first called Becquerel rays but later termed radioactive radiation (or simply *radioactivity*)‡, was similar to X-rays also in that it ionized air, as observed through the discharge of electroscopes.

Marie Curie (Fig. 1.1) subsequently showed that all uranium compounds produced *ionizing radiation* independent of the chemical composition of the salt. This was convincing evidence that the radiation was a property of the element uranium. Moreover, she observed that some uranium minerals such as pitchblende produced more ionizing radiation than pure uranium compounds. She wrote: "this phenomenon leads to the assumption that these minerals contain elements which are much more active than

† The Avogadro number ($N_A = 6.022 \times 10^{23}$) is defined as the number of atoms (or molecules) which are required to give the number of grams required by the gram atomic (or molecular) weight. This number is called a mole.

‡ The word radioactivity refers to the phenomenon *per se* as well as the intensity of the radiation observed.

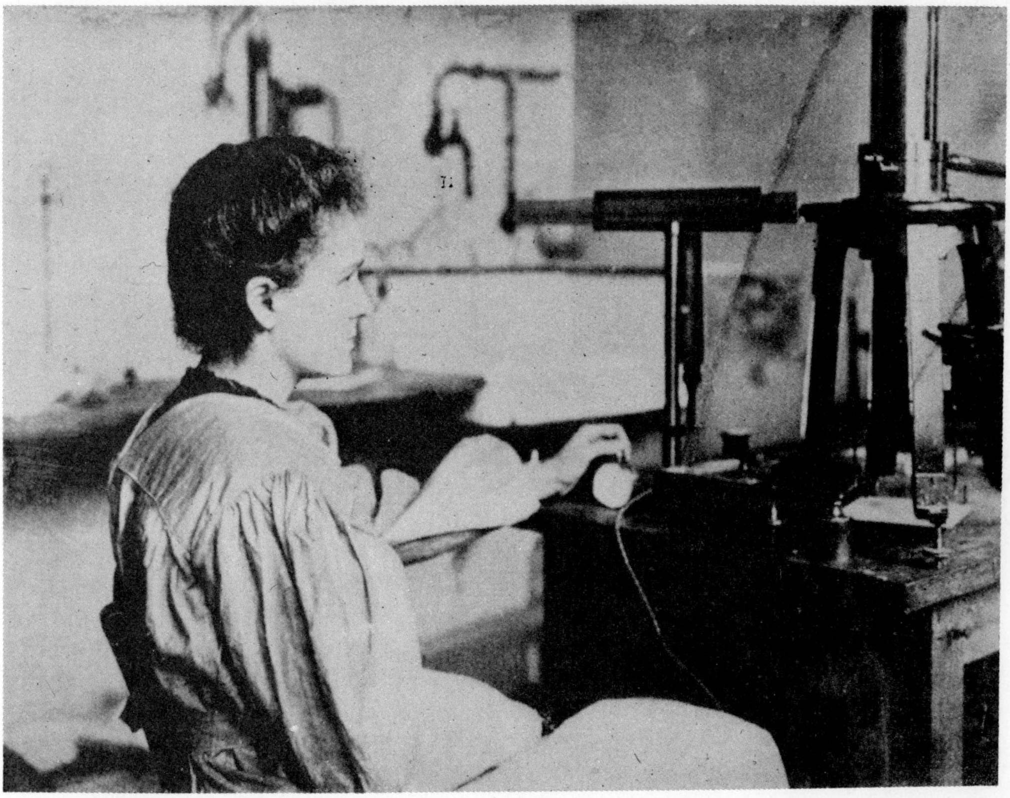

FIG. 1.1. Marie Curie in her laboratory in Paris using an electrometer for measuring the ionization caused by radium daughter products. She was born Sklodowska in 1867 in Warsaw, became professor at the Sorbonne in 1909, and Nobel laureate in physics (together with Henri Becquerel and her husband Pierre Curie) "in recognition of the extraordinary services they have rendered by their joint researches on the radiation phenomena discovered by Professor Henri Becquerel" in 1903. She was also awarded the Nobel prize in chemistry in 1911 "in recognition of her services to the advancement of chemistry by the discovery of the elements radium and polonium, by the isolation of radium, and the study of the nature and compounds of their remarkable element." (Photograph around 1905.)

uranium". She and her husband, Pierre Curie, began a careful purification of pitchblende, measuring the amount of radiation in the solution and in the precipitate after each precipitation separation step. These first *radiochemical* investigations were highly successful: "while carrying out these operations, more active products are obtained. Finally, we obtained a substance whose activity was 400 times larger than that of uranium. We therefore believe that the substance that we have isolated from pitchblende is a hitherto unknown metal. If the existence of this metal can be affirmed, we suggest the name *polonium*." It was in the publication reporting the discovery of polonium in 1898 that the word radioactive was used for the first time. It may be noted that the same element was simultaneously and independently discovered by W. Marckwald who called it "radiotellurium".

In the same year the Curies, together with G. Bemont, isolated another radioactive substance for which they suggested the name *radium*. In order to prove that polonium and radium were in fact two new elements, large amounts of pitchblende were processed, and in 1902 M. Curie announced that she had been able to isolate about 0.1 g of pure radium chloride from more than one tonne of pitchblende waste. The determination of the atomic weight of radium and the measurement of its emission spectrum provided the final proof that a new element had been isolated.

1.3. Radioactive decay

While investigating the radiochemical properties of uranium, W. Crookes and Becquerel made an important discovery. Precipitating a carbonate salt from a solution containing uranyl ions, they discovered that while the uranium remained in the supernatant liquid in the form of the soluble uranyl carbonate complex, the radioactivity associated with the uranium was present in the precipitate, which contained no uranium. Moreover, the radioactivity of the precipitate slowly decreased with time, whereas the supernatant liquid showed a growth of radioactivity during the same period (Fig. 1.2). We know now that this measurement of radioactivity was concerned with only beta- and gamma-radiations, and not with the alpha-radiation which is emitted directly by uranium.

Similar results were obtained by Rutherford (Fig. 1.3) and F. Soddy when investigating the radioactivity of thorium. Later, Rutherford and F. E. Dorn found that radioactive gases (*emanation*) could be separated from salts of uranium and thorium. After separation of the gas from the salt, the radioactivity of the gas decreased with time, while new radioactivity grew in the salts in a manner similar to that shown in Fig. 1.2. The rate of increase of the radioactivity with time in the salt was found to be completely independent of chemical processes, temperature, etc. Rutherford and Soddy concluded from these observations that radioactivity was due to changes within atoms themselves. They proposed that, when radioactive decay occurred, the atoms of the original elements (e.g. of U and Th) were transformed into atoms of new elements.

The radioactive elements were called *radioelements*. Lacking names for these radioelements, letters such as X, Y, Z, A, B, etc., were added to the symbol of the primary (i.e. parent) element. Thus, UX was produced from the radioactive decay of uranium, ThX from that of thorium, etc. These new radioelements (UX, ThX, etc.) had different

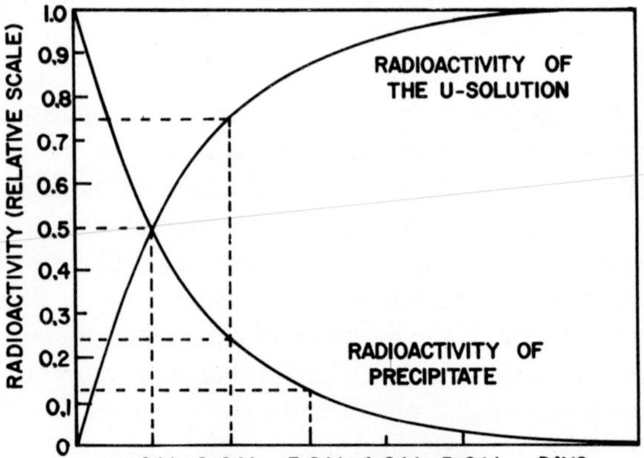

FIG. 1.2. Measured change in radioactivity from carbonate precipitate and supernatant uranium solution, i.e. the separation of daughter element UX (Th) from parent radioelement uranium.

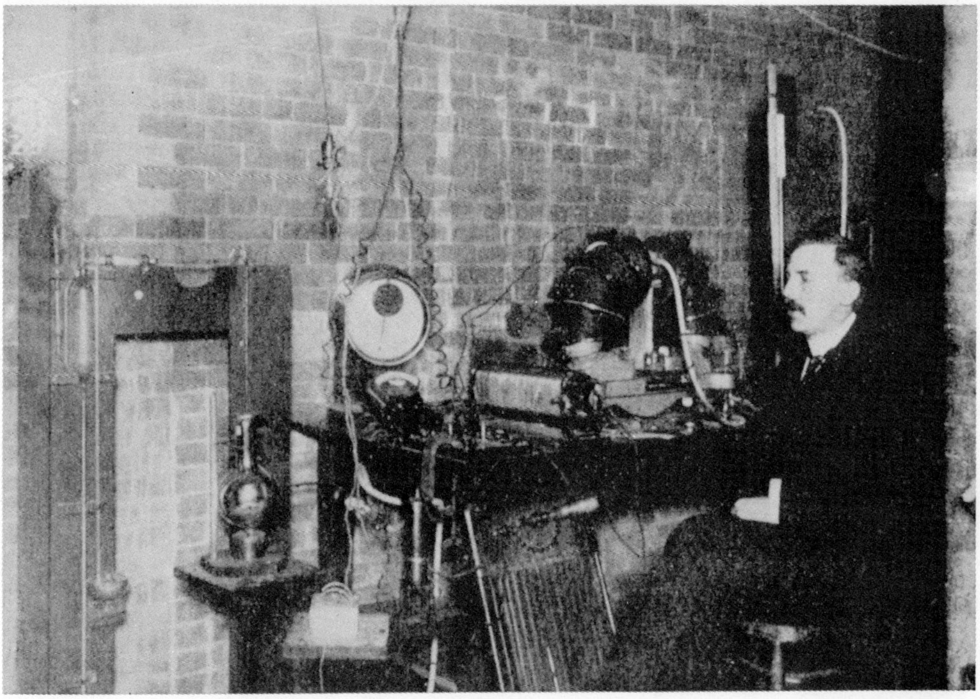

FIG. 1.3. Ernest Rutherford with α-measurement equipment in the basement of the University of Montreal in 1906. Rutherford was born in Nelson, New Zealand, in 1871. He became head of the Cavendish Laboratory, Cambridge, in 1919, and Nobel laureate in chemistry in 1908 "for his investigations into the disintegration of the elements, and the chemistry of radioactive substances".

chemical properties than the original elements, and could be separated from them through chemical processes such as precipitation, volatilization, etc. The radioactive daughter elements decayed further to form still other elements, symbolized as UY, ThA, etc. A typical decay chain could be written: Ra → Rn → RaA → RaB →, etc.

A careful study of the radiations emitted from these radioactive elements demonstrated that they consisted of three components which were given the designation alpha (α), beta (β), and gamma (γ). *Alpha-radiation* was shown to be identical to helium ions, whereas *beta-radiation* was identical to electrons. *Gamma-radiation* had the same electromagnetic nature as X-rays but was of higher energy. The rate of radioactive decay per unit weight was found to be fixed for any specific radioelement, no matter what its chemical or physical state was, though this rate differed greatly for different radioelements. The decay rate could be expressed in terms of a *half-life*, which is the time it takes for the radioactivity of a radioelement to decay to one-half of its original value. Half-lives for the different radioelements were found to vary from fractions of a second to millions of years; e.g. that of ThA is 0.1 of a second, of UX it is 24.1 days (Fig. 1.2), and of uranium, millions of years.

1.4. Discovery of isotopes

By 1910 approximately 40 different chemical species had been identified through their chemical nature, the properties of their radiation, and their characteristic half-lives.

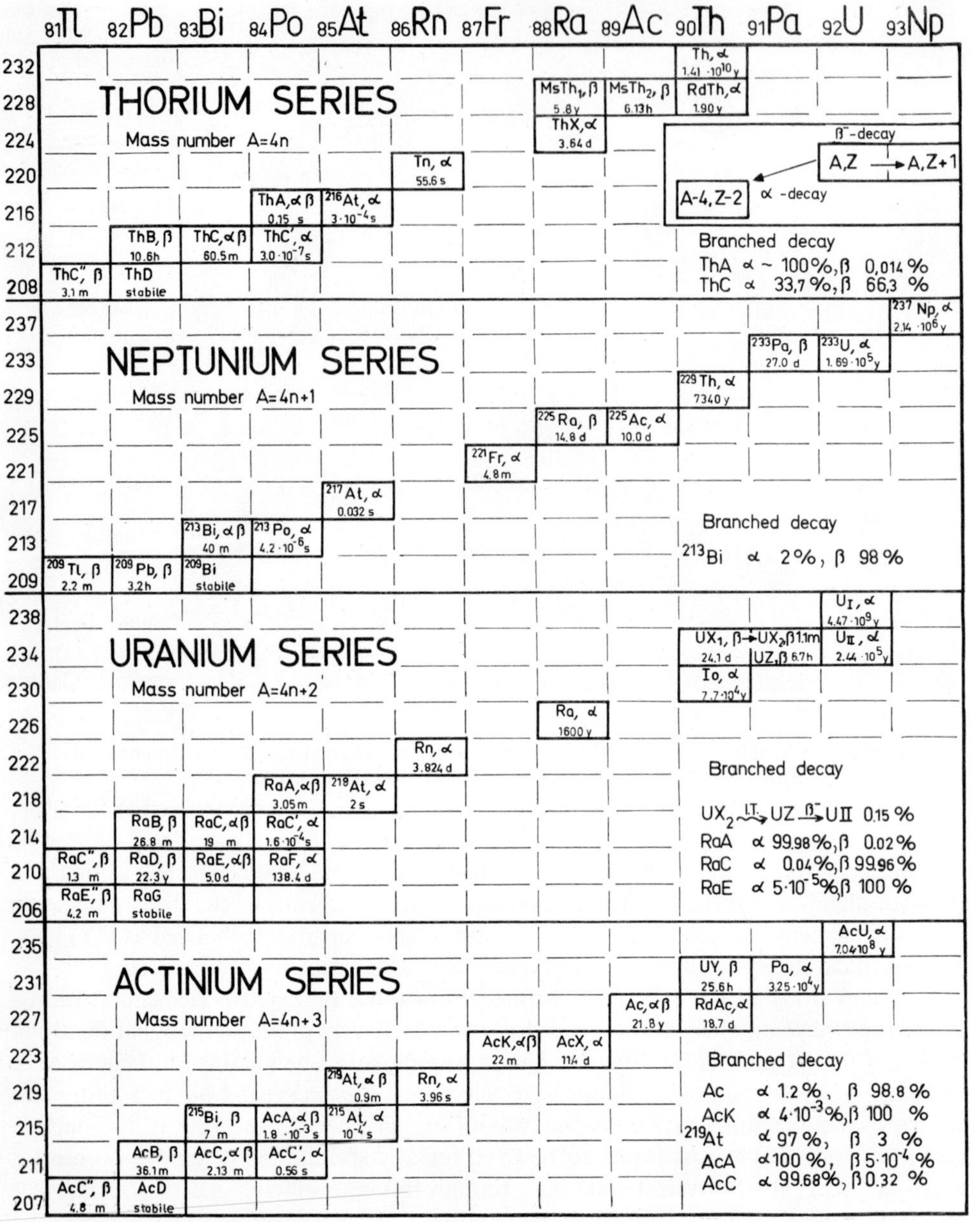

FIG. 1.4. The three naturally occurring radioactive decay series and the man-made neptunium series. Although ^{239}Pu (which is parent to the actinium series) and ^{244}Pu (which is parent to the thorium series) have been discovered in nature, the decay series shown here begin with the most abundant long-lived nuclides.

The study of the genetic relationships in the decay of the radioactive species showed that these radioelements could be divided into three distinct series. Two of these series originated in uranium and the third in thorium. B. Boltwood found that all three of the series ended in the same element—lead.

A major difficulty obvious to scientists at that time involved the fact that while it was known from the Periodic Table that there was space for only 11 elements between lead and uranium, approximately 40 radioelements were known in the decay series from uranium to lead. To add to the confusion was the fact that it was found that in many cases it was not possible to separate some of the radioelements from each other by normal chemical technique. For example, the radioelement RaD was found to be chemically identical to lead. In a similar manner, spectrographic investigations of the radioelement ionium showed exactly the same spectral lines that had been found previously to be due to the element thorium.

In 1913 K. Fajans and Soddy provided independently the explanation for these seemingly contradictory conditions. They stated that by the radioactive α-decay a new element is produced two places to the left of the mother element in the periodic system and in β-decay a new element is produced one place to the right of the mother element (Fig. 1.4, upper right insert). The radioelements that fall in the same place in the periodic system are chemically identical. Soddy proposed the name *isotopes* to account for different radioactive species which have chemical identity.

Research by J. J. Thomson soon provided conclusive support for the existence of isotopes. If a beam of positively charged gaseous ions is allowed to pass through electric or magnetic fields, the ions follow hyperbolic paths which are dependent on the masses and charges of the gaseous ions (see Fig. 2.1 and associated text). When these ion beams strike photographic plates, a darkening results which is proportional to the number of ions which hit the plate. By using this technique with neon gas, Thomson found that neon consists of two types of atoms with different atomic masses. The mass numbers for these two isotopes were 20 and 22. Moreover, from the degree of darkening of the photographic plate, Thomson calculated that neon consisted of about 90% of atoms with mass number 20, and 10% of atoms with mass number 22.

Thus a chemical element may consist of several kinds of atoms with different masses but with the same chemical properties. The 40 radioelements were, in truth, not 40 different elements but were isotopes of the 11 different chemical elements from lead to uranium.

1.5. Radioactive decay series

To specify a particular isotope of an element, the *atomic number* (i.e. order, number, or place in the Periodic Table of elements) is written to the left of the elemental symbol as a subscript and the mass number (i.e. the nearest integer value to the mass of the neutral atom, measured in atomic weight units) as a superscript. Thus the isotope of uranium with mass number 238 is designated $^{238}_{92}U$. Similarly, the isotope of protactinium with mass number 234 is designated $^{234}_{91}Pa$. For an alpha-particle we use either the Greek α or $^{4}_{2}He$. Similarly, the beta-particle is designated either by the Greek letter β or by the symbol $_{-1}^{0}e$. Later in the text we shall discuss a form of β-decay in which a positive electron is emitted; this is symbolized by $_{+1}^{0}e$ or β^{+}. Gamma-decay, which results in no change in either mass number or atomic number of the decaying nucleus, is symbolized by the Greek letter γ.

In radioactive decay both mass number and atomic number are conserved. Thus in the decay chain of $^{238}_{92}U$ the first two steps are written:

$$^{238}_{92}U \rightarrow {}^{234}_{90}Th + {}^{4}_{2}He \tag{1.1}$$

$$^{234}_{90}Th \rightarrow {}^{234}_{91}Pa + {}_{-1}^{0}e \tag{1.2}$$

Frequently, in such a decay chain, the half-life ($t_{1/2}$) for the radioactive decay is shown either above or below the arrow. A shorter notation is used commonly:

$$^{238}\text{U} \xrightarrow[4.5\times10^9\,\text{y}]{\alpha} {}^{234}\text{Th} \xrightarrow[24\,\text{d}]{\beta^-} {}^{234}\text{Pa} \xrightarrow[1.1\,\text{min}]{\beta^-} {}^{234}\text{U} \xrightarrow[2.5\times10^5\,\text{y}]{\alpha} {}^{230}\text{Th, etc.} \quad (1.3)$$

The three naturally occurring radioactive decay series, which are known as the *thorium series*, the *uranium series*, and the *actinium series*, are shown in Fig. 1.4. A fourth series, which originates in the synthetic element neptunium, is also shown. This series is not found on earth since all of the radioactive species in the series have decayed away long ago. Both the present symbolism of the isotope as well as the historical (i.e. "radioelement") symbolism are given in Fig. 1.4. Notice that the *rule of Fajans and Soddy* is followed in each series so that α-decay causes a decrease in atomic number by two units and mass number by four, whereas β-decay produces no change in mass number but an increase in atomic number by one unit. Moreover, we see a pattern occurring frequently in these series of an α-decay step followed by two β-decay steps.

1.6. Atomic models

Neither radioactive decay nor the discovery of isotopes provided information on the internal structure of atoms. Such information was obtained from scattering experiments in which a substance such as a thin metal foil was irradiated with a beam of α-particles and the intensity (measured by counting scintillations from scattered particles hitting a fluorescent screen) of the particles scattered at different angles measured (see Fig. 7.7). It was assumed that the deflection of the particles was caused by collisions with the atoms of the irradiated material. Some of the α-particles were strongly deflected through angles greater than 90°. While such large angle reflections were rare, only one in approximately 8000, they were quite unexpected and could not be explained by Thomson's "plum pudding" model of the atom. Consideration of these rare events led Rutherford in 1911 to the conclusion that the entire positive charge of an atom must be concentrated in a very small volume whose diameter was about 10^{-14} m. This small part of the atom he named the *nucleus*. Rutherford could calculate that concentration of all the positive charge and essentially all the atomic mass in the atomic nucleus could cause the scattering of α-particles through large angles when these particles passed close to the nucleus (see text to Fig. 7.7). The atomic electrons have much smaller mass and were assumed to surround this nucleus. The total atom with the external (circulating) electrons had a radius of approximately 10^{-10} m in contrast to the much smaller radius calculated for the nucleus.

It was soon shown that the positive charge of the atomic nucleus was identical to the atomic number assigned to an element in the periodic system of Mendeleev. The conclusion, then, is that in a neutral atom the small, positively charged nucleus was surrounded by electrons whose number was equal to the total positive charge of the nucleus. In 1913 N. Bohr, using quantum mechanical concepts, was able to propose such a model of the atom which in principle is still valid.

1.7. Summary

Our understanding of the nucleus has grown rapidly since Rutherford's scattering experiments. Some of the important steps in the history of nuclear science are listed

TABLE 1.1. *Historical survey of nuclear science*

1896	BECQUEREL discovers radiation from uranium (radioactivity). The intensity of the radiation is measured either through its ionization of air or through the scintillations observed when the radiation hits a fluorescent screen.
1896–1905	CROOKES, BECQUEREL, RUTHERFORD, SODDY, DORN, BOLTWOOD *et al.* Radioactive decay is found to be transformation of atoms leading to different radioelements which are genetically connected in radioactive decay series.
1898	P. and M. CURIE discover polonium and radium; the first radiochemical methods.
1898–1902	P. CURIE, DEBIERNE, BECQUEREL, DANLOS *et al.* discover that radiation affects chemical substances and causes biological damage.
1900	VILLARD and BECQUEREL propose that γ-radiation is electromagnetic in nature; finally proven in 1914 by RUTHERFORD and ANDRADE.
1900	BECQUEREL: β-rays are identified as electrons.
1902	First macroscopic amounts of a radioactive element (radium) isolated by M. and P. CURIE and DEBIERNE.
1903	RUTHERFORD: α-radiation is shown to be ionized helium atoms.
1905	EINSTEIN formulates the law of the equivalence between mass and energy.
1907	STENBECK makes the first therapeutic treatment with radium and heals skin cancer.
1911	RUTHERFORD, GEIGER, and MARSDEN conclude from measurement of the scattering of α-radiation against thin foils that atoms contain a very small positive nucleus.
1912	HEVESY and PANETH, in the first application of radioactive trace elements, determine the solubility of $PbCrO_4$ using RaD.
1912	WILSON develops the cloud chamber, which makes tracks from nuclear particles visible.
1913	HESS discovers cosmic radiation.
1913	FAJANS and SODDY explain the radioactive decay series by assuming the existence of isotopes. This is proven by J. J. THOMSON through deflection of neon ions in electromagnetic fields. ASTON separates the isotopes of neon by gas diffusion.
1913	N. BOHR shows that the atomic nucleus is surrounded by electrons in fixed orbitals.
1919	RUTHERFORD: first nuclear transformation in the laboratory, $^{4}He + {}^{14}N \rightarrow {}^{17}O + {}^{1}H$.
1919	ASTON constructs the first practical mass spectrometer and discovers that isotopic weights are not exactly integers.
1921	HAHN discovers nuclear isomers: $^{234m}Pa(UX_2) \xrightarrow[1,2\ \text{min}]{\gamma} {}^{234}Pa(UZ)$.
1924	DE BROGLIE advances the hypothesis that all moving particles have wave properties.
1924	LACASSAGNE and LATTES use radioactive trace elements (Po) in biological research.
1925–1927	Important improvements of the BOHR atomic model: PAULI exclusion principle, SCHRÖDINGER wave mechanics, HEISENBERG uncertainty relationship.
1928	GEIGER and MÜLLER construct the first GM tube for single nuclear particle measurements.
1931	VAN DE GRAAFF develops an electrostatic high voltage generator for accelerating atomic ions to high energies.
1932	COCKCROFT and WALTON develop the high voltage multiplier and use it for the first nuclear transformation in the laboratory with accelerated particles (0.4 MeV $^{1}H + {}^{7}Li \rightarrow 2\ {}^{4}He$).
1932	LAWRENCE and LIVINGSTON build the first cyclotron.
1932	UREY discovers deuterium and obtains isotopic enrichment through evaporation of liquid hydrogen.
1932	CHADWICK discovers the neutron.
1932	ANDERSSON discovers the positron, e^+ or β^+, through investigation of cosmic rays in a cloud chamber.
1933	UREY and RITTENBERG show isotopic effects in chemical reactions.
1934	JOLIOT and I. CURIE discover artificial radioactivity: $^{4}He + {}^{27}Al \rightarrow {}^{30}P + n$; $^{30}P \xrightarrow[2.5\ \text{min}]{\beta^+} {}^{30}Si$.
1935	YUKAWA predicts the existence of mesons.
1935	WEIZSÄCKER derives the semiempirical mass formulae.
1937	NEDDERMEYER and ANDERSSON discover μ-mesons in cosmic radiation using photographic plates.
1938	BETHE and WEIZSÄCKER propose the first theory for energy production in stars through nuclear fusion: $3\ He \rightarrow {}^{12}C$.
1939	HAHN and STRASSMAN discover fission after irradiation of uranium with neutrons.
1940	MCMILLAN, ABELSON, SEABORG, KENNEDY, and WAHL produce the first transuranium elements, neptunium (Np) and plutonium (Pu), and discover that ^{239}Pu is fissionable.
1940	Scientists in many countries show that ^{235}U is fissioned by slow neutrons, but ^{232}Th and ^{238}U only by fast neutrons, and that each fission produces two to three new neutrons while large amounts of energy are released. The possibility of producing nuclear weapons and building nuclear power stations is considered in several countries.

1942	FERMI and co-workers build the first nuclear reactor (critical on December 2).
1944	First gram amounts of a synthetic element (Pu) produced at Oak Ridge, USA. Kilogram quantities produced in Hanford, USA, in 1945.
1944	MCMILLAN and VEKSLER discover the synchrotron principle which makes it possible to build accelerators for energies > 1000 MeV.
1940–1945	OPPENHEIMER and co-workers develop a device to produce fast uncontrolled chain reactions releasing very large amounts of energy. First test at Alamagordo in the United States in 1945 produces an energy corresponding to 20,000 tonnes of TNT; this is followed by the use of atomic bombs in Japan.
1944–1947	Photo-multiplier scintillation detectors are developed.
1946	LIBBY develops the ^{14}C method for age determination.
1950	A nuclear shell model is suggested by MAYER, HAXEL, JENSEN, and SUESS.
1951	The first breeder reactor, which also produces the first electric power, is developed by Argonne National Laboratory, USA, and built in Idaho.
1952	The United States test the first device for uncontrolled large scale fusion power (the hydrogen bomb).
1953–1955	A. BOHR, MOTTELSON, and NILSSON develop the unified model of the nucleus (single particle effects on collective motions).
1955	CHAMBERLAIN, SEGRÉ, WIEGAND, and YPSILANTIS produce antiprotons.
1955	First nuclear powered ship (submarine *Nautilus*).
1954–1956	A 5 MWe nuclear power station starts at Obninsk, USSR, in 1954. First large civilian nuclear power station (45 MWe) starts at Calder Hall, England, in 1956.
1959	First civilian ship reactor used in the ice-breaker *Lenin*, launched in the USSR.
1961	A radionuclide (^{238}Pu) is used as power source in a satellite (Transit-4 A).
1961	Semiconductor detectors are developed.
1969	Soviet scientists obtain stable high density plasmas in Tokamak fusion reactor.
1972	Accelerator at National Accelerator Laboratory in Batavia, USA, achieves proton energies of 300 GeV (raised to 500 GeV in 1976).
1974	French scientists discover ancient natural nuclear reactor in Oklo, Gabon.

1955	Formation of United Nations Scientific Committee on the Effects on Atomic Radiation (UNSCEAR).
1957	Formation of the International Atomic Energy Agency (IAEA), with headquarters in Vienna.
1963	Partial Test Ban Treaty bans nuclear tests in the atmosphere, in outer space, and under water.
1968	Treaty on the Non-Proliferation of Nuclear Weapons (NPT) is signed by the "three depository governments" (USSR, UK, and USA), all nuclear weapons countries (NWC), and 40 other signatory states, all nonnuclear weapons countries (NNWC).
1971	The IAEA takes the responsibility for a safeguards system for control of fissile material in nonnuclear weapons countries.

in Table 1.1. Many of these discoveries and the practical consequences are discussed in the subsequent text.

We know now that the nucleus is composed of neutrons and protons, which can be described as existing in different energy levels—orbitals—analogous to the electronic orbitals of atoms. These neutrons and protons are held together in the nucleus by a very strong, short range attractive force whose origin is believed to lie in the exchange of subnuclear particles called mesons between the nucleons. Again, this model of the nuclear binding force has an analog in the covalent bond of molecules.

The discovery of fission resulted in nuclear weapons and created a whole new source of energy, while the use of radioisotopes led to important advances in almost every experimental area of scientific research. The variety of elementary particles and of antiparticles required new interpretations of the fundamental bases of nature extending from the nucleus to the universe as a whole. Yet with all this advance in knowledge, it should be evident in the chapters to come how much remains to be understood. We still must use several different nuclear models to explain different aspects of nuclear behavior. The nature of nuclear forces is not adequately described by present theories, nor do we have a satisfactory understanding of the nature and role of the subnuclear particles. In 80 years, nuclear science has created a revolution in our daily lives as well as in our knowledge of our physical world. It is not unlikely that in the next 80 years nuclear science will lead to even further advances in our understanding of nature and to even greater benefits to society.

1.8. Literature

1.8.1. *Recent general reading*

R. T. Beyer, *Foundations of Nuclear Physics*, Dover Publ. Inc., New York, 1949.
E. Farber, *The Evolution of Chemistry*, Ronald Press, New York, 1952.
L. Fermi, *Atoms in the Family*, University of Chicago Press, Chicago, 1954.
B. Russel, *Wisdom of the West*, McDonald, London, 1959.
R. G. Hewlett and O. E. Andersson Jr., *The New World, 1939/1946* (volume 1 of a *History of the U.S.A.E.C.*), Pennsylvania State University Press, University Park, 1962.
Nobel Lectures, Chemistry, and *Nobel Lectures, Physics*, Elsevier, Amsterdam, 1966 and later.
S. Glasstone, *Source Book on Atomic Energy*, van Nostrand, New York, 1967 (3rd edn.).
J. R. Partington, *A History of Chemistry*, Macmillan, London, 1970.
G. T. Seaborg, *Nuclear Milestones*, W. H. Freeman, San Francisco, 1972.

1.8.2. *Classics in nuclear chemistry*

M. Curie, *Traité de Radioactivité*, Paris, 1910; *Radioactivité*, Hermann, Paris, 1935.
S. Meyer and E. Schweidler, *Radioaktivität*, Leipzig, 1927.
E. Rutherford, J. Chadwick, and C. D. Ellis, *Radiations from Radioactive Substances*, Cambridge University Press, 1930 (reprinted 1951).
O. Hahn, *Applied Radiochemistry*, Cornell University Press, 1936.
G. Hevesy and F. A. Paneth, *A Manual of Radioactivity*, Oxford University Press, 1938.
G. T. Seaborg, *The Transuranium Elements*, Yale University Press, New Haven, 1958.
G. Hevesy, *Adventures in Radioisotope Research*, Pergamon Press, Oxford, 1962.
O. Hahn, *A Scientific Autobiography*, Charles Scribner's Sons, New York, 1966.

CHAPTER 2
Nuclei and Isotopes

Contents

For some time after Rutherford had demonstrated the presence of the atomic nucleus by his scattering experiments, it was believed that the constituent particles of the nucleus were electrons and protons. An oxygen atom of atomic number 8 and mass number 16 was believed to have a nucleus which consisted of 16 protons, each having approximately unit mass, and of 8 electrons. This would produce a resultant net nuclear charge of + 8 which was exactly balanced in the neutral atom by the 8 extranuclear electrons. By 1920, Rutherford and others were expressing doubt concerning this model of the nucleus. When J. Chadwick in 1932 demonstrated the existence of neutrons, he provided the basis for the present nuclear model wherein the nucleus is composed of only protons and neutrons. According to this model, the oxygen atom of mass number 16 has a nucleus which consists of 8 protons and 8 neutrons; since neutrons have no charge but are very similar to protons in mass, the net nuclear charge is + 8. There are 8 extranuclear electrons in the neutral atom of oxygen.

2.1. Species of atomic nuclei

The term *nucleon* is used to designate both protons and neutrons in the nucleus. The *mass number A* is the total number of nucleons. Thus

$$A = Z + N \qquad (2.1)$$

where Z is the number of protons (= the *atomic number*) and N is the number of neutrons. The elemental identity and the chemical properties are determined by the atomic number.

* An asterisk in the chapter title or chapter contents means that that chapter or section may be excluded for a minor course.

As we have seen in Chapter 1, an element may be composed of atoms that, while having the same number of protons in the nuclei, have different mass numbers and, therefore, different numbers of neutrons. Neon, for example, has an atomic number of 10, which means that the number of protons in the nuclei of all neon atoms is 10; however, 90% of the neon atoms in nature have 10 neutrons present in their nuclei while 10% of the atoms have 12 neutrons. Such atoms of constant Z but different A are called *isotopes*. The heavy hydrogen isotopes ^{2}H and ^{3}H are used so often in nuclear science that they have been given special names and symbols, deuterium (D) and tritium (T), respectively.

The word isotope is often misused to designate any particular nuclear species, such as ^{16}O, ^{14}C, ^{12}C. It is correct to call ^{12}C and ^{14}C isotopes of carbon since they are nuclear species of the same element. However, ^{16}O and ^{12}C are not isotopic since they belong to different elemental species. The more general word *nuclide* is used to designate any specific nuclear species; e.g. ^{16}O, ^{14}C, and ^{12}C are nuclides. The term *radionuclide* should be used to designate any radioactive nuclear species, although *radioisotope* is a common term used for the same purpose. To designate a nuclide, the mass number A should be written as a superscript to the left of the elemental symbol and the atomic number Z as a subscript to the left; however, the latter is often omitted since the symbol itself identifies the atomic number.

In addition to being classified into isotopic groups, nuclides may also be divided into groupings with common mass numbers and common neutron numbers. *Isotopes* are nuclides with a common number of protons (Z), whereas *isobar* is the term used to designate nuclides with a common number of nucleons (A), i.e. the same mass number. Nuclei with the same number of neutrons (N) but different atomic numbers are termed *isotones*. $^{40}_{19}K$ and $^{40}_{18}A$ are examples of isobars, while $^{3}_{1}H$ and $^{4}_{2}He$ are examples of isotones.

In some cases a nucleus may exist for some time in one or more excited states and it is differentiated on this basis. Such nuclei that necessarily have the same atomic number and mass number are called *isomers*. ^{60m}Co and ^{60g}Co are isomers; the ^{60m}Co nuclide exists in a high energy (*excited*) *state* and decays spontaneously by emission of a γ-ray with a half-life of 10.5 min to the lowest energy C, *ground state*, designated by ^{60g}Co.

$$^{60m}_{27}Co \xrightarrow[10.5\ \text{min}]{\gamma} {}^{60}_{27}Co$$

The symbol m stands for *metastable*, while g (or no symbol) refers to the ground state.

2.2. Atomic masses and atomic weights

The *universal mass unit*, abbreviated u (sometimes amu for atomic mass unit), is defined as one-twelfth of the mass of the ^{12}C atom which has been defined to be 12.000 00 u. The absolute mass of a ^{12}C atom is obtained by dividing the value 12.000 00 by the Avogadro number ($\mathbf{N}_A = 6.022\,04 \times 10^{23}$). The value for the mass of a ^{12}C atom, i.e. the nucleus plus the 6 extranuclear electrons, is thus $1.992\,68 \times 10^{-23}$ g. Atomic masses are expressed in units of u relative to the ^{12}C standard. This text uses M to indicate masses in units of u, and m in units of kilograms; $m = M/10^3\,\mathbf{N}_A$.

In nuclear science it has been found convenient to use the atomic masses rather than nuclear masses. The number of electrons are always balanced in a nuclear reaction, and the changes in the binding energy of the electrons in different atoms are insignificant within the degree of accuracy used in the mass calculations. Therefore the difference in atomic masses of reactants and products in a nuclear reaction gives the difference in the masses of the nuclei involved. In the next chapter, where the equivalence between mass and energy is discussed, it is shown that all nuclear reactions are accompanied by changes in nuclear masses.

The mass of the nucleus can be approximated by subtracting the sum of the masses of the electrons from the atomic mass. The mass of an electron is 0.000 549 u. In grams, this mass is 9.1165×10^{-28}. Since the neutral carbon atom has 6 electrons, the approximate mass of the nucleus is $1.99268 \times 10^{-23} - 6 \times (9.1165 \times 10^{-28}) = 1.99213 \times 10^{-23}$ g. This calculation has not included the difference in the mass of the 6 extra electrons attributable to the binding energy of these electrons. However, this binding energy has a mass equivalent which is smaller than the least significant figure in the calculation.

The mass of a neutron is 1.008 665 u while that of the hydrogen atom is 1.007 825 u. Since both neutrons and protons have almost unit atomic masses, the atomic mass of a nuclide should be close to the number of nucleons, i.e. the mass number. However, when a table of atomic masses is studied it becomes obvious that many elements have masses which are far removed from integral values. Chlorine, for example, has an atomic mass value of 35.453 u, while copper has one of 63.54 u. These values of the atomic masses can be explained by the effect of the relative abundances of the isotopes of the elements contributing to produce the observed net mass.

If an element consists of n_1 atoms of isotope 1, n_2 atoms of isotope 2, etc., the *atomic fraction* x_1 for isotope 1 is defined as:

$$x_1 = n_1/(n_1 + n_2 + \ldots) = n_1/\Sigma n_i \tag{2.2}$$

TABLE 2.1. *Isotopic data for some elements*

Element	*Z*	*N*	*A*	Atomic mass (u)	Abundance (%)	Atomic weight	Symbol
Hydrogen	1	0	1	1.007 825	99.985	1.007 97	^{1}H
	1	1	2	2.014 102	0.0155		^{2}H, D
	1	2	3	3.016 049	0		^{3}H, T
Helium	2	1	3	3.016 030	$\lesssim 0.0001$	4.0026	^{3}He
	2	2	4	4.002 603	100.00		^{4}He
Lithium	3	3	6	6.015 125	7.42	6.939	^{6}Li
	3	4	7	7.016 004	92.58		^{7}Li
Beryllium	4	5	9	9.012 186	100.00	9.0122	^{9}Be
Boron	5	5	10	10.012 939	~ 19.6	10.811	^{10}B
	5	6	11	11.009 305	~ 80.4		^{11}B
Carbon	6	6	12	12.000 000	98.892	12.0112	^{12}C
	6	7	13	13.003 354	1.108		^{13}C
Nitrogen	7	7	14	14.003 074	99.635	14.007	^{14}N
	7	8	15	15.000 108	0.365		^{15}N
Oxygen	8	8	16	15.994 915	99.759	15.999	^{16}O
	8	9	17	16.999 133	0.037		^{17}O
	8	10	18	17.999 160	0.204		^{18}O
Chlorine	17	18	35	34.968 851	~ 75.7	35.453	^{35}Cl
	17	20	37	36.965 898	~ 24.3		^{37}Cl
Uranium	92	143	235	235.043 915	0.724	238.029	^{235}U
	92	146	238	238.050 770	99.266		^{238}U

The *isotopic ratio* is the ratio between the atomic fractions (or *abundances*) of the isotopes. For isotopes 1 and 2, the isotopic ratio is

$$\xi_1 = x_1/x_2 = n_1/n_2; \quad \xi_2 = x_2/x_1 = n_2/n_1 \tag{2.3}$$

The *atomic mass of an element* (or *atomic weight*) M is defined as the average of the isotopic masses, i.e. M_i is weighted by the atomic fraction x_i of its isotope:

$$M = x_1M_1 + x_2M_2 + \ldots = \Sigma x_iM_i \tag{2.4}$$

As an example, natural chlorine consists of two isotopes of which one has an abundance of 75.7% and an atomic mass of 34.9689 u and the second has an abundance of 24.3% and a mass of 36.9659 u. The resultant average atomic mass for the element is 35.454. The atomic mass of copper of 63.55 may be contributed to the presence of an isotope in 69.09% abundance with a mass of 62.9296 u and of a second isotope of 30.91% abundance and 64.9278 u. Atomic masses and abundances of some isotopes are given in Table 2.1.

2.3. Determination of isotopic masses and abundances

The masses and relative abundances of different isotopes occurring naturally in an element can be determined with great exactness using the same technique Thomson employed to demonstrate the presence of isotopes in neon. The instrument used for this purpose is known as a mass spectrometer. The principles of construction of a *mass spectrometer* are shown in Figs. 2.1 and 2.2.

Let us first consider the movement of an ion in electric and magnetic fields, as shown in Fig. 2.1. The ion of mass m is assumed to have a charge q (C, coulomb), which is an integer (z) multiple of the electron charge $\mathbf{e}$ ($1.602\,19 \times 10^{-19}$ C): $q = z\mathbf{e}$. If it is accelerated from velocity zero to v (m s^{-1}) by an electric potential V (V, volts), it will acquire a kinetic energy E_{kin} corresponding to

$$E_{kin} = \tfrac{1}{2}mv^2 = qV \tag{2.5}$$

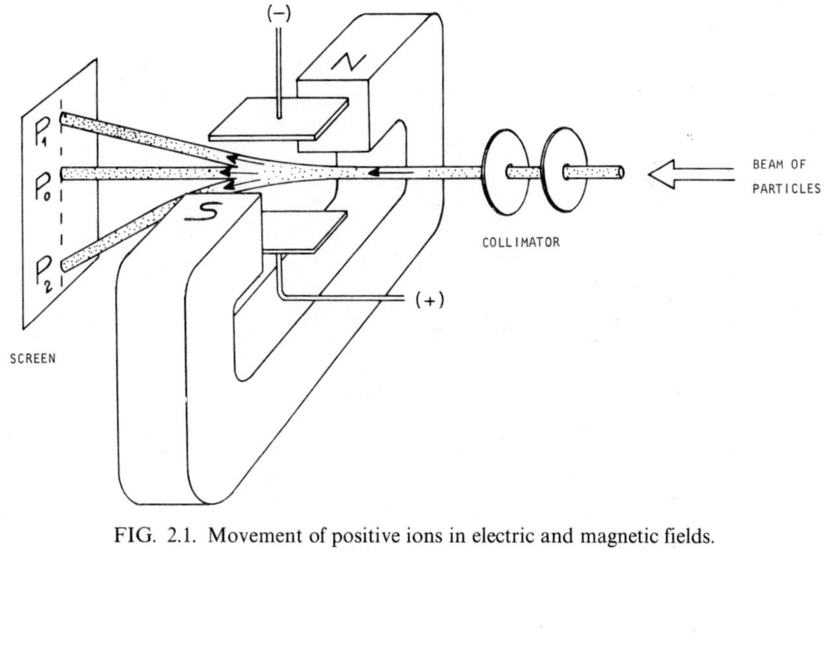

FIG. 2.1. Movement of positive ions in electric and magnetic fields.

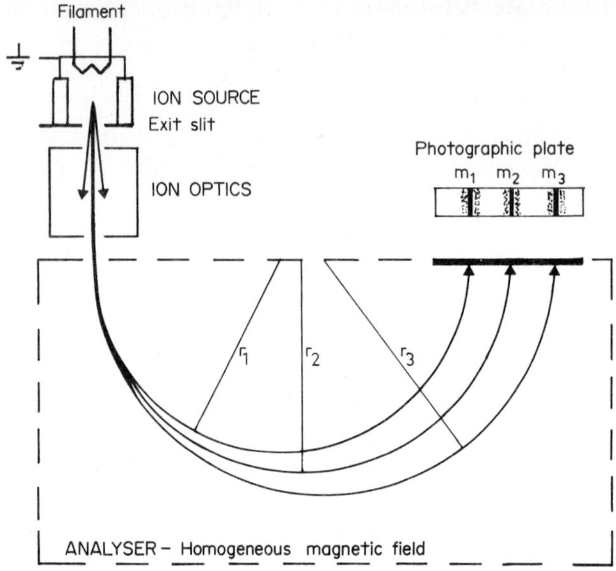

FIG. 2.2. The principle of the mass spectro*graph* (–meter, if the photographic plate is replaced by an ion current meter).

J, joule (or N m, newton meter), If q is given in units of the electronic charge, the kinetic energy will be in units of *electron volts* (eV). For transformation to other energy units, see Appendix IV.

Figure 2.1 shows the deviations of a positive ion in an electric field U (N/C, newton/coulomb) directed upwards, and a magnetic field B (T, tesla; or Wb m^{-2}, Weber m^{-2}) directed into the page. The force F (newton) acting on the ion in the electric field only is

$$F_e = qU \tag{2.6}$$

The ion will hit the screen at point P_1. In the magnetic field only the force is

$$F_m = qvB \tag{2.7}$$

The ion will hit the screen at point P_2. If the forces balance each other, i.e. when $F_e = F_m$, the ion hits the screen at point P_0. One obtains

$$v = U/B \tag{2.8}$$

In either one of the fields (F_e or F_m) the deviation is counteracted by the centrifugal force F_c, where

$$F_c = mv^2/r \tag{2.9}$$

and r is the radius of curvature. In the magnetic field only, the balance between F_m and F_c leads to

$$q/m = v/Br \tag{2.10}$$

where q/m is denoted as the *specific charge* of the ion.

In the mass spectrometer gaseous ions are produced in an *ion source*, e.g. by electron bombardment of the gas produced after heating the substance in a furnace (Fig. 2.2) or by electric discharge, etc. If positive ions are to be investigated, the ion source is given a high positive potential relative to the exit slits. This results in the ions being accelerated away from the source and into the ion optic system. The purpose of the ion optic system is to produce ions of exact direction and of constant velocity, which is achieved through the use of electrostatic and magnetic fields as described; cf. (2.8).

The spectrometer commonly consists of a homogeneous magnetic field which forces the ions to move in circular paths according to (2.10). Combining (2.5) and (2.10) gives

$$m = qr^2B^2/2V \tag{2.11}$$

where V is the ion acceleration potential. Mass spectrometers are usually designed so that of the three variables V, B, or r, two are constant and the third variable, which allows ions of different q/m value to be focused on the detector. The minimum value of q/m is always $\mathbf{e}/m$ because singly charged ions of the atomic or molecular species are almost always present. In order to avoid collisions between ions and gaseous atoms, which would cause scattering of the ion beam, the system is evacuated. The detector can be either a photographic plate or a charge collecting device.

The most common type of mass spectrometer (A. O. Nier 1940) uses a fixed magnetic field and a fixed radius of curvature for the ion beam. If the acceleration potential V is varied so that the masses m_1 and m_2 alternately are registered by the detector, producing ion currents I_1 and I_2, respectively, the abundance of each isotope can be calculated from the ratios $x_1 = I_1/(I_1 + I_2)$ and $x_2 = 1 - x$ when only two isotopes are present. The resolution of modern mass spectrometers is extremely high, as indicated by the values in Table 2.1.

For several decades, mass spectrometers were used primarily to determine atomic masses and isotopic ratios. Now they are applied to a large variety of chemical problems.

Some uses of mass spectrometry of direct interest to chemists involved in nuclear science are:

(a) *Molecular weight determination* can be made by mass spectrometry if gaseous ions can be produced with M/q values not exceeding about 400. This method is of great importance in radiation chemistry (Chapter 15) (where often a large number of products are produced which may be quite difficult to identify by other means) and in the analysis of organic compounds.

(b) The *study of chemical reactions* directly in the gas phase by mass spectrometry is possible. Using an ion source in which molecules are bombarded by a stream of low energy ($\lesssim 100$ eV) electrons, ionization and dissociation reactions can be studied, e.g.

$$C_8H_{18} + e^- \rightarrow C_8H_{18}^+ + 2e^-$$

$$\hookrightarrow C_4H_9^+ + C_3H_6^+ + CH_3\cdot + e^-$$

This technique has practical application in the petroleum industry for determining the composition of distillation and cracking products.

(c) The *method of isotopic dilution* can be used for reaction studies in condensed media. In this technique, the natural isotopic ratio in an element is changed by adding a sample enriched in one particular isotope. For example, a sample of nitrogen with 95% ^{15}N may be added to a reaction which involves nitrogen. The mass spectrometer is used to determine the isotopic ratios of the reaction products to discern how the added nitrogen has reacted. This essentially is a nonradioactive tracer method which has found great use in biochemistry, where the lack of suitable long-lived isotopes of a number of the light elements has precluded the normal radioactive tracer techniques.

(d) *Analysis of gas purity* (e.g. in gaseous diffusion plants in which ^{235}U and ^{238}U

are separated) is done conveniently by mass spectrometry. Not only are the ratios of the uranium isotopes determined, but also the air and water vapor which leaks into the system can be measured. This produces such ions as O_2^+, N_2^+, CO_2^+, and HF^+, which can be measured easily because mass spectrometry detects the presence of impurities in parts per million (ppm) concentration.

2.4. Isotopic ratios in nature

Mass spectrometric investigations of geologic material has shown that isotopic ratios vary slightly in nature with the largest variations observed for the heaviest and lightest elements.

For the heaviest elements the reason for the variation in the ratio of isotopes in an element can be related directly to the consequence of naturally occurring radioactive decay. For example, in a thorium mineral the decay sequence terminates in the formation of an isotope of lead with mass number 208 (see Fig. 1.2). By contrast, in uranium ore, in which ^{238}U is the most abundant isotope, the primary end product of the decay is an isotope of lead with mass number 206. This means that thorium minerals and uranium minerals contain lead with different isotopic composition. In fact, one of the best confirmations of the existence of the radioactive decay series came from the proof that lead had different net atomic weights in different minerals.

The isotopic ratios for the lightest elements depend in which natural material they are found. The ratio for $^7Li/^6Li$ varies from 12.20 to 12.48, while that of $^{12}C/^{13}C$ varies from 89.3 to 93.5. The isotopic ratio $^{18}O/^{16}O$ varies by almost 5% as shown in Fig. 2.3. However, since natural oxygen contains only 0.2% ^{18}O, even a 5% variation in the isotopic ratio has very little influence on the atomic weight of the oxygen. Table 2.2 shows the variation that has been measured for the $^{32}S/^{34}S$ ratio. The variation of

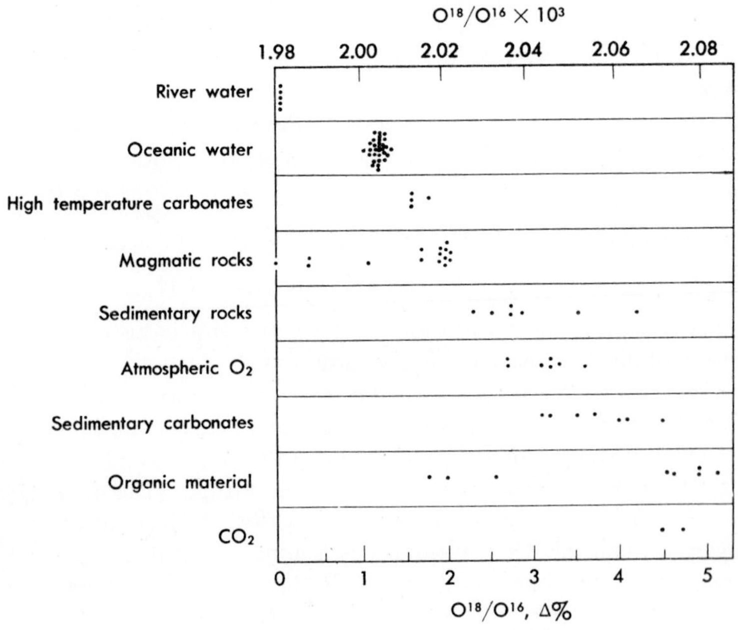

FIG. 2.3. Observed $^{18}O/^{16}O$ isotope ratios. (According to Vinogradov.)

TABLE 2.2. *Variation of isotopic ratio in natural sulfur*

Source of sulfur sample	$^{32}S/^{34}S$
Sulfate from Gulf Coast salt domes	21.42
Sulfate from sea water	21.78
Sulfate minerals from limestones	22.00
Native sulfur of volcanic origin	22.15
Native sulfur of biogenic origin	22.28
Hydrogen sulfide of biogenic origin	22.55

natural isotopic ratios for boron and chlorine causes the uncertainties in their abundances shown in Table 2.1.

2.5. Physicochemical differences for isotopes

Although the isotopic variations in the heaviest elements can be attributed to the consequences of radioactive decay, the variations observed in lighter elements are attributable to chemical behavior. The rates and equilibria of chemical reactions depend on the masses of the atoms involved in the reactions. This is further explained in Appendices B and C. As a consequence, isotopes may be expected to give somewhat different quantitative physicochemical values (the *isotope effect*). As examples of the isotope effect, we may note that the freezing point of H_2O is 0°C, while that for heavy water D_2O is 3.82°C. The boiling point of D_2O is 1.43° higher than that of H_2O. Similarly, while H_2 boils at 20.38 K (Kelvin), D_2 boils at 23.50 K. As a result of these differences in the boiling points, the vapor pressures at a particular temperature for H_2 and D_2 are not the same and distillation can be used to fractionate hydrogen isotopes. Other physical properties such as density, heat of vaporization, viscosity, surface tension, etc., differ similarly.

The optical emission spectra for isotopes are slightly different since electronic energy levels are dependent on the atomic masses. The light emitted when an electron falls from an outer orbit of main quantum number n_2 to an inner orbit of quantum number n_1 $(< n_2)$ is given by

$$\tilde{\nu} = \mathbf{R}_\infty Z^2 \frac{m_{\text{red}}}{\mathbf{m}_e}\left(\frac{1}{n_1^2} - \frac{1}{n_2^2}\right) \tag{2.12}$$

where $\tilde{\nu}$ is the wave number ($1/\lambda$ m^{-1}; $\tilde{\nu}$ is still most commonly given in cm^{-1}), $\mathbf{R}_\infty$ is the Rydberg constant (1.097×10^7 m^{-1}), $\mathbf{m}_e$ is the electron (rest) mass, and m_{red} the reduced mass, according to

$$m_{\text{red}}^{-1} = \mathbf{m}_e^{-1} + m_{\text{nucl}}^{-1} \tag{2.13}$$

where m_{nucl} is the nuclear mass. For the light hydrogen isotope, the H_α line† occurs at 656.285 nm, while the corresponding line for the deuterium isotope of hydrogen occurs at 656.106 nm. This difference could be predicted from (2.12) and its observation experimentally in 1932 provided the first evidence for the existence of a heavy hydrogen isotope. This spectral difference has some practical importance as it can be used for a spectroscopic analysis of the amount of heavy water in light water. Similar *isotopic line shifts* occur in the spectra of all elements, but are rarely as large as the shift of almost 0.2 nm observed for hydrogen. For the isotopes ^{235}U and ^{238}U, the isotopic shift is 0.025 nm.

The absorption spectra of molecules, which depend on transitions between different

†H_α-line refers to transition energy between $n_1 = 1$ and $n_2 = 2$.

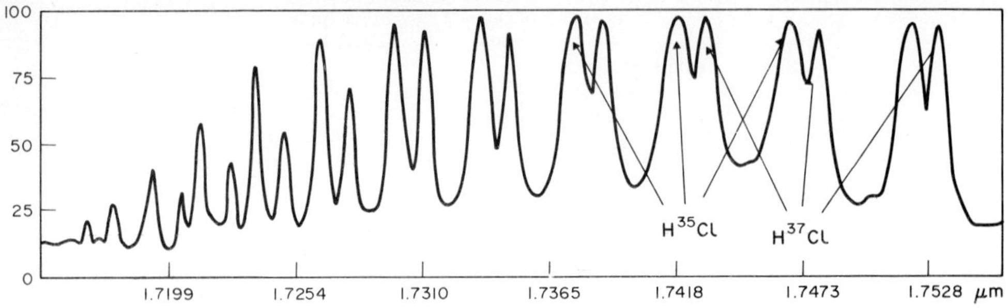

FIG. 2.4. Infrared spectrum of HCl showing one vibrational transition with 12 superimposed rotational transitions each for $H^{35}Cl$ and $H^{37}Cl$. (According to Hardy and Sutherland.)

rotational and vibrational energy states (see Appendix E), show similar isotopic line shifts (Fig. 2.4). The vibrational absorption bands for HDO at 2950 nm and 3830 nm are sufficiently different from the absorption bands of H_2O that they may be used for spectrophotometric analysis of heavy water in normal water.

2.6. Isotope effects in chemical equilibrium

Equilibrium constants can be calculated from the partition functions of the reactants and products provided the different translational, rotational, and vibrational energy states and their population in the molecules are known (see Appendix B). Because these three energy forms are mass dependent, slightly different partition functions are obtained for isotopic molecules. Consequently slightly different equilibrium constants are obtained in reactions of different isotopic composition. In isotopic mixtures, i.e. when two or more isotopes of one element are present at the same time, the situation becomes more complicated because of the *isotope exchange* taking place. Thus in the reaction

$$C^{16}O_2\,(g) + H_2{}^{18}O\,(l) \rightleftharpoons C^{18}O^{16}O\,(g) + H_2{}^{16}O\,(l)$$

the oxygen isotopes in CO_2 exchange with the oxygen isotopes in H_2O. The value of the equilibrium constant (mole fractions) $k = 1.046$ (0 °C) indicates that ^{18}O will be slightly enriched in the CO_2 by the reaction. Thus if carbon dioxide is bubbled through water, the emergent gas will be more enriched in the ^{18}O than the residual water. In this reaction, the *isotope effect* is said to be 4.6‰.

The following reaction occurs with carbonate ions in water:

$$C^{16}O_3^{2-} + H_2{}^{18}O \rightleftharpoons C^{18}O^{16}O_2^{2-} + H_2{}^{16}O \quad (k = 1.112\ (0°C))$$

In this reaction the isotope effect is 11.2‰, which is relatively large and results in enrichment of precipitated carbonate in ^{18}O compared to dissolved carbonate. The equilibrium constant for this reaction and, hence, the $^{18}O/^{16}O$ ratio, has a temperature dependence according to the equation

$$T(°C) = 18 - 5.47(\xi_{18} - \xi_T)$$

where ξ_T is the isotopic ratio (cf. (2.3)) at temperature T.

This effect leads to an interesting consequence. The isotopic ratios for sedimentary carbonate show variation of 2.04 to 2.07×10^{-3} (Fig. 2.3). If it is assumed that these differences are due to the precipitation of the carbonate at different temperatures, one can use

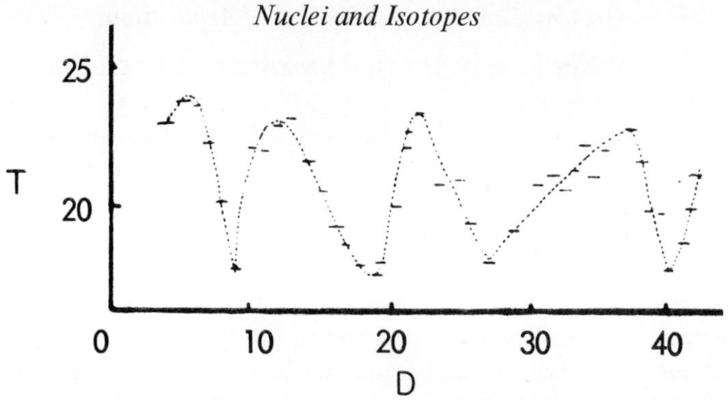

FIG. 2.5. Temperature calculated from the isotopic ratio $^{18}O/^{16}O$ in carbonate of a shell from the Jura period as a function of the distance from center of the shell. (According to Epstein.)

the isotopic ratios to obtain information on geologic temperature. In Fig. 2.5 the data for a shell from the Jura geologic period is shown. The oxygen ratio in the carbonate of the shell has been determined by starting from the center of the shell and measuring the isotopic ratio in each layer of annual growth. The result shows that the temperature at which the carbonate was formed varied during the life of the creature; further, the creature that inhabited the shell must have died during the spring of its fourth year since there was no further growth of the shell after the fourth period of increasing temperature.

Chemical isotope effects are particularly large for lighter elements in biological systems. The *chlorella algae* prefers deuterium over hydrogen, and tritium over deuterium. The enrichment factor depends on the conditions of growth; for deuterium to hydrogen an enrichment value of 1.6–3 has been found, while for tritium to hydrogen the enrichment factor is about 2.5. Bacteria behave similarly, with *coli bacteria* showing an enrichment factor for deuterium of 3.9. Inasmuch as some of the hydrogen atoms are not exchanged readily due to the inertness of their chemical bonds, the isotopic fractionation which involves the easily exchangeable hydrogen atoms in these biological processes must have even larger enrichment factors for deuterium and tritium than their measured values would indicate.

The peculiarity of biological material to prefer certain isotopes has led to studies of how biological material behaves in an isotopic environment which differs substantially from that found in nature. Normally it is found that the organisms wither away and lose their ability to reproduce. It has been possible to raise mice with only ^{13}C in their organism (fed on ^{13}C algae). Carp cannot survive a higher D_2O concentration than 30%, but, on the other hand, some organisms show a strong growth, and some microorganisms have been found to be able to live in pure D_2O or $H_2{}^{18}O$. Exchanging natural $^{14}NH_3$ for $^{15}NH_3$ seems to have little effect on biological systems.

In all of these investigations it should be noted that even when we characterize an isotopic effect as large, it is still quite small by normal reaction criteria except for hydrogen isotopes. For all but the very lightest elements we can assume in most normal chemical experiments that there is no isotope effect. This assumption forms a basis of the use of radioactive tracers to study chemical systems.

2.7. Isotope effects in chemical kinetics

The reason why higher organisms cannot survive when all light hydrogen atoms are replaced by deuterium is to be found not so much in a shift of chemical equilibria as in a shift in reaction rate leading to a fatal lowering of the metabolic rate when light isotopes are replaced by heavier.

In contrast to chemical equilibria, chemical reaction rates depend on the concentration of the reactants and transition states (see Appendix C) but not on the product. The concentration of the transition states depends on the activation energy for its formation and the frequency for its decomposition into the products. These factors can be derived from the partition function which, as mentioned above, differ slightly for molecules of different isotopic composition. For isotopes of the lighter elements, the activation energy term makes the main contribution to the reaction rate isotope effect, while for the heavier elements the vibrational frequency causing the decomposition into the products plays the larger role. Because the energy states usually are more separated for the isotopic molecules of the products and reactants than for the transition state, isotope effects are usually larger in reaction kinetics than in equilibria.

Studies of kinetic isotope effects are of considerable theoretical interest, particularly in organic chemistry. The practical applications are still meager, but this will not necessarily be so in the future. For example, by deuterating (i.e. replacing the hydrogen atoms by deuterium) lubricant oils for watches the reaction rate for oxidation of the oil considerably decreases, leading to up to five times longer lifetime for the oil (and, perhaps, correspondingly fewer visits to the watchmaker). Another example is the decrease in metabolic rate for ^{13}C compounds, which has led to the suggestion of its use for treatment of certain diseases, as e.g. porphyria.

2.8. Isotope separation

Many fields of fundamental science have found great advantage in using pure or enriched isotopes. The industrial use of nuclear power also requires the enrichment

TABLE 2.3. *Production of larger amounts of pure and enriched isotopes; for suppliers see Appendix F*

Isotope	Abundancy (%)	Enrichment (%)	Method	α	Quantities produced (to 1977)	Application
^2H, D	0.014	100	Electrolysis of H_2O	~5	$>10^3$ t	Moderator in reactors
			Chemical exchange: $HD + H_2O \rightleftharpoons H_2 + HDO$	2.7		Nuclear weapons?
^6Li	7.5	100	Electrolysis of LiOH	1.06	tonnes	For production of ^3T through $^6Li(n,\alpha)^3T$
^{10}B	20	95	Distillation of BF_3	1.0075	kg	For neutron detection through $^{10}B(n,\alpha)^7Li$
			Exchange distillation	1.028		
^{13}C	1.1	70	Distillation of CO	1.011	kg	Stable tracer for carbon
^{15}N	0.38	95	Distillation of NO	1.055	kg	Stable tracer for nitrogen
			Exchange $NH_3(g) - NH_4^+(l)$	1.035		
^{18}O	0.2	90	Exchange $CO_2(g) + H_2O(l)$	1.05	~100 g	Stable tracer for oxygen
^{20}Ne	91	100	Thermal diffusion		~100 g	Research
^{235}U	0.7	100	Gas diffusion of UF_6	1.0043	tonnes	Nuclear reactors, Nuclear weapons
		100	Electromagnetically		kg	
		>5	Gas centrifugation of UF_6	1.10	kg	
Most common nuclides	$\gtrsim 1$	~100	Electromagnetically		mg to g amounts	Research

of particular isotopes, primarily of the uranium fuel. The methods which have been developed to achieve isotopic fractionation may be divided into two groups (cf. Table 2.3).

(a) *Equilibrium processes.* These processes consume little energy, but the size of the isotope effect in normal chemical equilibrium limits their use to the isotope fractionation of very light elements, usually atomic number less than about 10.

(b) *Rate processes.* This includes processes which depend on such phenomena as ionic mobility, diffusion, electrolysis, electromagnetic separations, centrifugation, and kinetic processes of a chemical nature. While the isotopic effects in these processes are normally larger than for equilibrium processes, they require a large amount of energy and therefore have economic restrictions.

2.9. Isotope enrichment*

Figure 2.6 shows a flow scheme for an isotopic fractionation process. Each *stage* consists of a number of *cells* coupled in parallel; in the figure only one cell is shown for each separation stage, but in order to obtain a high product flow, the number of cells are usually high at the feed point and then decrease towards the product and waste stream ends, as can be inferred from Fig. 2.7. Each cell contains some physical arrangement, which leads to an isotope fractionation. Thus the atomic fraction for a particular isotope is different in the two outgoing streams from a cell; in the *product stream* the isotope is enriched (atomic fraction x'), while in the *waste stream* it is depleted (atomic

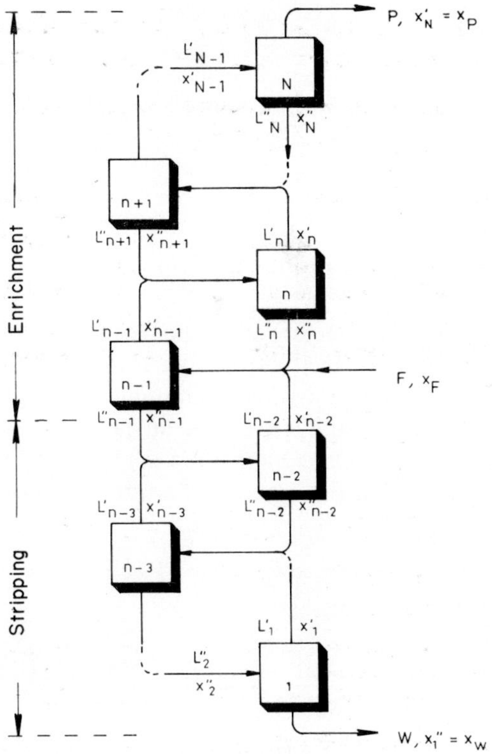

FIG. 2.6. Flow arrangement for an ideal cascade with total reflux for isotope separation.

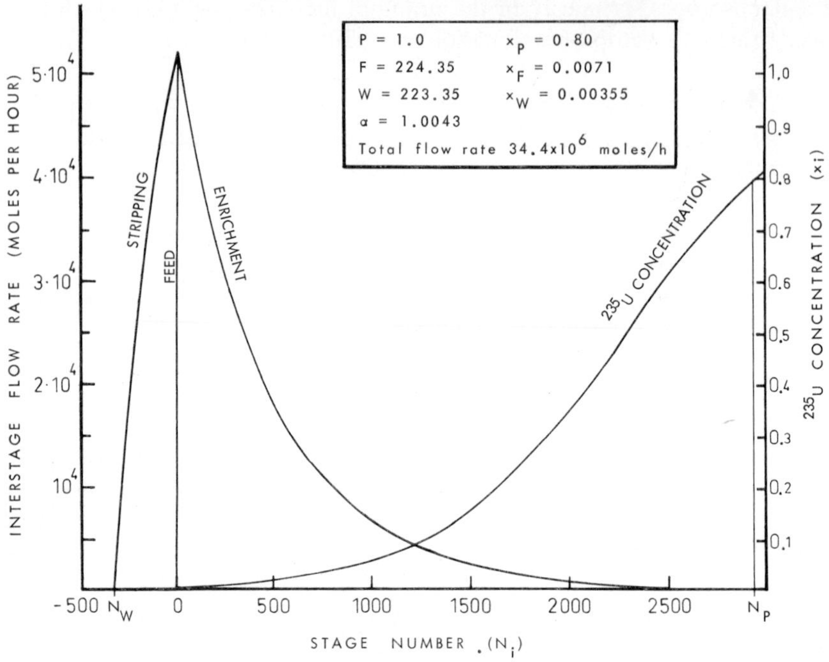

FIG. 2.7. Interstage flow rate and enrichment in an ideal cascade with total reflux having N_P enrichment and N_W stripping stages. The change in isotopic composition in the stripping part is not shown.

fraction x''). The *separation factor* α is defined as the quotient between the isotopic ratios of the product and waste streams for a single step, thus ((2.2) and (2.3)):

$$\alpha = \frac{\xi'}{\xi''} = \frac{x'/(1-x')}{x''/(1-x'')} \tag{2.14}$$

In most cases α has a value close to unity, as can be seen from Table 2.3, which lists the most common separation processes. The value of $\alpha - 1$ is commonly called the *enrichment factor*.

Since, in practice, the separation factors are often even smaller than these theoretically maximum values, to obtain a product with a high enrichment it is necessary to have a multistage process. The number of stages determine the degree of the enrichment of the product, while the number and size of cells in each stage determine the amount of product. This amount (P moles of composition x_P) is related to the amount of feed (F moles of composition x_F) and the amount of waste (W moles of composition x_W) by the equations

$$F = P + W \quad \text{and} \quad Fx_F = Px_P + Wx_W \tag{2.15}$$

From these equations we obtain

$$F = \frac{P(x_P - x_W)}{x_F - x_W} \quad \text{and} \quad W = \frac{P(x_P - x_F)}{x_F - x_W} \tag{2.16}$$

The number of stages required to separate feed into product and waste of specified composition is a minimum at *total reflux*, when $P = 0$. For this condition M. R. Fenske has derived a relation, which can be divided into one part for the enrichment:

$$N_P \ln \alpha = \ln \frac{x_P(1 - x_F)}{x_F(1 - x_P)} \tag{2.17a}$$

and one for the stripping part of the cascade:

$$N_W \ln \alpha = \ln \frac{x_F(1 - x_W)}{x_W(1 - x_F)} \tag{2.17b}$$

N_P and N_W are the minimum number of enrichment and stripping stages, respectively. In isotope separations α is often very close to one; $\ln \alpha$ can then be replaced by $(\alpha - 1)$. In practice some product flow is desired, the fraction withdrawn at the enrichment stage being known as "*the cut*" P/F. The number of stages required to produce the composition x_P then increases.

The most economic, and thus also the most common, type of cascade is the so-called *ideal cascade*. In this there is no mixing of streams of unequal concentrations, thus $x'_{n-1} = x''_{n+1}$ in Fig. 2.6. Although the number of stages required for a particular product is exactly twice the values given by (2.17) minus 1,

$$N_{\text{ideal}} = 2N - 1 \tag{2.18}$$

the interstage flow becomes a minimum, thus minimizing the inventory and work required. Also for the ideal cascade, the minimum number of stages are obtained at total reflux.

The interstage flow rate L_i that must be maintained in order that there be any enrichment in the ideal cascade at the point, where the concentration x_i occurs, is given by

$$L_i = \frac{2P}{(\alpha - 1)} \frac{(x_P - x_i)}{x_i(1 - x_i)} \tag{2.19}$$

To obtain the flow rate in the stripping part, P is replaced by W, and $x_P - x_i$ by $x_i - x_W$.

For the enrichment of natural uranium (0.71% in ^{235}U) to 80% in ^{235}U, the minimum number of enrichment stages is found to be almost 3.000 at an enrichment factor of 0.43%. The flow rate at the feed point becomes about 52.300 times the product flow P. Isotope separation plants therefore become very large even for small outputs. In Fig. 2.7 the interstage flow rate, as well as the isotopic enrichment, is given as a function of the stage number for N_P enrichment and N_W stripping stages.

Isotopic separation on a technical scale thus requires a large number of stages in order to obtain high enrichment and large amounts of materials in order to give a substantial amount of product. Small changes in the values of α can have a large economic impact in isotopic separation.

2.10. Isotope exchange between H_2O and H_2S*

As an example of an industrial isotope exchange process, let us consider the production of heavy water by the chemical reaction

$$H_2O\,(l) + HDS\,(g) \rightleftharpoons HDO\,(l) + H_2S\,(g) \begin{cases} k = 2.34\ (25°C) \\ k = 1.92\ (100°C) \end{cases}$$

From the values of the equilibrium constants k we see that the enrichment of deuterium

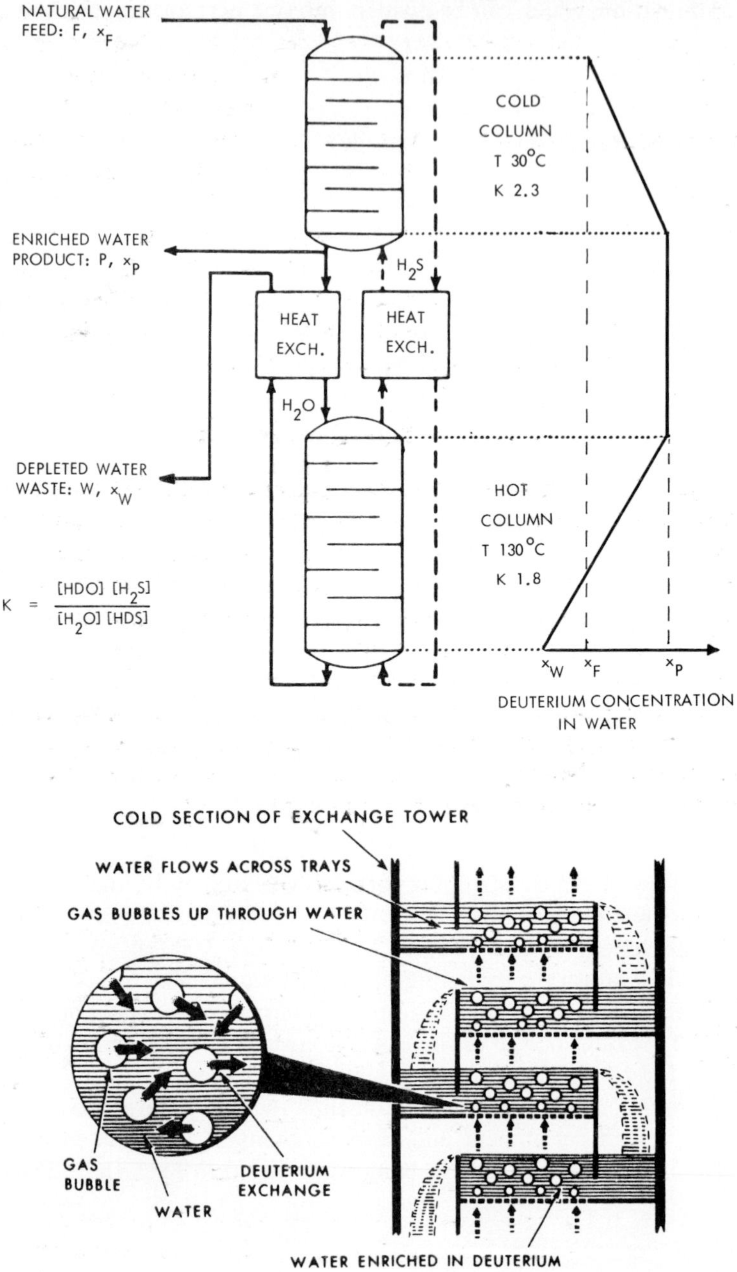

FIG. 2.8. The water–hydrogen sulphide dual-temperature distillation process for enrichment of heavy water.

in water increases with decreasing temperature. Use is made of this property in the two-temperature H_2O–H_2S exchange process, which is used in many countries to produce heavy water. A typical plant consists of several units as shown in Fig. 2.8. Through the upper distillation tower natural water flows downwards and meets hydrogen sulfide gas streaming upwards. As a result of the exchange between H_2O and H_2S, heavy hydrogen is enriched in the water. In the lower tower, which is operated at a higher temperature, the equilibrium conditions are such that deuterium is enriched in the hydrogen sulfide and moves with that gas to the upper tower. No catalyzer is required in order to achieve rapid equilibrium in this reaction. The product of the process is water which is enriched in deuterium from the top tower and water which is depleted in deuterium from the bottom tower. The hydrogen sulfide circulates through both towers with no net loss.

A plant to produce 400 tons annually has been constructed in Canada. The largest exchange towers are 60 m high and have a diameter of 6 m. In 5 units (only one unit is indicated in Fig. 2.8) the D_2O concentration is raised from 0.014% to 15%. Through distillation the concentration is raised further to 90%, followed by electrolysis to produce 99.97%. The 1976 cost of pure D_2O was ~US$100 per kg.

2.11. Electrolysis of H_2O*

Electrolysis of water produces hydrogen gas at the cathode, which contains a lower proportion of deuterium than the original water. The isotope effect stems from the differences in the rates of dissociation of a proton (H^+) and a deuteron (D^+) from water, and the rates of neutralization of these hydrated ions, and thus has a kinetic basis. Depending on the physical conditions α-values between 3 and 10 are obtained. For $\alpha = 6$ it is necessary to electrolyze 2700 l natural water (deuterium content 0.014%) to produce 1 l water containing 10% deuterium, mainly as HDO. In a multistage process the hydrogen gas is either burnt to recover energy, or used in a chemical process, e.g. in ammonia synthesis by the Haber–Bosch process. Although this technique has been used industrially in Norway to produce tonne amounts of pure D_2O, it is no longer considered economic except for final purification of D_2O.

2.12. Gaseous diffusion

In a gaseous sample the lighter molecules have a higher average velocity than the heavier molecules. In 1930, F. W. Aston in England showed that the lighter of the two neon isotopes, ^{20}Ne, diffused through the walls of porous vessels somewhat faster than the heavier isotope, ^{22}Ne. In the gas the average kinetic energy of isotopic molecules must be the same, i.e. $1/2M_L v_L^2 = 1/2M_H v_H^2$, where M_H and M_L are the masses of the molecules containing the heavy and light isotopes to be separated. The maximum theoretical separation factor in gaseous diffusion is given by

$$\alpha = v_L/v_H = \sqrt{M_H/M_L} \qquad (2.20)$$

The theory is more complicated, depending among other things upon the mean free path of the gaseous atoms, the pore length and diameter of the separating membrane, and the pressure difference over the membrane. If experimental conditions are carefully controlled, this theoretical maximum can be closely approached.

^{235}U is enriched through gaseous diffusion using the relatively volatile uranium compound UF_6. In addition to its volatility, UF_6 has the advantage that fluorine consists of only one isotope, ^{19}F. For the isotopic molecules $^{238}UF_6$ and $^{235}UF_6$, the value of 1.0043 is theoretically possible for α, cf. (2.20). The following conditions must be considered in the technical application of the separation.

(a) The cells are divided into two parts by a membrane which must have very small pores (e.g. 10–100 nm in diameter) in order to obtain isotopic separation. In order that large gas volumes can flow through the membrane, millions of pores are required for each square centimeter. Moreover, the membranes must have good mechanical stability to withstand the pressure difference across them.

(b) UF_6 sublimes at 64°C, which means that the separation process must be conducted at a temperature above this.

(c) UF_6 is highly corrosive and attacks most materials. The membrane must be inert to attack by UF_6. Water decomposes UF_6 according to the equation

$$UF_6\,(g) + 2H_2O\,(g) \longrightarrow UO_2F_2\,(s) + 4HF\,(g)$$

This reaction is quite undesirable as HF is highly corrosive and solid UO_2F_2 can plug the pores of the membranes. The tubing and cells of a plant are made principally of nickel and teflon to overcome the corrosion problems.

(d) In order to transport the large gas volumes and to keep the proper pressure drop across the membranes, a gaseous diffusion plant requires a large number of pumps. A large cooling capacity is needed to overcome the temperature rise caused by compression of the gas.

In §2.9 the number of stages and the interstage flow relative to the product flow was given for enrichment of ^{235}U from its natural isotopic abundance of 0.71% to a value of 80%. With a waste flow in which the isotopic abundance of ^{235}U is 0.35%, (2.15) shows that for each mole of product obtained 221 moles of feed are necessary. In more recent designs the concentration of ^{235}U in the waste is reduced even further, to ~0.2%. Isotope separation through gaseous diffusion is a very energy-consuming process due to the compression work and cooling required. An annual production of 10 Mkg

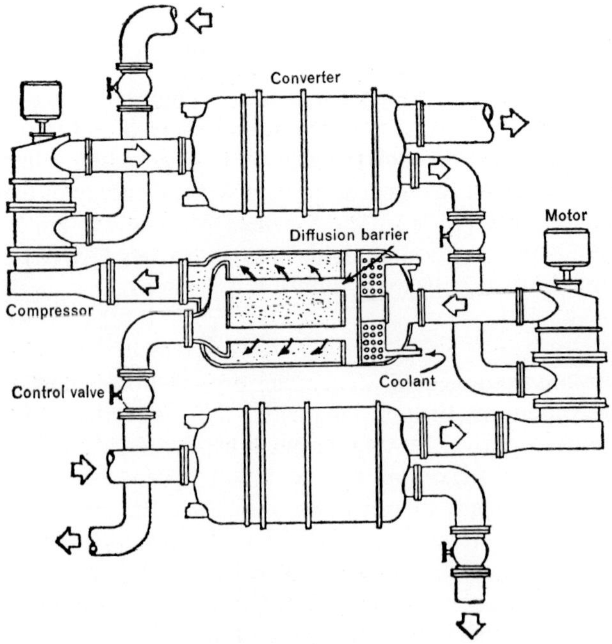

FIG. 2.9. Principle of gas diffusion barrier.

SWU requires an installed capacity of 3500 MW in present plants, or ~3 MWh kg^{-1} SWU. Improved technology may reduce this to ~ 2800 MW.

The work required to enrich uranium in ^{235}U increases rapidly with the ^{235}U content of the product. Because of varying domestic prices on natural uranium, as well as varying contents of ^{235}U in uranium obtained from used reactor fuel elements, so-called *toll enrichment* has been introduced. In this case, the purchaser himself provides the uranium feed into the separation plant and pays for the extra *separative work* required to make his desired product out of the regular flow of uranium in the plant. In Appendix D the cost calculation for the separative work is presented in terms of kg SWU (*separative work units*). A 1 GWe nuclear light water reactor station requires about 180×10^3 kg SWU in initial fueling, and then 70–90 $\times 10^3$ kg SWU for an annual reload.

Gaseous diffusion plants are known to exist in the United States, Great Britain, the Soviet Union, and France, and possibly in China. Figure 2.9 shows part of the plant of Oak Ridge, Tennessee. The US 1977 capacity of about 17 Mkg SWU will be upgraded to reach about 28 Mkg SWU in 1985. The French EURODIF plant at Tricastin is planned to reach a capacity of about 11 Mkg SWU by 1982. By 1986 the noncommunist world isotope enrichment capacity (*all* techniques) is expected to reach about 70 Mkg SWU.

2.13. Electromagnetic isotope separation

During the Manhattan Project of the United States, electromagnetic separation was used to obtain pure ^{235}U. This process is identical in theory to that described for the mass spectrometer. The giant electromagnetic separators were called *calutrons* and after the 1939–1945 war were used at Oak National Laboratory to produce gram amounts of stable isotopes of most elements up to a purity of 99.9% or more.

2.14. Gas centrifugation

Though gas diffusion has been the all dominating process for ^{235}U enrichment, its small separation factor, large energy consumption, and secrecy about the technique have encouraged interest in finding other and more advantageous processes. Most effort has been put into developing centrifugal processes because of the large separation factors achievable. Dynamic gas centrifugation, as described in this section, as well as "static" gas centrifugation (the nozzle technique, see next section) have now both reached industrial scale.

In a gas centrifuge, light molecules are enriched at the center and heavy molecules at the periphery (Fig. 2.11). It can be shown that the ratio of radial concentration to the axial concentration for isotopic molecules under equilibrium conditions and at low pressure (avoiding remixing due to the Brownian movement) is given approximately by

$$\frac{x_P}{x_F} \approx e^{\delta n} \qquad (2.21)$$

for n centrifuges (stages) connected in series. Here

$$\delta = (M_H - M_L)10^{-3} v_r^2 r^2 / 2\mathbf{R}T \qquad (2.22)$$

where r is the centrifuge radius and v_r the speed of rotation. The separation factor

FIG. 2.10. The gaseous diffusion plant in Oak Ridge, Tennessee. The erection of the plant cost about $600 million, and the power to run it amounted to about 1 GW. The left building is about 700 m long.

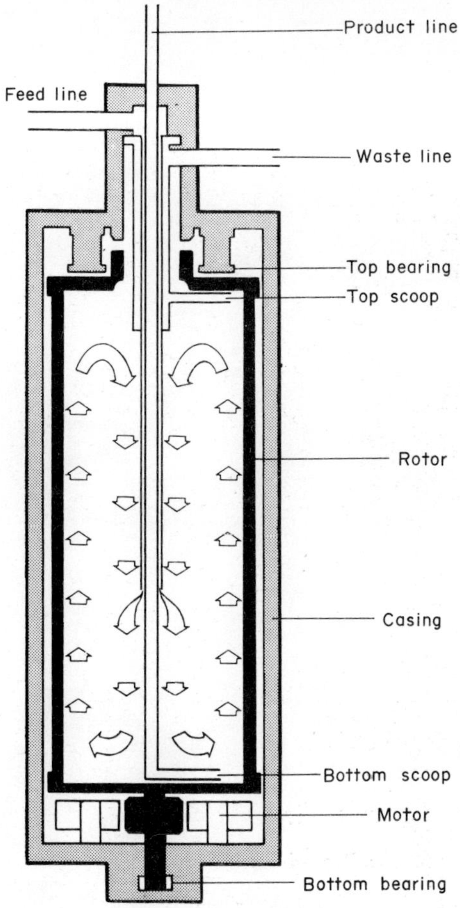

FIG. 2.11. Gaseous centrifuge for $^{235}UF_6$ enrichment.

$\alpha \approx 1 + \delta$. Estimated values for present centrifuges operating on UF_6 are: length 1 m; radius 0.2–0.4 m; rotational speeds of 50–80 × 10^3 rpm. The enrichment obtainable in one stage is limited by the material strength of the centrifuge bowl; the present materials limit $v_r r$ to $< 1000 \text{ m s}^{-1}$. Typical separation factors are 1.4–2.0 per stage, thus about 10 stages are required to enrich ^{235}U from 0.7% to 3% with a 0.2% tail. As compared, a diffusion plant would require ~ 1200 stages. Though rather few stages are required to upgrade natural uranium to reactor quality, a very large number of centrifuges are needed to produce large quantities of enriched material.

Centrifuge technology requires <7% of the power consumed by a diffusion plant, or <200 MW power for 10 Mkg SWU/y. This makes their environmental impact minimal, as compared to gas diffusion plants, which require substantial electric power installation and cooling towers with large water vapor effluent. Smaller plants of a few million kg SWU/y are already economical, and their output can be readily multiplied by installation of parallel processing trains. The very large number of centrifuges required due to their small size (each having a capacity of $\lesssim 15$ kg SWU/y) does not lead to excessive construction costs due to mass production. On the whole, centrifuge separation seems to lead to about the same enrichment cost as large scale diffusion plants.

FIG. 2.12. A pilot cascade with gas centrifuges for uranium isotope enrichment at Oak Ridge National Laboratory, USA.

Several demonstration plants are now running, the largest being at Capenhurst (England), Almelo (the Netherlands), and Oak Ridge (USA); Fig. 2.12 shows a "second generation" centrifuge cascade at Oak Ridge. The joint Dutch–German–British company URENCO/CENTEC intends to install a capacity of about 10 Mkg SWU by 1988, beginning with 2 Mkg SWU in 1982. In the United States DOE (Department of Energy) will build four centrifuge enrichment units with a total capacity of 8.8 Mkg SWU in Portsmouth, Ohio, to be in operation in 1988.

2.15. Other methods of isotope separation*

In theory all physicochemical procedures are capable of isotope separation. Some other methods which have been studied include distillation, molten salt electrolysis, solvent extraction, and ion exchange.

Tenths of kilograms of pure ^{13}C and ^{15}N have been produced at the Los Alamos

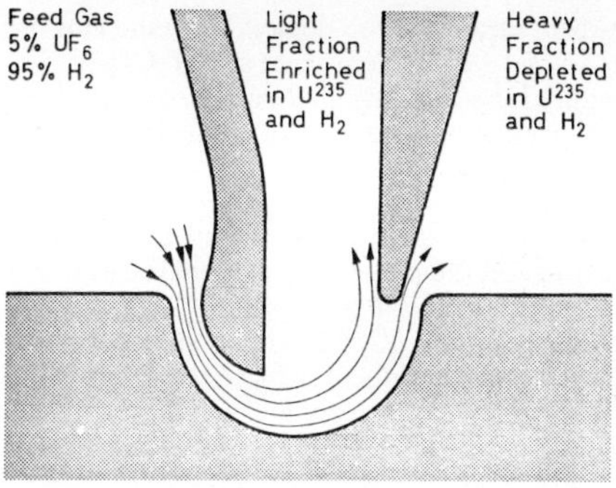

FIG. 2.13. Section through a separation nozzle arrangement showing the stream lines.

Scientific Laboratory through *distillation* of NO and CO at liquid air temperature. At the same time a fractionation between ^{16}O and ^{18}O occurs. ^{18}O has been suggested for production of $^{238}Pu^{18}O_2$ for use as an energy source in artificial hearts, because of the much lower secondary neutron emission (lower yield of $^{18}O(\alpha, n)$ than of $^{16}O(\alpha, n)$).

A novel method of separation, involving passage of a mixture of UF_6 and helium or hydrogen at very high velocities through a *nozzle*, as seen in Fig. 2.13, has been developed by E. W. Becker in Germany. The technique is sometimes referred to as "static" or "stationary-walled" gas centrifugation. The separation factor is typically 1.01–1.03 per stage, i.e. about three times better than in the gaseous diffusion process, and offers great possibilities for further improvements. Thus while the diffusion process requires about 1200 stages for a 3% ^{235}U enrichment (with 0.2% tails), the nozzle technique will require only about 500 stages. However, the power consumption is said to be larger, $\sim$ 4500 MW for an annual production of 10 Mkg SWU, as compared to 2800–3500 MW for a diffusion plant. The present operating cost is said to be higher than for both the diffusion and the centrifuge methods. The 5 Mkg SWU separation plant under construction at Valindaba, South Africa, uses the nozzle technique with hydrogen as a carrier gas for the UF_6.

Recently a number of *photoionization* processes have been investigated for isotopic separation, especially of uranium. In one such process UF_6 is irradiated by a laser beam, producing selective vibrational excitation in the $^{235}UF_6$ molecule (cf. § 2.5). By irradiation with ultraviolet (possibly, but not necessarily, by laser) light the excited molecule is brought to dissociation, leaving $^{238}UF_6$ undissociated. It is important that the ultraviolet-pulse follow quite rapidly after the laser pulse, so that the vibrationally excited $^{235}UF_6$ molecule does not lose its excitation energy through collision with surrounding molecules. It is obvious that this necessitates gas phase reactions. The $^{235}UF_n^{6-n}$ ion formed through the dissociation ($n < 6$) is then collected by the action of electromagnetic fields. This technique is not limited to UF_6; even pure uranium metal vapor and plutonium compounds have been separated into their isotopic constituents by strong laser light. Although research in this area has indicated a large scale feasibility of this and several similar processes, no predictions can yet be made of their technological value.

2.16. Exercises

2.1. How many atoms of ^{235}U exist in 1 kg of uranium oxide, U_3O_8, made of natural uranium?

2.2. What is the atomic fraction of deuterium in water with the mole fraction of 0.81 for H_2O, 0.18 for HDO, and 0.01 for D_2O?

2.3. The translational energy of one mole of a gas is given by 3/2 **R***T*, which corresponds to an average thermal molecular velocity v (the root mean square velocity), while the most probable velocity $v' = \sqrt{0.67}\,v$.

(a) What is the most probable velocity of a helium atom at 800° C?

(b) What voltage would be required to accelerate an α-particle to the same velocity?

2.4. A Nier type mass spectrometer has a fixed radius of curvature of 5 cm and a magnetic field of 3000 G; 1 G = 10^{-4} Wb m^{-2}. At what accelerating voltage will a Na^+ ion be brought to focus at the ion collector?

2.5. In a Dempster type (constant B and V) mass spectrograph utilizing 180° degree focusing, the ions $^{12}C^+$ and $^{11}BH^+$ are recorded simultaneously, the latter ion having a slightly larger orbit diameter. The separation between the lines recorded on the photographic plate is 0.0143 cm and the orbit diameter for the $^{12}C^+$ ion is 20 cm. What is the atomic mass of ^{11}B?

2.6. In an investigation the isotope ratio $^{18}O/^{16}O$ was found to be 2.045×10^{-3} for fresh water and 2.127×10^{-3} for carbon dioxide in the atmosphere. Calculate the equilibrium constant (mole fractions!) for the reaction

$$H_2{}^{18}O(l) + CO_2(g) \rightleftharpoons H_2O(l) + C^{18}O^{16}O(g)$$

2.7. How many ideal stages in an ordinary cascade are required at an $\alpha = 6$ to produce water in which 10% of the hydrogen is deuterium?

2.8. In a distillation column with total reflux, ^{10}B is enriched through exchange distillation of $BF_3O(C_2H_5)_2$ from the natural value of 20 atom % to a product containing 95% ^{10}B. The packed column has a length of 5 m and a diameter of 3 cm. What is the approximate height of a theoretical stage if the enrichment factor is 0.026?

2.9. A gaseous centrifuge plant is set up in order to enrich UF_6 of natural isotopic composition in ^{235}U. The centrifuges, which each have a length of 100 cm and a diameter of 20 cm, rotate at 40 000 rpm. The gas temperature is 70° C.

(a) Prove that the separation factor α in (2.14) can be approximated by ρ^δ according to (2.21) when the product flow is very small compared to the waste flow, and α is not far from 1.

(b) Using this approximation, what is the theoretical separation factor for one unit?

(c) Assuming that the enrichment factor obtained with the centrifuge is only 70% of the theoretical one, what number of units would be required in series in order to achieve UF_6 with 3% ^{235}U?

(d) How many centrifuges will be required to produce 10^6 kg SWU/y, i.e. about 6% of the total US 1973 capacity? Use eqn. (D.4) in Appendix D. $D\rho$ is 2.35×10^{-4} poise (g cm^{-1} s^{-1}).

2.17. Literature

G. H. Clewlett, Chemical separation of stable isotopes, *Ann. Rev. Nucl. Sci.* **1** (1952) 293.

J. Bigeleisen, Isotopes, *Ann. Rev. Nucl. Sci.* **2** (1953) 221.

T. F. Johns, Isotope separation by multistage methods, *Progr. Nucl. Phys.* (ed. O. R. Frisch) **6** (1957) 1.

M. Benedict and T. Pigford, *Nuclear Chemical Engineering*, McGraw-Hill, New York, 1957.

J. Kistemacher, J. Biegeleisen, and A. O. C. Nier (eds.), *Proceedings of the International Symposium on Isotope Separation*, North-Holland, Amsterdam, 1958.

J. H. Beynon, *Mass Spectrometry and its Application to Organic Chemistry*, Elsevier, Amsterdam, 1960.

L. Melander, *Isotope Effects on Reaction Rates*, Ronald Press, New York, 1960.

R. E. Weston Jr., Isotope effects in chemical reactions, *Ann. Rev. Nucl. Sci.* **11** (1961) 439.

H. London, *Separation of Isotopes*, G. Newnes, London, 1961.

A. E. Brodsky, *Isotopenchemie*, Akademie-Verlag, Berlin, 1961.

S. S. Roginski, *Theoretische Grundlagen der Isotopenchemie*, VEB Deutscher Verlag der Wissenschaften, Berlin, 1962.

IUPAC and IAEA, *Isotope Mass Effects in Chemistry and Biology*, Butterworths, London, 1964.

D. P. Herron, How toll enrichment will reduce fuel costs, *Nucleonics*, **22** (6) (1964) 62.

J. M. Blum, Le séparation isotopique, *Energie Nucléaire*, 1965, p. 207.

M. L. Smith (ed.), *Electromagnetically Enriched Isotopes and Mass Spectrometry*, Butterworths Sci. Publ., London, 1965.

F. A. White, *Mass Spectrometry in Science and Technology*, J. Wiley, 1968.

E. W. Becker, A comparison of three separation methods, *Nuclear News*, July 1969, p. 46.

R. W. Levin, Conversion and enrichment in the nuclear fuel cycle, in D. M. Elliot and L. E. Weaver (eds.), *Education and Research in the Nuclear Fuel Cycle*, University of Oklahoma Press, 1972.

R. N. Zare, Laser separation of isotopes, *Sci. Am.* Feb. 1977.

D. R. Olander, The gas centrifuge, *Sci. Am.* Aug. 1978.

CHAPTER 3

Nuclear Mass and Stability

Contents

3.1. Patterns of nuclear stability

There are approximately 275 different nuclei which have shown no evidence of radioactive decay and, hence, are said to be stable with respect to radioactive decay. When these nuclei are compared for their constituent nucleons, we find that approximately 60% of them have both an even number of protons and an even number of neutrons (*even–even nuclei*). The remaining 40% are about equally divided between those that have an even number of protons and an odd number of neutrons (*even–odd nuclei*) and those with an odd number of protons and an even number of neutrons (*odd–even nuclei*). There are only 6 stable nuclei known which have both an odd number of protons and odd number of neutrons (*odd–odd nuclei*); ${}^{2}_{1}H$, ${}^{6}_{3}Li$, ${}^{10}_{5}B$, ${}^{14}_{7}N$, ${}^{50}_{23}V$, and ${}^{180}_{73}Ta$. It is significant that the first 4 stable odd–odd nuclei are found in the very light elements ${}^{6}Li$, ${}^{10}B$, and ${}^{14}N$ in high abundance. The other 2 nuclei are found in quite low isotopic abundance (0.25 and 0.01%) and we cannot be certain that some of these nuclides are not unstable to radioactive decay with extremely long lifetimes.

Considering this pattern for the stable nuclei, we can conclude that nuclear stability is favored by even numbers of protons and neutrons. The validity of this statement can be confirmed further by considering for any particular element the number and types of stable isotopes; see for example, Table 3.1. Elements of even atomic number (i.e. even number of protons) are characterized by having a relatively sizable number of stable isotopes, usually more than 4. For example, the element tin, atomic number 50, has 10 stable isotopes while cadmium ($Z = 48$) and tellurium ($Z = 52$) each have 8. By contrast silver ($Z = 47$) and antimony ($Z = 51$) each have only 2 stable isotopes, and rhodium ($Z = 45$), indium ($Z = 49$), and iodine ($Z = 53$) have only 1 stable isotope. Many other examples of the extra stabilization of even numbers of nucleons can be found in Fig. 3.1 and from isotope charts. The guide lines of N and Z equal to 2, 8, 20, etc., have not been selected arbitrarily. These proton and neutron numbers represent unusually stable proton and neutron configuration, as will be discussed further in Chapter 6. The curved

TABLE 3.1. *Number of stable nuclides for some different proton (Z) and neutron (N) numbers*

Number of protons or neutrons	Number of stable nuclei with number of Protons (isotopes)	Neutrons (isotones)
19	2 (K)	0
20	6 (Ca)	5
21	1 (Sc)	0
27	1 (Co)	1
28	5 (Ni)	5
29	2 (Cu)	1
48	8 (Cd)	4
49	1 (In)	1
50	10 (Sn)	5
51	2 (Sb)	1
52	8 (Te)	4
81	2 (Tl)	1
82	4 (Pb)	8
83	1 (Bi)	1

line through the experimental points is calculated based on the liquid drop model of the nucleus which is discussed later in this chapter.

Typically, elements of odd Z usually have only one or two stable isotopes, and these have an even number of neutrons. This is in contrast to the range of stable isotopes of even Z, which includes nuclei of both even and odd N, although the former outnumber the latter. Tin ($Z = 50$), for example, has 7 stable even–even isotopes and only 3 even–odd

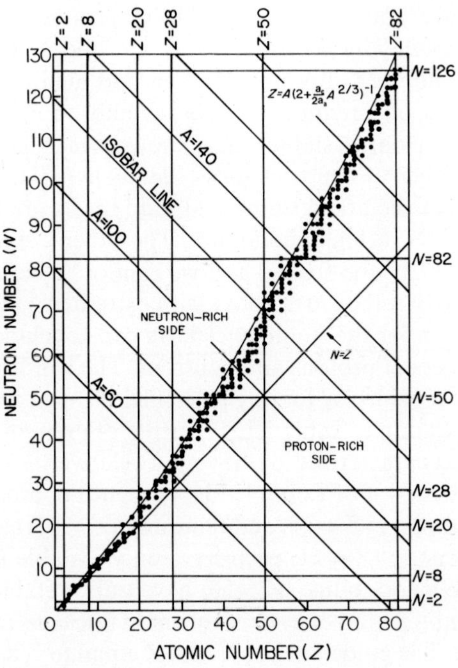

FIG. 3.1. Chart showing stable nuclides as a function of their proton (Z) and neutron (N) numbers. The numbers denoted, 2, 8, etc., are discussed in Chapter 6.

ones. We also find that nuclei of even N have only 1 or 2 stable isotopes.

The greater number of stable nuclei with even numbers of protons and neutrons is explained in terms of the energy stabilization gained by combination of like nucleons to form pairs, just as electrons of opposite spin pair in atoms and molecules. If a nucleus has, for example, an even number of protons, all these protons can exist in pairs. However, if the nucleus has an odd number of protons, at least one of these protons must exist in an unpaired state. The increase in stability resulting from complete pairing in elements of even Z is responsible for their ability to accommodate a greater range of neutron numbers as illustrated for the isotopes of tin relative to those of indium and antimony. The same pairing stabilization holds true for neutrons so that an even–even nuclide which has all its nucleons, both neutrons and protons, paired represents a quite stable situation. In the elements in which the atomic number is even, if the neutron number is uneven, there is still some stability conferred through the proton pairing. For elements of odd atomic number, unless there is stability due to an even neutron number, the nuclei are radioactive with rare exception. We should also note that the number of stable nuclear species is approximately the same for even–odd and odd–even cases. The pairing of protons and neutrons must thus confer approximately equal degrees of stability to the nucleus.

3.2. Neutron to proton ratio

If a graph is made (Fig. 3.1) of the relation of the number of neutrons to the number of protons in the known stable nuclei, we find that in the light elements stability is achieved when the number of neutrons and protons are approximately equal ($N = Z$). However, as we progress in the Periodic Table of the elements (i.e. along the Z-line), the value of the ratio of neutrons to protons, the N/Z ratio, for nuclear stability increases from unity to approximately 1.5 at bismuth. Thus pairing of the nucleons is not a sufficient criterion for stability: a certain ratio N/Z must also be prevalent. However, even this does not suffice for stability, because at high Z-values, a new mode of radioactive decay, α-emission, appears. Above bismuth the nuclides are all unstable to radioactive decay by α-particle emission, while some are unstable also to β-decay.

If a nucleus has a N/Z ratio too high for stability, it is said to be neutron-rich. It will undergo radioactive decay in such a manner that the neutron to proton ratio decreases to approach more closely the stable value. In such a case the nucleus must decrease the value of N and increase the value of Z, which can be done by conversion of a neutron to a proton. When such a conversion occurs within a nucleus, *negatron emission* is the consequence, with creation and emission of a negative β-particle designated by β^- or ${}_{-1}^{0}e$. For example:

$${}^{116}_{49}\mathrm{In} \rightarrow {}^{116}_{50}\mathrm{Sn} + {}_{-1}^{0}e$$

If the N/Z ratio is too low for stability, then radioactive decay occurs in such a manner as to lower Z and increase N by conversion of a proton to neutron. This may be accomplished through *positron emission*, i.e. creation and emission of a positron (β^+ or ${}_{+1}^{0}e$), or by absorption by the nucleus of an orbital electron (*electron capture*, EC). Examples of these reactions are:

$${}^{116}_{51}\mathrm{Sb} \rightarrow {}^{116}_{50}\mathrm{Sn} + {}_{+1}^{0}e \qquad \text{and} \qquad {}^{195}_{79}\mathrm{Au} \xrightarrow{\mathrm{EC}} {}^{195}_{78}\mathrm{Pt}$$

Positron and electron capture are competing Processes with the probability of the latter increasing as the atomic number increases. Beta decay is properly used to designate all three processes, β^-, β^+, and EC. (The term "beta decay" without any specification usually only refers to negatron emission.)

Thus in the early part of the Periodic Table, unstable neutron deficient nuclides decay by positron emission, but for the elements in the platinum region and beyond, decay occurs predominantly by electron capture. Both processes are seen in isotopes of the elements in the middle portion of the Periodic Table.

The alternative to positron decay (or EC) is proton emission, which, although rare, has been observed in about 40 nuclei very far off the stability line. These nuclei all have half-lives $\lesssim 30$ s. For example: ^{115}Xe, $t_{1/2}$ (p) 18 s; proton/EC ratio, 3×10^{-3}.

We can understand why the N/Z ratio must increase with atomic number in order to have nuclear stability when we consider that the protons in the nucleus all must experience a repulsive Coulomb force. The fact that stable nuclei exist means that there must be an attractive force tending to hold the neutrons and protons together, and that this attractive nuclear force must be sufficient in stable nuclei to overcome the disruptive Coulomb force. Conversely, in unstable nuclei there is a net imbalance between the attractive nuclear force and the disruptive Coulomb force. As the number of protons increases, the total repulsive Coulomb force must increase. Therefore, to provide sufficient attractive force for stability the number of neutrons increases more rapidly than that of the protons.

Neutrons and protons in nuclei are assumed to exist in separate *nucleon orbitals* just as electrons are in electron orbitals in atoms. If the number of neutrons is much larger than the number of protons, the neutron orbitals occupied extend to higher energies than the highest occupied proton orbital. As N/Z increases, a considerable energy difference can develop between the last (highest energy) neutron orbital filled and the last proton orbital filled. As a consequence, the stability of the nucleus can be enhanced when a neutron in the highest neutron orbital is transformed into a proton in a lower energy proton orbital; see the example for $A = 12$ in Fig. 3.2. These questions of nuclear forces and the energy levels of nucleons are discussed more extensively in Chapter 6.

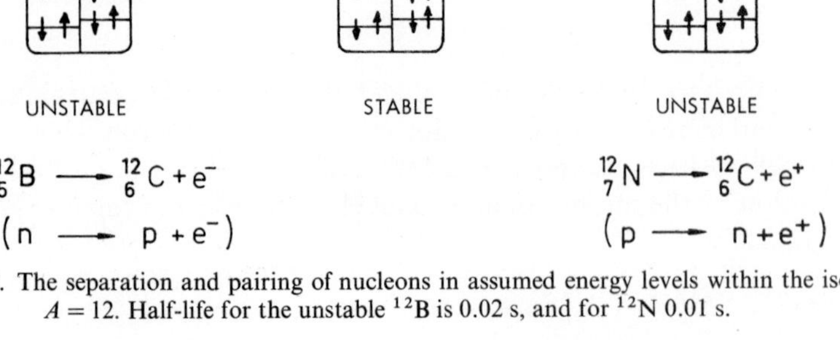

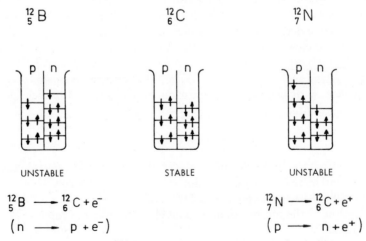

FIG. 3.2. The separation and pairing of nucleons in assumed energy levels within the isobar $A = 12$. Half-life for the unstable ^{12}B is 0.02 s, and for ^{12}N 0.01 s.

3.3. Mass defect

It was noted in Chapter 1 that the masses of nuclei are close to the mass number A. Using the mass of carbon-12 as the basis ($^{12}_{6}C = 12.000\,00$ u), the hydrogen atom and the neutron do not have exactly unit masses. We would expect that the mass M_A of an atom with mass number A would be given by the number of protons (Z) times the mass of the hydrogen atom ($\mathbf{M}_H$) plus the number of neutrons (N) times the mass of the neutron ($\mathbf{M}_n$), i.e.

$$M_A \approx Z\mathbf{M}_H + N\mathbf{M}_n \tag{3.1}$$

For deuterium with one neutron and one proton in the nucleus, we would then anticipate an atomic mass of

$$\mathbf{M}_H + \mathbf{M}_n = 1.007\,825 + 1.008\,665 = 2.016\,490 \text{ u}$$

When the mass of the deuterium atom is measured, it is found to be 2.014 102 u. The difference between the measured and calculated mass values, which in the case of deuterium equals $-0.002\,388$ u, is called the *mass "defect"* (ΔM_A):

$$\Delta M_A = M_A - Z\mathbf{M}_H - N\mathbf{M}_n \tag{3.2}$$

From the Einstein equation, $E = mc^2$, which is discussed further in Chapter 6, one can calculate that one atomic mass unit is equivalent to 931.5 MeV, where MeV is a million electron volts. The relationship of energy and mass would indicate that in the formation of deuterium by the combination of a proton and neutron, the mass defect of 0.002 388 u would be observed as the liberation of an equivalent amount of energy, i.e. $931.5 \times 0.002\,388 = 2.224$ MeV. The emission of this amount of energy (in the form of γ-rays) is observed when a proton captures a low energy neutron to form $^{2}_{1}H$. As a matter of fact, in this particular case, the energy liberated in the formation of deuterium has been used in the reverse calculation to obtain the mass of the neutron since it is not possible to determine directly the mass of the free neutron. With the definition (3.2) all stable nuclei are found to have negative ΔM_A values; thus the term "defect". (The student should not misinterpret this so that a negative mass defect is a mass excess!)

In nuclide (or isotope) tables the neutral atomic mass is not always given, but instead the *mass excess*, which we indicate as δ_A. It is defined as the difference between the measured mass and the mass number of the particular atom:

$$\delta_A = M_A - A \tag{3.3}$$

Mass excess values are either given in u (or, more commonly, in micromass units, μu) or in eV (usually keV). Table 3.2 contains a number of atomic masses, mass excess, and mass defect values, as well as some other information which is discussed in later sections.

When two elements form a compound in a chemical system, the amount of heat liberated is a measure of the stability of the compound. The greater this heat of formation (enthalpy, ΔH) the greater the stability of the compound. When carbon is combined with oxygen to form CO_2, it is found experimentally that 393 kJ of heat is evolved per mole of CO_2 formed. If we use the Einstein relationship, we can calculate that this would correspond to a total mass loss of 4.4×10^{-9} g for each mole of CO_2 formed (44 g). Although chemists do not doubt that this mass loss actually occurs, at present there are no instruments of sufficient sensitivity to measure such small changes.

TABLE 3.2. *Atomic masses and binding energies*

Element	Z	N	A	Atomic mass M_A (u)	Mass excess $M_A - A$ (μu)	Mass defect ΔM_A (μu)	Binding energy E_B (MeV)	E_B/A (MeV/nucleon)
n	0	1	0	1.008 665	8 665	0	–	–
H	1	0	1	1.007 825	7 825	0	–	–
D	1	1	2	2.014 102	14 102	– 2 388	2.22	(2.22)
T	1	2	3	3.016 049	16 049	– 9 106	8.48	2.83
He	2	1	3	3.016 030	16 030	– 8 285	7.72	2.57
He	2	2	4	4.002 604	2 604	– 30 376	28.29	7.07
He	2	4	6	6.018 893	18 893	– 31 417	29.26	4.87
Li	3	3	6	6.015 126	15 126	– 34 344	31.99	5.33
Li	3	4	7	7.016 005	16 005	– 42 130	39.24	5.60
Be	4	3	7	7.016 929	16 929	– 40 366	37.60	5.37
Be	4	5	9	9.012 186	12 186	– 62 439	58.16	6.46
Be	4	6	10	10.013 535	13 535	– 69 755	64.97	6.49
B	5	5	10	10.012 939	12 939	– 69 511	64.74	6.47
B	5	6	11	11.009 305	9 305	– 81 810	76.20	6.92
C	6	6	12	12.000 000	0 000	– 98 940	92.15	7.67
N	7	7	14	14.003 074	3 074	– 112 356	104.6	7.47
O	8	8	16	15.994 915	– 5 085	– 137 005	127.6	7.97
F	9	10	19	18.998 405	– 1 595	– 158 670	147.8	7.77
Ne	10	10	20	19.992 440	– 7 560	– 172 460	160.6	8.03
Na	11	12	23	22.989 771	– 10 229	– 200 284	186.5	8.11
Mg	12	12	24	23.985 042	– 14 958	– 212 838	198.2	8.25
Al	13	14	27	26.981 539	– 18 461	– 241 496	224.9	8.33
Si	14	14	28	27.976 929	– 23 071	– 253 931	236.5	8.44
P	15	16	31	30.973 765	– 26 235	– 282 250	262.9	8.48
K	19	20	39	38.963 710	– 36 290	– 358 265	333.7	8.56
Co	27	32	59	58.933 189	– 66 811	– 555 366	517.3	8.77
Zr	40	54	94	93.906 133	– 93 867	– 874 777	814.8	8.67
Ce	58	82	140	139.905 392	– 94 608	– 1 258 988	1172.7	8.38
Ta	73	108	181	180.948 007	– 51 993	– 1 559 038	1452.2	8.03
Hg	80	119	199	198.968 279	– 31 721	– 1 688 856	1573.0	7.90
Th	90	142	232	232.038 124	38 124	– 1 896 556	1766.6	7.62
U	92	144	236	236.045 637	45 637	– 1 922 202	1791.0	7.59
U	92	146	238	238.050 770	50 770	– 1 934 220	1801.7	7.57
Pu	94	146	240	240.053 882	53 882	– 1 946 758	1813.4	7.56

The energy changes in nuclear reactions are much larger. This can be seen if we use the relationship between electron volts and joules (or calories) in Appendix IV, and observe that nuclear reaction formulas and energies refer to single atoms (or molecules), while chemical reactions and equations refer to number of moles; we have:

$$1\ \text{eV/molecule} = \begin{cases} 1.6022 \times 10^{-19} \times 6.0220 \times 10^{23} = 96.48\ \text{kJ mole}^{-1}, \text{or} \\ 3.8268 \times 10^{-20} \times 6.0220 \times 10^{23} = 23.045\ \text{kcal mole}^{-1} \end{cases} \tag{3.4}$$

Thus, the formation of deuterium from a neutron and a hydrogen atom would lead to the liberation of 214.6×10^6 kJ (51.3×10^6 kcal) for each mole of deuterium atoms formed. By comparison, then, the nuclear reaction leading to the formation of deuterium is approximately half a million times more energetic than the chemical reaction leading to formation of CO_2.

It is not common practice to use mole quantities in considering nuclear reactions as the number of individual reactions under laboratory conditions is well below 6.02×10^{23}. Therefore, the energy and mass changes involved in the reaction of individual particles and nuclei are used in nuclear science.

3.4. Binding energy

The energy liberated in the formation of CO_2 from the elements, the heat of formation, is a measure of the stability of the CO_2 molecule. The larger the heat of formation the more stable the molecule since the more energy is required to decompose the molecule into its component atoms. Similarly, the energy liberated in the formation of a nucleus from its component nucleons is a measure of the stability of that nucleus. This energy is known as the *binding energy* (E_B) and has the same significance in nuclear science as the heat of formation has in chemical thermodynamics. We have seen that the binding energy of deuterium is 2.22 MeV. The ${}^4_2He$ nucleus is composed of 2 neutrons and 2 protons. The measured mass of the ${}^4_2He$ atom is 4.002 604 u. The mass defect is:

$$\Delta M_{He} = M_{He} - 2M_H - 2M_n = 4.002\,604 - 2 \times 1.007\,825 - 2 \times 1.008\,665 = -0.030\,376 \text{ u}$$

For the binding energy between the nucleons in a nucleus we write the simple relation

$$E_B\,(\text{MeV}) = -931.5\Delta M_A\,(\text{u}) \tag{3.5}$$

Thus the binding energy for ${}^4He$ is 28.3 MeV. It is quite unlikely that 2 neutrons and 2 protons would ever collide simultaneously to form a ${}^4He$ nucleus; nevertheless, this calculation is useful because it indicates that to break ${}^4He$ into its basic component nucleons would require at least 28.3 MeV.

A better indication of the relative stability of nuclei is obtained when the binding energy is divided by the total number of nucleons to give the binding energy per nucleon, E_B/A. For ${}^4He$ the value of E_B/A is 28.3/4 or 7.1 MeV, whereas for ${}^2H$ it is 2.22 for the single bond between the two nucleons. Clearly, the ${}^4He$ nucleus is considerably more stable than the ${}^2H$ nucleus. For most nuclei the values of E_B/A vary between the rather narrow range of 5–8 MeV. To a first approximation, therefore, E_B/A is relatively constant, which means that the total nuclear binding energy is roughly proportional to the total number of nucleons in the nucleus.

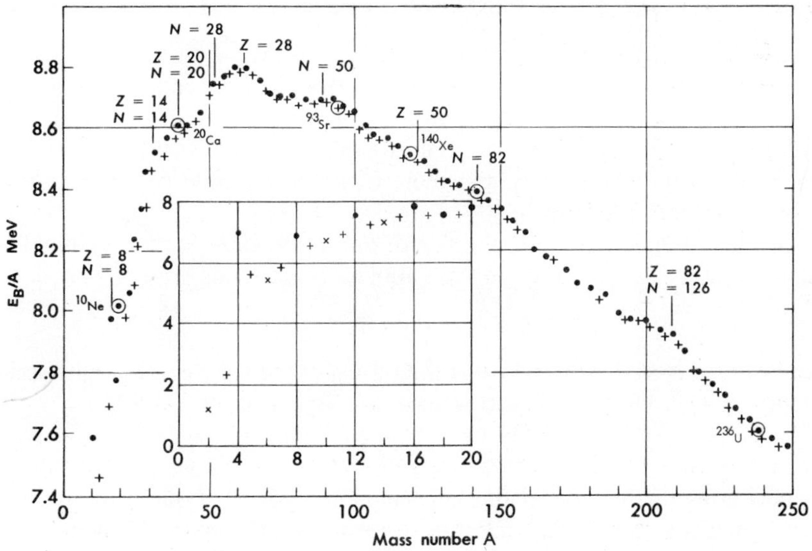

FIG. 3.3. Binding energy per nucleon (E_B/A) as a function of mass number (A). Each point is an average value for the nearest five isobars; ● refers to even–even, + odd–even and even–odd, and × to odd–odd isobars. (According to M. A. Preston.)

Figure 3.3 shows that the E_B/A values increase with increasing mass number up to a maximum around mass number 60 and then decrease. Therefore the nuclei with mass numbers in the region of 60, i.e. nickel, iron, etc., are the most stable. Also in this figure we have indicated the number of neutrons and protons which form especially stable configurations. This effect can be observed as small humps on the curve.

If two nuclides can be caused to react so as to form a new compound nucleus whose E_B/A value is larger than that of the reacting species, obviously a certain amount of binding energy would be released. The process which is called *fusion* is "exothermic" only for the nuclides of mass number below 60. As an example, we can choose the reaction

$$^{20}_{10}\text{Ne} + ^{20}_{10}\text{Ne} \rightarrow ^{40}_{20}\text{Ca}$$

From Fig. 3.3 we find that E_B/A for neon is about 8.0 MeV and for calcium about 8.6 MeV. Therefore, in the 2 neon nuclei $2 \times 20 \times 8.0 = 320$ MeV are involved in the binding energy, while $40 \times 8.6 = 344$ MeV binding energy are involved in the calcium nucleus. When 2 neon nuclei react to form the calcium nucleus the difference in the total binding energy of reactants and products is released, i.e. $344 - 320 = 24$ MeV.

Figure 3.3 also shows that a similar release of binding energy can be obtained if the elements with mass numbers greater than 60 are split into lighter nuclides with higher E_B/A values. Such a process, whereby a nucleus is split into two smaller nuclides, is known as *fission*. An example of such a fission process is the reaction

$$^{236}_{92}\text{U} \rightarrow ^{140}_{54}\text{Xe} + ^{93}_{38}\text{Sr} + 3\text{n}$$

The binding energy per nucleon for the uranium nucleus is 7.6 MeV, while those for the ^{140}Xe and ^{93}Sr are 8.4 and 8.7 MeV respectively. The amount of energy released in this fission reaction is approximately $140 \times 8.4 + 93 \times 8.7 - 236 \times 7.6 = 191.5$ MeV for each uranium atom fission.

3.5. Nuclear radius

Rutherford showed by his scattering experiments that the nucleus occupies a very small portion of the total volume of the atom. Roughly, the radii of nuclei vary from 1/10 000 to 1/100 000 of the radii of atoms. The common unit of atomic size is the Ångström (1 Å = 10^{-10} m), whereas the common unit of nuclear size is the fermi (1 fermi = 10^{-15} m). In the SI system these units are replaced by nanometer (1 nm = 10^{-9} m) and femtometer (1 fm = 10^{-15} m).

Experiments designed to study the size of nuclei indicate that the volumes of nuclei (V_n) are directly proportional to the total number of nucleons present, i.e.

$$V_n \propto A \qquad (3.6)$$

Since for a sphere $V \propto r^3$, where r is the radius of the sphere, for a spherical nucleus $r^3 \propto A$, or $r \propto A^{1/3}$. Using r_0 as the proportionality constant

$$r = r_0 A^{1/3} \qquad (3.7)$$

The implications of this is that the nucleus is composed of nucleons packed closely together with a constant density (about 0.2 nucleons fm^{-3}) from the center to the edge of the nucleus. This constant density model of the nucleus has been shown to be not completely correct, however. By bombarding nuclei with very high energy electrons or protons (up to $\gtrsim 1$ GeV) and measuring the scattering angle and particle energy,

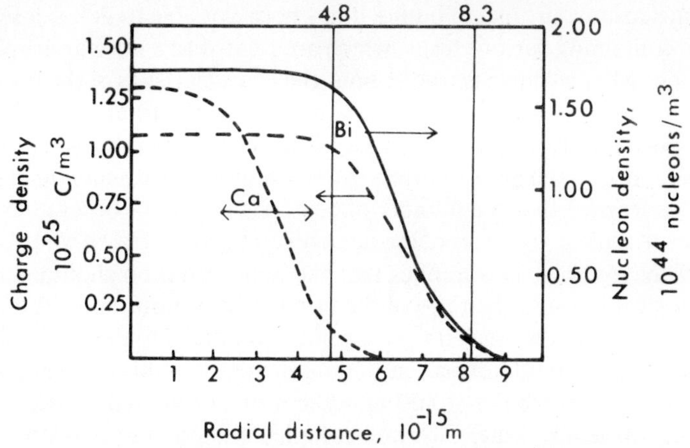

FIG. 3.4. Experimentally measured charge and nuclear density values for ^{40}Ca and ^{209}Bi as a function of the nuclear radius.

the charge and mass distribution density near the surface of the irradiated nucleus can be studied. These experiments have led to the conclusion that nuclei do not possess a uniform charge or mass distribution out to a sharp boundary, but rather have a fuzzy surface of decreasing density. The decreasing surface density is shown for a light and a heavy nucleus in Fig. 3.4. With an atomic number greater than 20 it has been found that a uniform charge and mass density exists over a short distance from the center of the nucleus, and this core is surrounded by a layer of decreasing density which seems to have a constant thickness of approximately 2.5 fm independent of mass number. In a bismuth nucleus, for example, the density remains relatively constant for approximately 5 fm then decreases steadily to one-tenth of that value in the next 2 fm (Fig. 3.4). It has also been found that not all nuclei are spherical, some being oblate and others prolate around the axis of rotation.

Despite the presence of this outer layer of decreasing density and the nonspherical symmetry, for most purposes it is adequate to assume a constant density nucleus with a sharp boundary. Therefore, use is made of the radius equation (3.7) in which the r_0 value may be assumed to be 1.4 fm. Using this relationship, we can calculate the radius of ^{40}Ca to be $r = 1.4 \times 10^{-15} \times 40^{1/3} = 4.79$ fm, and for ^{209}Bi to be 8.31 fm. These values are indicated in Fig. 3.4. For ^{80}Br a similar calculation yields 6.0 fm, while for ^{238}U the radius calculation is 8.7 fm. From these calculations we see that the radius does not change dramatically from relatively light nuclei to the heaviest.

3.6. Semiempirical mass equation

In preceding sections we have learned that the size as well as the total binding energy of nuclei are proportional to the mass number. These characteristics suggest an analogy between the nucleus and a drop of liquid. In such a drop the molecules interact with their immediate neighbors but not with other molecules more distant. Similarly, a particular nucleon in a nucleus is attracted by nuclear forces only to its adjacent neighbors. Moreover, the volume of the liquid drop is composed of the sum of the volumes of the molecules or atoms present since these are nearly incompressible. Again, as we learned above, this is similar to the behavior of nucleons in a nucleus. Based on the analogy of a nucleus to a droplet of liquid, it has been possible to derive a semiempirical mass equation containing various terms which are related to a nuclear droplet.

Let us consider what we have learned about the characteristics of the nuclear droplet. (a) First, recalling that mass and energy are equivalent, if the total energy of the nucleus is directly proportional to the total number of nucleons there should be a term in the mass equation related to the mass number. (b) Secondly, in the discussion of the neutron/proton ratios we learned that the number of neutrons could not become too large since the discrepancy in the energy levels of the neutron and proton play a role in determining the stability of the nucleus. This implies that the mass equation should contain a term which allows for variation in the ratio of the number of protons and neutrons. (c) Since the protons throughout the nucleus experience a mutual repulsion which affects the stability of nucleus, we should expect in the mass equation another term reflecting the repulsive forces of the protons. (d) Still another term is required to take into account that the surface nucleons, which are not completely surrounded by other nucleons, would not be totally saturated in their attraction. In a droplet of liquid this lack of saturation of surface forces gives rise to the effect of surface tension. Consequently, the term in the mass equation reflecting this unsaturation effect should be similar to a surface tension expression. (e) Finally, we have seen that nuclei with an even number of protons and neutrons are more stable than nuclei with an odd number of either type of nucleon and that the least stable nuclei are those for odd numbers of both neutrons and protons. This odd–even effect also must be included in a mass equation.

Taking into account these various factors, we can write a semiempirical mass equation. It is often more useful to write the analogous equation for the mass defect or binding energy of the nucleus, recalling (3.5). Such an equation, first derived by C. F. von Weizsäcker in 1935, would have the form:

$$E_B\,(\text{MeV}) = a_v A - a_a \frac{(N-Z)^2}{A} - a_c \frac{Z^2}{A^{1/3}} - a_s A^{2/3} \pm a_\delta A^{-1} \tag{3.8}$$

The first term in this equation takes into account the proportionality of the energy to the total number of nucleons (the volume energy); the second term, the variations in neutron and proton ratios (the asymmetry energy); the third term, the Coulomb force of repulsion for protons (the Coulomb energy); the fourth, the surface tension effect (the surface energy). In the fifth term, which accounts for the odd–even effect, a positive sign is used for even proton–even neutron nuclei and a negative sign for odd proton–odd neutron nuclei. For nuclei of odd A (even–odd or odd–even) this term has the value of zero. By comparison of this equation with actual binding energies of nuclei, one obtains a set of coefficients; such a set is

$$a_v = 14.0, \quad a_a = 19.3, \quad a_c = 0.59, \quad a_s = 13, \quad a_\delta = 130$$

With these coefficients the binding energy equations (3.2) and (3.5) give agreement within a few percent of the measured values for most nuclei of mass number greater than 40. Figure 3.5 shows the relative contribution of each term to E_B/A. The increase in E_B/A for light nuclei is seen to be due to the rapid decrease of the surface energy term at low A; also note the increasing effect of the disruptive Coulomb energy as A increases.

When the calculated binding energy is compared with the experimental binding energy, it is seen that for certain values of neutron and proton numbers, the disagreement is more serious. These numbers are related to the so-called "magic numbers", which

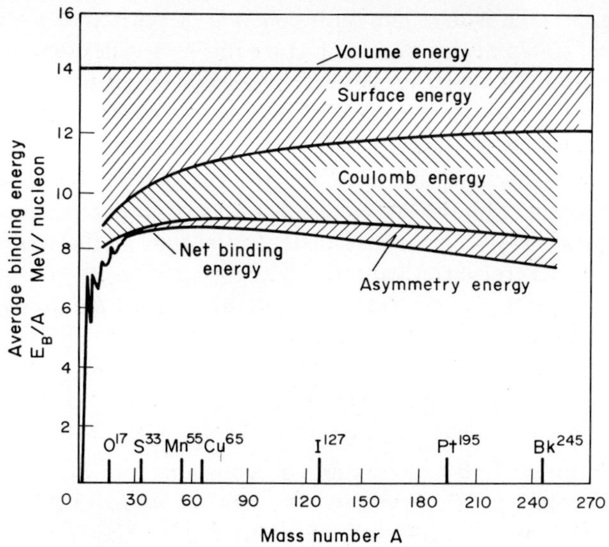

FIG. 3.5. The variation of the various terms in the semiempirical liquid drop binding energy equation as a function of A. (According to R. D. Evans.)

we have indicated in Figs. 3.1 and 3.3, whose recognition led to the development of the nuclear shell model described in a later chapter.

3.7. Valley of β-stability

If the semiempirical mass equation is written as a function of Z, remembering that $N = A - Z$, it reduces to a quadratic equation of the form

$$E_B = aZ^2 + bZ + c \pm dA^{-1} \tag{3.9}$$

where the terms a, b and c also contain A. This quadratic equation describes a parabola for constant values of A. Consequently, we would expect that for any family of isobars the masses should fall upon a parabolic curve. Such a curve is shown in Fig. 3.6. In

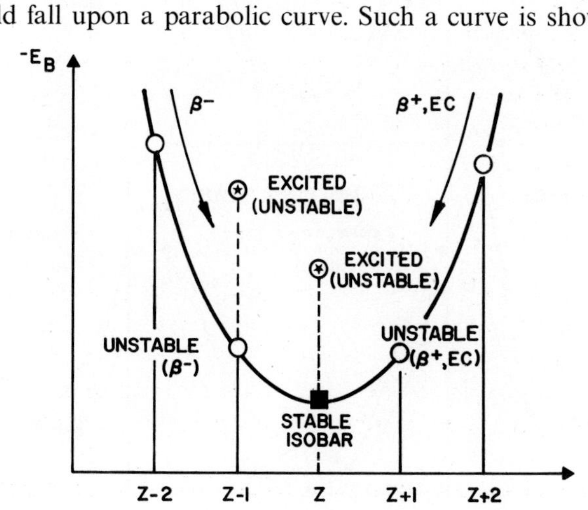

FIG. 3.6. Isobar cut across the stability valley showing schematically the position of different kinds of nuclei.

returning to Fig. 3.1, the isobar line with constant A but varying Z cuts diagonally through the line of stable nuclei. We can picture this as a valley, where the most stable nuclei lie at the bottom of it (cf. Figs. 3.1 and 3.6), while unstable nuclei lie up the valley sides as shown in Fig. 3.6. Any particular isobaric parabola can be considered as a cross-section of the valley of stability. The isobars located on the sides of the parabola (or slope of the valley) are unstable to radioactive decay to more stable nuclides lower on the parabola. Nuclides on the left hand side of the parabola (lower atomic numbers) are unstable to decay by β^- emission. Isobars to the right of the valley of stability are unstable to β^+ decay or electron capture.

At the bottom of the valley the isobars are stable against β^- decay. The maximum stability line can be calculated from (3.8) to be

$$Z = A\left(2 + \frac{a_c}{2a_a}A^{2/3}\right)^{-1} \tag{3.10}$$

and is shown in Fig. 3.1. By modifying the constants the fit can be improved in some areas but becomes poorer in others. For small A values (3.10) reduces to $Z = 2A$ or $N = Z$; thus the bottom of the stability valley follows the $N = Z$ line as indicated in Fig. 3.1 for the lighter nuclides.

A closer analysis of the theoretical relations for binding energy as a function of Z leads us to expect three different isobaric parabola depending on whether the nuclei are odd-A (even–odd or odd–even), odd–odd, or even–even (Fig. 3.7). In the first case, in which the mass number is odd, we find a single parabola (Fig. 3.7, (I)); whether β^-, β^+ or EC, beta decay leads to changes from odd–even to even–odd, etc. For even mass numbers one finds a double parabola as in Fig. 3.7 (II)–(V). As a conse-

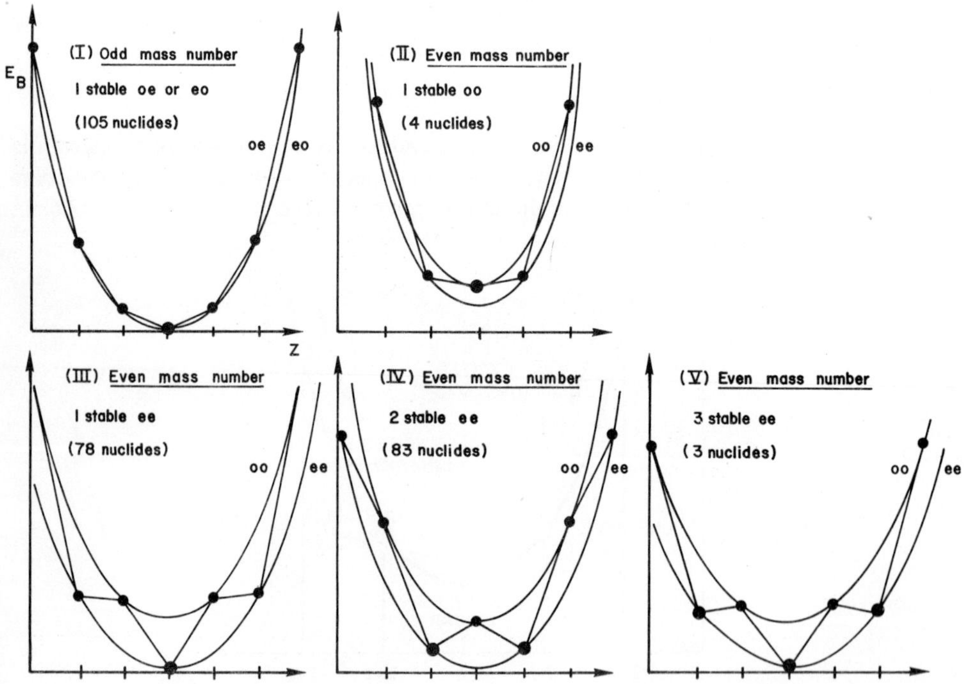

FIG. 3.7. Isobar parabolas for odd mass numbers (I: odd–even or even–odd nuclides) and for even mass numbers (cases II–V). The stable nuclides are indicated by heavier dots.

quence of the approximate nature of (3.8), where individual nuclear properties are not considered, the difference between the curves for the odd–odd and even–even nuclei may lead to alternatives with regard to the numbers of possible stable isobars. In fact it is possible to find three stable isobars (case V) although two (case IV) are more common. Although the odd–odd curve always must lie above the even–even curve, still an odd–odd nucleus may become stable, as is shown for case (II) in Fig. 3.7.

3.8. Other modes of instability

In this chapter we have stressed nuclear instability to beta decay. However, in §3.4 it was learned that very heavy nuclei are unstable to fission. There is also a possibility of instability to emission of α-particles in heavy elements and to neutron and proton emission. Figure 3.8 is an extension of Fig. 3.1 (axes are interchanged) to include all these properties.

Nuclei are unstable to forms of decay as indicated in Fig. 3.8. For example, making a vertical cut at $N = 100$, the instability from the top is: proton emission above the line, then, α-emission (for $N = 60$ it would instead be positron emission or electron capture, as these two processes are about equally probable), β-emission and, finally, neutron emission below the line. The field surrounding the stable nuclei shows the nuclei presently known (squares) and with calculated β-decay half-lives $\lesssim 10^{-8}$s (solid lines). For α-decay the figure indicates that for $A > 150$ ($Z \gtrsim 70$, $N > 80$) the nuclei are α-unstable, but in fact α-decay is commonly observed only above $A \gtrsim 200$. This is due to the necessity for the α-particle to pass over or penetrate the Coulomb barrier (cf. §6.7.3). Although neutron and proton emissions are possible energetically, they are not commonly observed as the competing β-decay processes are much faster.

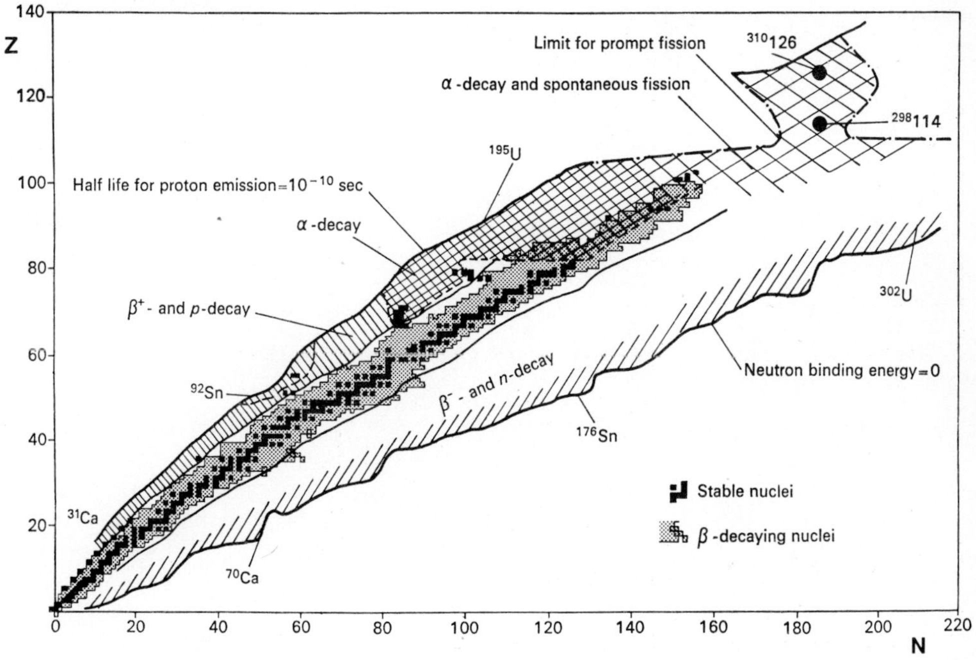

FIG. 3.8. Various modes of nuclear instability as a function of the number of protons (Z) and number of neutrons (N) in the nucleus. (According to I. Bergström.)

3.9. Exercises

3.1. Calculate the nucleon binding energy in ^{24}Mg from the atomic mass excess value in Table 3.2.

3.2. How many times larger is the nucleon binding energy in ^{24}Na than the electron binding energy when the ionization potential of the sodium atom is 5.14 V?

3.3. Assuming that in the fission of a uranium atom an energy amount of 200 MeV is released, how far would 1 g of ^{235}U drive a car which consumes 1 liter of gasoline (density 0.70 g cm^{-3}) for each 10 km? The combustion heat of octane is 5500 kJ $mole^{-1}$, and the combustion engine has an efficiency of 18%.

3.4. Estimate if fusion of deuterium into helium releases more or less energy per gram of material consumed than the fission of uranium.

3.5. When a neutron is captured in a nucleus, the mass number of the isotope increases one unit. In the following table mass excess values are given for three important isotope pairs:

^{235}U	40 906 keV	^{236}U	42 510 keV
^{238}U	47 291	^{239}U	50 579
^{239}Pu	48 573	^{240}Pu	50 190

If the average nucleon binding energy in this region is 7.57 MeV one can calculate the difference between this average binding energy and the one really observed in the formation of ^{236}U, ^{239}U, and ^{240}Pu. Calculate this difference. Discuss the possible significance of the large differences observed for $^{238}U/^{239}U$ pair as compared to the other pairs in terms of nuclear power.

3.6. With the semiempirical mass equation (3.8) estimate the binding energy per nucleon for ^{10}Be, ^{27}Al, ^{59}Co, and ^{235}U. Compare the results with the observed values in Table 3.2.

3.7. With eqn. (3.10) determine the atomic number corresponding to maximum stability for $A = 10, 27, 59$, and 239. Compare these results with the data in the isotope chart, Appendix D.

3.10. Literature

See Appendix A, §§ A.1 and A.2.

CHAPTER 4
Radioactive Decay

Contents

Introduction

Radioactive decay is a spontaneous nuclear transformation that has been shown to be unaffected by pressure, temperature, chemical form, etc. This insensitivity to extranuclear conditions allows us to characterize radioactive nuclei by their decay period and their mode and energy of decay without regard to their physical or chemical condition.

The period of radioactive decay is expressed in terms of the *half-life* ($t_{1/2}$), which is the time required for one-half of the radioactive atoms in a sample to undergo decay. In practice this is the time for the measured radioactive intensity (or simply, *radioactivity* of a sample) to decrease to one-half of its previous value (see Fig. 1.2). Half-lives vary from millions of years to fractions of seconds. While half-lives between a minute and a year are easily determined with fairly simple laboratory techniques, the determination of much shorter half-lives requires elaborate techniques with advanced instrumentation. The shortest half-life measurable today is about 10^{-14} s. Consequently, radioactive decay which occurs with a time period less than 10^{-14} s is considered to be instantaneous. At the other extreme, if the half-life of the radioactive decay exceeds 10^{15} y, the decay cannot be observed above the normal signal background present in the detectors. Therefore, nuclides which may have half-lives greater than 10^{15} y are considered to be stable to radioactive decay. It should be realized that 10^{15} y is about 10^5 times larger than the age of the universe.

Radioactive decay involves a transition from a definite quantum state of the original nuclide to a definite quantum state of the product nuclide. The energy difference between the two quantum levels involved in the transition corresponds to the decay energy. This decay appears in the form of electromagnetic radiation and as the kinetic energy of the products.

The mode of radioactive decay is dependent upon the particular nuclide involved. We have seen in Chapter 1 that radioactive decay can be characterized by α-, β-, and γ-radiation. *Alpha-decay* is associated with the emission of helium nuclei. *Beta-decay* is associated with the creation and emission of either negatrons or positrons and with the process of electron capture. *Gamma-decay* is the emission of electromagnetic radiation where the transition occurs between energy levels of the same nucleus. An additional mode of radioactive decay is that of *internal conversion* in which a nucleus loses its energy by interaction of the nuclear field with that of the orbital electrons, causing ionization of an electron instead of γ-ray emission. A mode of radioactive decay which is observed only in the heaviest nuclei is that of *spontaneous fission* in which the nucleus dissociates spontaneously into two roughly equal parts. This fission is accompanied by the emission of electromagnetic radiation and of neutrons. In the last decade also some unusual decay modes have been observed for nuclides very far off from the stability line, namely *neutron emission* and *proton emission.*

In the following, for convenience, we sometimes use an abbreviated form for decay reactions, as illustrated for the ^{238}U decay chain in §15:

$$^{238}\mathrm{U}(\alpha)^{234}\mathrm{Th}(\beta^-)^{234}\mathrm{Pa}(\beta^-)^{234}\mathrm{U}(\alpha), \text{ etc.,}$$

or, if half-lives are of importance:

$$^{238}\mathrm{U}(\alpha, 4.5 \times 10^9 \text{ y})^{234}\mathrm{Th}(\beta^-, 24 \text{ d})^{234}\mathrm{Pa}(\beta^-, 1.1 \text{ min})^{234}\mathrm{U}(\alpha, 2.5 \times 10^5 \text{ y}), \text{ etc.}$$

In the following chapter we discuss the energetics of the decay processes based on nuclear binding energy considerations and simple mechanics, then we consider the kinetics of the processes. In Chapter 6 where the internal properties of the nuclei are studied, the explanations of many of the phenomena discussed in this chapter are presented in terms of simple quantum mechanical rules.

4.1. Conservation laws

In radioactive decay—as well as in other nuclear reactions—a number of *conservation laws* must be fulfilled. Such laws place stringent limitations on the events which may occur.

Consider the reaction

$$X_1 + X_2 \rightarrow X_3 + X_4 \tag{4.1}$$

where X represents any nuclear or elementary particle. In induced nuclear reactions X_1 may be the bombarding particle (e.g. a ^{4}He atom in a beam of α-particles) and X_2 the target atom (e.g. ^{14}N atoms), and X_3 and X_4 the products formed (e.g. ^{1}H and ^{17}O).

Sometimes only one product is formed, sometimes more than two. In radioactive decay several products are formed; reaction (4.1) is then better written $X_1 \rightarrow X_2 + X_3$. For generality, however, we discuss the conservation laws for the case (4.1).

For the general reaction (4.1):

(a) The *total energy* of the system must be constant, i.e.

$$E_1 + E_2 = E_3 + E_4 \tag{4.2}$$

where E includes all energy forms: mass energy (§ 7.2), kinetic energy, electrostatic energy, etc.

(b) The *linear momentum*

$$p = mv \tag{4.3}$$

must be conserved in the system, and thus

$$p_1 + p_2 = p_3 + p_4 \tag{4.4}$$

The connection between kinetic energy E_{kin} and linear momentum is given by the relation

$$E_{\mathrm{kin}} = p^2/2m \tag{4.5}$$

(c) The *total charge* of the system must be constant, i.e.

$$Z_1 + Z_2 = Z_3 + Z_4 \tag{4.6}$$

where the charge is in electron units.

(d) The *mass number* in the system must be constant, i.e.

$$A_1 + A_2 = A_3 + A_4 \tag{4.7}$$

(e) The *total nuclear angular momentum* p_{I} of the system must be conserved, i.e.

$$(p_{\mathrm{I}})_1 + (p_{\mathrm{I}})_2 = (p_{\mathrm{I}})_3 + (p_{\mathrm{I}})_4 \tag{4.8}$$

Since there exist two types of angular momentum, one caused by orbital movement of the individual nucleons and the other due to the intrinsic spin of the nucleons (internal angular momentum), a more practical formulation of (4.8) is

$$\Delta I = \mathrm{I}_3 + \mathrm{I}_4 - \mathrm{I}_1 - \mathrm{I}_2 \tag{4.9}$$

where I is the (total) *nuclear spin quantum number*. The quantum rule is

$$\Delta I = 1, 2, 3, \ldots \tag{4.10}$$

i.e. the change of nuclear spin in a reaction must have an integral value.

The three first laws are general in classical physics; the last two refer particularly to nuclear reactions. In Chapters 5 and 6 other conservation laws are discussed for nuclear reactions, but these are less important in radioactive decay.

4.2. Alpha decay

4.2.1. *Detection*

Alpha particles cause extensive ionization in matter. If the particles are allowed to pass into a gas, the electrons released by the ionization can be collected on a positive electrode to produce a pulse of current. *Ionization chambers* and *proportional counters* are instruments of this kind, which permit the individual counting of each α-particle emitted by a sample. Alpha particles interacting with matter may also cause molecular excitation, which can result in fluorescence. This fluorescence—or *scintillation*—allowed the first observation of individual nuclear particles. The earlier detection screens, made of a thin layer of zinc sulphide on a glass plate, where the scintillations were observed by microscope, have been replaced by modern devices described in Chapter 17.

4.2.2. *Decay energy*

Alpha decay is observed for the elements heavier than lead and for a few nuclei as light as the lanthanide elements. It can be written symbolically as

$$ {}^{A}_{Z}X \rightarrow {}^{A-4}_{Z-2}X + {}^{4}_{2}He \tag{4.11}$$

Several examples are given in Chapter 1, and can be found in the radioactive decay series in Fig. 1.4.

The decay energy can be calculated from the known atomic masses, because the binding energy released (spontaneous decay processes are always exoergic) corresponds to a disappearance of mass, cf. eqns. (3.2) and (3.5). This energy is also called the *Q-value of the reaction*

$$Q\,(\mathrm{MeV}) = -931.5\Delta M\,(\mathrm{u}) \tag{4.12}$$

For α-decay we can define the Q-value as

$$Q_\alpha = -931.5(M_{Z-2} + M_{\mathrm{He}} - M_Z) \tag{4.13}$$

We always write the products minus the reactants within parentheses. A decrease in total mass in α-decay means a release of energy. The minus sign before the constant 931.5 is necessary to make Q positive for spontaneous decay.

An example will show the use of this equation. For the decay reaction $^{238}U \rightarrow {}^{234}Th + {}^{4}He$, the mass values for ^{238}U and ^{4}He are in Table 3.2; for ^{234}Th it is 234.043 583. Thus we obtain $Q_\alpha = -931.5\,(234.043\,583 + 4.002\,604 - 238.050\,770) =$ 4.269 MeV.

If the products are formed in their ground states, which is usually true for α-decay, the total decay energy is partitioned into the kinetic energies of the daughter nucleus (E_{Z-2}) and the helium nucleus (E_α):

$$Q_\alpha = E_{Z-2} + E_\alpha \tag{4.14}$$

Because of conservation of energy (4.2) and momentum (4.4)

$$E_{Z-2} = Q_\alpha M_\alpha / M_Z \tag{4.15}$$

and

$$E_\alpha = Q_\alpha M_{Z-2} / M_Z \tag{4.16}$$

From these equations we can calculate the kinetic energy of the ^{234}Th daughter to be 0.072 MeV, while that of the α-particle is 4.197 MeV. Because of the large mass difference between the α-emitting nucleus and the helium atom, almost all of the energy is carried away with the α-particle.

Although the kinetic energy of the daughter nucleus is small in comparison with that of the α-particle, it is large (72,000 eV) in comparison with chemical binding energies (< 5 eV). Thus the recoiling daughter easily breaks all chemical bonds by which it is bound to other atoms.

In 1904 it was observed by H. Brooks that measurements on ^{218}Po (RaA), obtained from radon, led to a contamination of the detection chamber by ^{214}Pb (RaB) and ^{214}Bi (RaC). This was explained by Rutherford as being due to daughter recoil in the α-decay of ^{218}Po in the sequence:

$$^{222}\text{Rn}(\alpha, 3.8\text{ d})^{218}\text{Po}(\alpha, 3.05\text{ min})^{214}\text{Pb}(\beta^-, 27\text{ min})^{214}\text{Bi}(\beta^-, 20\text{ min})\ldots$$

This recoil led to ejection of ^{214}Pb into the wall of the instrument. The use of the *recoil* of the daughter to effect its separation was employed by O. Hahn beginning in 1909 and played a central role in elucidating the different natural radioactive decay chains.

Alpha-decay energies are most precisely measured in magnetic spectrometers. From (2.5) and (2.10) it is calculated that

$$E_\alpha = \frac{2\mathbf{e}^2 B^2 r^2}{m_{\text{He}}} \tag{4.17}$$

From knowledge of the values of $\mathbf{e}, m_{\text{He}}$, B, and r, E_α can be calculated. A more common technique is to use *silicon surface barrier solid state detectors* combined with pulse height analyzers ("α-spectrometers", Chapter 17).

4.3. Beta decay

4.3.1. *Detection*

Energetic electrons cause ionization and molecular excitation in matter, although the effect is weaker and more difficult to detect than for α-particles. As a result the effect must be amplified for counting individual β-particle. Ionization is used in *proportional* and *Geiger counters.* Scintillation counting can also be used with various detector systems (Chapter 17).

4.3.2. *The β-decay process*

The radioactive decay processes which are designated by the general name of β-decay include negatron emission (β^- or ${}_{-1}^{0}$e), positron emission (β^+ or ${}_{+1}^{0}$e) and electron capture (EC). If we use the β-decay of ^{137}Cs as an example, we can write

$$^{137}\text{Cs} \rightarrow {}^{137\text{m}}\text{Ba} + \beta^-$$

This β-decay must occur between discreet quantum levels of the parent nuclide ^{137}Cs and the daughter nuclide $^{137\text{m}}$Ba.

The quantum levels of nuclei are characterized by several quantum numbers, an important one being the *nuclear spin.* The spin value for the ^{137}Cs ground state level is 7/2, while that of $^{137\text{m}}$Ba is 11/2. The electron emitted is an elementary particle of spin 1/2. In nuclear reactions the nuclear angular momentum must be conserved (4.8), which means that in radioactive decay processes the difference in total spin between reactant and products must be an integral value (4.10). Inspection of our example shows that this conservation of spin rule is violated if the reaction is complete as we have written it. The sum of the spin of the $^{137\text{m}}$Ba and of the electron is 11/2 + 1/2 or 6, while

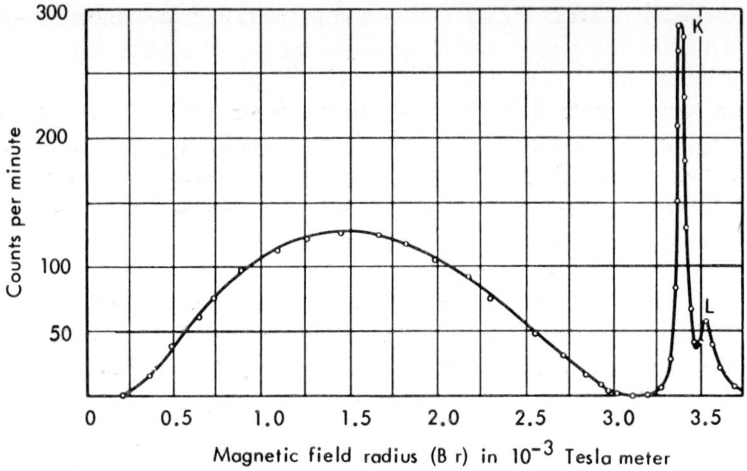

FIG. 4.1. Spectrum for electrons emitted by ^{137}Cs as observed with a magnetic spectrometer. *Br* (Gcm) is proportional to the square root of the energy (according to (4.18)). (From W. Gentner, H. Maier-Leibnitz, and W. Bothe.)

that of the ^{137}Cs is 7/2. Therefore, the change in spin (ΔI) in the process would seem to be 5/2 spin units. Inasmuch as this is a nonintegral value, it violates the rule for conservation of angular momentum. Before accounting for this discrepancy let us consider another aspect of β-decay which seems unusual.

Figure 4.1 shows the β-particle spectrum of ^{137}Cs as obtained by a magnetic spectrometer. The β-particle energy is calculated by the relation

$$E_\beta = e^2 B^2 r^2 / 2m_e \tag{4.18}$$

where m_e is the electron relativistic mass. The spectrum shows the number of β-particles as a function of *Br*, which is proportional to E_β through (4.18). We observe a continuous distribution of energies. This seems to disagree with our earlier statement that decay occurs by change of one nucleus in a definite energy state to another nucleus also in a definite energy state. The two sharp peaks designated K and L at the high energy end of the spectrum are not related to the beta spectrum itself and are discussed later in the chapter (§ 4.4).

4.3.3. *The neutrino*

This problem of "wrong" spin change and the continuous "nonquantized" spectrum led W. Pauli to the assumption that β-decay involves emission of still another particle which has been named the *neutrino* and given the symbol ν. The neutrino has a spin value of 1/2, an electrical charge of 0, and a mass of 0. It is therefore somewhat similar to the photon, which has neither mass, electric charge nor spin. However, while the photon readily interacts with matter, the neutrino does not. In fact the interaction is so unlikely that a neutrino has a very high probability of passing through the entire earth without reacting.

The spin attributed to the neutrino allows conservation of angular momentum; in our example, the total spin of the products would be $11/2 + 1/2 + 1/2$ or 13/2, and when the spin of ^{137}Cs, 7/2, is subtracted from this the result is 6/2 which is an acceptable integral value. Thus the decay reaction above is incomplete and must be written

$$^{137}\text{Cs} \rightarrow {}^{137\text{m}}\text{Ba} + \beta^- + \bar{\nu}$$

Notice we have replaced ν by $\bar{\nu}$, which is the designation of the *antineutrino*. Beta-decay theory has shown that antineutrinos $\bar{\nu}$ are emitted in negatron decay, and "regular" neutrinos ν in positron decay. We can consider the particles identical; cf. §5.4. Because of the extremely low probability of interaction of neutrinos with matter, they are often omitted in writing β-decay reactions.

The neutrino theory also explains the energy spectrum in β-decay. However, this necessitates the introduction of another important nuclear concept, that of *relativistic mass* and *rest mass*. In 1901 S. G. Kaufmann showed in experiments that the mass of an electron m seemed to increase when its velocity v approached that of the speed of light **c**. It was found that this increase followed an expression

$$m = m^0(1 - v^2/\mathbf{c}^2)^{-1/2} \tag{4.19}$$

based on H. Lorentz's studies of the relation between distance, speed of light, and time. m^0 is the rest mass of the particle (at velocity $v = 0$), while m is referred to as the relativistic mass. This relation is valid for any moving object, macrosopic or microscopic, whether it is a "particle" or a "wavepacket". Figure 4.2 shows $v/\mathbf{c}$ as a function of the kinetic energy of the particle, $E_{\text{kin}} = \frac{1}{2}mv^2$ (2.5).

If the parentheses in (4.19) is expanded by means of the binomial theorem of algebra, it approximates to

$$m = m^0 + \frac{\frac{1}{2}m^0v^2}{\mathbf{c}^2} \tag{4.20}$$

The last term is approximately the kinetic energy of the particle (2.5) divided by $\mathbf{c}^2$, and thus

$$m \approx m^0 + E_{\text{kin}}/\mathbf{c}^2 \tag{4.21}$$

The increase in mass, $\Delta m = m - m^0$, because of the kinetic energy of the particle,

$$E_{\text{kin}} = \Delta m\mathbf{c}^2 \tag{4.22}$$

was generalized by A. Einstein in the special theory of relativity, leading (after more

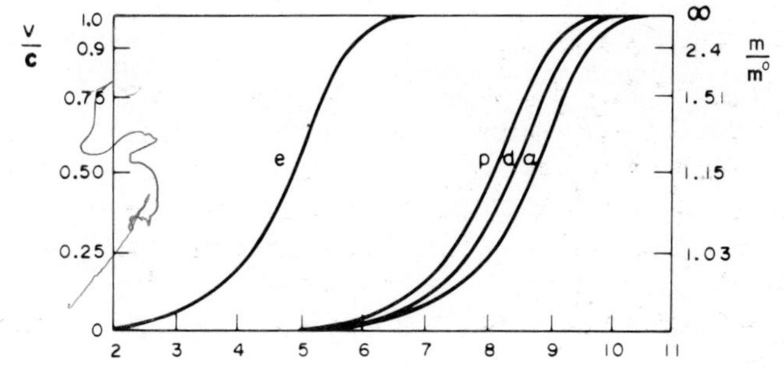

FIG. 4.2. Relativistic masses m for some common nuclear particles, divided by their rest masses $\mathbf{m}^0$, as a function of the kinetic energy of the particle.

detailed calculations) to the well known mass–energy relationship

$$E = m\mathbf{c}^2 \tag{4.23}$$

which we already have applied in the discussion of the nuclear binding energy (3.5).

When a neutrino is ejected from the nucleus it carries away energy, which may be considered to be kinetic in nature. Thus, according to (4.21) the neutrino can also be considered to have a mass > 0, and obviously also a momentum $p = mv$. Recoil studies of β-decay have proven this to be true.

In order to correctly apply (4.18) for the calculation of the β-decay energy, the relativistic electron mass must be used; as is seen from Fig. 4.2, already at 0.1 MeV, the relativistic mass of the electron is 15% larger than the rest mass $\mathbf{m}_e^0$. (In the following the rest masses of the electron, neutron, etc., will be denoted simply as $\mathbf{m}_e$, $\mathbf{m}_n$, etc.)

The energy released in β-decay is distributed between the neutrino, the electron, and the recoil of the daughter nucleus. This latter will be much smaller than the first two and can be neglected in a first approximation (§4.3.7). Therefore, the total β-decay energy can be considered to be distributed between the neutrino and the electron. For the decay $^{137}Cs \rightarrow {}^{137m}Ba$ it can be shown that the total decay energy Q_β is 0.514 MeV. This is also termed E_{max}. The neutrino energy spectrum is the complement of the β-particle energy spectrum. If the energy of the electron is 0.400 MeV, that of the neutrino is 0.114 MeV. If the electron energy is 0.114 MeV, the neutrino energy is 0.400 MeV.

In β^- decay the average value of the β-particle energy is approximately $0.3\,E_{max}$. In positron emission, the average energy of the positron particle is approximately $0.4\,E_{max}$.

4.3.4. *Negatron decay*

This process can be written symbolically as follows:

$$^{A}_{Z}X \rightarrow {}_{Z+1}^{\ A}X + {}_{-1}^{\ 0}\beta + \bar{\nu} \tag{4.24}$$

However, if we take the electrons into account, the neutral parent atom has Z orbital electrons, while the daughter atom, with a nuclear charge of $Z + 1$, must capture an electron from the surroundings, in order to become neutral:

$${}_{Z+1}^{\ A}X^+ + e^- \rightarrow {}_{Z+1}^{\ A}X \tag{4.25}$$

Moreover, since the negatron emitted provides an electron to the surroundings, the total electron balance remains constant. As a result, in the calculation of the decay energy it is not necessary to include the mass of the emitted β-particle as the use of the mass of the neutral daughter atom includes the extra electron mass. The equation for calculating the Q-value in negatron decay is thus:

$$Q_{\beta^-} = -931.5\,(M_{Z+1} - M_Z) \tag{4.26}$$

As an example we can take the decay of a free neutron in a vacuum; it transforms spontaneously with a half-life of 10.6 min to a proton.

$$^{1}_{0}n \rightarrow {}^{1}_{1}H + {}_{-1}^{\ 0}e$$

The Q-value for this reaction is

$$Q_{\beta^-} = -931.5(1.007\,825 - 1.008\,665) = 0.782 \text{ MeV}$$

4.3.5. *Positron decay*

Positron decay can be written symbolically as

$$^{A}_{Z}\text{X} \rightarrow {}_{Z-1}^{A}\text{X}^{-} + {}_{+1}^{0}\beta + \nu \rightarrow {}_{Z-1}^{A}\text{X} + {}_{-1}^{0}\text{e} + {}_{+1}^{0}\beta + \nu \qquad (4.27)$$

Here we must consider the net atomic charges. The daughter nucleus has an atomic number one less than the parent. This means that there will be one extra electron mass associated with the change in atomic number. Moreover, an electron mass must also be included for the positive electron emitted. When $^{22}_{11}$Na decays to $^{22}_{10}$Ne, there are 11 electrons included in the $^{22}_{11}$Na atomic mass but only 10 in the $^{22}_{10}$Ne atomic mass. Consequently, an extra electron mass must be added on the product side in addition to the electron mass associated with the positron particle. The calculation of the Q-value must therefore include two electron masses beyond that of the neutral atoms of the parent and daughter

$$Q_{\beta^+} = -931.5(M_{Z-1} + 2\mathbf{M}_{\text{e}} - M_Z) \qquad (4.28)$$

Each electron mass has an energy equivalent to 0.511 MeV, since 931.5 × 0.000 549 = 0.511.

Consider the calculation of the Q-value for the reaction

$$^{13}_{7}\text{N} \rightarrow {}^{13}_{6}\text{C} + \beta^+$$

For this reaction we have

$$Q_{\beta^+} = -931.5(13.003\,354 - 13.005\,738) - 2 \times 0.511 = 1.20\ \text{MeV}$$

4.3.6. *Electron capture*

The EC decay process can be written symbolically

$$^{A}_{Z}\text{X} \xrightarrow{\text{EC}} {}_{Z-1}^{A}\text{X} \qquad (4.29)$$

The captured electron comes from one of the inner orbitals of the atom. Depending on the electron shell from which the electron originates, the process is sometimes referred to as K-capture, L-capture, etc. The probability for the capture of an electron from the K-shell is several times greater than that for the capture of an electron from the L-shell, since the wave function of K-electrons is substantially larger at the nucleus than that of L-electrons. Similarly, the probability of capture of electrons in higher order shells decreases with the quantum number of the electron shell.

The calculation of the decay energy in electron capture follows the equation

$$Q_{\text{EC}} = -931.5(M_{Z-1} - M_Z) \qquad (4.30)$$

Note that like the case of the negatron decay, it is not necessary to add or subtract electron masses in the calculation of the Q-value in EC. An example of EC is the decay of ^{7}Be to ^{7}Li for which it is possible to calculate that the Q-value is 0.861 MeV. This reaction is somewhat exceptional since for neutron deficient nuclei with values of Z below 30, positron emission is the normal mode of decay. Electron capture is the predominant mode of decay for neutron deficient nuclei whose atomic number is greater than 80. The two processes compete to differing degrees for the nuclei between atomic numbers 30 and 80. Electron capture is observed through the emission of electrons from secondary reactions occurring in the electron shell because of the elemental change (see §4.7).

4.3.7. *Daughter recoil*

If the β-particle and the neutrino are emitted with the same momentum but in opposite direction, the daughter nucleus experiences no recoil. On the other hand, if they are both emitted in the same direction, or if all the energy is carried away with one of the particles, the daughter experiences maximum recoil. The daughter therefore recoils with kinetic energies from zero up to a maximum value (when the β-particle is emitted with maximum energy). We can therefore write

$$Q_\beta = E_d + E_{max} \tag{4.31}$$

where E_d is the recoil energy of the daughter nucleus. From the laws of conservation of energy and momentum, and taking the relativistic mass changes of the electron into account, one finds that the daughter recoil energy is

$$E_d = \frac{m_e \cdot E_{max}}{m_d} + \frac{E_{max}^2}{m_d c^2} \tag{4.32}$$

The recoil energy is usually ~ 100 eV, which still is sufficient for causing atomic rearrangements in surrounding molecules. In the decay of ^{14}C (to N), E_{max} is 0.155 MeV, which gives $E_d = 7$ eV. However, by labeling ethane, $^{14}CH_3\,^{14}CH_3$, with ^{14}C in both C positions, it was found that $^{14}CH_3NH_2$ was formed in 50% of the cases when one of the ^{14}C atoms in ethane had decayed, although the C ≡ N bond strength is only 2.1 eV. Most of the decays occur with less than the maximum recoil energy, which also can be averaged over the whole molecule. The small recoil also explains why decay reactions like

$$^{127}TeO_3{}^{2-} \longrightarrow {}^{127}IO_3{}^{-} + \beta^-$$

and

$$^{52}MnO_4{}^{-} \longrightarrow {}^{52}CrO_4^{2-} + \beta^+$$

are possible, even when E_d is tens of electron volts. However, secondary effects tend to cause the chemical bond to break following radioactive decay (see §4.7).

4.4. Gamma emission and internal conversion

Frequently, α- and β-decay leave the daughter nucleus in an excited state. This excitation energy is removed either by γ-ray emission or by a process called internal conversion.

The α-emission spectrum of ^{212}Bi is shown in Fig. 4.3. It is seen that the majority of the α-particles have an energy of 6.04 MeV, but a considerable fraction (about 30%) of the α-particles have higher or lower energies. This can be understood if we assume that the decay of parent ^{212}Bi leads to excited levels of daughter ^{208}Tl. This idea is supported by measurements showing the emission of γ-rays of energies which exactly correspond to the difference between the highest α-energy, 6.08 MeV, and the lower ones. For example, an ~ 0.32 MeV γ accounts for the 5.76 MeV α (6.08 − 5.76 = 0.32). The excited levels of ^{208}Tl are indicated in the insert of Fig. 4.3.

Gamma rays produce very low density ionization in gases so they are not usually counted by ionization, proportional, or Geiger counters. However, the fluorescence produced in crystals such as sodium iodide make scintillation counting of γ-rays efficient. Gamma ray spectra can be measured with very high precision using solid state semi-

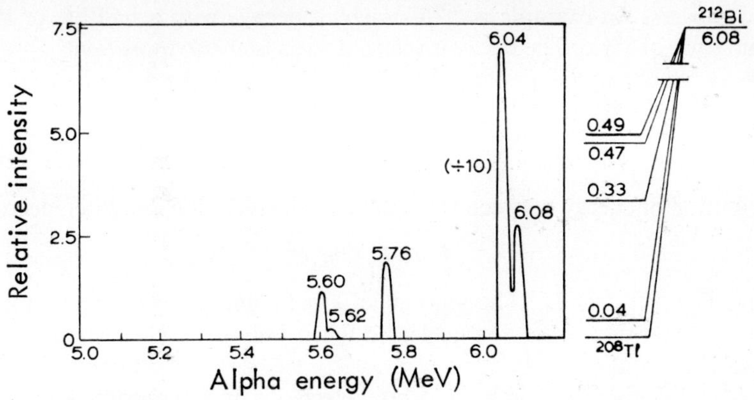

FIG. 4.3. Alpha energy spectrum from $^{212}Po \rightarrow {}^{208}Tl$. (According to E. B. Paul.)

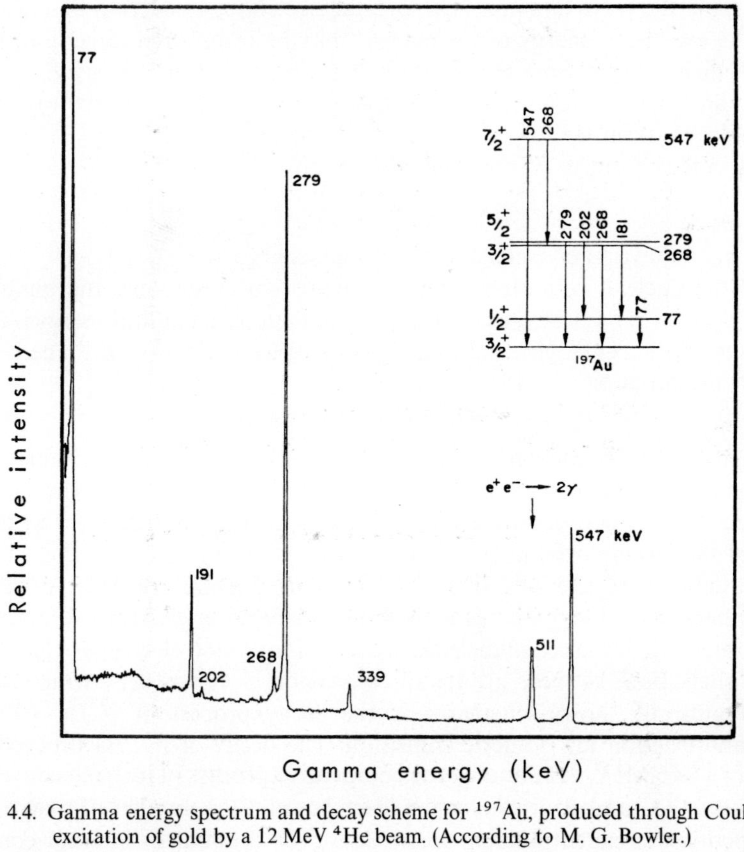

FIG. 4.4. Gamma energy spectrum and decay scheme for ^{197}Au, produced through Coulomb excitation of gold by a 12 MeV ^{4}He beam. (According to M. G. Bowler.)

conductor detectors (Chapter 16). Figure 4.4 shows such a spectrum for the decay of various excited states of ^{197}Au.

In the great majority of cases the emission of the γ-ray occurs immediately after α- or β-decay, i.e. within $\lesssim 10^{-12}$ s, but in some instances the nucleus may remain in the higher energy states for a measurable length of time. The longer-lived excited nuclei are called *isomers*. An example is ^{60m}Co, which decays with a half-life of 10.5 min to the ground state of ^{60}Co. The decay is referred to as *isomeric transition*.

The decay energy in γ-emission is distributed between the γ-ray quantum (E_γ) and the kinetic energy of the recoiling product nuclei (E_d). We can therefore write

$$Q_\gamma = E_d + E_\gamma \tag{4.33}$$

The distribution of energy between the γ-ray and the recoiling daughter, according to

$$E_d = E_\gamma^2/2m_d c^2 \tag{4.34}$$

shows that $E_d < 0.1\%$ of E_γ. The amount of kinetic energy of the recoiling nuclide is therefore so trivial that it may be neglected when only the γ-ray energy is considered; cf. exercise 4.4.

Gamma rays can interact with the orbital electrons of other atoms, so that the latter are expelled from that atom with a certain kinetic energy (see Chapter 14). A similar process, called *internal conversion*, can occur *within* the atom undergoing radioactive decay. Because the wave function of an orbital electron may overlap that of the excited nucleus, the excitation energy of the nucleus may be transferred directly to the orbital electron, which escapes from the atom with a certain kinetic energy E_e. No γ-ray is emitted in an internal conversion process; it is an alternate mode to γ-ray emission of de-excitation of nuclei.

Internal conversion can be represented symbolically as

$$^{Am}_{Z}X \longrightarrow {}^{A}_{Z}X^+ + {}^{0}_{-1}e \longrightarrow {}^{A}_{Z}X \tag{4.35}$$

where we again have to consider the net atomic charge.

Part of the nuclear excitation energy is required to overcome the binding energy E_{be} of the electron in its electronic orbital.† The remaining excitation energy is distributed between the recoiling daughter nucleus and the ejected electron E_e. The relationship is given by the equation

$$Q_\gamma - E_{be} = E_d + E_e \tag{4.36}$$

The ejected electron, known as the *conversion electron*, normally originates from an inner orbital, since their wave functions have greater overlap with the nucleus. It is to be noted that the conversion electrons are monoenergetic. Inasmuch as the binding energies of the atomic orbitals are different, the values of E_e reflect the differences in the electronic binding energies. In Fig. 4.1 two sharp peaks are observed just beyond E_{max}. The first peak, designated as K, is due to conversion electrons originating in the K atomic shell, while the peak labeled L is due to conversion electrons originating in the L atomic shell. Both of these groups of conversion electrons arise from the decay of ^{137m}Ba. Figure 4.5 shows schematically the decay process of $^{137}Cs \rightarrow {}^{137}Ba + \beta^-$; IT is an abbreviation for isomeric transition. The decay of ^{137m}Ba proceeds both by emission of a 0.66 MeV γ-ray and by the competitive process of internal conversion. The ratio between the number of conversion electrons and the number of γ-rays emitted in this competition is called the *conversion ratio*. The amount of internal conversion is not indicated in decay schemes like Fig. 4.5.

If we denote the conversion ratio as α_i, it is equal to the ratio of K-electrons ejected (which we may denote with I_{eK}) to that of gamma quantas emitted (I_γ):

$$\alpha_K = I_{eK}/I_\gamma \tag{4.37}$$

† Electron-binding energies are tabulated in standard physics tables.

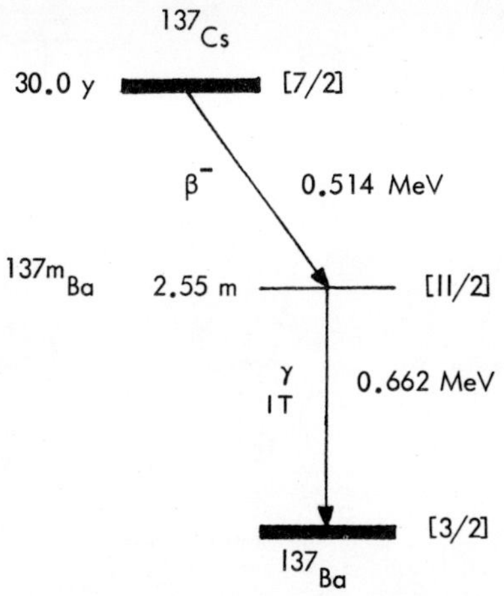

FIG. 4.5. Decay scheme of ^{137}Cs. IT stands for isomeric transition. Fractional figures within square brackets are nuclear spin values.

Usually $\alpha_K < 0.1$. Also $\alpha_K > \alpha_L > \alpha_M$, etc. For ^{137m}Ba the ratio of K-electron/L-electrons emitted is about 5 while the value of α_K is 0.094.

It can be shown that the energy of the recoiling nucleus (E_d, eqn. (4.36)) is much smaller than the kinetic energy of the ejected electron E_e and may be ignored. The mathematical expression to use is (4.32).

4.5. Spontaneous fission

As the nuclear charge increases to large values, nuclei become more unstable. This is reflected by decreasing half-lives for nuclei heavier than uranium. In 1940 K. Petrjak and G. Flerov found that ^{238}U in addition to α-decay also had a competing mode of radioactive decay termed spontaneous fission. In this mode two heavy fragments (*fission products*) are formed in addition to some neutrons. The reaction may be written

$$^{A}_{Z}X \rightarrow {}^{A_1}_{Z_1}X_1 + {}^{A_2}_{Z_2}X_2 + \nu n \qquad (4.38)$$

where ν is the number of neutrons, usually 2–3. The half-life for spontaneous fission of ^{238}U is very long, about 8×10^{15} y. This means that about 70 fissions occur per second in 1 kg of ^{238}U, which can be compared with the simultaneous emission of 45×10^9 α-particles.

With increasing Z, spontaneous fission becomes more common; i.e. the half-life for this decay process decreases. For $^{240}_{94}$Pu it is 1.2×10^{11} y; for $^{244}_{96}$Cm, 1.4×10^7 y; for $^{252}_{98}$Cf, 66 y; and for $^{256}_{100}$Fm, 3×10^{-4} y. In fact, spontaneous fission becomes the dominating decay mode for the heaviest nuclei (see Fig. 3.8).

Spontaneous fission is basically similar to fission induced by bombardment with low energy neutrons (§9.7).

4.6. Decay schemes and isotope charts

Information on the mode of decay, the decay energy, and the half-life are included in the *nuclear decay scheme*. A number of such schemes are shown in Fig. 4.6. The half-life is to the left of the line representing the energy level of the nuclide from which the

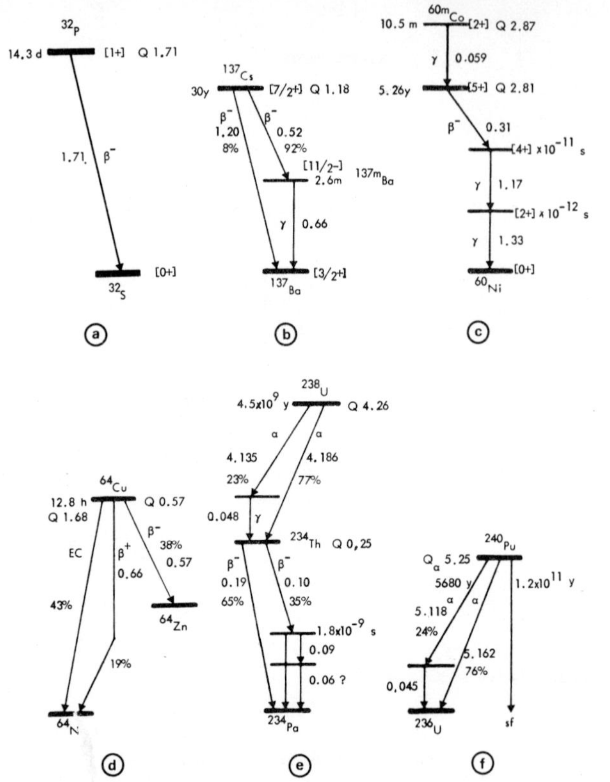

FIG. 4.6. Example. of different decay schemes; energies are in MeV. The schemes are explained in the text.

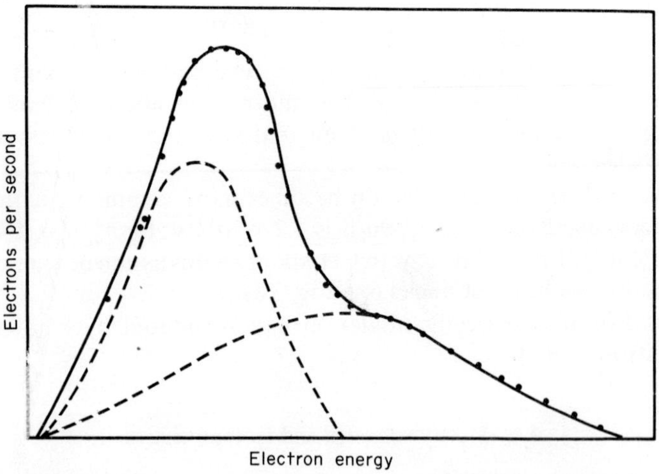

FIG. 4.7. Beta spectrum for the positron decay of ^{84}Rb showing two positron groups, one of 0.77 MeV (11%) and one of 1.05 MeV (10%) E_{max}. (^{84}Rb also decays through β^- emission.)

decay transition originates. The nuclear spin value of that state is given within brackets. It also gives the Q-value for a particular decay step. The mode of radioactive decay is characterized by the proper symbols such as α, β, γ, EC. This symbol also has next to it the decay energy for that particular transition, and the fraction decaying according to that mode if it is less than 100%.

Figure 4.6(a) shows the β^- decay of nuclide ^{32}P. The decay scheme of ^{137}Cs (Fig. 4.6(b)) differs somewhat from Fig. 4.5 and what we would expect from the curve in Fig. 4.1. The reason is that in the electron spectrum of Fig. 4.1, a small fraction of electrons (8%) emitted with an energy of 1.20 MeV could not be detected because of the insensitivity of the magnetic spectrometer used. It is common for nuclei to decay through different competing reactions, as in this case which involves different β-rays. If the higher energy β-decay had been as common as the lower one, we would have observed a mixed β-spectrum, as is indicated in Fig. 4.7. Figure 4.6(c) shows the decay of ^{60}Co, and also its isomeric *precursor* ^{60m}Co. The β^- decay is immediately followed by a cascade of two γ-rays. ^{64}Cu (Fig. 4.6(d)) decays through negatron (38%) and positron (19%) emission and electron capture (43%); this is referred to as *branched decay*. The vertical line in the angled arrow indicating the positron decay symbolizes the rest mass energy of the two electrons created, i.e. 1.02 MeV. Adding 1.02 MeV to 0.66 MeV (indicated at the line) gives 1.68 MeV, the Q-value for the decay from ^{64}Cu to ^{64}Ni. Figure 4.6(e) is a more complicated decay sequence for $^{238}U(\alpha)^{234}Th(\beta^-)^{234}Pa$. In the beginning of this chapter we pointed out that the decay of ^{238}U sometimes results in an excited state of the daughter ^{234}Th (in 23 out of 100 decays), although the excitation energy is comparatively small. Figure 4.6(f) shows how spontaneous fission competes with α-decay in ^{240}Pu. Instead of giving the percentage in the different decay branches, the half-life for that particular mode of decay may be given; conversion between half-lives and percentage is explained in §4.9.

Figure 4.6 gives simplified decay schemes. In practice they are much more detailed as shown for mass numbers 24 and 25 in Fig. 4.8. We return later to this figure for more detailed discussion. Three parabolas have been drawn through the figure, two for the odd–odd and even–even transitions ($A = 24$), and one for the odd–even, even–odd transitions ($A = 25$). These parabolas correspond to II and I, respectively, in Fig. 3.7.

Isotope charts (see Appendix N) can be considered as condensed isotope tables. Normally the nuclear charge Z is on the vertical axis, and the neutron number N on the horizontal one. The legend explains the information provided. Such isotope charts are very useful for rapid scanning of ways to produce a certain nuclide and to follow its decay modes.

4.7. Secondary processes in the atom

Once an electron is ejected from an atomic orbital due to internal conversion, electron capture, or some other process involved in radioactive decay, a vacancy is created in the electron shell which can be filled in several ways. Electrons from higher energy orbitals can occupy the vacancy. The difference in the binding energy of the two shells involved in the transition will be emitted from the atom as *X-rays*. This process is called *fluorescent radiation*.

If the difference in the binding energy for the transition is sufficient to exceed the binding energy of electrons in the L- or M-levels, emission of the energy as X-rays is not the predominant mode. Instead an *internal photoelectric* process can occur and the excess binding energy results in the emission of several low energy electrons which are called *Auger electrons.* The Auger electrons are much lower in energy than the electron from the *nuclear internal conversion* process, since the difference in electronic binding energies is in the eV range compared to the energies in the nuclear conversion process which are in the MeV range. The atom may be left in a state of high ionization by Auger emission; positive charges of 10–20 have been observed. When such high charges are neutralized, the energy liberated is sufficient to break chemical bonds (cf. § 4.3.2).

In isomeric decay the γ-energy is often so small that the daughter recoil is negligible. For example

$$^{80m}Br \xrightarrow[4.4h]{\gamma} {}^{80}Br$$

occurs through the emission of a γ-ray of 0.049 MeV, giving the daughter a recoil energy of only 0.016 eV. Still the decay leads to the emission of ^{80}Br from ethyl bromide, when the parent compound is $C_2H_5{}^{80m}Br$, even though the bond strength is 2.3 eV. The γ is highly converted and as the electron "hole" is filled, Auger emission occurs. Bromine ions from Br^+ to Br^{17+} have been observed through mass spectrometric analyses of the ethyl bromide gas phase.

4.8. Closed decay energy cycles

The masses for many short-lived nuclei are unknown although their decay modes and energies have been determined. From this the nuclear masses may be calculated, and consequently Q-values of different unknown decay modes can be obtained. This can be done through the use of closed decay energy cycles.

Suppose we need to know if ^{237}U can decay to ^{233}Th through α-emission. Of course this is a simple calculation if the masses of ^{237}U and ^{233}Th are known, but let us assume they are not. We have data that ^{237}U decays through β-emission (E_{max} 0.248 MeV) followed by γ-decay (E_γ 0.267 MeV). ^{233}Th decays through β-emission (E_{max} 1.230 MeV) directly to ^{233}Pa. ^{237}Np undergoes α-decay to ^{233}Pa with $E_\alpha = 4.79$ MeV.

We may construct a closed cycle including these decay energies

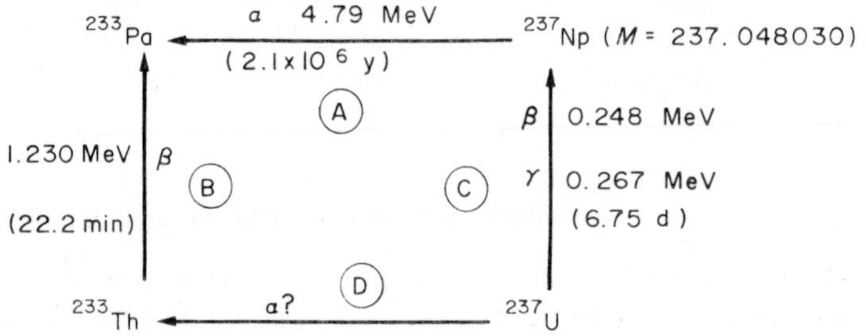

The Q-value for branch Ⓓ is $Q = -931.5\,(M_{^{233}Th} + M_{He} - M_{^{237}U})$. For branch Ⓐ we can calculate (4.16) $Q = \dfrac{E_\alpha M_Z}{M_{Z-2}} = 4.79\,\dfrac{237}{233} = -931.5\,(M_{^{233}Pa} + M_{He} - M_{^{237}Np})$. By introducing values for M_{He} and $M_{^{237}Np}$ we obtain $M_{^{233}Pa} = 233.040\,108$.

For branch Ⓑ we calculate

$$M_{^{233}Th} = M_{^{233}Pa} + 1.230/931.5 = 233.041\,428$$

For branch Ⓒ one obtains similarly

$$M_{^{237}\mathrm{U}} = M_{^{237}\mathrm{Np}} + (0.248 + 0.267)/931.5 = 237.048\,581$$

Thus all information is available for calculating branch Ⓓ. The Q-value is found to be 4.23 MeV, and the $E_\alpha = 4.23 \times 233/237$ or 4.16 MeV. Although spontaneous α-decay is energetically possible, it has not been detected. The systematics of α-decay (§4.15) indicates an expected half-life of $> 10^6$ y. Because the β-decay rate is much faster ($t_{1/2} = 6.75$ d), too few α's are emitted during the lifetime of ^{237}Np to be detected.

4.9. Kinetics of simple radioactive decay

Most radioactive isotopes which are found in the elements on earth must have existed for at least as long as the earth. The nonexistence in nature of elements with atomic numbers greater than 92 is explained by the fact that all the isotopes of these elements have lifetimes considerably shorter than the age of the earth.

Radioactive decay is a random process. Among the atoms in a sample undergoing decay it is not possible to identify which specific atoms will be the next to decay. We denote the *decay rate* by A. It is a measure of the number of disintegrations per unit time:

$$A = -\mathrm{d}N/\mathrm{d}t \tag{4.39}$$

The decay rate is proportional to the number of radioactive atoms N present: $A \propto N$. If 10^5 atoms show a decay rate of 5 atoms per second, then 10^6 atoms show a decay rate of 50 atoms per second. If the number of radioactive nuclei and the number of decays per unit time are sufficiently great to permit a statistical treatment, then

$$-\mathrm{d}N/\mathrm{d}t = \lambda N \tag{4.40a}$$

where λ is the proportionality constant known as the *decay constant*. If the time of observation Δt during which ΔN atoms decay is very small compared to t, one may simply write

$$A = \frac{\Delta N}{\Delta t} = \lambda N \tag{4.40b}$$

If the number of nuclei present at some original time $t = 0$ is designated as N_0, (4.40a) upon integration becomes the general equation for simple radioactive decay:

$$N = N_0 \mathrm{e}^{-\lambda t} \tag{4.41a}$$

In Fig. 4.9 the ratio of the number of nuclei at any time t to the original number at time $t = 0$ (i.e. N/N_0) has been plotted on both a linear (left) and logarithmic (right) scale as a function of t. The linearity of the decay curve in the semilogarithmic graph illustrates the exponential nature of radioactive decay. Since $A \propto N$, the equation can be rewritten as

$$A = A_0 \mathrm{e}^{-\lambda t} \tag{4.41b}$$

Commonly, log A is plotted as a function of t since it is simpler to determine the disintegration rate than it is to determine the number of radioactive atoms in a sample.

Instead of the decay constant λ, the average *lifetime* τ is sometimes used:

$$\tau = 1/\lambda \tag{4.42}$$

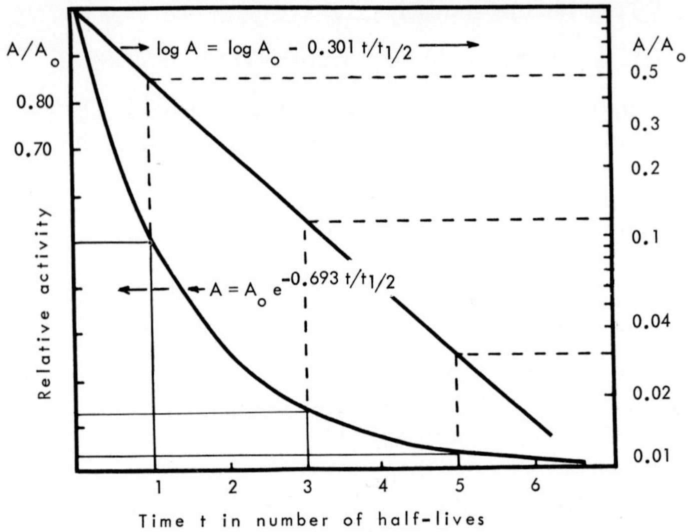

FIG. 4.9. Linear and logarithm plots of simple radioactive decay.

Even more common is the use of the *half-life*, $t_{1/2}$, which is the time needed to reduce the amount of radioactive material by a factor of 2. Thus

$$A/A_0 = 1/2 = e^{-\lambda t_{1/2}}$$

and thus

$$t_{1/2} = \ln 2/\lambda = 0.693/\lambda \tag{4.43}$$

$t_{1/2}$ is about 70% of the average lifetime τ.

The number of radioactive nuclei remaining at any time in a sample which at $t = 0$ had N_0 atoms can be calculated from the equation

$$N = N_0/2^n \tag{4.44}$$

where n is the number of half-lives which have passed. In radioactive work, 10 half-lives ($n = 10$) is usually considered as the useful lifetime for a radioactive species since $N = N_0/2^{10} = 10^{-3}\ N_0$; i.e. N, and hence A, is 0.001 of the original N_0 and A_0.

The decay rate is usually expressed as disintegrations per second (dps) or disintegrations per minute (dpm). In measuring radioactive decay experimentally it is very rare that every disintegration is counted. However, a proportionality exists for any particular detection system between the absolute disintegration rate A and the observed decay rate:

$$R = \psi A \tag{4.45}$$

where R is the *observed decay* or *count rate* and ψ the proportionality constant, known as the *counting efficiency*. This counting efficiency depends on many factors including the detector type, the geometry of the counting arrangement, and the type and energy of the radioactive decay. ψ commonly has a value between 0.1 and 0.5. Equation (4.45) is only valid provided $\Delta t \ll t_{1/2}$ (in which case $\Delta N \ll N$), where Δt is the time of measurement; this is the normal situation; cf. (4.40b).

Figure 4.10 shows the radioactivity of a ^{32}P sample measured every third day with a GM counter. It is seen that the activity decreases from about 8400 cpm at $t = 0$ to

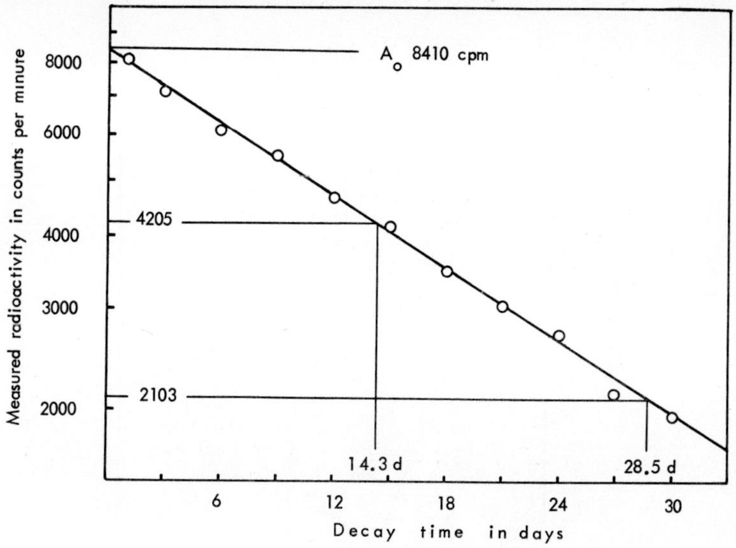

FIG. 4.10. Semilogarithmic plot of the measured decay of ^{32}P using a GM counter with a background of 20 cpm.

4200 cpm in 14.3 days, and to about 2100 cpm in 2 × 14.3 days. The uncertainty in the measurements is about the size of the circles, i.e. about ± 110 cpm at $t = 0$ and about ± 65 cpm at 30 days. In plots of this kind, the count rate measured in the absence of the sample (the *background*) must be subtracted from that obtained with the sample present to yield the correct radioactivity for the sample alone. In Fig. 4.10 the background is so small (i.e. about 20 cpm) that it has very little influence on the decay curve.

The half-life is such a definitive characteristic of a radioactive species that knowledge of it plus the decay energy is often sufficient to allow identification of a nuclide. A radioactive sample, which exhibits a half-life of 4.5×10^9 y with α-decay energies of 4.8 MeV (77%) and 4.3 MeV (23%), is almost certainly ^{238}U as there is no other nuclide known with this exact set of properties.

With (4.39), (4.43), and (4.45) one obtains

$$R = \psi N \ln 2/t_{1/2} \tag{4.46}$$

Knowing the counting efficiency ψ and the number of atoms N, the half-life can be calculated from measurement of R. For example, in a counting arrangement with $\psi = 0.515$ for α-particles, 159 cpm are observed from a ^{232}Th deposit of 1.27 mg (sample weight a). Thus $A = R/\psi = 309$ dpm, $N = a \cdot \mathbf{N}/M = 1.27 \times 10^{-3} \times 6.02 \times 10^{23}/232.0 = 3.295 \times 10^{18}$ ^{232}Th atoms, and $t_{1/2} = 0.693 \times 3.295 \times 10^{18}/309 = 7.40 \times 10^{15}$ min $= 1.41 \times 10^{10}$ y.

It is quite obvious that it would not be possible to determine such a long half-life by following a decay curve like the one in Fig. 4.10. Alternately, for short-lived nuclides such as ^{32}P one may use (4.46) with the known half-life and experimentally measured values of ψ and R to determine values of N.

4.10. Mixed decay

A radioactive sample may contain several different radioactive nuclides which are not genetically related. The decay of each nuclide follows the decay equations of the

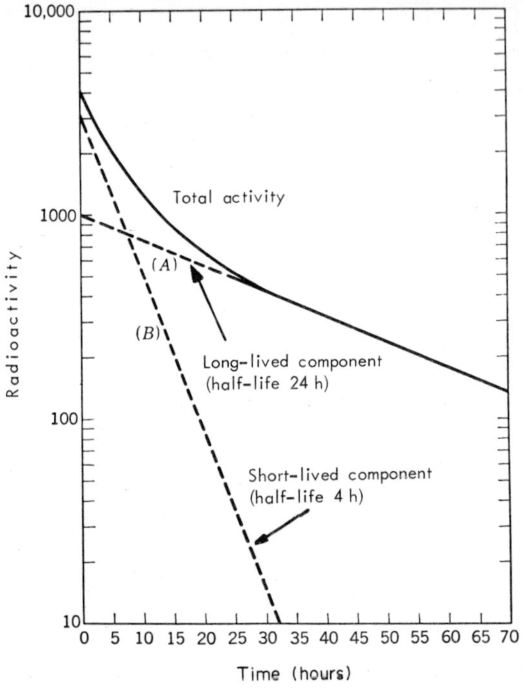

FIG. 4.11. Decay diagram of a mixture of two independently decaying nuclides with half-lives of 4 and 24 h.

previous section. The detector measures a certain amount of the radioactivity of each species so that

$$R = R' + R'' \tag{4.47a}$$

which with the introduction of (4.41b) and (4.45) gives

$$R = \psi' A'_0 \, e^{-\lambda' t} + \psi'' A''_0 \, e^{-\lambda'' t} \tag{4.47b}$$

Figure 4.11 shows the composite decay curve for the mixture of ^{71}Zn ($t_{1/2}$ 3.9 h) and ^{187}W ($t_{1/2}$ 23.8 h). If the half-lives of the species in the mixture differ sufficiently, as in this case, the decay curve can be resolved into the individual components. The long-lived nuclide (^{187}W, line A) can be observed to have linear decay at times long enough for the shorter-lived species to have died. The decay line related to this species can be extrapolated back to $t = 0$ and this line subtracted from the observed decay curve. The resulting curve should be the linear decay due to the shorter-lived species in the sample (line B). For more complex mixtures, this process may be repeated until the curve is completely resolved into linear components.

Sometimes it is possible in mixed decay to observe preferentially the decay of one species by proper choice of detection technique. For example, a proportional counter may be used at an operating voltage that allows detection of α-decay ($\psi_\alpha > 0$) but excludes detection of β-decay ($\psi_\beta = 0$). By contrast a typical Geiger counter can be used for β-decay but does not detect α-radiation since the α-particles do not penetrate the window of the Geiger tube. These problems are discussed more extensively in Chapter 17.

4.11. Radioactive decay units

The *Curie unit* (abbreviated Ci) was originally defined as the number of decays per second from one gram of ^{226}Ra. Based on the half-life of 1580 y which was accepted at the time of definition, the decay rate for one gram of ^{226}Ra would be

$$A = \frac{0.693 \times 6.02 \times 10^{23}}{226.1 \times 1580 \times 365 \times 24 \times 60 \times 60} = 3.70 \times 10^{10} \text{ dps}$$

The value of the half-life has been shown subsequently to be 1600 y, which would give a slightly different disintegration rate.

In the SI system the radiation unit is the Becquerel (Bq) and the activity is given in reciprocal seconds, s^{-1}:

$$1 \text{ Becquerel (Bq)} = 1 \text{ (disintegration) } s^{-1} \tag{4.48a}$$

Although the use of the Curie unit "is not recommended" it has been defined to be exactly

$$1 \text{ Curie (Ci)} = 3.7 \times 10^{10} \text{ } s^{-1} \text{ (Bq)} \tag{4.48b}$$

which corresponds to 222×10^{10} dpm. In this text we deviate slightly from the SI recommendations and use Ci as well as the abbreviations dpm and dps. It may be practical to use Ci at activities $\gtrsim$ M ($= 10^6$) dpm (60 MBq $\approx$ 2 mCi). In nuclear physics dps is commonly used, while in nuclear chemistry the unit dpm is preferred.

For the measured count rate the common unit in English is counts per minute (cpm), in German Impulse pro Minute (Ipm), and in French coups de minute (c/m); the smaller units are cps, etc.

The *specific radioactivity* S is defined as the decay rate A per unit amount W of an element or compound,

$$S = A/w \tag{4.49}$$

The SI unit of specific radioactivity is Bq kg^{-1} which however is impractically small. For practical purposes it is commonly defined in dpm g^{-1} or dpm $mole^{-1}$. *Activity concentration* (or "radioactive concentration") is given in Bq m^{-3}, or, more commonly, Ci m^{-3}. With the half-life of 1600 y the specific activity per gram of ^{226} Ra is 0.988 Ci or 3.65×10^{10} Bq or 2.19×10^{12} dpm. The specific activities of some of the longer-lived naturally occurring radioactive species are: K, 1.85×10^3 dpm g^{-1}; ^{232}Th, 0.243×10^6 dpm g^{-1}; ^{238}U, 0.746×10^6 dpm g^{-1}.

4.12. Branching decay

Several times in this chapter the possibility of competing modes of decay has been noted. In such competition, termed *branching decay*, the parent nuclide may decay to two or more different daughter nuclides: e.g.

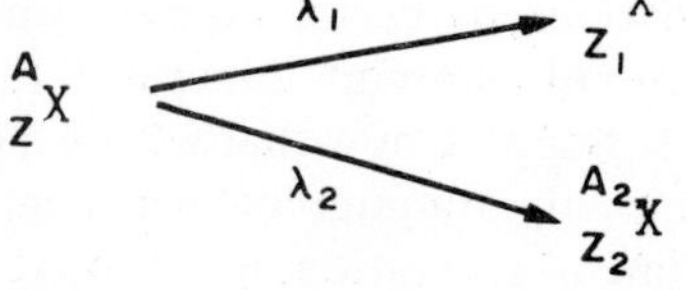

(4.50)

where for each branching decay a *partial decay constant* can be determined. These constants are related to the total observed decay constant for the parent nuclide as

$$\lambda_{\text{tot}} = \lambda_1 + \lambda_2 \ldots \tag{4.51}$$

Each mode of decay in branching may be treated separately; the decay in an individual branch has a half-life based on the partial decay constant. Since only the total decay constant (the rate with which the mother nuclide, ${}^{A}_{Z}X$ in (4.50), decays) is observable directly, partial decay constants are obtained by multiplying the observed total decay constant by the fraction of parent decay corresponding to that branch. ${}^{64}Cu$ decays 43% by electron capture, 38% by negatron emission, and 19% by positron emission. The observed total decay constant is equal to 0.0541 h^{-1} based on the half-life of 12.8 h. The partial constants are:

$$\lambda_{EC} = 0.43 \times 0.0541 = 0.0233\ h^{-1}$$

$$\lambda_{\beta^-} = 0.38 \times 0.0541 = 0.0206\ h^{-1}$$

$$\lambda_{\beta^+} = 0.19 \times 0.0541 = 0.0103\ h^{-1}$$

These partial decay constants correspond to partial half-lives of 29.7 h for electron capture decay, 33.6 h for β^- decay, and 67.5 h for positron decay.

4.13. Successive radioactive decay

There are many instances where a parent decays to a daughter which itself decays to a third species (i.e. a "grand-daughter"). The chains of radioactive decay in the naturally occurring heavy elements include as many as 10–12 successive steps.

$$X_1 \xrightarrow{\lambda_1} X_2 \xrightarrow{\lambda_2} X_3 \xrightarrow{\lambda_3} X_4 \tag{4.52}$$

The net rate of formation of the daughter atoms X_2 is the difference between the rate of formation of the daughter and her rate of decay, i.e.

$$dN_2/dt = N_1\lambda_1 - N_2\lambda_2 \tag{4.53}$$

where N_1 and N_2 are the number of parent and of daughter atoms, and λ_1 and λ_2, the decay constants of the parent and daughter, respectively. The solution of this equation is

$$N_2 = \frac{\lambda_1}{\lambda_2 - \lambda_1} N_1^0 (e^{-\lambda_1 t} - e^{-\lambda_2 t}) + N_2^0\, e^{-\lambda_2 t} \tag{4.54}$$

where N_1^0 and N_2^0 are the amounts of parent and daughter respectively at time $t = 0$. The first term in this equation tells us how the number of daughter nuclei vary with time as a consequence of the formation and subsequent decay of the daughter nuclei, while the second term accounts for the decay of those daughter nuclei that were present at $t = 0$.

Let us illustrate this relationship by an example among the naturally occurring radioactive decay series. In an old uranium mineral all the products in the decay chain can be detected (see Fig. 1.4). Suppose now that we use a chemical separation to isolate two samples, one containing only uranium and one containing only thorium (relation (1.3) and Fig. 1.2). At the time of separation, which we designate as $t = 0$, there are N_1^0 atoms of ${}^{238}U$ and N_2^0 atoms of ${}^{234}Th$. In the thorium fraction, which is free from uranium, $N_1^0 = 0$ and, therefore, the thorium atoms decay according to the last term in (4.54). This sample gives a simple exponential decay curve with a half-life of 24.1 d, as shown by the precipitate curve in Fig. 1.2. The uranium fraction at $t = 0$ is completely free of thorium; i.e. $N_2^0 = 0$. However, after some time it is possible to detect the presence of ^{234}Th. The change in the number of ^{234}Th with time follows the first term of (4.54); in fact, in Fig. 1.2 the measurements detect only the ^{234}Th nuclide (β-emitting), since the detection system is not sensitive to the α's from ^{238}U ($\psi_\alpha = 0$). The time of observation is much smaller than the half-life of the ^{238}U decay, so there will be no observable change in the number of atoms of uranium during the time of observation, i.e. $N_1 = N_1^0$. Further, since $t_{1/2}$ for ^{238}U $\gg t_{1/2}$ for ^{234}Th, i.e. $\lambda_1 \ll \lambda_2$, we can simplify (4.54) to

$$N_2 = \frac{\lambda_1}{\lambda_2} N_1 (1 - e^{-\lambda_2 t}) \tag{4.55}$$

According to this equation, N_2 increases with time with the half-life of ^{234}Th. In other words, after a period of 24.1 d there is 50% of the maximum value of ^{234}Th, after 48.2 d there is 75% of the final maximum value, etc. This is illustrated by the change in the uranium fraction activity in Fig. 1.2. Further, from the relationship (4.55) we can see that the maximum value of thorium ($t = \infty$) is given by

$$N_2 \lambda_2 = N_1 \lambda_1 \tag{4.56}$$

These equations, based on $\lambda_1 \ll \lambda_2$, show that the amount of daughter atoms becomes constant after some time. At that time the rate of decay of the daughter becomes equal to the rate of decay of the parent, i.e. $A_2 = A_1$, but the amounts of the parent and the daughter are not equal since N_2 is much smaller than N_1. This condition of $A_2 = A_1$ is known as *secular equilibrium*, which is a misnomer since this is a steady state and not a true equilibrium situation. It is also common to speak of *radioactive equilibrium* in referring to this steady state condition.

We can calculate that at secular equilibrium for each gram of ^{238}U there will be present 1.44×10^{-11} g of ^{234}Th and 4.9×10^{-16} g of ^{234}Pa. Since the specific radioactivity of 1 g of ^{238}U is 746 000 dpm/g, the decay rate of 4.9×10^{-16} g ^{234}Pa is also 746 000 dpm.

When the time of observation is very short compared to the half-life of the parent nuclide, as in secular equilibrium, no change in the decay rate of the parent is observed for many daughter half-lives. Our example of ^{137}Cs, which decays via the isomeric state $^{137\text{m}}$Ba to ^{137}Ba, presents another case of "secular equilibrium". If we have an "old" sample in which radioactive equilibrium has been reached (older than ~15 min, since the $t_{1/2}$ of the daughter is 2.6 min), and separate the cesium from the barium by precipitation and filtration of the $BaSO_4$, the activity measured from the precipitate will follow curve (1) in Fig. 4.12. In the filtrate solution the activity from ^{137}Cs, curve (2), is unchanged during our observation time. However, $^{137\text{m}}$Ba grows into the solution curve (3), so that the total activity of the solution, curves (2) plus (3), increases according to curve (4).

In many radioactive decay chains the half-life of the parent is longer than that of the daughter but it is short enough that a change in the disintegration rate of the parent is observable during the period of observation of the experiment. In such cases the system reaches the condition termed *transient equilibrium*. The length of time of observation of the activity of the sample may be the determining factor as to whether it appears to be transient or secular equilibrium. If a parent has a one month half-life and the observation of the change in decay rates of parent and daughter extends over an hour or even a few days, the data would follow the equation for secular equilibrium since the degree of change in the parent decay would be negligible. However, if the observation extends over a period of several weeks or months, then the change in the decay rate

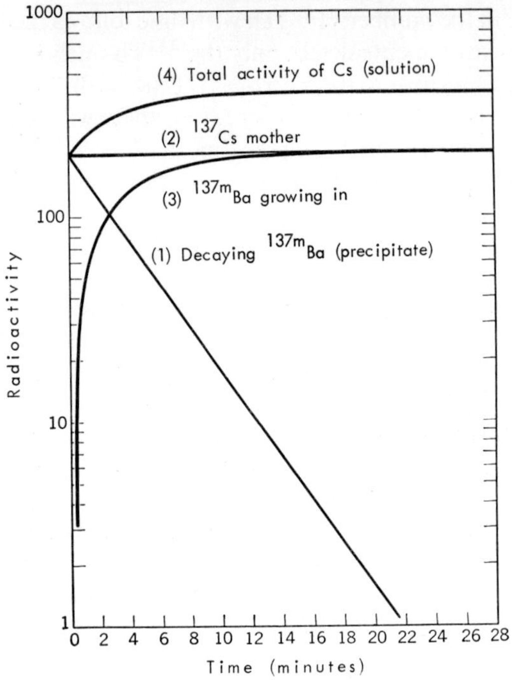

FIG. 4.12. Case of radioactive equilibrium: successive decay chain $^{137}Cs(t_{1/2}$ 30 y) → ^{137m}Ba ($t_{1/2}$ 2.6 min) → stable.

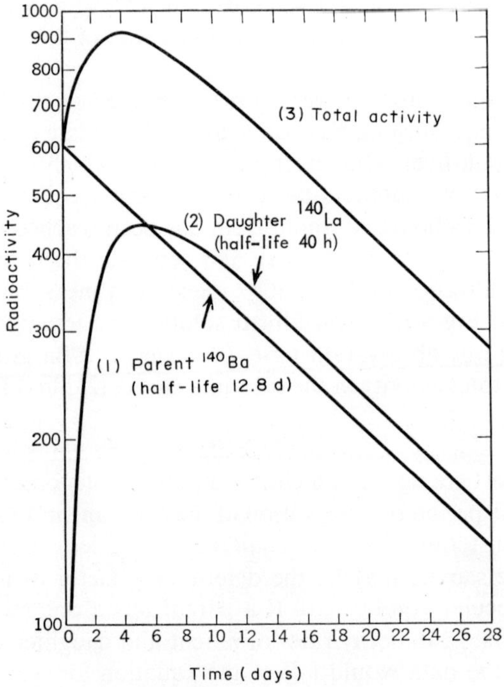

FIG. 4.13. Case of transient equilibrium: successive decay chain ^{140}Ba ($t_{1/2}$ 12.8 d) → ^{140}La ($t_{1/2}$ 40 h) → stable.

of the parent is significant and it would appear as transient equilibrium.

The case of transient equilibrium can be illustrated by an example, such as the decay chain

$$^{140}\mathrm{Ba}\,(\beta^-, 12.8\ \mathrm{d})^{140}\mathrm{La}\,(\beta^-, 40.2\ \mathrm{h})^{140}\mathrm{Ce}\ (\mathrm{stable})$$

^{140}Ba is one of the most important fission products. If we isolate barium, lanthanum grows into the sample. Figure 4.13 shows the decay of ^{140}Ba in curve (1), which follows the simple decay of (4.41). Curve (2) shows the activity of the daughter, for which the left half of eq. (4.54) is valid. Replacing decay constants by half-lives we can rewrite this equation as

$$A_2 = \frac{t_{1/2,2}}{t_{1/2,1} - t_{1/2,2}} A_1^0 (e^{-0.693t/t_{1/2,1}} - e^{-0.693t/t_{1/2,2}}) \tag{4.57}$$

At $t \ll t_{1/2,1} (t \ll 12.8\ \mathrm{d})$ the first exponential term is very close to 1 and A_2 increases proportional to $(1 - e^{-0.693t/t_{1/2,2}})$; this is the increasing part of curve (2). At $t \gg t_{1/2,2} (t \gg 40\ \mathrm{h})$, the second experimental term becomes much smaller than the first one, and A_2 decreases proportional to $e^{-0.693t/t_{1/2,1}}$. For this part of the curve we may write

$$N_2 = \frac{\lambda_1}{\lambda_2 - \lambda_1} N_1 \tag{4.58}$$

which is the relation valid for transient equilibrium. The total activity of the barium sample, curve (3), is the sum of curves (1) and (2).

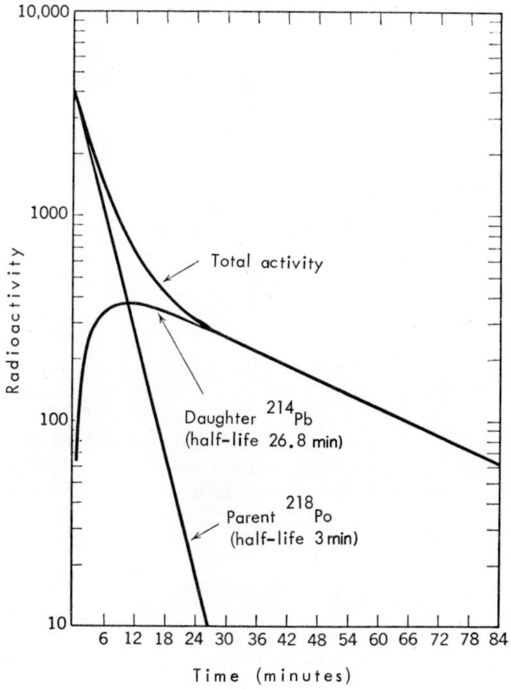

FIG. 4.14. Case of no equilibrium: successive decay chain ^{218}Po ($t_{1/2}$ 3 m) → ^{214}Pb (26.8 min)→ stable.

If the parent is shorter-lived than the daughter, the daughter activity grows to some maximum value and then decays with its own characteristic half-life. This contrasts to the case of transient equilibrium where the daughter has an apparent decay given by the half-life of the parent. An example of this is shown in Fig. 4.14 for the decay chain

$$^{218}\text{Po}(\alpha, 3\text{ m})^{214}\text{Pb}(\beta^-, 27\text{ m})^{214}\text{Bi}$$

The time necessary for obtaining the maximum daughter intensity in the nonequilibrium case of the shorter-lived parent is given by

$$t_{max} = \frac{1}{\lambda_2 - \lambda_1} \ln \frac{\lambda_2}{\lambda_1} \tag{4.59}$$

4.14. Radioisotope generators

The growth of radioactive daughters frequently has practical significance. For example, in radiation therapy and diagnostic medicine it is preferable to use short-lived nuclides. In fact, it is preferable to conduct tracer experiments with short-lived nuclides as this eliminates the problem of disposal of residual radioactive waste after completion of the experiment. It is convenient to have a long-lived mother in storage from which a short-lived daughter can be removed as required for use in tracer work. A few examples of uses of such mother–daughter pairs are discussed; others are included in Table 4.1. Such systems are called radioisotope generators.

^{222}Rn is sometimes used for the radiotherapeutic treatment of cancer. This product is isolated by separating it as a gas from the parent substance ^{226}Ra which is normally in the form of solid or a solution of $RaBr_2$. ^{222}Rn grows into the radium sample with a half-life of 3.8 d. After a 2-week period, following a separation of radon from radium, approximately 90% of the maximum amount of radon has grown back in the radium sample. Consequently, it is useful to separate ^{222}Rn each 2 weeks from the radium

TABLE 4.1. *Some common radioactive milking pairs. The decay properties include decay energy* (MeV) *mode of decay and half-life*

Mother nuclide	Decay properties	Daughter nuclide	Decay properties	Application
^{44}Ti	EC, γ; 47.3 y	^{44}Sc	1.5 β^+; 1.17; 4.0 h	Teaching
^{68}Ge	EC; 275 d	^{68}Ga	1.88 β^+, 1.08 γ; 1.14 h	Medical
^{90}Sr	0.55 β; 28.1 y	^{90}Y	2.27 β; 64 h	Heat source (large amounts),[a] calibration source
^{99}Mo	β, γ; 67 h	^{99m}Tc	0.14 γ; 6.0 h	Medical
^{113}Sn	EC, γ; 115 d	^{113m}In	0.40 γ; 1.66 h	Medical
^{132}Te	β, γ; 78 h	^{132}I	β, γ; 2.3 h	Medical
^{137}Cs	β, γ; 30.0 y	^{137m}Ba	0.28 γ; 2.55 min	Gamma radiography, radiation sterilization (large amounts)[a]
^{140}Ba	β, γ; 12.8 d	^{140}La	β, γ; 40.2 h	Lanthanum tracer
^{144}Ce	β, γ; 284 d	^{144}Pr	3.00 β; 17.3 min	Calibration source
^{210}Pb	β, γ; 21 y	^{210}Bi	1.16 β; 5.01 d	Calibration source
^{226}Ra	α; 1600 y	^{222}Rn	α; 3.82 d	Medical
^{238}U	α; 4.5 × 10^9 y	^{234}Th	β, γ; 24.1 d	Thorium tracer

[a] Mainly use of mother substance.

samples since further time provides very little additional radioactivity. The ^{222}Rn is an α-emitter; the therapeutic value comes from the irradiation of the tissue by the γ-rays of the decay daughters ^{214}Pb and ^{214}Bi which reach radioactive equilibrium extremely rapidly with the ^{222}Rn.

^{99m}Tc is used for diagnostic purposes for liver, spleen, and thyroid scanning. The ^{99}Mo parent is absorbed on a column of alumina and the daughter ^{99m}Tc removed by passage of saline solution at intervals governed by the equilibrium. The parent, when it is fixed in a semipermanent sample as on an adsorbent column, is often known as a *cow* and the removal of the daughter activity from the radioisotope generator (the "cow") is termed milking.

Another commonly used radioisotope generator is ^{132}Te from which ^{132}I may be milked. In this case ^{132}Te is adsorbed as radium tellurite on an alumina column, and the ^{132}I removed by passage of 0.01 M ammonia through the column. The ^{132}I is used both diagnostically and therapeutically for thyroid cancer.

4.15. Decay energy and half-life

It was observed early in both α- and β-decay that the longer the half-life the lower the decay energy. Although there are many exceptions to this observation H. Geiger and J. M. Nuttall formulated the law

$$\log \lambda_\alpha = a + b \log \hat{R}_{\text{air}} \tag{4.60}$$

for the natural α-active nuclides. Here a and b are constants, and $\hat{R}_{\text{air}}$ is the range of the α-particles in air which is directly proportional to E_α.

A similar relation was deduced by E. Fermi for the β-decay:

$$\log \lambda_\beta = a' + b' \log E \tag{4.61}$$

where a' is a constant related to the type of β-decay and $b' \approx 5$.

Although these rules have been superseded by modern theory and the enormous amount of nuclear data now available, they may nevertheless be useful as rough guides in estimates of half-lives and decay energies. In §6.7 more valid but more complicated relationships are discussed.

4.16. Exercises

For some of the problems necessary nuclear data are given in the tables or appendices.

4.1. ^{239}Pu emits α-particles of maximum 5.152 MeV. What is the recoil energy of the product formed?

4.2. Using a magnetic spectrometer the maximum energy of the electrons from ^{137}Cs was found in Fig. 4.1 to correspond to 3.15×10^{-3} Tesla m. Calculate the energy (a) assuming that the electrons are nonrelativistic, (b) with correction for relativistic mass increase.

4.3. ^{11}C decays through emission of positrons of a maximum energy of 1.0 MeV. Calculate the recoil energy of the daughter.

4.4. ^{16}N decays through β^- decay to ^{16}O with a half-life of 7.1 s. A number of very energetic γ's follow after the β-emission, the dominating one with an energy of 6.14 MeV. What is the ^{16}O recoil energy?

4.5. The binding energy of a K-electron in barium is 37 441 eV. Calculate from Fig. 4.1 the internal conversion energy for ^{137m}Ba (Fig. 4.5).

4.6. From the specific activity of potassium (1850 dpm/g K) and the fact that it all originates in the rare isotope ^{40}K(0.0119%), calculate the half-life of ^{40}K.

4.7. One may assume that when ^{238}U was formed at the genesis an equal amount of ^{235}U was formed. Today the amount of ^{238}U is 138 times the amount of ^{235}U. How long a time ago did the genesis occur according to this assumption?

4.8. The interior of the earth is assumed to be built up of a solid core (radius 1400 km) followed by a molten core (radius 3500 km) and a molten mantle (radius 6400 km) covered by a 35 km thick crust. One assumes that 2% by weight of the molten mantle is potassium; the mantle density is assumed to be 6000 kg m^{-3}. What energy outflow will the radioactive decay of this element cause at the earth's surface? The decay scheme of ^{40}K is given in eqn. (12.16). For the EC branch $Q = 1.505$ MeV, for the β^- branch 1.314 MeV.

Compare this energy output to the solar energy influx to the earth of 5.6×10^{26} J y^{-1}.

4.9. A hospital has a 1.5 Ci source of ^{226}Ra in the form of a $RaBr_2$ solution. If the ^{222}Rn is pumped out each 48 h, what is (a) the radon activity at that moment, (b) the radon gas volume at STP?

4.10. (a) Prove the correctness of eqn. (4.20) by using Newton's laws of motion. (b) Prove the correctness of eqn. (4.34).

4.11. A recently prepared ^{212}Pb sample has the activity of 10^6 dpm. (a) What is the activity 2 h later? (b) How many lead atoms are left in the sample at this moment? $t_{1/2}$ 10.6 h.

4.12. A radioactive sample was measured at different time intervals:

Time (h)	Activity (cpm)
0.3	11 100
5	5 870
10	3 240
15	2 005
20	1 440
30	1 015
40	888
50	826
100	625

Determine the half-lives of the two nuclides (not genetically related) in the sample and their activities (in Ci) at time $t = 0$. The background of the detection device was 100 376 counts per 1000 min; its counting efficiency was 17%.

4.13. The α-activity of a mixture of astantine isotopes was measured at different times after their separation, giving the following results:

t (min)	R (cpm)	t (min)	R (cpm)
12	756	121	256
17.2	725	140	215.5
23.1	638	161	178.5
30.0	600	184	150.7
37.7	545	211	127.3
47.5	494	243	101.9
59.5	435	276	84.9
73	380	308	68.2
87	341	340	55.0
102	288		

Calculate the half-lives of the isotopes and their activities at $t = 0$.

4.14. In the ion source of a mass spectrograph, UF_6 vapor is introduced which partly becomes ionized to UF_5^+. The ionic currents were measured at mass positions 333, 330, and 329. The ion current ratios were $I_{333}/I_{330} = 139$, and $I_{330}/I_{329} = 141.5$. What is the half-life of ^{234}U if that of ^{238}U is 4.5×10^9 y? Radioactive equilibrium is assumed to exist in the UF_6.

4.17. Literature

See Appendix A, §§ A.1 and A.2.

W. F. LIBBY, Chemistry of energetic atoms produced in nuclear reactions, *J. Am. Chem. Soc.* **69** (1947) 2523.

K. H. LIESER, Chemische Gesichtspunkte für die Entwicklung von Radionuklidgeneratoren, *Radiochim. Acta* **23** (2) (1976) 57.

CHAPTER 5

Cosmic Radiation and Elementary Particles

Contents

Introduction

The early workers in nuclear science found that their measuring equipment had a constant background level of radiation which could be eliminated only partially even with the aid of thick shielding walls of iron and lead. It was assumed initially that this radiation had its origin in naturally radioactive elements present in the materials in the laboratory. However, in 1911 V. F. Hess carried measuring equipment up into the atmosphere with the aid of balloons and learned that this background radiation increased with altitude. Obviously, at least a component of the laboratory background radiation had its origin in some extraterrestrial source. In recent years, test equipment carried outside of the earth's atmosphere by rockets has given us data which provide a fairly accurate picture of the composition of the radiation that comes to the earth from space.

The investigation of cosmic radiation has had a profound influence on nuclear science. When J. Chadwick in 1932 discovered the neutron, the picture of matter seemed complete: all matter appeared to be composed of four fundamental particles: protons, neutrons, electrons, and photons. However, through studies of the cosmic radiation C. D. Anderson discovered the *positron* (the first *antiparticle*) in the same year. Five years later Anderson and S. H. Neddermeyer discovered another new particle with a mass about one-tenth of a proton or about 200 times heavier than the electron. This particle is the *muon*, designated by μ. Since that time a large system of subnuclear particles has been discovered.

5.1. Primary cosmic radiation

A rather small fraction of the cosmic radiation consists of electromagnetic radiation and electrons. The former vary in energy from a small percentage of γ-rays to a considerable intensity of X-rays, to visible light and to radiation in the radiofrequency region. The types and intensities of this radiation have been of great importance to development of models of the formation and composition of the universe.

The major part of the cosmic radiation is nuclear particles with very high energy: approximately 70% protons, 20% α-particles, 0.7% lithium, beryllium, and boron ions, 1.7% carbon, nitrogen, and oxygen ions, the residual 0.6% ions of $Z > 10$ (see Table 11.4). These ions are bare nuclei prior to interaction since their kinetic energies exceed the binding energies of all of the orbital electrons.

Modern large radiotelescopes have demonstrated the existence of about 30 organic compounds in interstellar matter containing molecules as complicated as ethanol, methylamine and vinylcyanide (CH_2CHCN). However, these molecules are probably not contained in the cosmic radiation which reaches the earth.

The cosmic particle radiation can be divided by energy into two major groups (Fig. 5.1). One group has energies mainly below 1 GeV and consists primarily of protons. The source of most of this group is the sun. Its intensity varies in relation to solar eruptions since at the time of such an eruption a large amount of solar material, primarily hydrogen, is ejected into space.

The second group has energies up to 10^{10} GeV, although the intensity of the particles

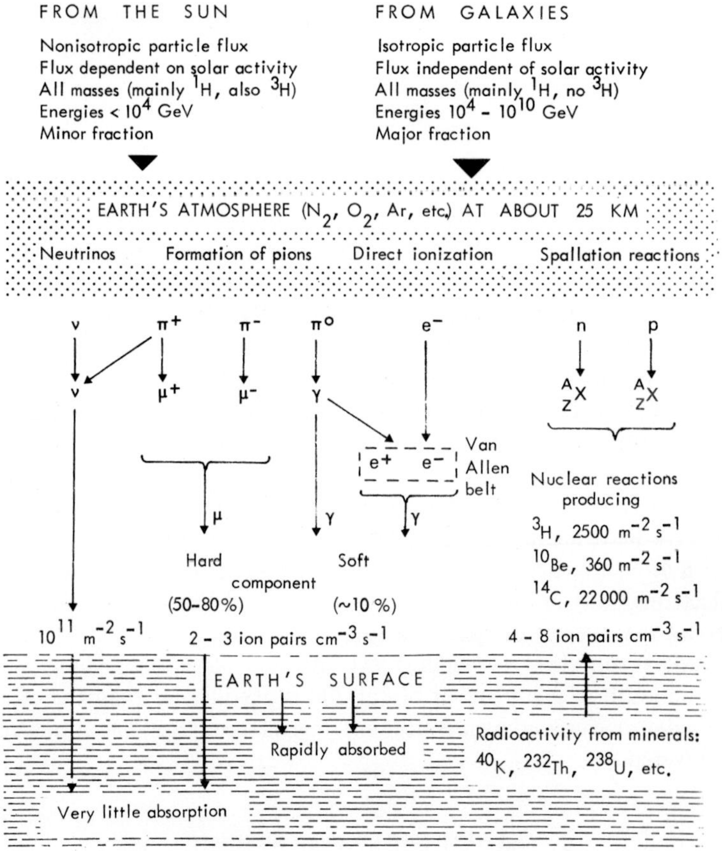

FIG. 5.1. Cosmic radiation consists of atoms and photons which reach the atmosphere from sources outside the earth, leading to the formation of numerous secondary particles, some (but not all) detectable at the earth's surface.

decreases with increasing energy, following the relation $N(E) \propto E^{-1.6}$, where $N(E)$ is the number of particles with energies in excess of E. Thus particles of 10^3 GeV have an intensity of about 10^{11} higher than particles of 10^{10} GeV. Within this high energy group the particles at the lower end of the energy spectrum are assumed to originate from sources within our galaxy (the Milky Way), while the particles of the higher energy end are assumed to come from sources outside of our galaxy. Different hypotheses, which for the most part are untested, suggest that the particles come from astronomical radio sources, super novae exploding, or colliding galaxies, etc. Moreover, some scientists suggest that at least a portion of this radiation is residue of the processes involved in the original formation of the universe. In all cases it is assumed that these high energy particles obtain their tremendous kinetic energies through acceleration in the magnetic field of galactic objects (synchrotron acceleration, §8.7).

When the primary cosmic particles enter the earth's atmosphere, they collide with the matter of that atmosphere and are annihilated. In this annihilation process a large number of new particles are formed whose total kinetic energy is less than that of the original primary radiation but whose total rest mass is larger than that of the primary particles. The main reaction products of the annihilation processes of the primary cosmic particles are particles which are known as *pions*, designated π. Since the pion rest mass is 0.147 u, the energy required to produce a pion is at least 137 MeV (in practice, in order to conserve momentum it exceeds 400 MeV (eqn. (4.4)). Figure 5.2 shows the

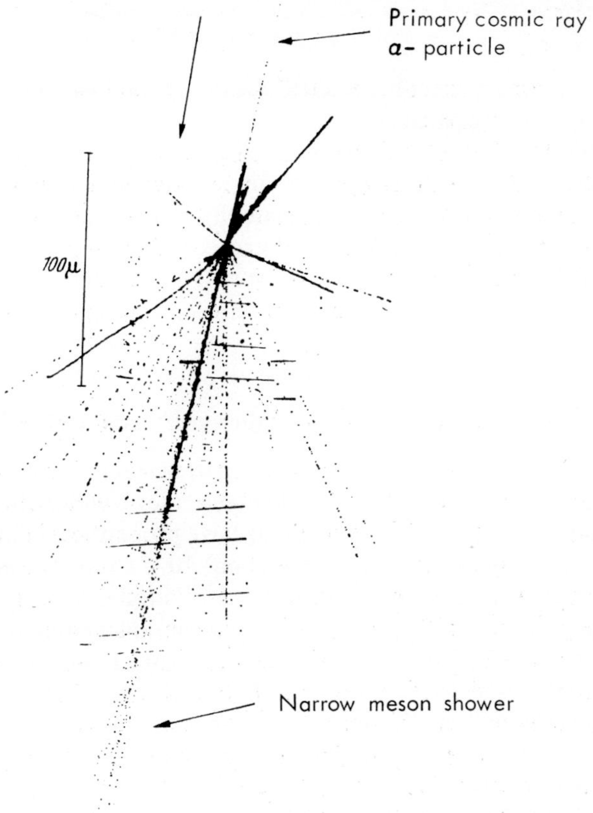

FIG. 5.2. Production of numerous secondary particles in a photographic emulsion caused by a 10^4 GeV helium atom. (According to Kaplan, Peters, and Bradt.)

FIG. 5.3. Holes in helmets of Apollo 12 astronauts caused by high energy cosmic rays. The holes have been made visible in an optical microscope through an etching technique. (According to *New Scientist*, April 22, 1973.)

reaction in which a primary helium ion of 10^4 GeV was annihilated to form a shower of over 60 new highly ionizing particles.

Figure 5.3 shows the effect of high energy cosmic rays hitting the helmets of Apollo 12 astronauts. It is probable that the cosmic ray intensity will put a limit to how long man can endure in outer space: it has been calculated that in a journey to the planet Mars about 0.1% of the cerebral cortex will be destroyed. The annihilation process occurs to such an extent that below an altitude of approximately 25 km above the earth the number of primary cosmic particles has been reduced to a quite small fraction of the original intensity.

5.2. Secondary reactions in the earth's atmosphere

Few of the pions formed in the annihilation process reach the earth's surface. They undergo radioactive decay to muons and neutrinos, or they collide with other particles in the atmosphere and are annihilated. The muons have properties similar to the electron, but are unstable, decaying with a lifetime of about 10^{-6} s to electrons and neutrinos. The collision reactions of the pions result in the formation of a large number of other particles such as electrons, neutrons, protons, and photons. Some of the electrons so formed are captured in a thick zone around the earth known as the inner van Allen belt.

The main part (50–80%) of the cosmic radiation which reaches the earth's surface consists of high energy muons. Muons have much less tendency to react with atomic nuclei than pions and, therefore, can penetrate the atmosphere and the solid material of the earth relatively easily. The remaining part, which is the lower energy component of the cosmic radiation that strikes the earth, consists of photons, electrons, and positrons. At sea level this part of the cosmic radiation gives rise to approximately 2–3 ion pairs s^{-1} cm^{-3} of air. It is this component of the cosmic radiation that gives rise to the cosmic ray portion of the natural background that is measured in nuclear detection devices in laboratories. The remainder of the natural background comes from naturally occurring radioactive elements in the laboratory materials and surrounding building. Some of the cosmic radiation interacts to make atmospheric radioactivity which is principally the nuclides ^{3}H and ^{14}C. This important radioactivity is treated in §12.1.

5.3. Elementary particles and forces of nature

Cosmic radiation contains a large number of the kind of particles which are called elementary particles. We have so far mentioned protons, electrons, neutrons, positrons, pions, muons, photons, and neutrinos. This group of elementary particles began to be considerably expanded about 1947 when physicists discovered the first of the so-called "strange" particles in cloud chamber pictures of cosmic rays. These new elementary particles were called strange because they lived almost a million million times longer than the scientists had any reason to expect at that time. Within a few years after the first such particle had been discovered, a total of 7 strange particles were known, and physicists began to suspect that the reason for their long lifetime was the different types of forces involved in their creation and annihilation. By 1958 there were still only 15 particles that were classified as elementary. However, the population of elementary particles literally exploded during the sixties, and it is difficult to present a complete list of such particles at the present time, partially because compiling such a complete list involves the question of definition of what is an elementary particle.

A reason for such a large increase in the number of elementary particles discovered during the decade of the sixties is the number of very high energy accelerators which began operation during that time. In these accelerators it was possible to impart sufficient kinetic energy to protons that interactions with nuclides transformed a large amount of the kinetic energy into matter. In order to produce a mass of about 1 u (i.e. the approximate mass of a proton or a neutron) about 1 GeV of energy must be transformed. Since the high energy accelerators built in the sixties could accelerate protons to tens of GeV, it was possible to produce elementary particles not only lighter but even heavier than protons.

The number of such particles now exceeds 100, and the study of these particles has created a new scientific area called *elementary particle physics*. This field is quite different from nuclear physics, which is concerned with composite nuclei only. A principal objective of elementary particle physics has been to group the particles together according to their properties to reflect a meaningful pattern which would indicate the basis for a theory of the nature of the elementary particles. This effort somewhat resembles the development of the Periodic Table, which led scientists to an understanding of atomic structure. Such a grouping is shown in Table 5.1, where the bases for classification are quantum physical properties ascribed to the particles. However, one of the most important properties, that of particle interaction (we have referred to this earlier as nuclear force), cannot easily be described in such terms.

The universe is an immense and incredibly diverse assembly. Yet in all this diversity scientists have been able to discover thus far only four basic ways in which objects can experience mutual attractions or repulsions. Before proceeding with further discussion of the elementary particles of nature, let us describe in a qualitative way these four basic *forces* of nature.

TABLE 5.1. *Classification and properties of some elementary particles*

	Class		Name	Symbol	Rest mass (MeV)	Rest mass (u)	Stability; lifetime (s)	Antiparticle
Fermions $s = 1/2$ Pauli princ. valid	Leptons (light) W, A – no, P – no		Neutrino	ν	0	0	Stable	$\bar{\nu}$
			Electron	e^-	0.5110	0.000 548 6	Stable	e^+, positron
			Muon	μ^-	105.7	0.1135	$\rightarrow e^- + \nu + \bar{\nu}$; 2×10^{-6}	μ^+
	Baryons (heavy) S A – yes P +	Nucleons	Proton	p^+	938.2	1.007 276	Stable	$\bar{p}^-$
			Neutron	n	939.5	1.008 665	$\rightarrow p + e^- + \bar{\nu}$; 1×10^3	$\bar{n}$
		Hyperons (hyper-nucleons)	Lambda	Λ	1115.4	1.1976	$\rightarrow p + \pi^-$; $\rightarrow n + \pi^+$; 3×10^{-10}	$\bar{\Lambda}$
			Sigma	$\Sigma^+, \Sigma^0, \Sigma^-$	1193 ± 4	1.281 ± 0.004	$\nrightarrow p, n, \Lambda, \pi^{\pm 0}$; 10^{-10}	$\bar{\Sigma}^-, \bar{\Sigma}^0, \bar{\Sigma}^+$
			Xi	Ξ^0, Ξ^-	1315; 1321	1.412; 1.418	$\rightarrow \Lambda + \pi^0$, $\rightarrow \Lambda + \pi^-$; 10^{-10}	$\bar{\Xi}^0, \bar{\Xi}^+$
Bosons	Photon W, A – no, P – no, $s = 1$		Photon	γ	0	0	Stable	same
$s = 0, 1, \ldots$ ~~Pauli princ.~~	Mesons (medium) S, A – no, P –, $s = 0$		pi meson	π^+, π^0, π^-	137 ± 3	0.147 ± 0.003	$\pi^0 \rightarrow 2\gamma$, 10^{-16}; $\pi^\pm \rightarrow \mu^\pm + \nu$, 3×10^{-8}	same
			K meson	K^+, K^0	496 ± 2	0.533 ± 0.002	$\nrightarrow \pi^{\pm 0}, u^{\pm 0}, \nu$; 10^{-8}–10^{-10}	$K^-, \bar{K}^0$

W = weak interaction; S = strong interaction; A – yes = baryon number must be conserved; P + = conservation of parity and particle group has positive or even parity; P – = negative or odd parity conservation; P – no = parity is not conserved; ~~Pauli principle~~ = Pauli principle not valid; s = spin quantum number.

The first and weakest force of nature is that of *gravity*. This is the force that causes all objects to attract one another and is responsible for the attraction of the planets to the sun in the solar system and of the solar system to the rest of the galaxy. It is also the force that holds us to the earth. It seems paradoxical that the weakest attraction of the four basic forces of nature is the one that is responsible for the assembly of the largest objects on the greatest scale.

A second force of nature with which we are all relatively familiar is that of the *electromagnetic force*. The electromagnetic force is expressed by Coulomb's law and is responsible for the attraction and repulsion of charged bodies. Just as the gravitational force holds the planets in their orbits about the sun and explains the stability of the planetary systems, so the electromagnetic force explains the attraction between electrons in atoms, atoms in molecules, and ions in crystals. It is the force that holds the atomic world together. It is approximately 10^{36} times stronger than the gravitational force. If gravity is the force underlying the laws of astronomy, electromagnetism is the force underlying the laws of chemistry and biology.

The third major force in nature has been discussed briefly in Chapter 3 where we called it the nuclear force. This force is also known as the *strong interaction force* and is the one responsible for holding nuclear particles together. Undoubtedly it is the strongest in nature but operates over the very short range of approximately 10^{-14} m. By contrast, although gravity is the weakest force, it operates over the immense distances involved in the universe. Electromagnetism binds electrons to nuclei in atoms with an energy corresponding to several electron volts, whereas the strong interaction force holds nucleons together in nuclei with energies corresponding to millions of electron volts. Particles that exert strong nuclear forces are called *hadrons* ("the strong ones"). They are subdivided into two classes (Table 5.1): *baryons* (heavy) which contain protons, neutrons, and heavier particles, and *mesons* (for medium) which include pions.

The fourth force is the one about which we know the least at the present time. It is the force that is involved in the radioactive β-decay of atoms and is known as the *weak interaction force*. Like the strong interaction, this weak interaction force operates over extremely short distances and is the force that is involved in the interaction of very light particles known as *leptons* (electrons, muons, and neutrinos) with each other and as well as their interaction with mesons, baryons, and nuclei. One characteristic of leptons is that they seem to be quite immune to the strong interaction force. The strong nuclear force is approximately 10^2 times greater than the coulombic force, while the weak interaction force is smaller than the strong attraction by a factor of approximately 10^{13}.

The strong interaction manifests itself in its ability to react in very short times. For example, for a particle which passes an atomic nucleus of about 10^{-15} m in diameter with a velocity of approximately 10^8 m s^{-1} (i.e. with a kinetic energy of $\sim$ 50 MeV for a proton and 0.03 MeV or an electron), the time of strong interaction is about 10^{-23} s. This is about the time of rotation of the atomic nucleus. The weak interaction force requires a much longer reaction time and explains why leptons such as electrons and photons do not react with atomic nuclei but do react with the electron cloud of the atom which has a diameter on the order of 10^{-10} m. There is sufficient time in passing this larger diameter for the weak interaction force to be effective. The long lifetimes of the "strange" particles are due to the fact that their decay involves the slow weak interaction force rather than the much faster strong interaction force.

5.4. Some general properties of elementary particles

Because macroscopic matter—as we see it around us—is ultimately built up of a few elementary particles, it is reasonable to expect that elementary particles possess the same properties that we observe in the macroscopic world. Three such properties—familiar to the chemist—are *mass*, *charge*, and *spin*. All elementary particles (except the photon) have one or several of these properties, but, in contrast to macroscopic matter, they are all quantized: the rest mass can only have fixed values, and so also for charge and spin. The combination of these three properties is usually sufficient to make an elementary particle unique, as can be seen from Table 5.1. In addition to these three fundamental properties a number of others can be ascribed to each elementary particle, as *baryon number*, *statistics*, *symmetry*, *parity*, *hypercharge*, *iso spin*, and *strangeness*, in addition to the strong and weak interaction discussed in the previous section.

We usually define a particle as something having *mass*. In Table 5.1 the neutrino and the photon have rest masses of zero. As has been tacitly assumed in discussing the β- and γ-decay, these particles show momentum and they behave as if they have a mass provided their energies exceed 0. We conclude that neutrinos and γ-rays have particle properties and a relativistic mass > 0 (for further discussion, see §7.1).

In §4.1 we mentioned the requirement of constant mass number in nuclear reactions. In particle physics, since all nuclei are baryons, this is referred to as the conservation of baryon number, and must be obeyed by all baryons (designated by A—yes in Table 5.1). However, this rule does not hold for the annihilation of a baryon through the interaction with its antiparticle. Figure 5.6 describes such a reaction and the subsequent elementary particle decay processes.

All leptons and baryons (this includes p^+, n, and e^-) have spin, i.e. they rotate around their own axes. Quantum mechanical calculations and experimental observations have shown that each particle has a fixed rotational energy which is determined by the *spin quantum number* s ($s = 1/2$ for leptons and elementary baryons). Particles of nonintegral spin are called *fermions* because they obey the quantum statistical rules devised by E. Fermi and P. A. M. Dirac, which state that two such particles cannot exist in the same closed system (nucleus or electron shell) having all quantum numbers the same (the Pauli principle). Photons and mesons have even spin: 0 for mesons and 1 for photons; even spin particles are called *bosons*.

Fermions share the property that they can be created and destroyed only in conjunction with an antiparticle of the same class. For example, if an electron is emitted in β-decay it must be accompanied by the creation of an antineutrino. Conversely, if a positron which is an antielectron is emitted in β-decay, it is accompanied by the creation of the neutrino. Bosons do not require the formation of an antiparticle in their reactions. Thus K-mesons can decay into two or three lighter mesons, none of which need to be antiparticles.

The *spin* of a charged particle will lead to the formation of a magnetic moment (μ_s) directed along the axes of rotation. The magnetic moment for the electron spin (9.27×10^{-24} JT^{-1}) is about 700 times larger than that of the proton spin (1.41×10^{-26} JT^{-1}), although theory predicts a larger ratio of 1836 ($= \mathbf{M}_e/\mathbf{M}_p$; see §6.2). The reason for this deviation is probably the fact that the proton is not an evenly charged rotating sphere. In fact the proton has been found to have an internal structure indicating layers of varying charge density (Fig. 5.4). This figure also indicates that the neutron has an uneven internal charge density, and consequently it also is expected to

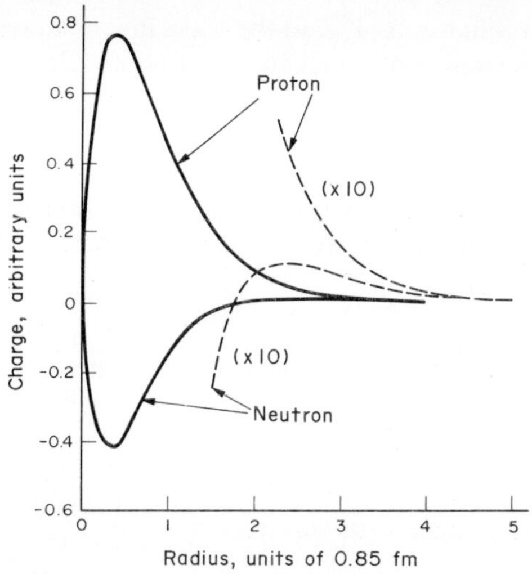

FIG. 5.4. Internal structure of the proton and neutron indicating varying charge densities. (According to R. Hofstadter.)

have a magnetic moment. Measurements have shown that it has a negative sign (as compared to the electron and proton) with a value of $0{\cdot}97 \times 10^{-26}$ J T^{-1}.

In an external magnetic field these particles may orient themselves parallel or antiparallel to the field. The free particles, however, have their axes (or rather, their spin vectors, see §6.2) pointing in all directions, as, for example, electrons in an X-ray tube, or protons in an accelerator beam, or neutrons in a nuclear reactor (see Fig. 6.3(a)). This spin is further discussed in §6.2.

In chemistry crystals are characterized by their *symmetry*, i.e. reflection, rotation, and/or inversion may or may not have the same pattern for two crystals. The simplest case of symmetry is a ball, which looks the same from all directions, whereas our left and right hands do not. Symmetry is an important physical concept. In elementary particle physics mirror symmetry requires that all particles should have their antiparticles with all properties the same (e.g. the mass, the spin, the lifetime, etc.) except for one which is the opposite. The antiparticles are listed in Table 5.1 (the photon and π-mesons are their own antiparticles).

We may well think about antimatter as consisting of antiprotons and antineutrons in the antinucleus surrounded by antielectrons (i.e. positrons). Superficially, there would be no way to distinguish such antimatter from our matter (sometimes called koino matter). It has been proposed that the universe is made up of matter and antimatter as a requirement of the principle of symmetry. In that case some galaxies, which we perhaps can observe, should be made up of antimatter. When such antimatter galaxies (or material expelled from them) collide with koino matter galaxies, matter is annihilated and tremendous amounts of energy are released.

Nature is not completely symmetric. If ^{60}Co atoms are oriented by a strong magnetic field such that their nuclear spin vectors are all parallel, we might expect that the β^- particles emitted in the radioactive decay would be expelled equally in all directions of space. They are not; the β-particles are always emitted along the direction opposite to the spin of the nucleus. Moreover, it has also been shown in such experiments that the antineutrinos always have their spin vector along the direction of flight.

Parity is associated with the properties of the wave function of the particle, and with symmetry. If all the coordinates X, Y, Z in the wave function are changed to $-X$, $-Y$, $-Z$, when the sign of the function is unchanged, the system has positive or even parity; when the sign is changed, the system has negative or odd parity. Parity must be conserved in nuclear reactions involving strong interaction but not in those involving weak interaction. From Table 5.1 it is seen that this law holds only for the strong interacting particles, baryons, and mesons. The experiment in the previous paragraph is not a violation of parity conservation as it involved β-decay which is a weak interaction.

5.5. Waves and particles

It is a daily experience that moving bodies have a kinetic energy which involves a mass and a velocity. Less familiar is the concept of M. Planck (1900) in which light moves in wave packets of energy:

$$E = \mathbf{h}\nu \tag{5.1}$$

Here ν is the frequency of light with wavelength λ

$$\mathbf{c} = \nu\lambda \tag{5.2}$$

and **h** the Planck constant, 6.63×10^{-34} J s.

Einstein in his theory of the photoelectric effect, and A. Compton in the theory on the scattering of photons (Chapter 14), showed that photons in fact have not only a discrete energy, but also a discrete momentum (cf. §4.33)

$$p_\nu = E_\nu/\mathbf{c} \tag{5.3}$$

Photons seem to collide with other particles as if they have a real mass and velocity as in the classical mechanical expression for momentum: $p = mv$ (4.3). If we put $v = \mathbf{c}$ and equate with (5.3) we obtain a hypothetical relativistic mass of the photon as

$$m_\nu = E_\nu/\mathbf{c}^2 \tag{5.4}$$

This is the mass–energy relation of Einstein, eqn. (4.23).

There are many examples of mass properties of photons. To the two mentioned above we may add the solar pressure (i.e. photons from the sun which push atoms away from the sun and into space), which has played a significant part in the formation of our planetary system, and measurements showing that photons are attracted by large masses through the gravitational force. Thus we see the evidence for the statement in the beginning that all elementary particles must have relativistic mass, even if the rest mass is zero.

It is reasonable to assume, as L. de Broglie did in 1924, that since photons can behave as moving particles, moving particles may show wave properties. From the previous equation, we can devise that the wavelength of such *matter waves* is

$$\lambda = \mathbf{h}/mv \tag{5.5}$$

This relation is of importance in explaining nuclear reactions, and has led to practical consequences in the use of electron diffraction and in the development of electron microscopy.

The wave and particle properties of matter complement each other (the complementarity principle; N. Bohr, 1928), and throughout this book we use models based on wave properties and sometimes on particle properties, depending on which more directly explain the particular phenomenon under discussion.

5.6. The Heisenberg uncertainty principle

The *uncertainty principle* (W. Heisenberg, 1927) states that it is impossible to measure simultaneously the exact position and the exact momentum of a particle. This follows directly from the wave properties of the particle. If, for example, we attempt to measure the exact position of an electron by observing the light emitted when it hits a scintillating screen, this act interferes with the movement of the electron causing it to scatter, which introduces some uncertainty in its momentum. The size of this uncertainty can be calculated exactly and is related to the Planck constant. If Δx denotes the uncertainty in position and Δp the uncertainty in momentum along the x axis, then

$$\Delta x \Delta p \geqq \frac{\mathbf{h}}{2\pi} \equiv \hbar \quad (5.6)$$

$\hbar$ is called "h-bar" and is 1.05×10^{-34} J s.

This principle holds for other *conjugate variables*, as angle θ and angular momentum p_θ

$$\Delta\theta\Delta p_\theta \geqq \hbar \quad (5.7)$$

and time and energy

$$\tau\,\Delta E \geqq \hbar \quad (5.8)$$

This latter equation relates the lifetime τ of an elementary (or nuclear) particle to the uncertainty in its energy (ΔE). For excited nuclear states this can be taken as the width of the γ-peak at half-maximum intensity (the "FWHM value") (Fig. 5.5). For example, if $\Delta E_\gamma = 1.6$ keV, then $\tau \geqq 1.05 \times 10^{-34}/1600 \times 1.60 \times 10^{-19}\,\text{s} = 4.1 \times 10^{-19}$ s. This is a long time compared to that of a nuclear rotation, which is about 10^{-23} s.

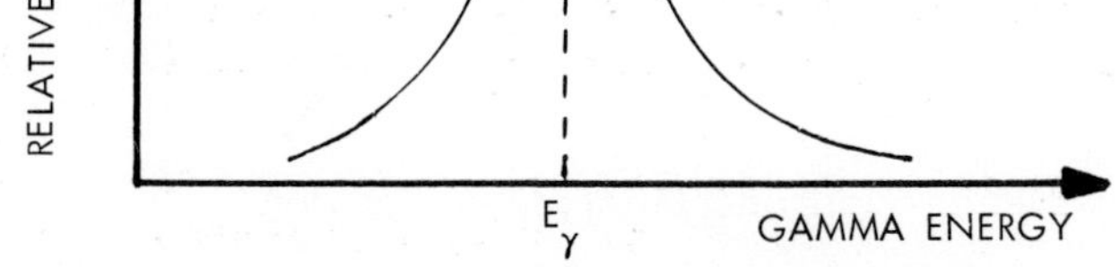

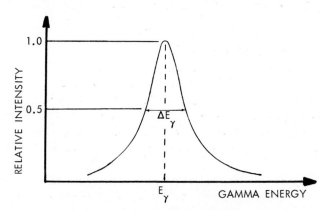

FIG. 5.5. The energy half-width value ΔE_γ is obtained as the width of the γ-peak at half-maximum intensity (FWHM value).

5.7. Some reactions of elementary particles*

The existence of the neutrino was predicted by W. Pauli in 1927 but it was not proven until in 1953 when F. Reines and C. L. Cowan conducted counting experiments close to a large nuclear reactor. Since in the β-decays following fission neutrinos are emitted, nuclear reactors are very intense neutrino sources. The detector consisted of a scintillating solution containing cadmium and was surrounded by photomultipliers to observe the scintillations occurring as a consequence of the following reactions:

$$\bar{\nu} + {}^1\text{H} \rightarrow \text{n (fast)} + \text{e}^+$$

$$\text{e}^+ + \text{e}^- \rightarrow 2\gamma_1$$

$$\text{n (thermal)} + {}^{113}\text{Cd} \rightarrow {}^{114}\text{Cd} + \gamma_2$$

The γ's emitted are of different energy; the γ_1 is 0.51 MeV, but γ_2 much higher. There is also a time lag between the γ's because of the time required for the fast neutrons to be slowed down to thermal energy. The detection system allowed a delay time to ascertain a relation between γ_1 and γ_2 (delayed coincidence arrangement). When the reactor was on, 0.2 cpm were observed, while it was practically zero a short time after the reactor had been turned off, thereby demonstrating the formation of neutrinos during reactor operation.

The sun is believed to be a very intense neutrino source due to the nuclear fusion processes (Chapter 11). About 5% of the energy output of the sun is in the form of neutrinos; at the earth about 10^{15} ν pass through our body every second (10^{11} ν s^{-1} cm^{-2} of earth surface), but only 1 neutrino is absorbed every 10 s. Attempts have been made to measure the sun's neutrino output—it should allow insight into the relative amounts of the different fusion reactions in the sun—by placing a tank of 25 m^3 tetrachloroethylene in a deep mine (to reduce other cosmic background). The neutrino detection method is based on the reaction

$$\nu + {}^{37}\text{Cl} \rightarrow {}^{37}\text{Ar} + \text{e}^-$$

The radioactive ^{37}Ar ($t_{1/2} = 35$ d) is removed from the liquid by a helium gas purge and placed in a small, low level proportional counter to observe the 2.8 keV Auger electrons emitted in the electron capture decay of ^{37}Ar. At present the results obtained indicate a solar neutrino output which is considerably less than expected.

The discovery of π-mesons (or *pions*) was reported by C. F. Powell and G. P. Occhialini in 1948 after they had analyzed tracks in photographic emulsions placed for some months on a mountain top to get a high yield of cosmic ray interactions. Pions are produced in large amounts in all high energy ($\gtrsim$ 400 MeV) nuclear reactions. In 1935 H. Yukawa suggested that the nucleons in a nuclide were held together through the exchange of a hypothetical particle, which we now recognize as the pion, just as hydrogen atoms in H_2^+ are held together through the exchange of an electron:

$$\text{p} \leftrightarrow \text{n} + \pi^+; \quad \text{n} \leftrightarrow \text{p} + \pi^-; \quad \text{p} \overset{\pi^0}{\leftrightarrow} \text{p}; \quad \text{n} \overset{\pi^0}{\leftrightarrow} \text{n}$$

Pions are the particles of strong interaction. In the reaction

$$^7\text{Li} + {}^1\text{H} \rightarrow {}^7\text{Be} + \text{n}$$

one may assume that the reaction occurs through the transfer of a π^+ from the proton to the neutron:

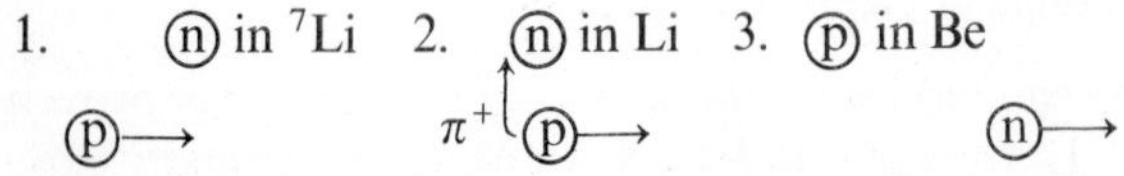

If the proton energy is 2 MeV it has a velocity of 2×10^7 m s^{-1}. The size of the lithium-nucleus is about 6×10^{-15} m. Thus the time for nuclear passage is $6 \times 10^{-15}/2 \times 10^7 = 3 \times 10^{-22}$ s, i.e. in the time range for strong interaction.

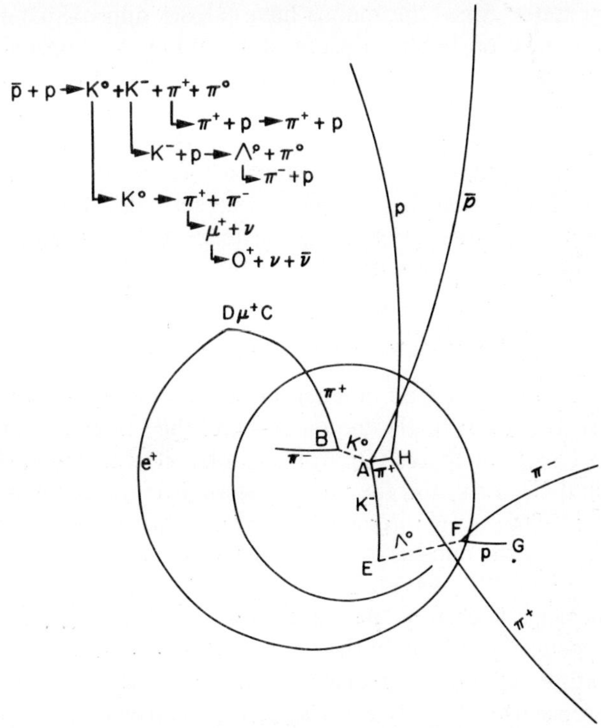

FIG. 5.6. The reaction products of an annihilated antiproton as seen in a liquid hydrogen bubble chamber. (According to *Annual Report 1961*, CERN.)

Figure 5.6 shows the reaction of an antiproton produced through the reaction

$$^{1}\mathrm{H}(\gtrsim 6\,\mathrm{GeV}) + {}^{1}\mathrm{H}\,(\text{liquid hydrogen}) \rightarrow {}^{1}\mathrm{H} + {}^{1}\mathrm{H} + {}^{1}\mathrm{H} + {}^{1}\bar{\mathrm{H}}$$

in a liquid hydrogen bubble chamber. The antiproton ($\bar{p}$) enters the bubble chamber along the path indicated by the arrow. At point A it is annihilated through the reaction with a proton, leading to the formation of four mesons K^0, K^-, π^+, and π^0. The uncharged particles give no tracks, but the K^0 decays at B into π^+ and π^-; π^+ decays at C into two leptons, μ^+ and ν, of which μ^+ in point D decays to e^+, ν, and $\bar{\nu}$. The neutrinos disappear without reaction, as did the π^0 originally formed. The K^- decays in point E to two unchanged particles, a Λ^0 hyperon and a π^0 meson; the π^0 is detected through its decay in point F to a proton, which is stopped at G, and a π^-. The originally formed π^+ at point A is scattered against a proton in point H. The information has been obtained through kinematic analyses: from the particle range, track density, and curvature (the bubble chamber is kept in a homogeneous magnetic field) one determines particle energy, mass, and charge.

5.8. Muonic atoms*

Muons interact only through the weak interaction force and, therefore, can spend a considerable length of time in the vicinity of the nucleus without interaction. As a result, negative muons can approach closely to atomic nuclei, and at sufficiently low velocities can be captured by a nucleus into fixed energy states, which are quite analogous to the electronic energy states. Since the muons have masses approximately 200 times that of the electron, the most probable positions of the muons are approximately 200 times closer to the nucleus than the analogous electronic states. Muonic atoms show X-ray excitation spectra quite analogous to electronic atoms and we can speak of the K, L, M, etc., shells of the muonic atoms in the same sense as we speak of those of electronic atoms. After a period of time of 10^{-16} s, which is long by atomic standards but quite short by our normal measurements, the muon interacts with the nucleus and is annihilated. Unfortunately, the rather brief time of existence of muonic atoms has not allowed the formation of chemical compounds.

5.9. Elementary vs. composite particles*

When two objects experience an interaction due to one of the four forces, the result is a composite entity. For example, the planets and the sun experience a gravitational attraction which results in the stability of the solar system. Similarly, the attraction between the orbital electrons and the nucleus of an atom results in a composite entity known as the atom. The strong interaction acts on the nucleons in the nucleus to form the composite entity of the nucleus. As physicists have been able to study neutrons and protons more intensively, they have found that they react as if they were made up of some more primitive objects known as *quarks*. Obviously, then, whether a particle is elementary or composite depends upon the point of view with which we are regarding the object and the nature of the force involved. To an astronomer a planet is an elementary particle in a real sense while to a chemist the nucleus is an elementary particle. However, to a nuclear physicist the nucleons are elementary particles, while to the high energy physicist it is doubtful what particles could be classified as elementary particles.

To circumvent this problem of what is elementary and what is composite, physicists have lately avoided the title elementary and have preferred to use a more general concept in which all particles are considered to be possible states in which matter can condense. These states are related to the force which forms them. In this sense the solar system is a state of the gravitational force, an atom is a state of electromagnetic force, and a nucleon is a state of the strong interaction force. This concept, which regards the various objects in the physical world as states of their interaction forces, is helpful in many ways. For example, not only is there a bewildering array of the so-called "elementary" particles but we have also learned that these particles have their opposites in an antiparticle. In the concept of states of matter, a particle can represent a positive energy state of a system while its analog antiparticle represents the negative state of the same system.

Some regular patterns have been found for the elementary particle. For example, if the spin of a certain class of particles is plotted against the square of the particle masses, all particles of the same "family" fall on a straight line (Regge trajectories). This indicates that many "new particles" in fact may only be excited states of the same particle, differing in quantum number such as spin (or "hyper charge"); in fact hundreds of such states are now known. For example, the neutron has a mass corresponding to 939 MeV and spin 1/2, and there is baryon with mass 1688 MeV and spin 5/2 with all other properties like the neutron; the heavier particle may thus be a highly excited neutron. Y. Ne'eman and M. Gell-Mann have shown that many of these particles can be combined in octets, as shown for baryons in Fig. 5.7.

Many attempts have been made to unify all particles in one simple theory. According to the "bootstrap theory", all particles are made up of one another in a perfect democracy,

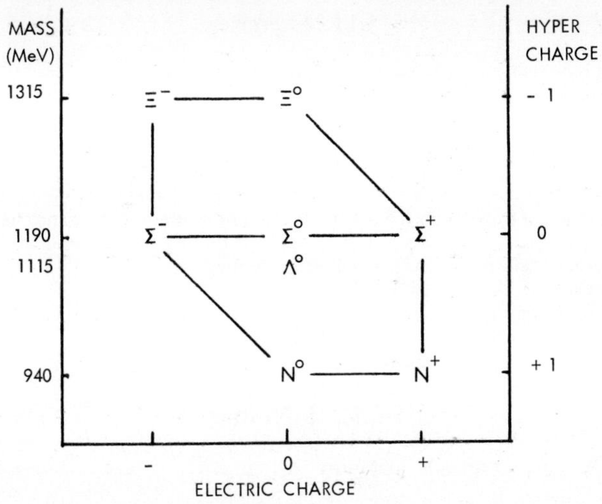

FIG. 5.7. Baryon octet (spin + 1/2). The N^0 and N^+ are the neutron (n) and proton (p), respectively, which are considered as two different states of the same nucleon (N).

and it is not possible to distinguish between composite systems, constituent elements, and binding elements. According to another theory all particles are made up of hypothetical particles called quarks. Quarks and antiquarks are assumed to have electric charges $+2/3$ and $-1/3$, and an atomic mass of 1/3 u. For example, a proton may consist of $+2/3$, $+2/3$, and $-1/3$ (mass 1, charge $+1$) and a neutron of $+2/3$, $-1/3$, and $-1/3$. Much research is presently underway in experimental particle physics to find such particles.

5.10. Particles and poets*

The amazing variety and interrelationships of elementary particles may seem confusing when we first study them. Such a sense of confusion as well as an appreciation of the excitement of this rapidly evolving field of science has inspired the well-known American poet and novelist, John Updike, to write the following:

News from the Underworld
(After Blinking One's Way Through "The detection of neutral weak currents," in *Scientific American*)

They haven't found the W
wee particle for carrying
the so-called "weak force" yet, but you
can bet they'll find some odder thing.

Neutrinos make a muon when
a proton, comin' through the rye,
hits in a burst of hadrons; then
eureka! γ splits from π

and scintillation counters say
that here a neutral lepton swerved.
Though parity has had its day,
the thing called "strangeness" is preserved.

(From *The American Scholar* **44** (1975) 584.)

5.11. Exercises

5.1. What proof exists that some cosmic rays do not come from the sun?

5.2. What is the primary cosmic radiation hitting earth's atmosphere? Does it penetrate to the earth's surface?

5.3. What intensity from cosmic radiation is expected for an unshielded laboratory γ-detector?

5.4. Which type of mesons are released in high energy particle interactions, and why?

5.5. What kinds of forces exist in nature? How does the weak interaction manifest its properties?

5.6. What are bosons and how do they differ from fermions? Does the difference have any practical consequence?

5.7. What proof exists that the photon has matter properties?

5.8. How can the neutrino be detected?

5.9. If the energy width of a peak in a γ-spectrum is known, what other property of the parent atom can be deduced? What is the name of the principle applied?

5.12. Literature

See Appendix A, § A.1 and Chapter 11 literature.

Cosmic radiation

S. Flügge (ed.), Kosmische Strahlung, *Handbuch der Physik*, Band XLV 1/1, Springer-Verlag, Berlin, 1961.
G. Burbidge, The origin of cosmic rays, *Sci, Am.* **215**, Aug. 1966.
V. F. Weisskopf, The three spectroscopies, *Sci. Am.* **218,** May 1968.
T. Gold, Pulsars—the key to kosmic rays?, *Sci. Res.*, June 9, 1969, p. 33.
A. W. Wolfendale (ed.) *Cosmic Rays at Ground Level*, The Institute of Physics, London, 1973.
A. Webster, The cosmic background radiation, *Sci. Am.* **231,** Aug. 1974.
D. D. Peterman and E. V. Benton, High LET particle radiation inside Skylab command module, *Health Physics* **29** (1975) 125.

Elementary particles

G. F. Chew, M. Gell-Mann, and A. H. Rosenfeld, Strongly interacting particles, *Sci. Am.* **210**, Feb. 1964.
K. W. Ford, *Die Welt der Elementarteilchen*, Springer-Verlag, 1966.
Y. Ne'eman, What's happening in particle physics? *Science and Technology*, March 1967.
L. C. C. Yvan, *Elementary Particles*, Academic Press, 1971.
C. E. Wiegand, Exotic atoms, *Sci. Am.* **227**, Nov. 1972.
B. P. Gregory, *Annual Report 1974*, CERN.
A. S. Goldhaber and M. M. Nieto, The mass of the photon, *Sci. Am.* May 1976.

CHAPTER 6

Nuclear Structure

Contents

Introduction

Throughout the ages and in every civilization, people have developed explanations of observed behaviors. These explanations are based on the principle of causality, i.e. every effect has a cause and the same cause produces always the same effect. We call these explanations *models*.

Scientists are professional model builders. Observed phenomena are used to develop a model, which then is tested through new experiments. This is familiar to every chemist: although we cannot see the atoms and molecules which we add into a reaction vessel, we certainly have some idea about what is going to happen. It is indeed our enjoyment in developing models which causes us to experiment in science. Of course, since man is fallible some models may turn out to be wrong, but as new data accumulate, wrong or naive models are replaced by better ones.

We have already shown how one model for the nuclear structure, the liquid drop model, has helped us to explain a number of nuclear properties, the most important being the shape of the stability valley. But the liquid drop model fails to explain other important properties. In this chapter we shall try to arrive at a nuclear model which takes into account the quantum mechanical properties of the nucleus.

6.1. Requirements of a nuclear model

Investigation of light emitted in atomic reactions led N. Bohr to suggest the quantized model for the atomic nucleus, which became the foundation for explaining the chemical properties of the elements and justifying their ordering in the Mendeleev periodic system. From studies of molecular spectra and from theoretical quantum and wave mechanical calculations based on Bohr's model, we are able to interpret many of the most intricate details of chemical bonding.

In a similar manner, nuclear reactions have yielded information which helps us develop a picture of nuclear structure. But the situation is more complicated for the nucleus than for the atom. In the nucleus there are two kinds of particles, protons and neutrons, packed close together, and there are two kinds of forces—the electrostatic force and the short range strong nuclear force. This more complex situation has caused slow progress in developing a satisfactory model, and no single nuclear model has been able to explain all the nuclear phenomena.

6.1.1. *Some general nuclear properties*

Let us begin with a summary of what we know about the nucleus, and see where that leads us.

In Chapter 3 we observed that the binding energy per nucleon is almost constant for the stable nuclei (Fig. 3.3) and that the radius is proportional to the cube root of the mass. We have interpreted this as reflecting fairly uniform distribution of charge and mass throughout the volume of the nucleus. Other experimental evidence supports this interpretation (Fig. 3.4). This information was used to develop the liquid drop model, which successfully explains the stability valley (Fig. 3.1). This overall view also supports the assumption of a strong, short range nuclear force.

A more detailed consideration of Figs. 3.1 and 3.3 indicates that certain mass numbers seem to be more stable, i.e. nuclei with Z- or N-values of 2, 8, 20, 28, 50, and 82 (see also Table 3.1). There is other evidence for the uniqueness of those numbers. For example, if either the probability of capturing a neutron (the neutron capture cross-section) or the energy required to release a neutron is plotted for different elements, it is found that maxima occur at these same neutron numbers, just as maxima occur for the electron ionization energy of the elements He, Ne, Ar, Kr, etc. (i.e. at electron numbers of 2, 8, 16, 32, etc.). The nuclear N- or Z-values of 2, 8, 20, 28, 50, and 82 are called "magic numbers".

It seems logical that these magic numbers indicate some kind of regular substructure in the nucleus. Moreover, since the same magic numbers are found for the neutrons and for the protons, we would further assume that the neutrons and the protons build their substructure independently of each other, but in the same way. Another fact that must be indicative of the nuclear substructure is the stability for nuclei with even proton or even neutron numbers. Since we know that the individual nucleons have spin, we could postulate that nucleons in the nucleus must pair off with opposed spins.

6.1.2. *Quantized energy levels*

The nucleus would thus seem to consist of independent substructures of neutrons and protons, with each type of nucleon paired off as far as possible. Further, the nucleons are obviously grouped together in the magic numbers. From the decay of radioactive nuclei we know that the total decay energy (Q-value) of any particular nuclide has a definite value. Moreover, γ-emission from any particular nucleus involves discrete, definite values. These facts resemble the quantized emission of electromagnetic radiation (X-ray, UV, visible light, etc.) from atoms. We may conclude a similar explanation for the nucleus: decay of radioactive nuclei, whether α, β, or γ, involves a transition between discrete quantized energy levels.

6.1.3. *The nuclear potential well*

In our development of a model of the nucleus we have so far not considered the nuclear binding energy. Let us imagine the situation wherein a neutron of low kinetic energy approaches a nucleus (Fig. 6.1). Since the neutron is uncharged it is not affected by the Coulomb field of the nucleus and approaches the nucleus with no interaction, until it is close enough to experience the strong nuclear force F_n, which is always attractive, i.e. F_n is a positive quantity. At the point r_n the neutron experiences the strong attraction to the nucleus and is absorbed. The surface of the nucleus is assumed to extend to r_s since this distance represents the radius of the nucleus over which the nuclear

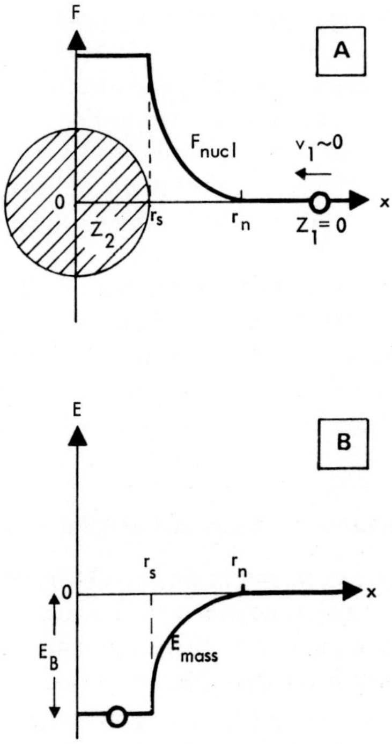

FIG. 6.1. The approach along the x-axes of a neutron (charge $Z_1 = 0$, velocity $v_1 \approx 0$) towards a nucleus (charge Z_2; Fig. A) leading to the neutron being trapped at a certain energy level, E_B (Fig. B). F_{nucl} is the nuclear force extending from the surface of the "solid core" at r_s to the distance r_n. E_B is the nuclear binding energy in the potential well.

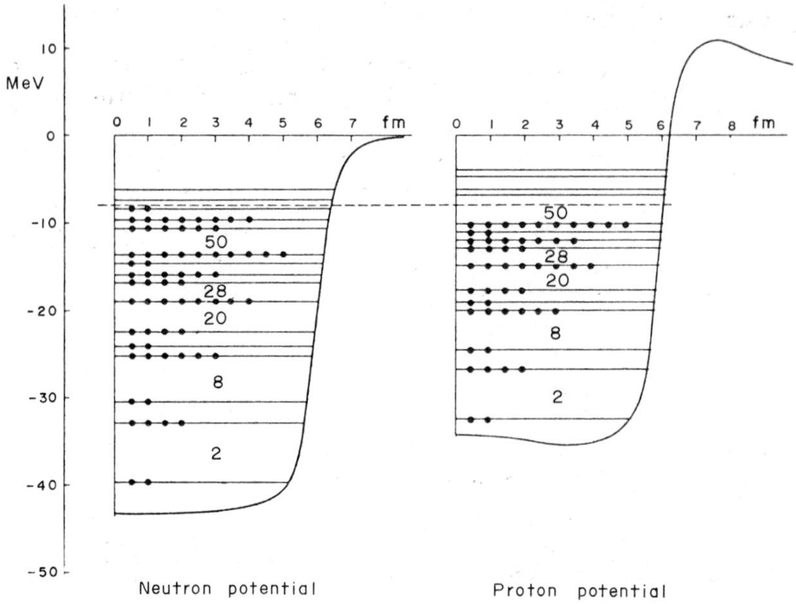

FIG. 6.2. The ^{116}Sn neutron and proton shell structure in the potential well. (According to S. G. Nilsson.)

force is constant. When the neutron is absorbed, energy is released and emitted in the form of a γ-quantum. The energy of the γ-ray can be calculated from the known masses of the reactants and product nuclides: $E_\gamma = -931.5(M_{A+1} - M_A - M_n)$.

The energy released is the (neutron) *binding energy* of the nucleus E_B. The total energy of the nucleus has thus decreased as is indicated in Fig. 6.1(B); it is common to refer to this decrease as a *potential well*. The nucleons can be considered to occupy different levels in such a potential well. The exact shape of the well is uncertain (parabolic, square, etc.) and depends on the exact mathematical form assumed for the interaction between the incoming particle and the nucleus.

Protons experience the same strong, short range nuclear force interaction as they contact the nucleus. However, they also experience a long range repulsive interaction due to the Coulomb force between the positive incoming protons and the positive charge of the nucleus. This repulsion prevents the potential well from being as deep for protons as for neutrons. Figure 6.2 shows the energy levels of the protons and of the neutrons in the nucleus of $^{116}_{50}$Sn.

6.2. Rotational energy and angular momentum

It is an intriguing fact that nucleons in nuclei, electrons in atoms, as well as large cosmic objects such as solar systems and even galaxies, are more dominated by rotational than by linear motion, although in our daily life the latter seems to be a more common phenomenon. In rotation there is a balance between two forces: the centrifugal force of inertia, which tries to move a body away from a center point, and an attractive force (gravitational, electrostatic, etc.), which opposes the separation. We can regard rotation as a constructive force, which builds atoms and planetary systems, while linear motion is destructive, moving bodies away from or against each other.

In the preceding section we assumed that the nucleus existed in some kind of a potential well. One may further assume that a nucleon moves around in this well in a way not too different from the way the electron moves around the atomic nucleus, i.e. with an oscillation between kinetic and potential energy. With some hypothesis about the shape of the nuclear potential well we can apply the *Schrödinger wave equation* to the nucleus. Without being concerned at this point with the consequences of assuming different shapes we can conclude that the solution of the wave equation allows only certain energy states. These energy states are defined by two quantum numbers: the *principal quantum number n*, which is related to the total energy of the system, and the *azimuthal (or radial) quantum number l*, which is related to the rotational movement of the nucleus.

6.2.1. *Rotational (mechanical) energy*

If a mass m circles in an orbit of radius r at a constant angular velocity ω (rad s^{-1}), the tangential velocity at this radius is

$$v_r = \omega r \tag{6.1}$$

The kinetic energy in linear motion is $E_{kin} = \frac{1}{2}mv^2$, and we therefore obtain for the (kinetic) rotational energy

$$E_{rot} = \tfrac{1}{2}m\omega^2 r^2 \tag{6.2a}$$

The angular velocity is often expressed as the frequency of rotation, $\nu_r = \omega/2\pi\ s^{-1}$. Equation (6.2a) can also be written

$$E_{rot} = \tfrac{1}{2}I_{rot}\,\omega^2 \tag{6.2b}$$

where

$$I_{rot} = \Sigma m_i r_i^2 \tag{6.3}$$

is the rotational *moment of inertia*. We consider as the rotating body a system of i particles of masses m_i, each individual particle at a distance r_i from the axis of rotation; then (6.2b) is valid for any rotating body. In nuclear science we primarily have to consider two kinds of rotation: the *intrinsic rotation* (or *spin*) of a body around its own axis (e.g. the rotation of the earth every 24 h), and the *orbital rotation* of an object around a central point (e.g. rotation of the earth around the sun every 365 days). Equation (6.2b) is valid for both cases, but (6.2a) only for the orbital rotation of a particle of small dimensions compared to the orbital radius, in which case $I_{rot} = mr^2$. For a spherical homogeneous spinning body (e.g. the earth's intrinsic rotation) of external radius r_{ex}, (6.2b) must be used, where $I_{rot} = \frac{2}{5}mr_{ex}^2$.

6.2.2. *Angular momentum*

Like linear motion, rotation is associated with a momentum, called the *angular momentum*. For the orbital rotation the *orbital angular momentum* (p_l) is

$$p_l = mv_r r \tag{6.4a}$$

while for spin the *spin angular momentum* (p_s) is

$$p_s = \omega I_{rot} \tag{6.4b}$$

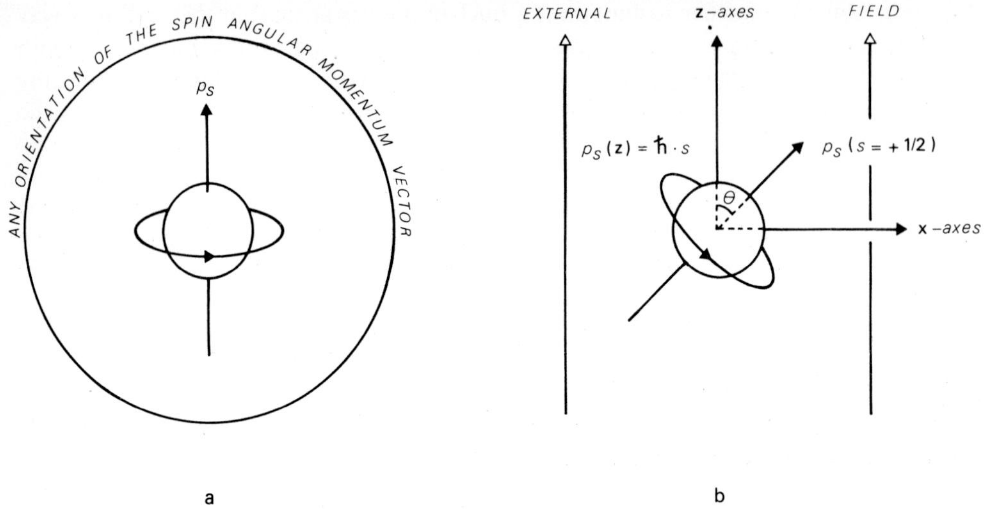

FIG. 6.3. A particle spinning in (a) field-free space and (b) in an external field along the z-axes.

Angular momentum is a *vector quantity*, which means that it has always a certain orientation in space, depending on the direction of rotation. For the rotation indicated in Figure 6.3(a), the vector can only point upwards. It would not help to turn the picture upside down because the coupling between rotational direction and its vector remains the same, as students south of the equator will agree.

Quantum mechanics prescribes that the spin angular momentum of electrons, protons, and neutrons must have the magnitude

$$p_s = \hbar\sqrt{s(s+1)} \tag{6.5}$$

here s is the *spin quantum number*. For a *single* particle (electron or nucleon) the spin s is always $\frac{1}{2}$. In addition to spin, the three atomic particles can have orbital movements. Again quantum mechanics prescribes that the magnitude of the orbital angular momentum of these particles

$$p_l = \hbar\sqrt{l(l+1)} \tag{6.6}$$

We shall refer to l as the *orbital* (*angular momentum*) *quantum number*. Only certain values are permitted for l, related to the main quantum number n:

for electrons: $0 \leqq l < n - 1$

for nucleons: $0 \leqq l \gtreqless n$

For nucleons but not for electrons l may (and often does) exceed n.

6.2.3. *Coupling of spin and orbital angular moments*

A rotating charge gives rise to a *magnetic moment* μ_s. The rotating electron and proton can therefore be considered as tiny magnets. Because of the internal charge distribution of the neutron (Fig. 5.4) it also acts as a small magnet. In the absence of any external magnetic field these magnets point in any direction in space (Fig. 6.3(a)), but in the presence of an external field they are oriented in certain directions determined by quantum mechanical rules. This is indicated by the angle θ in Figure 6.3(b), where we have the spinning particle in the center of a coordinate system. The quantum mechanical rule is that the only values allowed for the projections of spin angular momentum $p_s(z)$ on the field axes are:

$$p_s(z) = \hbar m_s \tag{6.7}$$

For composite systems, like an electron in an atom or a nucleon in a nucleus, the (*magnetic*) *spin quantum number* m_s may have two values, $+\frac{1}{2}$ or $-\frac{1}{2}$, because the spin vector has two possible orientations (up or down) with regard to the orbital angular momentum.

The orbital movement of an electron in an atom, or of a proton in the nucleus, gives rise to another magnetic moment (μ_l) which also interacts with external fields. Again quantum mechanics prescribes how the orbital plane may be oriented in relation to such a field (Fig. 6.4(a)). The orbital angular momentum vector $\vec{p}_l$ can assume only such directions that its projection on the field axes, $p_l(z)$, has the values

$$p_l(z) = \hbar m_l \tag{6.8}$$

m_l is referred to as the *magnetic orbital quantum numbers; m_l* can have all integer values between $-l$ and $+l$.

For a single-particle, nucleon or electron, the orbital and spin angular moments add vectorially to form a *resultant* vector (Fig. 6.4(b)),

$$\vec{p}_j = \vec{p}_l + \vec{p}_s \tag{6.9}$$

$\vec{p}_j$ will orient itself towards an external field so that only the projections

$$p_j(z) = \hbar m_j \tag{6.10}$$

are obtained on the field axes, m_j is the *total magnetic angular momentum quantum*

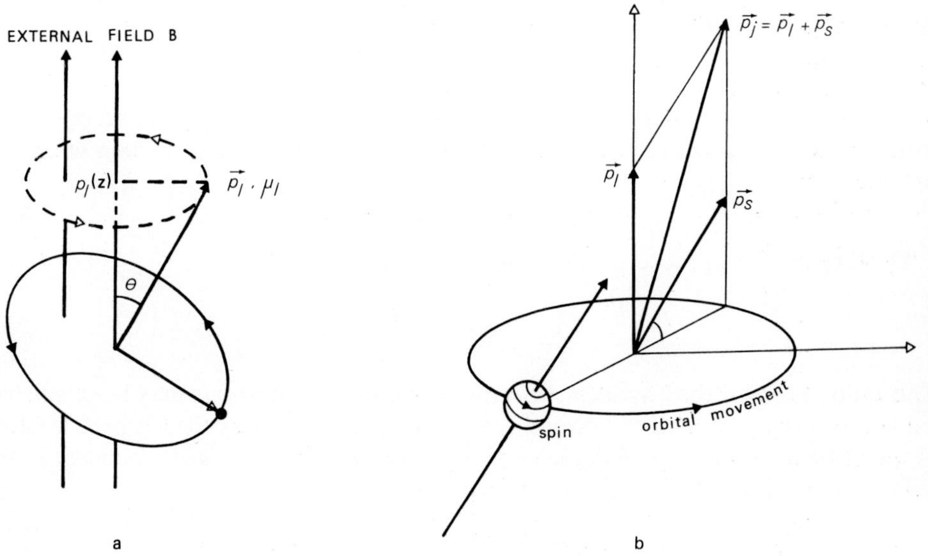

FIG. 6.4. (a) Angular momentum caused by an orbiting particle and permitted value projected on external field axes. The vector p_l precesses around the field axes as indicated by the dashed circle (ellipse in the drawing). (b) The l–s coupling of orbital angular momentum and spin leads to a resultant angular momentum p_j.

number; it can have all integer values between $-j$ and $+j$. The magnitude of p_j is

$$p_j = \hbar\sqrt{j(j+1)} \qquad (6.11)$$

where

$$j = l \pm s \qquad (6.12)$$

Here j is the *total (resultant) quantum number of the particle.*

6.2.4. *Magnetic moments*

a. Single particles P. A. M. Dirac showed in 1928 that the spin magnetic moment of the electron is:

$$\mu_s(\text{electron}) = \frac{\hbar \mathbf{e}}{2\mathbf{m}_e} = \mathbf{B}_e \qquad (6.13)$$

where $\mathbf{B}_e = 9.273 \times 10^{-24}$ joule per tesla ($\mathrm{J\,T^{-1}}$) (or $\mathrm{A\,m^2}$; in cgs units $\mathbf{B}_e$ is 9.273×10^{-21} erg per gauss). This value is referred to as one *Bohr magneton* (B.m.).

For the proton one would expect the spin magnetic moment μ_s to be

$$\mathbf{B}_e \cdot \mathbf{m}_e/\mathbf{m}_p = \mathbf{B}_n \qquad (6.14)$$

$\mathbf{B}_n$ (n for nucleon) has the value $5.051 \times 10^{-27}\ \mathrm{J\,T^{-1}}$ and is referred to as one *nuclear magneton* (n.m.). However, measurements show that

$$\mu_s(\text{proton}) = 14.1 \times 10^{-27}\ \mathrm{J\,T^{-1}} = 2.793\ \text{n.m.}$$

The reason for the higher value is found in the uneven charge distribution within the proton.

Recall that the neutron also has an uneven charge distribution. This gives rise to a neutron spin magnetic moment

$$\mu_s(\text{neutron}) = -1.913\ \text{n.m.}$$

The magnetic moment μ_l, caused by a charge q in *circular orbit*, can be calculated from classical physics: the current caused in the orbit, qv_r, times the area encircled, πr^2. Therefore

$$\mu_l = qv_r r/2 \qquad (6.15)$$

Dividing (6.15) by p_l according to (6.4a), gives

$$\gamma = \frac{\mu_l}{p_l} = \frac{q}{2m} \qquad (6.16a)$$

This ratio γ is called the *gyromagnetic ratio.* If γ and p_l are known, μ_l may be calculated. Equation (6.16a) is valid for electrons ($\mathbf{e}, \mathbf{m}_e$), but for protons only the left part. Because of the quantization of $p_l(z)$ (6.8) the component of the orbital magnetic moment in the field direction (Fig. 6.4(a)) is also quantized:

$$\mu_l(\text{electron}) = m_l \mathbf{B}_e \qquad (6.17)$$

Such a simple approach is not possible for nucleon in nucleus, because no nucleon moves completely independent of the other nucleons, nor is the orbital path always circular.

b. Atoms and nuclei In an atom with many electrons, the spin and angular moments of the electrons couple vectorially and separately to form resultant quantum numbers (using conventional symbolism)

$$S = \Sigma \vec{s}_i \tag{6.18a}$$

$$L = \Sigma \vec{l}_i \tag{6.18b}$$

which couple to form the total angular momentum quantum number of the atom

$$\vec{J} = \vec{L} + \vec{S} \tag{6.18c}$$

(see Table 6.2). Equations (6.18) are referred to as *Russell–Saunders coupling.* For the atom as a whole, the magnetic moment is

$$\mu(\text{atom}) = g_j \frac{\mathbf{e}}{2\mathbf{m}_e} p_j = g_j \mathbf{B}_e m_j \tag{6.19}$$

where g_j is the Landé factor and $+J \leqq m_j \leqq -J$. The Landé factor accounts for the effect of mutual screening of the electrons.

The *nuclear magnetic moment* depends on the spin and angular moments of the neutrons and protons. For the nucleus it is given by

$$\mu(\text{nucleus}) = g_I \frac{\mathbf{e}}{2\mathbf{m}_p} p_I \tag{6.20}$$

where g_I is the nuclear g-factor and p_I, the magnitude of the nuclear spin angular moment. Because this moment can have only the projections $\hbar m_I$ on the axes of a magnetic field where m_I is the nuclear magnetic angular momentum quantum number $(-I \leqq m_I \leqq I)$, we may write this equation

$$\mu(\text{nucleus}) = g_I\left(\frac{\mathbf{e}\hbar}{2\mathbf{m}_p}\right)m_I = g_I \mathbf{B}_n m_I \tag{6.21}$$

TABLE 6.1. *Summary of the properties of the atomic constituents (independent movements in a central potential field)*

Line	Property	Electron	Proton	Neutron
1	Mass (u)	0.000 5486	1.007 276	1.008 665
2	Charge (e units)	-1	$+1$	0
3	Spin quantum number	$s = 1/2$	$s = 1/2$	$s = 1/2$
4	Spin dipole moment	$\mu_e = 1$ B.m.	$\mu_p = 2.793$ n.m.	$\mu_n = -1.913$ n.m.
5	Orbital quantum number	$0 \leqq l \leqq n-1$	$0 \leqq l \lesseqgtr n$	
6	Permitted orbital field projections	$-\hbar l \ldots +\hbar l$	$-\hbar l \ldots +\hbar l$	
7	Total angular quantum number	$(j = l \pm s)$	$j = l \pm s$	
8	Total angular momentum	$(p_j = \hbar\sqrt{j(j+1)})$	$p_j = \hbar\sqrt{j(j+1)}$	
9	Particle symbolism	nl^i	nl_j	

Notes

2. The electron unit charge is $\mathbf{e} = 1.602 \times 10^{-19}$ C.
3. The spin angular momentum has the magnitude $\vec{p}_s = \hbar\sqrt{s(s+1)}$, with permitted projections on an external field axis $= \pm s\hbar$ (for e, p, and n $= \pm 1/2\hbar$).
7. l and s couple only in the one-electron system; see §6.2.4.
8. Permitted projections on an external field axis $\hbar m_j$, where $-j \leqq m_j \leqq j$.
9. n is the principal quantum number, l the azimuthal (orbital, radial) quantum number, i is the number of electrons in the particular n, l state, and j is the total angular momentum quantum number. For $l = 0$, 1, 2, etc., the symbols s, p, d, f, g, etc., are used.

TABLE 6.2. *Summary of atomic and nuclear properties associated with particle interactions (q.n. = quantum number)*

Line	Property	Electron–electron interaction	Nucleon–nucleon interaction
10	Spin–spin (*s–s*) coupling, q.n.	$S = \Sigma s_i$ (strong)	For each nucleon $j = l \pm s$
11	Total spin magnetic moment	$\mu_s = 2\sqrt{S(S+1)}$ B.m.	and $I = \Sigma j_i$
12	Orbit–orbit (*l–l*) coupling, q.n.	$L = \Sigma l_i$ (strong)	For ee nuclei: groundstate $l = 0$
13	Total orbital magnetic moment	$\mu_L = \sqrt{L(L+1)}$ B.m.	For eo, oe nuclei: groundstate $l = j$
14	Spin–orbit coupling, q.n.	$J = S + L$(weak)	For oo nuclei: groundstate varies
15	Resulting angular momentum	$p_J = \hbar\sqrt{J(J+1)}$	$p_I = \hbar\sqrt{I(I+1)}$
16	Resulting magnetic moment	$\mu_J = g_J\sqrt{J(J+1)}$ B.m.	$\mu_I = g_I\sqrt{I(I+1)}$n.m.
17	Transition selection rules	$\Delta L \pm 1$, $\Delta J = 0$ or ± 1	Δ_l = integer; $\Delta j > 0$
18	Multipole moment	Electric dipole	Electric or magnetic multipole
		Electron–nucleon interactions	
19	Grand atomic angular mom., q.n.	$F = J + I = S + L + I$	
20	Grand atomic angular mom.	$p_F = \hbar\sqrt{F(F+1)}$; field projections $0, \ldots \hbar F$.	
21	External field: none	Hyperfine spectrum (hfs): J levels split due to nuclear spin I	
22	External field: weak ($\lesssim 10^{-2}$ T)	Hfs F-levels split into $2F + 1$ levels (Zeeman effect)	
23	External field: average (~ 10 T)	Electron–nucleon q.n. decouple, producing $(2J + 1)(2I + 1)$ levels	
24	External field: very strong	Electron spin–orbit decouple, producing separate S- and L-levels	

TABLE 6.3. *Spin (I), parity (+for even, −for odd), nuclear magnetic moment (μ nuclear magnetons) and quadrupole moment $\hat{Q}(10^{-28}\ m^2)$ for some stable and radioactive nuclides* ^{A}X

Stable nuclides				Radioactive nuclides				
^{A}X	$I(\pm)$	μ_I	$\hat{Q}$	^{A}X	$I(\pm)$	μ_I	$\hat{Q}$	$t_{1/2}$
^{1}H	1/2+	2.793		$^{3}_{1}$T	1/2+	2.979		12.35 y
^{2}D	1+	0.854	0.003	$^{14}_{6}$C	0+			5736 y
^{10}B	3+	1.801	0.080	$^{24}_{11}$Na	5/2+	1.689		15.03 h
^{11}B	3/2−	2.689	0.040	$^{32}_{15}$P	1+	−0.252		14.3 d
^{12}C	0+			$^{36}_{17}$Cl	2+	1.285	−0.017	3.10^5 y
^{13}C	1/2−	0.702		$^{45}_{20}$Ca	7/2−			163 d
^{14}N	1+	0.404	0.010	$^{55}_{26}$Fe	3/2			2.7 y
^{16}O	0+			$^{60}_{27}$Co	5+	3.810		5.27 y
^{17}O	5/2+	−1.894	−0.026	$^{64}_{29}$Cu	1+	−0.216		12.70 h
^{19}F	1/2+	2.629		$^{95}_{40}$Zr	5/2+			64.0 d
^{23}Na	3/2+	2.218	0.110	$^{131}_{53}$I	7/2+	2.400	−0.400	8.04 d
^{31}P	1/2+	1.132		$^{137}_{55}$Cs	7/2+	2.838	0.050	30.1 y
^{33}S	3/2+	0.643	−0.055	$^{140}_{57}$La	3−			40.2 h
^{39}K	3/2+	0.391	0.090	$^{198}_{79}$Au	2−	0.580		2.70 d
^{59}Co	7/2−	4.649	0.400	$^{232}_{90}$Th	0+			1.41×10^{10} y(α)
^{87}Sr	9/2+	−1.093	0.360	$^{235}_{92}$U	7/2−	−0.350	4.100	7.04×10^{8} y(α)
^{141}Pr	5/2+	4.500	−0.060	$^{238}_{92}$U	$(1i^{11/2})$			4.47×10^{9} y(α)
^{197}Au	3/2+	0.115	0.580	$^{239}_{94}$Pu	1/2+	0.210		2.44×10^{4} y(α)

I is the *total nuclear spin*. We shall see in the next section how I can be determined.

In Table 6.1 we have summarized the most important properties of the atomic constituents, and in Table 6.2 their modes of interaction. In Table 6.3 some spin and magnetic moments are given for stable and radioactive nuclei.

6.2.5. *Precession*

Before going into the details of nuclear structure, there is one more property of the nucleon which must be considered. Both types of angular momenta, p_s and p_l, as well as their corresponding magnetic moments, are vector quantities. Quantum mechanics forbids $\vec{p}$ (and consequently μ) to be exactly parallel with an external field. At the same time the external field tries to pull the vector so that the plane of rotation becomes perpendicular to the field lines. The potential magnetic energy is

$$E_{\text{magn}} = \vec{B}\cdot\vec{\mu} = B\cdot\mu\cos\theta \tag{6.22}$$

(see Figs. 6.3 and 6.4) where B is the magnetic field strength. However, because of the inertia of the particle this does not occur until the particle has moved somewhat along its orbit, with the consequence that the vector starts to rotate around the field axes as shown in Fig. 6.4(a). The angular momentum vector therefore precesses around the field axes, like a gyroscope. We may rewrite (6.16a) as

$$\gamma = \vec{\mu}/\vec{p}_{\text{rot}} \tag{6.16b}$$

This form makes it more obvious why γ is referred to as the gyromagnetic ratio. Equation (6.16b) is valid both for angular momentum and spin, but of course the value is different for electrons and protons. The angular velocity ω_γ of the precession is found to be

$$\omega_\gamma = \mu\vec{B}/\vec{p}_{\text{rot}} \tag{6.23}$$

By replacing angular velocity with frequency ($\omega = 2\pi\nu$)

$$\nu_\gamma = \mu\cdot B/2\pi p_{\text{rot}} \tag{6.24a}$$

or

$$\nu_\gamma = \gamma\cdot B/2\pi \tag{6.24b}$$

where ν_γ is the Larmor precession frequency.

6.3. The single-particle shell model

6.3.1. *Quantum number rules*

Let us assume that a nucleon moves around freely in the nuclear potential well, which is spherically symmetric, and that the energy of the nucleon varies between potential and kinetic like a harmonic oscillator, i.e. the potential walls (see Figs. 6.1 and 6.2) are parabolic. For these conditions the solution of the Schrödinger equation yields:

$$E(\text{nucleon}) = \mathbf{h}\sqrt{2U_0/mr^2}\,[2(n-1)+l] \tag{6.25}$$

where U_0 is the potential at radius $r = 0$ and m is the nucleon mass. We have defined

n and l previously. The square root, which has the dimension s^{-1}, is sometimes referred to as the oscillator frequency (ω in Table 6.6). The following rules are valid for nucleons in the nuclear potential well:

(a) l can have all positive integer values beginning with 0, independent of n;
(b) the energy of the l state increases with increasing n as given by (6.25);
(c) the nucleons enter the level with the lowest total energy according to (6.25) independent of whether n or l is the larger;
(d) there are independent sets of levels for protons and for neutrons;
(e) the Pauli principle is valid, i.e. the system cannot contain two particles with all quantum numbers being the same;
(f) the spin quantum numbers must be taken into account (not included in (6.25)).

Following these rules, the nucleons vary greatly in energy and orbital motion. However, these rules do not exclude the existence of two nucleons with the same energy (so-called degenerate states), provided the quantum numbers differ.

6.3.2. *Nuclei without nucleon spin–orbit coupling*

If we calculate the sequence of energy levels on the assumption that n is constant, we obtain the pattern in Table 6.4. This is the sequence of electronic states in atoms but it does not agree with the observed magic numbers 2, 8, 20, 28, 50, etc., for nuclei.

Applying the rules of the previous section a new set of nucleon numbers is obtained: 2, 8, 20, 40, 70, etc. (Table 6.5, and left column Table 6.6). This level scheme allows a large amount of degeneracy. For example, for $n = 1$ and $l = 3$ (1f-state) we find from (6.25) that $[2(n-1)+l] = 3$, which value is obtained also for $n = 2$ and $l = 1$ (2p-state). Since the f-state can have 14 and the p-state 6 nucleons, the degenerate level of both states

TABLE 6.4. *Energy levels derived on bases of permitted values for the azimuthal quantum number l and spin quantum number s*

l	State	Possible quantum values	Number of states (including spin)	Accumulated nucleon number
0	s	0	$1 \times 2 = 2$	2
1	p	$-1, 0, +1$	$3 \times 2 = 6$	8
2	d	$-2, -1, 0, +1, +2$	$5 \times 2 = 10$	18
3	f	$-3, -2, -1, 0, +1, +2, +3$	$7 \times 2 = 14$	32
4	g	$-4, -3, -2, -1, 0, 1, 2, 3, 4$	$9 \times 2 = 18$	50
l	–	$-l \ldots +l$	$2(2l+1)$	–

TABLE 6.5. *Energy levels according to eqn.* (6.25)

Levels	Number of nucleons	Accumulated nucleons
1s	2	2
1p	6	8
1d and 2s	10 + 2	20
1f and 2p	14 + 6	40
1g, 2d, and 3s	18 + 10 + 2	70
1h, 2f, and 3p	22 + 14 + 6	112
...	...	...

TABLE 6.6. *Nuclear energy levels obtained for the single-particle model by solving the Schrödinger-equation (a) for a harmonic oscillator, and rounded-square well potentials (b) without and (c) with spin–orbit coupling. The $n\hbar\omega$ figures to the left are energies for model a, according to eqn. (6.25). Numbers in ( ) indicate orbital capacities and those in [ ] give cumulative capacity up to the given line.*
(According to G. E. Gordon and C. D. Coryell)

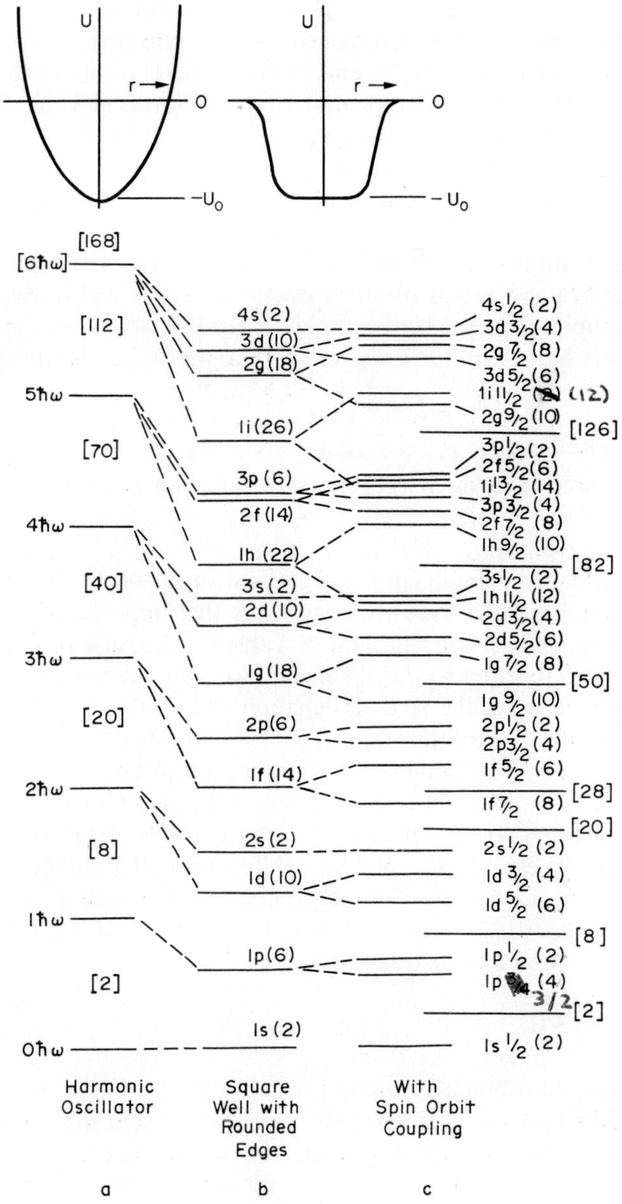

can contain 20 nucleons. However, the numbers still do not correspond to the experimental magic numbers.

A further refinement is possible if we assume that the nuclear potential well has straight walls, i.e. the potential energy $U(r)$ is $-U_0$ at $r < r_n$ while it is infinite at $r \geqq r_n$, where r_n is the nuclear radius. This assumption, when introduced (see Fig. 6.2 and figure in Table 6.6) in the Schrödinger-equation, leads to a splitting of the degenerate levels, so that the lowest energy is obtained for the state with lowest main quantum number n. For our example of the 1f- and 2p-states the 1f orbitals are lower in energy than the 2p. This refinement yields the middle row of levels in Table 6.6 but still does not lead to the correct magic numbers.

6.3.3. *Nuclear level scheme with nucleon spin–orbit coupling*

In multielectron atoms, the Russell–Saunders coupling is present in light atoms. However, in the heaviest atoms of many electrons and in highly charged nuclei, the *j-j* (spin–orbit) coupling better describes the systems. O. Haxel, J. J. O. Jensen, H. E. Suess, and M. Goeppert-Mayer in 1949 suggested that the nucleons always experience a strong spin–orbit coupling according to (6.12)

$$j = l + s \qquad (6.12)$$

and that the total spin of the nucleus I is the sum of the nucleon spins

$$I = \sum j \qquad (6.26)$$

In this way, all l levels are split into two levels with quantum values $l + \frac{1}{2}$ and $l - \frac{1}{2}$, of which the former has the lowest energy value (the opposite of the electron case). This yields the row of levels on the right in Table 6.6. Because of the energy splitting the new levels group together so they fit exactly with the experimental magic numbers.

As an example, consider the level designation $1i_{11/2}$. This has the following interpretation: the principal quantum number is 1; i indicates that the orbital quantum number l is 6; the angular momentum quantum number j is 11/2 ($j = l - 1/2$). The number of permitted nucleons in each level is $2j + 1$, thus 12 for $j = 11/2$.

The magic numbers correspond to sets of energy levels of similar energy just as in the atom the K, L, M, etc., shells represent orbitals of similar energy. The N-electronic shell contains the 4s, 3d, 4p sets of orbitals while the 4th nuclear "shell" contains the $1f_{5/2}$, $2p_{3/2}$, $2p_{1/2}$, and $1g_{9/2}$ sets of orbitals. Remember that there are separate sets of orbitals for protons and for neutrons.

6.3.4. *The nuclear spin*

The nuclear spin I is obtained from (6.12) and (6.26). Since j is always a half-integer, nuclides with odd number of nucleons (odd A) must have odd spin values (odd I), while those with even A must have even I. The nucleons always pair, so that even numbers of protons produce no net spin. The same is true for even numbers of neutrons. For nuclei of even numbers for both N and Z, the total nuclear spin I is always equal to zero. Some ground state nuclear spins are given in Table 6.3.

For odd A the nuclear spin is wholly determined by the single unpaired nucleon (single particle model). Let us take the nucleus ${}^{13}_{6}C$ as an example. It contains 6 protons and 7 neutrons. From Table 6.6 we conclude that there are 2 protons in the $1s_{1/2}$ level

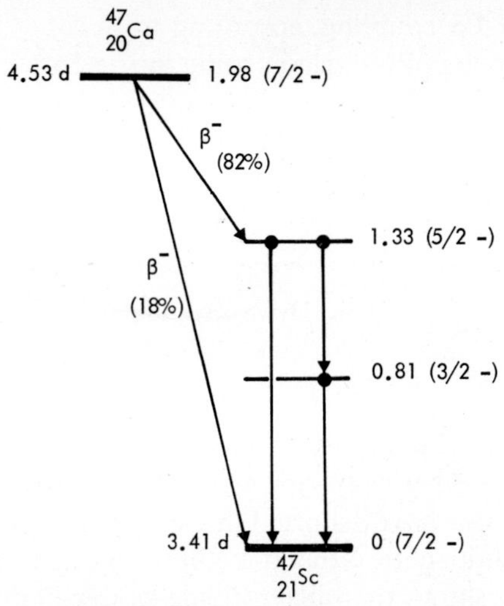

FIG. 6.5. Part of decay scheme for $A = 47$; the data are experimental values.

and 4 protons in the $1p_{3/2}$ level. A similar result is obtained for the first 6 neutrons but the 7th neutron must enter the $1p_{1/2}$ level. The value of I for ^{13}C is thus 1/2. Another example is $^{51}_{23}V$, which has 23 protons and 28 neutrons. Since $N = 28$, the neutrons do not contribute to I. From Table 6.6 we see that we can accommodate 20 protons in the orbitals $1s^2, 1p^6, 1d^{10}, 2s^2$. The next 3 protons must go into the $1f_{7/2}$ level ($1f^3_{7/2}$), where, however, 2 of them are paired. Therefore, the single unpaired proton has $j = 7/2$, leading us to predict $I = 7/2$, which also is the measured nuclear spin value.

When a nucleus is excited, either through interaction with other particles or in a decay process, for nuclei with N or Z values near magic numbers the paired nucleons seem not to be perturbed by the excitation (if it is not too large). As a result, we can associate the excitation with any unpaired nucleons. Let us choose the decay of ^{47}Ca to ^{47}Sc (see Fig. 6.5). ^{47}Ca has 27 neutrons and in the ground state the last 7 neutrons must occupy the $1f_{7/2}$ level. The ground state of (the unstable) ^{47}Sc has 21 protons, the unpaired proton can only be accommodated in the $1f_{7/2}$ level. Thus both of the ground states have $I = 7/2$. The next higher energy states for ^{47}Sc involve the levels $1f_{5/2}$ and $2p_{3/2}$. In the figure it is seen that these levels are observed although their order is reversed from that in Table 6.6. This reflects some limitation of the single-particle model, and is explained in §6.5.

For odd–odd nuclei the nuclear spin is given by

$$I\,(\text{odd–odd}) = j_p + j_n = (l_p \pm 1/2) + (l_n \pm 1/2) \tag{6.27a}$$

According to Nordheim's rule, if

$$j_p + j_n + l_p + l_n \text{ is even, then } I = |j_p - j_n| \tag{6.27b}$$

$$j_p + j_n + l_p + l_n \text{ is odd, then } I \leq |j_p + j_n|$$

The use of this rule can be illustrated by the case of $^{64}_{29}Cu$ which has its odd proton in the $1f_{5/2}$ orbital and its odd neutron in the $2p_{3/2}$ orbital. Thus for the proton, $j = 5/2$, $l = 3$; for the neutron, $j = 3/2$, $l = 1$. Because $l_n + l_p + j_n + j_p$ is even, we use $I = |5/2 - 3/2| = 1$, which is the observed spin value. The lightest nuclei are exceptions to this rule since they often exhibit LS coupling, according to (6.18). ^{10}B is an example; it has 5 protons and 5 neutrons, the fifth nucleon being in the $1p_{3/2}$ state. Thus $I = j_1 + j_2 = 3$, which is observed.

We mentioned in §6.2.4 that theoretical calculations of the nuclear magnetic moment, (6.21), are usually not satisfactory. The value of μ(nucleus) can have a number of values depending on m_I, with a maximum value of I.

6.4. Deformed nuclei

6.4.1. *Deformation index*

Both the liquid-drop model and the single-particle model assume that the mass and charge of the nucleus are spherically symmetric. This is true only for nuclei close to the magic numbers; other nuclei have distorted shapes. The most common assumption about the distortion of the nuclide shape is that it is ellipsoidal, i.e. a cross-section of the nucleus is an ellipse. Figure 6.6 shows the oblate (flying-saucer-like) and prolate (egg-shaped) ellipsoidally distorted nuclei; the prolate shape is the more common.

Deviation from the spherical shape is given by

$$\beta = \frac{2(a-c)}{a+c} \tag{6.28}$$

where a and c are the elliptical axes as shown on the prolate figure. For prolate shape $\beta > 0$, and for oblate shape $\beta < 0$. The maximum deformation observed is about $\beta = \pm 0.6$.

The deformation is related to the nuclear shell structure. Nuclei with magic numbers are spherical and have sharp boundary surfaces (they are "hard"). As the values of N and Z depart from the magic numbers the nucleus increases its deformation.

6.4.2. *Electric multipoles*

In a spherical nucleus we assume the charge distribution to be spherical and the nucleus acts as a monopole. In the deformed nuclei, the nuclear charge has a non-spherical distribution. The potential at a point $\bar{x}, \bar{y}, \bar{z}$ (Fig. 6.6) will be found to vary

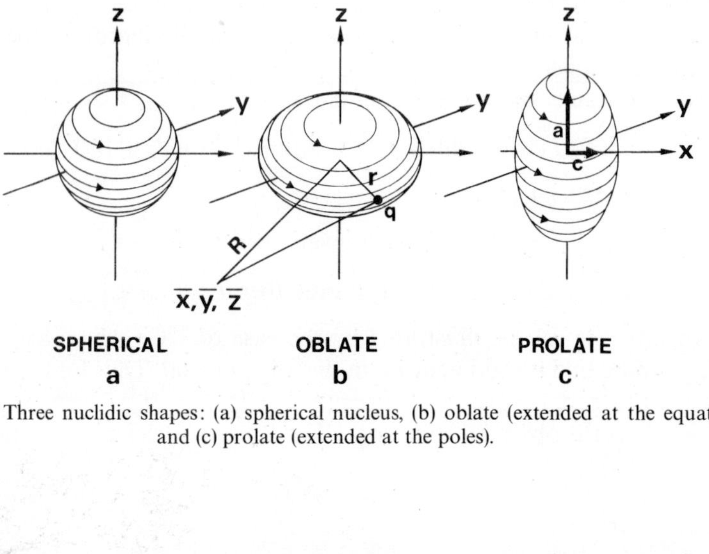

FIG. 6.6. Three nuclidic shapes: (a) spherical nucleus, (b) oblate (extended at the equator), and (c) prolate (extended at the poles).

depending on the charge distribution and mode of rotation of the nucleus. The nuclear charge may be distributed to form a dipole, a quadrupole, etc. Nuclei are therefore divided into different classes depending on their electrical multipolarity:

Class:	E0	Types: monopoles; multipolarity 0
	E1	dipoles; multipolarity 1
	E2	quadrupoles; multipolarity 2
	E3	octupoles; multipolarity 3

etc.

It has been found that nuclei with spin $I = 0$ have no multipole moment (class E0). According to theory, nuclei with $I = 1/2$ can have a dipole moment, but this has not yet been shown experimentally. Nuclei with $I = 1$ have quadrupole moments (class E2); they are fairly common. The quadrupole moment $\hat{Q}$ can be calculated for spheroidal (i.e. deformation not too far from a sphere) nuclei by

$$\hat{Q} = \frac{2z\mathbf{e}}{5}(a^2 - c^2) \tag{6.29a}$$

$\hat{Q}$ is usually referred to as the internal quadrupole moment, i.e. the expected value for a rotation around the z-axes. However, quantum mechanics makes this impossible and gives for the maximum observable quadrupole moment

$$\hat{Q}_{\text{obs}} = \hat{Q}\frac{I - 1/2}{I + 1} \tag{6.29b}$$

Thus $\hat{Q}_{\text{obs}} = 0$ for $I \leqq 1/2$. $\hat{Q}_{\text{obs}}$ is usually given in charge (C) times area (cm^2 or m^2). Most commonly 10^{-24} cm^2 (or 10^{-28} m^2) is used as unit and referred to as one barn, $\hat{Q}_{\text{obs}}$ is > 0 for the more common prolate shape, and < 0 for oblate. Some measured values are given in Table 6.3.

The rotation of nuclei with electric multipoles gives rise to formation of magnetic multipoles. Nuclei can therefore also be divided according to the magnetic multipolarity (M0, M1, etc.).

6.4.3. *The collective nuclear model*

In 1953 A. Bohr and B. Mottelsen suggested that the nucleus be regarded as a highly compressed liquid, undergoing quantized rotations and vibrations. Four discrete collective motions can be visualized. In Fig. 6.6 we can imagine that the nucleus rotates around the y-axes as well as around the z-axes. In addition it may oscillate between prolate to oblate forms (so-called *irrotation*) as well as *vibrate*, for example, along the x-axes. Each mode of such collective nuclear movement has its own quantized energy. In addition, the movements may be coupled (cf. coupling of vibration and rotation in a molecule).

The model allows the calculation of rotational and vibrational levels as shown in Fig. 6.7. If a ^{238}U nucleus is excited above its ground state through collision with a high energy heavy ion (Coulomb excitation), we have to distinguish between three types of excitation: (a) nuclear excitation, in which the quantum number j is changed to raise the nucleus to a higher energy level (according to Table 6.6); (b) vibrational excitation, in which case j is unchanged, but the nucleus is raised to a higher vibrational

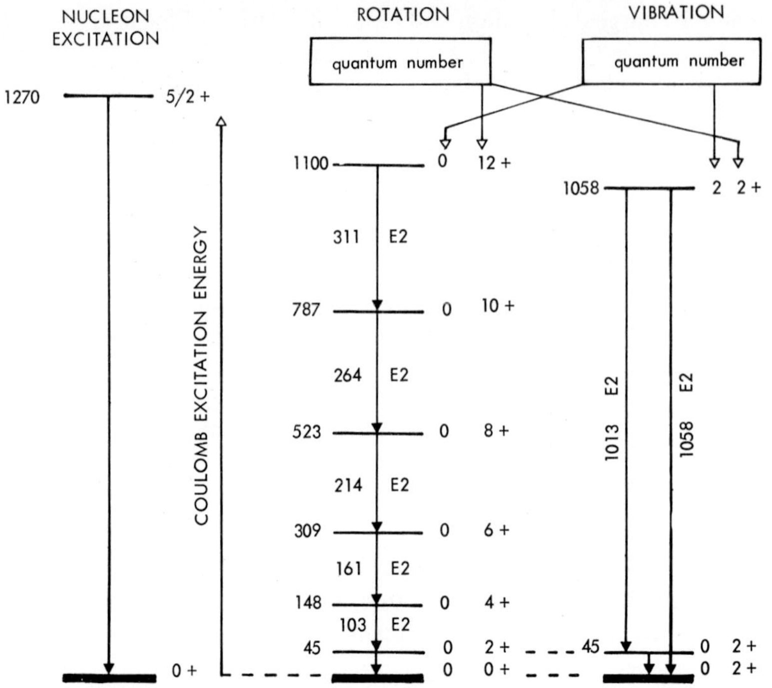

FIG. 6.7. Nucleonic, rotational, and vibrational levels observed in ^{238}U; energy in keV. (According to E. K. Hyde.)

level, characterized by a particular vibrational quantum number (indicated in the figure); (c) rotational excitation, also characterized by a particular rotational quantum number. The figure shows that the rotational levels are more closely spaced and thus transitions between rotational levels involve lower energies than de-excitation from excited nuclear or vibrational states.

The rotational energy levels can often be calculated from the very simple expression

$$E_{rot} = (\hbar^2/2I_{rot})n_r(n_r + 1) \tag{6.30}$$

where I_{rot} is the moment of inertia and n_r the rotational quantum number; this equation is identical to (B.16). The validity of this equation depends on whether the different modes of motion can be treated independently or not, which they can for strongly deformed nuclei like ^{238}U.

6.5. The unified model of deformed nuclei

The collective model gives a good description for even–even nuclei but cannot account for some of the discrepancy between observed spins and the spin values expected from the single-particle shell model. The latter was developed on the assumption of a nucleon moving freely in a symmetrical potential well, a situation which is valid only for nuclei near closed shells. The angular momentum of an odd-A deformed nucleus is due both to the rotational angular momentum of the deformed core and to the angular momentum of the odd nucleon. Consequently the energy levels for such a nucleus are different than for the symmetric shell model.

This situation was taken into account by S. G. Nilsson, who calculated energy levels

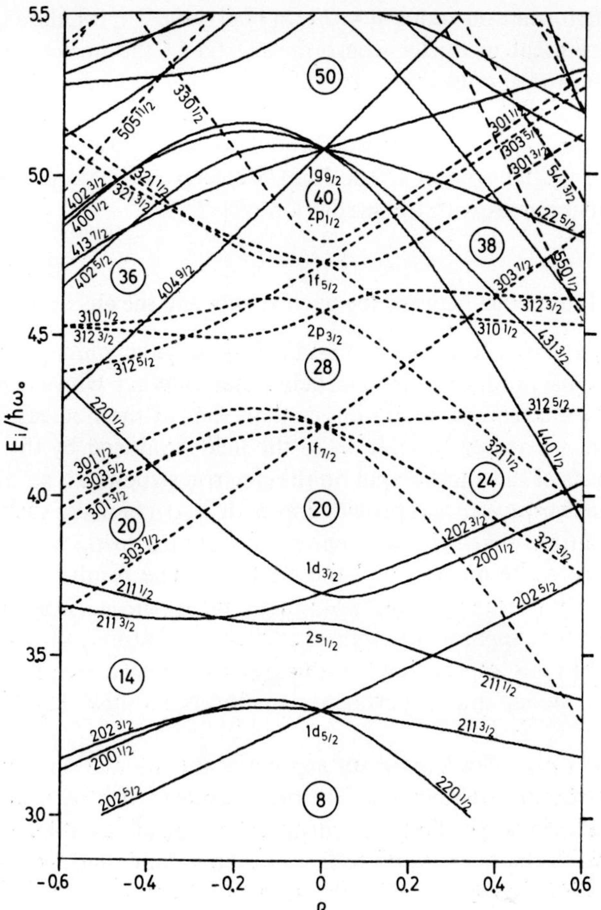

FIG. 6.8. Nilsson diagram of single-particle energy levels for deformed nuclei. The energy scale is given in units of $\hbar\omega_0$, where $\hbar\omega_0$ is approximately $41A^{-1/3}$ MeV. The figures along the center give the neutron or proton numbers and l–j values for the single nucleon. The figure combinations at the end of the lines are approximate quantum numbers; the first figure is the principal quantum number n. Even lines are for even parity, broken lines for odd. (According to S. E. Larsson, G. Leander, I. Ragnarsson, and N. G. Alenius.)

for odd nuclei as a function of the nuclear deformation β. Figure 6.8 shows how the energies of the Nilsson levels vary with the deformation β of the potential well. Each shell model level of angular momentum j splits into $j + 1/2$ levels (called *Nilsson levels* or *states*). Each level may contain up to two nucleons and form the ground state of a rotational band. In addition the undeformed levels ($\beta = 0$) appear in somewhat different order than for the symmetric shell model (Table 6.6). This leads to a reversal in order for some of the levels, e.g. 1f5/2 and 2p3/2 (this explains the observed level order for ^{47}Sc, Fig. 6.5). The Nilsson levels are quite different in all characteristics from the shell model states, and their prediction of energies, angular momenta, quantum numbers, and other properties agrees better with experimental data for deformed nuclei than those of any other model.

As an example we may choose $^{23}_{11}$Na, which has a quadrupole moment of 0.11 barn. Assuming the nuclear radius to be $1.1A^{1/3} = (a + c)/2$ (fm) we can use (6.28) and (6.29) to calculate a deformation index $\beta = 0.11$. (The value 1.1 for the constant r_0 in (3.7) gives better agreement in nuclei where the inertia of the nucleus is involved.) From Fig. 6.8 the 11th proton must enter the 3/2 level rather than the 1d5/2 level as the symmetric shell model indicates in Table 6.6. The experimental spin of 3/2 confirms the Nilsson prediction. Similarly, for the deformed $^{19}_{9}F$ and $^{19}_{10}Ne$, we expect from Table 6.8 the odd nucleon to give spin of 1/2 and not 5/2, as would be obtained from Table 6.6. Again, experiment agrees with the prediction of 1/2.

6.6. Interaction between the nuclear spin and the electron structure*

We have already seen how the spin and orbital angular momentum of the electrons and of the nucleus produce magnetic fields that interact with each other. The field produced by the electrons is much larger than that of the nucleus, and consequently the nuclear spin is oriented in relation to the field produced by the electron shell. By contrast the effects of the nuclear spin on the electron structure is so small that it usually is neglected. Nuclear physics has provided us with instruments of such extreme sophistication and resolution that there are many ways of measuring with great accuracy the interaction between the nucleus and the electrons. The result has been new research tools of utmost importance, most prominent being the *nuclear magnetic resonance* (nmr) techniques. The separate disciplines of chemistry, atomic physics, nuclear physics, and solid state physics approach each other closely in such techniques, and an understanding of the theory and experimental methods requires knowledge of all these subjects.

In this section only a few important aspects of the interaction between nuclear spin and electronic structure are reviewed. The methods described are usually not considered to fall within the framework of nuclear chemistry, but in all scientific fields it is important to be able to reach the border and look at developments and techniques used on the other side. Such information is often the seed to further scientific development.

6.6.1. *Hyperfine spectra*

In §6.3.4 we mentioned that the electrons in the atomic shell have Russell–Saunders coupling (6.18), $\vec{J} = \vec{L} + \vec{S}$, where $\vec{J}$ is referred to as the *internal quantum number*. The magnetic field created by the electrons interacts with that caused by the nuclear spin to yield the grand atomic angular momentum vector

$$\vec{F} = \vec{J} + \vec{I} \tag{6.31a}$$

The magnitude of this momentum is

$$p_F = \hbar\sqrt{F(F+1)}, \tag{6.31b}$$

but only projections $p_F(z) = 0, \ldots \hbar F$ are permitted on an external (to the atom) field axis. This leads to a large number of possible energy levels, although they are limited by certain selection rules.

The nuclear spin can orient itself in relation to $\vec{J}$ in

$2I + 1$ directions if $I \leq J$
$2J + 1$ directions if $J \leq I$

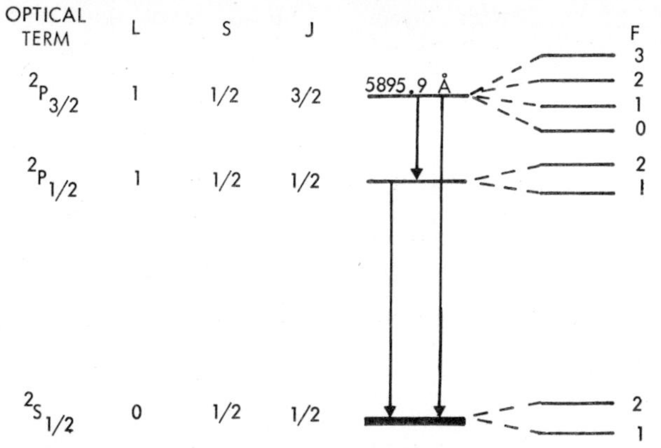

FIG. 6.9. The development of hyperfine lines in an optical spectrum of sodium due to the nuclear spin I, which enters through the coupling $\vec{F} = \vec{J} + \vec{I}$.

Consider ^{23}Na as an example (Fig. 6.9);† it has a nuclear spin $I = 3/2$. The yellow sodium line of 5896 Å is caused by de-excitation of its electronically excited P state to the ground state $2S_{1/2}$.

The difference between the two P-states is very small, only 0.0022 eV (or about 6 Å units), and can only be observed with high resolution (*fine spectra*). To each of these three levels the nuclear spin I has to be added, yielding the quantum number F according to (6.31). It is easy to determine that the level number rule holds, e.g. for $I = J$ we must have $2I + 1 = 4$ possible levels. These levels can be observed in optical spectrometers only at extremely high resolution (*hyperfine spectrum*, hfs). The energy separation between the hfs lines depends on the nuclear magnetic dipole μ_I, the spin value I, and the strength of the magnetic field produced at the nucleus, as discussed in §6.2.4 (see also §6.6.3). The energy separation is very small, on the order of 10^{-5} eV, corresponding to a wavelength difference of about 1/100 of an Å. Even if there is great uncertainty in the energy determination of the levels, simply counting the number of hyperfine lines appearing for a certain electronic $\vec{J}$-level gives the value of the nuclear spin, because $-I \leqq m_I \leqq I$ (number of m_I values $= 2I + 1$).

At very high resolutions it may be possible to determine μ_I from hfs, and from this to calculate I. The hyperfine splitting and shifting of optical lines have yielded important information not only about nuclear spins and magnetic moments, but also about the electric charge distribution and radius of the nucleus.

6.6.2. *Atomic beams*

The hyperfine spectra are obtained from light sources in the absence of any external (to the atom) magnetic field. If the source is placed between the poles of a magnet, whose strength B is progressively increased, the following sequence of changes takes place (see Table 6.2, electron–nucleon interaction).

Suppose B is increased slowly to 10^{-2} T (100 gauss). Each hyperfine level F is found to split into $2F + 1$ levels, since the quantum mechanical rule of permitted projections

†The letters S and P stand for $\vec{L} = 0$ (and thus $\Sigma l_i = 0$) and 1, respectively. The superscript 2 refers to the number of possible S-values; thus $2S + 1 = 2$, i.e. $\vec{S} = \Sigma s_i = 2$. The subscript gives the $\vec{J}$-value; with (6.18c) the ground state $\vec{J} = 1/2$ means $L = 0$ and $S = 1/2$ ($S = -1/2$ not possible). For the excited state both $S + 1/2$ and $S = -1/2$ are possible, giving rise to the two P-states.

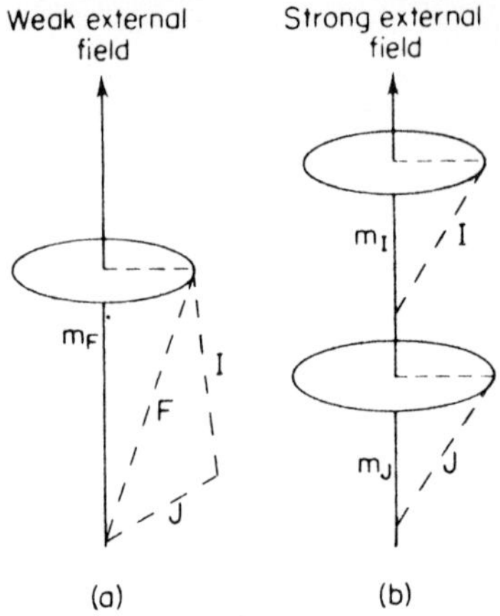

FIG. 6.10. (a) Coupling of electronic and nuclear angular moments J and I in a very weak magnetic field. Their sum F forms the quantum number m_F along the direction of the magnetic field, the maximum magnitude of which is $p_F(z) = \hbar F$. In a strong magnetic field the electronic and nuclear angular moments decouple, so that I and J act independently, giving projections m_j and m_I on the field vector.

on an external field vector comes into operation. These permitted projections can vary from zero to a maximum value of $\hbar F$; for $F = 3$ seven new lines are obtained.

Further increases in B to 10 T (10^5 gauss) leads to a decoupling of F into its components J and I (Fig. 6.10). For each projection of J on the field vector there are $2I + 1$ (lines $(-I \ldots 0 \ldots +I)$, giving altogether $(2J + 1)(2I + 1)$ lines. The splitting of the spectral lines in a weak magnetic field is called the Zeeman effect.

The decoupling of the angular momentum of the atom into its electronic and nuclear components is used in the *atomic beam apparatus* to determine nuclear magnetic moments and spin values. A beam of atoms, produced in an oven, is allowed to enter a tube along which a series of magnets (usually three) have been placed. The magnetic field splits the atomic beam into several component beams, each containing only atoms which have the same values for all the quantum numbers. Between the magnets there is a small coil connected to a high frequency oscillator, which produces a weak oscillating magnetic field. When this oscillator is tuned to an energy $\mathbf{h}\nu$, which exactly matches the energy difference between two quantum states of the atom, energy may be absorbed producing a transition from one of the states to another. Subsequently the atoms are deflected differently by the magnetic field than they were before the energy absorption. By a combination of homogeneous and heterogeneous magnetic fields the atomic beam apparatus allows only those atoms which have absorbed the energy quantum $\mathbf{h}\nu$ to reach the detector. From the properties of the magnetic fields and the geometric dimensions and frequency of the instrument, the magnetic moment of the atom in different quantum states can be determined. This allows calculation of the magnetic moment and spin of the nucleus. This technique is of interest to the nuclear chemist because spin values can be obtained for short-lived nuclei in submicroscopic amounts. In contrast the hfs technique requires macroscopic amounts of atoms.

6.6.3. *Nuclear magnetic resonance*

When an atom is placed in an external magnetic field (strength B) so that J and I decouple (cf. (6.31)), the nuclear magnetic moment vector $\vec{\mu}_I$ must precess around

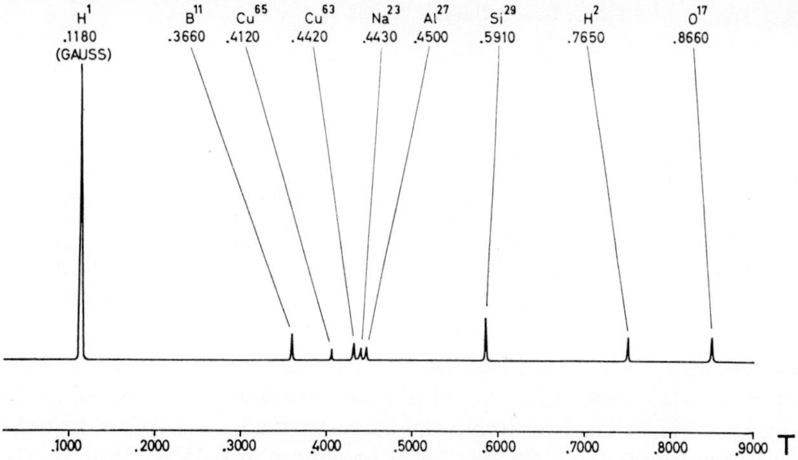

FIG. 6.11. Nuclear magnetic resonance spectrum for a glass vessel containing water and some copper. The magnetic field is given in tesla (T).

the field direction, with the components in the direction of the field restricted to

$$\mu_I = g_I \mathbf{B}_n m_I \tag{6.21}$$

In the external field the states with different m_I have slightly different energies. The potential magnetic energy of the nucleus is

$$E_{\text{magn}} = -\vec{\mu}_I \vec{B} = -g_I \mathbf{B}_n m_I B \tag{6.32}$$

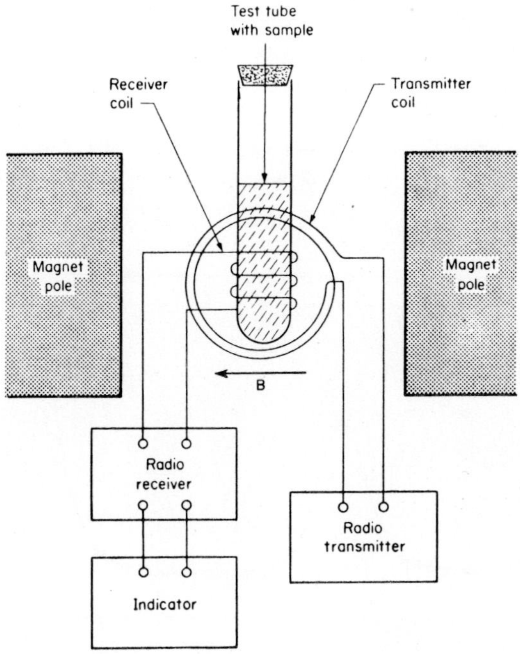

FIG. 6.12. Arrangement for nmr measurements. (According to W. J. Moore.)

The energy spacing between two adjacent levels is:

$$\Delta E = g_I \mathbf{B}_n B \tag{6.33}$$

because $\Delta m_I = \pm 1$. For example, consider the case of the nucleus ^{19}F for which $g_I = 5.256$ in a field of 1 T; $\Delta E = 5.256 \times 5.0505 \times 10^{-27} \times 1 = 2.653 \times 10^{-26}$ J. The frequency of electromagnetic radiation corresponding to this energy is 4.0×10^7 s^{-1} or 40 MHz. This lies in the short wave region, $\lambda = 7.5$ m. This frequency is also the same as that of the Larmor precession, as given by (6.24b).

These relations can be used to calculate the nuclear magnetic moment, if I is known, or vice versa. Figure 6.11 shows the results of an experiment in which a sample has been placed in a variable magnetic field containing two coils (Fig. 6.12), one connected to a radio receiver operating at 5 MHz and the other a detector coil connected to an amplifier. The sample is a glass tube (B, O, Na, Al, Si atoms) containing a piece of copper alloy (Cu, Al) in water (H, D, O). By varying the magnetic field a number of resonances are observed. For example, for ^{23}Na one has a resonance at 0.443 T. With (6.33) we can calculate $g_I = 1.48$, and, consequently, the magnetic moment of ^{23}Na is 2.218 (Table 6.3). Further, for ^{23}Na, $m_I = 3/2$ and $I = 3/2$.

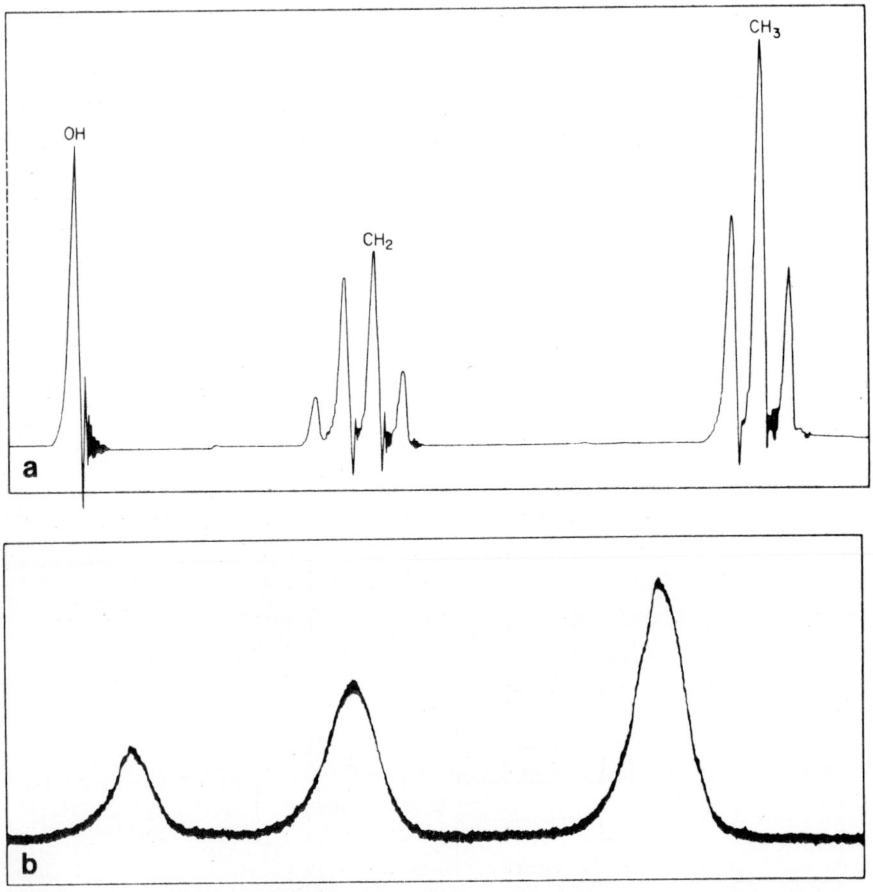

FIG. 6.13. Nmr spectrum for alcohol. The chemical shift is in ppm of frequency ($\nu_{meas} \times 10^6$ per 60 Mc). The solid line (b) is the unresolved spectrum.

The magnetic field experienced by the nucleus is not exactly equal to the external field because of the shielding effect of the electron shell, even if I and J are decoupled. Although this shielding of the nucleus is very small, about $10^{-5}B$, it can still easily be detected with modern equipment. The shielding effect depends on the electronic structure. In ethanol CH_3CH_2OH, the proton nuclei are in three slightly different electronic environments (the CH_3, the CH_2, or the OH group). Therefore, a high resolution nmr spectrum for proton resonances, pmr, looks like that in Fig. 6.13; TMS (tetramethylsilane), $(CH_3)_4$ Si, is the reference sample. The magnetic field axis is usually replaced by δ, which may be the change either in frequency (if ν is varied) or in the magnetic field (if B is varied) relative to the value for the reference sample. This change, usually given in parts per million (ppm), is called the *chemical shift* because it depends on the chemical environment of the nucleus. In the figure the resolution is very high, about 0.01 ppm.

The structural information that can be provided by this method is very detailed, and a new and deeper insight in chemical bonding and molecular structure is provided. The nmr technique has therefore become a central tool for the investigation of chemical structures in solids, liquids, and the gaseous state.

6.7. Radioactive decay and nuclear structure

In the preceding section we have described three methods of determining nuclear spin—one optical and two magnetic. The nuclear spin plays a central role in forming the nuclear energy states. It is therefore to be expected that it also should be of importance in nuclear reactions and in radioactive decay. Let us consider some rules for the lifetimes of unstable nuclei, for their permitted modes of decay, and for the role of nuclear spin. Knowing these rules, it is, for example, possible from a decay scheme to predict the spin states of levels which have not been measured.

6.7.1. *Gamma-decay*

Photons are emitted in the transition of a nucleus from a higher energy state (level) to a lower

$$E_\gamma = \mathbf{h}\nu = E_\mathrm{f} - E_\mathrm{i} \tag{6.34}$$

where f and i refer to the final and initial states. Because the photon has spin 1, de-excitation through γ-emission is always accompanied by a spin change $\Delta I \geqq 1$. This leads to a change in the charge distribution of the nucleus and consequently also to a change in its magnetic properties. Depending on the type of change occurring through the γ-emission, the radiation is classified as electric or magnetic (§6.4) according to the scheme in the left part of Table 6.7. Although parity need not be conserved in γ-decay, the parity change associated with the different types of γ-transitions is listed in that table.

Based on the single-particle model, J. Blatt and V. F. Weisskopf have calculated probable lifetimes for excited states assuming a model nucleus with a radius of 6 fm. For decay involving electric multipole transitions the average lifetime τ is proportional to $r^{-2\Delta l}$, and for decay involving magnetic multipoles, to $r^{-2(\Delta l-1)}$; in the decay the nuclear spin quantum number s does not change. The calculated values are included in Table 6.7.

TABLE 6.7. *Classification of radiation emitted in γ-decay and lifetime calculations of the excited states for given γ-energies* (According to J. Blatt and V. F. Weisskopf)

Type of radiation	Name of transition	Spin change $\Delta I = \Delta l$	Parity change	Average lifetime (τ) in seconds for γ-energy level		
				1 MeV	0.2 MeV	0.05 MeV
E1	Electric dipole	1	Yes	4×10^{-16}	5×10^{-14}	3×10^{-12}
M1	Magnetic dipole	1	No	3×10^{-14}	4×10^{-12}	2×10^{-10}
E2	Electric quadrupole	2	No	2×10^{-11}	6×10^{-8}	6×10^{-5}
M2	Magnetic quadrupole	2	Yes	2×10^{-9}	5×10^{-6}	5×10^{-3}
E3	Electric octupole	3	Yes	2×10^{-4}	2	70 h
M3	Magnetic octupole	3	No	2×10^{-2}	180	200 d

The decay of ^{24m}Na ($I = 1+$) to ^{24}Na ($I = 4+$) provides a useful example. The transition involves $\Delta l = 3$, no (i.e. there is no parity change), so it is designated as M3. The γ-energy is 0.473 MeV and the nuclear radius of ^{24}Na is about 3.7 fm. In order to compare this energy with those given in Table 6.7, a radius correction from the assumed 6 fm to the observed 3.7 fm must be made; accordingly $E_{corr} = E_{obs}(r_i/6)^{-2(\Delta l - 1)}$ for M-type radiation. This gives a hypothetical energy of $0.473\,(3.7/6)^{-4} = 3$ MeV. According to Table 6.7 the lifetime should be > 0.02 s; the observed value is 0.035 s. The agreement between experiment and calculation is sometimes no better than a factor of 100.

The Blatt–Weisskopf relationship between energy and lifetime is only applicable for excited nucleonic states, not for rotational states. In the decay of these states the rotational quantum number always changes by two units and the lifetime of the states is proportional to $E_\gamma^{-5}\hat{Q}^{-2}$. The lifetimes are so short that no rotationally excited isomers have been observed as yet.

6.7.2. *Beta-decay*

Beta-decay theory is quite complicated and involves the weak nuclear interaction force, which is even less understood than the strong interaction. The theory for β-decay derived by E. Fermi in 1934 leads to the expression

$$\lambda = G|M|^2 f \tag{6.35a}$$

for the decay constant. G is a constant, $|M|$ is the nuclear matrix element for the β-transformation (i.e. of a proton into a neutron for positron emission, or the reverse for negatron emission), f is a function of E_{max}, and Z. $|M|$ depends on the wave function and gives the "order" of decay. Since $t_{1/2} = \ln 2/\lambda$

$$\ln 2 = G|M|^2 f t_{1/2} \tag{6.35b}$$

we see that the product $ft_{1/2}$ should be constant for a decay related to a certain $|M|$. The ft value (omitting index 1/2) is often referred to as the comparative β-half-life, and nomograms for its calculation are given in nuclear data tables and decay schemes. The lower the ft value the higher is the probability for decay, and the shorter is the half-life. G. Gamow and E. Teller have given selection rules for β-decay which are useful for estimating decay energy, half-life, or spin in a certain decay process, if two of their properties are known. These rules are summarized in Table 6.8.

For example, the decay of ^{24}Na occurs 99% through β-emission (with an E_{max} = 1.4 MeV) to an excited state of ^{24}Mg (Fig. 4.8). The log ft value of the transition is 6.1.

TABLE 6.8. *Gamow–Teller selection rules for β-decay*

Transition type	ΔI	Parity change	Log ft
Super allowed	0	No	3
Allowed	0, $\pm 1^{(a)}$	No	4–6
First forbidden	0, ± 1, ± 2	Yes	6–9
Second forbidden	± 2, $\pm 3^{(b)}$	No	10–13

[a]Not $0 \to 0$, [b]Also $0 \to 0$.

The ground state of ^{24}Mg is 0+ ; the excited state has positive parity. Thus the selection rules indicate an allowed transition for which the only spin changes permitted are 0 and ± 1. The ground state of ^{24}Na is 4+, and the observed excited state (at the 4.12 MeV level) is 4+, in agreement with the rule.

6.7.3. *Alpha-decay theory*

If we calculate the Q-value for the α-decay reaction (4.11) from the mass formulae (4.12) we find that $Q \gtrsim 0$ for all nuclei with $A > 150$, which means that we would expect all elements heavier than the rare earths to be unstable with respect to α-decay. Although α-decay is observed for some rare earth elements, it occurs frequently only for $A \geqq 210$, i.e. nuclides heavier than ^{209}Bi.

In §4.15 we mentioned the discovery by Geiger and Nuttall that the lower the α-energy the longer was the half-life of the α-decay; doubling the decay energy may reduce the half-life by a factor of 10^{20}. Alpha-decay has been observed with energies from slightly greater than 2.5 MeV (e.g. ^{150}Gd, $E_\alpha = 2.73$ MeV, $t_{1/2} = 2.1 \times 10^6$ y) to about 9 MeV (e.g. ^{114}At, $E_\alpha = 9.2$ MeV, $t_{1/2} = 2$s). The isotopes of the actinide elements typically have α-energies between 5 and 7 MeV.

The observed stability against α-decay for nuclei in the range $150 \leqq A \leqq 210$ can be explained by assuming the α-particle exists as a (preformed) entity inside the nucleus but with insufficient kinetic energy to overcome the "internal Coulomb barrier". This barrier is assumed to be of the same type, although of somewhat different shape, as the external Coulomb barrier, which is discussed in some detail in §7.4.

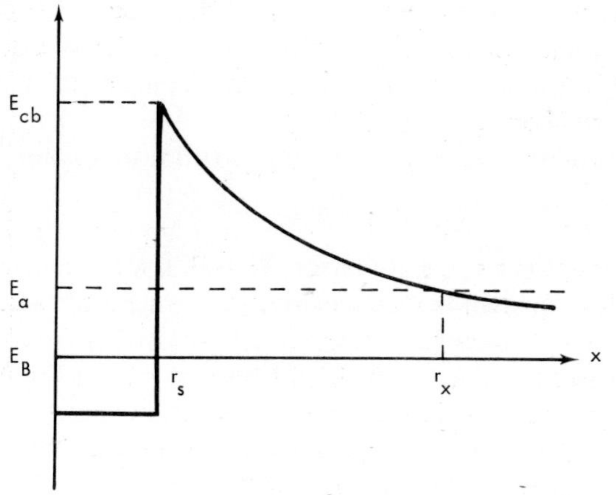

FIG. 6.14. Alpha-penetration through the potential wall.

Assume that the average kinetic energy is at the level marked E_α in Fig. 6.14. If the particles in the nucleus have a Boltzmann energy distribution, any particular α-particle can acquire sufficient kinetic energy through collisions to overcome the barrier (E_{cb}). It would be emitted with an energy E_{cb} which for an element like uranium is 26 MeV. However, the observed α-energy is only 4.2 MeV.

This contradiction was explained by G. Gamov, and independently by R. W. Gurney and E. U. Condon, in 1928, by using a quantum mechanical model, which retained the feature of the "one-body model" with a preformed α-particle inside the nuclear potential wall.

Consider α-decay in which the emitted α-particle has an energy of E_α. The kinetic energy in the nucleus (i.e. before emission) of this α-particle would be

$$E_\alpha^0 = 1/2M_\alpha^0(v_\alpha^0)^2 = E_\alpha + E_B \tag{6.36}$$

The α-particle bounces with velocity v_α^0 repeatedly between the potential walls. The time between two successive impacts is

$$\tau_\alpha = 2r_s/v_\alpha^0 \tag{6.37}$$

As we have noted in α-decay the α-particles escape the nucleus with lower energy than the potential barrier, like passing through a tunnel in a mountain. The probability for such tunneling must be directly related to the number of impacts per unit time

$$\lambda = T_0/\tau_\alpha = T_0 v_\alpha^0/2r_s \tag{6.38}$$

$T_0(\leqq 1)$ is called the *barrier transmission factor*, which is the probability of tunneling per impact:

$$T_0 = 4\left(\frac{E_{cb} - E_\alpha}{E_B + E_\alpha}\right)^{1/2} e^{-2G} \tag{6.39a}$$

where

$$G = \frac{2Z_1 Z_\alpha e^2}{\hbar v}(\cos^{-1} u - u\sqrt{1 - u^2}) \tag{6.39b}$$

and

$$u^2 = r_s/r_x = E_\alpha/E_{cb} \tag{6.39c}$$

The same relation holds—with appropriate substitutions—for any charged particle trying to enter the nucleus from outside the potential barrier (cf. §7.4), where, however, only one impact is possible.

Combining (6.38) and (6.39) and taking the logarithm we obtain

$$\log \lambda = \text{const} - 2G \tag{6.40}$$

which is very similar to the empirical Geiger–Nuttall law.

The observed decay constant is very sensitive to E_α: a 1 MeV increase in E_α increases λ (and decreases $t_{1/2}$) by a factor of about 10^5. It is also very sensitive to the nuclear radius: a 10% increase in r_s (or the Coulomb radius r_c; see Fig. 7.4) which means a corresponding decrease of the Coulomb barrier height, increases λ by a factor of 150.

The α-decay theory was the first successful (quantum mechanical) explanation of radioactive decay, and as such played a major role in further development of nuclear theories and models. Although its simplicity causes it to fail considerably for nonspherical

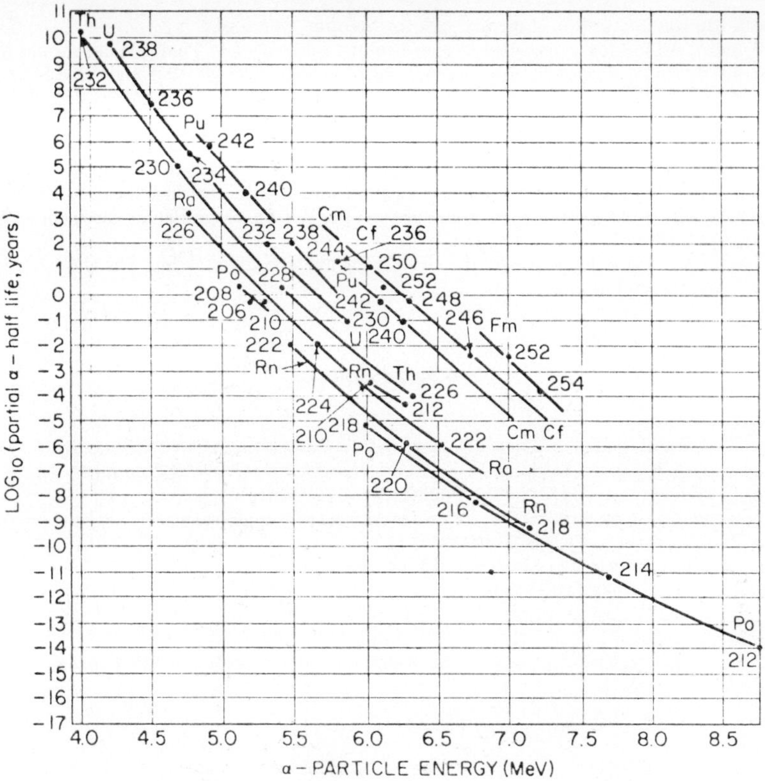

FIG. 6.15. Alpha-decay half-lives as a function of decay energy.

nuclei as well as those near closed shells, such effects can be taken into account in more advanced Nilsson-type calculations.

In order to predict α-decay energies and half-lives we use correlations usually referred to as α-systematics. Figure 6.15 shows an example, where E_α and $t_{1/2}$ are correlated with the Z and A of α-emitting nuclei. Such correlations have been very useful in the identification of isotopes of actinides and transactinide elements.

6.8. Exercises

6.1. The quantum numbers $s = 1/2$ and $l = 2$ are assigned to a particle. (a) If spin and orbital movements are independent, how many space orientations (and thus measured spectral lines if no degeneration of energy states occur) are possible in an external field of such a strength that both movements are affected? (b) How many lines would be observed if spin and orbital movements are coupled?

6.2. In the hydrogen atom the K-electron radius is assumed to be 0.529×10^{-10} m (the Bohr radius). (a) Calculate the orbital velocity of the electron assuming its mass to be $\mathbf{m}_e$. (b) How much larger is its real mass because of the velocity? Does this affect the calculations in (a)?

6.3. A beam of protons pass through a homogeneous magnetic field of 0.5 T. In the beam there is a small high frequency coil which can act on the main field so that the proton spin flips into the opposite direction. At what frequency would this occur?

6.4. Calculate the nuclear Landé factor for ^{11}B.

6.5. In Table 6.5 the first degenerate levels have been given. Using the same assumptions, what states will be contained in the next level and how many nucleons will it contain?

6.6. How deep is the nuclear well for ^{116}Sn if the binding energy of the last nucleon is 9 MeV?

6.7. (a) Calculate the spins and nuclear g factors for (a) ^{45}Ca, (b) ^{60}Co, and (c) ^{141}Pr, using data in Table 6.3.

6.8. The observed quadrupole moment of ^{59}Co is 0.40 barn. (a) What is the deformation value β? (b) What spin value is expected from the Nilsson diagram?

6.9. Which neutron and proton states account for the spin value I of ^{14}N?

6.10. A γ-line at 0.146 MeV is assigned to a + 4 rotational level of ^{238}Pu. (a) What should the energy of the + 2 and + 6 rotational levels be? Compare with the measured values of 0.044 and 0.304 MeV. (b) If ^{238}Pu is considered to be a homogeneous sphere, what will its apparent radius be? Compare with that obtained using relation (3.7).

6.11. A ^{239}Pu compound is placed in a test tube in a 40 MHz nmr machine. At what field strength does resonance occur with the nuclear spin? Is the measurement possible? Relevant data appear in Table 6.3.

6.12. Using the Gamow theory the probability for tunneling of an α-particle in the decay of ^{238}U is $1:10^{38}$, and the α-particle hits the walls about 10^{21} times per second. What average lifetime can be predicted for ^{238}U from this information?

6.13. Estimate with the Gamow theory the half-life for the α-decay of ^{147}Sm. The Q_α is 2.314 MeV. E_B in (6.36) is the α-particle binding energy.

6.9. Literature

See Appendix A, § A.1.

I. PERLMAN, A. GHIORSO, and G. T. SEABORG, Systematics of alpha-radioactivity, *Phys. Rev.* **77** (1950) 26.
I. PERLMAN and J. O. RASMUSSEN, Alpha radioactivity, *Handbuch der Physik* **42**, Springer-Verlag, 1957.
W. J. MOORE, *Physical Chemistry*, 3rd edn., Prentice-Hall, I. 1962.
G. E. GORDON and C. D. CORYELL, Models for nuclear structure of spherical nuclei, *J. Chem. Ed.* **44** (1967) 636.
E. K. HYDE, Nuclear models, *Chemistry* **40** (1967) 12.
D. R. INGLIS, Nuclear models, *Physics Today*, **22** (6) (1969) 29.
M. BARANGER and R. A. SORENSEN, The size and shape of atomic nuclei, *Sci. Am.* **221** (Aug. 1969) 59.
U. M. PARIHH, *Absorption Spectroscopy of Organic Molecules*, Addison–Wesley, 1973 (application of NMR and mass spectrometry).
D. ISABELLE and G. RIPHA, La Structure de noyau atomique, *La Recherche*, June 1974, p. 543.

CHAPTER 7

Nuclear Reactions: I. Energetics

Contents

Introduction

Reactions between an atomic nucleus and another particle are called induced nuclear reactions. In some such reactions, new nuclei are formed (*nuclear transmutations*); in others the original nucleus is excited to a higher energy state (*inelastic scattering*); in a third class, the nucleus is unchanged (*elastic scattering*). Spontaneous nuclear reactions, which are involved in the radioactive decay of unstable nuclei, have been discussed in Chapter 4. In this chapter the emphasis is on the mass and energy relationships in induced reactions.

7.1. Mechanics of induced nuclear reactions

We have discussed a number of laws governing nuclear reaction processes. A summary of these laws provides us with a picture of the mechanics of nuclear reactions occurring below 1 GeV, in which energy range the formation of new elementary particles can be neglected.

First, let us consider a generalized nuclear reaction as shown in Fig. 7.1, in which the *projectile* is indicated by subscript 1 and the *target* atom by subscript 2; the reaction leads to the formation of two new species, which are designated by subscripts 3 and 4.

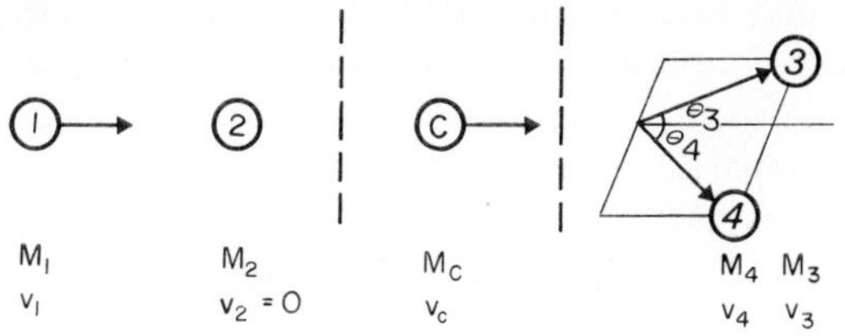

FIG. 7.1. Schematic picture of a nuclear reaction.

The velocity of the projectile v_1 is greater than zero, but the velocity of the target atom v_2 is made to be zero by using the target nucleus as the origin of the system of reference. In nuclear reactions it is common to assume that an intermediate product, the *compound nucleus*, is formed. This is designated in Fig. 7.1 by the subscript c. The compound nucleus may have a very short lifetime (less than 10^{-13} s) prior to decay to the products.

All the *conservation laws* derived in §4.1 are applicable to induced nuclear reactions. These are:

(a) the conservation of total energy: $\Delta E = 0$ (4.2)
(b) the conservation of linear momentum: $\Delta p = 0$ (4.4)
(c) the conservation of total charge: $\Delta Z = 0$ (4.6)
(d) the conservation of mass number: $\Delta A = 0$ (4.7)
(e) the conservation of spin: $\Delta I = 0$ (4.9)

"Conservation of total energy" means that the total energy of the products must equal the total energy of the reactants, i.e. $E_{\text{products}} - E_{\text{reactants}} = \Delta E = 0$. For (b) it should be remembered that linear momentum is a vector property; thus

$$p_1 + p_2 = p_3 \cos\theta_3 + p_4 \cos\theta_4 \tag{7.1a}$$

$$p_3 \sin\theta_3 = p_4 \sin\theta_4 \tag{7.1b}$$

Some of these conservation laws (e.g. (d) and (e)) are not always obeyed in high energy reactions ($\geqq 1$ GeV) in which new elementary particles are formed. The kinetic energy equation $E_{\text{kin}} = \frac{1}{2}mv^2$ (2.5) was deduced by I. Newton in 1687 assuming that the mass of the particle was independent of its velocity. We have seen that this is not true at high particle velocities. For example the relativistic mass increase of a bombarding proton is about 1% at kinetic energy of 10 MeV. In all dynamic relations involving moving particles, the relativistic mass must be used. This is particularly important in the next chapter, where we discuss the effect of the acceleration of charged particles to high energies in nuclear particle accelerators.

7.2. The mass energy

As for radioactive decay, the energy of a nuclear reaction is given by its Q-value (3.5, 4.12):

$$Q\,(\text{MeV}) = -931.5\,\Delta M^0\,(\text{u}) \tag{7.2}$$

where

$$\Delta M^0 = M_3^0 + M_4^0 - M_1^0 - M_2^0 \tag{7.3}$$

The reaction energy may thus be computed from the rest masses of the reactants and the products. If mass disappears in the reaction ($\Delta M^0 < 0$), energy is released: the reaction is said to be *exoergic*, and Q is positive. For $Q < 0$ the reaction is *endoergic* and $\Delta M^0 > 0$. (For comparison, in chemistry a negative value of ΔH is associated with exothermic spontaneous reactions.) Tables of Q-values, especially for light projectiles, can be found in several literature sources. The most comprehensive one is given by F. Everling, L. A. Koenig, J. H. E. Mattauch, and A. H. Wapstra. Part of such a table is reproduced in Table 7.1.

TABLE 7.1. *Different ^{14}N-reactions*

(From F. Everling, L. A. Koenig, J. H. E. Mattauch and A. H. Wapstra, "Consistent set of energies liberated in nuclear reactions". *1960 Nuclear Data Tables*, US Government Printing Office, Washington 25 DC, 1961)

TARGET	REACTION (IN	, OUT)	END-PRODUCT	Q - VALUE KEV	ERROR KEV
14 N	(GAMMA	, 1N)	13 N	-10553.2	1.3
	(	, 1H)	13 C	-7549.4	0.7
	(	, 2D)	12 C	-10271.76	0.16
	(	, 3T)	11 C	-22735.0	3.0
	(	, 3HE)	11 B	-20734.48	0.49
	(	, 4HE)	10 B	-11613.4	0.8
	(	,1N+1N)	12 N	-30880	90
	(	,1N+1H)	12 C	-12496.48	0.43
	(	,1H+1H)	12 B	-25082.8	1.0
14 N	(1N	,GAMMA)	15 N	10834.3	0.9
	(	, 1H)	14 C	626.55	0.39
	(	, 2D)	13 C	-5324.7	0.8
	(	, 3T)	12 C	-4014.13	0.44
	(	, 3HE)	12 B	-17364.9	1.0
	(	, 4HE)	11 B	-157.3	0.7
	(	,1N+1N)	13 N	-10553.2	1.3
	(	,1N+1H)	13 C	-7549.4	0.7
	(	,1H+1H)	13 B	-20200	50
14 N	(1H	,GAMMA)	15 O	7291.1	1.7
	(	, 1N)	14 O	-5931	5
	(	, 2D)	13 N	-8328.5	1.3
	(	, 3T)	12 N	-22400	90
	(	, 3HE)	12 C	-4778.61	0.20
	(	, 4HE)	11 C	-2922.3	3.1
	(	,1N+1N)	*	*	*
	(	,1N+1H)	13 N	-10553.2	1.3
	(	,1H+1H)	13 C	-7549.4	0.7
14 N	(2D	,GAMMA)	16 O	20735.40	0.26
	(	, 1N)	15 O	5066.4	1.8
	(	, 1H)	15 N	8609.6	0.8
	(	, 3T)	13 N	-4295.6	1.3
	(	, 3HE)	13 C	-2056.3	0.7
	(	, 4HE)	12 C	13573.85	0.40
	(	,1N+1N)	14 O	-8155	5
	(	,1N+1H)	14 N	-2224.71	0.40
	(	,1H+1H)	14 C	-1598.17	0.36

TARGET	REACTION (IN	, OUT)	END-PRODUCT	Q - VALUE KEV	ERROR KEV
14 N	(3T	,GAMMA)	17 O	18619.8	0.9
	(	, 1N)	16 O	14477.76	0.47
	(	, 1H)	16 N	4852	6
	(	, 2D)	15 N	4576.6	0.9
	(	, 3HE)	14 C	-137.93	0.36
	(	, 4HE)	13 C	12263.3	0.8
	(	,1N+1N)	15 O	-1191.3	1.9
	(	,1N+1H)	15 N	2351.9	0.9
	(	,1H+1H)	15 C	-6638.1	1.1
14 N	(3HE	,GAMMA)	17 F	15840.0	2.3
	(	, 1N)	*	*	*
	(	, 1H)	16 O	15242.24	0.27
	(	, 2D)	15 O	1797.9	1.7
	(	, 3T)	14 O	-5166	5
	(	, 4HE)	13 N	10024.0	1.4
	(	,1N+1N)	*	*	*
	(	,1N+1H)	15 O	-426.8	1.8
	(	,1H+1H)	15 N	3116.4	0.9
14 N	(4HE	,GAMMA)	18 F	4403.9	4.0
	(	, 1N)	17 F	-4737.2	2.3
	(	, 1H)	17 O	-1192.9	0.9
	(	, 2D)	16 O	-3110.22	0.38
	(	, 3T)	15 O	-12521.6	1.8
	(	, 3HE)	15 N	-9742.9	0.9
	(	,1N+1N)	*	*	*
	(	,1N+1H)	16 O	-5334.9	0.5
	(	,1H+1H)	16 N	-14961	6

TARGET 14 N (N = 7 , Z = 7)

MASS EXCESS	ERROR			BIND.EN.	ERROR	
2863.60	0.16	KEV	12C=0	104656.9	3.1	KEV
3074.38	0.17	MICRO-MU	12C=0			
7007.98	0.16	KEV	16O=0			
7526.19	0.17	MICRO-MU	16O=0			

The relativistic mass equation (4.21)

$$E_{\text{kin}} = \mathbf{c}^2(m - m^0) \tag{7.4}$$

can be separated into two terms if we define

$$E^0_{\text{mass}} \equiv m^0\mathbf{c}^2 \tag{7.5}$$

and

$$E_{\text{tot}} \equiv m\mathbf{c}^2 \tag{7.6}$$

Then

$$E_{\text{tot}} = E_{\text{kin}} + E^0_{\text{mass}} \tag{7.7}$$

We will call E^0_{mass} the *mass energy*, which is independent of the kinetic energy of the particle. E^0_{mass} is mass *potential* energy, and can be converted into any other energy form because (7.5) is a form of Einstein's mass–energy relation. E^0_{mass} is closely related to the nuclear binding energy (cf. 3.5).

In the previous section we pointed out that for $Q > 0$ the reaction is exoergic and, as a consequence of that, mass has to disappear ($\Delta M^0 < 0$). If the total energy of the system is constant (conservation rule (a)), it becomes obvious from (7.7) that when E^0_{mass} decreases E_{kin} must increase. Thus the products of the nuclear reaction have a higher kinetic energy than the reactants.

7.3. Dissecting a nuclear reaction

As an example of a nuclear transmutation reaction, let us consider the following:

$$^{14}_{7}\text{N} + ^{4}_{2}\text{He} \rightarrow [^{18}_{9}\text{F}]^* \rightarrow ^{17}_{8}\text{O} + ^{1}_{1}\text{H}$$

The intermediate compound nucleus $^{18}_{9}F$ is in square brackets to indicate its transitory nature and marked with an asterisk to indicate that it is excited. Induced nuclear reactions are often written in an abbreviated manner indicating first the target and then, in parentheses, the projectile and the smaller product, followed by the major product outside the parentheses. In the case of the sample reaction, we would write $^{14}N(\alpha, p)\,^{17}O$. The abbreviations used for ^4He, ^1H, $^2H(=D)$, $^3H(=T)$, etc., are α, p, d, t, etc.; see also Table 7.1. The reactions may be classified by the particles in parentheses; the sample reaction is called an (α,p) type.

The sample reaction is of historical interest since this is actually the reaction studied by Lord Rutherford in 1919 when he produced the first induced nuclear transformation in the laboratory. A cloud chamber photograph of the reaction is shown in Fig. 7.2.

From the atomic rest masses we can calculate that the change in mass for this reaction is

$$\Delta M^0 = (M^0_3 + M^0_4 - M^0_1 - M^0_2) = (16.999\,133 + 1.007\,825 - 14.003\,074 - 4.002\,604) = 0.001\,280\ (\text{u})$$

This increase in mass corresponds to a Q-value of -1.19 MeV (7.2). The energy required for this endoergic reaction can only be obtained through the kinetic energy of the projectile.

In the collision between the projectile and the target nucleus, which we have assumed to be stationary, the compound nucleus formed always acquires a certain kinetic

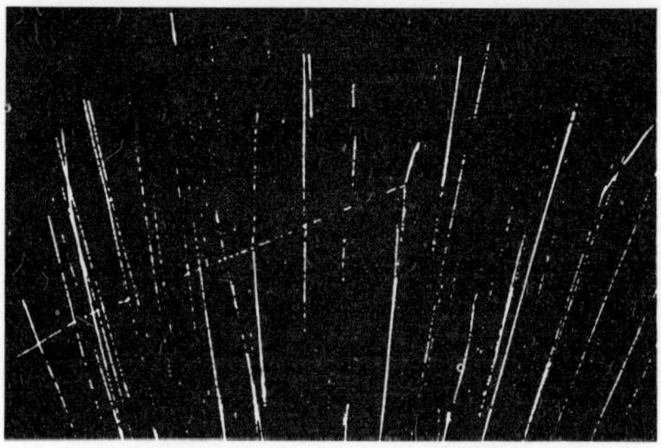

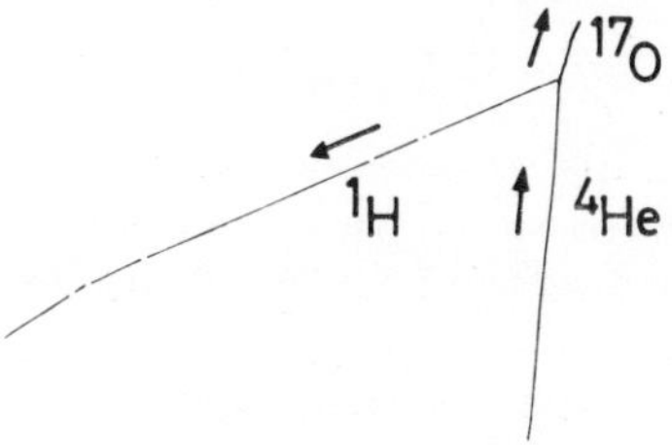

FIG. 7.2. Cloud chamber photograph of the Rutherford reaction $^4He + {}^{14}N \rightarrow {}^{17}O + {}^1H$. (According to P. M. S. Blackett and D. S. Lees.)

energy which can be calculated from the conservation laws to be

$$E_{kin,c} = \frac{M_1}{M_1 + M_2} E_{kin,1} \approx \frac{A_1}{A_c} E_{kin,1} \tag{7.8}$$

The mass numbers can be substituted for the atomic masses since we are carrying the calculation of the energy to three significant figures only. In order for the reaction to occur, the kinetic energy of the projectile $E_{kin,1}$ must exceed the kinetic energy of the compound nucleus $E_{kin,c}$ by the value of Q:

$$E_{kin,1} = E_{kin,c} - Q \tag{7.9}$$

The minimum energy necessary for a reaction to occur is called the *threshold energy* E_{tr}, which according to (7.8) and (7.9) is

$$E_{tr} = -Q\left(\frac{M_1 + M_2}{M_2}\right) \approx -Q\frac{A_c}{A_2} \quad (\text{if } Q < 0) \tag{7.10}$$

For our sample reaction, the threshold energy is $1.19 \times 18/14 = 1.53$ MeV.

If the projectile has a higher kinetic energy than the threshold energy, the products would have a correspondingly higher combined kinetic energy, assuming no transformation into other forms of energy. Rutherford used α-particles from the radioactive decay of ^{214}Po, which have kinetic energies of 7.68 MeV. The products, therefore, had considerable kinetic energy, which is shown by the thick tracks in the cloud chamber picture of Fig. 7.2.

In Fig. 7.3 the energy pattern for our reaction is summarized. The Q-value for the first step of the reaction ($^4He + {}^{14}N \rightarrow {}^{18}F$) is $+4.40$ MeV (Q_1). The kinetic energy of the ^4He is 7.68 MeV, so conservation of momentum requires that the ^{18}F nucleus

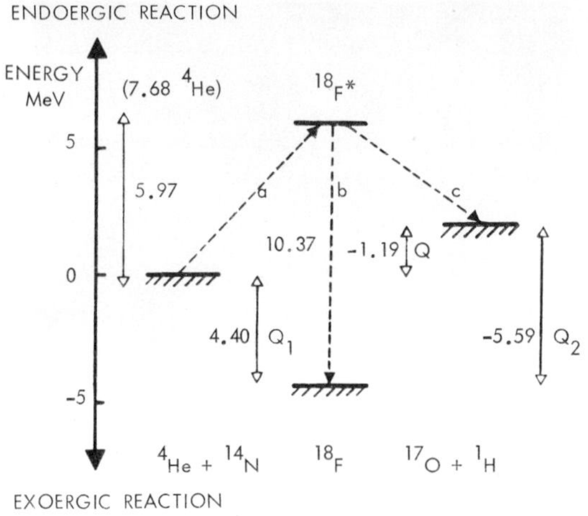

FIG. 7.3. Energy diagram of the transmutation $^4He + ^{14}O \rightarrow ^{18}F^* \rightarrow ^{17}O + ^1H$ caused by bombarding ^{14}N by 7.68 MeV α-particles from ^{214}Po.

have $\frac{4}{4+14} 7.68 = 1.71$ MeV. Thus, the internal excitation energy of [18F]* is $4.40 + 7.68 - 1.71 = 10.37$ MeV. For the second step, $^{18}F \rightarrow ^{17}O + ^1H$, the Q_2-value is -5.59 MeV. This requires that the total excitation energy of the products be $10.37 + 1.71 - 5.59 = 6.49$ MeV.

7.4. The Coulomb barrier

Equation (7.7) is a special case of a more general equation applicable to all systems of particles:

$$E_{tot} = E_{kin} + E_{pot} \tag{7.11}$$

where E_{kin} = translational, rotational, vibrational, etc., energy and E_{pot} = mass energy, gravitational, electrostatic energy, surface energy, chemical binding energy, etc. This is the total energy referred to in (4.2).

In nuclear reactions we would include the mass energy for the atomic masses in their ground state E^0_{mass}, the excitation energy of the nucleus above its ground state E_{exc}, the absorption or emission of photons in the reaction E_ν, and—in reactions between charged particles—the electrostatic potential (Coulomb) energy E_{coul}. Since we are concerned only with reactions induced by neutral or positively charged particles, the Coulomb energy is zero or positive (i.e. repulsive). The incoming projectile must possess sufficient kinetic energy to overcome any repulsion. During the compound nucleus stage, this appears as excitation energy. In the second step of the reaction the repulsion of charged products results in greater kinetic energy. Thus, generally,

$$E_{tot} = E_{kin} + E_{coul} + E^0_{mass} + E_{exc} + E_\nu \tag{7.12}$$

In nuclear reactions the total energy must be conserved, although the distribution of this energy in the different forms of (7.12) usually changes during the course of the reaction. For example, any decrease in the mass energy term must be balanced by a complimentary increase in one or more of the other energy terms. An example of this occurs in the explosion of nuclear weapons where a fraction of the rest mass is trans-

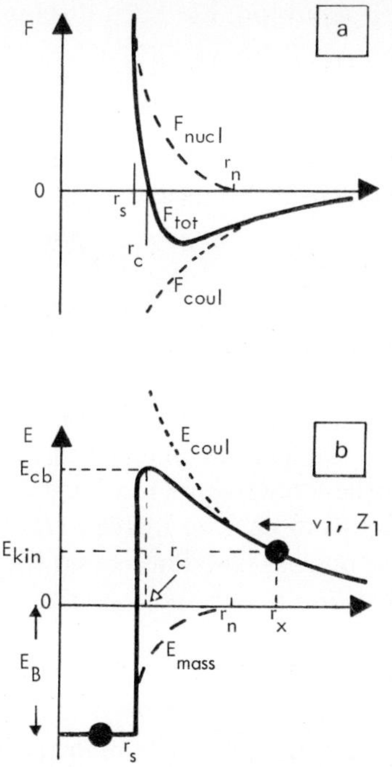

FIG. 7.4. Forces (a) and energy (b) conditions when a charged projectile (Z_1, v_1) reacts with a target nucleus.

formed to other forms of energy. The opposite reaction, the transformation of kinetic energy to mass, occurs in the production of elementary particles and in high energy acceleration of particles in cyclotrons, synchrotons, etc.

Let us, as an example of the use of (7.12), consider a reaction of a positively charged particle (M_1, Z_1, v_1) with a target atom ($M_2, Z_2, v_2 \approx 0$). We make two simplifying assumptions: the target nucleus is in the center of the coordinate system (i.e. no recoil, $v_c = 0$) and relativistic mass corrections can be neglected.

Because both projectile and target have positive charge they must repel each other according to the Coulomb law:

$$F_{coul} = \frac{eZ_1 eZ_2}{x^2} k \qquad (7.13)$$

where k is 8.99×10^9 (N m^2 C^{-2}). This force is shown as a function of the distance between the particles in Fig. 7.4a. At a distance greater than r_n only the Coulomb repulsive force is in operation; however, for distances less than r_n both the attractive nuclear force F_{nucl} and the repulsive Coulomb act upon the system. The total force is given by $F_{tot} = F_{coul} + F_{nucl}$. This is shown by a solid line in the figure. At some particular distance designated r_c, the forces balance each other and at shorter distances ($x < r_c$), the attractive nuclear force dominates. The distance r_c is known as the Coulomb radius; to be more exact, r_c is the sum of the projectile and target radii.

In Fig. 7.4b the variation in the value of the different terms of (7.12) is indicated as a function of the distance x between the two particles. For convenience we can separate the course of the reaction into a sequence of three steps.

(1) At long distances from the target nucleus the kinetic energy of the projectile is

decreased due to the Coulomb repulsion. For such distances the nuclear force can be neglected and

$$E^0_{kin} = E_{kin} + E_{coul} \tag{7.14}$$

where E^0_{kin} is the original kinetic energy of the projectile and E_{coul} is the electrostatic (Coulomb) potential energy, which at distance x is

$$E_{coul} = -\frac{Z_1 Z_2 \mathbf{e}^2}{x} k \tag{7.15}$$

As the projectile approaches the target nucleus, the Coulomb repulsion causes the potential energy to increase as the kinetic energy of the particle decreases. If this decrease in kinetic energy of the particle is such that the kinetic energy reaches a value of zero at any distance greater than r_c, the particle is reflected away from the nucleus before it is close enough to experience the attractive nuclear force. The projectile is thus hindered by a *Coulomb potential barrier* from causing nuclear reaction. A necessary condition for charged projectiles to cause nuclear reactions is that E^0_{kin} must exceed the Coulomb barrier height E_{cb}:

$$E_{cb} = \frac{Z_1 Z_2 \mathbf{e}^2}{r_c} k \tag{7.16}$$

This equation is useful for determining the Coulomb radius ("distance of closest approach") of nuclei; for conservation of momentum see (7.18 and 7.19).

(2) If $E^0_{kin} > E_{cb}$ the attractive nuclear force dominates and the particle is absorbed by the target nucleus. Assuming $Q > 0$, the E^0_{mass} decreases. This means that E_{exc} increases, and the system is transformed into an excited compound nucleus. Moreover, since the projectile had a certain kinetic energy, the excitation of the compound nucleus is

$$E_{exc} = Q + E^0_{kin} \tag{7.17}$$

This equation is based on the assumption that the compound nucleus is the center of reference. It should be noted that the height of the Coulomb barrier does not influence the excitation energy of the nucleus in any way other than that a projectile must have a kinetic energy greater than that value before the reaction can occur.

(3) In reaction all E_{kin} and E_{coul} are transferred into E_{exc} and E^0_{mass} for the compound nucleus. The excitation of the compound nucleus can be removed either through the emission of γ-rays, or by the decay of the compound nucleus into different product nuclei. In the former case, $E_{exc} = E_\gamma$. In the latter case we again have to bring in (7.12).

Thus far we have discussed three different kinds of nuclear radii—r_s, r_c, and r_n. In addition we have treated the target and the projectile as points in space. We learned in Chapter 3 that, experimentally, the nuclear radius is given by $r = r_0 A^{1/3} \times 10^{-15}$ m. Experiments have indicated that the values of r_0 are approximately 1.1 for the radius r_s of constant nuclear force, 1.3 for the Coulomb radius r_c, and 1.4 for the nuclear force radius r_n, which includes surface effects.

To calculate the value of the energy of the Coulomb barrier we can use a model in which the target nuclei and the projectile are just touching so that r_c is taken as the sum of the radii of the projectile and the target nucleus. Moreover, we must now consider that the compound nucleus does not remain immobile in the collision but receives a certain kinetic energy determined by the conservation of momentum. With this model

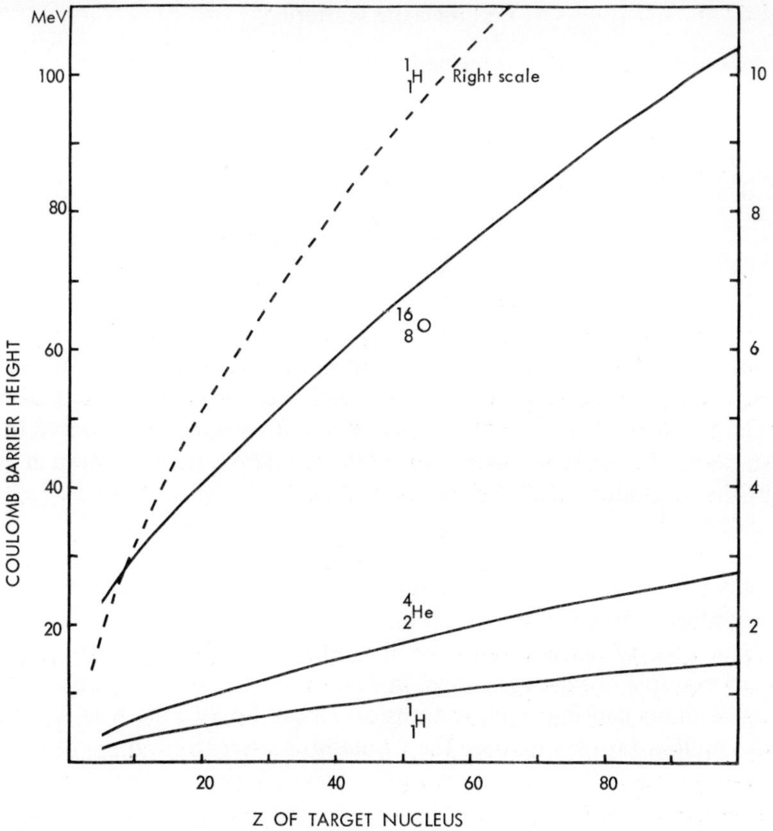

FIG. 7.5. The Coulomb barrier height, E_{cb} (min) according to (7.18), for reactions between a target element Z (of most common A) and projectiles of ^{1}H, ^{4}He, and ^{16}O.

the Coulomb barrier energy is given by the equation

$$E_{cb\,(min)} = 1.109\left(\frac{A_1 + A_2}{A_2}\right)\left(\frac{Z_1 Z_2}{A_1^{1/3} + A_2^{1/3}}\right) \quad \text{(MeV)} \tag{7.18}$$

$E_{cb\,(min)}$ is the minimum energy that a projectile of mass A_1 and charge Z_1 must have in order to overcome the Coulomb barrier of a target nucleus of mass A_2 and charge Z_2. For the reaction between an α-particle and the nucleus ^{14}N, $E_{cb\,(min)}$ has a value of 4.95 MeV; it is obvious that Rutherford's α-particles of 7.68 MeV from the decay of ^{210}Po had sufficient kinetic energy to cause reaction.

In Fig. 7.5 the values of $E_{cb\,(min)}$ for proton, α-, and oxygen-16 particles are shown as a function of the atomic number of the target nucleus. The A-values of the target are of the most stable isotope of the element Z.

In most cases the projectile–target collision is not head-on. Considering conservation of momentum, the distance x_{min} of closest approach becomes

$$x_{min} = \frac{2Z_1 Z_2 \mathbf{e}^2}{m_{red}(v_1^0)^2} k\left(1 + \sin^{-1}\left(\frac{\theta}{2}\right)\right) \tag{7.19}$$

θ is the scattering angle $m_{red}^{-1} = m_1^{-1} + m_2^{-1}$; v_1^0 is the projectile maximum velocity.

7.5. Inelastic scattering

In the class of nuclear reactions termed inelastic scattering, part of the kinetic energy of the projectile is transferred to the target nucleus as excitation energy without changing the values of A or Z of either target or projectile. However, the collision of the projectile and target and the formation of compound nucleus does result in a value of Q different than zero. The process can be written:

$$X_1 + X_2 \rightarrow X_c^* \begin{cases} \xrightarrow{a} X_1 + X_2 + \gamma & \\ \xrightarrow{b} X_1 + X_2^* & Q<0 \end{cases} \quad \begin{matrix} (7.20a) \\ (7.20b) \end{matrix}$$

The reaction path a indicates that the energy Q is emitted as a γ-ray when the compound nucleus decays. In reaction path b the Q is retained in the decay of the compound nucleus as excitation energy of the target nuclide. The latter exists in an excited state and may transform to the ground state quite rapidly or may exist for a measurable time as an isomer. An example of inelastic scattering reaction (7.20b) is the formation of an isomer of silver by the irradiation of ^{107}Ag with neutrons

$$^{107}\mathrm{Ag(n, n')}^{107\mathrm{m}}\mathrm{Ag} \rightarrow {}^{107}\mathrm{Ag} + \gamma$$

The half-life of ^{107m}Ag is 44 s. The energy relationships in inelastic scattering are the same as for induced nuclear transformations.

A technique called *Coulomb excitation* is used to induce rotationally excited states in nuclei; an example of this was shown in Fig. 6.7. In order to impart high energy to the nucleus without causing nuclear transformation, heavy ions with kinetic energy below that required for passing over the Coulomb barrier ($E_1 < E_{\mathrm{cb\,(min)}}$) may be used as projectiles.

7.6. Elastic scattering

In elastic scattering energy is transferred from the projectile to the target nucleus but the value of Q is zero:

$$\mathrm{X}_1(v_1) + \mathrm{X}_2(v_2 \approx 0) \rightarrow \mathrm{X}_1(v_1') + \mathrm{X}_2(v_2' > 0) \quad Q = 0 \tag{7.21}$$

From the conservation of momentum and energy one finds that for a collision in which $\theta_3 = \theta_4 = 0$ (a head-on collision):

$$v_1' = \frac{M_1 - M_2}{M_2 + M_1} v_1; \quad v_2' = \frac{2M_1}{M_2 + M_1} v_1 \tag{7.22}$$

v_1 is the projectile velocity, and v_1' and v_2' are the velocities of the scattered species.

An important elastic scattering reaction in nuclear reactors involves the slowing down of neutrons from the high kinetic energies, which they possess when emitted in nuclear fission to very low energies at which they have much higher reaction probabilities with the nuclear fuel. The neutrons are slowed to energies comparable to those of a neutron gas at the temperature of the material in which they are moving and, hence, they are known as *thermal neutrons*. The most probable kinetic energy for particles at thermal equilibrium at the temperature T is given by the relation $E_{\mathrm{kin}} = \mathbf{k}T$. At 25°C the most probable energy for thermal neutrons is 0.026 eV, while the average energy is approximately 0.040 eV. In nuclear reactions it is necessary to consider the most probable kinetic energy rather than the average energy (cf. Chapter 11).

TABLE 7.2. *Distance from neutron source for maximum thermal flux*

Neutron source	Neutron energy (at source)	Moderator, H_2O	Moderator, D_2O
RaBe	Average 4 MeV	10 cm	
TD reaction	14 MeV	15 cm	
Fission	Average 2 MeV	7 cm	21 cm

Equation (7.22) shows that for the case of head-on (or central) collisions ($\theta = 0$) between energetic neutrons and protons at thermal energies ($v_1 \approx 0$), the velocity of the neutron is reduced in a single step to thermal energy as a result of the approximately equal masses of the neutron and proton. This effect can be seen in a head-on collision between equally heavy billiard balls. If the target atom is ^{12}C, the velocity of a neutron decreases approximately 15% in a single collision, while the energy of the neutron decreases 28%. If the target atoms are as heavy as uranium, the decrease in velocity and in kinetic energy is on the order of about 1% per collision.

The slowing down of energetic neutrons to low kinetic energies is called *moderation*. Obviously, the lightest atoms are the best *moderators* for fast neutrons and heavier atoms are poorer moderators. Conversely, heavier atoms will be better *reflectors* for energetic neutrons since they will scatter them back with little loss in energy. For thermal neutrons,

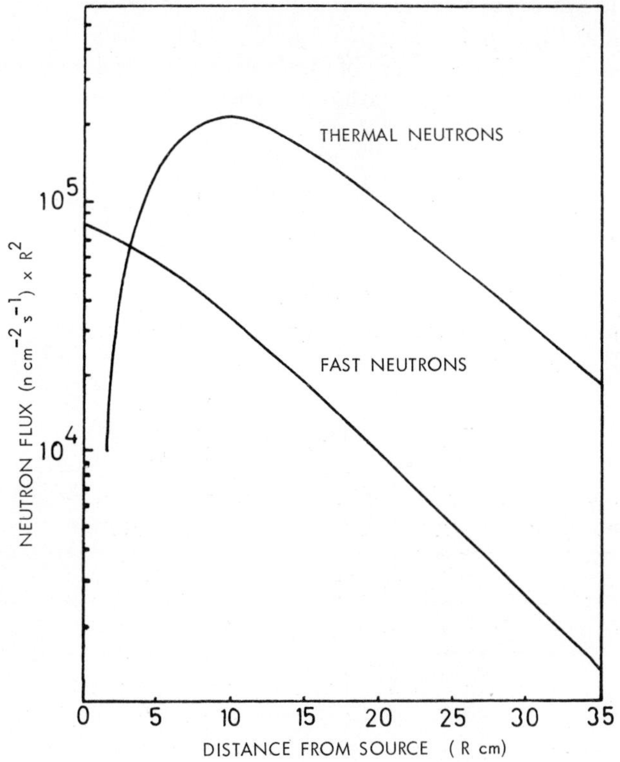

FIG. 7.6. The thermalization in water of fast neutrons from a RaBe point neutron source. Source intensity is 10^6 n s^{-1}.

however, heavy atoms are not good reflectors because of their greater tendency to absorb the neutrons. Noncentral collisions ($\theta > 0$) are more common than central ones, so the average decrease in neutron energy per collision is less than the values calculated for central collisions (see further §19.5).

In a point neutron source, surrounded by a moderator, the thermal neutron flux (particles per unit area and unit time) as a function of the distance from the source is found initially to increase and then to decrease; see Fig. 7.6. The distance from the source at which the flux is at maximum is given in Table 7.2. This is the optimum position for thermal neutron induced reactions.

7.7. Rutherford scattering*

If a collimated beam of particles ($Z_1, A_1, E_{\mathrm{kin},1}$) strikes a foil ($Z_2, A_2$) so that most of the particles pass through the foil without any reduction in energy, it is found that many particles are scattered away from their incident direction (Fig. 7.7). We can neglect multiple scattering and nuclear transformations, since they are many times less than the number of scattering events.

Geiger and Marsden found that scattering follows the relation

$$\frac{d\sigma}{d\Omega} = \left(k\frac{Z_1 Z_2 \mathbf{e}^2}{2m_{\mathrm{red}}(v_1^0)^2}\right)^2 \frac{1}{\sin^4 \theta/2} \tag{7.23}$$

where $d\sigma/d\Omega$ is the differential cross-section (reaction cross-section in barns or m^2 per steradian), θ is the scattering angle (Fig. 7.7) and v_1^0 incident particle velocity. The differential cross-section is a measure of the probability σ of a scattering event per unit solid angle Ω (1 steradian covers $1/4\pi$th of the area of a sphere):

$$\frac{d\sigma}{d\Omega} = \frac{n}{n_o N_v x \Delta\Omega} \tag{7.24}$$

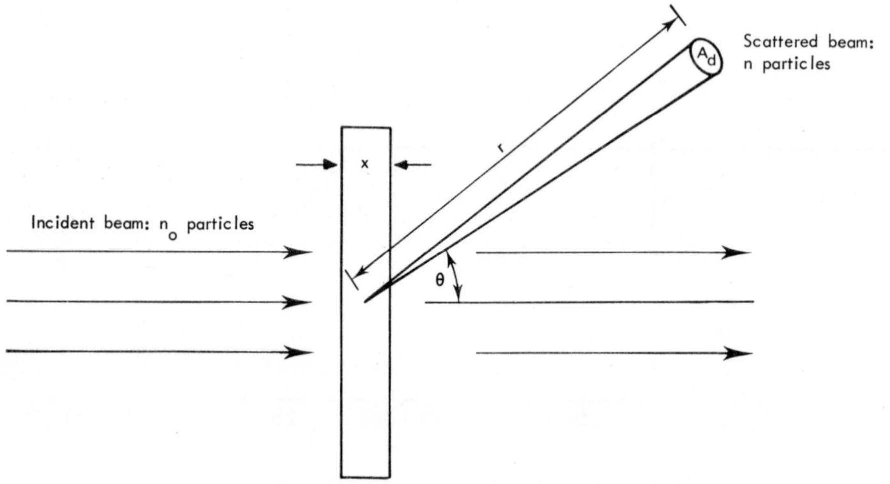

FIG. 7.7. Geometry for measuring the α-scattering by a thin foil.

where n is the number of projectiles scattered into the detector which subtends a solid angle

$$\Delta\Omega = A_d r^{-2} \tag{7.25}$$

with respect to the center of the target; A_d is the detector area and r is the distance between target and detector; n_o is the number of projectiles hitting the target, which is x thick and contains N_v scattering atoms per unit volume ($N_v = \rho N_A/M$, where ρ is the density).

By bombarding foils made of different metals with α-particles from radioactive elements, (7.23) can be used to prove that Z_2 is identical to the atomic number of the metal. This implies that the atom has a central core with a positive charge of that magnitude. Such experiments showed some particles scattered almost directly back (i.e. $\theta > 90°$), a fact which baffled Lord Rutherford: "It was quite the most incredible event that ever happened to me in my life. It was almost as incredible as if you fired a fifteen inch shell at a piece of tissue paper and it came back and hit you." From the scattering equations he concluded that the scattering center, the "nucleus", had a diameter which was 1000 times smaller than the atom. This same scattering technique was used 55 years later in the US lunar explorer Surveyor 5 (1967) to determine the composition of the lunar surface.

7.8. Exercises

7.1. (a) What kinetic energy must be given to a helium atom in order to increase its mass by 1%? (b) What are the mean velocity and the mean kinetic energy of a helium atom at STP?

7.2. Calculate the distance of closest approach for 5 MeV α-particles to a gold target.

7.3. In a Rutherford scattering experiment ^{2}D atoms of 150 keV are used to bombard a thin ^{58}Ni foil of a linear density of 67×10^{-6} g cm^{-2}. The detector subtends a solid angle of 1.12×10^{-4} sr and detects 4816 deuterons out of a total of 1.88×10^{12} incident on target. Calculate (a) the differential cross-section (in barns). (b) What is the distance between target and the solid state detector, which has a surface area of 0.2 cm^2?

7.4. In Rutherford scattering on a silver foil using α-particles from a thin-walled radon tube, the following data were observed: $d\sigma/d\Omega$ 22(θ 150°), 47(105°), 320(60°), 5260(30°), 105 400(15°) barns per steradian. Calculate the energy of the incident α-particles.

7.5. Alpha-particles from ^{218}Po(E_α 6.0 MeV) are used to bombard a gold foil. (a) How close to the gold nucleus can these particles reach? (b) What is the nuclear radius of gold according to the radius–mass relation ($r_0 = 1.3$)?

7.6. What is the Q-value for the reactions: (a) ^{11}B(d, α)^{9}Be; (b) ^{7}Li(p, n)^{7}Be?

7.7. What is the maximum velocity that a deuteron of 2 MeV can impart to a ^{16}O atom?

7.8. Calculate the mass of an electron accelerated through a potential of 2×10^8 V.

7.9. ^{12}C atoms are used to irradiate ^{239}Pu to produce an isotope of berkelium. What is the Coulomb barrier height?

7.10. Measurements made on the products of the reaction ^{7}Li(d, α)^{5}He have led to an isotopic mass of 5.0137 for the hypothetical nuclide ^{5}He. Show that this nuclear configuration cannot be stable by considering the reaction $^5\text{He} \rightarrow {}^4\text{He} + \text{n}$.

7.11. In an experiment one hopes to produce the long-lived (2.6 y) ^{22}Na through a d, 2n-reaction on neon. What is (a) the Q-value, (b) the threshold energy, (c) the Coulomb barrier height, and (d) the minimum deuteron energy for the reaction? The mass excesses (in keV) are − 5182 for ^{22}Na and − 8025 for ^{22}Ne.

7.9. Literature

See Appendix A, §§ A.1 and A.7.

CHAPTER 8

Accelerators and Neutron Sources[a]

Contents

Introduction

In this chapter we are concerned with the sources—accelerators, reactors, radioactive nuclides, etc.—that can be used to obtain charged particles, neutrons, or photons that can induce nuclear reactions or chemical changes. Thus we include various accelerators for producing beams of charged particles as well as neutrons. Although accelerators produce neutrons, a more copious neutron source is the nuclear reactor. Smaller neutron sources, of the spontaneous fissioning nuclide ^{252}Cf, or of mixtures of elements like radium and beryllium, in which nuclear transmutations (mostly (α, n)-reactions) produce neutrons, are also useful to the nuclear chemist. Radioactive sources, producing high intensities of α-, β-, or γ-radiation, have several practical applications, e.g. power generators in heart pacemakers, for γ-radiographic pictures, and for sterilization of food. This chapter covers a very wide range of different types of radiation and particle sources. The connection between the main sections is more in the application than in the equipment. For example, γ-rays can be used for a number of different purposes and can be produced in several different ways. The user must judge which type of source is the most suitable for any particular purpose.

8.1. Charged particle accelerators

In order to induce nuclear reactions with positively charged projectiles such as protons, α-particles or oxygen ions, it is necessary that the projectile particles have sufficient kinetic energy to overcome the Coulomb barrier created by the repulsion

[a] Largely optional in minor course.

between the positive charges of the projectile and the nucleus. While there is some probability that a positive projectile can tunnel through the Coulomb barrier at kinetic energies lower than the maximum value of the barrier, this probability is quite small until the kinetic energy is close to the barrier maximum.

As discussed in Chapter 7, in 1919 Lord Rutherford caused artificial transmutation of one element into another by using as projectiles the α-particles emitted in the radioactive decay of ^{214}Po. These α-particles had sufficient kinetic energy (7.7 MeV) upon emission from the polonium nucleus to overcome the Coulomb barrier and react with nitrogen nuclei. However, the kinetic energy of α-particles emitted in radioactive decay is insufficient to overcome the Coulomb barrier to react with nuclei of higher atomic numbers. Consequently, means had to be devised for acceleration of the charged projectiles to kinetic energies sufficient to achieve reaction.

The principle to be used to achieve the higher kinetic energies was obvious. The projectile particles would have to be ionized to obtain positively charged ions. If these ions could be accelerated through a potential difference of 1000 V, they would acquire 1000 eV additional kinetic energy, per unit of charge. If an α-particle of a + 2 charge was to be accelerated through a potential difference of 10^6 V, it would acquire an additional kinetic energy of 2 MeV (eqn. (2.5)). The problem of obtaining the desired kinetic energy involved two aspects: first, the production of the charged particles; second, the acceleration through the necessary potential difference.

8.2. Ion source

The problem of producing the positive ions was in principle relatively simple. If a gas is bombarded by energetic electrons, the atoms of the gas are ionized and positive ions produced. Figure 8.1 shows a typical type of an ion source used in an accelerator.

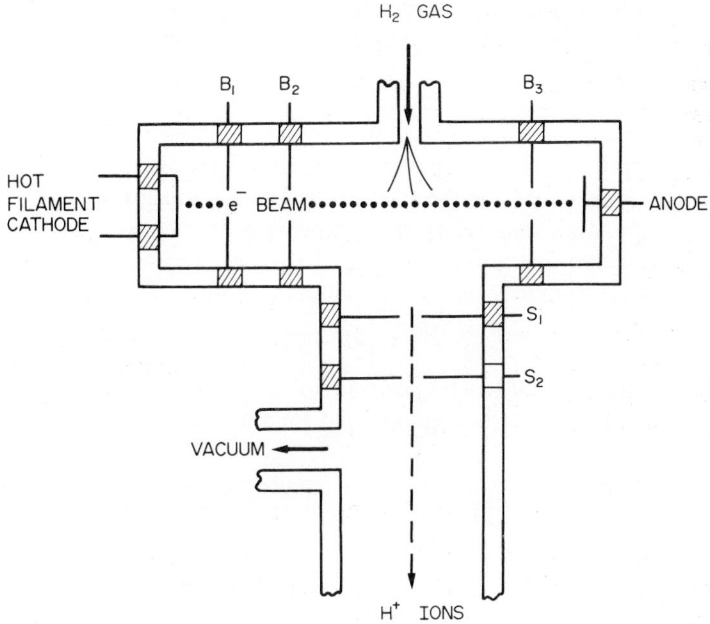

FIG. 8.1. A schematic representation of an ion source producing protons by electron bombardment of hydrogen gas.

As hydrogen gas flows into the region above the filament, the electrons being emitted by the filament are accelerated to an anode (a typical voltage drop over the electrodes B_1–B_2 may be 100 V) and in their passage through the gas cause ionization. The positive ions are extracted by attraction to a negative electrode (the voltage drop over S_1–S_2 may be 1–10 kV) into the accelerator region. The vacuum at the beam extraction is of the order 10^{-4} Pa ($\sim 10^{-6}$ mmHg) but in the ionization compartment it is normally 10^{-2} Pa ($\sim 10^{-4}$ mmHg). The basic principle is the same in the various types of ion sources that have been developed to meet the demands of the variety of the ions and accelerator types in use.

8.3. Single-stage accelerators

The first successful accelerator for causing nuclear transmutations was developed by J. D. Cockcroft and E. T. S. Walton in 1932 when they developed a means of applying a high voltage across an acceleration space. Gamov's development of the tunneling theory for α-decay led F. G. Houtermans to suggest an opposite reaction, a high energy proton tunneling through the Coulomb barrier into a target nucleus. Calculations showed that such tunneling could occur even at about 100 keV. Cockcroft, who was an electrical engineer, realized that it would be possible to build a high tension generator capable of accelerating protons to such energies (although even Rutherford first thought it technically impossible).

Cockcroft–Walton accelerators are still used for obtaining "low" kinetic energies, up to 4 MeV for protons. Presently, such accelerators are used in many installations as the first stage of acceleration in a more complex machine designed to produce high energy beams (see Fig. 8.8).

In a single-stage accelerator, such as a Cockcroft–Walton type, the total potential produced from a high voltage generator is imposed across the accelerator, i.e. between the ion source and the target. The kinetic energy, E_{kin}, of the projectile is:

$$E_{kin} = 1/2mv^2 = nqV \tag{8.1}$$

where q is the charge of the accelerated ions (Coulomb) and V is the imposed potential across the acceleration gap. This equation is identical to (2.5) except that n is the number of accelerating stages ($n = 1$ for the Cockcroft–Walton machine).

In recent years small and relatively inexpensive accelerators have come into use based on the Cockcroft–Walton principle. These are known as *transformer–rectifier accelerators* and are primarily used for production of neutrons through the reaction:

$$^3_1H + ^2_1H \rightarrow ^4_2He + n$$

Tritium targets are bombarded by accelerated deuterons. Tunneling of the Coulomb barrier (see Fig. 8.10) results in a good yield for this reaction even for energies of 0.1 MeV.

Figure 8.2 illustrates an inexpensive transformer–rectifier type accelerator. Deuterium molecules leak through a heated palladium foil into the vacuum of the ion source, where the high frequency electron discharge decomposes the deuterium molecules to form D^{+1} ions. The deuterium ions are extracted from the ion source with a relatively low negative potential to enter the acceleration tube with approximately 2.5 keV kinetic energy. The voltage difference of 100 kV across the acceleration space is obtained from a transformer coupled to a rectifier connected to a set of cylindrical electrodes.

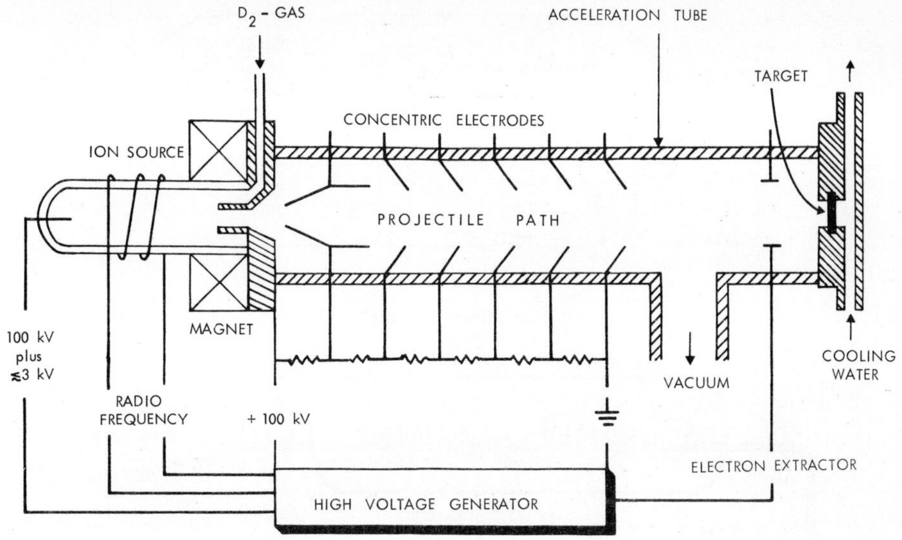

FIG. 8.2. Principle of a single-stage linear accelerator. The D_2 gas enters the vacuum through a palladium membrane.

The cylindrical electrodes serve the purpose both of acceleration and of focusing the ion beam into the proper path. As the beam particles exit from the last electrode, they strike the target, which consists of a metal foil covered with titanium on which tritium has been absorbed. Approximately 5 Ci of tritium can be absorbed as titanium tritide on an area of 5 cm^2. The target is cooled by water to minimize tritium evaporation. The passage of the beam through the target results in the emission of electrons which are removed by attraction to an anode to avoid interference with the beam.

With a voltage of 100 kV and an ion current of 0.5 mA, this accelerator can produce approximately 10^{10} ns^{-1} with an energy of 14 MeV. It is often desirable to reduce the kinetic energy of these neutrons to thermal values since thermal neutrons have a much larger probability of interacting with target nuclei. Moderation of the neutrons to thermal energies ($\sim$0.025 eV) can be accomplished by placing water or paraffin around the target (see §7.6). The flux of the thermal neutrons is relatively low, on the order of 10^8 n cm^{-2} s^{-1}. The flux of both higher energy and thermal neutrons increases as the beam current increases.

More advanced designs of this type of accelerator use coupling of transformers, rectifiers, and condensers to increase the voltage. With the larger single-stage accelerators it is possible to obtain beams of protons and deuterons which have energies of several MeV and currents of about 10 mA.

8.4. van de Graaff accelerators

A type of electrostatic accelerator that has been of great value to nuclear science since its invention is that designed by R. J. van de Graaff in 1931. The van de Graaff accelerator (VdG) can provide beams of higher energy than the single-stage Cockcroft–Walton accelerators with a very precisely defined energy. In a modification known as the tandem–VdG beams of 20 MeV protons and 30 MeV α-particles can be achieved. Positive ions of higher Z and electrons can also be accelerated in VdG machines.

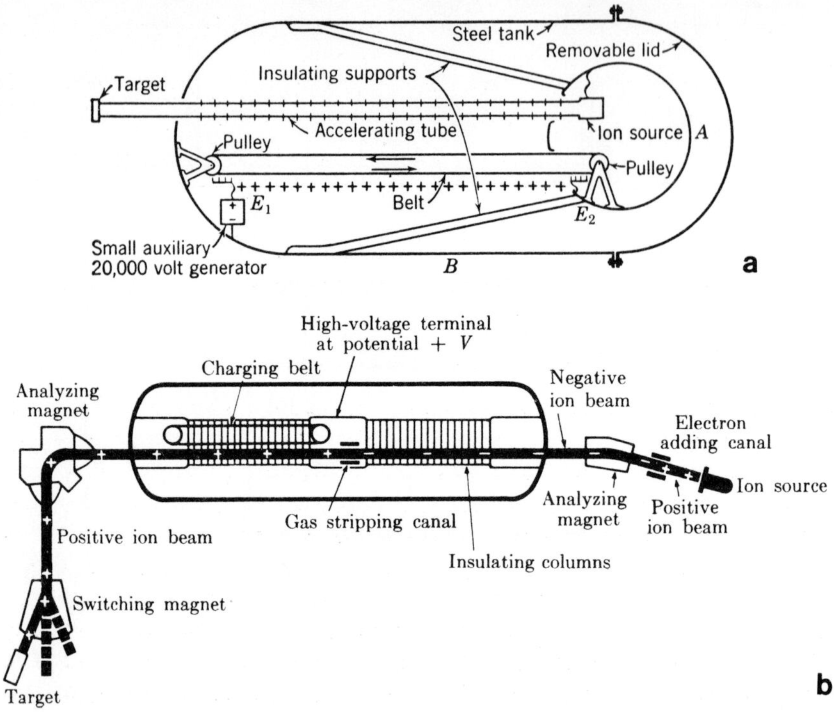

FIG. 8.3. (a) The main principle of the van de Graaff accelerator. (According to R. S. Shankland.) (b) The principle of the tandem–VdG accelerator. (According to R. V. van de Graaff.)

The principle of the VdG accelerator is illustrated in Fig. 8.3. A rapidly moving belt of a nonconducting material such as rubberized fabric accumulates positive charge as it passes an array of sharp spray points which transfer electrons from the belt to the spray points. The positive charge produced on the belt is continuously transferred by the movement of the belt to another set of spray points at which electrons are transferred to the belt from a hollow metal cylinder sphere. The result is the transfer of the positive charge to the sphere. By a continuous process of transfer of positive charge to the sphere via movement of the belt a high potential relative to ground can be built up on the sphere. The limit of the voltage that can be accumulated on the hollow sphere is determined by its discharge potential to the surrounding housing. If it is insulated by some nonconducting gas such as N_2, CO_2, or SF_6, potentials of ~ 10 MV can be accumulated. This potential can be transferred to an electrode providing the possibility of accelerating protons to energies of 10 MeV in a single stage.

The single-stage VdG has only one accelerating tube. In the tandem VdG modification, hydrogen atoms are first given a negative charge by bombardment with electrons in the ion source, and then accelerated from ground to the positive potential at the center electrode. At this stage, in the center of the accelerator, they are stripped of electrons by passage through a metal foil or a gas volume to become protons. In the second stage of acceleration the positive ions are accelerated as they move from the positive potential on the center electrode back to ground. The emergent beam of protons has a total energy of 20 MeV if the accelerating potential is 10 MV. Beams of helium, nitrogen, oxygen, and heavier ions can be accelerated similarly. The highest energy so far achieved in a tandem VdG accelerator is 60 MeV corresponding to M^- ions in the first stage and M^{5+} in the second ($1 \times 10 + 5 \times 10 = 60$ MeV).

Although the energy of the beams produced by VdG generators is extremely precise, the current intensities, 10–50 μA, are somewhat less than those of other accelerators.

The beam current i (amperes) is given by the relation

$$i = qI_0 = \mathbf{e}zI_0 \tag{8.2}$$

where q is the projectile charge (C), I_0 the incident particle current (particles s^{-1}), $\mathbf{e}$ 1.6×10^{-19}, and z the net charge (in electron units) of the beam particle (ion).

8.5. Multiple-stage linear accelerators

The potential obtained from a high voltage generator can be used repeatedly in a multiple-stage accelerator process ($n > 1$ in (8.1)). The linear accelerator operates on this principle. The accelerator tube consists of a series of cylindrical electrodes called drift tubes. The electrodes are coupled to a radiofrequency generator in the manner shown in Fig. 8.4. The high voltage generator gives a maximum voltage V which is applied to the electrodes by the radiofrequency so that the electrodes alternate in the sign of the voltage at a constant high frequency. If the particles arrive at the gap between electrodes in proper phase with the radiofrequency such that the exit electrode has the same charge sign as the particle and the entrance electrode the opposite, the particles are accelerated across the gap. Each time the particles are accelerated at an electrode gap they receive an increase in energy of qV; for n electrode gaps the total energy acquired is nqV (8.1). Inside the drift tubes no acceleration takes place since the particles are in a region of equal potential.

V. Veksler and E. McMillan independently demonstrated the principle of phase stability, which is based on an effect which tends to keep the particles in phase with the radiofrequency oscillation of the potential and allows beams of sufficient intensity for use in research. In the lower part of Fig. 8.4 the variation of potential at point A with time is shown. In operation, a particle in proper phase should arrive at the acceleration gap at time t_0. If a particle has less energy, it takes longer to traverse the drift tube and arrives late at the gap, e.g. at time t_2. Since the potential is now higher, it receives a greater acceleration and accelerates, thereby becoming more in phase. If a particle is

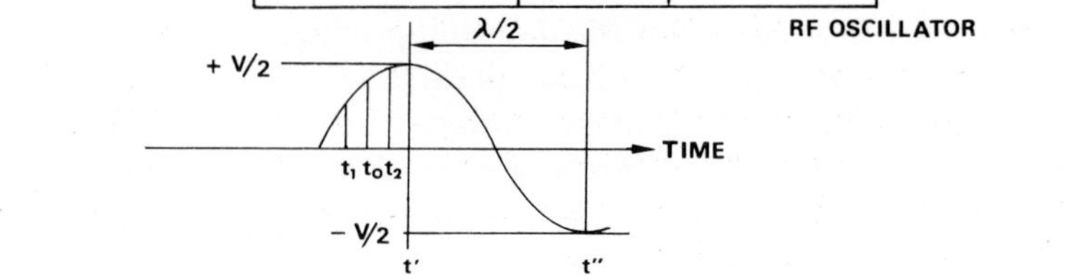

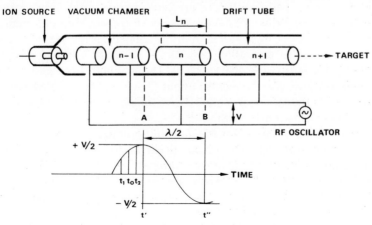

FIG. 8.4. Principle of a multiple-stage high voltage linear accelerator

too energetic, it arrives too early at the gap t_1 and receives a smaller acceleration, thereby putting it better in phase. This bunching of the ions not only keeps the ions in phase for acceleration but also reduces the energy spread in the beam.

Since the velocity of the beam particles are increasing as they progress down the accelerator but the oscillator frequency remains constant, it is necessary to make the drift tubes progressively longer so as to allow the beam particles to arrive at the exit of the tube in phase with the oscillation of the potential. Acceleration to high energies requires relatively long acceleration tubes. This increases the risk that during the time they are in the equal potential volume of the acceleration tube, the projectile will be defocused and lost from the beam by collisions with the walls of the acceleration tube. This can be minimized by the use of electrostatic and magnetic lenses which focus the beam into the center line of the acceleration tube.

For protons and deuterons the increase in energy below 10 MeV is directly related to the increase in the velocity of the particles. However, as the kinetic energy exceeds 10–20 MeV the relativistic mass becomes important since the velocity is approaching the velocity of light (Fig. 4.2). At projectile energies above 100 MeV, the relativistic mass increase is greater than the increase in velocity and at very high energies, the GeV range, the energy increase is practically all reflected by an increase in the relativistic mass (at 1.9 GeV, the proton mass is double that of the rest mass).

The phase requirement (Fig. 8.4) is that the time from point A to B (i.e. $t''-t'$) must be

$$L_n/v_n = \lambda/2\mathbf{c} \tag{8.3}$$

Taking the relativistic mass increase into consideration it can then be shown that the length of the drift tube at the nth stage is

$$L_n = \lambda/2\sqrt{1-(nk+1)^{-2}} \tag{8.4a}$$

where

$$k = qV/m^0\mathbf{c}^2 \tag{8.4b}$$

When $v \rightarrow \mathbf{c}, nk$ becomes $\gg 1$ and thus $L_n \rightarrow \lambda/2$. The length of the drift tubes therefore becomes constant at high particle energies.

A linear accelerator in use at the Lawrence Berkeley Laboratory of the University of California is the Super-HILAC (for HIgh energy Linear ACcelerator), which has a 52 m long acceleration path (excluding projectile injector and target arrangement). The machine is designed primarily for synthesis of transuranium elements through bombardment of targets of high atomic number elements with heavy projectile ions. The beam current is in the range 10^{12} particles s^{-1}. Through a special stripping device the heavy ions get a fairly high charge, $\gtrsim 10$. The maximum energy is 8.5 MeV per atomic mass unit, which corresponds to about 2 GeV for accelerated uranium ions. The Super-HILAC is the successor of the HILAC which was used to produce a number of transuranium elements during 1957–1971.

Several projects are in progress for the (hopeful) production of very heavy (transactinide) elements, where both linear and nonlinear accelerators (or a combination of them) are to be used. The projectile mass numbers and energies for these projects are summarized in Fig. 8.5: the USSR (Dubna), Federal Republic of Germany (UNILAC), Great Britain (Daresbury), France (ALICE and GANIL), in addition to the Super-HILAC and its further extension, the Bevalac (this is a hook-up of the Super-HILAC to the Bevatron synchrocyclotron; see next section).

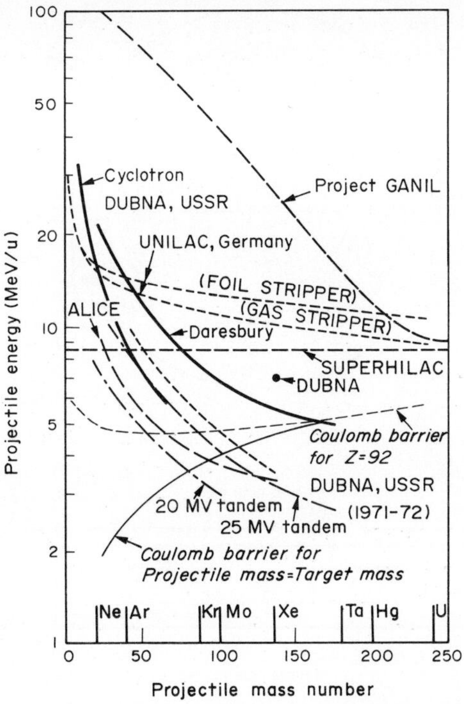

FIG. 8.5. Projects (in 1976) for accelerating heavy ions to high energies.

The world's largest linear accelerator is located at Stanford University in California. In this accelerator electrons are raised to energies of 20 GeV and used to study elementary particles, the distribution of charges within nuclei, etc. The accelerator tube is 3000 m long; the diameter is about 0.1 m. The maximum beam intensity is 8 μA. At 20 GeV the electrons have a velocity of 0.999 999 999 7**c** and the electron mass is 40,000 times greater than rest mass (approximately equivalent to that of a sodium atom).

8.6. Cyclotrons

The difficulties inherent in the length of high energy linear accelerators were ingeniously overcome by E. O. Lawrence and M. S. Livingston in their invention of the cyclotron in 1930. The basic operating principles of the cyclotron are shown in Fig. 8.6. The particles are accelerated in spiral paths inside the two semicircular flat evacuated metallic cylinders called dees (D's). The two dees are coupled to a high frequency alternating voltage. The volume within the dees corresponds to an equipotential condition just as does the volume within the drift tubes of linear accelerators. The dees are placed between the two poles of a magnet so that the magnetic field operates upon the ion beam to constrain it within flat circular paths inside the dees. At the gap between the dees the ions experience acceleration due to the imposition of the potential difference. The beam particles originate at the ion source at the center of the cyclotron, and as they spiral outwards in the dees they acquire an increase in energy for each passage across the gap of the dees. The target is located either inside the vacuum chamber (internal) or can be taken out through a thin "window" (external beam) after extraction from the circular path by a deflector.

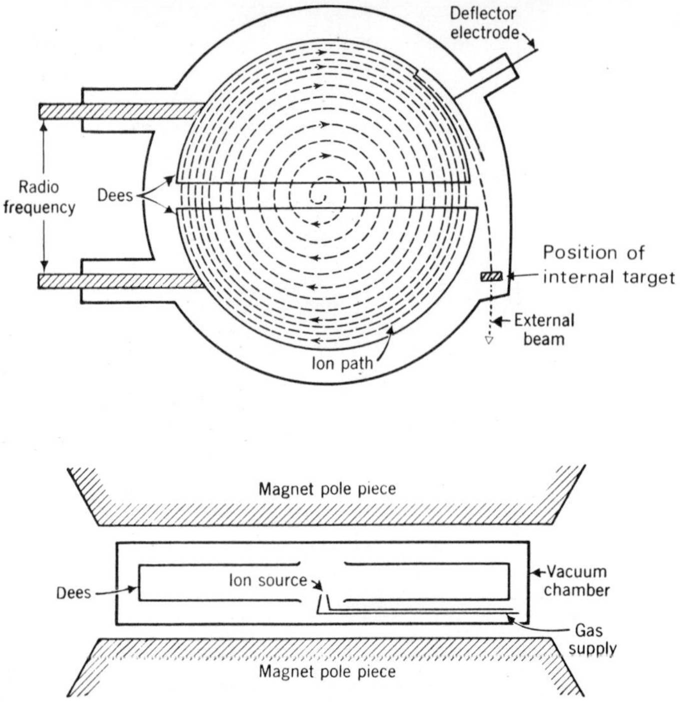

FIG. 8.6. Principle of a cyclotron. (According to R. S. Shankland.)

As the projectiles acquire energy, the radius of their path within the dees increases. The longer pathlength in the dees is related to the energy of the projectile by the equation (§2.3)

$$E_{\text{kin}} = \frac{q^2 r^2 B^2}{2m} \tag{8.5}$$

where r is the radius of the beam path at the energy E_{kin} and B is the value of the magnetic field. This equation is valid for particles of nonrelativistic velocities. The successful operation of the cyclotron is based on the principle that the increase in the radius of the path in the dee exactly compensates for the increase in particle velocity. As a result, the particles as they spiral through the dees always arrive at the gap when the potential difference is of the right polarity to cause acceleration.

The frequency (in Hz) requirement is

$$\nu = qB/2\pi m \tag{8.6a}$$

or

$$\nu = v/2\pi r \tag{8.6b}$$

where v is the particle velocity at radius r. From (8.6a) it follows at constant ν that B/m must be constant also.

The maximum energies available in conventional cyclotrons of the constant frequency type are about 10 MeV for protons, 20 MeV for deuterons, and 40 MeV for α-particles. The shape of the electrostatic field at the gap between the dees as well as the design of the magnetic field to produce a slight nonuniformity at the outer edges of the dees

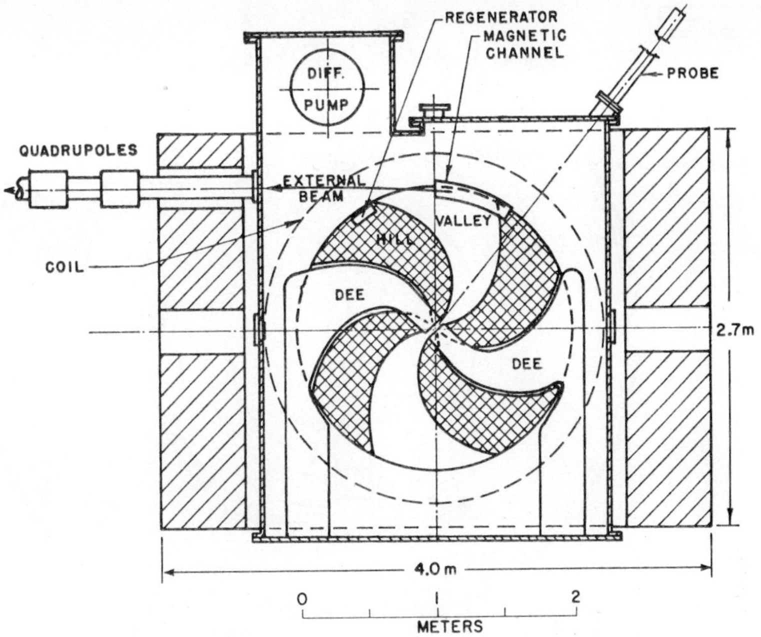

FIG. 8.7. A 150 MeV sector-focused proton cyclotron in Berkeley, USA, showing the unconventional magnet and (1/4-wave) D-designs. Maximum internal beam current 0.1 mA, ext. 0.1 μA. Average magnetic field 2.0 T. D-voltage 60 kV, RF frequency 30 MHz. The cyclotron is especially designed for therapy, radiography, and isotope production. (According to University of California, LBL-5075, 1975.)

produce a focusing effect on the beam particles. The ion currents are usually $\lesssim 0.5$ mA for the internal beam, and about a factor 10 less for an external beam.

As the energies increase, so does the relativistic mass. In order to have the beam particles arrive at the gap in phase, from (8.5) we can see that either the frequency or the magnetic field must be modified to compensate for the increasing mass. The former is done in synchrocyclotrons (next section) while the latter is used in sector focused cyclotrons. In the latter, the magnet has spiral ridges which provide sufficient modification of the field to allow the acceleration of particles to much higher energies than is possible with a conventional cyclotron (Fig. 8.7). Although there is a limit to the energies achievable with cyclotrons of constant frequency and magnetic field, they have the great advantage of high beam currents.

The magnetic bending of the beam corresponds to a centrifugal acceleration of the charged particles. The acceleration or deceleration of electric charges is always accompanied by the emission of electromagnetic radiation. Although only a small part of the energy input of a cyclotron is lost through this *synchrotron radiation*, it is still sufficient to require considerable radiation shielding (usually concrete walls). Synchrotron radiation is much less of a problem for linear accelerators.

8.7. Frequency modulated cyclotrons and synchrotrons

The frequency of a cyclotron can be modulated to take into account the variation in velocity and mass as relativistic effects increase in importance. At very high energies the radius becomes very large. This has led to two somewhat different accelerator

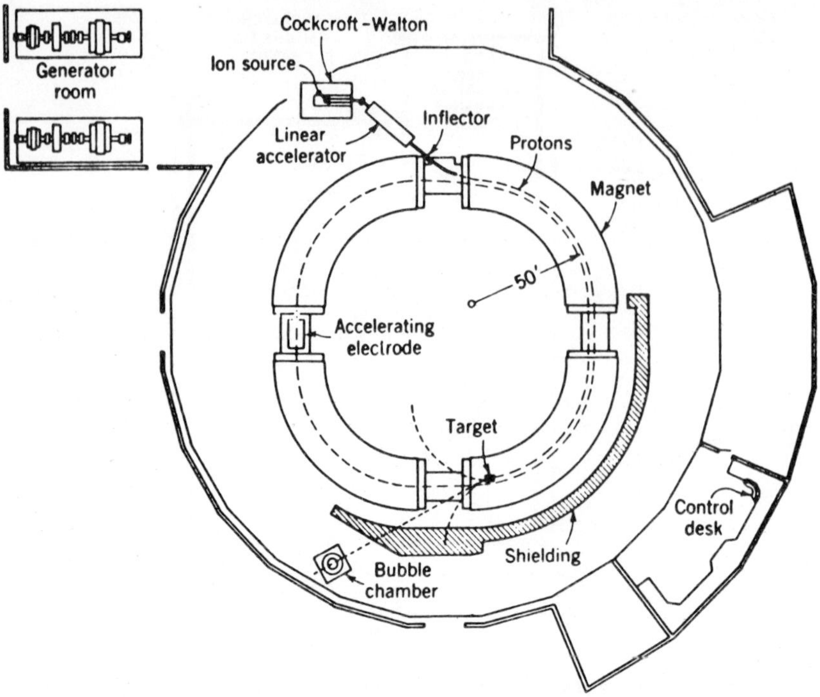

FIG. 8.8. Principle of a modulated multistage circular accelerator.

designs. The *frequency modulated* (FM) or *synchrocyclotron* maintains the original cyclotron principle with a spiral particle path, while in synchrotrons the particle path is fixed in a circular orbit; Fig. 8.8 is an example of the latter.

In a synchrotron there is a balance between frequency and the magnetic field, one or both being varied during the acceleration (cf. (8.6a)). From (8.5) and (8.6) one obtains

$$E_{\text{kin}} = \nu \pi r^2 B q \tag{8.7}$$

by eliminating the mass from the relation for the kinetic energy of the projectile. In these machines the particles are accelerated in bursts, since the frequency of the accelerating potential must be modulated throughout the traversal of one beam burst in the machine, and only the particles in that burst would be in resonance with the frequency change of the accelerating gap. As a result, beam currents are much smaller in FM cyclotrons and synchrotrons than in constant frequency cyclotrons.

Synchrotrons which accelerate positive particles use smaller accelerators, usually linear, for injecting the beam into the circular synchrotron. Figure 8.8 shows a typical synchrotron design, although it specifically refers to the Bevatron in Berkeley, which provides acceleration of protons to 6.5 GeV (the mass increase of a 6.5 GeV proton is 6.9 u), and heavier ions like carbon or neon to about 2 GeV u^{-1}. The Bevatron injectors are a 0.5 MeV Cockcroft–Walton machine followed by a 10 MeV multistage linear accelerator. The energy increase over the accelerating electrode is only 1500 V. As the particles accelerate, the magnetic field continues to increase in such a way that the radius of the particle orbit remains constant. The frequency of the oscillating voltage applied to the accelerating electrode must also increase, since the time required for a complete circuit becomes shorter and shorter until the velocity of the particles approaches the velocity of light, after which the circuit time remains constant. The whole accelerating cycle requires several seconds, during which time the particles make many millions of orbits. Each burst of particles contains about 10^{12} particles, and in the Berkeley machine there is one burst every 6 s. The successful operation of the synchrotron depends on the existence of phase stability.

The Bevatron is a rather old machine, put into operation in 1954. Larger and more recent (proton) accelerators are the 28 GeV machine of CERN and the 76 GeV machine of Serpukhov (USSR). The present largest machine is at Batavia in the USA (500 GeV); it contains 1000 magnets along its 6.3 km orbit, each magnet weighing 10 t.

At CERN, a machine producing 400 GeV protons is designed to produce 10^{13} protons s^{-1}; the acceleration diameter is 2.2 km. The injection time (with 10 GeV protons) is 23 s (to fill the acceleration ring), while the acceleration time is 9 s; for each cycle 2.5 MeV is added in energy. The field strength of the 200 bending magnets is 1.8 T and the power consumption is 50 MW. Its cost is estimated to be about US \$ 2×10^9 (1975).

Still higher energies have been achieved in collisions between particles circulating in two beam trains in opposite direction by using intersecting storage rings (ISR). In this system the energy loss due to momentum conservation in the reaction between high speed projectiles and stationary target atoms is eliminated. At CERN proton collision energies of 1700 GeV have been obtained (Fig. 8.9), and several machines of this type are under development.

Very compact synchrotrons have been built in which electrons are accelerated in a fixed orbit (radius $\lesssim 1$ m) in a strong magnetic field ($\gtrsim 1$ T) to energies in the GeV range. The intense electromagnetic (synchrotron) radiation has usually a maximum intensity at $\lesssim 1$ keV and is used for studies of photon induced reactions (e.g. for meson production), for radiation chemistry use, etc.

8.8. Neutron sources

Many radioactive isotopes are most conveniently produced by irradiation of a target with neutrons. There are three main ways of obtaining useful neutron fluxes:

(a) through accelerator-induced reactions (Table 8.1);
(b) through nuclear reactors (Table 8.2);
(c) through nuclear reactions induced by radioactive isotopes (Table 8.3).

8.8.1. *Accelerators*

Table 8.1 gives the most common accelerator-induced nuclear reactions leading to

TABLE 8.1. *Neutron yields for some accelerator produced reactions (see also Fig. 8.10)*

Reaction	Q (MeV)	Target	Yield[(a)]
$^7Li(p,n)^7Be$	− 1.64	Li salt	10^7 n s^{-1} μA^{-1} for 2.3 MeV H^+
$^2D(d,n)^3He$	3.27	D_2O ice	7×10^7 n s^{-1} μA^{-1} for 1 MeV D^+
$^3H(d,n)^4He$	17.6	^3H in Ti	2×10^7 n s^{-1} μA^{-1} for 0.1 MeV D^+
$^9Be(d,n)^{10}B$	3.79	Be metal	2×10^{10} n s^{-1} μA^{-1} for 7 MeV D^+
$^9Be(\gamma,n)^8Be$	− 1.67	Be metal	10^9 n s^{-1} μA^{-1} for 10 MeV e^-
$U(\gamma,fp)$	− 5.1	U metal	10^{11} n s^{-1} μA^{-1} for 40 MeV e^-

[(a)] 1 $\mu A = 6.24 \times 10^{12}$ single-charged ions per second.

FIG. 8.9. One of the colliding points of the CERN ISR machine. The enormous radius of curvature of the beam paths should be observed. The foreground shows one of the bending magnets.

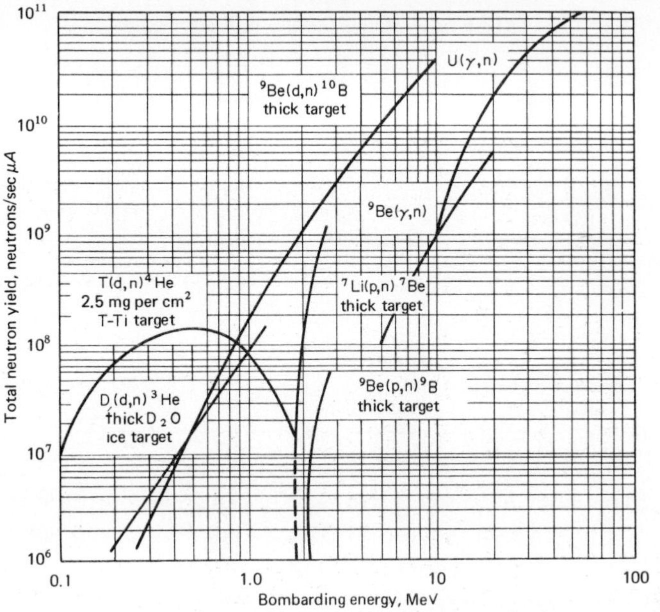

FIG. 8.10. Neutron yield as a function of projectile energy.

the production of neutrons. In general, the neutron yield increases with increasing beam energy up to some maximum value of the latter after which competition of other nuclear reactions reduces the probability of the initial neutron producing reaction (Fig. 8.10). The least expensive way to get a neutron flux of sufficient intensity for producing radioactive nuclides in easily measurable quantities (e.g. for activation analysis) is through the T + D reaction using a Cockcroft–Walton accelerator.

The reaction

$$^{7}Li + {}^{1}H \rightarrow {}^{7}Be + {}^{1}n$$

is used for the production of monoenergetic neutrons (cf. §9.3). Higher neutron yields can be obtained by the reaction of ^{9}Be with deuterons. Neutrons can also be obtained from ^{9}Be by irradiation in electron accelerators. The passage of the electrons in the target results in the production of high energy Bremsstrahlung γ-rays (§14.4.2) which react with the beryllium to produce neutrons. An even higher neutron flux is obtained for the same reason when uranium is bombarded by electrons.

8.8.2. *Nuclear reactors*

The energy of the neutrons produced in nuclear fission averages 1–2 MeV. In thermal reactors, the neutrons are moderated to energies of $\lesssim$0.1 eV. Such low energy neutrons mainly cause reactions in which neutrons are captured and γ-rays emitted whereas more energetic neutrons cause reactions in which neutrons, protons, or α-particles may be emitted after the initial neutron capture.

The neutron flux of some nuclear reactors is indicated in Table 8.2; it contains reactors of various types, mainly in order of increasing costs. The first reactor ("Slowpoke") is designed for low budget university use in areas of reactor education, limited research, small scale isotope production, and activation analyses. The TRIGA reactors are useful for the same purposes, but are more powerful and have a higher flux; thus they permit

TABLE 8.2. *Nuclear reactors as neutron sources*

Name	Location	Use	First critically	Power (MW th)	Thermal flux ($n\ cm^{-2}\ s^{-1}$)
Slowpoke 2	Ottawa, Canada	Research and education	1971	0.020 0.002	1.0×10^{12} (for $\lesssim 2.5$ h) 1.0×10^{11} (for > 24 h)
TRIGA Mark II	Mainz, Fed. Rep. Germany	Research	1965	0.10 250[a]	5×10^{11} $10^{14}\ n\ cm^{-2}$ per pulse
R2	Studsvik, Sweden	Materials testing	1963	50	2.4×10^{14}
High Flux Reactor	Grenoble, France	Research	1971	60	$> 5 \times 10^{14}$
HFIR	Oak Ridge, USA	Isotope production	1965	100	4×10^{15}
Biblis	Biblis, Fed. Rep. Germany	Power production	1975	3700	$\sim 10^{14}$

[a]In one pulse; pulse width at half-maximum value, 30 ms.

more advanced research, but also require more resources in control, shielding, operation, etc. The isotope production reactor, HFIR in Oak Ridge, and similar reactors, are designed primarily for production of isotopes of the heaviest elements, which are otherwise hard to obtain. Other research and testing reactors are described in §19.13.

8.8.3. *Radioactive sources*

Neutrons were discovered through the reaction between α-particles, emitted by radioactive substances, and light elements: ^{4}He (from Po) + $^{9}Be \rightarrow {}^{12}C + n$ (J. Chadwick 1932). All early neutron research was conducted with sources of this kind, the most popular being the RaBe mixture. A number of radioactive n-sources are listed in Table 8.3.

The nuclide ${}^{252}_{98}Cf$ emits neutrons through spontaneous fission decay. All the other neutron sources listed involve a radioactive nuclide whose decay causes a nuclear reaction in a secondary substance which produces neutrons. For example, ${}^{124}_{51}Sb$ produces neutrons in beryllium powder or metal as a result of the initial emission of γ-rays. Radium, polonium, plutonium, and americium produce neutrons by nuclear reactions induced in beryllium by the α-particles from their radioactive decay. For the neutrons produced either by spontaneous fission in californium or by the α-particle reaction

TABLE 8.3. *Radioactive neutron sources*

Material	Half-life	Neutron yield ($n\ s^{-1}\ Ci^{-1}$)	γ-dose rate[a] ($mRhm\ Ci^{-1}$)
$^{226}Ra + Be$	1 620 y	13×10^{6}	850
$^{239}Pu + Be$	24 400 y	9×10^{6}	4
$^{239}Pu + {}^{18}O$	24 400 y	$\lesssim 2.9 \times 10^{5}$	[b]
$^{241}Am + Be$	458 y	2.5×10^{6}	2.5
$^{210}Po + Be$	138 d	2.5×10^{6}	$\lessdot 0.3$
$^{124}Sb + Be$	60 d	$\sim 0.2 \times 10^{6}$	1000
^{252}Cf	2.6 y	$5 \times 10^{12}\ n\ s^{-1}\ g^{-1}$	[c]

[a]mRhm common abbreviation for milliroentgen per hour at 1 m from the source.
[b]This is a variable source of $Pu^{16}O_2 + H_2{}^{18}O$; by transferring the water through heating into the porous oxide, the reaction $^{18}O(\alpha, n)^{21}Ne$ is initiated; by cooling, water is returned to storage and n output decreases to < 0.01 maximum output.
[c]Gamma dose rate increases with time due to formation of γ-emitting daughters.

with beryllium, the energy is between 0 and 10 MeV, while for the neutrons emitted in the α, n-reaction involving ^{124}Sb, the energy is 0.02 MeV. Neutron sources are commercially available; the list price of ^{252}Cf was US$10 per μg in 1975.

Recently neutron multipliers have come into use. These usually consist of a tank containing highly enriched ^{235}U. The uranium releases 100 n for each trigger neutron sent in; the common source is ^{252}Cf. These units are used for neutron radiography (§14.11.2), activation analysis (§18.2.3), etc.

Although the neutron flux from radioactive sources is comparatively small, they are still quite useful for specific purposes, due to their extreme simplicity, reliability (they cannot be shut off), and small size. The γ-radiation which is also present is, however, a disadvantage. From this point of view, ^{252}Cf is the most preferable neutron source, but, on the other hand, it has a relatively short half-life compared with some of the other sources. ^{252}Cf is further discussed in §13.3.4.

8.9. Intense radioactive radiation sources

Before the development of nuclear reactors, the strongest radioactive radiation sources available were the radioactive elements isolated from uranium ore. The amount of radium so far isolated is of the order of kilograms (i.e. $\sim 10^3$ Ci).

Millions of curies of radioactive nuclides are produced in each nuclear power station. Because of the expected demand for strong radioactive sources for agricultural, medical, military, etc., purposes, several MCi of ^{90}Sr, ^{137}Cs, ^{144}Ce, and ^{147}Pm have been isolated, and plans for production of these nuclides in the 100 MCi scale had been announced. However, the demand has not been as high as expected, primarily because of the high costs for these sources. It is likely that the price will decrease as a consequence of improved reactor reprocessing methods (see Chapter 20).

By means of (n, γ)-reactions in special target materials placed in reactors, other nuclides can be produced. For example, several MCi of ^{60}Co have been produced in the United States, the United Kingdom, and France for γ-radiographic, medical, or sterilization purposes. Small amounts of highly intensive radioactive sources of many elements have been produced for medical purposes (primarily cancer treatment).

8.10. Areas of application for accelerators

Let us consider a few uses of accelerators. In §8.8 we have discussed the use of accelerators for neutron production. Electron as well as positive particle beams from these accelerators can be used directly for the study of radiation effects in materials (see Chapter 15) or to produce X-rays. Accelerators are used in medicine for producing intense γ- or X-ray beams for cancer treatment, or charged particle beams for similar purposes (the "proton-knife"). Accelerators in the 10 MeV range are very useful in producing radionuclides, which cannot be made through (n, γ) reactions in nuclear reactors. The precise definition of energy available in van de Graaff accelerators makes them particularly useful for the study of the energy levels of nucleons in nuclides. van de Graaff cyclotrons, linear accelerators, and synchrotrons have been extensively used for the study of nuclear reactions.

Very high energy accelerators are used for the production and study of elementary particles. Once the beam energy exceeds a couple of hundred MeV, mesons and other elementary particles are produced in nuclear reactions and can be studied directly or they can be used to cause various types of nuclear reactions.

8.11. Exercises

8.1. In a small linear accelerator containing 30 stages, He^{2+} ions are accelerated by a 150 kV, 100 MHz RF source. The ions are used to bombard a metal target to induce a specific reaction. (a) What is the proper length of the last drift tube? (b) What is the maximum projectile energy achieved? (c) What is the heaviest target in which a nuclear transformation can be induced (no tunneling)?

8.2. In order to propel a space vehicle to high velocities after its exit from the earth's gravitational pull, "ion rockets" might be used. These can be considered as simple accelerators for charged particles. The electroneutrality of the vehicle is conserved through emission of electrons from a hot filament. The gain in linear velocity Δv is given by the "rocket formula":

$$\Delta v = v_e \ln \frac{m_{RO}}{m_R}$$

where v_e is the exhaust velocity and m_{RO} and m_R are the initial and final mass of the vehicle respectively. (a) Calculate the propelling power for a rocket, which emits 10 keV protons at a current of 1 A. (b) What is the final velocity gain of such vehicle with 2000 kg initial mass after 1 year's operating time? (c) For the same net available power and operating time, as in (b), calculate the final velocity gain of a 2000 kg vehicle which emits Cs^+ ions at a current of 1 A.

8.3. The TD reaction is used to produce 14 MeV neutrons, which are considered to be emitted isotropically from the target. What is the fast neutron "flux" at 5 cm from target when the ion current is 0.2 mA and the acceleration voltage is 300 kV? Use Fig. 8.10.

8.4. Protons are accelerated to 12 GeV in a synchrotron in which the bending magnets have a maximum field strength of 14.3 T. What is the radius of curvature of the proton orbit?

8.5. In a linear accelerator for protons the first drift tube is 1 cm long and the accelerating potential is 25 kV. (a) Since the speed of light is not approached, how long should the second and third drift tubes be? (b) What should the full-wave frequency of acceleration be?

8.6. Calculate the maximum energy (a) for protons, deuterons, and helium ions in a cyclotron, whose maximum orbit diameter is 1.25 m and whose frequency is 12 MHz. (b) What magnetic fields would be required in each case?

8.7. The belt of a van de Graaff generator is 50 cm wide and it moves with a linear velocity of 10 m s^{-1}. Calculate how much current it can carry to the high voltage terminal when the surface charge is limited by gas breakdown. The gas (at 15 atm pressure) can withstand a field strength of 2×10^7 Vm^{-1}; the dielectric constant of the gas is 8.8×10^{-12} F m^{-1}. The distance between the belt and nearest wall is 1 m; disregard end effects.

8.8. In a VdG generator a 100 μA beam of He^{2+} ions is accelerated to an energy of 5 MeV before striking a target. How many grams of radium are required to provide the same number of α-particles?

8.12. Literature

See Appendix A, § A.1.

S. Flügge (ed.) *Handbuch der Physik*, Vol. XLIV: *Instrumentelle Hilfsmittel der Kernphysik*, Springer-Verlag, 1952.
M. S. Livingston and J. P. Blewett, *Particle Accelerators*, McGraw-Hill, 1962.
P. H. Rose and A. B. Wittkower, Tandem Van de Graaff accelerators, *Sci. Am.* **223** (Aug. 1970) 24.
CERN/1050, *The 300 GeV Programme*, CERN, Geneva 1972.
R. R. Wilson, The Batavia accelerator, *Sci. Am.* **231** (Febr. 1974) 72.
Californium-252 Progress, US ERDA, Savannah River Operations Office, PO Box A, Aiken, South Carolina 29801 (current publication).

CHAPTER 9

Nuclear Reactions: II. Mechanisms and Models

Contents

Introduction

The variety and complexity of nuclear reactions make this a fascinating area of research quite apart from the practical value of understanding fusion and fission. From studies of such properties as the relative amounts of formation of various competing products, the variation of the yields of these with bombarding energy, the directional characteristics and kinetic energies of the products, etc., we may formulate models of nuclear reaction mechanisms. Such models lead to systematics for nuclear reactions and make possible predictions of reactions not yet investigated.

9.1. The reaction cross-section

The probability for a nuclear reaction is expressed in terms of the *reaction cross-section*. The geometric cross-section that a nucleus presents to a beam of particles is πr^2. If we use 6×10^{-15} m as an average value for the nuclear radius, the value of πr^2 becomes $3.14\ (6 \times 10^{-15})^2 \approx 10^{-28}\ \mathrm{m}^2$. This average geometric cross-section of nuclei is reflected in the unit of reaction probability which is the *barn*, where $1\ \mathrm{b} = 10^{-28}\ \mathrm{m}^2$ (or $10^{-24}\ \mathrm{cm}^2$).

Consider the bombardment of a target containing N_v atoms per m^3 by a homogeneous flux ϕ_0 of particles (Fig. 9.1). The flux is expressed in units of particles $\mathrm{m}^{-2}\mathrm{s}^{-1}$. The target atoms N_v refer only to the atoms of the species involved in the nuclear reaction. If a Li–Al alloy is bombarded to induce reactions with the lithium, N_v is the number of lithium atoms per m^3 in the alloy, not the total of lithium and aluminum atoms. The

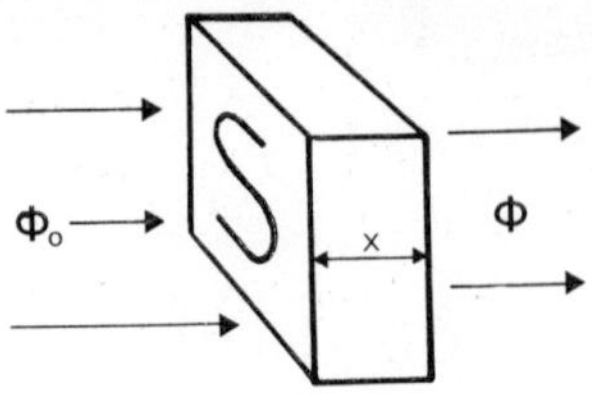

FIG. 9.1. Reduction of particle flux by absorption in a target.

change in the flux, $d\phi = \phi - \phi_0$, may be infinitesimal as the particles pass through a thin section of target thickness dx. This change depends on the number of target atoms per unit area (i.e. $N_v\,dx$), the flux ($\phi_0 \approx \phi$), and the reaction probability σ.

$$-d\phi = \phi\sigma N_v\,dx \tag{9.1}$$

The negative sign indicates that the flux decreases upon passing through the target due to reaction of the particles with the target atoms: thus $-d\phi$ is the number of reactions. Integration gives:

$$\phi = \phi_0 e^{-\sigma N_v x} \tag{9.2}$$

where ϕ_0 is the projectile flux striking the target surface. For targets which have a surface area of S (m^2) exposed to the beam, for the irradiation time t, the total number of nuclear reactions ΔN is:

$$\Delta N = (\phi_0 - \phi)St \equiv \phi_0 St(1 - e^{-\sigma N_v x}) \tag{9.3}$$

For a thin target in which the flux is not decreased appreciably upon passage through the target, this expression can be reduced to:

$$\Delta N = \phi_0 St\sigma N_v x = \phi_0 \sigma t N_v V = \phi_0 \sigma t N_t \text{ (thin target)} \tag{9.4}$$

where $V = Sx$ is the target volume, and $N_t = N_v V$ is the number of target atoms. Notice that, as a result of the product Sx, which equals the volume of the target, the relationship on the right of (9.4) is independent of the geometry of the target and involves only the total number of atoms in it.

Equation (9.4) can be used only when particle fluxes are homogeneous over the whole irradiated sample. In nuclear reactors, where the area of the sample is much smaller than the area of the flux, it is convenient to express the flux in terms of neutrons $m^{-2}\,s^{-1}$ and the target in terms of total number of atoms, as above.

By contrast, in an accelerator the target surface is often larger than the cross-section of the ion beam, and (9.4) cannot be used without modification. For accelerators the beam intensity (i.e. particle current) I_0 (particles s^{-1}) is given by (8.2); 1 A corresponds to $6.24 \times 10^{18}\,z^{-1}$ charged particles s^{-1}, where z is the charge in electron units on the particle, e.g. 2 for He^{2+}. In (9.4) $\phi_0 S$ must be substituted by $6.24 \times 10^{18}\,iz^{-1}$, where i (A) is the electric current.

In (9.4) $N_v x$ has the dimensions of atoms m^{-2}. This is a useful quantity in many calculations and for a pure elemental target is equal to

$$N_v x = \frac{N_A \rho x y_i}{10^{-3} M} \tag{9.5}$$

where ρ is the density of the target (kg m^{-3}), M the atomic weight (in *gram* mole) of the element, and y_i the isotopic fraction of reactive atoms i in the target.[†]

As an example of the use of these equations, consider the irradiation of a gold foil by thermal neutrons. Assume the foil is 0.3 mm thick with an area of 5 cm^2 and the flux is 10^{13} n cm^{-2} s^{-1}. The density of gold is 19.3 g cm^{-3} while the cross-section for the capture of thermal neutrons by ^{197}Au is 99 b. Transferring these units into SI and introducing them into (9.3) for an irradiation time of 10 min yields a value of 4.6×10^{15} for the number of ^{198}Au nuclei formed. If the thin target equation is used (9.4), the value of 5.0×10^{15} nuclei of ^{198}Au is obtained. If the same gold foil is bombarded in a cyclotron with a beam of protons of 1 μA when the cross-section is 1 b and $t = 10$ min, the number of reactions is 6.6×10^{12}.

Frequently, the irradiated target consists of more than one nuclide which can capture bombarding particles to undergo reaction. The *macroscopic* cross-section, which refers to the total decrease in the bombarding particle flux, reflects the absorption of particles by the different nuclides in proportion to their abundance in the target as well as to their individual reaction cross-sections. Assuming that the target as a whole contains N_v atoms m^{-3} with individual abundances designated by y_1, y_2, etc., for nuclides 1, 2, etc., the individual cross-sections are σ_1, σ_2, etc. The macroscopic cross-section $\hat{\Sigma}$ (m^{-1}) is

$$\hat{\Sigma} = N_v \sum_{1}^{n} y_i \sigma_i \qquad (9.6)$$

For a target which is x m thick one obtains

$$\phi = \phi_0 e^{-\hat{\Sigma} x} \qquad (9.7)$$

The value $\hat{\Sigma}^{-1}$ is the average distance a projectile travels between successive collisions with the target atoms (the *mean free path*).

9.2. Partial reaction cross-sections

The irradiation of a target may lead to the formation of a number of different products. For example, the irradiation of ^{63}Cu with protons can produce the nuclides ^{62}Zn, ^{63}Zn, ^{62}Cu, all of which are radioactive:

$$^{63}_{29}Cu + ^{1}_{1}H \xrightarrow{(p,n)} ^{63}_{30}Zn + n$$

$$^{63}_{29}Cu + ^{1}_{1}H \xrightarrow{(p,2n)} ^{62}_{30}Zn + 2\,n$$

$$^{63}_{29}Cu + ^{1}_{1}H \xrightarrow{(p,pn)} ^{62}_{29}Cu + ^{1}H + n$$

The formation probability of each product corresponds to a *partial reaction cross-section.* The total reaction cross-section σ_{tot} is the sum of all the partial cross-sections and measures the probability that the projectile causes a nuclear reaction independent of the products formed. Thus, the decrease in intensity of the particle flux is proportional to σ_{tot}. The amount of an individual product formed is proportional to σ_i, where σ_i corresponds to the partial reaction cross-section for the formation of the ith product.

Some partial cross-sections have their own names such as the *scattering cross-section* for elastic and inelastic scattering (σ_{scat}), the *activation cross-section* (σ_{act}) for the formation

[†] The dimensions used here for M (gram $mole^{-1}$, or 10^{-3} kg $mole^{-1}$) is not in accordance with the SI system, which requires M in kg $mole^{-1}$; this unit has not yet been accepted by IUPAC (International Union of Pure and Applied Chemistry). We therefore continue with the gram mole.

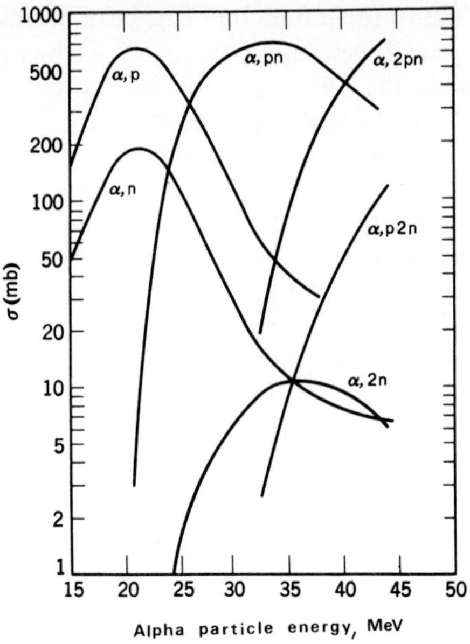

FIG. 9.2. Excitation functions for reactions between energetic ^4He ions and ^{54}Fe target nuclei. The abscissa is the kinetic energy of the projectile in the laboratory system. (According to F. S. Houck and J. M. Miller.)

of radioactive products, the *fission cross-section* (σ_f) for fission processes, and *absorption* or *capture cross-sections* (σ_{abs} or σ_{capt}) for the absorption or capture of particles. If all of these processes take place, one obtains (with caution to avoid overlapping reactions)

$$\sigma_{tot} = \sigma_{scat} + \sigma_{act} + \sigma_f, \text{ etc.} \tag{9.8}$$

In the irradiation of ^{235}U with thermal neutrons, σ_{scat} is about 10 b, σ_{act} (for forming ^{236}U) is approximately 107 b, and σ_f is 582 b.

The reaction cross-section depends on the projectile energy as shown in Fig. 9.2. The curves obtained for the partial reaction cross-section as a function of projectile energy are known as *excitation functions* or *excitation curves.*

9.3. Resonances and tunneling*

Experimentally, it is found that nuclear reactions sometimes occur at energies less than that required by the Coulomb barrier. This behavior is related to the wave mechanical nature of the particles involved in a nuclear reaction.

As a projectile approaches a target nucleus in a nuclear reaction, the probability that there will be overlap and hence interaction in their wave functions increases. This concept was used in 6.7.3 to explain the emission of α-particles with energies less than that required by the Coulomb barrier height. Such tunneling may also occur for projectiles approaching the nucleus from the outside. An example is provided by the reaction of protons with lithium (Fig. 9.3).

$$^7_3Li + ^1_1H \rightarrow {}^8_4Be^* \rightarrow \begin{cases} 2\,^4_2He & Q = 17.4\text{ MeV} \quad \text{(a)} \\ ^8_4Be + \gamma & Q = 17.2\text{ MeV} \quad \text{(b)} \\ ^7_4Be + n & Q = -1.64\text{ MeV} \quad \text{(c)} \end{cases}$$

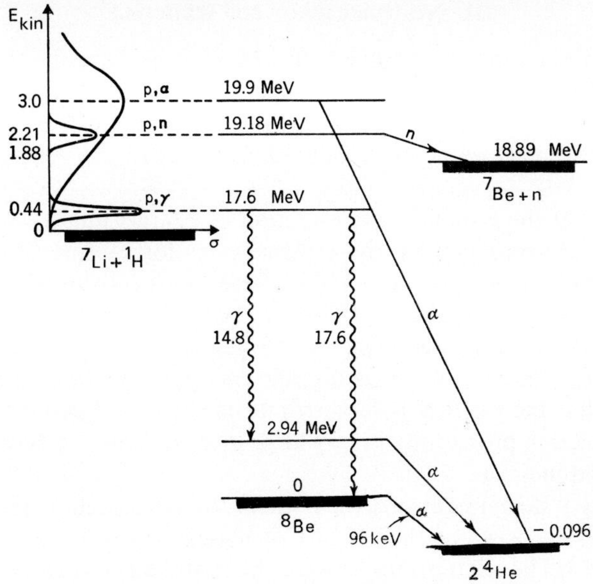

FIG. 9.3. Yield curves for the reaction between protons and ^{7}Li, leading to different excited levels in ^{8}Be, followed by decay to stable end products.

For this reaction the value of E_c is 1.3 MeV. However, due to tunneling the reactions begin to occur at lower proton energies. At an energy of 0.15 MeV about 0.1% of the protons penetrate the Coulomb barrier, at 0.3 MeV about 1%, and at 0.6 MeV about 20%.

The reaction cross-section is closely related to the excited energy states of the compound nucleus. Four such levels are shown for ^{8}Be* in Fig. 9.3. To the left of the figure the (p, γ), (p, n) and (p, α) partial cross-sections (excitation functions) are shown as a function of the proton kinetic energy. The maximum cross-section for reaction (b) occurs at a proton kinetic energy of 0.44 MeV, which, together with the Q-value, 17.2 MeV, of the reaction $^7\text{Li} + {}^1\text{H} \rightarrow {}^8\text{Be}$, leads to an excitation energy of 17.6 MeV, which exactly matches an excited level of the same energy in ^{8}Be*. At an excitation energy of 19.18 MeV another energy level is reached in the compound nucleus leading to its decay into $^7\text{Be} + \text{n}$, reaction (c). The excitation energy is achieved from release of nuclear binding energy (17.25 MeV) and from the proton kinetic energy. The amount needed is $19.18 - 17.25 = 1.93$ MeV. In order to conserve momentum, the proton must have $\frac{8}{7} \times 1.93 = 2.21$ MeV in kinetic energy. The increase in cross-section when the total excitation energy matches an excited energy level of the compound nucleus is known as a *resonance*.

This particular reaction is of interest for several reasons. It was the first nuclear reaction that was produced in a laboratory by means of artificially accelerated particles (Cockcroft and Walton 1932; cf. §8.3). Reaction (b) is used still for the production of high energy γ-radiation (17 MeV), while reaction (c) is used as a source of mono-energetic neutrons. The energy of the neutrons in reaction (c) is a function of the proton energy and the angle between the neutron and the incident proton beam. A necessary requirement, however, is that the threshold energy ($1.64 \times 8/7 = 1.88$ MeV) must be exceeded, the Q-value for reaction (c) being -1.64 MeV.

9.4. Neutron capture and scattering

Unlike charged particles, no Coulomb barrier hinders neutrons from reaching the target nucleus. This leads to generally higher reaction cross-sections for neutrons, particularly at very low energies. Moreover, since neutrons can be produced in very high fluxes in nuclear reactors, neutron-induced processes are among the more important nuclear reactions.

We have seen that the geometric cross-section of a target nucleus is on the order of 1 b, or 10^{-28} m^2. Experimentally, the cross-sections for capture of energetic ("fast") neutrons ($\gtrsim 1$ MeV) are often close to 1 b. However, for neutrons whose kinetic energy is in the 1–100 eV region, some nuclei show very large cross-sections—as high as 10^5 b. Such values can be explained as being due to neutron capture where the compound nucleus is excited exactly to one of its discrete energy levels (resonance capture). This does not mean that the nucleus is larger than its calculated geometric cross-section but that the interaction probability is very large in such cases—greater than the calculation of πr^2 would indicate.

For low energy ("slow") neutrons ($\ll 1$ MeV) the cross-section is also larger than the πr^2 value, and decreases as the velocity increases; this relation, $\sigma \propto v^{-1}$, is shown for boron in Fig. 9.4. The relationship between the cross-section and the neutron velocity can be understood in wave mechanical terms since the wavelength associated with the neutron increases with a decrease in velocity. According to the matter-wave hypothesis

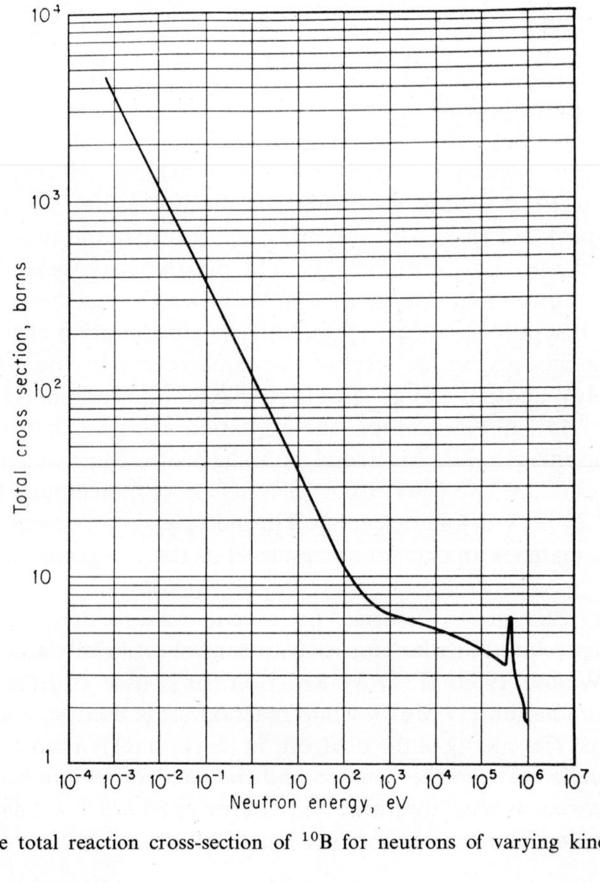

FIG. 9.4. The total reaction cross-section of ^{10}B for neutrons of varying kinetic energies.

(§5.5) the wavelength associated with a moving particle is $\lambda = \mathbf{h}/mv$ which can be written:

$$\lambda = \mathbf{h}(2mE_{\text{kin}})^{-1/2} = 0.286 \times 10^{-8}\,(mE_{\text{kin}})^{-1/2} \tag{9.9}$$

While the wavelength of a slow neutron is about 1 Å (Ångström), that of a fast neutron is less than 1/1000 of that. Since the reaction probability increases with increasing particle wavelength, a slow-moving particle has a higher probability of reaction than a faster one of the same kind.

In effect, the wave properties make neutrons appear much larger than their geometric size and increase the probability of interaction with the nucleus. From (9.9) it follows that

$$\sigma_{\text{capt}} \propto \lambda_{\text{n}} \propto E_{\text{n}}^{-1/2} \propto v_{\text{n}}^{-1} \tag{9.10}$$

where λ_{n}, E_{n}, and v_{n} are the wavelength, kinetic energy, and velocity, respectively, of the neutron. This relation is known as the *1/v law*. This law is valid only where no resonance absorption occurs. In Fig. 9.4 $\sigma_{\text{tot}} = \sigma_{\text{capt}} + \sigma_{\text{scat}}$, but σ_{scat} is approximately constant. Since σ_{capt} decreases as E_{n} increases, at higher energies ($\gtrsim 500$ eV) σ_{scat} dominates over σ_{capt}, except for the resonance at 0.3 MeV. The capture of neutrons in ^{10}B leads to the formation of ^{7}Li and ^{4}He.

9.5. Neutron diffraction*

From (9.9) we calculated that for neutrons the wavelength at thermal energies is on the order of 1 Å, i.e. of the same order of magnitude as the distance between atomic planes in a crystal. Thermal neutrons can, therefore, be scattered by crystals in the same manner as X-rays. For studies of crystal structures by neutron diffraction, a source of monoenergetic neutrons can be obtained from the spectrum of neutron energies in a reactor by the use of monochromators.

The probability of scattering of neutrons without energy change (*coherent scattering*) is approximately proportional to the area of the nucleus. As a consequence, coherent scattering of neutrons is less dependent on the atomic number than the scattering of X-rays which is proportional to the electron density (i.e. $\propto Z^2$). As a result of this difference, for a compound consisting of both heavy and light atoms, the position of the lighter atoms can be more easily determined using neutron diffraction, while X-ray diffraction is better for locating the heavier atoms. Neutron diffraction is, therefore, valuable for complementing the information obtained on the position of heavy atoms by X-ray diffraction. Neutron diffraction is particularly valuable in the location of hydrogen atoms in organic and biological materials.

9.6. Models for nuclear reactions*

No single model is successful in explaining all the aspects of the various types of nuclear reactions.

Let us consider three models which have been proposed for explaining the results of nuclear reaction studies.

9.6.1. *The optical model*

In the process of elastic scattering the direction of the particles is changed but none of the kinetic energy is converted to nuclear excitation energy. This would indicate that the reaction is independent of the internal structure of the nucleus and behaves much like the scattering of light from a crystal ball. Consequently, a model has been developed based on mathematical techniques used in optics. Light shining on a transparent crystal ball is transmitted with some scattering and reflection but no absorption. Light shining on a black crystal ball is all absorbed and there is no transmission or scattering. In nuclear reactions the incoming particles are scattered in elastic scattering and are absorbed in induced transmutations. Therefore, if the nucleus is to act as a crystal ball it can be neither totally transparent nor totally black. The optical model of the nucleus is also known as the *cloudy ball model*, indicating that nuclei both scatter and absorb the incoming particles.

The nucleus is described as a potential well containing neutrons and protons. The equation for the nuclear potential includes terms for absorption and scattering. This potential can be used to calculate the probability for scattering of incident particles and the angular distribution of the scattering. The model is in excellent agreement with experiments for scattering. Unfortunately, this model does not allow us to obtain much information about the consequences of the absorption of the particles which lead to inelastic scattering and transmutation.

9.6.2. *Liquid-drop model*

As the excitation energy of an excited nucleus increases, the energy levels get closer together. Eventually, a continuum is reached where the density of nuclear levels is so great that it is no longer possible to identify individual levels. (This is similar to the case for electronic energy levels of atoms.) When the excited nucleus emits a proton or neutron while in the continuum energy, the resultant nucleus may be still sufficiently energetic that it remains in the continuum region.

N. Bohr has offered a mechanism to explain nuclear reactions in nuclei which are excited into the continuum region. When a bombarding particle is absorbed by a nucleus, the kinetic energy of the bombarding particle plus the binding energy released by its capture provide the excitation energy of the compound nucleus. In this model, the *compound nucleus* becomes uniformly excited in a manner somewhat analogous to the warming of a small glass of water upon addition of a spoonful of boiling water. As the nucleons move about and collide in the nucleus, their individual kinetic energies vary with each collision just as those of molecules in a liquid change in molecular collisions. As this process continues, there is an increase in the probability that at least one nucleon will gain kinetic energy in excess of its binding energy. That nucleon is then evaporated (i.e. leaves the nucleus) analogously to the evaporation of molecules from liquid surfaces.

The evaporation of the nucleon decreases the excitation energy of the residual nucleus by an amount corresponding to the binding energy plus the kinetic energy of the released nucleon. The evaporation process continues until the residual excitation energy is less than the binding energy of a nucleon. The excitation energy remaining at this point is removed from the nucleus by emission of γ-rays.

Assume that the compound nucleus $^{188}_{76}Os$ is formed with a total excitation energy of 20 MeV. If the average binding energy of a neutron is 6 MeV and if each neutron leaves with 3 MeV of kinetic energy, evaporation of a neutron de-excites the nucleus by 9 MeV. Therefore, evaporation of two neutrons would leave the residual ^{186}Os nucleus with an excitation energy of only 2 MeV. Since this is below the binding energy of a neutron, further evaporation is not possible and γ-ray emission removes the final 2 MeV. If the ^{188}Os compound nucleus was formed by α-bombardment of ^{184}W, the reaction is represented as $^{184}_{74}W \xrightarrow{+\alpha} [^{188}_{76}Os]^* \xrightarrow{-2n} {}^{186}_{74}Os$.

9.6.3. *Lifetime of the compound nucleus*

An important assumption of the Bohr compound nucleus theory is that the time it takes for the accumulation on one nucleon of enough energy to allow evaporation is long by nuclear standards. This time is of the order of 10^{-14} s as compared to a time of 10^{-20} s required for a nucleon to cross the nuclear diameter once. Since the time is so long and there are so many internucleon collisions, the nucleus retains no pattern ("no memory") of its mode of formation, and the mode of decay should be, therefore, independent of the mode of formation. For example, ^{150}Dy ($t_{1/2} = 7.2$ min) can be formed in the following two ways:

$$\left.\begin{array}{l} ^{12}_{6}C + {}^{144}_{60}Nd \\ ^{20}_{10}Ne + {}^{136}_{56}Ba \end{array}\right\} \rightarrow [^{156}_{66}Dy^*] \rightarrow {}^{150}_{66}Dy + 6n$$

both form the excited $^{156}Dy^*$. To a first approximation, the probability for the ^{150}Dy formation is dependent only on the excitation energy of the compound nucleus $^{156}Dy^*$ but not on the manner in which this compound nucleus is formed (Fig. 9.5). This assumption of no memory of the mode of formation is not valid if very different amounts of angular momenta are involved in different modes of formation. For example, for proton induced reactions, the excitation energy is essentially all available for internal (nucleon) excitation. By contrast, as heavier bombarding particles are used, the average angular momentum

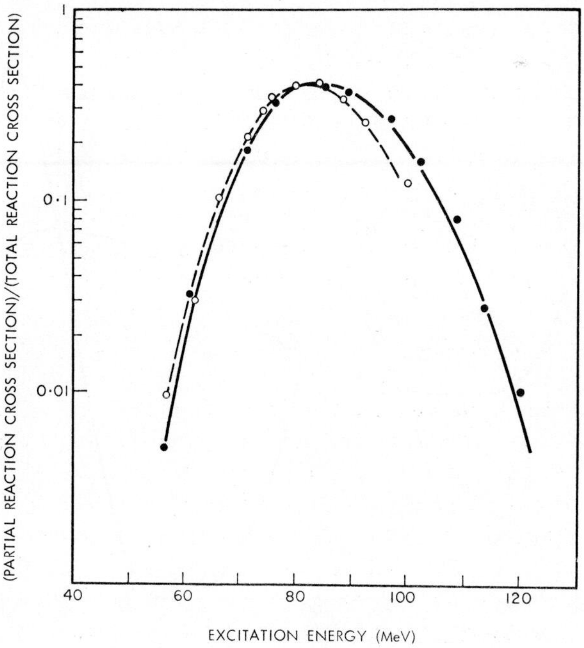

FIG. 9.5. Excitation functions for the formation of ^{150}Dy, through bombardment of either ^{144}Nd with ^{12}C (curve ○) or ^{136}Ba with ^{20}Ne (curve ●). (According to M. Lefort.)

of the compound system increases and the excitation energy is divided between the rotation and internal (nucleon) excitation. The modes of subsequent decay of the compound nucleus is affected by the amount of excitation energy that was involved in the angular momentum of the compound nucleus.

In general, neutron emission is favored over proton emission for two reasons. First, since there are usually more neutrons than protons in the nucleus, a neutron is likely to accumulate the necessary evaporation energy before a proton does. Second, a neutron can depart from the nucleus with a lower kinetic energy—the average neutron kinetic energy is 2–3 MeV. On the other hand, evaporating protons must penetrate the Coulomb barrier, so they often need about 5 MeV above their binding energy. It takes a longer time for this amount (*ca.* 12 MeV vs. 9 MeV for neutrons) to be concentrated on one nucleon.

Obviously, such a simple picture ignores a large number of complicating effects that can, in particular cases, reverse the order of these cross-sections. Nevertheless, despite its simplicity, the compound nucleus theory has been of great value in explaining many aspects of medium energy nuclear reactions (i.e. $\lesssim$ 10 MeV per nucleon of the bombarding particle).

9.6.4. *Direct interaction model*

The compound nucleus theory considers that the bombarding particle interacts with the nucleus as a whole. The nucleus is excited uniformly and evaporation of low energy nucleons follows. This model fails to explain some of the observed phenomena

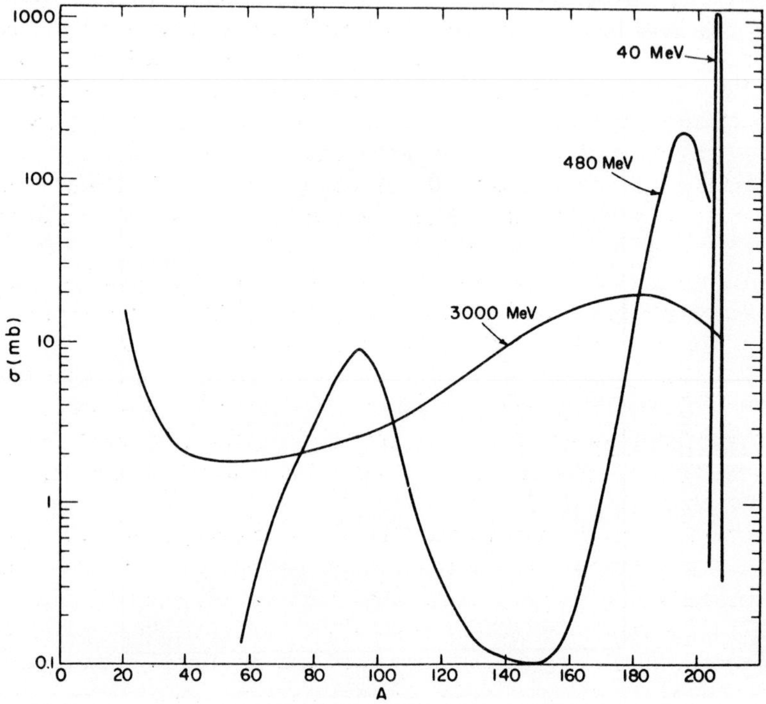

FIG. 9.6. Mass yield curve obtained by bombardment of lead with high energy protons. (According to J. M. Miller and J. Hudis.)

as the kinetic energy of the bombarding particle increases. One such observation is the occurrence of high energy neutrons and protons in the emitted particles. Another is the large cross-sections for reactions such as $X_1(p, pn)X_2$ at energies where 6 or 7 nucleons are evaporated in order to de-excite the nucleus.

Figure 9.6 shows the cross-section for production of nuclides of $A = 20$–200 when ^{208}Pb is bombarded with protons of energies 40, 480, and 3000 MeV. Such reactions, yielding a large number of products at high projectile energies, have been extensively studied by nuclear chemists, partly because the mixture of products required separation by sophisticated radiochemical techniques. Although some ambiguity exists in its use, often the term *spallation* is used for reactions in which a number of particles are emitted as a result of a direct interaction. At very high energies, not only is a broad range of products formed but the probability for the formation of these products is, within an order of magnitude, similar for every mass number. Further, studies of bombarding energies above 100 MeV show that high energy protons, neutrons, and heavier particles are emitted from the nucleus in a forward direction. Compound nucleus evaporation would be expected to be isotropic (i.e. show no directional preference in the center-of-mass system).

R. Serber has suggested a mechanism that satisfactorily accounts for most features of nuclear reactions at bombardment energies above 50 MeV for protons, deuterons, and α-particles. He proposed that high energy reactions occur in two stages.

(i) During the first stage the incoming particle undergoes direct collision with individual nucleons. In these collisions the struck nucleon often receives energy much in excess of its binding energy. Consequently, after each collision both the initial bombarding particle and the struck nucleon have some probability of escaping the nucleus since their kinetic energies are greater than their binding energies. If both particles escape, the nucleus is usually left with only a small amount of excitation energy. This explains the high cross-section for (p, pn) reactions. In support of this explanation, both the emitted proton and neutron have large kinetic energies. Either one or both of the original pair may collide with other nucleons in the nucleus rather than escape. During this initial stage, known as the *knock-on-cascade process*, the total number of direct collisions may be one or many. After a period lasting about 10^{-19} s, some of the struck nucleons have left the nucleus.

(ii) In the remaining nucleus the residual excitation energy is uniformly distributed. The reaction then enters its second and slower stage, during which the residual excitation energy is lost by nucleon evaporation. This stage resembles the compound nucleus process very closely.

In Fig. 9.6 at 40 MeV only the second, evaporation stage is observed as seen by the narrow mass distribution curve. The curves for 480 and 3000 MeV reflect the increased importance of the first, direct interaction stage which leads to a broad spectrum of product mass numbers.

The experimental data have been successfully reproduced using a calculation technique known as the Monte Carlo method and assuming a Fermi gas model for the nucleus. This model treats the nucleons like molecules of a very cold ideal gas in a potential well. The nucleons do not follow the Pauli exclusion principle and fill all vacant orbitals.

In the *Monte Carlo calculation*, the history of a single incident proton is studied in all its collisions in the nucleus. Each collision with a nucleon is characterized by probability distributions for occurrence, for energy, for angular distribution, and for pion formation. The calculations are repeated for different impact points. Numbers are picked from

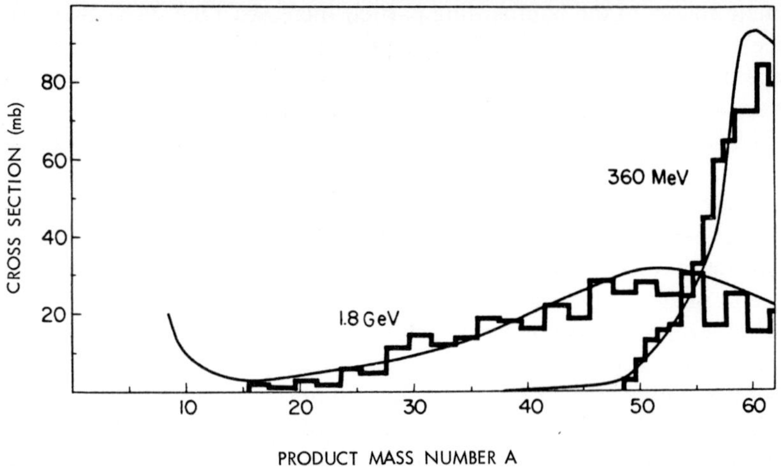

FIG. 9.7. Comparison of mass yield curves for proton interaction with copper as predicted by Monte Carlo calculations (histogram) with experimental results (curve). (According to G. Friedlander.)

a collection of random numbers for each case to ensure that equal weight is given to all possible impacts, etc. It is the use of random numbers that is the basis of the Monte Carlo technique. Modern computers allow so many random number calculations that a reliable pattern of events emerges.

Figure 9.7 shows the quite satisfactory agreement for two bombardment energies between experimental yields of A in proton bombardment of copper (solid lines) and Monte Carlo calculations (histograms).

9.7. Nuclear fission

Nuclear fission is unique among nuclear reactions since the nucleus divides into roughly equal parts with the release of a large amount of energy, about 200 MeV per fission. It is probably no overstatement to say that fission is the most important nuclear process, both for its potential to destroy civilization through the use of weapons and its potential through reactors to supply abundant power for all people.

An aspect of fission which has been studied extensively is the distribution of mass and charge among the fragments formed in fission. No matter how nuclei are made to undergo fission, fragments of various masses are formed, which result in production of chemical elements as light as zinc (atomic number 30) and as heavy as gadolinium (atomic number 64), with half-lives from fractions of seconds to millions of years. Approximately 400 different nuclides have been identified as products in the fission of ^{235}U by neutrons. Study of these fission products has required extensive radiochemical work and continues as a very active field of research—for example, in the measurement of very short-lived products. Although fission is an extremely complicated process, and still challenges theorists, satisfactory models for most of the fission phenomena have been developed.

9.7.1. *Mass and charge distribution*

If fission was symmetric, i.e. if two fragments of equal mass (and charge) were formed,

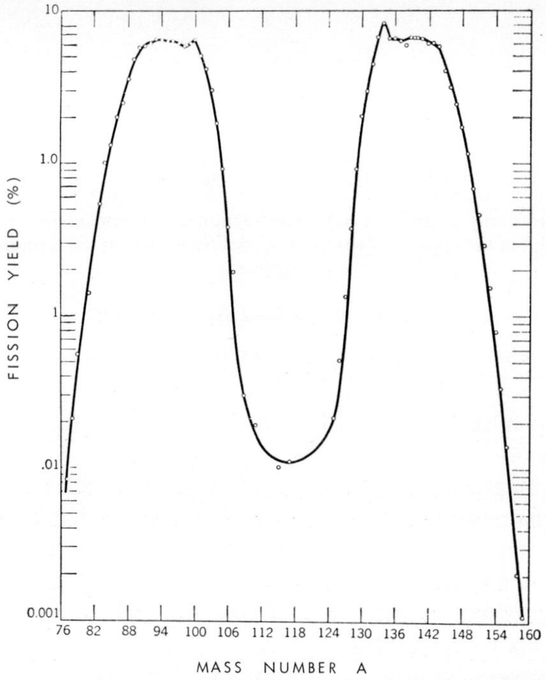

FIG. 9.8. Mass yield curve for fission of ^{235}U with thermal neutrons. (According to E. K. Hyde.)

the thermal neutron fission of ^{235}U would lead to the production of two $^{118}_{40}Pd$ nuclides,

$$^{235}U + n \rightarrow 2\,^{118}_{46}Pd$$

However, a plot showing the amount of different masses formed is a curve with two maxima—one near mass number 97 (close to the n-shell $N = 50$) and a second near mass number 137 (close to the n-shell $N = 82$) (Fig. 9.8). These two masses are formed together in the most probable split which is asymmetric ($A_1 \neq A_2$). As Fig. 9.8 shows, symmetric fission ($A_1 = A_2$) is rare in fission of ^{235}U by thermal neutrons—the yield for $A = 115$ is only 0.01% compared to 6% for $A_1 = 97$ (or $A_2 = 137$). Since two fission products are formed in each fission event, mass yield curves like that in Fig. 9.8 must total 200%.

From the mass yield curve in Fig. 9.8 we learn that complementary fission products (i.e. the two products A_1 and A_2 with identical yield values symmetrically located around the minimum) add up to about 234, not 236. Direct neutron measurements reveal that on the average 2.5 neutrons are emitted in fission. This number increases as the Z of the target and as the bombarding energy increase.

Because the n/p ratio for $^{235}_{92}U$ is 1.6, while the ratio necessary for stability is 1.2–1.4 in the elements produced in fission, fission fragments always have too large a value for the n/p ratio. This is compensated partially by the emission of several neutrons in the act of fission. However, the number of neutrons emitted is not sufficient to lower the n/p ratios to stable values. To achieve further lowering, the fission fragments, after neutron emission, successively undergo radioactive decay steps in which β-particles are emitted. Since β-decay occurs with no change in A, the successive β-decay steps follow the isobar parabola of the stability valley (see Fig. 3.1; cf. also the nuclide chart,

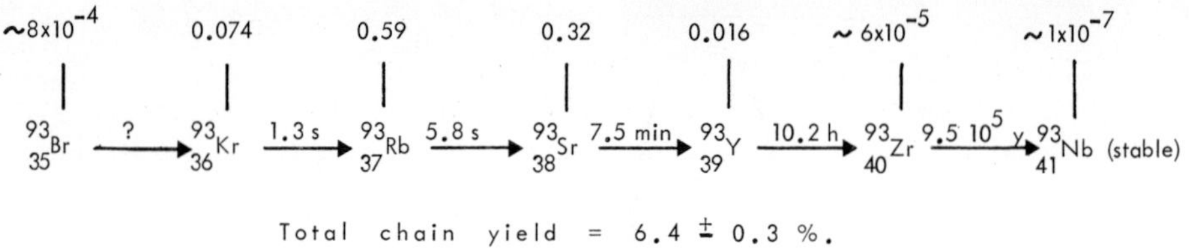

FIG. 9.9. Fission decay chain of $A = 93$. The fractional independent yields (i.e. fractions of the cumulative yield) in the upper row refer to nuclides believed to be formed directly in the fission process.

Appendix N). For $A = 137$, ${}^{137}_{52}Te$ is the first nuclide measured. The chain sequence is

$$ {}^{137}_{52}\mathrm{Te} \xrightarrow{3,5\ \mathrm{s}} {}^{137}_{53}\mathrm{I} \xrightarrow{24\ \mathrm{s}} {}^{137}_{54}\mathrm{Xe} \xrightarrow{3,8\ \mathrm{min}} {}^{137}_{55}\mathrm{Cs} \xrightarrow{30,1\ \mathrm{y}} {}^{137}_{56}\mathrm{Ba}\ \text{(stable)} $$

According to Appendix N, thermal fission of ${}^{235}U$ leads to a yield of 6.26% for the $A = 137$ chain.

In addition to measuring the variation of mass yield, the variation of fission yield in isobaric mass chains as a function of the proton number has been studied. In Fig. 9.9, "individual" yield data are presented for the $A = 93$ chain. In general, the charge distribution yields follow a Gaussian curve with the maximum displaced several units below the value of Z for stable nuclides. For $A = 93$ the yield is largest for $Z = 37$ and 38 compared to the stable value of $Z = 41$.

Fission of heavy elements other than uranium can be made to occur by bombardment, particularly if we use charged particles such as protons accelerated to high energies. The mass distribution curve for this type of fission is interesting. At low bombarding energies it is asymmetric as it is with low energy neutrons. However, as the energy of bombardment is increased the valley between the peaks of the curve becomes more shallow and, at high energies, a single-humped symmetric curve is obtained (Fig. 9.10).

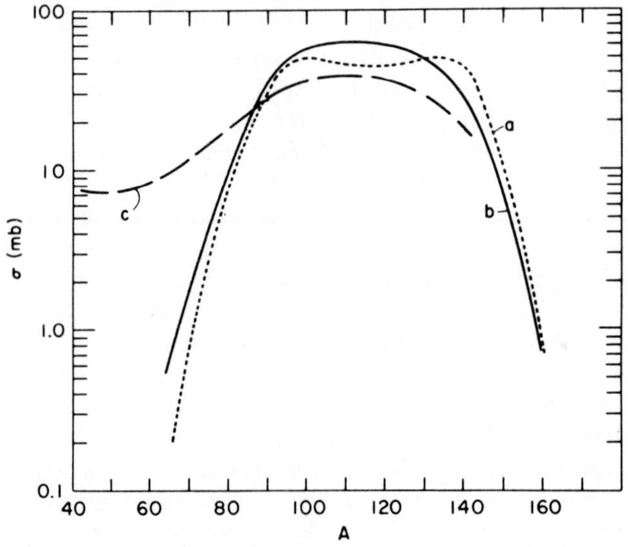

FIG. 9.10. Mass yield curves for fission products from uranium irradiated with protons: (a) 100 MeV, (b) 170 MeV, and (c) 2.9 GeV. (According to G. Friedlander.)

Thus, the most probable mode of mass split changes from asymmetric at low energies to symmetric at high energies. As the energy is increased the fission yield curve $\sigma_{\text{fiss}}(A)$ becomes indistinguishable from yield curves of the type in Fig. 9.6 ascribed to direct interaction mechanisms.

9.7.2. *Energy of fission*

From the curve of the binding energy per nucleon (see Fig. 3.3), we calculated in §3.3 that about 200 MeV would be released in the fission of a heavy element. In this section we consider how this fission energy is partitioned.

The neutrons emitted have an average kinetic energy of ~2 MeV. For the average of 2.5 neutrons emitted in fission of ^{235}U by thermal neutrons, about 5 MeV of the fission energy is required. The emission of γ-rays at the time of fission (*prompt* γ-rays) accounts for another 6–8 MeV. The largest part of the fission energy is observed as the kinetic energy of the fission products. We can estimate this by calculating the Coulombic repulsion energy of a probable fission product pair, $^{93}_{37}$Rb and $^{143}_{55}$Cs. The model used is two touching spherical nuclei with a charge center distance $d = 12.7$ fm (Fig. 9.11a). From (7.15) we calculate that the repulsive Coulomb energy (and, hence, the kinetic energy of separation) is 175 MeV. If the model is modified slightly to include a neck between the two nuclei, increasing the distance between the charge centers to 13.5 fm, the kinetic energy would be 165 MeV, as experiment requires. Such an elongated shape with a small neck at the time of separation in fission (the "scission" shape) is supported by several types of evidence.

The remaining ~23 MeV of fission energy is retained in the fission product nuclei as internal excitation energy. This energy is released in a sequence of β-decay steps

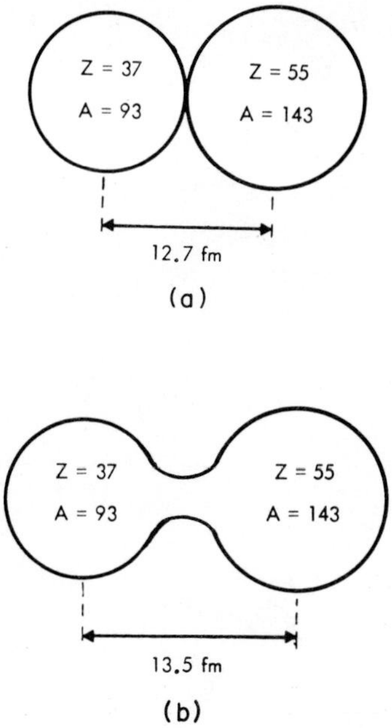

FIG. 9.11. Asymmetric fission of ^{236}U showing the distance between the effective centers of charge in the case (a) of touching spheroids, (b) of a scissioning dumbbell. The nuclear radii are calculated from $r = 1.3A^{1/3}$.

TABLE 9.1. *Recent data for energy distribution in thermal fission of* ^{235}U *in MeV*

Prompt energy		176.5 ± 5.5
of which the kinetic energy of fission products	164.6 ± 4.5	
kinetic energy of 2.5 neutrons (prompt)	4.9 ± 0.5	
γ-energy (prompt)	7.0 ± 0.5	
Delayed energy from fission products decay		23.5 ± 5.0
of which the kinetic energy of β's	6.5 ± 1.5	
neutrino radiation	10.5 ± 2.0	
γ-energy	6.5 ± 1.5	
Total		200 ± 6

in which the n/p values are adjusted to stability. Table 9.1 summarizes the distribution of fission energy.

9.7.3. *Fission models*

The analogy between nuclei and liquid droplets was found to be useful in deriving the semiempirical mass formula (§3.6) when N. Bohr and J. Wheeler explained fission just months after its discovery by using the same model. The surface tension of a liquid causes a droplet to assume a spherical shape, but if energy is supplied in some fashion, this shape is distorted. If the attractive surface tension force is greater than the distorting

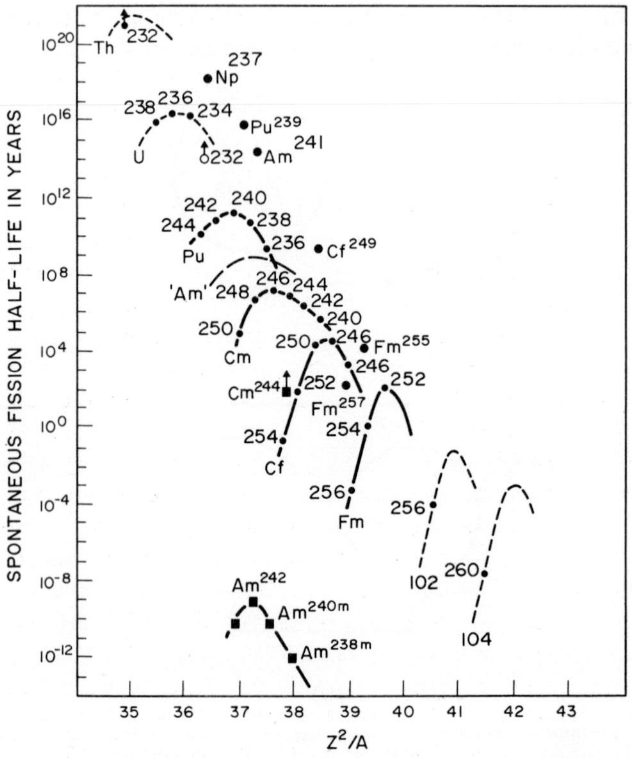

FIG. 9.12. Spontaneous fission lifetimes. (According to V. M. Strutinsky and S. B. Bjørnholm.)

force, the drop oscillates between spherical and elongated shapes. If, however, the distorting force becomes larger than the attractive force, the drop elongates past a threshold point and splits (fission).

In §3.6 we described how the repulsive forces between the protons in the nucleus could be expressed by a term a_c proportional to $Z^2/A^{1/3}$, and the surface tension attraction by another term a_s proportional to $A^{2/3}$. The repulsive Coulomb force tends to distort the nucleus in the same way a distorting force does a droplet, while the surface tension tries to bring it in to a spherical form. The ratio between the two opposing energies should measure the instability to fission of the nucleus:

$$\text{Instability} \propto \frac{\text{energy of repulsion}}{\text{energy of attraction}} = \frac{(Z^2/A^{1/3})}{A^{2/3}} = \frac{Z^2}{A} \tag{9.11}$$

This Z^2/A is known as the *fissionability parameter* because the liquid drop model predicts that the probability of fission should increase with increase in Z^2/A. Of all naturally occurring nuclides only ^{235}U can be fissioned by thermal neutrons, while ^{238}U fission requires energetic neutrons ($\gtrsim 2$ MeV). With increasing Z (>92) the fission probability with thermal neutrons increases and the half-life of the radioactive decay by spontaneous fission decreases. Both of these processes are more probable for even Z-elements than for odd Z-elements. The half-life for spontaneous fission decay is given in Fig. 9.12 as a function of the fissionability parameter.

In the semiempirical mass equation a spherical shape is assumed. If N and Z are kept constant and the potential energy of the nuclear liquid drop is calculated as a function of deformation from spherical to prolate, the curve in Fig. 9.13a is obtained. The nucleus exists normally in the ground state level of the potential well. In order to undergo fission it must be excited above the fission barrier which is about 5–6 MeV. As the diagram shows, this means excitation of the nucleus into the continuum level region if the nucleus retains the shape associated with the potential well. However, if the nucleus deforms, some excitation energy goes into deformation energy. At the top of the barrier, the nucleus is highly deformed and has relatively little internal excitation energy. It exists in well-defined vibrational levels, and fission occurs from such a level. This is known as the "saddle point" (the top of the barrier) of fission.

It has been recognized that the liquid-drop model semi-empirical mass equation cannot calculate the correct masses in the vicinity of neutron and proton magic numbers. More recently it was realized that it is less successful also for very deformed nuclei midway between closed nucleon shells. An additional complication with the simple liquid-drop model arose when isomers were discovered which decayed by spontaneous fission. Between uranium and californium a number of nuclides were found to decay by spontaneous fission with half-lives of 10^{-2} to 10^{-9} s, millions of times slower than prompt fission which occurs within 10^{-14} s but millions of times faster than normal spontaneous fission (Fig. 9.12). For example, ^{242}Cm has a ground state half-life to spontaneous fission of 10^6 years, while an isomeric state of ^{242}Cm has been found to fission with $t_{1/2}$ of 10^{-7} s.

V. M. Strutinsky has developed a model which satisfactorily explains the fission isomers. For such short half-lives the barrier must be only 2–3 MeV. Noting the manner in which the shell model levels vary with deformation (§6.5, the "Nilsson levels"), Strutinsky incorporated shell corrections in the liquid-drop model and obtained the

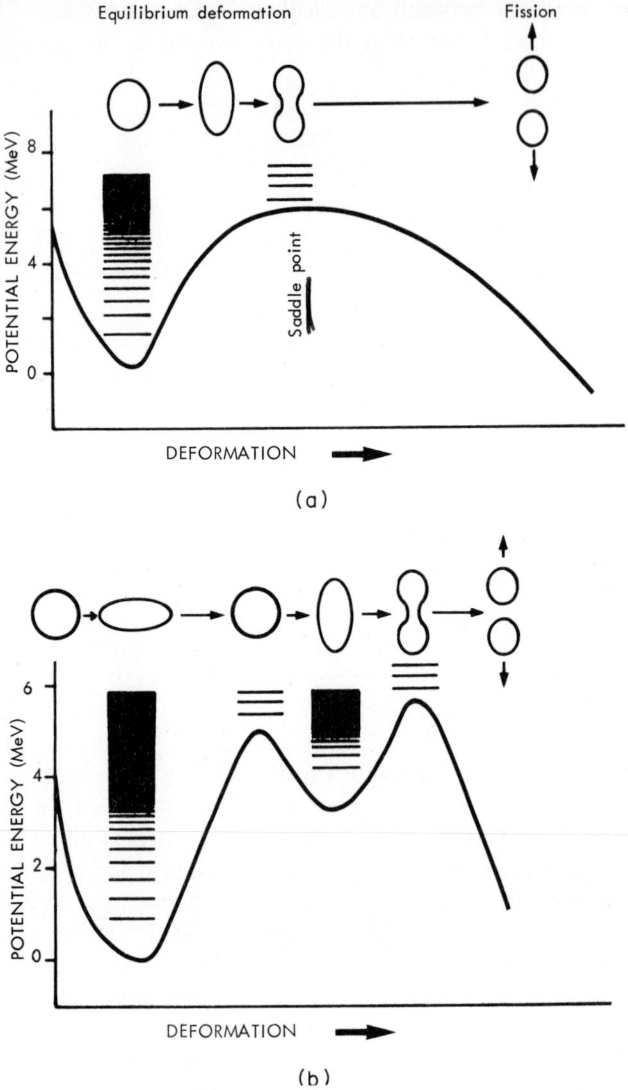

FIG. 9.13. (a) The liquid-drop model potential energy curve. (b) Same, but modified by the shell model.

"double-well" potential energy curve in Fig. 9.13b. In the first well the nucleus is a spheroid with the ratio of the major axis about 25% larger than the minor. In the second well, the deformation is much larger, the axis ratio being about 1.8. A nucleus in the second well is metastable (i.e. in isomeric state) as it is unstable to γ-decay to the first well or to fission. Fission from the second well is hindered by a 2–3 MeV barrier, while from the first well the barrier is 5–6 MeV, accounting for the difference in half-lives.

The single-well curve in Fig. 9.13a predicts symmetric fission whereas the double-well curve (Fig. 9.13b) leads to the correct prediction of asymmetric fission. Incorporation of shell effects in fission also leads to the prediction that the half-lives of very heavy nuclides ($Z \gtrsim 106$) must be longer than the simple liquid-drop model would indicate. This has led to an interest in "super heavy" elements of Z about 110–118.

9.8. Photonuclear reactions*

If a photon transfers sufficient energy to a nucleus to excite it to a higher state, three possibilities for de-excitation exist: (a) the same energy is immediately re-emitted isotropically, (b) a long-lived isomer may be formed which decays through emission of one or more γ-rays, and (c) the nucleus disintegrates. The first process (a) is referred to as the Mössbauer effect and is discussed in Chapter 14. The second process (b) of nuclear excitation has been discussed earlier. The third process (c) is referred to as *photonuclear disintegration.* The energy transferred to the nucleus must be sufficient to excite it above the dissociation energy for a proton, neutron, or other particle.

The simplest photodisintegration process is that of the deuteron, whose binding energy is 2.23 MeV. If the γ-ray energy absorbed exceeds this value, a neutron and a proton are formed. This is a common reaction in nuclear reactors using heavy water as moderator, because the fission γ-ray energy is often several MeV. However, γ-rays of such energy rarely occur in the radioactive decay of nuclides (i.e. with half-lives of hours or longer). The cross-section for photodisintegration of ^2H has a maximum value of 2.4 mb at 4.3 MeV E_γ.

The energy necessary for photodisintegration of a nucleus is calculated from known nuclear masses. It is obviously easier to remove one particle than several from a nucleus. As a result we find that E_γ must be $\gtrsim 5$ MeV for photodisintegration of heavier nuclei.

For $10 \leq E_\gamma \leq 40$ MeV the photon wavelength is comparable to the nuclear size. It is therefore easily absorbed, which causes collective nuclear vibrational motions (so-called dipole vibrations, because the neutrons and protons are assumed to vibrate in separate groups). De-excitation occurs through γ-emission. This is known as the "giant resonance" region, because the total cross-section for heavier nuclides goes up to hundreds of millibarns. For higher E_γ, nucleons may be expelled, the main reactions being (γ, n), $(\gamma, 2n)$ and (γ, np) in descending importance. As the γ-energy increases and the wavelength decreases to nucleon dimensions, interaction with nuclear groups (e.g., deuterons) or single nucleons takes place. Below 550 MeV one pion plus a nucleon may be emitted in the de-excitation following the photon absorption. At higher energies, several pions may be emitted. Very little is known about the details of these processes.

9.9. Exercises

9.1. A 0.01 mm thick gold foil, 1 cm^2 in area, is irradiated with thermal neutrons. The (n, γ) cross-section is 99 b. What is the transformation rate at a n-flux of 10^{15} n cm^{-2} s^{-1}?

9.2. Assume the irradiation time of the gold foil in the previous problem is one week. What percentage of the original gold atoms in the target have undergone transformation?

9.3. A water-cooled copper foil (0.1 mm thick) is irradiated by the internal beam of a sector focused cyclotron with 1.2 mA H^+ ions of 24 MeV for 90 min. The reaction $^{63}Cu(p, pn)^{62}Cu$ occurs with a probability of 0.086 b. Copper consists 69% of ^{63}Cu. The proton beam has a cross-section of only 15 mm^2. (a) How many ^{62}Cu atoms have been formed? (b) What fraction of the projectiles have reacted to form ^{62}Cu? (c) What cooling effect is required (kW) at the target?

9.4. Calculate the macroscopic capture cross-section for reaction of natural uranium with thermal neutrons. See nuclide chart.

9.5. Calculate the kinetic energy of the ^4He ion formed through thermal neutron capture in ^{10}B.

9.6. What minimum photon energy is required for the reaction $^{11}B(\gamma, n)^{10}B$?

9.7. A 2 MeV neutron collides elastically with an iron atom (^{56}Fe). What is the average temperature (corresponding to the maximum velocity) which can be ascribed to the iron nucleus after the collision?

9.10. Literature

See Appendix A, §§ A.1 and A.7.

P. M. Endt and M. Demeur, *Nuclear Reactions*, North-Holland, 1959.
V. Goldanskii and A. M. Baldwin, *Kinetics of Nuclear Reactions*, Pergamon Press, Oxford, 1961.
E. K. Hyde, I. Perlman, and G. T. Seaborg, *Nuclear Properties of the Heavy Elements: III, Fission Phenomena*. Prentice-Hall, 1964.
Physics and Chemistry of Fission, IAEA, Vienna, 1965, 1969, and 1973.
V. S. Fraser and J. C. D. Milton, Nuclear fission, *Ann. Rev. Nucl. Sci.* **16** (1966) 379.
M. Lefort, *Nuclear Chemistry*, van Nostrand, 1968.
R. Vandenbosch and J. R. Huizinga, *Nuclear Fission*, Academic Press, 1973.
R. Vandenbosch, Spontaneously fissioning isomers, *Ann. Rev. Nucl. Sci.*, **27** (1977) 1.
Nuclear Structure Physics, ed. S. J. Hall and J. M. Irvine, Proc. 18th Scottish U. Sum. Sch. in Physics, Edinburgh, 1977.

CHAPTER 10

Production of Radionuclides

Contents

Introduction

The production of radionuclides can be considered at three different levels: at about or below the curie level for research work; at the kilocurie level for biological and industrial uses; and at the megacurie level in nuclear power stations. The production principles are the same with the main steps being target preparation, irradiation, and its secondary (recoil and hot atom) effects, target processing, and isolation and specification (labeling, purity, etc.) of the radionuclides. The handling of radioactive samples, especially at the higher levels, requires special techniques and care, and are of such importance that they are discussed in several chapters (Chapters 15, 16, and 21).

In this chapter we emphasize the principles of the production of pure radionuclides by different techniques without special consideration of the scale of production.

The annual sales of radionuclides for use in research and in various applications exceed US $100 million, reflecting their importance.

10.1. General considerations

Radioactive nuclides may be prepared by a wide variety of particle accelerators and by nuclear reactors; however, only cyclotrons or reactors of at least moderate particle flux are able to produce sources of sufficiently high specific radioactivities to be of practical interest. These two methods of production supplement each other in that in general they do not produce the same isotopes of an element.

In cyclotrons charged particles such as protons, deuterons, and α-particles bombard the target nuclei, and after emission of one or more particles to remove the excess excitation energy, a radioactive product nuclide may result. The capture of positively charged particles and the subsequent emission of neutrons result in isotopes which are *neutron deficient* compared to the stable isotopes of the element. A second important point in cyclotron bombardments is that normally the product is not isotopic with the target. As a result, after chemical separations a product of high specific activity is obtained since it is not diluted by the target material. This contrasts with the production in a nuclear reactor where product and target are usually isotopic. Even if the same radioactive nuclide can be made in a nuclear reactor, cyclotron production may be favored if high specific activity is necessary for a particular experiment. An example of an important cyclotron-produced radionuclide is sodium-22 formed by the reaction:

$$^{24}_{12}\mathrm{Mg} + ^{2}_{1}\mathrm{H} \rightarrow ^{22}_{11}\mathrm{Na} + ^{4}_{2}\mathrm{He}$$

Nuclear reactors maintain themselves by chain fission reactions propagated by neutrons (Chapter 19). If other materials are introduced into the reactor, neutrons may be absorbed producing *neutron rich* isotopes, frequently radioactive. A typical reaction is

$$^{23}\mathrm{Na}(\mathrm{n}, \gamma)^{24}\mathrm{Na}$$

A second mode of production in reactors is by the fission process itself as the majority of fission products are radioactive and cover a wide range of atomic numbers of varying abundances. These radioactivities may be isolated with high specific activity in contrast to the neutron capture products. However, in fission products the isolation of any particular radioactive element can require a fairly lengthy separation from the large number of other radioactivities present.

Radioactive nuclides may also be obtained as daughters of parent activities which have been previously made either by cyclotron or reactor radiation. Several examples are discussed in §4.14 of such radioactive isotope generators.

Which mode of production is utilized depends on several factors. The distance from the production site and the shipping time, the length of time for separation and purification processes, and of the experiment itself, determine the half-life with which it is practical to work. The type of experiment and equipment available for handling and measurement determines the preferable decay type and intensity. All these factors must be considered in determining which nuclide is to be used. For example, if in an experiment requiring sodium tracer these aspects indicate the necessity of a time of at least several weeks, the use of reactor-produced ^{24}Na ($t_{1/2}$ 15.0 h) is excluded and the cyclotron-produced ^{22}Na ($t_{1/2}$ 2.6 y) is required. If the experiment requires the use of very high specific activities, either cyclotron or fission product activities may be necessary. Reactor activities are usually much cheaper and easier to obtain on a routine basis, because many targets can be irradiated simultaneously in the reactor, whereas in a cyclotron only one target at a time can be bombarded.

Although these considerations are important for choosing a radionuclide to suit a particular purpose, the possibilities of production may be limited either by the availability of facilities (reactors, different accelerator beams) or simply by the reaction cross-sections. Figure 10.1, which represents a small section of an isotope chart

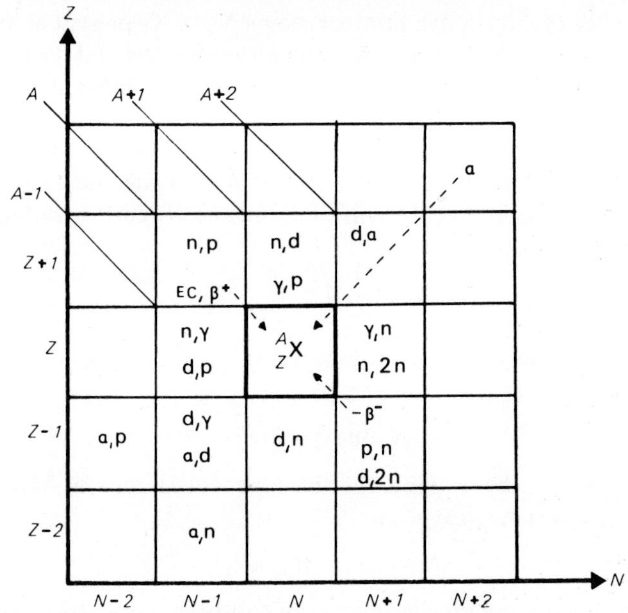

FIG. 10.1. Part of isotope chart showing modes of formation of element ${}^{A}_{Z}X$ through various (x, x') irradiation processes and through radioactive decay (dashed arrows).

(Appendix N), shows a *product* nuclide, ${}^{A}_{Z}X$, and paths for its formation. For example, it can be produced through an (n, γ) or a (d, p) reaction on ${}^{A-1}_{Z}X$, or a (d, α) reaction on ${}^{A+2}_{Z+1}X$, etc. The choice of reaction depends on the particular nuclides available, projectile energy, and the cross-sections (available in tables of nuclear data, Appendix A.7). As pointed out in §9.2, the same projectile may yield several different products: e.g. in Fig. 10.1 we have indicated five different deuteron-induced reactions.

It should also be realized that an induced nuclear reaction may be followed by rapid radioactive decay, leading to the desired radionuclide. Thus an n, 2n reaction may yield ${}^{A}_{Z}X$, which rapidly decays to ${}_{Z+1}{}^{A}X$ through β^- emission, e.g.

$$^{238}\text{U(n, 2n)}^{237}\text{U}$$

followed by

$$^{237}\text{U} \xrightarrow[6.75\text{ d}]{\beta^-} {}^{237}\text{Np}$$

yielding ^{237}Np for investigation of neptunium chemical behavior.

10.2. Irradiation yields

The production of a radioactive nuclide by irradiation in a reactor or an accelerator can be written as:

$$X_{\text{proj}} + X_{\text{target}} \xrightarrow{k} X_1 \xrightarrow{\lambda_1} X_2 \xrightarrow{\lambda_2} X_3 \tag{10.1}$$

We are here mainly concerned with the product X_1 (produced at rate k), which is assumed to be radioactive, decaying in a single step or via a chain of decays to a final stable nuclide. The number of radioactive product atoms N_1 of X_1 present at any time t is the difference between the total number of product atoms formed and the number of radioactive decays that have occurred in time dt. This leads to the differential equation

$$dN_1 = k\,dt - \lambda_1 N_1\,dt \qquad (10.2)$$

Integration between the limits of $N_1 = 0$ (this assumes that no radioactive product nuclides X_1 exist at the beginning of the irradiation) and N_1 corresponding to an irradiation time t_{irr}:

$$N_1 = \frac{k}{\lambda_1}(1 - e^{-\lambda_1 t_{irr}}) \qquad (10.3)$$

N_1 increases exponentially with time towards a maximum value, which is reached when $t_{irr}\lambda_1 \gg 1$ (i.e. $t_{irr} \gg t_{1/2}$):

$$N_1(\max) = k/\lambda_1 \qquad (10.4)$$

However, for $t_{irr} \ll t_{1/2}$, by expanding the exponential into a McLauren's series, $e^{-\lambda_1 t_{irr}} \approx 1 - \lambda_1 t_{irr}$ which in (10.3) gives

$$N_1 = kt_{irr} \quad (t_{irr} \ll t_{1/2}) \qquad (10.5)$$

If we measure the irradiation time in number of half-lives, $a = t_{irr}/t_{1/2}$, we obtain

$$N_1 = N_1(\max)\,(1 - 2^{-a}) \qquad (10.6)$$

This equation shows that irradiation for one half-life ($a = 1$) produces 50% of the maximum amount, two half-lives give 75%, etc.

After the end of the irradiation time, the radioactive product nuclides continue to decay according to their half-lives. We indicate this decay time after the end of irradiation (termed the *cooling time*) as t_{cool}. The number of product nuclides, which is a function of the irradiation time t_{irr} and cooling time t_{cool}, is given by

$$N_1 = \frac{k}{\lambda_1}(1 - e^{-\lambda_1 t_{irr}})e^{-\lambda_1 t_{cool}} \qquad (10.7)$$

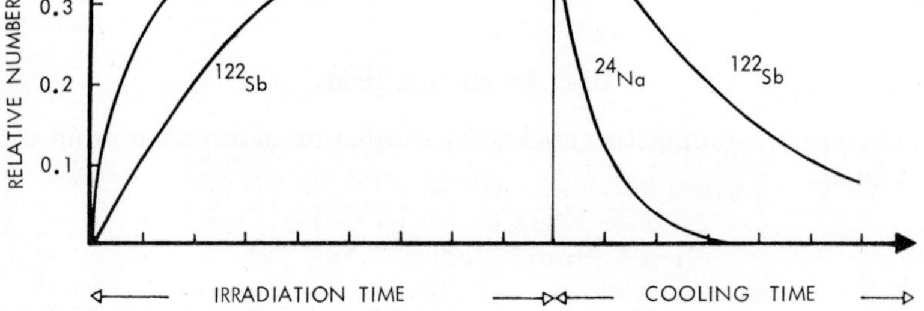

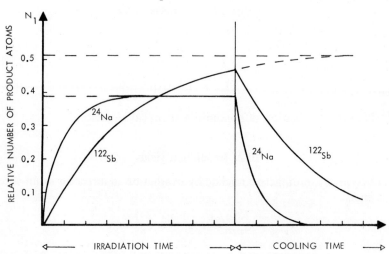

FIG. 10.2. Increase in number of product nuclei during irradiation time, and subsequent decrease after the end of irradiation (cooling time).

The effect of the cooling time is shown in Fig. 10.2 for two sample nuclides.

The production rate $k(\approx \Delta N/t_{irr})$ is given in §9.1:

$$k = \phi\sigma N_t \text{ (thin target in neutron flux cf. (9.4))} \tag{10.8a}$$

$$k = 6.24 \times 10^{18}\, i\sigma N_v x z^{-1} \text{ (thin target in accelerator)} \tag{10.8b}$$

(see (8.2)) where N_t is the total number of target nuclides with cross-section σ, and $N_v x$ is the number of target atoms per unit area. If a fission product is isolated, (10.8a, b) has to be multiplied by the nuclide fission yield y_i.

For (10.8) to be valid, the projectile energy must be maintained constant in order that the reaction cross-section remains constant. Moreover, the number of nuclear reactions must remain small compared to the total number of target atoms to justify the assumption of constant N_t or N_v. Finally, (10.8) assumes that the flux is not decreased in passing through the target (this is true only for a thin target).

Usually the radioactive decay rate A_1 is measured, rather than the number of atoms N_1. Recalling $A = \Delta N/\Delta t = \lambda N$, (10.3) and (10.8a) yield

$$A_1 = k(1 - e^{-\lambda_1 t_{irr}})e^{-\lambda_1 t_{cool}} \tag{10.9}$$

This equation represents the basic relationship for the production of radionuclides; in a reactor k is calculated according to (10.8a); for accelerator irradiation (10.8b) is used.

Let us take the production of radioactive sodium by reactor irradiation and cyclotron bombardment as examples of the application of these equations: Na_2CO_3 ($M = 106$) is irradiated in a reactor to produce ^{24}Na, which has a half-life of 15.0 h and emits β- and γ-rays. The reaction cross-section of ^{23}Na (100% in natural sodium) is 0.53 b for thermal neutrons. Assuming a 5 g sample, 60 h irradiation time, and a thermal flux of 10^{12} n cm^{-2} s^{-1}, we calculate:

$$A = 10^{12} \times 0.53 \times 10^{-24} \times \frac{2 \times 5}{106} \times 6.02 \times 10^{23}(1 - e^{-0.693 \times 60/15}) = 2.8 \times 10^{10} \text{ dps}$$

at the end of the irradiation time.

^{24}Na may be produced through accelerator irradiation of magnesium by the reaction ^{26}Mg(d, α)^{24}Na. With 22 MeV D$^+$ ions the reaction cross-section is about 25 mb. Assuming a 0.1 mm thick magnesium foil ($M = 24.3$, $\rho = 1.74$ g cm^{-3}) with an area larger than the projectile beam, and a 2 h irradiation at 100 μA, the activity of ^{24}Na produced (natural magnesium contains 11.0% ^{26}Mg) is:

$$A = 6.24 \times 10^{18} \times 100 \times 10^{-6} \times 25 \times 10^{-27}\left(0.11\frac{1.74}{24.3}6.02 \times 10^{23} \times 0.01\right)$$
$$\times (1 - e^{-0.693\, 2/15}) = 6.53 \times 10^{7} \text{ dps}$$

at end of irradiation.

10.3. Second-order reactions*

For large cross-sections, short half-lives, and long bombardments, second-order capture products may be formed. If the first-order product is radioactive, then its concentration at any time is dependent on (i) the decay constant, (ii) the cross-section for production of the second-order product, as well as (iii) the cross-section for its own

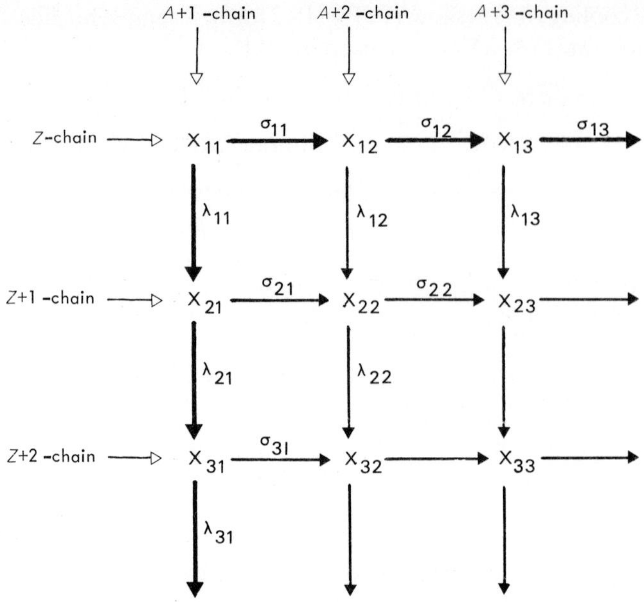

FIG. 10.3. Formation of higher nuclides through multiple neutron capture and associated decay chains.

production. Taking these possibilities into account we arrive at a scheme according to Fig. 10.3. For simplicity, we may assume that the induced nuclear transformations only involve single-neutron capture, and that only β^- decay occurs. However, it should be obvious that the scheme in Fig. 10.3 is applicable to all kinds of nuclear formation and decay reactions.

For the (n, γ) β^- case the upper horizontal row of Fig. 10.3 represents the successive formation of higher isotopes of the target element (the constant Z-chain) and the vertical rows the isobaric decay chains of each of these isotopes (the constant A-chains). The first of these two rows is indicated by heavy arrows. Chains which involve both induced transformations and radioactive decay play a central role in theories about the formation of the elements in the universe, in the thermonuclear reactions in the stars (Chapter 11), and in the synthesis of transuranium elements (Chapter 13).

For each nuclide in Fig. 10.3 we can write a differential expression for the change in concentration with time as a function of the irradiation and the radioactive decay:

For X_{11}: $dN_{11}/dt = -(\phi\sigma_{11} + \lambda_{11})N_{11} = -\Lambda_{11}N_{11}$ (10.10a)

X_{21}: $dN_{21}/dt = \lambda_{11}N_{11} - (\phi\sigma_{21} + \lambda_{21})N_{21} = \Lambda^*_{11}N_{11} - \Lambda_{21}N_{21}$ (10.10b)

X_{31}: $dN_{31}/dt = \lambda_{21}N_{21} - (\phi\sigma_{31} + \lambda_{31})N_{31} = \Lambda^*_{21}N_{21} - \Lambda_{31}N_{31}$ (10.10c)

... ...

X_{12}: $dN_{12}/dt = \phi\sigma_{11}N_{11} - (\phi\sigma_{12} + \lambda_{12})N_{12} = \Lambda^*_{11}N_{11} - \Lambda_{12}N_{12}$ (10.10d)

X_{13}: $dN_{13}/dt = \phi\sigma_{12}N_{12} - (\phi\sigma_{13} + \lambda_{13})N_{13} = \Lambda^*_{12}N_{12} - \Lambda_{13}N_{13}$ (10.10e)

... ...

X_{22}: $dN_{22}/dt = \lambda_{12}N_{12} + \phi\sigma_{21}N_{21} - (\phi\sigma_{22} + \lambda_{22})N_{22}$ (10.10f)

Let us now consider three different cases:

Case (i): Successive radioactive decay This occurs after formation of the first radioactive (parent) member X_{11}. Thus we only consider the left vertical chain, and assume σ_{11}, σ_{21}, etc. $= 0$. In this $A+1$ chain the second index is constant, and we omit it for case (i). This case is valid for all natural decay chains and for fission product decay chains. For each member of the decay chain we may write the differential equation

$$\frac{dN_i}{dt} = \lambda_{i-1} N_{i-1} - \lambda_i N_i \quad (i > 1) \tag{10.11}$$

We have already solved this relation in §10.2 for the formation of X_1 (the radioactive mother), and in §4.13 we showed how the amount of X_2 (the radioactive daughter) varies with time.

The general solution to the case with many successive decays is usually referred to as the *Bateman equations* (H. Bateman 1910):

$$N_n(t) = \lambda_1 \lambda_2 \ldots \lambda_{n-1} N_1^0 \sum_1^n C_i e^{-\lambda_i t} \tag{10.12a}$$

where $N_n(t)$ is the number of atoms for the nth species after time t, N_1^0 is the number of atoms of X_1 at $t = 0$, t is the *decay time*, and

$$C_i = \prod_{j=1}^{j=n} (\lambda_j - \lambda_i)^{-1} \quad \text{for} \quad j \neq i \tag{10.12b}$$

For $i = 1$, eqn. (4.41a) for simple radioactive decay is obtained. It is assumed that $N_i^0 = 0$ for $(i > 1)$. If that is not the case, additional chains, starting with each $N_i > 0$, require a new series of Bateman equations which are added together. For $i = 3$ and $N_2^0 = N_3^0 = 0$ at $t = 0$ one obtains

$$N_3(t) = \lambda_1 \lambda_2 N_1^0 (C_1 e^{-\lambda_1 t} + C_2 e^{-\lambda_2 t} + C_3 e^{-\lambda_3 t}) \tag{10.13}$$

where

$$C_1 = (\lambda_2 - \lambda_1)^{-1} (\lambda_3 - \lambda_1)^{-1}$$
$$C_2 = (\lambda_1 - \lambda_2)^{-1} (\lambda_3 - \lambda_2)^{-1}$$
$$C_3 = (\lambda_1 - \lambda_3)^{-1} (\lambda_2 - \lambda_3)^{-1}$$

Case (ii): Successive neutron capture This is the upper horizontal row in Fig. 10.3 (the constant Z-chain). Because the first index is constant in this chain, we omit it for case (ii). We assume $\lambda_{11}, \lambda_{12}$, etc. $= 0$. This is a valid approximation often used for the formation of transuranium elements in a nuclear reactor, which are long-lived in relation to the time of irradiation and observation. For each member of the formation chain we may write

$$\frac{dN_i}{dt} = \phi\sigma_{i-1} N_{i-1} - \phi\sigma_i N_i \tag{10.14}$$

By replacing λ_i with $\phi\sigma_i$ this equation becomes identical with (10.11). Thus the solution of case (ii) is obtained by using the Bateman eqn. (10.12). It should be noted that in case (ii) t is the *irradiation time* (t_{irr}).

Case (iii): Combined induced transformation and radioactive decay These combined effects can be taken into account by using the Bateman equations with some modifications, as developed by W. Rubinson.

We introduce the abbreviations

$$\Lambda = \phi\sigma + \lambda \tag{10.15}$$

and

$$\Lambda^* = \phi\sigma^* + \lambda^* \tag{10.16}$$

Λ_i is referred to as the disappearance constant of nuclide i, while Λ_i^* is the partial formation constant, and σ^* and λ^* the partial reaction cross-sections and decay constants for nuclide $i+1$; σ, σ^*, λ and λ^* may be equal to zero in some cases. The exact meaning Λ and Λ^* is explained below with examples. Introducing these abbreviations for the formation of the X_{i1} species of the constant $A+1$-chain, and dropping the second index, which is $= 1$ in the left vertical row in Fig. 10.3, we obtain the equation:

$$\frac{dN_i}{dt} = \lambda_{i-1}N_{i-1} - \Lambda_i N_i = \Lambda^*_{i-1}N_{i-1} - \Lambda_i N_i \tag{10.17}$$

Similarly, we find for the formation of the X_{1i} species (the constant Z-chain)

$$\frac{dN_i}{dt} = \phi\sigma_{i-1}N_{i-1} - \Lambda_i N_i = \Lambda^*_{i-1}N_{i-1} - \Lambda_i N_i \tag{10.18}$$

where the first index ($= 1$) has been dropped for simplicity. These two equations turn out to be identical, provided the partial formation constant Λ^* (10.16) for each nuclide (i.e. each index set) is used with the correct σ^* and λ^*; in (10.17) $\sigma^* = 0$, and in (10.18) $\lambda^* = 0$. With this reservation in mind the Bateman equations become

$$N_n(t) = \Lambda_1\Lambda_2^* \ldots \Delta_{n-1}^* N_1^0 \sum_1^n C_i e^{-\Lambda_i t} \tag{10.19a}$$

where

$$C_i = \prod_{j=1}^{j=n} (\Lambda_j - \Lambda_i)^{-1} \quad (j \neq i) \tag{10.19b}$$

The time t is the irradiation time t_{irr}.

The application of this expression and the use of Λ^* are best illustrated with a practical example.

If 1.0 mg of Tb_2O_3 ($M = 366$) is bombarded for 30 days at a flux of thermal neutrons of 10^{14} cm^2 s^{-1}, the following processes occur:

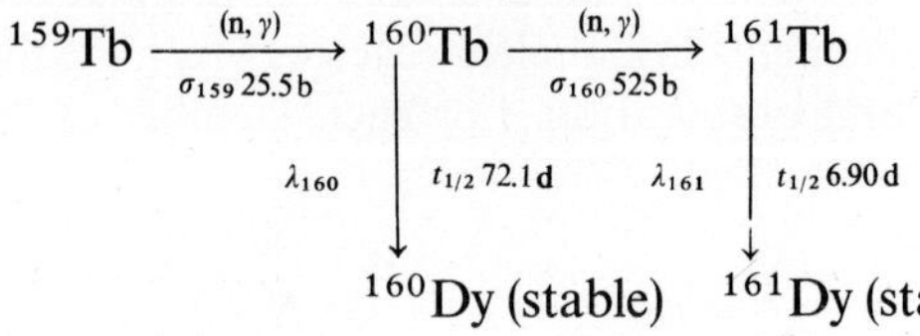

The appropriate (10.19) equation for the formation of ^{161}Tb (cf. also (10.13)) is:

$$N_{161} = \Lambda_{159}\Lambda_{160}^* N_{159}^0 \left(\frac{e^{-\Lambda_{159} t}}{(\Lambda_{160} - \Lambda_{159})(\Lambda_{161} - \Lambda_{159})} + \frac{e^{-\Lambda_{160} t}}{(\Lambda_{159} - \Lambda_{160})(\Lambda_{161} - \Lambda_{160})} + \frac{e^{-\Lambda_{161} t}}{(\Lambda_{159} - \Lambda_{161})(\Lambda_{160} - \Lambda_{161})} \right) \tag{10.20}$$

By considering the different reaction paths, we obtain

$$(\Lambda_1 =):\ \Lambda_{159} = \phi\sigma_{159} = 10^{14} \times 25.5 \times 10^{-24} = 2.55 \times 10^{-9}\ \mathrm{s}^{-1}$$

$$(\Lambda_2 =):\ \Lambda_{160} = \phi\sigma_{160} + \lambda_{160} = 10^{14} \times 25 \times 10^{-24} + \frac{0.693}{72.1 \times 24 \times 3600} = 1.64 \times 10^{-7}\ \mathrm{s}^{-1}$$

$$(\Lambda_3 =):\ \Lambda_{161} = \lambda_{161} = \frac{0.693}{6.9 \times 24 \times 3600} = 1.162 \times 10^{-6}\ \mathrm{s}^{-1}$$

$$(\Lambda_3^* =):\ \Lambda_{160}^* = \phi\sigma_{160} = 10^{14} \times 525 \times 10^{-24} = 5.25 \times 10^{-8}\ \mathrm{s}^{-1}$$

$$(N_1^0):\ N_{159}^0 = 2\ \frac{6.02 \times 10^{23}}{366} \times 10^{-3} = 3.28 \times 10^{18}\ \text{atoms}$$

$$(t =):\ t = t_{\mathrm{irr}} = 30 \times 24 \times 3600 = 2.59 \times 10^{6}\ \mathrm{s}^{-1}$$

Note that Λ_{159}^* is the rate of formation of ^{160}Tb, and Λ_{160}^* the rate of formation of ^{161}Tb, while Λ_{159}, Λ_{160}, and Λ_{161} are the rates of disappearance of ^{159}Tb, ^{160}Tb, and ^{161}Tb. Introducing the experimental values yields $N_{161} = 5.68 \times 10^{14}$ atoms. The radioactive decay is $A_{161} = \lambda_{161} N_{161} = 1.162 \times 10^{-6} \times 5.68 \times 10^{14} = 6.60 \times 10^{8}$ dps or 4×10^{10} dpm.

10.4. Target considerations*

The success of an irradiation experiment depends to a large extent on the target considerations. In many nuclear experiments as much time has to be devoted to the target preparation as to the rest of the experiment. If the purpose is to produce a radionuclide for a simple tracer experiment, the consideration in this section may be sufficient. If, however, the requirements are a product of extreme purity, very high specific activity, or very short half-life, special techniques must be used, which are discussed in §§10.5–10.7.

10.4.1. *Physical properties*

In thin targets the flux and energy of the projectile do not change during passage through the target. When the projectile energy is considerably changed (in a thick target) there is no simple way to calculate the yield.

The exact specification of when a target is thin depends on the experiment; in most cases thin means a target with a surface weight of not more than a few mg/cm^2 or a thickness of a few μm or less. The *surface weight* (or *linear density*) is $x\rho$, where x is the target thickness and ρ its density. Thin targets are usually used in accelerator irradiations and are made of a solid material, either metal or compound. In order to have sufficient mechanical strength the target is usually supported on a *backing material* such as aluminum. The backing material may be part of a solid target holder which has provisions for cooling to dissipate the heat developed in the bombardment. Thin targets are fixed on the backing material through electrolytic precipitation, vacuum deposition, mass spectrometric collection (for isotopically pure targets), or a number of other techniques.

In reactor irradiations it is more common to use thicker targets in order to obtain higher yields. In traversing the target the effective flux (i.e. $\phi\sigma$) does not decrease as rapidly for neutrons as for charged particles and, therefore, thicker targets can still be considered "thin" in neutron irradiations. However, if the product σN is large, the neutron flux may decrease appreciably at the center of the target, although for thermal neutrons there is no energy change in such a situation.

The target material can be a solid, liquid, or gas. For reactor irradiations it is common to place the target material in a container of polypropylene or of some metal of relatively low neutron capture cross-section. Pyrex and sodaglass are not good materials because of the reactions $^{10}B(n,\alpha)^7Li$ (which has a large σ_{capt}) and $^{23}Na(n,\gamma)^{24}Na$ (which yields a long-lived γ-product). In silica some γ-active ^{31}Si ($t_{1/2} = 2.6$ h) is produced ($\sigma_{capt} = 0.11$ b for thermal n). When metal sample holders, usually aluminum and magnesium are used, some γ-activity is produced; however, the cross-sections for the formation of the products ^{28}Al and ^{27}Mg are small, and these nuclides decay rapidly ($t_{1/2} = 2.3$ and 9.5 min, respectively). The primary requirement for the container is that it should be leak-tight and dissipate heat energy at a rate sufficient to avoid melting.

Sometimes of greater importance than whether the target is thin or thick is the extent of *burn-out* of the target by the irradiation. If the number of target atoms at the beginning of the irradiation is N^0_{target} and at time t_{irr} is N_{target}, then the burn-out can be estimated with the relation

$$N_{target} = N^0_{target}(1 - \sigma_{tot}\phi t_{irr}) \qquad (10.21)$$

where σ_{tot} includes all processes whereby the target atoms are reduced. When $\sigma_{tot}\phi t_{irr} \ll 1$, burn-out can be neglected.

In the example of the irradiation of the 0.3 mm thick gold foil in §9.1, from (9.2) we can calculate that the flux reduction is 17%; for a 0.01 mm thick foil it would be 0.62%, but for a 1 mm foil, 44%. The target thickness does not enter into the burn-out, eqn. (10.21). For a 10 min irradiation of 10^{13} n cm^{-2} s^{-1}, the burn-out is negligible. However, a one-month irradiation at the same flux yields 0.26% burn-out, while a month at a flux of 10^{15} gives about 20% burn-out.

10.4.2. *Chemical properties*

There are two ways to produce a pure radionuclide not contaminated with any other radioactivity. An extremely pure target can be used with a reaction path which is unique. Alternatively, the radioactive products can be purified after the end of the bombardment. For example, a 10 g sample of zinc irradiated for one week with 10^{13} n cm^{-2} s^{-1} yields a sample of ^{65}Zn ($t_{1/2}$ 244 d) with 7.1×10^9 dps. If, however, the zinc target is contaminated with 0.1% of copper, in addition to the zinc activity, 3.0×10^9 dps of ^{64}Cu ($t_{1/2}$ 12.7 h) is formed. In another example element $_{102}No$ was believed to be discovered initially in a bombardment of a target of curium by carbon ions. The observed activity, however, was later found to be due to products formed due to the small amount of lead impurity in the target. Similarly, neutron activation of samarium must be very free of europium contamination because of the larger europium reaction cross-sections. Handbooks of activation analysis often contain information on the formation of interfering activities from impurities.

Even a chemically pure target may yield products of several elements, particularly in cyclotron irradiation, where many reaction paths are often possible. In the bombardment of magnesium with deuterons, the following reactions occur:

$$^{24}Mg\,(79\%)(d,\alpha)\,^{22}Na(t_{1/2}\ 2.6\ y)$$

$$^{25}Mg\,(10\%)(d, n)\ ^{26}Al\ (t_{1/2}\ 7 \times 10^5\,y)$$

$$^{26}Mg\,(11\%)(d, \alpha)\,^{24}Na\ (t_{1/2}\ 15\,h)$$

A short time after the irradiation ^{24}Na is the predominant activity; after longer cooling times the ^{22}Na is always contaminated by ^{26}Al. In such a case chemical purification after the irradiation is required to yield a pure product.

A suitable target material may be unavailable, or very expensive because of a low natural abundance. ^{45}Ca, which is the most useful calcium radioisotope ($t_{1/2}$ 163 d), is produced through n, γ capture in ^{44}Ca; unfortunately, ^{44}Ca only occurs to 2.1% in nature. While irradiation of natural calcium may yield a specific activity of 5–10 mCi g^{-1}, isotope enrichment before irradiation to essentially pure ^{44}Ca yields $\lesssim$ 500 mCi g^{-1}.

When the target is a chemical compound its radiation stability becomes important. In general organic compounds are not stable towards either neutron or charged particle bombardment. However, with neutrons a very particular and useful kind of reaction, known as the Szilard–Chalmers reaction (see § 10.6), may take place. Simple inorganic compounds like NaCl are stable, and for some inorganic compounds like Na_2CO_3 the radiation decomposition (to Na_2O and CO_2) are of little importance.

10.5. Product specifications

Commercially available radioactive products (*radiochemicals*) come in a number of chemical forms. In Appendix F we have listed a number of organizations which sell stable or radioactive nuclides in one form or another; their catalogs list the type of radioactive sources and compounds available, purity of the products, maximum and specific activities, radiation decay characteristics and accuracy of standards, labeling positions, etc.

10.5.1. *Radiochemical processing*

It may be a long process between the irradiated target to a pure radiochemical compound. Separation procedures for radiochemical purification are similar whether used for tracer levels in the laboratory or for kilocurie levels in reactor fuel reprocessing. Most chemical techniques can be applied in the processing: precipitation, ion exchange, solvent extraction, electrodeposition, electrophoresis, distillation, etc. The basic purposes are to eliminate radioactive contaminants and to avoid diluting the radionuclide by isotopic stable atoms. The purification may have to be carried out behind lead or concrete shielding. If the radionuclide or radiochemical is to be used in medical and biological work, the sample may have to be sterile. A number of publications about radiochemical separation procedures are given in the literature list, § 10.9. We particularly call attention to the US NAS-NRC monograph series on the radiochemistry of the elements.

10.5.2. *Specific activity*

The specific activity or concentration of a radionuclide, i.e. radioactivity per gram, mole, or volume, is a very important factor in radionuclide work, e.g. in chemical work with substances of low solubility, in microbiology, etc.

Radionuclides of high specific activity are produced either through accelerator irradiation or through secondary reactions in the target (§ 10.6) in a reactor. Maximum specific activity is obtained when the radioactive nuclide is the only isotope of the element. This is not possible to achieve in regular reactor irradiation through (n, γ) capture processes. For example, reactor-produced ^{24}Na may be obtained in specific activities of $\lesssim 5$ Ci g^{-1}, while the specific activity of accelerator-produced ^{24}Na may exceed 10^3 Ci g^{-1}; however, the total activities available are usually the inverse.

A *carrier-free* radioactive sample is usually one in which the radionuclide is not diluted with isotopic atoms. In reactor production of ^{24}Na from target ^{23}Na, each ^{24}Na is diluted with a large number of ^{23}Na atoms. ^{24}Na cannot be made carrier-free in a reactor. If a carrier-free radionuclide has been produced, e.g. through accelerator irradiation, which then must be purified, its concentration is so low that it may not follow the normal chemical rules. A macroscopic amount of carrier, either isotopic or not, may have to be added to carry the radionuclide through the proper chemical purification steps. We discuss this further in Chapter 17.

10.5.3. *Labeling*

A radioactively *labeled* (or "*tagged*") compound is one in which one of the atoms is radioactive. Preparation of labeled compounds may involve lengthy chemical synthesis starting with the radioactive nuclides in elementary form or in a simple compound.

Most syntheses of ^{14}C-labeled organic or biochemical compounds follow conventional methods with appropriate modifications to contain the radioactivity. $^{14}CO_2$ gas liberated by action of an acid on labeled $Ba^{14}CO_3$ is used commonly as a starting point in the syntheses. Three examples can illustrate how compounds can be labeled at different sites of the molecule:

1. Acetic acid-2-^{14}C:

$$Ba^{14}CO_3 \xrightarrow{HCl} C^{14}O_2 \xrightarrow[\text{catalyst}]{H_2} {}^{14}CH_3OH \xrightarrow{PI_3} {}^{14}CH_3I$$

$$\xrightarrow{KCN} {}^{14}CH_3CN \xrightarrow{H_2O} {}^{14}CH_3CO_2H$$

2. Acetic acid-1-^{14}C:

$$Ba^{14}CO_3 \xrightarrow{HCl} {}^{14}CO_2 \xrightarrow{CH_3MgBr} CH_3{}^{14}CO_2H$$

3. Acetic acid-1, 2-^{14}C:

(a) $$Ba^{14}CO_3 \xrightarrow[\text{heat}]{Mg} Ba^{14}C_2 \xrightarrow{H_2O} {}^{14}C_2H_2 \xrightarrow[\text{catalyst}]{H_2O} {}^{14}CH_3{}^{14}CHO \xrightarrow{[O]}$$

$$ {}^{14}CH_3{}^{14}CO_2H$$

(b) $$K^{14}CN \xrightarrow{^{14}CH_3I} {}^{14}CH_3{}^{14}CN \xrightarrow{H_2O} {}^{14}CH_3{}^{14}CO_2H$$

In these examples labeling is said to be *specific* in that only one position in the compound contains radioactive nuclides. Alternatively, labeling may be *general* when several positions are labeled, which often is the case for tritium-labeled compounds (cf. §10.6.2).

Complicated organic compounds, e.g. penicillin, can be labeled through *biosynthesis*. For example, if penicillin is grown in a substrate containing a simple ^{35}S compound, ^{35}S is incorporated into the penicillin mold. Since other ^{35}S-containing products may be formed also, the ^{35}S penicillin must be purified, e.g. through solvent extraction or paper chromatography (§18.8).

Thousands of labeled organic compounds are commercially available and some organizations offer labeling service on request. See Appendix F.

10.5.4. *Radiochemical purity*

An important consideration in the use of radionuclides is their radiochemical purity since, should several radionuclides of different elements be present in a tracer sample used in an experiment, the result could be ambiguous and misleading. In a radiochemically pure sample, all radioactivity comes from a single radioactive element. If the radioactivity comes from a single isotope, the sample may be said to be radioisotopically pure.

The radioisotopic purity of a sample may be ascertained by measuring the half-life. This method obviously can be used only with radionuclides that show sufficient decay during the time of observation for a reliable half-life to be calculated. For radioactivity of longer half-lives the radioisotopic purity can be checked by measurement of the type and energy of the emitted radiations (Chapter 17) or by processing the tracers through one or more chemical steps characteristic of the element of which the radioactivity is presumably an isotope. For example, in the use of ^{90}Sr (half-life 28 y), the radioactivity can be processed through a number of typical reactions characteristic of strontium. If the radioactivity follows the expected chemistry, then the experimenter can be rather certain that it is strontium. This chemical check does not preclude the possibility that there might be several isotopic nuclides in the sample. However, the presence of several isotopes need not be a handicap in tracer studies concerned with the characteristic properties of a particular element.

When a compound labeled with a radioactive nuclide is used to study molecular behavior, it is necessary to ascertain the *radiochemical purity of the compound* and not just the purity of the radioactive isotope. This means it is necessary to be certain that the radioactivity comes only from a specific element and only from a specific compound incorporating that element. This can be determined through selective chemical operations in which it is frequently desirable that the compound should not be destroyed. Examples of such methods involve preparative gas chromatography, thin-layer chromatography, paper chromatography, dialysis, ion exchange, and solvent extraction. Checking that an element or compound is radiochemically pure is important not only for sources of samples prepared in the laboratory but for commercial samples as well. For example, ^{14}C-tagged organic compounds obtained commercially may have been prepared at varying times prior to purchase. Frequently, the decomposition in the sample due to the radiolysis that accompanies the radioactive decay results in the presence of products that interfere with the course of the experiment for which the labeled compound is to be used.

In some cases the radioactive product after a period of storage will have grown a sufficient amount of radioactive daughter that it is no longer sufficiently pure radiochemically. In these cases, the daughter activity must be separated prior to use. For example, the ^{90}Y normally present in a ^{90}Sr sample must be eliminated if the tracer solution is to be used to study strontium chemistry.

10.6. Recoil separations*

10.6.1. *Target recoil products*

When a projectile of high energy reacts with a target atom, the reaction products are emitted in the direction of the projectile, measured in laboratory coordinates. We use the term *recoil* to describe this. The products, whose kinetic energy can be calculated by the conservation laws of Chapter 7, are slowed and eventually stopped through collision with other atoms. Consequently, if the target is relatively thin, there is a certain probability that the recoil range of the product exceeds the thickness of the target and the target atoms can therefore escape the target. In such cases the recoiling products can be caught on special foils behind the target (catcher foils). These catcher foils can be stationary or they can be a moving band of metal or plastic, or even a gas, which sweeps past the target material.

The recoil technique was used in the discovery of the first transuranium element, neptunium, by E. McMillan and P. H. Abelson in 1939. A uranium salt, $(NH_4)_2U_2O_7$, was fixed on a paper and surrounded by more layers of thin paper to form a "sandwich" which was irradiated by thermal neutrons. At the time of fission the fission products separated with high kinetic energy and were caught in the layers of paper. From the number of layers of paper penetrated by individual fission products, their energy and mass could be calculated. A strong β^- activity was found in the first layers of paper, showing that it had relatively little recoil energy. From chemical experiments it was concluded that this activity was not a fission product but was due to ^{239}U which decayed into ^{239}Np:

$$^{238}U(n,\gamma)^{239}U \xrightarrow{\beta^-} {}^{239}Np$$

Such recoil techniques have been used extensively in the synthesis of higher transuranium elements. As an example, consider the formation of element 103, lawrencium (Fig. 10.4). The $_{103}Lr$ nuclides, which were formed in the reaction in the target between ^{252}Cf and projectiles of ^{11}B, recoiled out of the thin target and were slowed down in

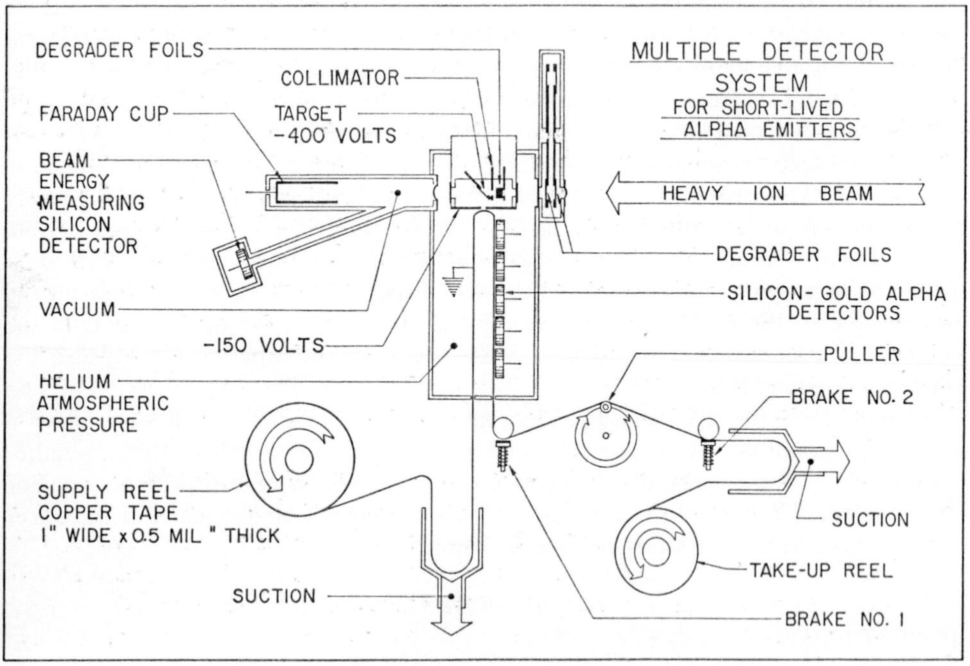

FIG. 10.4. Multiple detector system used in the production of short-lived α-decaying heavy actinides. (According to Lawrence Berkeley Laboratory.)

the helium gas by atomic collisions. The low energy recoil species existed as positive ions which were attracted to the moving, negatively charged, metal-coated plastic band. This band transported the lawrencium atoms on its surface to a number of α-particle detectors where the α-decay of the lawrencium was measured. From the velocity of movement of the plastic band and the measurement of the radioactivity by energy sensitive nuclear detectors, the half-lives and α-energies of the product nuclei could be determined. Comparison of these data with predictions from theory allowed assignment of the mass number of the nuclide.

Target recoil products may have a detrimental effect also. In reactor irradiation of biological samples to determine their trace metal concentration it was found that n, γ products formed in the material of the sample container recoiled out of the container wall into the sample, contaminating it. Such contamination can be eliminated by using a container which forms a minimal amount of such products.

10.6.2. *Hot atom reactions*

The momentum imparted to a nucleus in a nuclear reaction with a charged particle or a fast neutron almost invariably is sufficient to result in the rupture of chemical bonds holding the atom in the molecule. The same is true for the recoil energy imparted to a nucleus by the emission of an α-particle as well as in β-decay, γ-decay, etc.; the recoil energies in radioactive decay are discussed in §§4.2–4.4.

These recoiling atoms are known as *hot atoms* since their kinetic energy is much in excess of thermal equilibrium values. A hot atom may move as much as several hundred atomic diameters after the rupture of the bond before being stopped even though this takes only about 10^{-10} s. Initially the hot atom is highly ionized (e.g. for $Z = 50$ ionic charges up to $+20$ have been observed). As the hot atom is slowed down to thermal energies, it collides with a number of other particles in its path producing radicals, ions, and excited molecules and atoms. At thermal energies it may become a neutral atom (particularly if the matrix material is metallic), an ion, possibly in a different oxidation state (a common situation in inorganic material), or it may react to form a compound with molecules in its path (the prevalent situation in matrices of covalent material).

Chemical effects of nuclear transformations (*hot atom chemistry*) have been extensively studied in connection with induced nuclear transformations, both in the gas phase, in solution and in the solid state. In the latter cases the dissipation of the kinetic energy and neutralization of the charge within a small volume produces a high concentration of radicals, ions, and excited molecules in the region where the recoiling (frequently radioactive) atom is slowed down to energies where it can enter into stable combination. Usually the product molecule is neither very specifically nor completely randomly labeled. We shall discuss only one aspect of hot atom reactions; the topic is treated mainly in Chapter 15.

Hot atom reactions can be used for synthesis of labeled compounds. So far this technique has been used primarily for ^{3}H and ^{14}C labeling. For example, an organic gaseous compound is mixed with ^{3}He and the sample is irradiated with neutrons. The reaction $^{3}He(n, p)^{3}H^{\rightarrow}$ produces tritium as hot atoms (indicated by the small arrow) with kinetic energies of approximately 0.2 MeV. These hot atoms react with the organic molecules to produce labeled compounds. An example is the formation of labeled butyl alcohol

$$C_4H_9OH + {}^{3}H^{\rightarrow} \rightarrow C_4H_9O^{3}H + {}^{1}H^{\rightarrow}$$

The reaction is referred to as a *hot atom displacement* reaction. In an alternate method the solid organic compound may be mixed with Li_2CO_3, and this solid mixture irradiated with neutrons to cause the reaction: $^6Li(n, \alpha)^3H^{\rightarrow}$. Again the hot tritium atom combines with the organic species to cause labeling. In a third and simpler method, the organic compound is simply mixed with tritium gas. The β-radiation from tritium decay causes excitation and decomposition which leads to capture by the organic radicals of some of the remaining tritium. These reactions are not selective because reactions with decomposition products also occur. Therefore the tritium labeled product must be purified. Gas phase labeling usually results in incorporation of tritium in several sites of the molecule, e.g. toluene has about 9% of the total label in the CH_3 group, 54% in the *ortho* sites of the ring, 24% in the *meta* sites, and the remaining 13% at the *para* site.

^{14}C-labeling can be achieved in a similar manner by using the $^{14}N(n, p)\,^{14}C^{\rightarrow}$ reaction. The recoiling ^{14}C energy is about 0.6 MeV. A suitable labeling mixture is NH_4NO_3 + the organic compound; cf. ref. A. P. Wolf (1960).

10.6.3. *The Szilard–Chalmers process*

In 1934 L. Szilard and T. A. Chalmers discovered that bond breaking could occur for atoms following nuclear reaction or radioactive decay even though the recoil energy in the initial process is not sufficient to overcome the bonding energy. In the case of thermal neutron capture the processes involved in the emission of the γ-ray, which removes the nuclear excitation energy, impart recoil energy to the atom to break most chemical bonds. If, after rupture of the bonds, the product atoms exist in a chemical state different and separable from that of the target atoms, the former may be isolated from the large mass of inactive target. This provides a means of obtaining high specific activities in reactions where target and product are isotopic.

This process is known as the *Szilard–Chalmers reaction* and was discovered when, following the irradiation of ethyl iodide with thermal neutrons, it was found that radioactive iodide could be extracted from the ethyl iodide with water. Moreover, when iodide carrier and silver ions were added to this aqueous phase, the radioactive iodide precipitated as silver iodide. The obvious interpretation of these results is that the neutron irradiation of the ethyl iodide, which caused the formation of ^{128}I, ruptured the bonding of this atom to the ethyl group. The bond energy of iodine to carbon in C_2H_5I is about 2 eV. Since this exceeds the recoil energies of neutron capture, the bond breakage must have resulted from the γ-emission which followed neutron capture and not the capture process itself. The reaction can be written:

$$C_2H_5\,^{127}I + n \rightarrow [C_2H_5\,^{128}I]^* \rightarrow C_2H_5 + {}^{128}I^{\rightarrow} + \gamma$$

The $^{128}I^{\rightarrow}$ loses its kinetic energy and is stabilized as an iodine atom or iodide ion; it can be recaptured by the C_2H_5 radical (*retention* of activity in C_2H_5I). In addition to the necessity that the recoiling species have sufficient energy to rupture the bond, it is also necessary for a successful enrichment of specific activity by the Szilard–Chalmers process that there is no rapid exchange at thermal energies between the active and inactive iodine atoms in ethyl iodide:

$$C_2H_5\,^{127}I + {}^{128}I \overset{\text{slow}}{\rightleftharpoons} C_2H_5\,^{128}I + {}^{127}I$$

After the rupture of the bond, the hot atoms may also replace iodine or even hydrogen atoms from ethyl iodide molecules, similar to the thermal exchange:

$$^{128}I^{\rightarrow} + C_2H_5{}^{127}I \longrightarrow C_2H_5{}^{128}I + {}^{127}I$$

Again, the result is retention of the activity within the organic molecule. Retention is decreased by diluting the ethyl iodide with alcohol which reduces the probability of collision and exchange with C_2H_5I molecules as the hot atoms of $^{128}I^{\rightarrow}$ are being slowed.

Isomeric transitions which proceed by emission of γ-rays may not provide sufficient recoil energy to break covalent bonds. However, in these cases of very low energy isomeric transitions, the extent of internal conversion is large. This results in vacancies in the lower electron orbitals (§4.7). When electrons in higher orbitals move to fill the vacancies, the difference in electron-binding energies is sufficient to cause some ionization, resulting in relatively high charge states for the atom, leading to bond rupture. An example of such a process is seen in the reaction whereby an alkyl bromide is irradiated with thermal neutrons. During neutron irradiation the reaction is

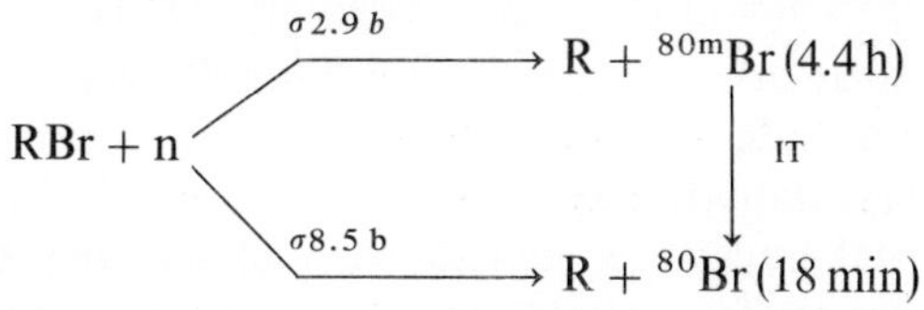

Extraction by water of this bromide produces an aqueous sample containing both ^{80}Br and ^{80m}Br. Some ^{80}Br and ^{80m}Br are retained as alkyl bromide in the organic phase. If, after a period of an hour, a new extraction is carried out on the organic phase, only the ^{80}Br is found in the aqueous sample. This ^{80}Br is the result of the reaction

$$R\,^{80m}Br \xrightarrow{IT} R + {}^{80}Br + \gamma$$

In addition to the organic halides a number of inorganic systems undergo similar reactions. From neutron irradiation of acid or neutral solutions of permanganate most of the ^{56}Mn activity is removable as MnO_2. From solid or dissolved chlorates, bromates, iodates, perchlorates, and periodates, active halide samples may be isolated in relatively high yields. Systems with Te, Se, As, Cu, and several Group VIII elements have been shown to be applicable to the Szilard–Chalmers process. The great advantage in all these cases is that a relatively small amount of activity may be isolated from a large mass of isotopic inactive target. Because of retention there is also some labeling of the target molecules.

10.7. Fast radiochemical separations*

Of the approximately 2000 known nuclides, about one-third have half-lives $\lesssim$ 10 min. The chemical identification and the assignment of the mass number of the short-lived nuclides, some with half-lives $\lesssim$ 0.1 s, require special techniques, especially when the irradiation yields a mixture of different nuclides.

These short-lived nuclides are either located on isobar lines far away from the bottom of the stability valley or at the transuranium end of the valley. In the efforts to extend our knowledge about the chemical elements as far as possible, Z-values of $\sim$ 106 have been reached. However, with increasing Z the half-lives become very short, such that the longest-lived isotope of $Z = 105$ has $t_{1/2} \approx 40$ s. Short-lived nuclides are of interest

TABLE 10.1. *Production methods for radio-isotopes of the 13 lightest elements*

Element	Production[(a)]	Half-life
H	$^{6}Li(n,\alpha)^{3}T$	12.3 y
He	$^{9}Be(n,\alpha)^{6}He$	0.80 s
Li	$^{7}Li(d,p)^{8}Li$	0.84 s
Be	$^{6}Li(d,n)^{7}Be$	53.4 d
B	$^{11}B(d,p)^{12}B$	0.02 s
C	$^{14}N(n,p)^{14}C$	5736 y
N	$^{12}C(d,n)^{13}N$	9.7 min
O	$^{14}N(d,n)^{15}O$	2.03 min
F	$^{19}F(n,2n)^{18}F$	109.7 min
Ne	$^{22}Ne(n,\gamma)^{23}Ne$	38 s
Na	$^{23}Na(n,\gamma)^{24}Na$	15.0 h
Mg	$^{26}Mg(n,\gamma)^{27}Mg$	9.5 min
Al	$^{27}Al(n,\gamma)^{28}Al$	2.3 min

[(a)]Many of the products may be obtained through alternative reactions.

because knowledge of their decay properties substantially adds to our picture of the stability surface for the nuclides. These nuclides also give valuable information about nuclear vibrational and rotational states.

Beside the fundamental interest in short-lived nuclides, they may be of practical value. For example, there are few suitable long-lived isotopes of the lightest elements (Table 10.1). By use of "on-line" techniques which use a scheme of automated, continuous irradiation–separation–experimentation–measurement systems, radiochemical experi-

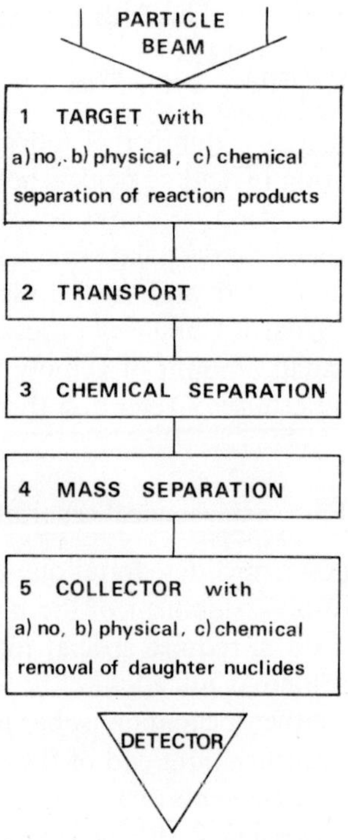

FIG. 10.5. Sequence of steps in production and study of very short-lived nuclides.

ments can be conducted on a time scale as short as tens of seconds. Not only does this allow tracer studies with very short-lived nuclides but it offers the additional advantage that the radioactivity is all gone in a very short time, thereby eliminating concern over contamination.

A primary requirement for work with short-lived nuclides (i.e. nuclides with $t_{1/2} \lesssim$ a few min) is that the site of production and the site of experimentation be close together. With half-lives of a few seconds it is still possible to use rapid manual procedures consisting of (a) target dissolution, followed by (b) chemical isolation of the radionuclide through solvent–extraction–stripping or high pressure/temperature ion exchange, and/or (c) precipitation or electroplating before (d) measurement. In this scheme the eventual chemical studies or other uses must be included, usually between steps (b) and (c). For very short-lived nuclides, all steps must be automated.

Figure 10.5 shows a block diagram of the possible steps in these fast experiments. The transport time must be short. Although some fast cars and runners were useful in earlier days, transportation now is by a pneumatic system for solid targets or by a liquid or gasflow line. With such systems transportation rates may be up to 100 m s^{-1}.

There are a number of nonchemical techniques involving time-of-flight and magnetic mass analysis of target recoil products, energy sensitive detectors, etc., with which half-lives down to 10^{-12} s can be measured. However, of more interest to us are the fast chemical systems. Let us review a few representative types.

(i) *Rapid off-line chemistry*. Figure 10.6 shows a set-up to recover fission product

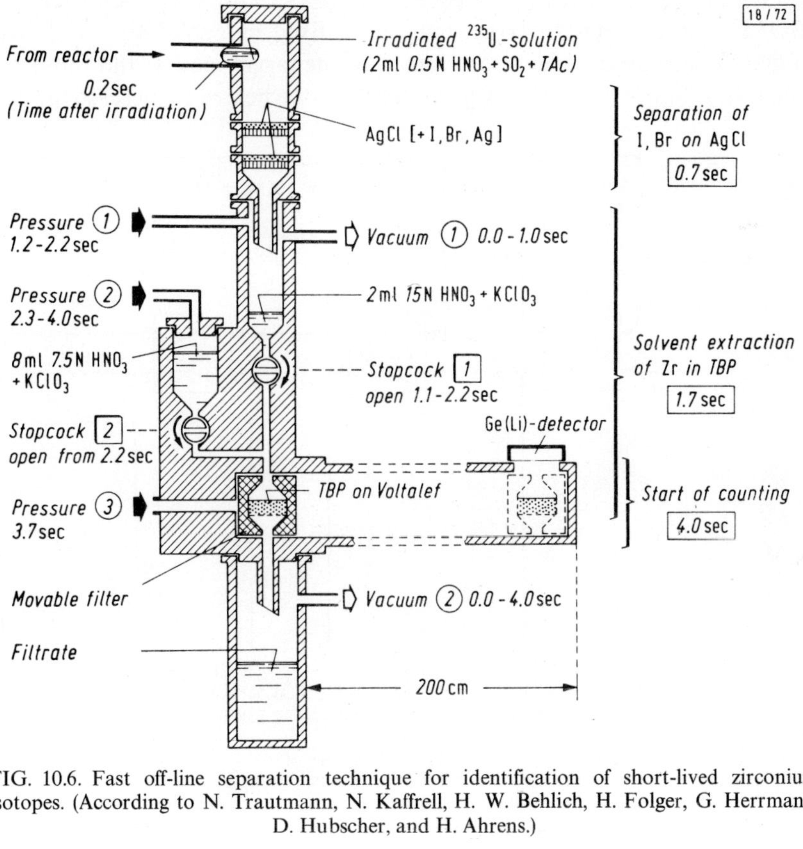

FIG. 10.6. Fast off-line separation technique for identification of short-lived zirconium isotopes. (According to N. Trautmann, N. Kaffrell, H. W. Behlich, H. Folger, G. Herrmann, D. Hubscher, and H. Ahrens.)

zirconium isotopes. (a) A solution of ^{235}U sealed in a plastic capsule and irradiated with thermal neutrons is projected pneumatically into the apparatus where the capsule gets smashed. (b) The solution is filtered through two layers of preformed, inactive silver chloride, which, by heterogeneous exchange, rapidly takes up the fission products I and Br. (c) The next step involves solvent extraction, where the organic phase containing the reagent tributyl phosphate (TBP) has been adsorbed on the surface of small plastic grains. The filtrate from (b) is mixed with strong $HNO_3 + KBrO_3$ and drawn through the stationary organic phase (marked "movable filter" in the figure), to remove zirconium. (d) After washing the organic phase with nitric acid, taken from the reservoir at left, the filter layer containing the zirconium is transferred pneumatically to the detector at the right. (e) Counting is started 4 s after the end of irradiation. The figure shows the time sequences for the operations in numbers. Stopcocks are operated pneumatically and valves magnetically by signals from electronic timers. The chemical yield of zirconium amounts to about 25%. This experiment led to the identification of ^{99}Zr with a half-life of 1.8 s.

In this system proper modification of conventional techniques provided very fast separations. Precipitation (in the preformed precipitate through heterogeneous exchange), solvent extraction (in the reversed phase chromatography version), and ion exchange (in the heated pressurized mode) have all been used in identification and study of short-lived transuranides (elements 102 and higher).

(ii) A rather different technique is illustrated by the ISOLDE collaboration (*Isotope Separation On Line DEvelopment*) at CERN, and OSIRIS project at Studsvik, Sweden. In the ISOLDE experiment the target material is fixed in a solid but still fairly open chemical matrix. The products formed react with the target matrix, and those with the "right" chemistry diffuse out of the target (which may be heated to increase the diffusion

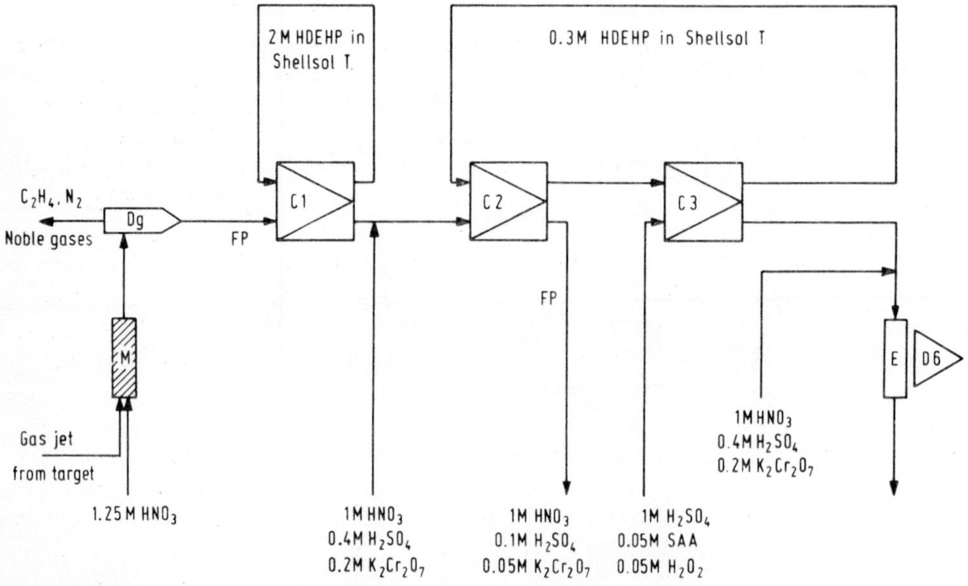

FIG. 10.7. On-line technique using liquid flow centrifugal separators (symbolized by C) and multiple detectors (D) for identification of short-lived rare earth isotopes. (According to N. Trautmann, P. O. Aronsson, T. Bjørnstad, N. Kaffrell, E. Kvale, M. Skarestad, G. Skarnemark, and E. Stender.)

FIG. 10.8. A SISAK battery of four centrifugal separators with static mixer and flow controls (above) and electric drives (below the centrifuges). The height of the rack is about 0.5 m. The maximum flow rate through each centrifuge is about $150\,l\,h^{-1}$ at 20 000 rpm; the hold-up time for absolute phase separation is then 0.3 s.

rate) and are swept away by a gas stream into the ionization chamber of a mass spectrometer where the products are separated according to their q/m ratio (2.10). In the OSIRIS project additional separation is carried out with a *thermochromatographic* column placed after the mass separator. The gas stream passes through a tube or column containing a partly selective adsorbent (silica gel, KCl, $BaCl_2$, K_2CrO_7, etc.). The temperature gradient along the tube leads to deposition of elements (or compounds) of different vapor pressure at different positions. It has been shown that fairly selective deposition and separation can be achieved, although the thermochromatographic technique is not absolutely element specific. The technique yields isobaric nuclides within a certain Z-range for decay measurements. Half-lives down to 0.1 s have been determined. It has been used for identification of spallation products, fission products, and transuranium elements. In the latter case, G. N. Flerov, I. Zvara *et al.* in Dubna, USSR, did not use mass separation; the mass was estimated from decay systematics and decay energy measured with solid state track detectors.

(iii) In the SISAK (Short-lived Isotopes Studied by the AKufve technique) collaboration, a target containing an element (e.g. uranium) on a thin foil is bombarded with projectiles, while a gas stream behind the target continually sweeps away the products recoiling out into the gas stream (cf. Fig. 10.4). The "gas jet", which contains catcher molecules (e.g. C_2H_4), is thoroughly mixed with an aqueous solution (M, Fig. 10.7), which afterwards is degassed (Dg). The aqueous solution containing the radioactive products is passed through a series of fast and selective solvent extraction loops, where continuous flow mixers and centrifugal separators are used (the "AKUFVE-technique") (Fig. 10.8). The solution from the final separation passes through an absorbent E which retains the element of interest. The adsorbent is measured by a detector D. The procedure is highly specific and takes about 3 s from target to final detector. Because it is an on-line technique, very detailed decay schemes can be measured for nuclides with half-lives down to < 1 s.

10.8. Exercises

10.1. ^{24}Na is produced through the reaction ^{26}Mg(d, α)^{24}Na. A 0.2 mm thick magnesium foil is irradiated for 1 h by a current of 130 μA of 22 MeV D^+ ions in a cyclotron. The foil has a much larger area than the cross-section of the beam. What is the specific activity of ^{24}Na if the magnesium foil (3 cm^2) contains 0.003% Na and σ for the reaction is assumed to be 25 mb?

10.2. Oxygen can be determined through the reaction $^{16}O(n, p)^{16}N \xrightarrow[7s]{\beta,\gamma}$; σ_{np} at 14 MeV n is 49 mb. 3.982 g of a fatty acid were irradiated for 20 s in 4×10^8 n cm^{-2} s^{-1}. After the irradiation the sample was rapidly transferred with a rabbit system to a scintillation detector which had an efficiency of 1.1% for the ^{16}N γ-rays (~ 6 MeV). Exactly 8 s after the end of the irradiation, the sample was counted for 1 min, yielding 13.418 counts above background. What was the oxygen fraction of the sample?

10.3. ^{199}Au can be formed through two successive n, γ-reactions on ^{197}Au (100% in nature). If 1 g ^{197}Au is irradiated with 10^{14} n cm^{-2} s^{-1} in 30 h, what will the disintegration rate of ^{199}Au be at the end of the irradiation? The chain of events to be considered is

$$^{197}Au \xrightarrow[99\ b]{} {}^{198}Au \xrightarrow[26\,000\ b]{} {}^{199}Au.$$

with ^{198}Au decaying with $t_{1/2}$ 2.7 d and ^{199}Au decaying with $t_{1/2}$ 3.15 d.

10.4. ^{246}Cm has a half-life of 4000 y. It can be obtained through neutron capture in ^{245}Cm, which has a half-life of 11 000 y; the reaction cross-section is 200 b. Both isotopes are also fissioned by thermal neutrons, σ_{245} 1800 b and σ_{246} 15 b. Because one does not want to lose too much ^{245}Cm, the irradiation is timed to give a maximum yield of ^{246}Cm. If the neutron flux is 2×10^{14} n cm^{-2} s^{-1}, when does the ^{246}Cm concentration reach its maximum?

10.5. A sample containing 50 g ethyl iodide was irradiated with neutrons from a RaBe source of 1×10^5 n cm^{-2} s^{-1} for 2 h. If there is a 20% retention and separation occurs 5 min after end of irradiation, what will be the activity of the AgI sample 10 min after separation? The detector has an 8% efficiency for the emitted γ-radiation. Assume the reaction cross-section for $^{127}I(n,\gamma)^{128}I \xrightarrow[25\ min]{\beta}$ to be 6.2 b.

10.6. ^{59}Fe of high specific activity can be produced through thermal neutron irradiation of a solution of 1 ml 0.1 M potassium hexacyanoferrate. The recoiling (free) iron atoms, which are produced with a 40% yield, are quantitatively extracted into an organic solvent. How long must the irradiation be in a reactor of 3×10^{14} n cm^{-2} s^{-1} to obtain 1 mCi ^{59}Fe? The reaction cross-section for $^{58}Fe(n, \gamma)^{59}Fe \xrightarrow[45\ d]{\beta,\gamma}$ is 1.1 b.

10.9. Literature

H. BATEMAN, The solution of a system of differential equations occurring in the theory of radio-active transformations, *Proc. Cambridge Phil. Soc.* **15** (1910) 423.

W. RUBINSON, The equations of radioactive transformations in a neutron flux, *J. Chem. Phys.* **17** (1949) 542.

J. PRAWITZ and J. RYDBERG, Composition of products formed by thermal neutron fission of 235U, *Acta Chem. Scand.* **12** (1958) 369, 377.

I. J. GRUVERMAN and P. KRUGER, Cyclotron-produced carrier-free radioisotopes: thick target yield data and carrier-free separation procedures, *Int. J. Appl. Rad. Isotopes* **5** (1959) 21–31.

A. P. WOLF, Labeling of organic compounds by recoil methods, *Ann. Rev. Nucl. Sci.* **10** (1960) 259.

IAEA, *Production and Use of Short-lived Radioisotopes from Reactors*, Vienna, 1963.

B. J. WILSON, *The Radiochemical Manual*, UKAEA, The Radiochemical Centre, Amersham, 1966.

J. KLEINBERG (ed.), *Collected Radiochemical Procedures*, USAEC Rept. LA-1721 (3rd edn.), CFST, 1967.

A. K. LAVRUKHINA, T. C. MALYSHEVA, and A. K. PAVLOTSKAYA, *Chemical Analysis of Radioactive Materials*, translated by Scripta Technica Ltd., Cleveland, Ohio, CRC Press, 1967 (International Scientific Series).

S. AMIEL, Modern rapid radiochemical separation, in *Nuclear Chemistry* (L. YAFFE ed.), Vol. II, p. 251, Academic Press, 1968.

I. ZVARA, YU. T. CHUBURKOV, R. CALETKA, and M. R. SHALAEVSKII, *Radiokhimiya* **11** (1969) 163 (= *Sov. Radiochem.* **11** (1969) 161).

G. HERRMANN and H. O. DENSCHLAG, Rapid chemical separations, *Ann. Rev. Nucl. Sci.* **19** (1969) 1.

NAS-NS, *Source Material for Radiochemistry*, No. 42, 1970.

S. BORG, B. RYDBERG, L. E. DE GEER, G. RUDSTAM, B. GRAPENGIESSER, E. LUND, and L. WESTGAARD, On-line separation of isotopes at the reactor in Studsvik (OSIRIS) *Nucl. Instrum. Meth.* **91** (1971) 109.

D. DeSoete, R. Gijbels, and J. Hoste, *Neutron Activation Analysis*, Wiley-Interscience (1972).

P. O. Aronsson, B. E. Johansson, J. Rydberg, G. Skarnemark, J. Alstad, B. Bergersen, E. Kvale, and M. Skarestad, SISAK—a new technique for rapid, continuous (radio) chemical separations, *J. Inorg. Nucl. Chem.* **36** (1974) 2397.

W. Müller and R. Lindner (eds.), *Transplutonium 1975*, North-Holland/Elsevier, 1976.

N. Trautmann, Rapid chemical separations, In *3rd Int. Conf. Nuclei Far from stability*, CERN, 1976.

H. L. Rawn, L. C. Karraz, J. Denimal, E. Kugler, M. Skarestad, S. Sundell, and L. Westgaard, New techniques at ISOLDE-II *Nucl. Instr. Methods*, 1976.

NAS-NS, Monograph Series on the "Radiochemistry of Elements" (covers almost all the elements).

G. Rudstam, The on-line mass separator OSIRIS and the study of short-lived fission products, *Nucl. Instr. Meth.* **139** (1976) 239.

Radiopharmaceuticals and other compounds labeled with short-lived radionuclides, *Int. J. Applied Rad. Isotopes* **28** (1977) Nos. 1/2.

CHAPTER 11

*Thermonuclear Reactions and Nucleogenesis**

Contents

Introduction

In a system of particles at thermal equilibrium the temperature is defined by the average kinetic energy, or velocity, of the particles, as described in Appendix B, § B.2. A star is an example of a system of particles enclosed in a gravitational vessel in a fashion similar to a solution of compounds in a glass beaker. The velocities of particles in the star are high enough to cause nuclear reactions and transformation of elements, just as the energy of the molecules in a glass vessel may be high enough to cause chemical reactions leading to new compounds. In both examples, energy is lost from the system through electromagnetic radiation (visible light, heat, etc.). The necessity of overcoming the chemical activation energy also has an analogy in thermonuclear reactions, where the reaction cross-sections are highly energy dependent. Even though the *average energy* may be insufficient to cause a reaction, the *Boltzmann* high energy "tail" (see Fig. App. B.1) may contribute a sufficient number of high energy particles to cause an observable reaction rate. Thus the chemical reactions at earthly temperatures have a close parallel in thermonuclear reactions at ultra high temperatures of millions of degrees or more.

The fusion of light nuclei into heavy ones releases large amounts of binding energy and is the main stellar energy source. So far we have been able to reproduce such systems only in thermonuclear weapon explosions, but considerable research is directed to discover ways to harness the fusion energy in a more controlled manner. Development of controlled nuclear fusion will require the solution of a number of nuclear and chemical problems. In this chapter we discuss these problems.

11.1. Energy production in the sun

Our sun is located in the Milky Way galaxy about 30 000 light-years away from the center. The Milky Way is a lens-shaped spiral galaxy with an estimated diameter of about 100 000 light-years. It has about 10^{11} stars, most of them larger than our sun. In space there are some 10^9 galaxies similar to ours.

The sun, which contains about 99.8% of the mass of the solar system (sun + planets),† consists of approximately 73% hydrogen, 25% helium, and 2% carbon, nitrogen, oxygen,

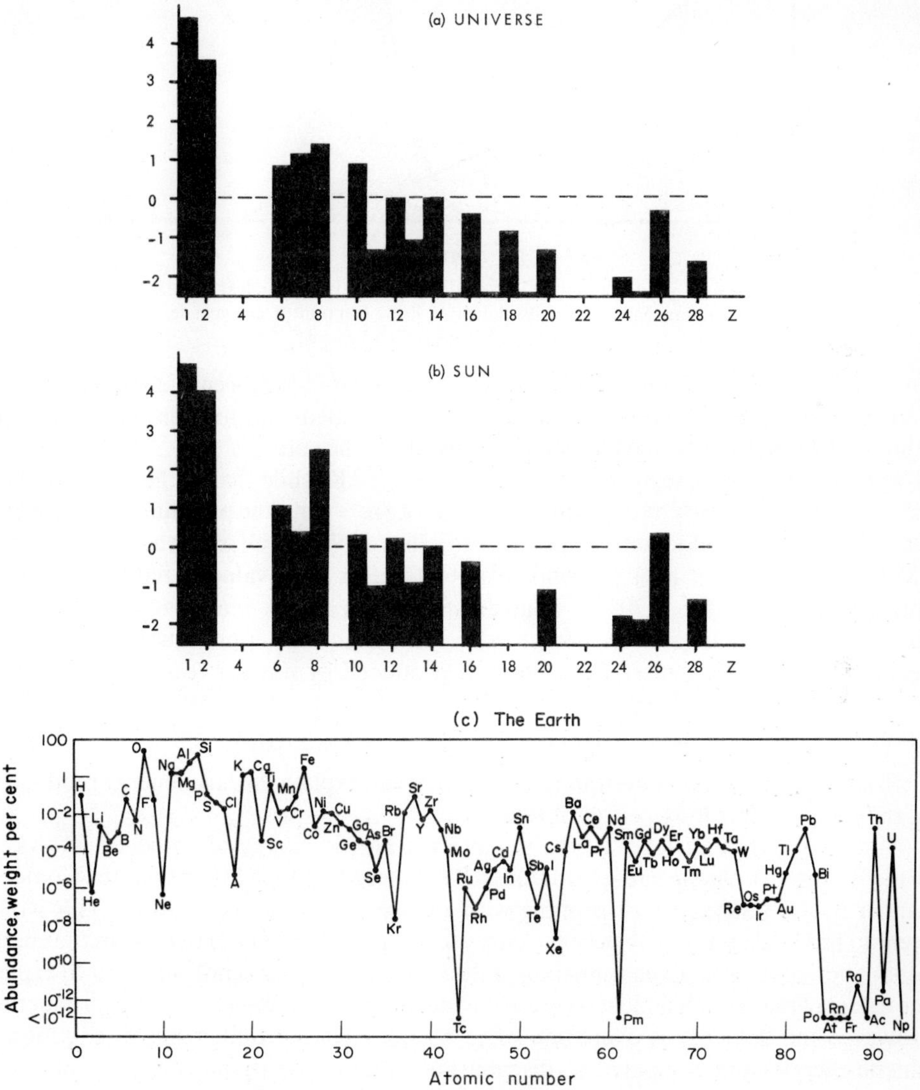

FIG. 11.1. The most abundant elements in (a) the universe, (b) the sun, and (c) the earth. For (a) and (b) the amount is logarithmic with regard to silicon (abundancy = 1) and for (c) it is in percentage. All other elements are below the base line. The reliability is not better than a factor of 2. (Figure 11.1c according to B. Mason.)

† The mass and density of the sun are 1.99×10^{30} kg (this mass is usually symbolized by $\mathbf{m}_\odot$) and 1410 kg m^{-3}, the corresponding values for the earth are 5.98×10^{24} kg and 5520 kg m^{-3}.

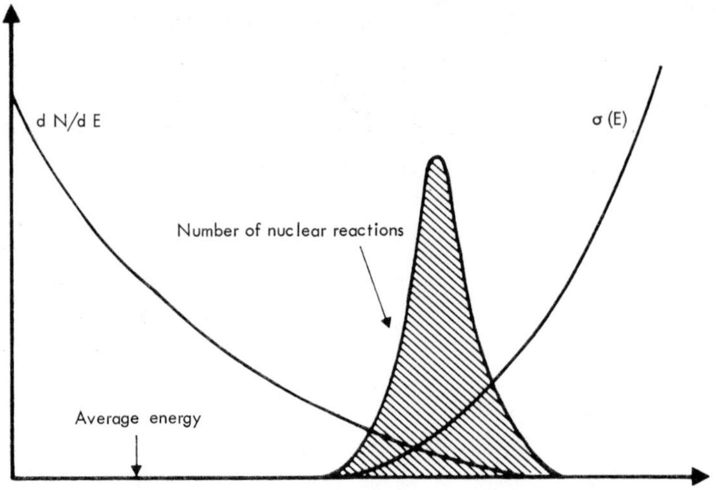

FIG. 11.2. Reaction yield $\sigma(E)\mathrm{d}N/\mathrm{d}E$ (shaded area) for thermonuclear processes. The left curve is the Boltzmann particle distribution; $\mathrm{d}N$ is the fraction of all particles which have a kinetic energy between E and $E + \mathrm{d}E$.

and other elements. In all, approximately 70 elements have been detected in the sun and there is no reason not to believe that all the elements up to uranium are present. Figure 11.1 b shows the most abundant elements in the sun.

The average surface temperature of the sun is 5780 K while that of the center is about 1.5×10^7 K. The density at the core is about 10^5 kg m^{-3}, and the pressure about 2×10^{16} Pa. The energy production of the sun is 3.76×10^{26} J s. Of this the earth receives 1353 J m^{-2}s^{-1} at an area perpendicular to the flux (this value is referred to as the *solar constant*); of this ~ 30% is immediately reflected back into space. It is assumed that this production rate has been relatively constant for approximately five billion years of the sun's existence. This energy is produced by nuclear reactions in the core of the sun, whereby hydrogen is converted into helium

$$4\,^1\mathrm{H} \rightarrow {}^4\mathrm{He} + 2\beta^+ + 2\nu \qquad (Q = 25\ \mathrm{MeV}) \tag{11.1}$$

Such a nuclear fusion process is the only one that can explain the large energy production of the sun over this long period of time.

At a temperature of 1.4×10^7 K, the average kinetic energy of a nuclide is 1.8 keV, while the most probable energy is 1.2 keV. This energy is much lower than the Coulomb barrier to the interaction of protons with each other, which is 1.1 MeV. This would seem to provide a serious objection to the use of the reaction (11.1) to explain the energy production of the sun. Even tunneling of the Coulomb barrier is not sufficient to explain the extent of reaction that must be present to account for the observed energy production. However, the difficulty is overcome when we consider the Boltzmann distribution of kinetic energies and compare this distribution with the variation of the reaction cross-section with energy. Figure 11.2 shows, on a linear scale, the energy distribution of protons at a given average temperature as well as the dependence of the cross-section on energy. As we see, these two curves $\mathrm{d}N/\mathrm{d}E$ and $\sigma(E)$ overlap. A small fraction of the particles do have sufficient energies to cause reactions ($= \sigma(E)\,\mathrm{d}N/\mathrm{d}E$) as shown by the shaded area. Since the nuclear reactions are a consequence of the high average temperature of the reaction particles, the process is called *thermonuclear*.

11.2. Abundance of elements in the universe

It is a most difficult task to attempt to assess the relative abundance of the elements in something as large as our universe. Nevertheless, scientists have assembled what is probably a fairly accurate assessment. In Fig. 11.1a the most abundant elements are given so that they can be compared with the abundance in the sun and the earth. They are also given by the smoothed curve in Fig. 11.3.

Approximately 90% of the total number of atoms in the universe are hydrogen and the remaining 10% are almost all helium. The rest of the elements account for less than 1% of all the atoms in the universe. It is interesting to observe that in most stars (including the sun) the fraction of helium is about twice as high as for the universe. Another interesting observation is that there is a steady decrease in the abundances of the elements from the lightest, hydrogen, to the heaviest ones with the notable exception of the iron group, which seems to be in much higher abundance than would be expected. Any theory for the formation of the elements would seem to start with hydrogen as a basis and proceed to explain how the heavier elements were successively formed from hydrogen. Moreover, it should explain why iron is so abundant and why certain very light elements such as lithium and beryllium occur very rarely.

Before beginning a consideration of the possible schemes for the synthesis of the elements, there are some astrophysical observations that should be noted. The age of the Milky Way is thought to be approximately 15×10^9 y; the values suggested vary

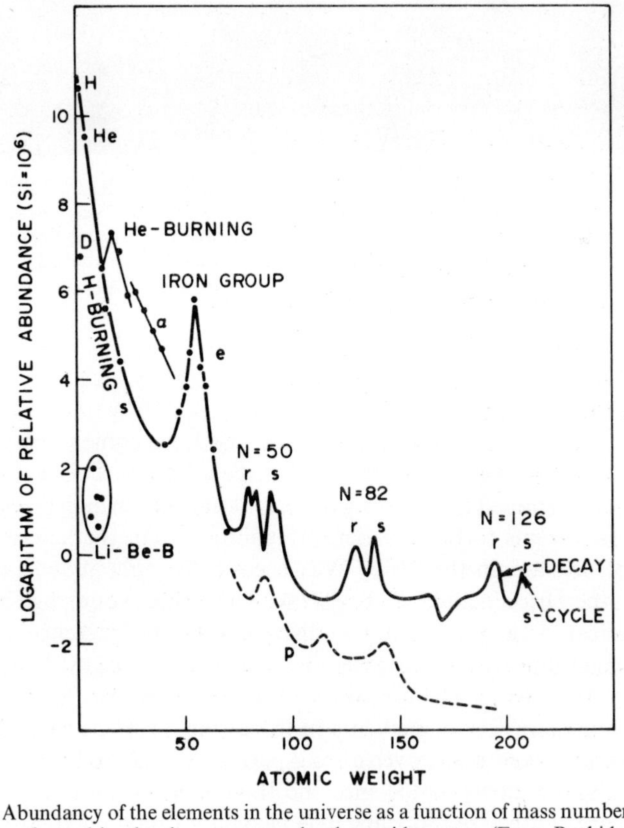

FIG. 11.3. Abundancy of the elements in the universe as a function of mass number; s indicates mass numbers formed by the slow process, r by the rapid process. (From Burbidge, Burbidge, Fowler, and Hoyle.)

FIG. 11.4. The Orion nebula is made up of a gas cloud that glows through excitation by four extremely hot young stars, which are partly hidden behind the cloud. These stars are very young, probably less than a million years. Star formation is probably going on right now in the gas whose density is about 1000 times that of interstellar space (in space about 1 atom cm^{-3}). The gas temperature is about 10 000 K. (Lick Observatory photograph.)

between 12 and 18×10^9. In the following we shall call 10^9 y one *eon*. The oldest rocks found on the moon and the dates of meteorites that have come to earth would seem to establish an age of approximately 4.65 eons for these bodies. This is less than the age of the earth's crust determined from dating of the oldest rocks, which gives a value of about 2.6 eons. Our sun seems to be considerably younger than the age of our galaxy and, in fact, some of the stars in the Milky Way seem to be only about a million years old (e.g. T-Tauri in the Orion nebula; cf. Fig. 11.4). This rather young age has been deduced from astrophysical data and from the discovery of the presence of technetium in the stars. Technetium is an unstable element whose longest-lived isotope has a half-life of only 4.1×10^6 y. The discovery of a variety of ages for the stars as well as for the existence of a variety of elements in these stars leads us to assume that the formation of stars has taken place over a long period of time and is still in process. This concept must also be incorporated into the theory of the formation of the elements, if the elements are synthesized in stars. The elements which make up our solar system seem to have an age of 7–13 eons.

11.3. The big-bang beginning of stellar evolution

Our information about the universe comes from spectral measurements at all wave-lengths. In the optical region the continuous spectrum tells us about the surface temperature of the stars which combined with the luminosity gives information on the size of the star. The study of the spectral lines yields not only the elemental surface composition but also its density and the movement of the star. Study of microwave spectra from space reveals the composition of material in interstellar space; so far about 50 molecules as complex as formic acid and methylamine have been discovered. High-energy cosmic radiation (Chapter 5) and investigation of pulsars, quasars, etc., add further to the cosmological picture of the universe and how it was formed. For our own solar system, the radiochemical analyses of meteorites and the moon provide a valuable source of data for constructing a model of how our planetary system was formed. Also the studies of so-called nucleochronometers, i.e. age determinations based on radioactive decay chains, are an essential source of information for cosmologists.

One theory of the origin of the elements originated with G. Gamov and R. A. Alpher and incorporates proposals of H. Bethe, and is known as the Alpher–Bethe–Gamov or *big-bang theory*. They originally postulated that in the beginning of our universe all matter in it existed as neutrons in some super giant nucleus (the "ylem") which exploded. During the explosion neutrons were converted through β-decay ($t_{1/2}$ 11 min) to protons, and the protons combined with neutrons and other protons to start the process of forming heavier elements. In this fashion all the elements were formed in a matter of less than an hour by a process of neutron capture. Out of this expanding cloud the stars and planets were formed through condensation processes.

The theory has later been modified, so that the ylem may have contained some particles heavier than the neutron, and also—and more important—the radiation density (i.e. number of photons) in the ylem must have been much higher than the particle density, ~ 18 per cm^3. It is believed that this process leads to the formation of nuclides up to ^4He, but not higher. According to this approach about 25% of the expanding cloud was transformed into helium, the rest was hydrogen. It is a great triumph of this theory that astronomical observations fully support this composition. There is evidence for the expansion of the universe from a more or less central point, though this point cannot be localized. Because of the Doppler effect, the light emitted by stars moving away from us is shifted towards larger wave-lengths (red-shift phenomenon). All galaxies are found to be receding from us and from each other (the expansion of the universe), and the red-shift can be used to calculate the velocity of recession. If this velocity is assumed to be constant, the time it would take to move the galaxies back to the starting point is found to be about 13 eons. Corrections for the decrease in expansion velocity of a galaxy due to decreasing gravitational pull from the other galaxies increases the value to 15–20 eons.

Recently I. E. Segal has found that the red-shift observed is proportional not to our distance from the galaxies (Hubble's rule), but to the square of the distance. This would have the effect that the galaxies (especially so-called quasars) are evenly distributed in time and space. A conclusion is that one of the strongest arguments for the big-bang hypothesis disappears, and that instead the universe may have existed for eternity.

The temperature in the ylem has been assumed to be 0.1–1 MeV (~ 2–20×10^9 K). When such a gas expands adiabatically, the temperature must decrease. Thus from the size of the universe Alpher, R. C. Herman, and Gamov calculated that the universe today should be filled with a blackbody radiation of about 5 K. Such a radiation, though with a temperature of 2.7 K, has recently been discovered to be arriving from all directions in space. This is among the strongest evidence for the big-bang hypothesis.

From a consideration of the abundance of the elements the original big-bang theory seemingly encounters some almost insurmountable difficulties. The main one is that, because all nuclides with mass numbers 5 and 8 are extremely short-lived, it would seem that if the process of elemental formation occurs through successive neutron or proton reactions, helium would be the heaviest element that could be formed, and the matter in the universe would consist of only hydrogen and helium. Since this is obviously not true (even if hydrogen and helium do constitute 99%), the big-bang theory must be modified.

11.4. Stellar formation

A more complete explanation of the universe considers that the elements of $Z > 2$ in their present abundance were formed by nucleogenesis in individual stars as they proceed through their individual life cycles.

The stars are assumed to form from the condensation of interstellar matter, which would consist mostly of hydrogen and helium isotopes. The interstellar gas density varies considerably; in our part of space it is estimated to amount to a few hydrogen atoms per cubic centimeter. Nevertheless, it is believed that the total mass of matter involved as interstellar gas is of the same order of magnitude as the total mass of all the stars and their planets. As the gas collects more densely in an area, at some critical density the hydrogen degenerates into H^+ ions and nonlocalized ("free") electrons (this is the *plasma state*), which leads to a considerable pressure reduction. The gravitational forces cause an increased rate of condensation and, as a result of the release of gravitational energy, the temperature of the gas rises. When the temperature in the condensed core reaches approximately 5×10^6 K, thermonuclear reactions leading to conversion of hydrogen to helium begin to occur with significant intensity. The temperature of the gas further increases, leading to a rather hot core but a much cooler outer atmosphere, which we have noted is the situation in the sun.

Hydrogen burning in the core does not proceed directly as in reaction (11.1), since a four-body simultaneous collision is a very unlikely process. Rather, for stars in which there is initially only hydrogen in the core, the process must occur through the interaction of two protons to form first a deuteron (Table 11.1). This process involves the weak interaction force in the positron formation which makes it very slow measured on the

TABLE 11.1. *The proton–proton chain for* ^{4}He *formation*
The mean reaction time refers to conditions in the center of our sun. About 90% of our sun's energy production is assumed to come from this chain

Reaction steps	Q-value (MeV)	Mean reaction time	
$^1H + {}^1H \rightarrow {}^2H + \beta^+ + \nu$	0.44	1.4×10^{10} y	(11.2)
$^2H + {}^1H \rightarrow {}^3He + \gamma$	5.49	6 s	(11.3)
$^3He + {}^3He \rightarrow {}^4He + 2{}^1H$	12.86	9×10^5 y	(11.4)
Overall $4{}^1H \rightarrow {}^4He + 2\beta^+ + 2\nu$	24.7[(a)]	1.4×10^{10} y	(11.5)

[(a)]Because the neutrinos leave the sun with an energy of 0.4 MeV, the energy production in the sun is 24.3 MeV/He formed.

nuclear time scale. In the sun only about one $^1H + {}^1H$ collision in 10^{22} leads to reaction. If the reaction rate had been much higher or lower, the difference in solar energy would have made life impossible on earth, but perhaps possible on some other planet in our solar system. The deuteron reacts with an additional proton to form ^3He. Laboratory experiments have shown that ^3He does not interact with protons; however, two ^3He nuclei can fuse to form ^4He with the liberation of two protons.

If the energy production in the sun has lasted for 4.6 eons, according to this cycle, only about 6% of the hydrogen content of the sun has been consumed.

A number of alternate reactions could be involved, for example:

$$^2H + {}^3H \longrightarrow {}^4He + n \qquad Q = 17.59 \text{ MeV} \tag{11.6}$$

$$^2H + {}^2H \longrightarrow {}^3He + n \qquad 3.27 \tag{11.7a}$$

$$^2H + {}^2H \longrightarrow {}^3H + p \qquad 4.03 \tag{11.7b}$$

$$^2H + {}^3He \longrightarrow {}^4He + p \qquad 18.35 \tag{11.8}$$

$$^3H + {}^4He \longrightarrow {}^6Li + n \qquad -2.73 \tag{11.9}$$

$$^1H + {}^6Li \longrightarrow {}^3He + {}^4He \qquad 4.02 \tag{11.10}$$

$$^3He + {}^4He \longrightarrow {}^7Be + \gamma \qquad 1.59 \tag{11.11a}$$

$$^7Be + e^- \longrightarrow {}^7Li + \nu \qquad 1.37 \tag{11.11b}$$

Some of these have been suggested as alternate steps in the helium production of hydrogen. In Fig. 11.5 the reaction cross-sections for the first three processes are plotted as a function of projectile energy. They are also of interest in controlled thermonuclear reactions (§11.10).

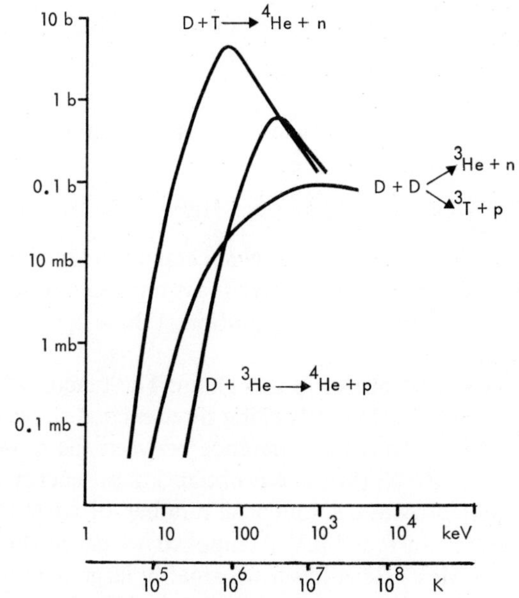

FIG. 11.5. Cross-section for heavy hydrogen atom reactions as a function of the particle kinetic energy (center-of-mass system). The temperature scale is calculated with the relation $E = \mathbf{k}T = 8.6 \times 10^{-5}\, T\,(\text{eV})$.

11.5. Formation of elements heavier than helium

Since helium has a greater mass than hydrogen, there is greater gravitational attraction between helium nuclei. As the burning in the core proceeds, the helium that is formed collects in the center of the core of the star while the hydrogen burning moves to the outer layers. Moreover, as the hydrogen fuel in the core is consumed, the temperature of the core decreases somewhat. The net result is that as the amount of helium is increased and the hydrogen depleted, the core contracts through the increased gravitational attraction of the helium, and the temperature of the core rises. This rise of the core temperature heats up the outer layers of hydrogen and results in an expansion of the outer mantle of the star. This in turn results in a cooler surface and the star appears to irradiate more red light. The star at this period of its life becomes known as a *red giant*. We can expect that in time our sun will pass through a red giant stage at which time its diameter should expand sufficiently to engulf the inner planets of the solar system; this is not expected to occur until about five eons from now.

As the density and the gravitational attraction of the helium in the core of the red giant increases, the temperature rises proportionately until the helium nuclei acquire sufficient kinetic energy to undergo fusion reactions. This produces a sudden heating at the center, and in fact the core can be considered to detonate (the *helium flash*) though this detonation is dampened by the large outer masses. It is at this point that the problem of mass numbers 5 and 8 is circumvented, although such a circumvention is indicated already in reactions (11.9) and (11.11). Two helium nuclei combine to form ^{8}Be which exists for a very short period of time ($t_{1/2}$ 2×10^{-16} s). Under the conditions of the high densities and temperature in the helium core of a red giant even though ^{8}Be has such a short lifetime, there would always be a small equilibrium amount present. It has been calculated that in some red giants ($\rho = 10^7$ kg m^{-3}, $T = 1.5 \times 10^8$ K) one ^{8}Be atom is in equilibrium with about 10^9 ^{4}He. This amount is sufficient to allow some probability of capture of a third helium nucleus by the ^{8}Be to form ^{12}C. The reaction then is

$$3\,^{4}He \rightarrow {}^{8}Be + {}^{4}He \rightarrow {}^{12}C + \gamma \tag{11.12}$$

Once ^{12}C has been formed, further reactions with helium can explain the formation of oxygen, neon and higher elements

$$^{12}C(^{4}He, \gamma)^{16}O(^{4}He, \gamma)^{20}Ne(^{4}He, \gamma)^{24}Mg(^{4}He, \gamma)^{28}Si \tag{11.13}$$

(Fig. 11.6). As we can see from Fig. 11.1, the elemental abundance of the lighter elements shows higher values for those elements where mass numbers are divisible by four. This correlates well with the proposal that the formation of these light elements occurs mainly in the helium capture chain.

There are some interesting features in this chain. The reaction ^{8}Be (ground state) + $^{4}He \rightarrow {}^{12}C$ has a Q-value of 7.37 MeV. The first three excited states of ^{12}C have energies of 4.43, 7.66, and 9.64 MeV. Thus the difference between the Q-value and the middle state is 0.29 MeV. At 1.5×10^8 K, the most probable kinetic energy is only 13 keV; however, the Boltzmann tail provides a sufficient number of particles (about 1 in 10^{10}) that the additional energy of 0.29 MeV is supplied to cause the resonance reaction $^{8}Be + {}^{4}He \rightarrow {}^{12}C^*$ (7.66 MeV). This has a reasonably large reaction cross-section (the width of the state is 8 eV). In the absence of this level, carbon production would be very much smaller, and a universe with a much lower carbon content would be quite unlikely to develop life.

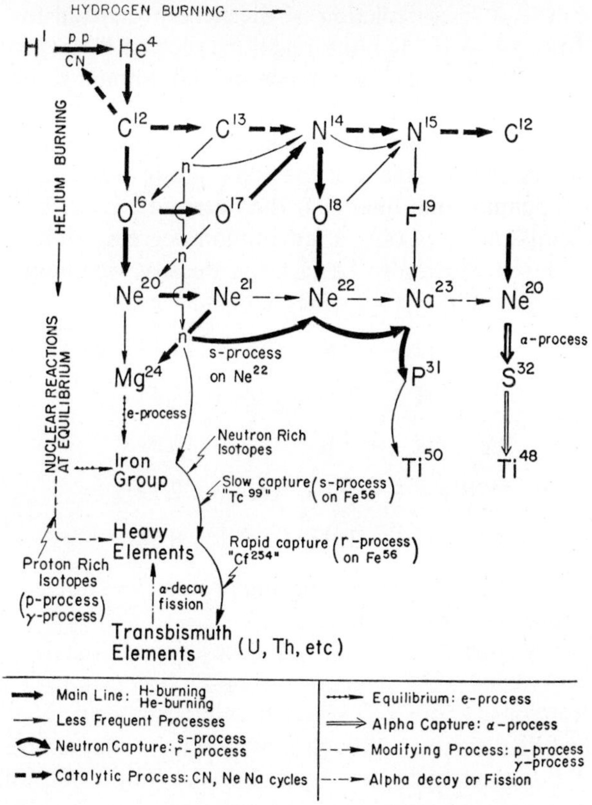

FIG. 11.6. Schematic diagram of the nuclear process by which the synthesis of the elements in stars takes place. (From Burbidge, Burbidge, Fowler, and Hoyle.)

The short half-life of ^{8}Be plays a significant role. If ^{8}Be had been much more stable, the helium-consuming chain would have proceeded much more rapidly. In fact after the helium ignition the energy production rate would have increased enormously and the red giant would have exploded as a supernova. However, the burning would not have gone much further than to ^{12}C. We would then have had a world mainly consisting of carbon ("graphite planets") and unused helium and hydrogen. On the other hand, had ^{8}Be been even less stable, the fusion synthesis would never have been able to bridge the mass 8, and no higher elements would have been formed.

The Q-value for the reaction $^4\text{He} + {}^{12}\text{C} \rightarrow {}^{16}\text{O}$ is 7.16 MeV. The nearest excited levels of ^{16}O are 7.12 and 8.87 MeV. The Q-value surpasses the first of these levels and is 1.71 MeV below the second. The latter is not useful for ^{16}O production since the Boltzmann distribution provides too small a fraction of particles with the needed kinetic energy of 1.71 MeV. However, the resonance width of the 7.12 MeV level makes the reaction possible for the low energy distribution tail. Had it been easier to make ^{16}O, we would have had less carbon and more oxygen in the universe. This probably would have been a hindrance in the development of life, since it is believed that life must start in a reducing (low oxygen-containing) atmosphere.

Helium burning, which only occurs for stars with $m \gtrsim 3.5\mathbf{m}_\odot$, leads to the formation of heavier elements which accumulate in the core of the star. Because the convection in stars with $m \geqq \mathbf{m}_\odot$ is small, successive layers of heavier elements are formed. The poor convection can lead to a depletion of the helium demand for the synthesis of higher elements. However, at 10^9 K photoreactions such as $^{20}Ne(\gamma, \alpha)^{16}O$ can proceed and produce helium atoms of high enough energy and amount to produce or capture heavier nuclei. As the pressure and temperature in the star increase, elements up to iron may be formed. A star at this stage may be characterized by a central core of iron group elements surrounded by layers of the silica group elements magnesium, neon, carbon plus oxygen, helium, and, finally, at the outermost layer, hydrogen. The outer layer may have a temperature of only a few million degrees, while the center may be 10^9 K. In this outer layer, which also has a lower density, we might expect to find Li, Be, and B. However, the conditions are such that reactions like

$$^{2}H + ^{2}H \longrightarrow ^{4}He \qquad 23.85 \text{ MeV} \tag{11.14}$$

$$^{7}Li + ^{1}H \longrightarrow 2\,^{4}He \qquad 17.35 \tag{11.15}$$

$$^{9}Be + ^{1}H \longrightarrow ^{10}B \qquad 6.59 \tag{11.16a}$$

$$^{10}B + n \longrightarrow ^{7}Li + ^{4}He \qquad 2.79 \tag{11.16b}$$

$$^{11}B + ^{1}H \longrightarrow 3\,^{4}He \qquad 8.68 \tag{11.17}$$

consume the Li, Be, and B. Moreover the burning process does not lead to the formation of appreciable amounts of these nuclides.

The possibility of the helium-burning sequence depends on the mass of the star. For a star of mass $<0.5\mathbf{m}_{\odot}$ the temperature never becomes high enough to ignite helium, while for a mass $<3.5\mathbf{m}_{\odot}$ no carbon burning can occur. The latter stars eventually evolve into white dwarves, although gravitational collapse and convection may lead to ignition of the outer hydrogen layer. The resulting explosions blow some of the matter in the outer layer into space (*nova explosion*).

For stars of masses $>3.5\mathbf{m}_{\odot}$ helium burning becomes the important energy source. As the core continues to heat, reactions such as

$$^{12}C + ^{12}C \rightarrow ^{24}Mg + \gamma \tag{11.18}$$

and

$$^{16}O + ^{16}O \rightarrow ^{32}S + \gamma \tag{11.19}$$

take place at $T > 10^9$ K; however, carbon burning requires a star mass $\gtrsim 7.5\mathbf{m}_{\odot}$. This leads to the production of photons which in turn can cause emission of nucleons from nuclei with masses between 12 and 32. The result is a "soup" of nuclei, nucleons, α-particles and photons, in which increasing temperature leads to reactions between the lighter particles and the heavier nuclei to form elements as heavy as the iron group. This is the origin of the relatively high abundancy of the iron group element.

Our galaxy has stars much larger and much smaller than our sun, and of widely different ages. Thus the processes we have described are occurring at present all around us in the Milky Way. Moreover, the stars emit considerable amounts of matter into space. As a result, interstellar gas, out of which new stars are formed, contains atoms heavier than hydrogen, although hydrogen is the most abundant element.

11.6. The carbon cycle

For stars which have been formed from interstellar material containing a significant

TABLE 11.2. *The CNO cycle for helium formation as devised by H. Bethe and C. F. von Weizsäcker*
The mean reaction time refers to conditions in the center of our sun. About 10% of our sun's energy production is assumed to come from this cycle

Reaction steps	Q-value (MeV)	Mean reaction time	
$^{12}C + {}^{1}H \rightarrow {}^{13}N + \gamma$	1.94	10^7 y	(11.20a)
$\downarrow$ $^{13}C + \beta^+ + \nu$	1.20	7 min	(11.20b)
$^{13}C + {}^{1}H \rightarrow {}^{14}N + \gamma$	7.55	3×10^6 y	(11.21)
$^{14}N + {}^{1}H \rightarrow {}^{15}O + \gamma$	7.29	3×10^8 y	(11.22a)
$\downarrow$ $^{15}N + \beta^+ + \nu$	1.74	2 min	(11.22b)
$^{15}N + {}^{1}H \rightarrow {}^{12}C + {}^{4}He$	4.86	10^5 y	(11.23)
Overall $4{}^{1}H \rightarrow {}^{4}He + 2\beta^+ + 2\nu$	24.6	3×10^8 y	(11.5)

amount of carbon, a different process has been postulated for the conversion of hydrogen to helium. This process is summarized in Table 11.2. Branches to this sequence have been suggested, leading to other nuclides, such as fluorine isotopes. The net result of the basic reactions is still the conversion of four hydrogen atoms into a helium atom, and the star proceeds through the same sequence of evolution that we have outlined for the normal hydrogen-burning stars. At temperatures $< 1.5 \times 10^7$ K the direct proton burning is more important than the CNO chain.

11.7. Formation processes from stable isotope abundancies

So far we have only discussed formation of elements up to iron, outlining the processes of formation of these elements through helium capture, although considerable proton capture also must have occurred. Such processes produce nuclei with N/Z close to 1 (see Fig. 3.1). When such a "proton rich" nucleus decays along an isobar line, if several stable isobars exist the "first" one reached is the one with the lowest N/Z value. We should expect that this isobar would be formed in a higher abundancy than other stable isobars; this also implies that, in most cases, the isotope of a given element with the lowest N-number (or with the highest Z-value) would be most abundant. The isotope chart (Appendix N) shows this to be correct. A simple comparison is made in Table 11.3. For mass number $A \lesssim 72$ it is obvious that the isobar with the highest Z-value is more abundant. An exception occurs at $A \sim 50$ and $N \sim 28$ which can be explained as due to the closed nuclear shells at these values.

For $A \gtrsim 72$ obviously the isobar with the highest N/Z is the most abundant. Exceptions are again found at the magic numbers 50 and 82. It is a reasonable conclusion that these nuclei must have been formed in a process very different from that for the lighter elements. In fission of heavy elements neutron rich products are obtained which decay through β^- emission until a stable nucleus is reached which is a relatively high N/Z isotope. A similar situation holds for elements produced through very high flux neutron irradiation in reactors. The theories for the formation of elements above iron assume that these elements have been formed through processes similar to these although the details are still unclear, as we learn in the next section.

TABLE 11.3. *Comparison of isobars of highest abundance for different values of A*

Mass number of most abundant isobars			
Highest Z	Highest Z $\approx$ Lowest Z	Lowest Z	Abundances (%)
36			
	40		
46, 48			
	50		Ti = 5.34, V = 0.25, Cr = 4.35 ($N \sim 28$)
54, 58, 64, 70			
		74	
	76		
		78, 80	
	82		
		84, 86	
	92		Zr = 2.8, Mo = 16.7, Ru = 5.5 ($N \sim 50$)
		94, (96), 98	
	100		
		102, 104, 106, 108	
	110		
		112, 113, 114, 115	
			Cd = 7.6, Sn = 14.3 ($Z = 50$)
		120, 122, 123	
	124		
		126, 128, 130, 132 134, 136, 138	
142			Ce = 11.1, Nd = 27.1 ($N \sim 82$)
		144	
148			
	150		
		152, 154, 156, 158 160, 162, 164, 168 170, 174, 176, 180 184, 186, 188, 190 192, 196, 198, 204	

11.8. The slow process

Through hydrogen and helium burning, neutrons are formed. A few of the reactions have been indicated ((11.6), (11.7a), (11.9)) and many others certainly occur. As the heavier elements form in the star, the neutron production increases considerably, since such reactions become more prevalent the heavier are the elements involved in the charged particle and γ-induced reactions. The mode of production of the elements changes from that of helium capture to that of neutron capture, so that the elements from iron to bismuth can be formed by a slow process of neutron capture (n, γ reactions), interrupted by β-decay whenever this is faster than the next capture step.

Such a process is known as the slow, or *s-process*. The reaction probability for the capture of neutrons increases with the atomic number of the element. Even though the abundances of the elements decrease with increasing atomic number, the probability of the formation of heavier elements increases with atomic number. Consequently, a leveling in the overall probability of elemental formation explains the flattening of the abundance curve for $A > 100$.

The formation of ^{104}Pd from ^{100}Ru can serve as an example of the steps in the s-process of elemental formation

$$\begin{array}{lllllll} ^{100}_{44}\text{Ru} \xrightarrow{(n,\gamma)} {}^{101}_{44}\text{Ru} \xrightarrow{(n,\gamma)} {}^{102}_{44}\text{Ru} \xrightarrow{(n,\gamma)} & {}^{103}_{44}\text{Ru} & \\ & \downarrow \beta & \\ & {}^{103}_{45}\text{Rh} \xrightarrow{(n,\gamma)} & {}^{104}_{45}\text{Rh} \\ & & \downarrow \beta \\ & & {}^{104}_{46}\text{Pd} \end{array} \quad (11.24)$$

The discovery of the element promethium (for which the longest-lived isotope has a half-life of only 18 y) in a star (HR 465) in the Andromeda constellation shows that an s-process must be occurring. A possible reaction path is $^{146}\text{Nd}(n,\gamma)^{147}\,\text{Nd}(\beta^-, 11\ \text{d})$ $^{147}\text{Pm}(\beta^-, 2.6\ \text{y})$. The s-process is believed to last for $\sim 10^7$ y, which is a short period in the total lifetime of a star.

As the star proceeds through the s-process stage we can expect that the fusion reactions decrease and the gravitational contraction of the star continues, leading to the release of even more gravitational energy. The contraction in the size of the star increases the rotational velocity and results in the ejection of some of the outer mantle into space, thereby exposing the inner, hotter core. This turbulence may become so violent that hydrogen from outer layers is mixed in with deeper layers of higher temperature, leading to instantaneous hydrogen burning and a very rapid rise in energy production. This is observed as a sudden light increase from the star (nova). After the nova stage the star would continue to cool until it becomes a white dwarf. The density of such a body would be enormous, about 10^8 kg m^{-3}.

11.9. Formation of the heaviest elements and supernova explosions

The s-process cannot explain the formation of the elements heavier than bismuth as the transbismuth elements have a number of short-lived species which would prevent the possibility of the formation of thorium and uranium in the amounts observed in nature. The abundance of the heaviest elements are believed to occur through another process.

In a star in which heavier elements are accumulated in its center, the energy production is carried on in an envelope surrounding the core. For a red giant with an original mass $>7.5\mathbf{m}_\odot$ (or a mass $>3.5\mathbf{m}_\odot$ at the end of the helium-burning period) the energy loss through photon and, especially, through neutrino emission is very large. This has several consequences: (i) the emission of energy into space cools the core and the giant begins to contract, (ii) the release of gravitational energy increases temperature and pressure in the envelope. Under the development of these conditions, the intense photon field begins to cause photodisintegration in elements like iron, releasing helium and neutrons. The helium immediately fuses and the intense heat developed spreads as a heat shock, which passes out to the cool shell of hydrogen and helium, initiating in this mantle new thermonuclear reactions. As a result the whole star may explode via the phenomenon known as a *supernova*. It has also been suggested that the intense pressure causes the reaction

$$p + e^- \rightarrow n + \nu$$

This process reduces the effective number of the particles in the stellar core, which there-

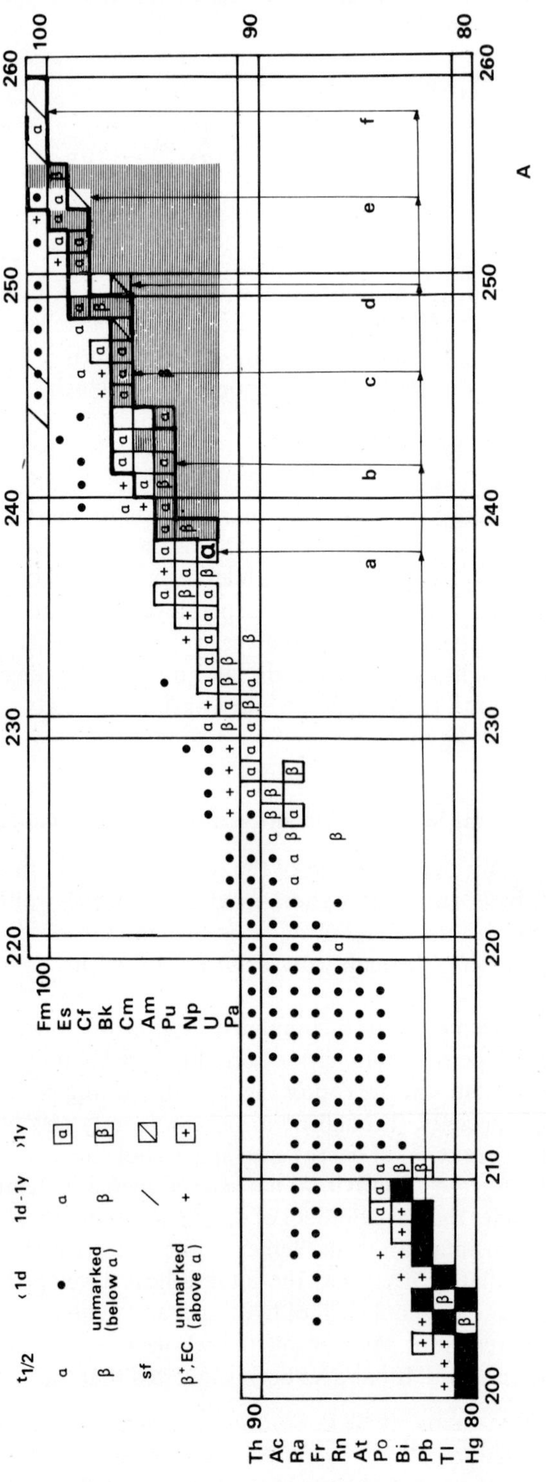

FIG. 11.7. Isotope chart of the heaviest elements showing, in the shaded area, nuclides produced in the Mike thermonuclear weapons test through successive neutron capture in ^{238}U followed by β^- decay. The arrow beginning with lead indicates successive n-capture and a series of β^- decay chains, of which a, b and c all terminate in ^{238}U (b and c after α-decay) and d, e and f are interrupted by sf-decaying nuclides. The heavy-framed zone shows build-up of transuranium elements through thermal neutron capture in ^{238}U followed by β^- decay.

fore collapses inwards until a new pressure sets up, when the interior region of the star matter mainly becomes a neutron fluid.

The supernova stage is a very short-lived one during which neutron production is extremely intense and provides a method whereby the earlier barrier of the short-lived isotopes between polonium and francium is no longer effective and the heaviest elements are rapidly synthesized. This mode of element formation is known as the rapid or *r-process.*

In Fig. 11.7 the build-up of ^{238}U from neutron capture in lead, and successive β^- decay, is illustrated by the sequence via arrow (a) as well as similar processes in ^{238}U leading to higher transuranic nuclides. The n-capture in the r-process has been suggested to go up to $Z \sim 100$ and $N \approx 184$. In the intense neutron field a considerable amount of (mainly fast) fission of the newly synthesized heavy elements probably also occurs. This may partly explain the peaks at N 50 and 82 in Fig. 11.3, which also correspond to maximum yields at A 95 and 140 in thermal fission (see Fig. 9.8). Some stars are unique in that they have an unusually high abundance of fission products; in such stars (e.g. HD 25354) also spectral lines from heavy actinides, like americium and curium, have been observed.

Astrophysicists were interested in the observation that the explosion of thermonuclear hydrogen bombs containing uranium resulted in the formation of elements 99 and 100. These elements were synthesized in the extremely short time of the explosion by the intense neutron fluxes bombarding the uranium (shaded area in Fig. 11.7). The explosion of the hydrogen bomb duplicated in a very small way what is believed to be the process of the formation of the heaviest elements in supernovae. The neutron fluxes and exposures in the s- and r-processes as compared to those in a nuclear explosion and a reactor are given in the following table:

	Flux (n cm^{-2} s^{-1})	× time	=	exposure (n cm^{-2})
s-process	$\sim 10^{16}$	~ 1000 y		$\sim 3 \times 10^{27}$
r-process	$\gtrsim 10^{27}$	1–100 s		$> 10^{27}$
Nuclear explosion	$> 10^{31}$	< 1 μs		$\sim 10^{25}$
Nuclear reactor	$\sim 10^{14}$	~ 1 y		$\sim 10^{21}$

The intensity of the neutron flux as well as the very short time preclude the possibility of β-decay as a competitor to neutron capture in the r-process. This can result in a different isotopic distribution of the elements for the r-process compared to that formed in the s-process. The following reaction sequence illustrates the r-process in which β-decay can occur only after the explosion has terminated and the intense neutron fluxes decreased (compare with the sequence (11.24)):

$$^{100}\mathrm{Ru}(n,\gamma)\,^{101}\mathrm{Ru}(n,\gamma)\,^{102}\mathrm{Ru}(n,\gamma)\,^{103}\mathrm{Ru}(n,\gamma)\,^{104}\mathrm{Ru}(n,\gamma)\,^{105}\mathrm{Ru}(n,\gamma)\,^{106}\mathrm{Ru}$$

$$^{103}\mathrm{Ru} \xrightarrow{\beta} {}^{103}\mathrm{Rh}$$

$$^{105}\mathrm{Ru} \xrightarrow{\beta} {}^{105}\mathrm{Rh} \xrightarrow{\beta} {}^{105}\mathrm{Pd}$$

$$^{106}\mathrm{Ru} \xrightarrow{\beta} {}^{106}\mathrm{Rh} \xrightarrow{\beta} {}^{106}\mathrm{Pd}$$

After completion of the r-process, ^{103}Ru, ^{105}Ru, ^{106}Ru undergo β-decay to isotopes of Rh and Pd. In this r-sequence ^{104}Ru–^{106}Ru are formed, but in the s-sequence beginning with ^{100}Ru, the heaviest ruthenium isotope is $A = 103$.

TABLE 11.4. *Relative abundance of elements in cosmic rays as compared with their cosmic abundance*

Element	Cosmic abundance	In cosmic rays
Hydrogen	10 000	10 000
Helium	1 500	700
Lithium, boron, beryllium	10^{-5}	15
Carbon, nitrogen, oxygen, fluorine	1.5	40
Neon through potassium	0.2	14
Elements with atomic number greater than 20	0.1	5

In the supernova explosion large masses of materials are ejected into interstellar space. This contributes to the existence of heavy elements in space. Table 11.4 shows the high fraction of elements heavier than helium in cosmic rays as compared with the cosmic abundancy. In fact, even uranium has been observed in cosmic rays and in our sun. Since our sun is undergoing the simplest type of hydrogen-burning cycle, it is not possible for the heavier elements to have been synthesized by the sun. Consequently, their presence indicates that the sun has been formed as a later generation star from material that was initially formed in an earlier star which has exploded or it has accumulated matter from such a star. It also shows that our solar system still is being showered by debris from some distant supernova explosion.

The carbon cycle stars must also correspond to second generation stars because of the presence of sufficient amounts of ^{12}C in their core to allow the carbon cycle to start. The same star may pass through several supernovae explosions whereby it loses large amounts of the lighter elements from the outer mantle in each explosion. The chemical composition of a star thus not only indicates its age but also tells us to which generation of stellar evolution it belongs.

A supernova explosion leaves a residue at the center of the expanding gas cloud generated at the explosion. This residue is believed to contract into a *neutron star*. At a density of $>10^{17}$ kg m^{-3} matter consists of relatively closely packed nuclei and degenerate electrons. It now becomes favorable for the protons to capture the electrons

$$p + e^{-} \rightarrow n + \nu \tag{11.25}$$

The neutrinos are emitted and only neutrons are left (cf. the ylem). If our sun would develop into a neutron star of density 10^{18} kg m^{-3}, its radius would become only 1% of that of the earth. However, such an evolution will not occur; for stars $\lesssim 1.44\mathbf{m}_{\odot}$ (the *Chandrasekhar limit*) the final destiny is to end up as a white dwarf without ever undergoing a supernova explosion.

The neutron star has a considerable energy reservoir in its rotation, whose frequency is about one rotation per second. Particles ejected from this rapidly rotating object would be caught in the rotating magnetic field and accelerated to relativistic tangential velocities. They then emit high intensities of bremsstrahlung. This would appear as pulses of radiation with the frequency of the rotation. Such radiation pulses have been observed from cosmic sources known as *pulsars*. The discovery of a pulsar at the center of the expanding Crab Nebula, which is the remnants of a supernova explosion observed by the Chinese in the year 1054, has provided strong evidence for the scheme outlined above.

The pulsars have been suggested to be the source of the high energy cosmic ray particles observed in our atmosphere.

The different elemental formation processes are summarized in Fig. 11.6, taken from the original paper by Burbidge, Burbidge, Fowler and Hoyle.

11.10. The origin of the earth

Much of the information about the formation of our solar system and the planets is based upon radiochemical analyses of meteorites and moon samples, and more recently from photographs and analysis of samples from the other planets, especially Venus and Mars and its moons. This has led to a picture of the formation of the earth which still is very uncertain.

Figure 11.1c shows the best estimates available today on the bulk composition of the earth. It is obvious that the composition varies considerably from that of the cosmic abundances. We see that the more volatile elements such as hydrogen and helium are almost absent from the earth. This is because the mass of the earth is too small to provide sufficient gravitational attraction for these light atoms. By the same reasoning we can understand why the moon has not been able to retain an atmosphere of nitrogen and oxygen since its mass is even smaller than that of the earth.

Radiochemical investigations of meteorites and of minerals from the earth and the moon have given us an idea not only of the age of these samples (Chapter 12) but also of the temperature and pressure conditions which once existed on their parent bodies. From these data some reasonable conclusions can be drawn about the processes whereby the earth was formed. According to the most generally accepted model, the sun, planets, and our moon were formed 4.6 eons ago by accumulation and condensation of cosmic dust. How this happened is not quite clear.

The combination of gravitational and centrifugal force in the rotating cloud of condensing matter led to collection of the lighter elements in the center of the system. The fact that the elemental composition of the earth and inner planets is quite different from that of the sun (which we believe is a better representation of the primary cosmic matter of our solar system) is probably due to the fact that those lighter elements, which were not attracted to the inner core of the condensing planetary matter, were forced through the increasing radiation pressure of the primeval sun to the outer part of the evolving solar system. Thus the earth became less rich with respect to the inert gases as well as hydrogen, ammonia, and methane, which are found in large amounts in the outer planets. However, these gases, which also have been discovered in interstellar space, are believed to have been the main constituents of the primordial ("reducing") atmosphere of the earth, although recent information about the composition of the planet Mars throws some doubts on this hypothesis. The earth seems to have been molten after its primary condensation and solidified some 3.0–3.5 eons ago. By some poorly understood processes, carbon dioxide and water were formed subsequently in larger amounts, and—with the development of living species (so-called "archaeo-bacteria", producing methane from hydrogen and carbon dioxide, believed to be as old as 3.5 eons; the earliest protozoa appeared about two eons ago)—also increasing amounts of oxygen ("oxidizing atmosphere"). The oxygen content, which is now about 20%, is believed to have been only 1% about 600 million years ago.

TABLE 11.5. *Reactions for controlled fusion reactors*
The kinetic energy (MeV) of some of the products are given in parentheses, and the minimum thermonuclear reaction energy in square brackets

	Q-value (MeV)	
The D–T reaction		
$^{2}H + ^{3}H \rightarrow ^{4}He(3.5) + n(14.1)$; [4 keV][(a)]	17.6	(11.6)
n (slow) + $^{6}Li \rightarrow ^{4}He + ^{3}H$	4.8	(11.26)
Overall: $^{2}H + ^{6}Li \rightarrow 2\,^{4}He$	22.4	(11.27)
The D–D reaction		
$^{2}H + ^{2}H \rightarrow ^{3}H(1.0) + ^{1}H(3.0)$; [35 keV][(b)]	4.0	(11.7a)
$^{2}H + ^{2}H \rightarrow ^{3}He(0.8) + n(2.5)$; [35 keV][(b)]	3.3	(11.7b)
$^{2}H + ^{3}H \rightarrow ^{4}He + n$	17.6	(11.6)
$^{2}H + ^{3}He \rightarrow ^{4}He(3.6) + ^{1}H(14.7)$; [30 keV]	18.3	(11.8)
Overall: $6\,^{2}H \rightarrow 2\,^{4}He + 2\,^{1}H + 2n$	43.2	(11.28)

[(a)]Corresponding to a temperature of 4.5×10^{7} K.
[(b)]Corresponding to a temperature of 4×10^{8} K.

11.11. Controlled thermonuclear reactors

For over 20 years an increasing amount of research has been put into the development of controlled thermonuclear reactors (CTR). The obvious reason is that they could be the long term answer to the energy demand of the world (cf. §21.1). If the heavy hydrogen in the oceans could be made to fuse as in reaction (11.3) under controlled conditions, this source of energy would last for millions of years.

Because fusion reactions are most easily achieved with hydrogen atoms due to the low Coulomb barrier and favorable wave mechanical transmission factor, only hydrogen atom reactions are being considered for controlled fusion. The reactions, which are believed to be of most importance, are summarized in Table 11.5.

In this section we consider briefly some aspects of fusion reactions.

11.11.1. *Physical principles*

Two factors determine the possibility of achieving a thermonuclear fusion reaction: the particle temperature, which must be $\gtrsim 10^{8}$ K, and the product $n\tau$, which must be $\gtrsim 10^{20}$ particles $s^{-1}\,m^{-3}$ (the *Lawson limit*), where n is the particle density and τ the confinement time. Research is being directed to obtaining these necessary conditions in either steady state or pulsed operation.

The steady state reactor is limited in power density by heat transfer and other considerations to about $n = 10^{20}$–10^{21}. Since each collision involves two particles, the fusion power density varies as the square of the particle density. At 1 Pa (i.e. 3×10^{20} particles m^{-3}) the power density would be tens of MW per m^{3}; at atmospheric densities it would be 10^{10} larger. This leads to a required confinement time of about 0.1–1s.

A large number of machines based on *magnetic confinement* of hydrogen ions have been built. The best results so far (1974) obtained are in the Tokamak machine in Moscow: $T \sim 10^{7}$ K, $n \sim 5 \times 10^{19}$, and $\tau \sim 0.02$ have been achieved simultaneously. In other experiments higher temperatures ($\sim 10^{8}$ K) have been achieved, but under much shorter confinement times. Alternatively, longer confinement times, ~ 1 s, have been produced but with n only $\sim 10^{17}\,m^{-3}$.

Pulsed operation is being studied (*inertia confinement*). Small pellets of solid D_2 and/or T_2 are dropped into the middle of a chamber where they are irradiated by intense beams of photons from lasers or from electrons from accelerators located around the chamber. The surface of the pellet immediately vaporizes, leading to a jet-stream of particles away from the pellet and an impulse (temperature–pressure wave) traveling into the pellet, increasing the central temperature to $> 10^8$ K. This causes a small explosion, whose energy can be collected with moderate difficulty. Because the particle density is high the pulse time can be very short and still meet the Lawson criterion. With electron beams considerable progress has been reported from the USSR and the USA in which temperatures of $\sim 10^9$ K have been reached and fusion neutrons produced. Larger machines of this type are now being developed. For photon beams the lasers must be of higher energy (10^5–10^6 J pulse^{-1}) than is presently available. An experimental 100 kJ pulse^{-1} CO_2 laser is expected to be in operation in the US in 1983. This should be compared with the expected energy output, 10^7–10^8 J pulse^{-1}. With a repetition frequency of 100 pellets s^{-1}, a power output of 1–10 GW would be achieved.

11.11.2. *The magnetic confinement reactor*

At $T = 10^5$ K the average kinetic energy of a hydrogen atom is 13 eV. Since the electron-binding energy is 13.0 eV, the hydrogen at $T \gtrsim 10^6$ is a plasma of H^+ ions and free electrons. No construction material can withstand a plasma of this energy. Thus the particles are kept away from the walls of the vessel containing the fusion reaction by strong magnetic fields. Figure 11.8 shows the principle of magnetic confinement in the Tokamak machine, which is an example of one of a number of magnetic configurations tested; temperatures above 6×10^7 K have been achieved.

Independent of the particular confinement design, a number of problems have to be solved before a thermonuclear reactor can function.

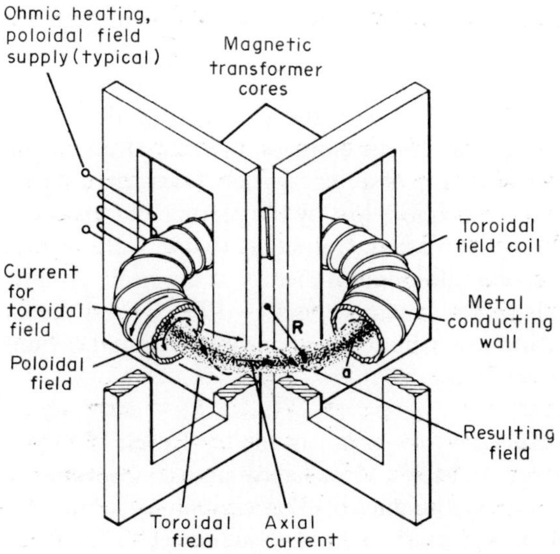

FIG. 11.8. The Tokamak plasma confinement scheme. The plasma major (toroidal) radius R will be 3.0 m and the minor radius $a \sim 1.3$ m, in the JET project (Joint European Torus), which will be ready around 1980. (According to Post and Ribe.)

(a) *Particle injection and withdrawal* In a steady state machine, whether it operates in a pulsed mode or not, fuel (i.e. T and D) must be injected and after consumption reaction products (^{3}He, ^{4}He) must be withdrawn. Because of the strong magnetic field, injection of ions is an extremely difficult problem. Consequently, systems with injection of high energy neutral particles are being studied. This could be accomplished through particle charge exchange:

$$D^+ \text{ (high energy)} + D \text{ (low energy)} \rightarrow D \text{ (high energy)} + D^+ \text{ (low energy)}$$

or

$$D_2^+ \text{ (high energy)} \rightarrow D + D^+ \text{ (both high energy)}$$

In the latter case dissociation is achieved through collision with other atoms. All ions are removed magnetically, leaving a beam of uncharged D-particles, which can enter the fusion volume.

The withdrawal problem is even more difficult if it is to be achieved by distortion of the magnetic field (in the plasma all particles become ionized) because this may lead to instabilities of the fusion plasma as a whole. As an alternative, the fusion reactor may be shut down when sufficient large amounts of "helium-ash" have built up and then purged and filled with fresh D–T mixture: the repetition cycle would be $\lesssim 20$ per h.

If the high temperature particles were to strike the walls of the vessel the walls would rapidly erode. This has two serious consequences: (i) some ionized wall material will sputter into the plasma and poison the fusion reaction, and (ii) the wall material must be replaced repeatedly. The solutions to these problems require a large amount of research in high temperature chemistry.

(b) *Plasma confinement and heating* The particles injected must be accelerated to sufficient energies and/or the density must be increased to meet the Lawson criterion. Many methods have been suggested based on the use of high frequency, of very high energy (relativistic) electrons, of laser light beams, or of increasing magnetic field, in addition to ohmic heating, which occurs naturally in the Tokamak machine. Another problem appears in connection with the particle injection, withdrawal, and heating because these processes lead to the emission of bremsstrahlung and synchrotron radiation. The radiation is mainly of an energy much less than that corresponding to the fusion temperature. It is therefore lost by the plasma and instead absorbed in the parts of the vessel closest to the plasma. Consequently the walls of the vacuum vessel must be effectively cooled. An additional difficulty is the heat insulation required between the very warm walls of the vacuum vessel ($\sim 1000°C$) and the current-carrying coils. The effect of radiation upon the coolant (probably liquid helium) and upon the coil materials is not known.

(c) *Energy extraction and fuel cycle* In the D–T cycle 80% of the energy appears as neutron kinetic energy (14 MeV). Most steady state concepts involve the capture of this energetic neutron in a surrounding blanket containing lithium as a metal or salt (e.g. Li_2BeF_4) in which new tritium is bred (reaction (11.27)). This develops additional kinetic energy, which is converted into blanket heat. Most of the energy of the fusion process appears therefore in the hot blanket, very similar to the heating of fuel rods of a fission reactor. If the blanket is in the form of molten lithium metal or salt, it can simply be pumped through a heat exchanger, which in the secondary circuit produces steam for a turbine. The similarity with the liquid-metal-cooled fast breeder reactor (LMFBR, Chapter 19) is obvious, and the materials problem is therefore similar. The tritium so produced must be recovered for recycling. This is somewhat difficult for the metal system because of the formation of lithium–tritium compounds. The problem is reduced if the blanket is LiF (or a salt mixture thereof) in which case TF would be expected to form. TF could be recovered easily, e.g. by the purging of the melt.

Instead of the more conventional steam turbine/electric generator design it may be possible to produce electricity directly from the flowing lithium metal. However, because of the high temperature of the melt, probably $\gtrsim 1000°C$, a satisfactory technique is not yet available.

As indicated from Fig. 11.5, the T–D reaction is the most probable starting point for a fusion reactor. This means that hundreds of grams of tritium would be needed initially. These amounts are available through the same processes used to produce tritium for hydrogen weapons. A likely technique involves the irradiation of a ^{6}Li–Al alloy in a high flux thermal fission reactor. In this alloy both tritium and ^{4}He are produced. They can be separated on the basis of their different vapor pressures, or through their different chemical reactivities. The consequence is that fusion reactors will demand considerable amounts of lithium. This may set limits to the use of fusion reactors on a large scale, although it has been estimated that the usable lithium reserves are ten times the expected usable fossil reserves.

(d) *Construction materials* The construction material close to the plasma must be heat resistant, radiation resistant, and have a low neutron capture cross-section. For the inner vacuum chamber refracting material and metals like vanadium and niobium have been suggested. High energy heavy particles and electrons, which will be abundant, must lead to a minimum of corrosion and vaporization. Fast neutrons cause considerable atom displacement (and thus structural changes) in the construction material. In fact, about every atom of the wall material will be displaced once a day. Annealing due to the high wall temperature should considerably reduce the damage. Through (n, p) and (n, α) reactions, fast neutrons also produce significant quantities of hydrogen and helium within the wall material. The helium may collect in voids, forming bubbles, and the hydrogen may form hydrides, both considerably altering the properties of the structural material. Although experience from the intense neutron fluxes in LMFBR will be of great value to fusion reactor research, the fission neutrons are of much lower ($\lesssim 2$ MeV) energy and thus present less of a problem.

11.11.3. *Commercial fusion reactor design*

Although controlled fusion devices are still in the future, the general principles of a commercial fusion reactor of the steady state type have been studied, primarily to uncover those aspects where research is needed. We have already discussed some of the problems involving nuclear chemistry and materials sciences.

Figure 11.9 shows a design study. The circular doughnut-shaped tube with its concentric vacuum chamber (cf. Fig. 11.8) may have an overall external diameter of about 20 m, while the cross-section of the vacuum chamber is about 4 m in diameter. The blanket would consist of a LiNb alloy, about 1 m thick. It may also contain a moderator, e.g. of graphite (not shown in the figure), to increase the yield of the ^{6}Li(n, α)^{3}T reaction.

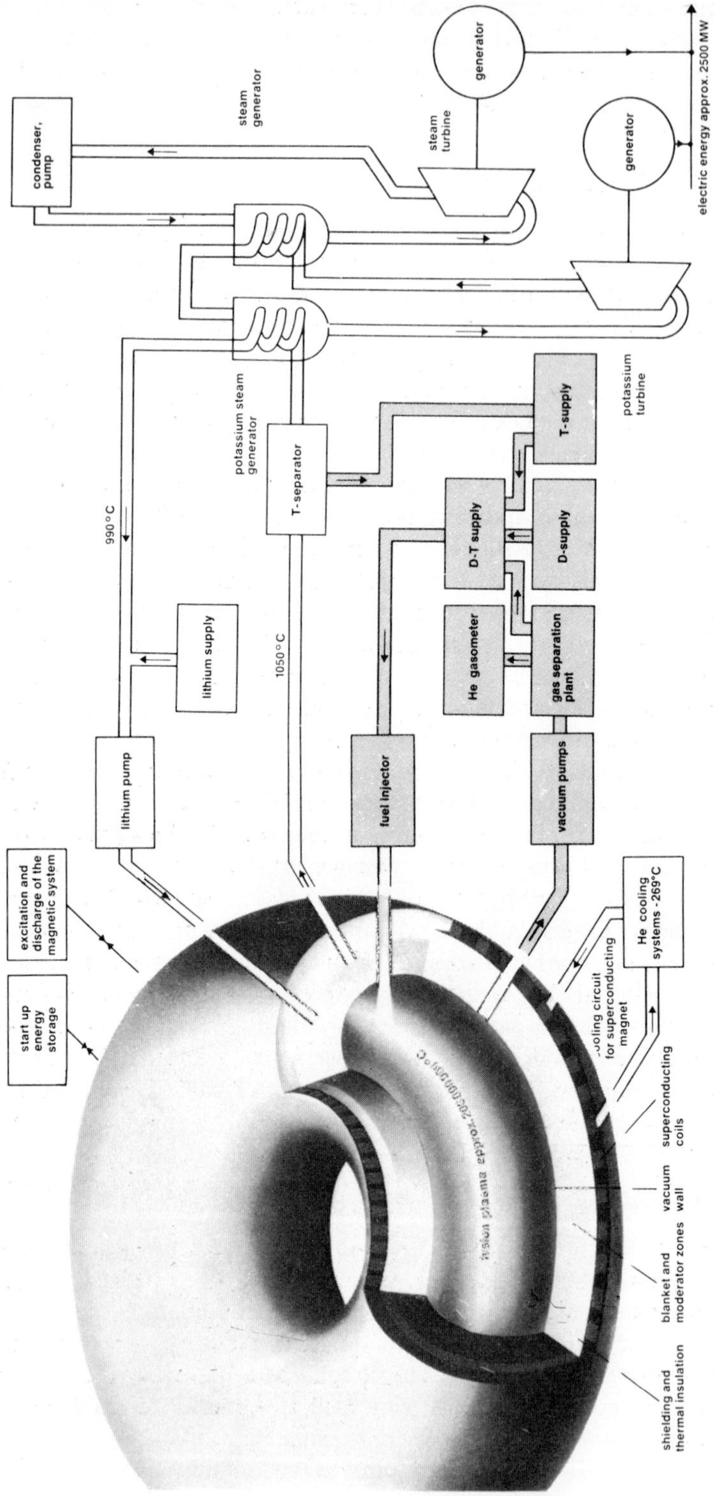

FIG. 11.9. A conceptual design of a fusion reactor system. (According to R. Gerwin.)

The structural material is niobium. The magnetic field strength is ~ 8 T (~ 80 kilogauss). With a tritium inventory of about 10 kg, and a consumption of about 1 kg d^{-1}, it would produce 5 GW_e (GW electric power) with a plasma temperature of $\gtrsim 10$ keV and a fast neutron flux of 4×10^{15} n $cm^{-2} s^{-1}$.

11.11.4. *Health and environmental aspects*

The design period of fusion reactors must also be used to evaluate environmental hazards, e.g. thermal pollution, operational safety, and radioactive wastes.

Because the present design of fusion power stations operates on the same principle as fission reactors, the thermal efficiency should be about the same. The necessarily larger power output will lead to a larger amount of energy removed by the coolant. Thus the thermal pollution is expected to be larger than for present fission power stations. Large scale development of magneto hydrodynamic systems may minimize this undesirable condition.

An operating reactor would contain several kilograms of tritium (10 kg of ^{3}T is about 10^8 Ci of radioactivity) which has a hazard corresponding to the noble gas fission products of the fission reactors. However, tritium is more difficult to contain because of its ability to pass through many metals (e.g. niobium), and more easily so at high temperatures. Moreover, tritium can exchange with hydrogen atoms in water, and thus become an ingestion hazard. Tritium is already a problem for fission reactors; for fusion reactors the problem is at least a factor of 1000 greater. The fusion reactor must therefore be made extremely tight against tritium leakage; the fraction permitted to leak out should be $< 10^{-7}$, which poses a serious technical problem.

The fusion reactor, like the fast fission breeder reactor, would contain large amounts of liquid metal. Very high activities are induced in the structural material. The preferred material at the present seems to be niobium or vanadium. Niobium has better structural properties and lower neutron capture cross-section than vanadium, but in niobium long-lived activities are induced: ^{94}Nb, formed through ^{93}Nb (100%) (n, γ), has 2×10^4 y, and ^{95}Nb, formed through (2n, γ), 35.1 d half-lives. In addition the fast neutrons may excite an isometric state of ^{93}Nb with a half-life of 13.6 y; this will be the dominating activity after the 35 d nuclide has died. If niobium is used, once the fusion reactor is stopped, the radiation field would quickly reduce to about 0.2‰, at which level it would persist for decades. Probably forced cooling would be necessary after reactor turn-off to avoid structural damage. This hazard will be less if vanadium is selected as construction material, although considerable amounts of ^{49}V ($t_{1/2}$ 330 d), and possibly also some very long-lived ^{53}Mn, are formed. The effects of fast neutrons on vanadium are not well known. Under all circumstances the induced activity will cause a maintenance hazard.

Very much magnetic energy would be stored in the fusion reactor, something like 10^{11} J. It is not clear what the consequences would be if this energy cannot decay under controlled conditions (which would take something like 6h).

The health and environmental problems of the fusion and the fission reactor are of similar kind, though smaller amounts of long-lived radioactivity are produced in the former case (i.e. no actinides). The environmental aspects of the fission reactor are treated in Chapters 19–21.

11.12. Exercises

11.1. (a) What is the most probable kinetic energy of a hydrogen atom at the interior of the sun ($T = 1.5 \times 10^7$ K)? (b) What fraction of the particles would have energies in excess of 100 keV?

11.2. Consider a power reactor in which microspheres ($r = 0.3$ mm) of frozen 1 : 1 T–D mixture (density 170 kg m^{-3}) are fused by laser irradiation. The laser compresses the spheres to $N_v = N_v^0 10^4$, where N_v^0 is the number of atoms per m^3 at ordinary pressure, and also heats it to a temperature corresponding to almost 20 keV. The energy developed through the T–D fusion reaction leads to expansion of the spheres, which occurs with the velocity of sound ($v_s = 10^8$ m s^{-1}). This leads to no more than 25% of the particles fusing. (a) By what factor will the Lawson criterion be exceeded? (b) What power is produced if the fusion micro-explosions occur at a rate of 30 per s?

11.3. If the deuterium in sea water is extracted and used to produce energy through reaction (11.7), what amount of gasoline (heat of combustion is assumed to be 35 MJ l^{-1}) would be equivalent to 1 l of sea water?

11.4. What amount (kg) of hydrogen is consumed per second in the fusion reaction (11.1)?

11.5. Deuterium is to be injected into a fusion reactor at a density of 10^{20} D^+ and 10^{20} e^- m^{-3} and an energy of 100 keV. How much of the deuterium must fuse to compensate for the ionization and injection energy? The ionization energy of the deuterium atom is 13 eV.

11.6. In a water power station with a fall height of 20 m and a water flow of 500 m^3 s^{-1}, the electric power output at 100% efficiency can be calculated. If the heavy water in the fall could be extracted and converted in a fusion reactor according to (11.7) with a 25% efficiency, which of the two power sources would yield more energy?

11.13. Literature

E. M. Burbidge, G. R. Burbidge, W. A. Fowler, and F. Hoyle, Synthesis of the elements in stars, *Rev. Mod. Phys.* **29** (1957) 547.

H. Craig, S. L. Miller, and G. J. Wasserburg (eds.), *Isotopic and Cosmic Chemistry*, North-Holland, 1963.

W. D. Arnett, C. J. Hansen, J. W. Truran, and A. G. W. Cameron (eds.), *Nucleosynthesis*, Gordon & Breach, 1965.

F. J. E. Peebles and D. T. Wilkinson, The primeval fireball, *Sci. Am.* **216** June (1967) 28.

D. D. Clayton, *Principles of Stellar Evolution and Nucleosynthesis*, McGraw-Hill, 1968.

A. O. J. Unsold, Stellar abundances and the origin of the elements, *Science* **163** (1969) 1015.

M. A. Ruderman, Solid stars, *Sci. Am.* **224** (Feb. 1971) 24.

F. Reines (ed.), *Cosmology, Fusion and Other Matter*, Colorado Ass. University Press, 1972.

D. N. Schramm and W. D. Arnett (eds.), *Explosive Nucleosynthesis*, University of Texas Press, Austin and London, 1973.

D. N. Schramm, The age of the elements, *Sci. Am.* **230** (Jan. 1974) 69.

V. Trimble, The origin and abundances of the chemical elements, *Rev. Mod. Phys.* **47** (1975) 877.

J. E. Ross and L. H. Aller, The chemical composition of the sun, *Science* **191** (1976) 1223.

V. Trimble, Cosmology: man's place in the universe, *Am. Sci.* **65** (1977) 76.

J. Rydberg and G. Choppin, Elemental evolution and isotopic composition, *J. Chem. Ed.* **54** (1977) 742.

D. N. Schramm and R. V. Waggoner, Element production in the early universe, *Ann. Rev. Nucl. Sci.* **27** (1977) 37.

C. Rolfs and H. P. Trautvetter, Experimental nuclear astrophysics, *Ann. Rev. Nucl. Sci.* **28** (1978) 115.

R. N. Clayton, Isotopic anomalies in the early solar system, *Ann. Rev. Nucl. Sci.* **28** (1978) 501.

Fusion power

D. J. Rose, Controlled nuclear fusion: status and outlook, *Science* **172** (1971) 797.

H. Postma, Engineering and environmental aspects of fusion power reactors, *Nuclear News*, April 1971.

D. Gruen, *The Chemistry of Fusion Technology*, Plenum Press, 1972.

R. F. Post, Prospects for fusion power, *Physics Today* **6** (1973) 30.

R. F. Post and F. L. Ribe, Fusion reactors as future energy sources, *Science* **186** (1974) 397.

CHAPTER 12

Naturally Occurring Radioactive Elements and Extinct Radioactivity

Contents

Introduction

Approximately one-third of the 90 elements occurring in nature have been found to have one or more isotopes which exhibit radioactivity in measurable quantities. If we add to this number the 15 or so synthetic elements, approximately one-half of the elements in the Periodic Table exhibit radioactivity in their normal state. This is indicated by the circles in the Periodic Table (inside front cover). However, with the exception of the synthetic elements technetium ($Z = 43$) and promethium ($Z = 61$), the elements lighter than bismuth which have naturally occurring radioactive isotopes exhibit such weak radioactivity (because of the very long half-lives) that extremely sensitive measuring devices are required to detect their radioactivity.

The naturally occurring radioactive nuclides can be divided into three groups: (i) those formed from cosmic radiation, (ii) those with lifetimes comparable to the age of the earth, and (iii) those that are part of the natural decay chains.

12.1. Radionuclides formed through cosmic radiation

Cosmic irradiation in the atmosphere produces neutrons and protons which can react with the atoms of the atmosphere (N_2, O_2, Ar, etc.) to result in the production of

TABLE 12.1. *Light radionuclides discovered in rain water*

Nuclide	Decay	Atmospheric production rate (atoms m^{-2} s^{-1})	Nuclide	Decay	Atmospheric production rate (atoms m^{-2} s^{-1})
^{3}H	β^-, 12.35 y	2 500	^{32}P	β^-, 14.3 d	
^{7}Be	EC, 53.4 d	81	^{33}P	β^-, 25.3 d	
^{10}Be	β^-, 1.6×10^6 y	360	^{35}S	β^-, 87.5 d	14
^{14}C	β^-, 5736 y	22 000	^{38}S	β^-, 2.83 h	
^{22}Na	β^+, 2.60 y	0.6	^{34m}Cl	γ, β^+, 32 min	
^{24}Na	β^-, 15.0 h		^{36}Cl	β^-, 3.0×10^5 y	11
^{28}Mg	β^-, 21.1 h		^{38}Cl	β^-, 37.2 min	
^{26}Al	β^+, 7.16×10^6 y	1.7	^{39}Cl	β^-, 56 min	16
^{31}Si	β^-, 2.62 h		^{39}Ar	β^-, 269 y	
^{32}Si	β^-, 280 y				

radioactive nuclides, some of which are listed in Table 12.1. These nuclides are brought from the atmosphere to the earth by rain water. The longer-lived ones have been discovered in meteorites also.

These radionuclides are formed in extremely low concentrations and their detection has only recently been accomplished through development of ultrasensitive counting techniques. The nuclides with half-lives of months to years have been used in studying atmospheric mixing processes, while the shorter-lived ones such as ^{38}Cl or ^{38}S have been used as natural tracers for atmospheric aerosols and precipitation processes. The longer-lived ones (T, ^{14}C, and ^{10}Be) serve as tracers of geological processes and in dating biological and geological material. However, only T and ^{14}C are of sufficient importance to deserve some detailed discussion.

12.1.1. *Tritium*

Satellite measurements have shown that the earth receives some of the tritium ejected from the sun. However, by far the larger amount of tritium found in nature comes from bombardment of atmospheric nitrogen by fast neutrons

$$n(\text{fast}) + {}^{14}N \rightarrow {}^{12}C + {}^{3}H \qquad (12.1)$$

The yield for this reaction is about 2500 atoms tritium per second per square meter of the earth's surface; the global inventory is therefore about 3.5 kg. The average residence time of the tritium in the atmosphere is 1.6 y which is a small fraction of its half-life of 12.35 y.

The tritium concentration in natural hydrogen is measured in TU or tritium units, 1 TU = 1 atom ^{3}H per 10^{18} atoms ^{1}H. A tritium unit corresponds to approximately 7.1 dpm/l of water. Prior to 1952 the concentration of tritium in water was normally 1–10 TU. However, after the hydrogen bomb tests conducted in the atmosphere during the decade of the fifties and early sixties such a large amount of tritium was added to the atmosphere that by 1962 the tritium content in organisms and in water was between 100 and 15000 TU, and in the atmosphere as high as 10^6 TU; in 1968 the atmospheric value had decreased to 30000 TU. In 1977 rain and surface water in Europe contained $\sim$ 150 TU ($\sim$ 500 pCi l^{-1}); the Rhine contained about twice as much due to releases from nuclear power stations. The ^{3}H content in food in the USA in 1961–1963 was approximately 400 dpm/g of hydrogen, still well below the accepted biological level ($\sim 10^5$ dpm g^{-1}). It will take approximately 100 years of no further atmospheric testing of hydrogen bombs before the tritium content in nature is reduced again to its normal pretesting level.

Water exposed to the atmosphere is constantly having its tritium content renewed so that a balance exists between the rate of renewal and the rate of decay. Consequently, it is possible by measurement of the tritium content to date water or aqueous solutions (e.g. wine) that have been isolated from contact with the atmosphere for some time.

The circulation of waters in lakes and oceans as well as the exchange of water in the atmosphere has been measured by this technique. In such investigations the principal difficulty lies in the extremely low concentrations of the tritium, although it can be increased through isotope enrichment. For example, electrolysis of alkaline solutions enriches ^3H in the water phase by a factor 100–1000.

12.1.2. 14-*Carbon*

The formation of tritium occurs via the bombardment of atmospheric nitrogen with energetic neutrons from cosmic radiation. However, if the nitrogen is struck by a neutron which has been thermalized, the reaction leads to the formation of ^{14}C:

$$n(\text{slow}) + {}^{14}N \rightarrow {}^{14}C + {}^{1}H \qquad (12.2)$$

This reaction occurs with a yield of approximately 2.2×10^4 atoms $^{14}C\ s^{-1}\ m^{-2}$ of the earth's surface; from this a global annual production rate of 0.042 MCi is calculated. The half-life of ^{14}C is 5736 y, although 5570 is commonly used as a ^{14}C standard year. The global equilibrium inventory has been calculated to be about 63 t, or 280 MCi. It is reasonable to assume that the production of ^{14}C in the atmosphere has been constant for at least a million years, which means that equilibrium exists between the rates of formation and decay of the ^{14}C in the atmosphere. Moreover, the half-life of ^{14}C is sufficient to allow equilibrium between the ^{14}C in the atmosphere and the exchangeable carbon in natural materials. In 1952 the specific activity due to ^{14}C in natural carbon was 15 dpm g^{-1} of carbon, which is not too difficult a level of activity to detect. Like tritium through the test of nuclear weapons the amount of ^{14}C has been considerably increased; the amount added to the atmosphere is estimated to be 1500 kg ^{14}C. In 1963 investigations on food showed that the specific activity of ^{14}C was 15–25 dpm g^{-1} while the concentration of ^{14}C in atmospheric CO_2 had increased to about 22 dpm g^{-1}. Many hundreds of years will be required before the specific activity of ^{14}C in carbon is reduced to the values which predated atomic weapons testing; the reduction is due to exchange with carbon in the oceans and precipitations to ocean bottoms, and dilution of indigenous carbon. On the other hand, the combustion of fossile fuels yields CO_2, which increases the amount of inactive carbon in the atmosphere. This dilution is about 3% for the period 1900–1970.

12.1.3. *Dating by* ^{14}C

The discovery that all living organic material has a certain specific radioactivity due to ^{14}C led W. Libby to a new method for determination of the age of biological material. This method, which has been of great importance to archeology, is based on the assumption that after the organism incorporating the biological material has died, no exchange takes place between the carbon atoms of the material and those of the surroundings. Thus, the number of ^{14}C atoms in such material decreases with time according to the half-life of ^{14}C. The equation for this dating procedure is

$$S\,(\text{dpm g}^{-1}) = 15\,e^{-0.693t/5736} \qquad (12.3a)$$

or

$$t\,(\text{y}) = \frac{(\log 15 - \log S)\,5736}{0.301} \qquad (12.3b)$$

For example, if the specific activity of a sample is measured to be 0.1 dpm g^{-1}, then (12.3b) gives a value of 41 300 years as the time since the material ceased to exchange its carbon. Only with extreme care and very sophisticated equipment can a specimen this old be determined with reliability, but shorter times can be measured more accurately since the specific activities are larger.

There are many interferences that must be considered in the use of carbon dating such as the possibility that exchange of carbon with the surroundings occurred after the specimen died, the possibility of isotopic effects between ^{14}C and ^{12}C in the biological material, and whether the increase in the ^{14}C content in the atmosphere in the last two decades has affected the measured specific activity.

A large number of important age determinations have been made with the ^{14}C dating method. It was believed that a large ice cap had covered parts of the North American continent until about 35 000 years ago. Dating of wood and peat by the ^{14}C method has shown that the ice must have lasted until about 11000 years ago. Moreover, several hundred pairs of sandals found in a cave in Oregon have been shown to be about 9000 years old, indicating that tribes with significant cultures had developed soon after the withdrawal of the ice cap from North America. By analyzing inorganic carbon in bone apatite, it has been determined that hunters in Arizona killed mammoths there 11 260 years ago. Another example of the use of ^{14}C dating which has attracted widespread attention involves the Dead Sea scrolls. There was considerable controversy about their authenticity until ^{14}C dating showed that their age was slightly more than 1900 years. This proved them to be about a thousand years older than any other documents connected with the Old Testament.

The assumption that cosmic radiation has been constant for long geological periods has been confirmed through ^{14}C dating of archeological materials whose ages can be ascertained through other methods. For example, by counting the annual rings on old trees such as the *Sequoia gigantea*, which can be almost 4000 years old, and by determining the ^{14}C content of each ring, the variation of cosmic radiation through several thousand years can be studied accurately.

12.2. Long-lived naturally occurring radionuclides

As the detection technique for radioactivity has been refined, a number of long-lived radionuclides have been discovered in nature. The lightest ones have been mentioned in §12.1. The heavier ones up to $_{82}Pb$ are listed in Table 12.2. The even heavier radionuclides of the natural decay series are presented in Fig. 1.4. In addition to these, both $_{93}Np$ and $_{94}Pu$ have been discovered on earth. ^{237}Np, with $t_{1/2}$ 2.14 × 10^6 y and ^{239}Pu with 2.44 × 10^4 y are too short-lived to have survived the 4 eons since the solar system was formed. These nuclides are always found in minerals containing uranium and

TABLE 12.2. *Naturally occurring radioactive nuclides between* $_{18}$Ar *and* $_{82}$Pb

Nuclide	Isotopic abundance (%)	Decay mode	Half-life (years)
^{40}K	0.011 9	β^-, EC	1.3×10^9
^{50}V	0.24	EC	6×10^{15}
^{87}Rb	27.85	β^-	4.7×10^{10}
^{113}Cd	12.3	β^-	9×10^{10}
^{115}In	95.72	β^-	6×10^{14}
^{123}Te	0.87	EC	1.24×10^{13}
^{138}La	0.089	β^-, EC	1.3×10^{11}
^{144}Nd	23.85	α	2.1×10^{15}
^{147}Sm	15.0	α	1.1×10^{11}
^{148}Sm	11.2	α	7×10^{15}
^{152}Gd	0.20	α	1.1×10^{14}
^{156}Dy	0.06	α	2×10^{14}
^{176}Lu	2.6	β^-	3×10^{10}
^{174}Hf	0.18	α	2×10^{15}
^{187}Re	62.60	β^-	5×10^{10}
^{186}Os	1.6	α	•2×10^{15}
^{190}Pt	0.012 7	α	6×10^{11}

thorium, and it is assumed that the neutrons produced in these minerals through (α, n), (γ, n), and spontaneous fission of ^{238}U form the neptunium and plutonium through n-capture and β-decay processes. The n-production rate in the uranium mineral pitchblende (containing 50% U) is about 50 n kg^{-1} s^{-1}.

The typical value for the ^{239}Pu/^{238}U ratio in minerals is 3×10^{-12}. However, values as high as 10^{-5} to 10^{-7} have been found in volcanic samples, which cannot have been contaminated with atmospheric ^{239}Pu debris from nuclear weapons testing. Two explanations have been suggested: (1) the ^{239}Pu is a daughter product of unknown super-heavy long-lived elements in the volcanic interior; (2) a much higher neutron flux exists (or existed in recent geologic time) in the deep geologic formations. The Oklo-type natural reactor (see Chapter 19) is an example of this possibility.

The long-lived plutonium isotope ^{244}Pu, which decays through α-emission and spontaneous fission (0.13%) with a total half-life of 8.26×10^7 y, was discovered in rare earth minerals in 1971. If this is a survival of primeval ^{244}Pu, only 10^{-15}% of the original can remain. An alternate possibility is that this ^{244}Pu is a contaminant from cosmic dust (e.g. from a supernova explosion in more recent times than the age of the solar system).

In Table 12.2 ^{50}V is the nuclide of lowest elemental specific activity (0.006 dpm g^{-1}) while the highest is ^{87}Rb (5.4×10^4 dpm g^{-1}) and ^{187}Re (5.3×10^4 dpm g^{-1}). The latter specific activities are not difficult to measure, but the specific activity of ^{50}V is so low that there is serious doubt that ^{50}V is radioactive. As our ability to make reliable measurements of low activities increases, the number of elements between potassium and lead with radioactive isotopes in nature is expected to increase. Because of their long half-lives, many of the nuclides in Table 12.2 play an important role in dating geologic samples.

12.3. Elements in the radioactive decay series

In Chapter 1 the existence of four series of radioactive nuclides was discussed briefly. The first series, known as the *thorium series*, consist of a group of radionuclides related through decay in which all the mass numbers are evenly divisible by four (the 4*n series*). This series has its origin in the radionuclide ^{232}Th (abundancy $x = 100\%$; specific activity $S = 2.4 \times 10^5$ α-dpm g^{-1}), which undergoes α-decay with a half-life of 1.41×10^{10} y. The *terminal nuclide* in this decay series is the stable species ^{208}Pb, which is also known as ThD. The transformation from the original parent ^{232}Th to the final product ^{208}Pb requires 6 α- and 4 β-decays. The longest-lived intermediate is 6.7 y ^{228}Ra.

The second series consist of a group of radionuclides whose mass number when divided by 4 has a remainder of 1 (the 4*n* + 1 *series*). This series is known as the *neptunium series* since the longest-lived, and therefore parent, species is ^{237}Np, which undergoes α-decay with a half-life of 2.1×10^6 y. Inasmuch as this half-life is considerably shorter than the age of the earth, ^{237}Np can no longer exist on earth, and, therefore, the neptunium series is not found as a natural occurrence. However, it has been discovered in the spectra of some stars. ^{237}Np is formed in nuclear reactors from uranium by thermal neutrons

$$^{235}\text{U}(\text{n},\gamma)^{236}\text{U}(\text{n},\gamma)^{237}\text{U} \xrightarrow{\beta^-} {}^{237}\text{Np} \tag{12.4}$$

and by fast neutrons

$$^{238}\text{U}(\text{n},2\text{n})^{237}\text{U} \xrightarrow{\beta^-} {}^{237}\text{Np} \tag{12.5}$$

Tens of kilograms of ^{237}Np have been produced (the total production rate in the present nuclear power plants approaches 1 kg d^{-1}) and all the members of the neptunium decay series have been isolated and identified. The end product of the neptunium series is ^{209}Bi, which is the only stable isotope of bismuth. Seven α-decays and 4 β-decays are required in the sequence from the parent ^{237}Np to ^{209}Bi. An important nuclide in the neptunium decay series is the uranium isotope ^{233}U, which has a half-life of 1.8×10^5 y (the most stable intermediate) and, like ^{235}U, is fissionable by slow neutrons.

The *uranium decay series* consist of a group of nuclides that, when their mass number is divided by 4, have a remainder of 2 (the 4*n* + 2 *series*). The parent of this series is ^{238}U ($x = 99.3\%$), which undergoes α-decay with a half-life of 4.47×10^9 y. The specific activity of uranium, including ^{235}U and ^{234}U, is 1.522×10^6 α-dpm g^{-1}. The stable end product of the uranium series is ^{206}Pb, which is reached after 8 α- and 6 β-decay steps. This is a particularly important series in nature since it provides the more important isotopes of the elements radium, radon, and polonium, which can be isolated in relatively large amounts in the processing of uranium minerals. Each tonne of uranium is associated with 0.340 g of ^{226}Ra. Since the annual uranium production is of the order of 30 000 t y^{-1}, considerable amounts of radium are present in the mine wastes.

Freshly isolated ^{226}Ra reaches radioactive equilibrium with all its daughter products to ^{210}Pb in several weeks (see Fig. 1.4). Many of the daughters emit energetic γ-rays. Therefore radium has been used as a γ-source in medical treatment of cancer (radiation therapy), either directly or as a source for the short-lived ($t_{1/2}$ 3.8 d) daughter ^{222}Rn. Prior to the atomic energy projects of World War II, the primary purpose of processing uranium ores was the isolation of radium, and more than a kilogram has been produced. However, the medical importance of radium has declined greatly since the introduction of other radiation sources. At present the largest use of radium is as a small neutron source (see Table 8.5).

Although the chemistry of radium is simple (like barium), the fact that it produces a radioactive gas (radon) complicates its handling. The decay of radon produces "airborne" radioactive atoms of At, Po, Bi, and Pb which may cause contamination of the surroundings. Work with radium compounds should be carried out within enclosures to avoid this contamination.

The inert gas radon diffuses from minerals of thorium and uranium, and adds radioactivity to the atmosphere by both its own presence and that of its daughters. The average concentration of radon near ground level is normally 400 dpm m^{-3}, almost all in the form of ^{222}Rn. This concentration is higher near large uranium or thorium deposits. In many places in the world (so-called "spas") radioactive wells are considered beneficial to health both for bathing and for drinking. The primary source of this radioactivity is dissolved radium. For example, famous spas in Europe have a radium content in the water of > 0.1 μCi m^{-3} ($> 10^5$ dpm m^{-3}) and a radon concentration of > 100 μCi m^{-3} ($> 10^8$ dpm m^3).

The fourth radioactive decay series, known as the *actinium series*, consists of a group of nuclides whose mass number divided by 4 leaves a remainder of 3 (*the* $4n + 3$ *series*). This series has its start with the uranium isotope ^{235}U (previously known as actinouranium), which has an abundance of 0.71% and a half-life of 7.04×10^8 y for α-decay; the specific activity of ^{235}U is 4.80×10^6 dpm g^{-1}. The stable end product of the series is ^{207}Pb, which is formed after 7 α- and 4 β-decays. The actinium series includes the most important isotopes of the elements protactinium, actinium, francium, and astatine. Inasmuch as ^{235}U is a component of natural uranium, these elements can be isolated in the processing of uranium minerals. The longest-lived protactinium isotope, ^{231}Pa ($t_{1/2}$ 3.4×10^4 y) has been isolated on the 100 g scale, and is the main isotope for the study of protactinium chemistry. ^{227}Ac ($t_{1/2}$ 21.8 y) is the longest-lived actinium isotope.

12.4. Age determination of geological materials

Prior to the discovery of radioactivity, geologists could obtain only poor estimates of the time scale of the evolution of the earth. The oldest geologic materials were assumed to be some 10 million years old, and it was believed that this represented the age of the earth. However, with the discovery of radioactivity early in this century, geologists developed more objective methods for such age determination ("nuclear clocks"). Thus in 1906 Rutherford estimated the ages of uranium and thorium minerals from their helium content and found them to be 400×10^6 y old. A year later B. B. Boltwood obtained a value of 2.2×10^9 y from the assumption that all uranium and thorium ultimately decayed to lead. When we realize how few isotopes in the chains had been discovered at the time, the calculation is surprisingly good. Nuclear clocks have given us a completely new picture of the age and evolution of the earth as well as the formation of elements and of the universe. These new areas, which are considered briefly below, are called *nuclear geochronology* and *cosmochronology*.

The techniques used are selective chemical isolation, mass spectrometric isotope analyses, absolute radioactivity counting, isotope dilution, fission track counting, thermoluminescence, etc. These are all described elsewhere in this book.

Nuclear clocks fall into two categories: "decay clocks", like ^{14}C and T, which follow simple exponential decay laws, and "accumulation clocks", which are based on the growth of decay products. Only the latter are used in geochronology and cosmochronology.

12.4.1. *Helium content of minerals*

In the uranium decay series 8 α-particles are emitted in the decay from ^{238}U to ^{206}Pb. Thus for every 8 helium atoms found in a uranium mineral, one atom of ^{238}U must have decayed to ^{206}Pb. If we designate the number of original uranium atoms as N^0, the number which has decayed with time t would be $N^0 - N$, where N is the number of uranium atoms present now. This leads to the relationship $N^0 - N = N_{He}/8$, where N_{He} is the number of helium atoms. Thus

$$N_{He}/8 = N^0_{238} - N_{238} = N^0_{238} - N^0_{238}e^{-\lambda_{238} t} = N^0_{238}(1 - e^{-\lambda_{238} t}) \tag{12.6}$$

and dividing by N_{238}

$$\frac{N_{He}}{8N_{238}} = \frac{N^0_{238}(1 - e^{-\lambda_{238} t})}{N^0_{238}e^{-\lambda_{238} t}} = e^{\lambda_{238} t} - 1 \tag{12.7}$$

From a determination of the amounts of helium and ^{238}U it is possible to calculate the time t, i.e. the age of the mineral.

Unfortunately, the method is not quite as simple as it first seems since uranium contains both ^{235}U as well as ^{238}U and many uranium minerals contain small amounts of ^{232}Th. It is, therefore, necessary to correct for the formation of helium from the decay of the ^{235}U and ^{232}Th series. Further, if the mineral has lost any helium through diffusion or other processes during its existence, the helium content would be abnormally low, leading to erroneously small values of t. This method, therefore, can only give lower limits of the ages of minerals.

12.4.2. *Lead content of minerals*

Lead has four stable isotopes of which three are end products of radioactive decay series. The fourth lead isotope, ^{204}Pb, is found in lead minerals in about 1.4% isotopic abundance and has no radiogenetic origin. At the time of formation of the earth, all the ^{204}Pb in nature must have been present mixed with unknown amounts of the other lead isotopes. If a lead-containing mineral lacks ^{204}Pb, it can be assumed that presence of the other lead isotopes together with uranium and/or thorium must be due to their formation in the decay series. In such ^{204}Pb-free minerals if it is possible to determine the amount of the parent nuclide ^{238}U and of the end product ^{206}Pb, the age of the mineral can be determined from the general equation

$$t = \frac{1}{\lambda} \ln\left(1 + \frac{N_d}{N_p}\right) \tag{12.8}$$

where N_p are the number of parent atoms (e.g. ^{238}U) and N_d the number of *radiogenic* daughter atoms (e.g. ^{206}Pb), and λ is the decay constant of the parent.

This method is more reliable than the helium method since there is far less possibility of any of the lead that has been formed by radioactive decay having diffused or been leached from the mineral during its geologic age. Mineral samples from the earth have yielded values as great as 3×10^9 y by the lead content method. Unfortunately, there is a problem with this method also. The decay series all pass through isotopes of the inert gas radon, and if the radon is lost from the mineral the ^{206}Pb content leads to an erroneously low age.

Since the lifetimes of the uranium isotope ^{238}U and ^{235}U are different, the isotopic

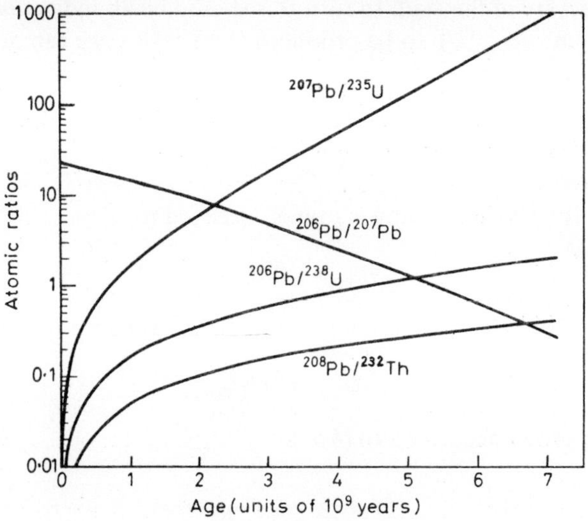

FIG. 12.1. Atomic ratios at the time of measurement as a function of the age of the minerals. (According to E. K. Hyde.)

ratio between their end products ^{206}Pb and ^{207}Pb can also be used for age determination. One can derive the relationship

$$N_{207}/N_{206} = \frac{1}{138}\frac{(e^{\lambda_{235}t} - 1)}{(e^{\lambda_{238}t} - 1)} \tag{12.9}$$

where the factor 1/138 is the present isotopic abundancy ratio of the uranium isotopes, i.e. N_{235}/N_{238}. This method has given values of 2.6×10^9 y for uranium and thorium minerals. When applied to stony meteorites, a somewhat longer age of $4.55(\pm 0.07) \times 10^9$ y is obtained. Figure 12.1 shows the variation of different lead, uranium, and thorium atomic ratios with time. If several such nuclear clocks agree, the age determination can be considered to be fairly reliable.

12.4.3. *Art dating**

Art forgery is unfortunately a common criminal activity. The authenticity of an art object can only be determined by the artist, who may no longer be available as most valuable art objects are rather old.

Nuclear chemical techniques have been of substantial aid in determining the age of such objects. We have already mentioned the carbon-14 dating, and in Chapter 18 we give another example in the use of activation analyses for art identification. Let us consider a specific case of dating paintings from measurements of their radioactive ^{210}Pb content. Lead is used in paintings as "white lead", i.e. an oil emulsion of lead sulfate.

When lead is extracted from the ore it is in secular equilibrium with its precursors radium and uranium. The radium and most of its descendants are removed during processing while the ^{210}Pb accompanies the other lead isotopes. The separation of radium from lead is not always complete. Because of the long half-life of ^{226}Ra (1600 y) as compared to the short-lived ^{210}Pb (22.3 y), the determination of the excess ^{210}Pb over the equilibrium amount received from ^{226}Ra decay provides a scale for the time since the manufacture of the lead. In turn this should be an indication of the age of the object. If we assume all ^{210}Pb to be produced from ^{226}Ra, we can derive the following relations:

At the time the lead was isolated, there was a secular equilibrium between the isotopes

$$\lambda_{226}N^0_{226} = \lambda_{210}N^0_{210} \tag{12.10}$$

The N^0_{210} atoms of ^{210}Pb have during time t decayed to

$$N_{210} = N^0_{210}e^{-\lambda_{210}t} \tag{12.11}$$

At the time of the lead isolation, a fraction x of the ^{226}Ra accompanied the lead

$$N'_{226} = xN^0_{226} \tag{12.12}$$

From this ^{210}Pb grows according to (4.54)

$$N'_{210} = \frac{\lambda_{226}}{\lambda_{210} - \lambda_{226}} xN^0_{226}(e^{-\lambda_{226}t} - e^{-\lambda_{210}t}) \tag{12.13}$$

N^0_{226} can be calculated from

$$N_{226} = N^0_{226}e^{-\lambda_{226}t} \tag{12.14}$$

Combining these equations, and putting $\lambda_{210} - \lambda_{226} \approx \lambda_{210}$, we get

$$\sum N_{210} = N_{226}\lambda^{-1}_{210}e^{\lambda_{226}t}(x\,e^{-\lambda_{226}t} - (1-x)\,e^{-\lambda_{210}t}) \tag{12.15}$$

N_{226}, the amount of ^{226}Ra, and N_{210}, the amount of ^{210}Pb, can be measured with great accuracy (to 0.001 cpm for α-active samples). Because ^{210}Pb is a β^- emitter, the measurements are preferentially done on the α-emitting daughter–daughter ^{210}Po. This leaves us with two unknowns, x and t. It is found experimentally that x is 0.1–0.001; the average of a large number of paintings < 80 y old gives $x \approx 0.01$. By comparison with other objects of the same artist, for which t is known with certainty, an acceptable x-value can be obtained, which then permits a reliable estimate of t for the object in doubt.

For example, paintings by the world-famous Dutch artist Vermeer van Delft (1632–1975) had a value of $(1 - R_{Ra}R^{-1}_{Pb})$ of ~ 0.10, while the forgeries during 1930–1940 by van Meegeren gave a value of 0.97. R_{Ra} is the radioactivity measured from ^{226}Ra. It is obvious that when $R_{Ra} \ll R_{Pb}$, the paintings must be very recent. The paintings by van Meegeren were so excellent in style that every authority accepted his forgeries as authentic works. However, threatened by the death penalty for selling "Vermeer" paintings to the Germans during World War II, van Meegeren confessed the forgery. These particular forgeries have since become valuable objects for art collectors in their own right.

12.4.4. *Potassium and rubidium content of minerals**

^{40}K$(1.3 \times 10^9$ y) and ^{87}Rb$(6.2 \times 10^{10}$ y), which occur with their respective elements in nature, have lifetimes of the same magnitude as the age of the earth. Their decay equations are

$$^{40}\text{K} \begin{cases} \xrightarrow{\text{EC, }11\%} {}^{40}\text{Ar} & (99.6\% \text{ of natural Ar}) \\ \xrightarrow{\beta^-,\ 89\%} {}^{40}\text{Ca} & (96.8\% \text{ of natural Ca}) \end{cases} \qquad (12.16)$$

$$^{87}\text{Rb} \xrightarrow{\beta^-} {}^{87}\text{Sr} \quad (7.02\% \text{ of natural Sr}) \qquad (12.17)$$

Considering these decay schemes, it can be shown that for the K–Ar clock (12.16) must be written

$$t = \lambda_{tot}^{-1} \ln\left(\frac{N_{Ar}}{0.11 N_K} + 1\right) \qquad (12.18)$$

where λ_{tot} is the total decay constant (see §4.12), N_{Ar} is the number of radiogenic atoms of ^{40}Ar, and N_K is the present atomic abundance of ^{40}K. Some uncertainty is associated with the use of the ^{40}K/^{40}Ar ratio because of the possibility of the loss of gaseous argon from minerals. It is probable that practically all the ^{40}Ar in our atmosphere has been formed through the radioactive decay of ^{40}K.

The measurement of the ^{87}Rb/^{87}Sr ratio by mass spectrometry is one of the most reliable methods for geologic age determinations; meteorite values as high as 4.6×10^9 y have been obtained. For the Rb–Sr clock a relation of type (12.9) is valid. Corrections must be made for nonradiogenic ^{87}Sr; it is believed that the primordial ^{87}Sr/^{86}Sr ratio was 0.70.

12.4.5. *Fission track dating**

The spontaneous fission rate of ^{238}U (the sf $t_{1/2}$ of 8×10^{15} y corresponds to 0.42 dpm/g ^{238}U) leads to the formation of fission tracks in all geological samples containing uranium. Because of the rather general occurrence of uranium, such minerals are common. Figure 17.2e shows such fission tracks in a sample from the Oklo mine. The tracks are about 1.7×10^9 y old.

If the U-concentration of a sample is known (N_{238}) and the concentration of fission tracks determined (N_{sf}), the age of the sample may be calculated from

$$N_{238} = (N_{238} + N_{sf})\, e^{-\lambda_{sf} t} \qquad (12.19)$$

where λ_{sf} is the spontaneous fission decay rate. This is the basis for *fission track dating*, which has become an important tool in cosmology, archeology, and geology. The absolute age of a sample can be determined over an enormous span of time, from as far back as the birth of the solar system up to only a few decades.

Three requirements must be fulfilled for the application of this technique (cf. §17.1):

(1) The fission tracks must be retained with 100% efficiency. Thus annealing due to time or heat causes erroneous results. Heat anneals the tracks, but normally the temperature must exceed the crystallization temperature. In many crystals, which have not been exposed to excessive heat since their formation, the tracks are retained. If there has been annealing due to high temperatures the tracks can be used to date the time of annealing. When pottery is fired, the crystals in the clay are "reset to zero". In Japan, fission tracks have dated pottery from 300 BC to 700 years ago.

(2) The concentration of uranium must not have changed during the years, nor should any other spontaneous fission decaying material ever have been present. While

correction for sf decay of ^{235}U is small (only about 1%), it has been found in meteoritic samples that have been dated with several techniques that fission track dating gives a higher age (the tracks are too numerous). It has been possible to deduce that some tracks have been formed by sf decay of ^{244}Pu, which has a half-life of only 8.3×10^7 y. No plutonium tracks are found in terrestrial samples: all ^{244}Pu was gone when the earth solidified. Some meteorites never melted, and those with ages of $4.6–4.0 \times 10^9$ y show that the original matter out of which they condensed contained a $^{244}Pu/^{238}U$ ratio of about 0.01 (also much higher values have been suggested).

12.4.6. *Thermoluminescent dating**

Most geological minerals such as quartz have the ability to emit visible light at temperatures below that at which red glow appears (*thermoluminescence*). This is due to the fact that the crystal contains imperfections in its lattice produced by high energy radiation during the ages, and which are "frozen" at normal temperatures (cf. §15.9) but released upon heating. The amount of light emitted is proportional to the radiation dose received. If the annual dose rate is constant and known, the age of the crystal can be calculated.

Within the clay of almost all low-fired earthenware there are many inclusions of free quartz, felspars, etc., of sizes from a few to several hundred microns. The quartz may be free of radioactivity, but the surrounding clay matrix often contains trace amounts of U, Th, and ^{40}K, which, along with cosmic radiation, produce an absorbed dose of the order of 0.1–1.0 rad y^{-1}. Very accurate radioactivity determinations of the matrix material are required for good results.

As an example, the age of a piece of Roman pottery found in England had been estimated by other means to be 1970y; thermoluminescence dating gave 1850 ± 30 y. This technique has become a valuable archeological tool. When used for authenticity testing of the famous Chinese glazed ceramics from the early 6th dynasties (about 1500 years ago), many forgeries were discovered. Many of these were made in China with the same kind of clay and glaze soon after the discovery of the true ceramics in the beginning of this century. It was not possible to distinguish them from the true antiques by chemical means, but thermoluminscence studies made the difference obvious.

12.5. Cosmochronology*

Estimates of the age of our galaxy must come mainly from theoretical calculations. However, radiochemical investigations can yield information about at least three important steps in the development of our planetary system: the formation of the elements in our solar system, the condensation of the cosmic dust into our planets, and the final solidification of the earth's crust. The latter information is obtained from the dating of geologic samples as already discussed. As long as the earth was molten, the different chemistry of mother and daughter elements caused them to separate. The dates obtained from these clocks give the age of the mineral; the oldest known minerals on earth are $\sim 3.7 \times 10^9$ y.

Some glassy materials in meteorites and from the moon are assumed to be original condensed matter, which never has undergone melting. The oldest ages seem all to converge at 4.6×10^9 y. Thus we conclude that this was the time when our planetary system began to form.

Further information about this event has been obtained by studying tracks which nuclear decay processes leave in certain minerals, as described in § 12.4.5. Fission tracks can only persist in minerals that have not been heated because heating above 600°C causes the tracks to disappear. The fact that ^{244}Pu fission tracks have been found in iron meteorites and in lunar samples shows that ^{244}Pu existed when the planetary system formed. Because of the short half-life of ^{244}Pu (8×10^7 y) it can be concluded that the mineral sample must have formed within a few hundred million years after the nuclide ^{244}Pu itself was formed. This is probably also the time for planetary formation. The existence of primordial plutonium indicates that a nucleogenetic r-process took place during the formation of the planets. Our solar system was showered by debris from a recent but distant supernova explosion; this explosion might have disturbed the "peaceful" gas cloud in our part of the universe and in fact initiated the formation of our solar system.

The spontaneous fission decay of ^{244}Pu yields ^{129}I, which decays as

$$^{129}\mathrm{I} \xrightarrow[1.6\times 10^7\ \mathrm{y}]{\beta^-} {}^{129}\mathrm{Xe}\ \text{(stable)}$$

The fact that meteorites containing about 100 times excess fission tracks also contain excess ^{129}Xe is strong support both for the existence of ^{244}Pu and that the meteorites condensed quite rapidly out of freshly produced new elements. The study of xenon isotope ratios and their relation to uranium and plutonium concentrations and the age of the elements is known as "xenology".

It is possible to go a step further back in the history of the universe and calculate the time when the elements in the solar system were formed assuming that all elements were not formed at the same time as the ^{244}Pu. Let us assume two radioactive nuclides X_i and X_j (e.g. ^{235}U and ^{238}U), which are formed through a neutron capture process during the time t_{ir}. When $t_{ir} = 0$ the elemental synthesis begins (the formation of our galaxy, or of the universe, depending on the hypothesis; cf. Chapter 11), and when $t_{ir} = t$ all elements have been formed. Some time passes, which we (analogous with § 10.2) call the cooling time t_c. At the end of this time the elements begin to condense, so that the formation of the oldest meteorites date from the end of the cooling time. It is obvious that we can apply (10.7), which for the nuclide ratio becomes

$$\left(\frac{N_i}{N_j}\right)_{t_c} = \frac{k_i\lambda_j(1 - e^{-\lambda_i t_{ir}})e^{-\lambda_i t_c}}{k_j\lambda_i(1 - e^{-\lambda_j t_{ir}})e^{-\lambda_j t_c}} \tag{12.20}$$

λ_i and λ_j are known, and N_i and N_j can be deduced at time t_c for the samples, e.g. through radioactivity measurements, fission track counting, mass spectrographic analyses of accumulated daughter products, etc.

However, the production rates must be estimated. We shall consider the simplest way to do this, which is with the aid of Fig. 11.7. The successive neutron capture in the r-process leads to addition of a large number of neutrons to the target nucleus. This is followed by successive β^- decay, until stability or some decay reaction (α or sf) interrupts the sequence. For example, ^{238}U can be assumed to be formed through the reaction (a) $^{208}_{82}$Pb(30n,γ)$^{238}_{92}$Pb (10β^-)$^{238}_{92}$U. However, the reaction (b) $^{208}_{82}$Pb(34n,γ)$^{242}_{82}$Pb(12β^-) $^{242}_{94}$Pu also occurs, and—to a first approximation—in about equal yield. The half-life of the α-active ^{242}Pu is only 4×10^5 y, so it decays rapidly to ^{238}U. Similarly, ^{246}Cm (c) is formed, and decays into ^{238}U. The β^- decays (d), (e), and (f) all end in nuclides decaying through spontaneous fission, so there is no contribution to the ^{238}U. Since three chains

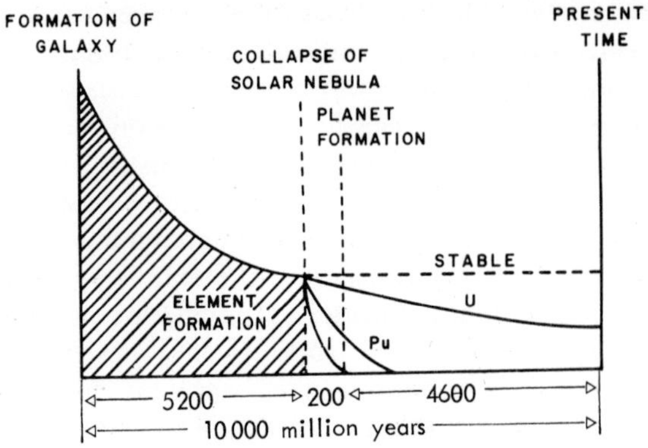

FIG. 12.2. Schematic showing (on the left) the rate of element production, which declines with the age of the galaxy, and (on the right) the concentrations of elements within the solar system. Isotopes such as ^{129}I and ^{244}Pu were present at the beginning of the collapse of the solar nebula. At this time newly formed elements stopped being added to the solar system from the galaxy. Sufficient amounts of ^{129}I and ^{244}Pu remained after the formation of solid materials to produce characteristic isotope anomalies and fission track excesses in meteoritic and lunar materials. (According to Fleischer, Price, and Walker.)

yield ^{238}U in equal amounts, the production rate of ^{238}U is 3.0. For other nuclides, by the same reasoning, one gets for ^{244}Pu 2.9, for ^{232}Th 5.9, and for ^{235}U 6.0. The production ratio for ^{235}U/^{238}U will be 6.0/3.0 = 2.0. A more refined calculation, considering the known neutron capture cross-sections, reduces this value to 1.65.

It may be doubted if as many as 30 neutrons can be added to ^{208}Pb to produce a ^{238}Pb nuclide. However, this is not a serious objection because it is immaterial if β-decay steps occur before $A = 238$ is reached. As can be seen from the thermonuclear weapons experiment shown in the shaded part of Fig. 11.7, the general principle holds.

In order to solve (12.20) for the two unknowns t_{ir} and t_c, two nuclear clocks are needed. A common pair, in addition to those already mentioned, is ^{129}I/^{127}I, which has been very thoroughly investigated. The production rate ratio is assumed to be 1.3. For the ^{235}U/^{238}U–^{129}I/^{127}I pairs, values of $t_{ir} = 9.9 \times 10^9$ y and $t_c = 80 \times 10^6$ y have been obtained. Adding the age of the sample, 4.6×10^9 y, to t_{ir} gives 14.5 eons for the time of the formation of the elements. This is a high value for the age of our galactic system, and more refined calculations (by C. M. Hohenberg, assuming several successive r-processes, but of different intensity) have modified the value to 8–9 eons. The t_c value has also been modified to about 0.18 eons.

On the basis of investigations of this kind and on theories of elemental formation discussed in Chapter 11, a picture has emerged of the formation of the elements, our galaxy and solar system, as shown in Fig. 12.2. The existence of ^{129}I and ^{244}Pu at the time of the planet formation indicates that some elements were formed later than when the galaxy formed. According to some theories, our solar system was repeatedly showered in earliest times by debris from supernovae.

12.6. Uranium

The only important use of uranium is as fuel in the nuclear energy industry and in weapons. However, the high density of metallic uranium (18.8 g cm^{-3}) makes depleted uranium (i.e. the tails from isotope enrichment plants), which is less valuable than natural uranium, useful for compact radiation shielding (§ 14.10).

12.6.1. *Isotopes*

Natural uranium consists of 99.3% ^{238}U and 0.72% ^{235}U. It has already been noted that these radioactive isotopes are the parent nuclides for the uranium and actinium decay series, respectively. The specific radioactivity of uranium is sufficiently low that in amounts of less than a kilogram, uranium can be treated without any particular radiological precautions. However, as with all of the heavy elements, it is a toxic substance: the threshold limit, 0.20 mg/m^3 air, is about the same as for lead (0.15 mg/m^3 air).

Both ^{238}U and ^{235}U are fissionable by fast neutrons but only ^{235}U is fissioned by thermal neutrons. A third important uranium isotope is ^{233}U, which does not exist in nature but can be produced through the irradiation of natural thorium with thermal neutrons.

$$^{232}\text{Th} + \text{n} \xrightarrow{\sigma 7.6\text{ b}} {}^{233}\text{Th} \xrightarrow[22\text{ min}]{\beta^-} {}^{233}\text{Pa} \xrightarrow[27\text{ d}]{\beta^-} {}^{233}\text{U} \xrightarrow[1.6\times10^5\text{ y}]{\alpha} \qquad (12.21)$$

It is estimated that >100 kg ^{233}U have been produced synthetically by this process. ^{233}U is also fissionable with thermal neutrons and represents the mode whereby thorium can be converted in a breeder nuclear reactor to fissionable material.

12.6.2. *Occurrence*

Uranium cannot be considered a very rare element. It is found in a large number of minerals (at least 60 are known) spread all over the globe. The earth's crust contains about 2.7 ppm U, which makes it about as abundant as arsenic or boron. The concentration of uranium in sea water is ~3.3 mg m^{-3}, with a variation of 0.5–5 mg m^{-3}, depending on the location.

The reason for the broad distribution of uranium is the relatively high solubility of UO_2^{2+} ions in carbonate, sulfate, and chloride solutions. The solubility of U^{4+} is much less; hence, when U(VI) is reduced to U(IV) (e.g. by organic matter), the U(IV) is sorbed or precipitates. In most minerals uranium is in the tetravalent state. The most important mineral is uraninite (a U(IV) + (VI)-oxide), in which the uranium concentration is 50–90%; it is found in Western Europe, Central Africa, and Canada. In the USA and the USSR carnotite (a K + U vanadate) is the most important mineral and contains 54% uranium. The historically famous mineral pitchblende is a variety of uraninite.

In the high grade ores the mineral is mixed with other minerals so the average uranium concentration in the crushed ore is much less: e.g. ≲0.5% on the Colorado Plateau. Uranium is often found in lower concentration, of the order of 0.01–0.03%, in association with other valuable minerals such as apatite, and in carbon-containing materials like shale, peat, etc. If these minerals are recovered for other purposes, the uranium may be produced as a byproduct for a price competitive with that for richer uranium ores.

The world reserves of uranium are estimated as about 3.1 Mt of uranium in the range ≲US $80 per kg,† and 1.1 Mt in the range US $80–130 per kg, about equally

† Prices refer to fall of 1978.

distributed as known and estimated reserves. The latter price range includes reserves containing down to ~ 300 g uranium per tonne ore (300 ppm). The largest low-priced reserves (i.e. highest grade) are found in the USA, South Africa, Australia, and Canada, while the largest reserves in the higher price group are found in Brazil, Spain, and Sweden. It has repeatedly been claimed that the reserves of very high price uranium (up to $\lesssim$\$500 per kg) are quite large, up to 50 Mt U_3O_8. At this high price it is said to be economical to recover uranium from sea water. It has been estimated that in mining ore with 50 ppm uranium, only 10% of the uranium energy content will be lost in the mining and recovery procedure. Mining of extremely low-grade ore for uranium (e.g. ore containing $\lesssim$ 50 ppm uranium) would most certainly constitute a severe environmental disturbance. The demand up to year 2000 is estimated as > 2.0 Mt natural uranium, if plutonium and uranium are returned to the fuel cycle (§20.3.2), or otherwise as > 2.3 Mt, for the nuclear program in Appendix I. A uranium price of US\$ 110 per kg corresponds to about 3 mills kWh^{-1} electricity produced (1 mill = US\$1/1000).

12.6.3. *Production*

Uranium ores differ widely in composition, containing a variety of other elements which must be removed. As a result the production methods differ considerably depending on the particular ore to be processed although in every case very selective processes must be used. The following is the common scheme:

The ore is mined in open pits or under ground. In the latter case, the mine must be well ventilated or face masks must be worn in order to avoid inhalation of radon and its radioactive daughter products.

Most commonly, the uranium ore is crushed and, if in a suitable mineral, concentrated through flotation. If the uranium exists in the tetravalent state it is oxidized to the hexavalent state by piling the uranium in large heaps and letting it oxidize in the air, sometimes with the aid of bacteria. The material is subsequently leached with sulfuric acid during which the uranium is dissolved as the sulfate complex, mainly $UO_2(SO_4)_2^{2-}$. The radioactive tailings containing radium and daughter products constitute a special problem; these radioactive wastes must be fixed so that they do not enter into the biosphere (see Chapter 21). The uranyl sulphate complex can be selectively removed from the aqueous solution by means of anion exchange resins or by extraction into an organic solvent (*solvent extraction*). When anion exchange resins are used the resin is usually contained in baskets, which are sunk into the pulp (i.e. crushed ore–H_2SO_4 mixture). More frequently a tertiary amine like tri-octyl amine in kerosene serves as a *liquid anion exchanger* in the solvent extraction separation. Denoting the amine liquid ion exchanger as R_3N, the extraction step may be written

$$2(R_2NH)_2SO_4(\text{org}) + UO_2(SO_4)_2^{2-}(\text{aq}) \rightleftharpoons (R_3NH)_4UO_2(SO_4)_3(\text{org}) + SO_4^{2-}(\text{org})$$

where org and aq refer to the liquid phase in which the compound exists.

A flow-sheet is illustrated in Fig. 12.3. The main part is a series of mixer–settler tanks (this type of chemical process equipment is described in §20.5.4)where the extraction of the uranium into the organic phase and subsequent washing out into a new aqueous phase ("stripping") takes place. In this process the uranyl complex can be removed from the organic phase with an aqueous solution of sodium carbonate, which transforms uranium into a nonextractable carbonate complex, mainly $UO_2(CO_3)_3^{4-}$. By adding ammonia or sodium hydroxide, uranyl precipitates out as ammonium diuranate or

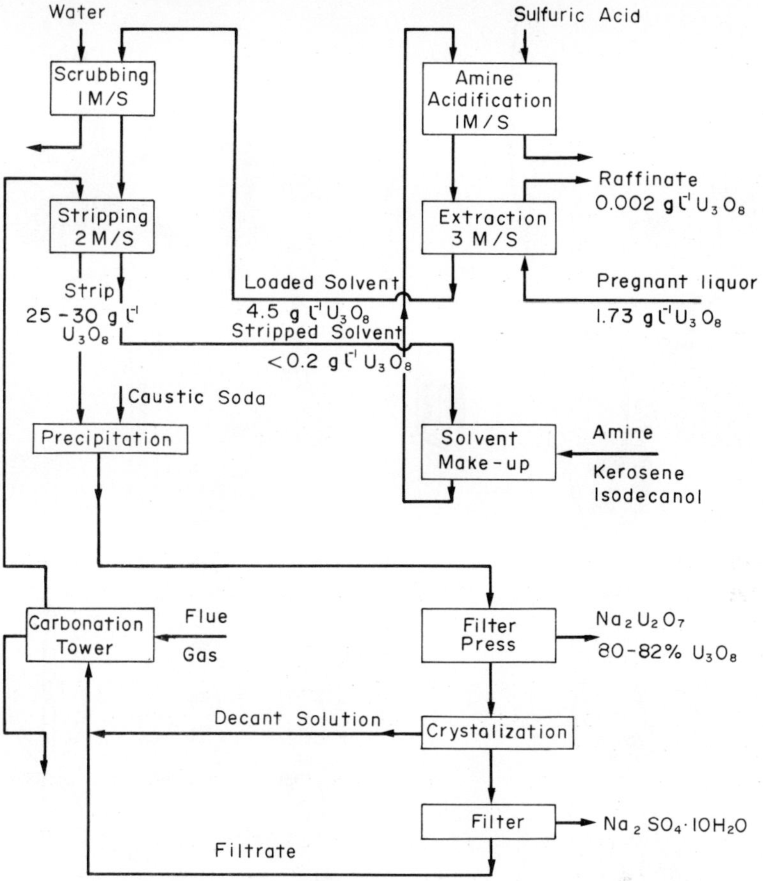

FIG. 12.3. General flow-sheet for the extraction of uranium with tertiary amine. 3 M/S means a battery of three mixer–settler stages. (According to S. O. Andersson, A. Ashbrook, D. S. Flett, G. M. Ritchey, and D. R. Spink.)

sodium diuranate, $Na_2U_2O_7$, which is filtered and dried. This intermediate product is referred to as "yellow cake", and contains 65–70% U. The ammonium salt can be heated to produce UO_3 ("orange oxide"). A uranium plant layout is pictured in Fig. 12.4. In this particular plant molybdenum also is recovered. Stripping is therefore with NaCl rather than caustic.

The yellow cake must be further treated in order to obtain a final pure product. This is usually done by an additional solvent extraction process of uranyl nitrate using tributyl phosphate, diethyl ether or methyl isobutyl ketone (MIBK or hexone). After precipitation of the uranium in the final strip solution, drying and heating forms UO_3, which is reduced by hydrogen gas to UO_2. This in turn can be contacted with HF to form UF_4 ("green salt").

Recently, a more direct process has been suggested in which uranium is obtained in an aqueous chloride solution after the extraction. Using a divided electrolytic cell with ion-exchanging membranes, the uranyl is reduced to U(IV) chloride and treated with HF in a liquid phase to yield UF_4.

The uranium tetrafluoride may be reduced through a thermite process with calcium

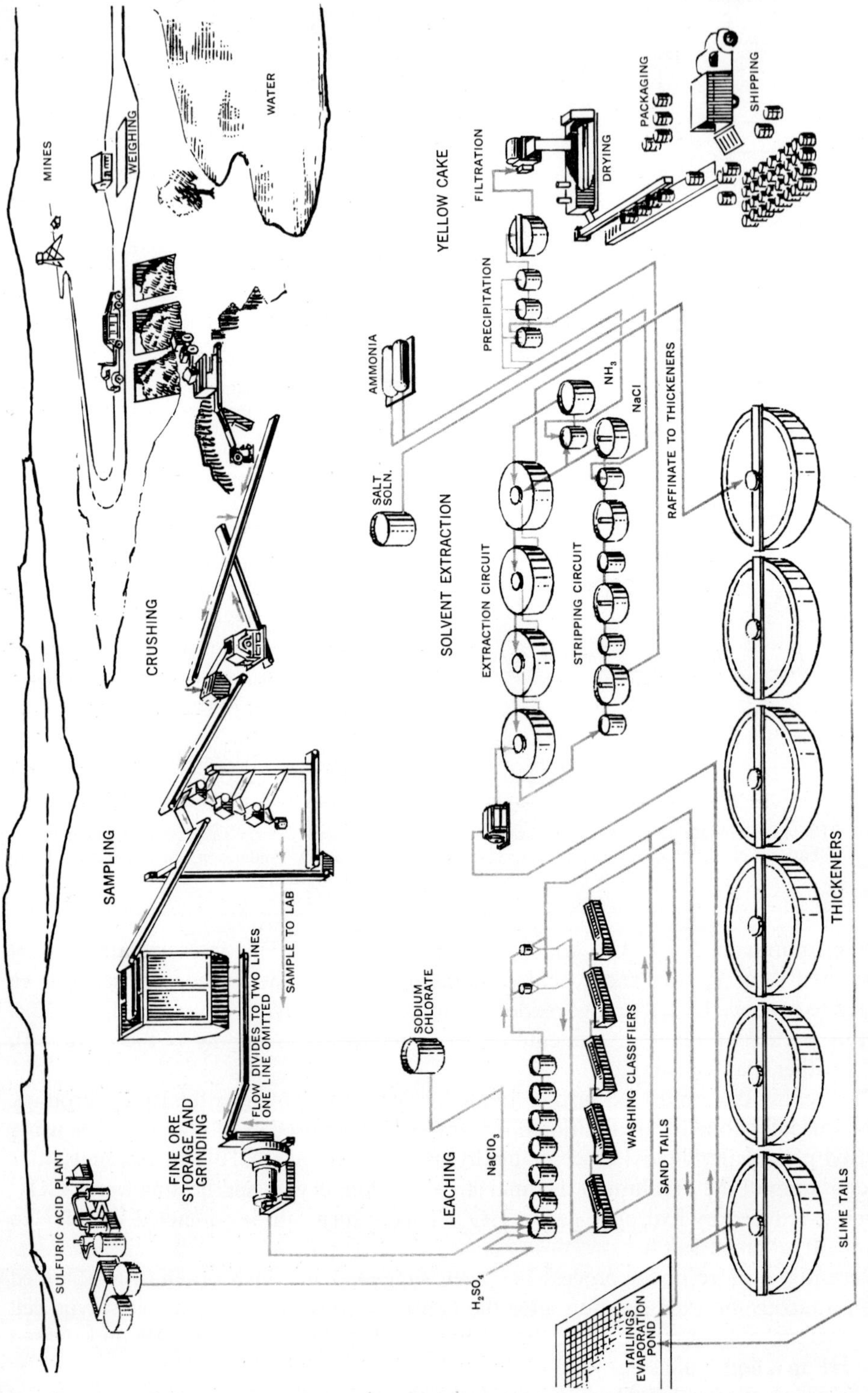

FIG. 12.4. Layout of the Kerr McGee uranium mill in Grants, New Mexico. Annual production is about 5000 t U_3O_8, which requires a mining of about 5000 t of ore per day.

metal to uranium metal, or treated with fluorine gas to form UF_6. The UO_2 may be heated with carbon at 1800° to produce uranium carbide, UC. Thus, while UO_3 and UF_4 are intermediate products, UO_2, UC, UF_6, and uranium metal are the final products. The UF_6 is used in gaseous diffusion processes for separation of uranium isotopes while the other final products are used as reactor fuel materials.

This process produces very pure uranium, which is a prerequisite for use as a fuel material in nuclear reactors. Commercial reactor uranium is usually better than 99.98% pure and its content of neutron poisons (nuclides which have high capture cross-sections for neutrons such as B, Cd, Dy) is less than 0.00002%.

Massive uranium metal oxidizes slowly in air at room temperature. A golden yellow color appears, which eventually darkens to black. This oxidation can be avoided by nickel plating. As a powder, uranium metal is very reactive at elevated temperatures and can be used to remove practically all impurities from rare gases. The metal is attacked slowly by water. Rapid reaction occurs with hydrochloric acid, and moderately rapid dissolution takes place with nitric acid. Only very slow reactions occur with sulfuric acid and hydrofluoric acid.

The chemistry of aqueous uranium is discussed in Chapter 13 together with the chemistry of the other actinides.

The world production was 23 kt U_3O_8 in 1977, with the industry running at about two-thirds of its capacity. Much more than this may soon be needed: the estimated annual demand in year 1985 is 60 kt and in the year 2000 about 150 kt U_3O_8. The discrepancy between demand and capacity has increased uranium prices from the long time (up to 1973) level of about \$6 per lb (\$13 per kg) to about \$40 per lb in September 1976, and contracted prices of \$54 per lb for delivery in 1980. In comparing capacity with demand, it should be realized that the lead time of the uranium industry is about 8 y, and that nuclear power producers have to contract uranium and enrichment services long before the true demand at the power station. The lead time is the time necessary for the uranium industry to react to a demand and expand output to satisfy it.

12.7. Thorium

The practical value of thorium is limited mainly to the nuclear energy industry as a fuel in high temperature gas-cooled reactors and possibly in the future in special thorium-breeder reactors.

12.7.1. *Isotopes*

Natural thorium consists 100% of the isotope ^{232}Th which is the parent nuclide for the thorium decay series. The specific radioactivity for thorium is even lower than that of uranium, and it is normally treated as a nonradioactive element. The principal interest for thorium in the atomic energy industry is its use in forming the nuclide ^{233}U. For radioactive tracer studies the nuclide ^{234}Th ($t_{1/2}$ 24.1 d) is used after separation from natural uranium: this separation is commonly and most simply done by shaking a concentrated aqueous acid uranyl nitrate solution repeatedly with diethyl ether, which removes the uranium to the organic phase, leaving the thorium in the aqueous solution.

12.7.2. *Occurrence and production*

Thorium is somewhat more common in nature than uranium, with an average content in the earth's crust of 10 ppm (by comparison the average abundance of lead is about 16 ppm in the earth's crust). In minerals it occurs only as oxide. The content of thorium in sea water has been found to be $< 0.5 \times 10^{-3}$ g m^{-3}, which is lower than that of uranium because of the lower solubility of Th^{+4} compounds.

The most common thorium mineral is monazite, a golden brown rare earth phosphate containing 1–15% ThO_2 and usually 0.1–1% U_3O_8. It is also found in small amounts in granite and gneiss. The largest deposits of monazite are found in India, Egypt, South Africa, the USA, and Canada, with 200–400 kt ThO_2 in each country. The total reserves at a price (1973) of $\lesssim$ \$20 per kg are estimated to about 2 Mt ThO_2. Because thorium often occurs with other valuable metals, in addition to the lanthanides, such as niobium (in Brazil), uranium (in Canada and the USA), and zirconium (in Egypt and India), it can be produced as a byproduct. The demand for ThO_2 has been so limited that the price has increased very little in past years.

The following procedure is used for producing thorium from monazite sand. The sand is digested with hot concentrated alkali which converts the oxide to hydroxide. The filtered hydroxide is dissolved in hydrochloric acid and the pH adjusted between 5 and 6, which precipitates the thorium hydroxide but not the main fraction of the lanthanide elements. The thorium hydroxide is dissolved in nitric acid and selectively extracted with methyl isobutyl ketone or tributyl phosphate in kerosene. This gives a rather pure organic solution of $Th(NO_3)_4$. The thorium is stripped from the organic phase by washing with alkali solution. No significant thorium requirements are expected before 1985 and no large thorium market exists at present.

12.8. Exercises

12.1. Cosmic-ray irradiation of the atmosphere yields 0.036 ^{10}Be atoms cm^{-2} s^{-1}. If this ^{10}Be is rapidly carried down into sea water, which is assumed to have a volume of 1.4×10^{18} m^3, what will the equilibrium radioactivity of ^{10}Be in 1 m^3 sea water be? The earth's surface area is 510×10^6 km^2.

12.2. In Greenland ice the ^{10}Be radioactivity has been measured to be 0.0184 dpm m^{-3}. How old is this ice if it was formed out of water in equilibrium with cosmic-ray ^{10}Be (see previous question)?

12.3. The CO_2 in the atmosphere is in exchange with carbon in living organisms, humus, dissolved organic compounds, and carbonate in the sea water, the latter being the main reservoir (88%) of all exchangeable carbon. The amount of such exchangeable carbon is estimated to be 7.9 g cm^{-2} of the earth. When cosmic-ray produced ^{14}C is mixed into this exchangeable carbon, what will the specific activity become?

12.4. On the label of a bottle of cognac bought in 1976 it is stated that the cognac is over 20 years old. An analysis showed a tritium content of 80 TU. Discuss the trustworthiness of the statement.

12.5. 0.11 cm^3 helium gas at NTP was isolated from 100 g of uranium mineral containing 5 ppm uranium. How old is the mineral?

12.6. A mineral was found to contain 39.1 g K and 872×10^{-6} l Ar at NTP. How old is the mineral?

12.7. A uranium mineral was found to contain the lead isotopes ^{204}Pb, ^{206}Pb, and ^{207}Pb in the ratio 200:830:280, as determined with a mass spectrometer. Estimate the age of the mineral.

12.8. The heat flow from the earth's crust is 0.060 J m^{-2} s^{-1}. The mean thickness of the crust is 17 km. The average concentration of uranium, thorium, and potassium in granite is estimated to be 4 ppm (by weight), 18 ppm, and 3.6%, respectively. Assuming that 70% of the crust is made up of granite (feldspar + quartz, density 2.6 g cm^{-3}), what will the heat flow at the earth's surface be from each of these elements? Assume β-heat as $1/3\ E_{max}$; for α-decay assume $E_\alpha = Q_\alpha$. Discuss the results.

12.9. A 1 GWe nuclear power station uses annually about 30 t uranium enriched to 3% ^{235}U. (a) How much natural uranium has been produced to keep it running? Assume waste stream from isotope separation plant to contain 0.3% ^{235}U. (b) How much low grade ore (assume 0.06% uranium) must be mined, if the uranium recovery efficiency in the process is 70%?

12.10. The assumed uranium resources in Japan are 4 kt, in Argentina 12 kt and in France 48 kt U_3O_8. How many 1 GWe reactor years can these uranium amounts sustain in each country at the uranium consumption rate (a) of the previous exercise?

12.9. Literature

W. F. Libby, *Radiocarbon Dating*, University of Chicago Press, 1956.

G. G. Marvin and E. F. Greenleaf, Methods of uranium recovery from ores, *Progr. in Nucl. Chem., Process Chemistry*, Pergamon Press, 1956.

M. Smutz, The recovery of thorium from monazite, ibid.

C. E. Junge, *Air Chemistry and Radioactivity*, Academic Press, 1963.

H. Craig, S. L. Miller, and G. J. Wasserburg (eds.), *Isotopic and Cosmic Chemistry*, North-Holland, 1963.

E. I. Hamilton and L. H. Ahrens, *Applied Geochronology*, Academic Press, 1965.

B. Keisch, Dating works of art through their natural radioactivity: improvements and applications, *Science* **160** (1968) 413.

C. M. Hohenberg, Radioisotopes and the history of nucleosynthesis in the galaxy, *Science* **166** (1969) 212.

E. E. Alexander *et al.*, ^{244}Pu-confirmation as an extinct radioactivity, *Science* **172** (1971) 837.

C. Keller, *The Chemistry of the Transuranium Elements*, Verlag Chemie GmbH, 1971.

D. York and R. M. Farquhar, *The Earth's Age and Geochronology*, Pergamon Press, 1972.

M. Eisenbud, *Environmental Radioactivity*, 2nd edn., Academic Press, 1973.

OECD, *Uranium Resources, Production and Demand*, OECD, NEA, and IAEA, Paris, 1973.

S. H. U. Bouie, Natural sources of nuclear fuel, *Phil. Trans. R. Soc. Lond.* A, **276** (1974) 495.

D. N. Schramm, The age of the elements, *Sci. Am.* **230** (Jan. 1974) 69.

J. Schirak, Fission tracks in the Allende chondrite evidence for ^{244}Pu, *Earth Planet Sci. Lett.* **23** (1974) 308.

P. K. Kurda, Fossil nuclear reactor and plutonium-244 in the early history of the solar system. I in IAEA, *The Oklo-phenomenon*, Vienna, 1976.

T. Cairns, Archeological dating by thermoluminescence, *Anal. Chem.* **48** (1976) 266A.

R. L. Fleischer, P. B. Price, and R. M. Walker, *Nuclear Tracks in Solids*, University of California Press, 1975.

CHAPTER 13

Synthetic Elements

Contents

Introduction

The hypothesis of Aristoteles (384–322 BC) that all substances in nature were made of one basic substance, though with different qualities (cold or warm, wet or dry), and that one substance could be changed into another by changing the qualities (as was observed, e.g. in metal making by heating various "earths") led many philosophers—particularly in the Middle Ages—to attempts to transform less noble metals into the most valuable one—gold. Although they failed for reasons that we understand today, in 1935 E. Amaldi, O. D'Agostino, E. Fermi, B. Pontecorvo, R. Rasetti, and E. Segré in Rome succeeded in producing gold atoms from another element (the reaction chain was $^{198}Pt(n,\gamma)^{199}Pt(\beta^+, 31\ min)^{199}Au(\beta^-, 3.1\ d)$).

The first *new element* made by man was $_{43}Tc$, discovered by C. Perrier and E. Segré in Rome in 1937 in molybdenum, which had been bombarded with deuterons. Since then 14 more elements have been synthesized. Although some of these have been found to exist in stars (deduced from spectra), they are not found on earth.

13.1. Technetium and promethium

All the isotopes of the elements technetium ($Z = 43$) and promethium ($Z = 61$) are radioactive with relatively short half-lives. Consequently these elements are not found in nature and their isotopes can only be produced through nuclear reactions in the laboratory.

The β-stability valley for $Z = 43$ has a minimum in the neighbourhood of $N \approx 55$; thus, for $Z = 43$, A-values around 97 and 99 are the most likely to be stable (recall in Chapter 3 that odd–odd nuclei are less stable than odd–even). If we consider all the isobars for $A = 95$ through 102 we find that for each mass number in this range there is already at least one stable nuclide for the elements with $Z = 42$ and 44 (^{94}Mo, ^{95}Mo, ^{96}Mo, ^{97}Mo, ^{98}Mo, ^{100}Mo, ^{96}Ru, ^{98}Ru, ^{99}Ru, ^{100}Ru, ^{101}Ru, ^{102}Ru). Since adjacent isobars cannot both be stable, the existence of these stable species excludes the possibility of stable odd–even isotopes for technetium. The longest-lived isotopes of technetium are those of $A = 97$ (2.6×10^6 y), $A = 98$ (4.2×10^6 y), and $A = 99$ (2.1×10^5 y).

For promethium we can show in an analogous manner that all the possible mass numbers for stable promethium isotopes are already occupied by stable isotopes of neodymium and samarium. The longest-lived promethium isotope is ^{145}Pm, which has a half-life of only 18 y.

^{99}Tc is formed in the fission of ^{235}U. Tens of kilograms of this nuclide have been isolated from used reactor fuel elements, and it is estimated that, were it desirable to do so, approximately 100 kg y^{-1} could be obtained. The chemical properties of technetium are similar to those of its homologs in the Periodic Table—manganese and rhenium. ^{99}Tc is a widely used radionuclide in nuclear medicine.

Small yields of the nuclide ^{147}Pm (2.6 y) are produced in fission. Promethium lies between neodymium and samarium in the Periodic Table and shows the close relationship in chemical properties of all the lanthanide elements.

13.2. Production of transuranic elements*

The discovery of the earlier transplutonium elements ($Z = 95$–100) has been intimately associated with the atomic energy program in the United States. It is of some interest to consider the general scientific background that led to the synthesis of a few of these elements, as well as their role in the nuclear energy and weapons technology, which have such important consequences for our society.

13.2.1. *Historical background*

After the discovery of the neutron by Chadwick in 1932, the group led by Fermi in Italy bombarded a large number of elements with neutrons to study the radioactivity induced through (n, γ) reactions. The decay characteristics of the radioactivity in the neutron-irradiated uranium were interpreted to indicate that some of the products were probably transuranium elements, i.e. elements with atomic numbers greater than uranium. However, many anomalies existed in the chemical properties of these "transuranium elements", and in 1939 O. Hahn (Fig. 13.1) and F. Strassman in Germany conducted a series of extremely careful chemical investigations which showed that these "transuranium elements" were in fact isotopes of elements in the middle of the Periodic Table such as Sr, La, Ba, etc. The explanation of this unexpected result was provided by L. Meitner and O. Frisch, who postulated that the uranium atoms were caused to split by thermal neutrons into two approximately equal fragments. Further investigation showed that it was the isotope ^{235}U which had undergone fission and that the fission cross-sections were several hundred times larger for thermal neutrons than for fast neutrons. Moreover, very large amounts of energy, approximately 200 MeV per fission, and an average of 2.5 neutrons per fission were released.

FIG. 13.1. Lise Meitner and Otto Hahn at the Kaiser-Wilhelm Institute in Berlin in 1913. Hahn was born 1879 in Frankfurt am Main and became Nobel laureate in chemistry in 1944 "for his discovery of the fission of heavy nuclei".

It occurred to a number of scientists that if the neutrons released in fission could be captured by other uranium atoms to cause further fission, a chain reaction should be the consequence. It was calculated that the minimum (i.e. critical) amount of uranium, assuming pure ^{235}U, for such a chain reaction was approximately 100 kg. In such a *critical mass* the chain reaction would spread with explosive velocity, and 1 kg of ^{235}U could be calculated to give the same explosive force as 20,000 t of TNT. As early as the beginning of 1941 these considerations led to plans to isolate isotopically pure ^{235}U for use in a nuclear weapon. Moreover, if the neutrons could be slowed to thermal values in uranium there was a possibility that the rate of increase of the chain reaction could be regulated and even kept constant. In 1940 F. Joliot in France, with this latter consideration in mind, applied for a patent for a nuclear reactor for controlled energy production.

13.2.2. *Neptunium*

Early in 1940 McMillan and Abelson in the United States synthesized and identified a new element with atomic number 93 to which they gave the name neptunium. The reaction used in the synthesis was

$$^{238}_{92}\mathrm{U}(\mathrm{n}, \gamma)\,^{239}_{92}\mathrm{U} \xrightarrow[23\ \mathrm{min}]{\beta^-} {}^{239}_{93}\mathrm{Np} \xrightarrow[2.3\ \mathrm{d}]{\beta^-} ({}^{239}_{94}\mathrm{Pu}) \tag{13.1}$$

The experimental recoil technique for separating the fission products from the target material and neutron capture products is described in §10.6.1. Chemical experiments showed that the product with the 2.3 d half-life could be reduced by SO_2 to a lower valency state (presumably + 4), which could be precipitated out as a fluoride (carrier: LaF_3). This distinguished the element from uranium. In its oxidized state (using BrO_3^- as oxidant) it showed the same chemistry as hexavalent uranium. Because no fission product is expected to have this behavior, the assumption that the element with the 2.3 d half-life was a transuranic element was verified.

13.2.3. *Plutonium*

At the end of 1940 an isotope of element 94 (plutonium) was synthesized by G. T. Seaborg (Fig. 13.2), E. M. McMillan, J. W. Kennedy, and A. C. Wahl by the bombardment of uranium with deuterons in a cyclotron.

$$^{238}_{92}\mathrm{U}(\mathrm{d}, 2\mathrm{n})\,^{238}_{93}\mathrm{Np} \xrightarrow[2.1\ \mathrm{d}]{\beta^-} {}^{238}_{94}\mathrm{P}\mu \xrightarrow[88\ \mathrm{y}]{\alpha} ({}^{234}_{92}\mathrm{U}) \tag{13.2}$$

It was identified as a new element, which was distinctly different from both uranium and neptunium in its redox properties (e.g. the + 3 and + 4 oxidation states were more stable). A second isotope of element 94, ^{239}Pu, was synthesized very shortly afterwards as an α-radioactive daughter of ^{239}Np with a half-life of 24000 y.

Experiments with ^{239}Pu proved that theoretical predictions that it would exhibit high fissionability with both thermal and fast neutrons were correct. This meant that ^{239}Pu in sufficient quantity would also experience an instantaneous nuclear explosion like ^{235}U. If controlled nuclear fission could be accomplished in a nuclear reactor, it would be possible to produce large amounts of plutonium by neutron bombardment of ^{238}U. The ^{239}Pu could be isolated by chemical methods which were expected to be simpler than the isotopic separation required to obtain pure ^{235}U. As a consequence, the production of ^{239}Pu became a major project of the atomic bomb program of the United States during World War II.

Even though plutonium may have been formed in considerable amounts in the r-process (§11.8), the most stable isotope, ^{244}Pu, has a half-life of only 8.3×10^7 y, which hardly permits any of it to have survived on earth to our time. However, traces of this ^{244}Pu have recently been discovered in cerium-rich rare earth minerals. If the amount found is extrapolated, considering the enrichment of plutonium in the mineral, one gram of the earth's crust will contain 3×10^{-25} g ^{244}Pu. This would leave us with about 10 g of natural plutonium left from the genesis of the earth, at which time it existed in parts per billion of matter.

As discussed in Chapter 19, ^{239}Pu has been formed in natural uranium reactors at a later stage of the earth's evolution. Recently man has synthesized hundreds of tonnes of ^{239}Pu; in addition, by 1985 the accumulated amount of other higher actinides within the

FIG. 13.2. Glenn T. Seaborg with ion exchange equipment for separation of transuranium elements at the University of California, Berkeley, in 1951. Seaborg was born in 1912 in Ishpenning, USA, and became Nobel laureate in chemistry in 1951 (together with Edwin M. McMillan) "for their discoveries in the chemistry of the transuranium elements".

European community is expected to be > 3 t ^{237}Np, > 1 t ^{238}Pu, > 3 t ^{241}Am, and > 100 kg ^{244}Cm; the production in the United States will be of the same magnitude.

13.2.4. *The Manhattan Project*

This was the code name for the wartime nuclear weapons program in the United States. It had as its objectives the isolation of large amounts of pure ^{235}U, the development of controlled nuclear fission in a nuclear reactor, the production and isolation of ^{239}Pu in the reactor, and, finally, the development of the use of these nuclides in atomic bombs.

The first large amounts of pure ^{235}U were obtained in 1944–1945 through electromagnetic and gaseous diffusion separations in a plant built in Oak Ridge, Tennessee, which had a production capacity of several hundreds of kilograms of pure ^{235}U per year. These plants, still in use, are described in Chapter 2.

At the same time, large plutonium-producing reactors were constructed under the direction of Fermi. The first test reactor became critical (self-sustaining) on December 2, 1942, in Chicago. In 1945 three large plutonium-producing reactors started operation in Hanford, Washington, with production capacities of hundreds of kilograms of plutonium per year which were isolated in chemical processing facilities also at Hanford.

The chemical isolation of plutonium was complicated because of the large amounts of radioactive fission products simultaneously produced in the irradiated uranium; this process is the subject of Chapter 20. In the study of the chemistry of these fission products many new isotopes, including several of technetium and promethium, were discovered.

13.2.5. *Americium and curium*

It was found that successive neutron capture in ^{239}Pu produced isotopes of plutonium which underwent β-decay forming new transplutonium elements with atomic numbers 95 and 96, named americium and curium. The reaction sequence was

$$\begin{array}{l} {}^{239}_{94}\mathrm{Pu}(\mathrm{n},\gamma)\,{}^{240}_{94}\mathrm{Pu}(\mathrm{n},\gamma)\,{}^{241}_{94}\mathrm{Pu} \\ \qquad\qquad\qquad\qquad\quad \downarrow \beta^- \\ \qquad\qquad\qquad\qquad\quad {}^{241}_{95}\mathrm{Am}(\mathrm{n},\gamma)\,{}^{242}_{95}\mathrm{Am} \\ \qquad\qquad\qquad\qquad\qquad\qquad\qquad \downarrow \beta^- \\ \qquad\qquad\qquad\qquad\qquad\qquad\qquad {}^{242}_{96}\mathrm{Cm} \end{array} \tag{13.3}$$

These elements had been discovered slightly earlier by Seaborg and co-workers in 1944 in bombardments of uranium and plutonium with α-particles in a cyclotron.

$$ {}^{238}_{92}\mathrm{U}(\alpha,\mathrm{n})\,{}^{241}_{94}\mathrm{Pu} \xrightarrow[15\,\mathrm{y}]{\beta^-} {}^{241}_{95}\mathrm{Am} \xrightarrow[433\,\mathrm{y}]{\alpha} \tag{13.4}$$

$$ {}^{239}_{94}\mathrm{Pu}(\alpha,\mathrm{n})\,{}^{242}_{96}\mathrm{Cm} \xrightarrow[163\,\mathrm{d}]{\alpha} \tag{13.5}$$

All of the isotopes so far described had half-lives of sufficient length for ordinary chemistry although the small amounts made special techniques necesary (§18.7).

13.2.6. *Berkelium and californium*

By 1949 Seaborg's group had synthesized a few milligrams of ${}^{241}_{95}$Am from reactor bombardment of plutonium. This material was used as the target in a cyclotron bombardment. Immediately following irradiation, the target was dissolved and the target and products separated by passage through a column of ion exchange resin using an eluting solution of ammonium citrate (§18.8.2). An α-emitting species with a half-life of 4.5 h was identified as the isotope of element 97 of mass number 243; the name berkelium was proposed.

Later the same technique was used with a target of a few micrograms of ^{242}Cm and the first isotopes of californium ($Z = 98$) were discovered.

$$^{241}_{95}\text{Am}(\alpha, 2n)\,^{243}_{97}\text{Bk} \xrightarrow[4.5\ \text{h}]{\text{EC}} \tag{13.6}$$

$$^{242}_{96}\text{Cm}(\alpha, n)\,^{245}_{98}\text{Cf} \xrightarrow[44\ \text{min}]{\alpha} \tag{13.7}$$

The last mentioned four actinides (Am, Cm, Bk, and Cf) have +3 as their most stable valency state, just as the rare earth elements, and their chemistry is very similar. This similarity is used for identification of the particular actinide; Fig. 13.3 shows elution curves for lanthanides and actinides from a cation exchange column (cf. §18.8.2).

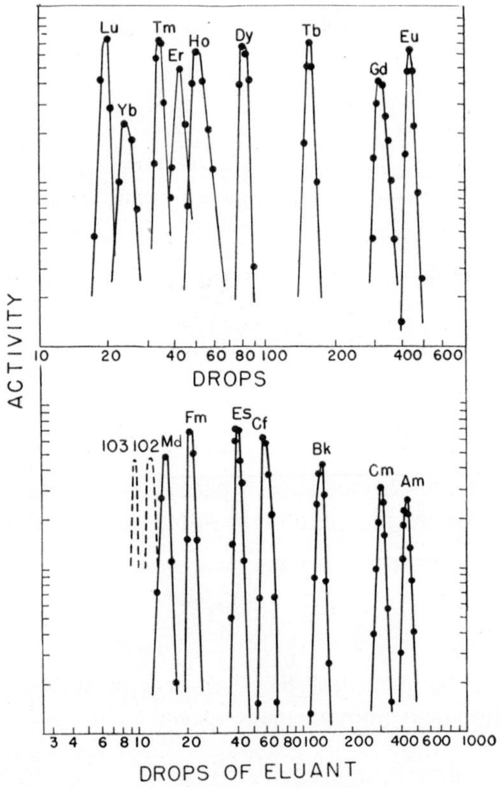

FIG. 13.3. The elution of tripositive actinide and lanthanide ions. Dowex-50 ion exchange resin was used with ammonium-α-hydroxyisobutyrate as the eluant. The positions predicted for elements 102 and 103 are indicated by broken lines. (According to Katz and Seaborg.)

13.2.7 *Einsteinium and fermium*

In 1952 the United States set off the first test thermonuclear explosion (code name "Mike"). The early analysis of debris from Mike showed that a heavy isotope of plutonium, ^{244}Pu, had been made by multineutron capture in ^{238}U which had been part of the device. More extensive chemical purification of some of the radioactive coral from the test site proved that isotopes of elements 99 (einsteinium) and 100 (fermium) had been made in the explosion. The neutron flux during the very brief burning time of Mike had been so intense (§11.9) that it resulted in capture by ^{238}U nuclei of as many as 17 neutrons. After this multineutron capture, which ended when the device blew itself apart, a sequence of β-decays occurred. The reaction sequence is shown as the shaded area in Fig. 12.2.

Soon after the initial discovery of these elements in the debris from the thermonuclear explosion, they were isolated as products of reactor irradiation (cf. Fig. 12.2). In this case

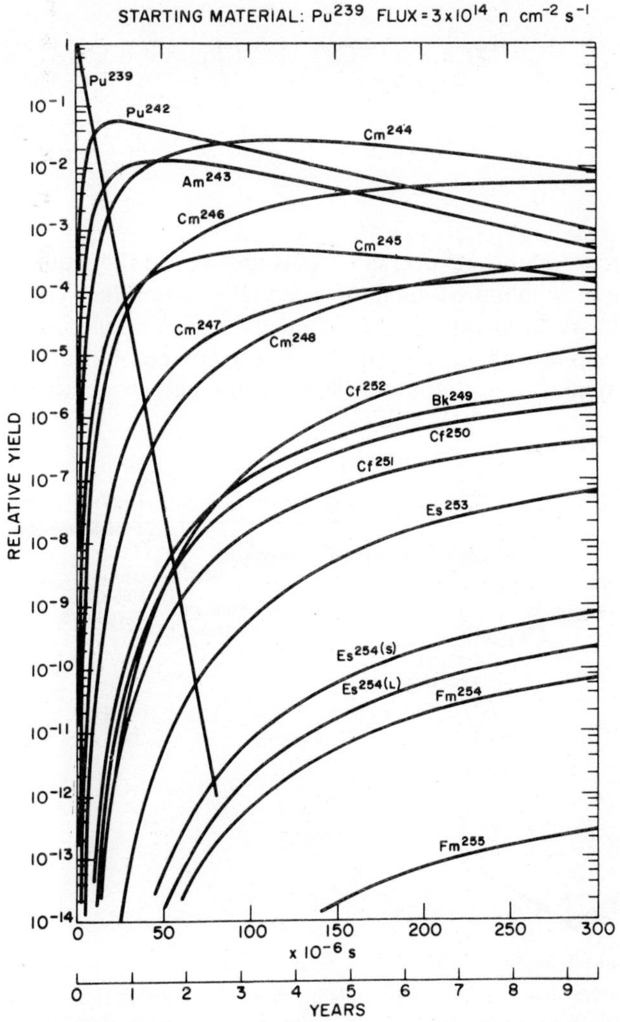

FIG. 13.4. Production of higher actinide isotopes through irradiation of ^{239}Pu in a thermal neutron flux of 3×10^{14} n cm^{-2} s^{-1}. (According to J. Milsted, P. Fields and D. N. Metta.)

the neutron capture occurs over a long time and β-decay processes compete with neutron capture depending on the $t_{1/2}(\beta^-)$ and the neutron flux. The reaction sequence is:

$$\begin{array}{l} ^{239}_{94}\text{Pu}(n,\gamma)\,^{240}\text{Pu}(n,\gamma)\,^{241}\text{Pu}(n,\gamma)\,^{242}\text{Pu}(n,\gamma)\,^{243}\text{Pu} \\ \qquad\qquad\qquad\qquad\qquad \downarrow \beta^- \\ \qquad\qquad\qquad\qquad\quad ^{241}_{95}\text{Am}(n,\gamma)\,^{242}\text{Am}(n,\gamma)\,^{243}\text{Am}(n,\gamma)\,^{244}\text{Am} \\ \qquad\qquad\qquad\qquad\qquad\qquad\quad \downarrow \beta^- \qquad\qquad\qquad\quad \downarrow \beta^- \\ \qquad\qquad\qquad\qquad\qquad\qquad ^{242}_{96}\text{Cm}(n,\gamma)\,^{243}\text{Cm}(n,\gamma)\,^{244}\text{Cm}(n,\gamma)\,^{245}\text{Cm}(n,\gamma)\ldots\ldots \end{array} \tag{13.8}$$

$$\begin{array}{l} \ldots\ldots\ldots\ldots\ldots\ldots\ldots\ldots\ ^{249}\text{Cm} \\ \qquad\qquad\qquad\qquad\quad \downarrow \beta^- \\ \qquad\qquad\qquad\qquad ^{249}_{97}\text{Bk}(n,\gamma)\,^{250}\text{Bk} \\ \qquad\qquad\qquad\qquad\quad \downarrow \beta^- \qquad\quad \downarrow \beta^- \\ \qquad\qquad\qquad\qquad ^{249}_{98}\text{Cf}(n,\gamma)\,^{250}\text{Cf}(n,\gamma)\,^{251}\text{Cf}(n,\gamma)\,^{252}\text{Cf}(n,\gamma)\,^{253}\text{Cf} \\ \qquad\qquad\qquad\qquad\qquad\qquad\qquad\qquad\qquad\qquad\qquad\qquad \downarrow \beta^- \\ \qquad\qquad\qquad\qquad\qquad\qquad\qquad\qquad\qquad\qquad\qquad ^{253}_{99}\text{Es}(n,\gamma)\,^{254}\text{Es} \\ \qquad\qquad\qquad\qquad\qquad\qquad\qquad\qquad\qquad\qquad\qquad\qquad\qquad\quad \downarrow \beta^- \\ \qquad\qquad\qquad\qquad\qquad\qquad\qquad\qquad\qquad\qquad\qquad\qquad\qquad ^{254}_{100}\text{Fm} \end{array} \tag{13.9}$$

Figure 13.4 shows the production of various nuclides in the sequence as a function of time for a reactor with a predominantly thermal neutron flux. The amount of product decreases as we proceed through the sequence since fission competes with the (n,γ) reactions for even–odd nuclei. The capture–fission competition is depicted in Fig. 13.5.

The production of transplutonium elements in highly irradiated reactor fuels creates

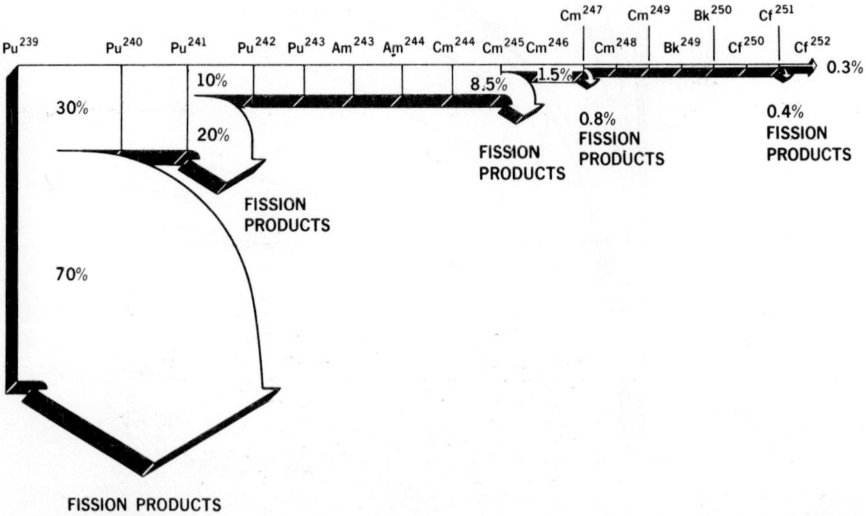

FIG. 13.5. The major paths of transuranium element production in a nuclear reactor. Both fission and neutron capture take place: in ^{239}Pu about 70% fissions, the remainder is transmuted to ^{240}Pu. The split between fission and capture is repeated at all even–odd isotopes up the chain, so that only 0.3% remains as transuranium elements when ^{252}Cf is reached.

a long term radioactive waste problem which is perhaps more troublesome than the fission products discussed in Chapters 20 and 21.

13.2.8. *The heaviest actinides: mendelevium, nobelium, lawrencium*

By 1955 transmutation of ^{239}Pu had produced 10^9 atoms of ^{253}Es. This was electroplated on gold and bombarded with helium ions to give the reaction

$$^{253}_{99}\text{Es} + ^{4}_{2}\text{He} \rightarrow ^{256}_{101}\text{Md} + ^{1}_{0}\text{n} \qquad (13.10)$$

A technique was used that allowed the atoms of ^{256}Md to recoil from the very thin target onto a "catcher" foil. Thirteen atoms of ^{256}Md made in 9 h of irradiation were isolated by elution from a column of ion exchange resin and identified by the spontaneous fission of the daughter $^{256}_{100}$Fm

$$^{256}_{101}\text{Md} \xrightarrow[1.3\text{ h}]{\text{EC}} {}^{256}_{100}\text{Fm} \xrightarrow[2.6\text{ h}]{\text{sf}} \qquad (13.11)$$

In 1957 Russian scientists claimed to have synthesized element 102 by irradiating ^{241}Pu with ^{16}O ions, using nuclear emulsions for α-energy determination. They assigned

$$^{241}_{94}\text{Pu} + ^{16}_{8}\text{O} \rightarrow ^{252}_{102}\text{No} + 5^{1}_{0}\text{n}$$

an 8.8 ± 0.5 MeV α with a half-life of 2–40 s, to element 102. However, their technique and results were disputed by scientists at the Radiation Laboratory in Berkeley, California (where all previous actinides had first been identified), who produced and identified definitively element 102 in 1958 by use of a double-recoil technique. The reaction scheme was

$$^{246}_{96}\text{Cm} + ^{12}_{6}\text{C} \xrightarrow[\text{1st recoil}]{} {}^{254}_{102}\text{No} + 4^{1}_{0}\text{n}$$
$$\downarrow \alpha\text{. 2nd recoil} \qquad (13.12)$$
$$^{250}_{100}\text{Fm}$$

A schematic of the technique is shown in Fig. 13.6.

Lawrencium, element 103, was synthesized at Berkeley in 1961 by the reactions

$$^{250,1,2}_{98}\text{Cf} + ^{10,11}_{5}\text{B} \rightarrow ^{258}_{103}\text{Lr} + (2-5)^{1}_{0}\text{n} \qquad (13.13)$$

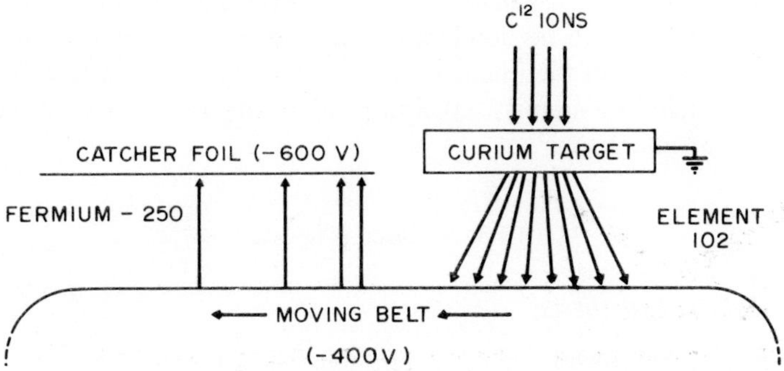

FIG. 13.6. Double-recoil technique used for identification of element 102. (According to A. Ghiorso.)

by the technique shown in Fig. 10.4. The half-life of the product (later determined as ~ 4 s) was too short to allow any chemistry. It was the first actinide to be identified through purely instrumental methods.

13.2.9. *The transactinide elements*

The very short half-lives and the small amounts of product in the initial experiments have not allowed chemical identification of the transactinide elements. Identification by nuclear properties is less conclusive as it depends usually on correlation with predicted energies and half-lives obtained by extrapolation of systematic trends. The controversies resulting from the uncertainties inherent in this approach have been a feature of the history of the transactinide elements, just as they were for nobelium and lawrencium. In fact, it is still uncertain which claims represent the actual discovery for these elements. The Russian group at Dubna and the American group at Berkeley continue to contest the other's claims to the initial synthesis of elements 104, 105, 106, and 107, though both groups undoubtedly have produced isotopes of at least elements 104–106. Partially the disagreements are due to the different techniques used. In Dubna irradiations have often been made with heavier ions (^{50}Ti, ^{51}V, etc.) than at Berkeley. The Russian group has used one-atom-at-a-time separations with thermochromatography (cf. §10.7) and solid state track detectors (§17.1), while the American group has continued with fast ion exchange separations and energy sensitive solid state detectors (§17.5). In all cases, heavy ions such as ^{12}C and heavier were used to bombard heavy element targets. The product radioactivity was isolated by recoil techniques which sometimes allowed measurement of the half-lives and decay energies. In some cases, identification was based on measurement of daughter activities.

The two groups have proposed different names for these elements, thereby emphasizing their claim to the discoveries; for 104 Ku (Kurchatovium) and (Rutherfordium) and for 105 Ns (NielsBohrium) and Ha (Hahnium). In the Periodic Table we have used $_{104}$Ku and $_{105}$Ha but imply no judgement of the conflicting claims.

In the 1960s a transuranium production program was initiated in the United States and located at Oak Ridge National Laboratory, where three new facilities were built: the High Flux Isotope Reactor (HFIR), Oak Ridge Isochronous Cyclotron (ORIC) and the Transuranium Processing Facility (TRU). Starting with ^{239}Pu irradiated to high burn-up ($\gtrsim 10^{22}$ n cm^{-2}), ^{242}Pu, ^{243}Am, and ^{244}Cm were produced in considerable amounts, isolated at the TRU and refabricated into new targets for irradiation in the HFIR and ORIC. A 25 MeV Tandem–VdG has been added so that heavy ions up to ~ 18 MeV$/A$ at $A < 50$ can be accelerated. Subsequent new reprocessing, which is a formidable task because of the intensity of radioactivity from the fission products and heavy elements (including n-emission), and irradiations would produce heavy nuclides up to californium and higher.

13.3. Actinide properties

13.3.1. *The actinide hypothesis*

Neptunium can be considered as a possible homolog of the transition elements manganese, technetium, and rhenium. However, the first investigators were surprised to learn that neptunium was more similar in its chemical behavior to uranium than to these elements. Later investigations on plutonium showed that it was also more similar chemically to uranium than to the homologs iron, ruthenium and osmium. This led to the suggestion that uranium, neptunium, and plutonium constituted a new group called the uranides with a characteristic multiplicity of valence states. The initial attempts to synthesize and isolate americium and curium were based on the assumption that they would have a similar variety of valence states. However, after an initial period of failure, Seaborg and co-workers tried a new attack in which it was assumed that elements 95 and 96 exhibit a rather constant $+3$ state. Separation schemes based on a $+3$ state soon led to the isolation and identification of these elements. Americium and curium had properties not only different from their possible transition element homologs in the Periodic Table but also rather different from the uranide group of elements.

As early as 1923 N. Bohr had suggested that there might exist a group of 15 elements at the end of the Periodic Table that would be analogous in their properties to the 15 lanthanide or rare earth elements. This idea, combined with the increasing stability of the $+3$ oxidation state for the transuranium elements as one proceeded from $Z = 93$ to 96, led Seaborg to the conclusion that these new elements constituted a second rare earth series whose initial member was actinium. As the atomic number increases from 90, electrons would be added in the 5f subshell similar to the occupation of the 4f subshell in the lanthanides. This series would be terminated with element 103 since this would correspond to the addition of 14 electrons for a completed 5f subshell.

Seaborg's actinide hypothesis was a subject initially of considerable objection since it seemingly predicted that the trivalent oxidation state would be the most stable in aqueous solution for all of the elements between $Z = 89$ and 103 just as the lanthanides all exhibit a stable $+3$ state in solution. However, in aqueous solution the most stable oxidation states are $+4$ for thorium, $+5$ for protactinium, $+6$ for uranium and neptunium, and $+4$ for plutonium. Only for the elements beginning with americium is the $+3$ the most stable state in solution. In the synthesis of elements heavier than curium, the actinide hypothesis was confirmed inasmuch as all the elements have a stable $+3$ state in aqueous solution. Recent investigations have shown that the last three elements mendelevium, nobelium, and lawrencium also have a divalent state in

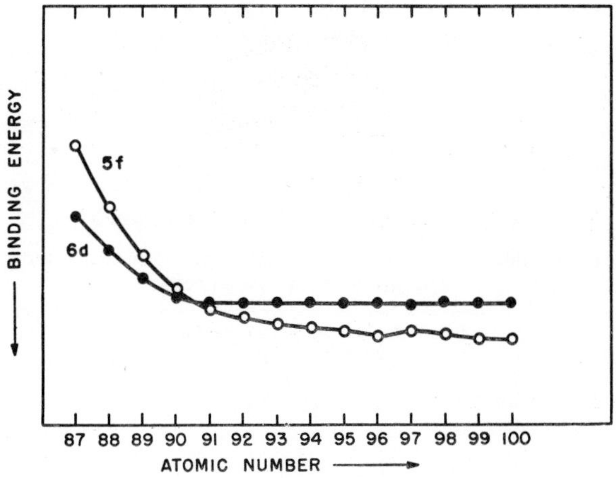

FIG. 13.7. Qualitative description of binding energies for electrons in 5f and 6d shells for elements from $_{87}$Fr to $_{100}$Fm. (According to Katz and Seaborg.)

solution (which probably is the most stable for nobelium). This corresponds to the divalent state observed for ytterbium in the lanthanide elements. Later studies indicated that, indeed, curium ($Z = 96$) is the midpoint of the actinide series inasmuch as its electronic configuration is $5f^7$. However, for $Z = 90$ to $Z = 94$ the 5f and 6d orbitals are very close in energy and the electronic occupation is variable (Fig. 13.7).

The recognition of the similarity in chemical properties between the actinide and lanthanide elements was an important contributing factor in the synthesis and isolation of the transcurium elements. Most of the chemical identification was carried out by eluting the elements from columns of cation exchange resin. The pattern of the elution behavior from the resin bed of the lanthanide elements made it possible to predict with good accuracy the expected elution position for a new actinide element (Fig. 13.3). This technique constituted the most definitive chemical evidence in the discovery experiments for the elements from atomic numbers 97 through 101.

13.3.2. *Chemistry of actinides*

All 15 actinide elements are now known. Table 13.1 lists their valency states. In Fig. 13.8 redox diagrams are given for the most important actinides; for comparison, standard potentials are included for some useful redox reagents. Any particular actinide can be obtained in a desired valency state by the use of proper oxidizing or reducing agents (Table 13.2; cf. also Fig. 13.8).

The pentavalent state of the actinides is less stable than the other states (except for Pa) and normally undergoes disproportionation. Plutonium is particularly interesting in the variety of oxidation states that can coexist in aqueous solutions (Fig. 13.8, at 1000 mV). For example, a plutonium solution in 0.5 M HCl of 0.0003 M Pu concentration at 25°C, which is initially 50% Pu(IV) and 50% Pu(VI), will equilibrate within a few days via disproportionation reactions to an equilibrium system that is 75% Pu(VI), 20% Pu(IV), and few percent each of Pu(V) and Pu(III).

The reactions are:

$$2PuO_2^+ + 4H^+ = Pu^{4+} + PuO_2^{2+} + 2H_2O \qquad (13.14a)$$

and

$$PuO_2^+ + Pu^{4+} = PuO_2^{2+} + Pu^{3+} \qquad (13.14b)$$

The chemical properties are different for the different valency states (Table 13.3), while in the same valency state the actinides closely resemble each other. The compounds formed are normally quite ionic.

TABLE 13.1. *Oxidation states of the actinides*

The most stable oxidation states are in bold figures; those not known in solution are within parentheses

89 Ac	90 Th	91 Pa	92 U	93 Np	94 Pu	95 Am	96 Cm	97 Bk	98 Cf	99 Es	100 Fm	101 Md	102 No	103 Lr
						(2)		(2)	(2)	2	2	2	**2**	
3	(3)	(3)	3	3	3	**3**	**3**	**3**	**3**	**3**	**3**	**3**	3	**3**
	4	4	4	4	**4**	4	4	4						
		5	5	**5**	5	5								
			6	6	6	6								
				7	7									

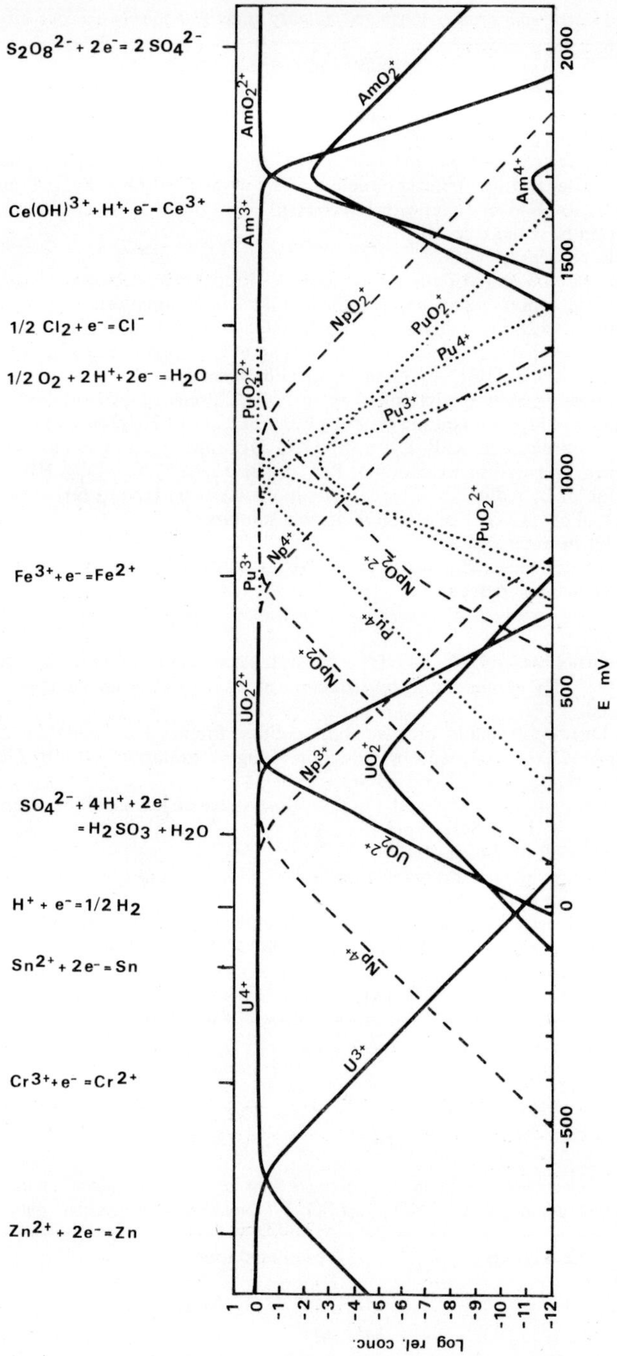

FIG. 13.8. Redox diagrams for U, Np, Pu, and Am in 1 M $HClO_4$ at 25°C. Each actinide is assumed to be isolated in the solution at a total concentration of 1 M. The ordinate is the relative concentration of a specific valency state at a particular redox potential of the solution (abscissa). The concentration lines cross at the standard potential for the redox reaction, e.g. $U^{4+} + e = U^{3+}$ at -0.63 V. (According to J. O. Liljenzin.)

TABLE 13.2. *Preparation methods and stability of actinide ions in aqueous solutions*

Ion	Stability and method of preparation
U^{3+}	Slow oxidation by water, rapid oxidation by air, to U^{4+}. Prepared by electrolytic reduction (Hg cathode).
Np^{3+}	Stable to water, rapid oxidation by air to Np^{4+}. Prepared by electrolytic reduction (Hg cathode).
Pu^{3+}	Stable to water and air. Oxidizes by action of its own α-radiation to Pu^{4+}. Prepared by reduction by SO_2, Zn, U^{4+} or H_2(g) with Pt catalyst.
Md^{3+}	Stable. Can be reduced to Md^{2+}
No^{3+}	Unstable, reduces to No^{2+}.
Pa^{4+}	Stable to water. Rapid oxidation by air to Pa(V). Prepared by electrolytic reduction (Hg cathode) and by the action of Zn amalgam, Cr^{2+} or Ti^{3+} in hydrochloric acid.
U^{4+}	Stable to water. Slow oxidation by air to UO_2^{2+}. Oxidation in nitrate medica catalyzed by UV light. Prepared by oxidation of U^{3+} by air or electrolytic reduction of UO_2^{2+} (Hg cathode) and reduction of UO_2^{2+} by Zn or H_2(g) with Ni catalyst.
Np^{4+}	Stable to water. Slow oxidation by air to NpO_2^+. Prepared by oxidation of Np^{3+} by air or reduction of higher oxidation states by Fe^{2+}, SO_2, I^- or H_2(g) with Pt catalyst.
Pu^{4+}	Stable in concentrated acids, e.g. 6 M NHO_3. Disproportionates to Pu^{3+} and PuO_2^+ at lower acidities. Prepared by oxidation of Pu^{3+} by BrO_3^-, Ce^{4+}, $Cr_2O_7^{2-}$, HIO_3 or MnO_4^- in acid solution or by reduction of higher oxidations states by HNO_2, NH_3OH^+, I^-, 3 M HI, 3 M HNO_3, Fe^{2+}, $C_2O_4^{2-}$ or HCOOH in acid solution.
Am^{4+}	Not stable in water.
Bk^{4+}	Stable to water. Slow reduction to Bk^{3+}. Prepared by oxidation of Bk^{3+} by $Cr_2O_7^{2-}$ or BrO_3^-.
PaO^{3+} or PaO_2^+	Stable. Difficult to reduce.
UO_2^+	Disproportionates to U^{4+} and UO_2^{2+}. Most stable at pH 2.5. Prepared by electrolytic reduction of UO_2^{2+} (Hg cathode) and by reduction of UO_2^{2+} by Zn amalgam or H_2(g). pH around 2.5 used.
NpO_2^+	Stable. Disproportionates only at high acidities. Prepared by oxidation of lower oxidation states by Cl_2 or ClO_4^- and by reduction of higher oxidation states by NH_2NH_2, NH_2OH, HNO_2, H_2O_2/HNO_3, Sn^{2+} or SO_2.
PuO_2^+	Disproportionates to Pu^{4+} and PuO_2^{2+}. Most stable at low acidities. Prepared by reduction of PuO_2^{2+} by I^- or SO_2 at pH 2.
UO_2^{2+}	Stable. Difficult to reduce.
NpO_2^{2+}	Stable. Easy to reduce. Prepared from lower oxidation states by oxidation by Ce^{4+}, MnO_4^-, Ag^{2+}, Cl_2 or BrO_3^-.
PuO_2^{2+}	Stable. Fairly easy to reduce. Reduces slowly under the action of its own α-radiation. Prepared by oxidation of lower oxidation states by BiO_3^-, HOCI or Ag^{2+}.
AmO_2^{2+}	Stable. Reduces fairly rapidly under action of its own α-radiation. Prepared by electrolytic oxidation (Pt anode) in 5 M H_3PO_4 or by $S_2O_8^{2-}$ in the presence of Ag^+.

TABLE 13.3. *Characteristic reactions of actinide ions of different valency states with some important anions.* Cl^-, NO_3^-, *and* SO_4^{2-} *do not precipitate actinide ions*

Reagent	Conditions	Precipitated ions	Not precipitated
OH^-	pH $\gtrsim$ 5	$M^{3+}, M^{4+}, MO_2^+, MO_2^{2+}$	
F^-	4M H^+	M^{3+}, M^{4+}	MO_2^+, MO_2^{2+}
IO_3^-	0.1M H^+	M^{4+} (M^{3+} may oxidize)	MO_2^+, MO_2^{2+}
PO_4^{3-}	0.1M H^+	M^{4+} (Ac^{3+} partly)	M^{3+} (Pu^{3+} and higher An)
CO_3^{2-}	pH > 10	M^{3+}, M^{4+} (as hydroxide)	MO_2^{2+} (anionic complex)
CH_3COO^-	0.1M H^+	MO_2^{2+}	M^{3+}, M^{4+}, MO_2^+
$C_2O_4^{2-}$	1M H^+	M^{3+}, M^{4+}	MO_2^+, MO_2^{2+}

TABLE 13.4. *Metallic and ionic radii of the actinides and the interatomic distances in the actinyl(V) and actinyl(VI) ions (Å)*

Element	M^0	M^{3+}	M^{4+}	M^{5+}	M^{6+}	M(V) – O	M(VI) – O
Ac	1.88	1.076					
Th	1.80		0.984				
Pa	1.63		0.944	0.90			
U	1.56	1.005	0.929	0.88	0.83		1.71
Np	1.55	0.986	0.913	0.87	0.82	1.98	
Pu	1.60	0.974	0.896	0.87	0.81	1.94	
Am	1.74	0.962	0.888	0.86	0.80	1.92	
Cm	1.75	0.946	0.886				
Bk		0.935	0.870				

According to the usual conventions, the metallic radii refer to the coordination number 12, the ionic radii to the coordination number 6.

The ionic radii of the actinide elements of the different valency states decrease with increasing atomic number (the *actinide contraction*) (Table 13.4). Consequently the charge density of the actinide ions increases with increasing atomic number and, therefore, the probability of formation of complexes and of hydrolysis increases with atomic number. For example, the pattern of stabilities of complexes in the tetravalent states is

$$Th^{4+} < U^{4+} < Np^{4+} < Pu^{4+} \qquad (13.15)$$

For the same element, the stability of the complexes varies with the oxidation state in the series

$$M^{4+} > MO_2^{2+} > M^{3+} > MO_2^{+} \qquad (13.16)$$

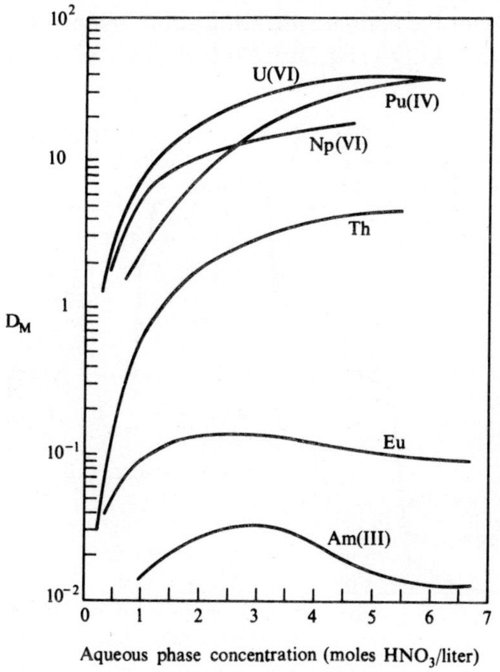

FIG. 13.9. Distribution of actinide species and Eu(III) between 30 vol. % TBP in kerosene and aqueous HNO_3. (According to J. T. Long.)

The reversal between M^{3+} and MO_2^{2+} reflects that the hexavalent metal atom in the linear $[OMO]^{2+}$ is only partially shielded by the two oxygen atoms; thus the metal ion MO_2^{2+} has a higher charge density than M^{3+}.

The pattern in Fig. 13.3, where the heavier actinides are eluted before the lighter ones, means that the α-hydroxyisobutyrate eluant forms stronger complexes as the cation radius decreases (Table 13.4). The complexes, which have a lower positive or even negative charge, are less strongly adsorbed on the resin.

In a similar way the extraction pattern shown in Fig. 18.17 can be explained. The ionic radii increase in the order $Cm^{3+} < Am^{3+} < Ac^{3+}$, and consequently the ease of extraction is in the same order (i.e. Cm^{3+} is extracted at a lower pH than Am^{3+}, etc.). In general, the ease of solvent extraction follows the sequence (13.16). This is further illustrated in Fig. 13.9, which shows the extraction of some actinide ions from HNO_3 solutions by the reagent tributylphosphate (TBP) dissolved in kerosene. In this case the extracted species are $MO_2(NO_3)_2(TBP)_2$, $M(NO_3)_4(TBP)_2$, and $M(NO_3)_3(TBP)_3$.

The chemistry involved in the isolation and purification of the actinide elements from irradiated reactor fuel elements is discussed in Chapter 20.

13.3.3. *The solid state*

The actinide metals can be produced by heating the tri- or tetrafluoride with metallic alkaline earth or alkali elements:

$$PuF_4 + 2Ca \longrightarrow Pu + 2CaF_2 \qquad (13.17)$$

The metals exhibit several allotropic modifications: 3 for uranium and 6 for plutonium between room temperature and the melting point (1130° for U and 640° for Pu) (Fig. 13.10). The density of the actinide metals at room temperature shows an unusual variation: Th 11.8; Pa 15.4; U 19.1; Np 20.5; Pu 19.9; Am 13.7; Cm 13.5; Bk 14.8 g cm^{-2}.

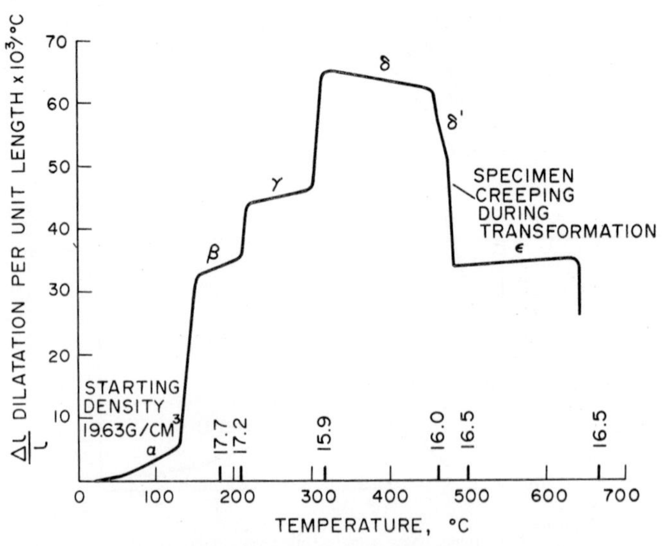

FIG. 13.10. Dilatation curve and densities of high purity plutonium. (According to M. B. Waldron, J. Garstone, J. A. Lee, P. G. Mardon, J. A. C. Marples, D. M. Poole, and K. G. Williamson.)

All the metals are very electropositive and attacked by water vapor with production of hydrogen. They are slowly oxidized in air and at higher temperatures, in the form of small chips, they are pyrophoric. The oxides, nitrides, and halides are produced most easily by heating the metals in the appropriate elemental gas. The fluorides are among the most important solid actinide compounds since they are the starting material for the production of the metals and of the volative hexafluorides (the hexafluoride of uranium is used in isotopic enrichment in gaseous diffusion plants, Chapter 2). The solid metal, oxide, and carbide of uranium are used in reactor fuel elements. The preparation of uranium compounds is described in § 12.6; the corresponding plutonium compounds are prepared similarly.

13.3.4. *Applications of actinides*

The spontaneous decay of the transuranium elements by α-decay and/or fission results in energy releases. Since very small amounts of some nuclides (e.g. ^{238}Pu, ^{244}Cm, ^{252}Cf) can be sources of appreciable energy, these radionuclides can be used in small power generators. This use of actinide elements is discussed to a greater extent in § 14.11.3.

^{241}Am emits a 60 keV γ-ray and has been used as a γ-radiation source to measure thickness of metal sheets and of deposited metal coatings, the degree of soil compaction, sediment concentration in flowing streams, and to induce X-ray fluorescence in chemical analysis (§ 14.8). As an α-particle emitter, ^{241}Am has been mixed with beryllium to make intense neutron sources for logging oil wells and for measuring water content in soils and even in process streams in industrial plants. It is extensively used for elimination of static electricity and in smoke detectors where its use depends on the ionization of air (see also § 14.11).

A nuclide for which many new uses recently have been found is ^{252}Cf. The characteristics of this important radioisotope are given in Table 13.5. ^{252}Cf is the only nuclide that

TABLE 13.5. *Nuclear properties of* ^{252}Cf

Mode of decay	
α-emission	96.9%
Spontaneous fission	3.1%
Half-life	
α-decay	2.731 ± 0.007 y
Spontaneous fission	85.5 ± 0.5 y
Effective (*a* and sf)	2.646 ± 0.004 y
Neutron emission rate	2.31×10^{12} n s^{-1} g^{-1}
Neutrons emitted per spontaneous fission	3.76
Average neutron energy	2.348 MeV
Average α-particle energy	6.117 MeV
Gamma emission rate (exclusive of internal conversion X-rays)	1.3×10^{13} photons s^{-1} g^{-1}
Dose rate at one meter in air	
Neutron	2.2×10^{3} rem h^{-1} g^{-1}
Gamma	1.6×10^{2} rads h^{-1} g^{-1}
Decay heat	
From α-decay	18.8 W g^{-1}
From fission	19.7 W g^{-1}
Source volume (excluding void space for helium)	< 1 cm^{3} g^{-1}

can provide a useful neutron intensity over a sufficiently long half-life to make it a useful neutron source. The low rates of heat emission, γ-radiation, and helium evolution allow fabrication of simple, small ^{252}Cf sources that require no external power supply nor any maintenance but that can provide moderately high, neutron fluxes. Among the applications of ^{252}Cf we may list the following:

(a) process control by a variety of on-stream nondestructive analytical techniques;
(b) medical diagnosis by activation analysis;
(c) production of short-lived radioisotopes at locations where they will be used, thus avoiding decay during transportation from an accelerator or reactor at another site;
(d) industrial neutron radiography, which images low density materials—especially hydrogenous materials (better than X-rays);
(e) possible medical treatment of tumors with ^{252}Cf sources that can be implanted in the body;
(f) petroleum and mineral exploration in which the compactness and portability of ^{252}Cf neutron sources facilitate testing for valuable deposits, particularly in inaccessible places such as deep wells and the sea floor;
(g) moisture measurements;
(h) hydrology studies to locate sources of water;
(i) nuclear safeguards tests, e.g. for criticality control in reactor fuel storage areas and for nuclear materials accountability (detection and recovery of fissionable material; implementation of nuclear agreements).

13.4. Superheavy elements

The controversies surrounding the discovery claims of the elements above atomic number 101 are the result of the very short half-lives of these elements. In fact, simple extrapolation of the half-lives as a function of atomic number would suggest that elements 107 or 108 might be the last that can be synthesized. Not only are the half-lives to α-decay shortening (Fig. 6.15) but those for spontaneous fission (Fig. 9.12) decrease even more rapidly as Z increases. However, in 1966 extra stability was predicted for a new, heavier group of nuclides $Z \gtrsim 114$ (cf. Fig. 3.8). These predictions have led to vigorous research efforts to synthesize these superheavy elements in the laboratory and to find them in nature.

Some years ago it was generally assumed that the next proton magic number after $Z = 82$ would be $Z = 126$ since this was the neutron pattern. However, more careful theoretical study indicates that the proton shell closure occurs probably for $Z = 114$ due to the effect of proton–proton repulsion. The most probable neutron magic number is $N = 184$, although some calculations indicate $N = 164$ as a possibility. These calculations are rather sensitive to the exact type and shape of the potential, so the proton magic number may be slightly smaller or larger than 114 while the same is true for the neutron number of 184. Nevertheless, the superheavy nucleus $^{298}114$ seems the best bet for maximum stability in this region.

Figure 13.11 shows the variation in nuclear deformation calculated for the fission barrier $^{298}114$. Of particular interest are the small local fluctuations at small deformation. The minimum of 8 MeV at zero deformation constrains the nucleus to a spherical shape. Spontaneous fission is a very slow process in this situation since it involves

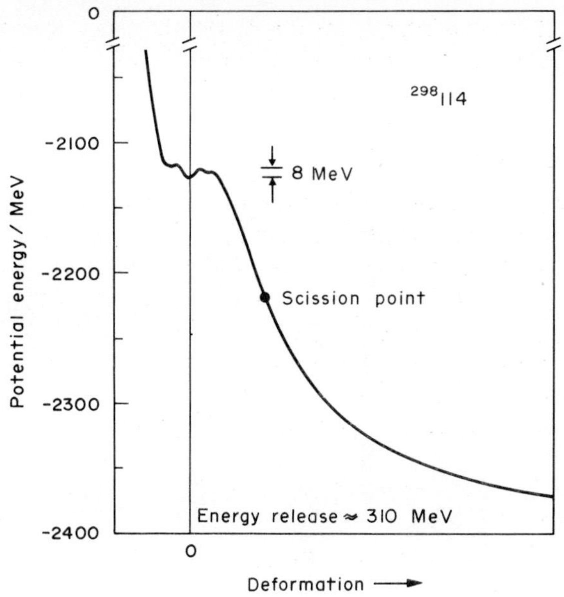

FIG. 13.11. Dependence of potential energy on deformation for a nucleus like $^{298}_{184}114$. (According to R. Nix *Proc. Int. Conf. on Nuclear Reactions Induced by Heavy Ions, Leysin 1970.*)

tunneling the 8 MeV barrier. These local fluctuations in the potential energy curve in Fig. 13.11 result from adding corrections for shell effects to a liquid drop model. The resistance to deformation associated with closed shell nuclei produces much longer half-lives to spontaneous fission than would be expected from calculations based on a liquid drop model solely.

The result of these calculations is the prediction of an "island of stability" for nuclides of about $Z = 114$ and $N = 184$. Figure 13.12 shows this pictorially. The expected half-lives are very sensitive to the parameters used, and various groups have differed by 10^5 in their estimates. However, lifetimes of years to perhaps millions of years are possible

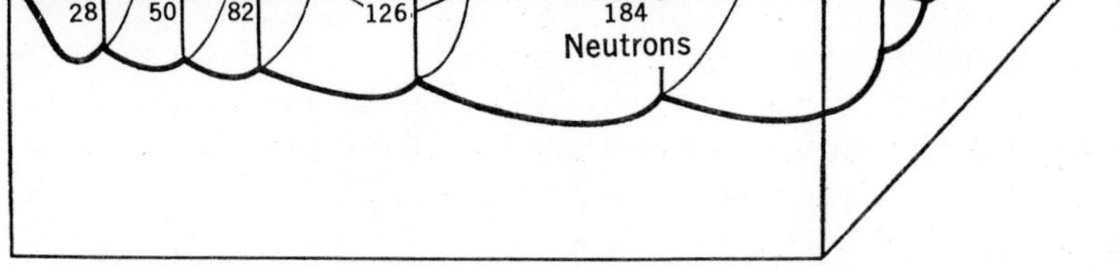

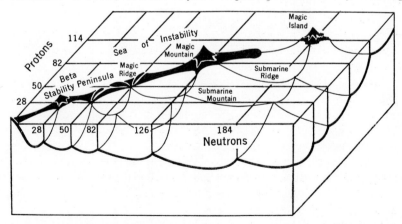

FIG. 13.12. Nuclear stability illustrated in a scheme that shows a peninsula of known elements and an island of predicted stability (around $Z = 114$ and $N = 184$) in a sea of instability. Stable region on the "mainland peninsula" is represented by a mountain ridge. (According to S. G. Thomson and C. F. Tsang.)

TABLE 13.6. *Some predicted properties of superheavy elements*

	Element 113 (eka-thallium)	*Element* 114 (eka-lead)	*Element* 117 (eka-astatine)	*Element* 118 (eka-radon)	*Element* 119 (eka-francium)	*Element* 120 (eka-radium)
Chemical group	III	IV	VII(halogen)	VIII(noble gas)	I(alkali metal)	II(alkaline earth)
Atomic weight	297	298	311	314	315	316
Atomic volume (cm^3 $mole^{-1}$)	18	21	45	50	80–90	45
Density (g cm^{-3})	16	14			3	7
Most stable oxidation state	+1	+2	−1	0	+1	+2
Oxidation potential (V)	$M \rightarrow M^{+} + e^{-}$ −0.6	$M \rightarrow M^{2+} + 2e^{-}$ −0.8	$2M^{-} \rightarrow M_2 + 2e^{-}$ +0.25–0.50		$M \rightarrow M^{+} + e^{-}$ +2.9–3.0	$M \rightarrow M^{2+} + 2e^{-}$ +2.9
First ionization potential (eV)	7.4	8.5	9.3	9.8	3.4–3.8	5.4
Second ionization potential (eV)		16.8	16	15	23	10
Ionic radius (Å)	1.48	1.31	2.3		1.8–1.9	1.5–1.7
Metallic radius (Å)	1.75	1.85			2.9	2.3
Melting point (°C)	430	70	350–550	−15	0–30	680
Boiling point (°C)	1100	150	610	−10	630	1700

for $Z = 110–114$. Given the uncertainty in the calculations, this must be extended to an estimate that nuclei between $Z = 110$ and 126 could be sufficiently stable to isolate and measure if synthesized. The possible properties of some of these superheavy elements have been summarized in Table 13.6.

The predictions that the half-lives of some superheavy elements may be very long have led to searches for evidence of their existence in nature. Several reports of tentative evidence for such elements have appeared but subsequent experimentation has explained the observations as based on known elements with only a few exceptions. The xenon isotopes in some meteorites indicate a volatile progenitor element, possibly of $Z = 111–116$. Fissioning and α-emitting species leave tracks and haloes in mica samples. The existence of "giant" haloes has been interpreted as reflecting the very high energy spontaneous fission of superheavy elements.

Similarly, attempts to synthesize superheavy elements by bombardment of targets with heavy ions so far have not been successful. The attempts to synthesize these elements continues in many countries as there is general agreement that the superheavy island of stability is there waiting for its Columbus.

13.5. Exercises

13.1. What nuclear reactions would be suitable to make gold out of mercury?

13.2. What fraction of neptunium is in the + 4 state in a 0.1 M $Fe(SO_4)_2$ solution of acidity 1 M H^+?

13.3. Irradiation of ^{238}U with deuterium yields ^{238}Pu. Will any other plutonium isotope be produced?

13.4. What are the decay products of ^{252}Cf, of ^{253}Cf, and of ^{254}Cf?

13.5. ^{244}Pu decays through spontaneous fission with a half-life of 2.5×10^{10} y. Estimate the number of neutrons emitted per fission if the measured n-emission rate is 6235 n s^{-1} g^{-1}.

13.6. Flerov bombarded ^{207}Pb with ^{54}Cr and obtained a product which within 4–10 ms decayed by spontaneous fission. Suggest a product nucleon.

13.7. In Fig. 13.9, Eu(III) is more easily extracted than its homolog Am(III). Suggest an explanation.

13.8. A cardiac pacemaker contains 150 mg ^{238}Pu. What is its heat output? Use data in the isotope chart.

13.9. What electronic configuration would you ascribe to nobelium considering its place in the periodic system and its chemistry?

13.6. Literature

J. Katz and G. T. Seaborg, *The Chemistry of the Actinide Elements*, Methuen/Wiley, 1957.

E. K. Hyde and G. T. Seaborg, The transuranium elements, *Handbuch der Physik*, Band XLII, Springer, 1957.

G. T. Seaborg, *The Transuranium Elements*, Yale University Press, 1958.

G. T. Seaborg, *Manmade Transuranium Elements*, Prentice-Hall, 1963.

G. H. Coleman, *The Radiochemistry of Plutonium*, NAS-NS 3058, Washington, 1965.

A. D. Gel'man, A. I. Moskvin, L. M. Zaitsev, and M. P. Mefod'eva, *Complex compounds of transuranides*, Israel Program for Scientific Translations, Jerusalem, 1967.

G. T. Seaborg and J. L. Bloom, The synthetic elements: IV, *Sci. Am.* **220**, No. 4 (April 1969).

J. M. Cleveland, *The Chemistry of Plutonium*, Gordon & Breach, 1970.

D. C. Hoffman, F. O. Lawrence, J. L. Mewherter, and F. M. Rourke, Detection of Plutonium-244 in nature, *Nature* **234** (1971) 132.

S. G. Thompson and C. F. Tsang, Superheavy elements, *Science* **178** (1972) 1047.

K. W. Bagnall (ed.), Lanthanides and actinides, *Inorganic Chemistry*, Series one, Vol. 7, Butterworths, 1972.

Gmelin Handbuch, Ergänzungswerk zur 8. Auflag: Vol. 4, *Transuranium Elements*, Part C, *The Compounds* (1972); Vol. 7a, *Transuranium Elements*, Part A1, II, *The Elements* (1974); Vol. 7b, *Transuranium Elements*, Part A2, *The Elements* (1973); Vol. 20, *Transuranium Elements*, Part D1, *Solution Chemistry* (1975).

Papers on Lanthaniden und Actiniden by B. Kanellakopulos, C. Keller, W. Müller, G. Koch, A. Brusdeylins, H. L. Scherff, and K. Maas in *Chemiker-Zeitung* **97** (1973) 513 ff.

G. A. Burney and R. M. Harbour, *The Radiochemistry of Neptunium*, NAS-NS-3060, Washington, 1974.

A. J. Freeman and J. B. Darby, Jr. (eds.), *The Actinides: Electronic Structure and Related Properties*, Academic Press, 1974.
M. Taube, *Plutonium—A General Survey*, Verlag-Chemie GmbH, 1974.
V. I. Spitzyn and J. J. Katz (eds.), *Chemistry of Transuranium Elements*, Pergamon Press, 1976.
W. Muller and R. Lindner (eds.), *Transplutonium 1975*, North-Holland/Elsevier, 1976.
W. Muller and H. Blank (eds.), *Heavy Element Properties*, North-Holland/Elsevier, 1976.
H. Blank and R. Lindner (eds.), *Plutonium 1975 and Other Actinides*, North-Holland/Elsevier, 1976
R. Walgate, New world beyond uranium, *New Scientist* **24** (June 1976), 696.
G. A. Cowan, A natural fission reactor, *Sci. Am.* July 1976.

CHAPTER 14

Absorption of Nuclear Radiation

Contents

Introduction

Our understanding of the nature of nuclear particles is based on their mode of interaction with matter. Knowledge about this interaction is essential in a variety of areas of nuclear science, such as the proper utilization and construction of detection and measuring devices for radiation, the design of radiation shielding, the medical and biological applications of radiation, radiochemical synthesis, etc.

The term *nuclear radiation* is used to include all elementary (including photons) and charged particles having velocities in excess of approximately 100 eV whether the particles have been produced through nuclear reactions (spontaneous or induced) or have acquired their energy in electrostatic accelerators. This lower energy limit is very high in comparison to ionization energies (usually < 15 eV) and to the energies involved in chemical bonds (normally 1–5 eV). Therefore, nuclear radiation can cause ionization, directly or indirectly in its passage through matter; this is reflected in the common name *ionizing radiation*. Neutrons of energies < 100 eV are included because their absorption (capture) by nuclei results in emission of nuclear radiation with energies $\gg 100$ eV.

The passage of such high energy radiation through matter results in the transfer of energy from the radiation to the atoms and molecules of the absorber material. This transfer of energy continues until the impinging particle has reached the same average kinetic energy as the atoms comprising the material, i.e. until thermal equilibrium is obtained.

In considering the absorption of nuclear radiation it is appropriate to view the overall process from two aspects: (1) processes occurring to the nuclear particles themselves as their energies are reduced to thermal equilibrium value; such *absorption processes* are the principal consideration of this chapter; (2) processes in the absorbing material due to the effect of the transfer of energy. This transfer results initially in excitation and ionization which cause physical and chemical changes. The study of these effects is the domain of *radiation chemistry* and is considered in the next chapter.

14.1. Survey of absorption processes

The reduction in the intensity of a beam of ionizing particles can be caused either by reaction with the nuclei of the absorbing material (nuclear reactions) or with the atomic electrons (electron collision). In Table 14.1 the most important processes involved in the absorption of nuclear radiation in matter are listed along with the probability for each process. Comparison shows that the probability of interactions with electrons is considerably greater than that of a nuclear reaction; the only exception to this is the case of neutron absorption. In fact the principal mode of interaction between the particle and the atoms of the absorbing material involves the electromagnetic fields of the particle and the atomic electrons. Since neutrons are neutral particles, we do not expect such electromagnetic interaction to occur; in order for neutrons to transfer energy it is necessary that they experience a collision with a nucleus. Consequently for all particles except neutrons, nuclear reactions can be neglected in considering the processes involved in the reduction of the intensity of the particle beam.

As nuclear radiation passes the atoms of an absorber it can transfer some of its energy to the atoms. If the amount of energy transferred is sufficient, ionization results. The positive ion and the electron thus formed are known as an *ion pair*. Frequently the electrons from this primary ionization have sufficiently high kinetic energy to cause secondary ionization in other atoms. The number of electrons produced in secondary ionization is often larger than that of the primary ionization but the average kinetic energies of the secondary electrons are lower than those of the primary electrons (§ 14.4). In many interactions the initial radiation transfers insufficient energy for ionizations; instead an electron is raised to a higher, excited energy level of the atom. These *excited atoms* rapidly return to lower energy states by emission of electromagnetic radiation such as X-rays, visible light, etc. For neutrons the absorption process involving the capture of the neutron (cf. §§ 4.4 and 10.6) imparts sufficient recoil energy to cause ionization and excitation.

14.2. Absorption curves

In order to measure the absorption of nuclear radiation, the experiments must be performed in such a manner as to eliminate as many of the interfering factors as possible. Usually a well-collimated beam is used. This is illustrated in Fig. 14.1 for a point radioactive source. The relation between the disintegration rate A and the count rate R is

TABLE 14.1. *Survey of nuclear radiation absorption processes*
The reaction cross-sections (σ) give only order of magnitude at about 1 MeV in $Z \approx 20$

	Reacting particles and fields	Type of reaction	σ (barns)	Name of process
1	*Protons and heavier ions* react with			
1a	orbital electrons	Particle energy loss through atomic excitation and ionization	$\gtrsim 10^5$	Ionization, (atomic) excitation
1b	atomic nucleus	Particle elastically scattered	$\lesssim 10$	Nuclear scattering
1c	atomic nucleus	Particle inelastically scattered	< 1	Nuclear (coulomb) excitation
1d	atomic nucleus	Particle captured, formation of compound nucleus ($E_p > E_c$ (min))	$\lesssim 0.1$	Nuclear transmutation
2	*Electrons* (e^-, β^-, β^+) react with			
2a	orbital electrons	Particle energy loss through atomic excitation and ionization	$\gg 10^2$	Ionization, (atomic) excitation
2b	orbital electrons	Slow β^+ annihilated, 2–3 photons formed	(100%)	Positron annihilation
2c	electric field of nucleus	Particle scattered with energy loss, continuous emission of $h\nu$ ($E_e \gg 1$ MeV)	$\gg 1$	Bremsstrahlung
3	*Photons* (γ) react with:			
3a	field of orbital electrons	γ scattered without energy loss	$\lesssim 0.01$	Coherent scattering
3b	free (outer) electrons	γ scattered with energy loss, ionization	$\lesssim 10$[a] (3b–3d)	Compton effect
3c	bound (inner) electrons	γ completely absorbed, one electron knocked out		Photo effect
3d	field of nuclear force	γ annihilated, formation of positron–negatron pair ($E_\gamma > 1.02$ MeV)		Pair formation
3e	atomic nucleus	γ scattered without energy loss	$\lesssim 10^{-3}$ (3e–3g)	Mössbauer effect
3f	atomic nucleus	γ scattered with energy loss		Nuclear excitation
3g	atomic nucleus	γ absorbed by nucleus, nuclear transmutation ($E_\gamma > 5$ MeV)[b]		Nuclear photo effect
4	*Neutrons* react with			
4a	atomic nucleus	n scattered with energy loss	$\lesssim 10$	Neutron moderation
4b	atomic nucleus	n captured, nuclear transformation	$\lesssim 10^4$	Neutron capture

[a]See Fig. 14.16; σ increases strongly with decreasing energy.
[b]Threshold energy for $Be(\gamma, \alpha)\,^4He$ 1.6 MeV, $D(\gamma, n)H$ 2.2 MeV.

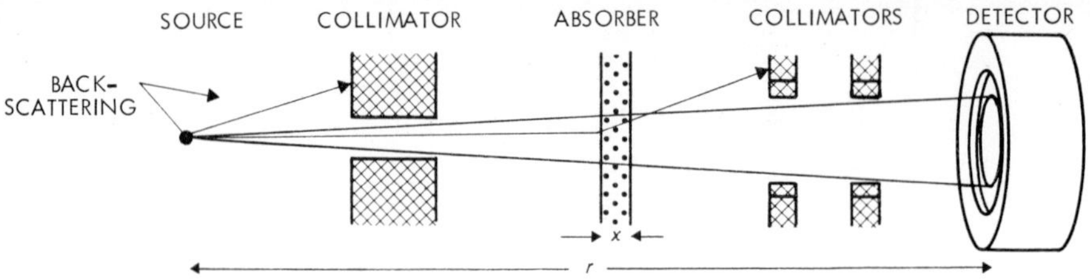

FIG. 14.1. Geometrical arrangement for measuring absorption curves.

given by (4.45):

$$R = \psi A$$

The *counting efficiency* ψ includes a number of factors:

$$\psi = \psi_{\text{sample}} \psi_{\text{abs}} \psi_{\text{det}} \psi_{\text{geom}} \tag{14.1}$$

If conditions were ideal we would have no self-absorption or scattering in the sample (in which case $\psi_{\text{sample}} = 1$), no absorption of radiation between the sample and the detector window ($\psi_{\text{abs}} = 1$), and the detector would have a 100% efficiency (sensitivity) to a "count" for each particle reaching its window ($\psi_{\text{det}} = 1$).

The geometric efficiency ψ_{geom} is 1 for 4π-geometry, i.e. for a spherical detector subtending a 360° solid angle about the sample. Although such detectors exist (§17.8), more commonly the sample is counted outside the detector at some distance r, as indicated in Fig. 14.1. If the detector window offers an area of S_{det} perpendicular to the radiation, the geometrical efficiency is approximated by (for small ψ_{geom})

$$\psi_{\text{geom}} \approx S_{\text{det}} (4\pi r^2)^{-1} \tag{14.2}$$

(The geometry for nonpoint sources is given in Appendix G.)

The radioactivity measured is proportional to the particle flux ϕ reaching the detector

$$R = k\phi \tag{14.3}$$

where $k = S_{\text{det}} \psi_{\text{det}}$ and

$$\phi = \psi_{\text{abs}} \phi_0 \tag{14.4}$$

and

$$\phi_0 = \psi_{\text{sample}} nA/4\pi r^2 \tag{14.5}$$

ϕ_0 is the flux of particles (particles $m^{-2} s^{-1}$) from the source at the detector with no absorption between source and detector ($\psi_{\text{abs}} = 1$), and n is the number of particles emitted per decay ($n > 1$ only for γ following some α- and β-decays). Thus if every β-decay yields 2γ (in cascade) and γ is the measured radiation, then $n_\gamma = 2$, and $\phi_0 = n_\gamma A_\beta / 4\pi r^2$ γ-quantas $m^{-2} s^{-1}$. In branched decay n will not be an integer. Equation (14.5) is the so-called $1/r^2$-*law* since the measured flux varies as the inverse of the square of the distance to the source.

These equations are valid as long as the conditions at the source and at the detector, as well as r, are kept constant. When an absorber is inserted between the source and detector (Fig. 14.1), ψ_{abs} depends on the absorber thickness x (m^{-1}). For an absorber thickness $x = 0$, $\psi_{\text{abs}} = 1$ in accurate measurements. There is a small absorption due to the air between the sample and detector unless the measurement is done in a vacuum.

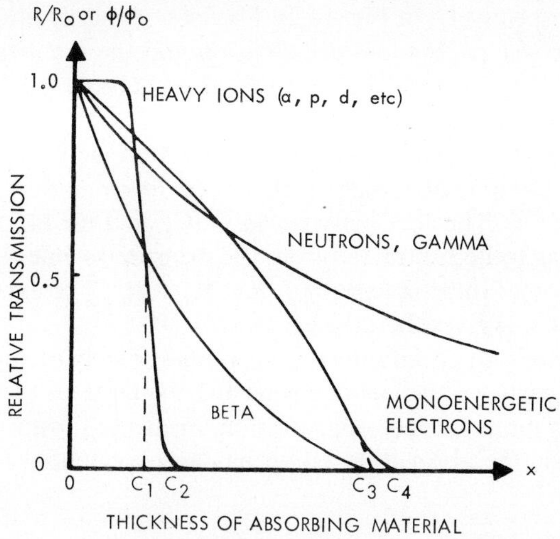

FIG. 14.2. Absorption curves showing relative transmission ϕ/ϕ_0 (or R/R_0) as a function of the absorber thickness x. C_1 and C_3 are average range, C_2 and C_4 maximum (or extrapolated) range.

Absorption curves relate the variation of either R or ϕ to the thickness of the absorbing material. In Fig. 14.2 the relative transmission ϕ/ϕ_0 is plotted as a function of absorber thickness for different kinds of radiation. For charged particles, i.e. electrons, protons, and heavier ions, ϕ/ϕ_0 reaches zero at a certain x-value (x_{max}); this is referred to as the maximum range $\hat{R}$ of the particles. The range can be expressed by either the *average range* ($x = C_1$ for heavy ions and C_3 for electrons) or the *maximum* (or extrapolated)

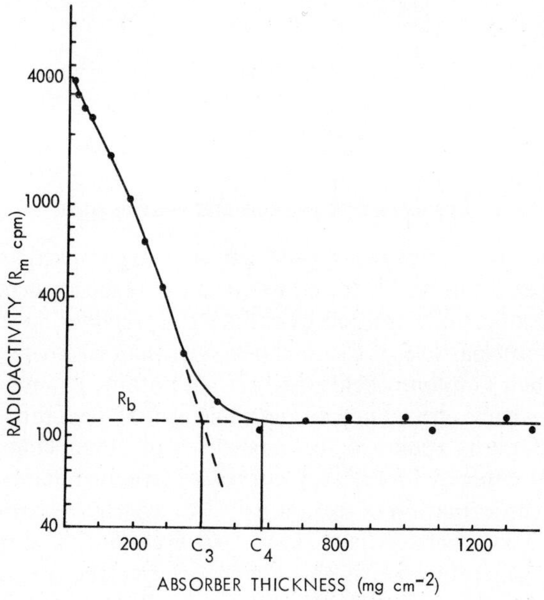

FIG. 14.3. Absorption curve for ^{32}P β-radiation showing the extrapolated range (C_4) and the average range (C_3). The dashed curve is obtained when the background is subtracted from the measured curve. The absorber thickness is measured in linear density, ρx.

range (C_2 and C_4, respectively, in Fig. 14.2). The loss of energy involves collisions with atomic electrons, and the energy loss per collision and the number of collisions varies from one ionizing particle to the next, resulting in a slight *straggling* in the range. The average range is the meaningful one.

Figure 14.3 shows an absorption curve for ^{32}P. The radioactivity R has been measured as a function of aluminum absorber thickness in *linear density*, kg m^{-2}, or more commonly in mg cm^{-2}. The low activity "tail" (C_4 in Fig. 14.3) is the background activity R_b, which has to be subtracted from the measured value R_m to obtain the true value for the radiation of interest (e.g. ^{32}p): $R = R_m - R_b$. The extrapolation of R to a value equal to R_b (i.e. C_3) gives the range.

Whereas it is possible to specify maximum ranges for charged particles, this is not possible for neutral particles such as neutrons and γ-quanta. If the absorber is not too thick, these particles undergo only one collision, or at the most a few, before they are absorbed. As a result the absorption curve has an exponential form similar to that described in §9.1.

$$\phi = \phi_0 e^{-\mu x} \tag{14.6}$$

where μ is the *total attenuation coefficient*. Thus for n and γ we have

$$\psi_{abs}(x) = e^{-\mu x} \tag{14.7a}$$

The reduction in intensity of a beam can occur by two mechanisms. One involves the deflection or scattering of the particles from the direct line of path between the source and the detector and is described by the *scattering coefficient* μ_s. The second mode of reduction is the complete transfer of the projectile energy to the absorbing material (the particles are "captured") and is designated by the (*energy*) *absorption coefficient* μ_a. The (total) attenuation coefficient in (14.6) is the sum of both these modes.

$$\mu = \mu_s + \mu_a \tag{14.7b}$$

Both μ_s and μ_a can be measured independently. The (total) attenuation coefficient is of primary interest in radiation shielding, while the (energy) absorption coefficient is essential in considering radiation effects on matter.

14.3. Absorption of protons and heavier ions

The mode of interaction of protons and heavier charged particles with the atoms of the absorbing material can be illustrated by considering the absorption of α-particles. With rare exception, α-particles emitted by radioactive nuclides have energies between 4 and 9 MeV. Since the α-particles are so much heavier than electrons, they are deflected very slightly when their Coulomb fields interact with atoms or molecules to form ion pairs. As a result, α-particles travel in a straight line as they pass through matter, which explains the straight paths observed for α-particles in cloud chamber photographs (Fig. 14.4). This is in contrast to the very curved or irregular paths of the secondary electrons emitted in the formation of the ion pair. For a 5 MeV α-particle the maximum energy of the secondary electrons is 2.7 keV. However, only a small fraction of the secondary electrons actually receive this much energy; the average energy of the secondary electrons is closer to 100 eV. The ionization caused by more energetic secondary electrons is usually referred to as δ-tracks.

In solids and liquids the total path length for α-particles from radioactive decay is

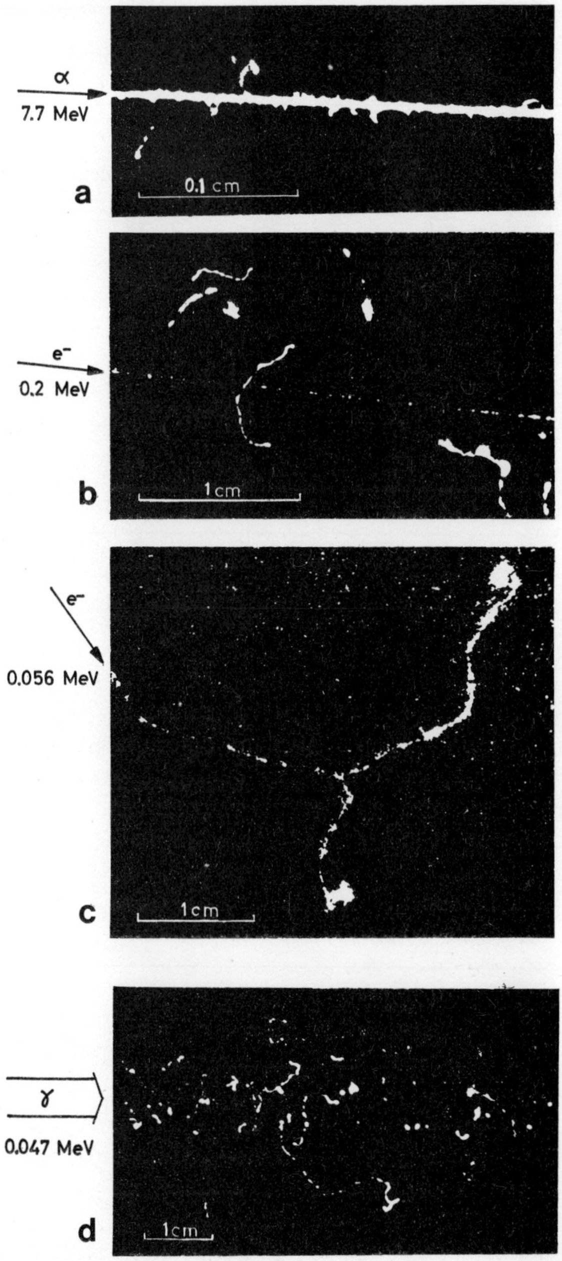

FIG. 14.4. Cloud chamber tracks of α, β, (e⁻) and γ-rays at 1 bar in air ((a), (b), and (c)) and in methane (d). All figures show secondary electrons. (b) A fast electron produces a straight line in addition to two δ-tracks of more energetic secondary electrons (> 1 keV). Also tracks of secondary electrons are shown at some distance from the main track; they are caused by absorption of the soft X-rays produced in the absorption of the primary electrons (c) Collision of a 56 keV electron leading to two δ-rays, forming ion clusters at the end of their tracks. (d) Shows secondary electrons of all energies $\lesssim$ 10 keV. (According to W. Gentner, H. Maier-Leibnitz, and H. Bothe.)

TABLE 14.2. *Range in air and water, and average linear energy transfer (LET) values for different radiation*
Upper half refers to monoenergetic (accelerated) particles. For β-decay $E_{abs} = 1/3\ E_{max}$

Radiation	Energy (MeV)	Maximum range: cm air	Maximum range: mm water	Average LET value in water (keV μm^{-1})
Electron	1	405	4.1	0.24
	3	1400	15	0.20
	10	4200	52	0.19
Proton	1	2.3	0.023	43
	3	14	0.014	21
	10	115	1.2	8.3
Deuteron	1	1.7	—	—
	3	8.8	0.088	34
	10	68	0.72	14
Helium	1	0.57	0.005 3	190
	3	1.7	0.017	180
	10	10.5	0.11	92
Fiss. fragment	100	2.5	0.025	3300
^{226}Ra (α)	E_α 4.80	3.3	0.033	145
^{210}Po (α)	E_α 5.30	3.8	0.039	136
^{222}Rn (α)	E_α 5.49	4.0	0.041	134
^{3}T (β)	E_{max} 0.018	0.65	0.005 5	1.1
^{35}S (β)	E_{max} 0.167	31	0.32	0.17
^{90}Sr (β)	E_{max} 0.544	185	1.8	0.10
^{32}P (β)	E_{max} 1.71	770	7.9	0.07
^{90}Y (β)	E_{max} 2.25	1.020	11	0.07
^{137}Cs (γ)	E_γ 0.66	$x_{1/2}$ = 8.1 cm H_2O		0.39
^{60}Co (γ)	E_γ 1.20–1.30	$x_{1/2}$ = 11.1 cm H_2O		0.27

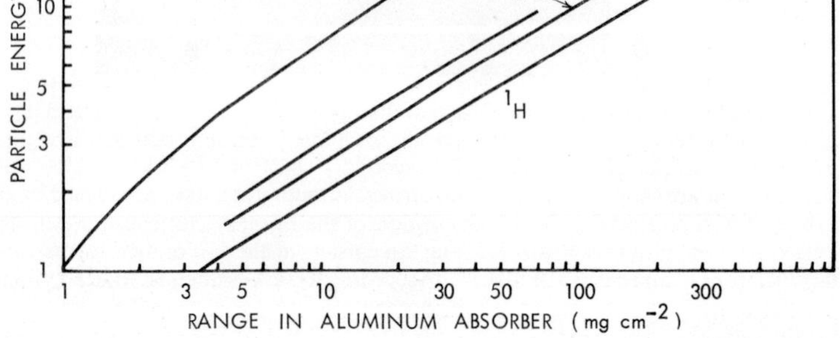

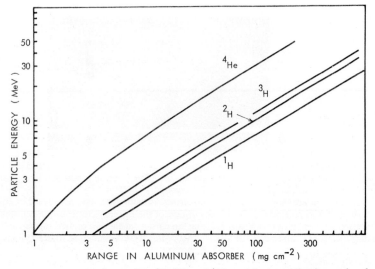

FIG. 14.5. Range of energetic ^4He, ^3H, ^2H, and ^1H particles in an aluminum absorber.

quite short. However, in gases at standard temperature and pressure the paths are several centimeters in length (Table 14.2). The range in air for α-particles with an initial energy E_α MeV can be calculated by the empirical equation ($\rho_{air} = 1.293$ kg m^{-3}):

$$\hat{R}_{air} = 0.31\,E_\alpha^{3/2}(\text{cm}) = 0.40\,E_\alpha^{3/2}\,(\text{mg cm}^{-2}) \qquad (14.8)$$

The range $\hat{R}_z$ in other materials can be approximated roughly by

$$\hat{R}_z = 0.173 E_\alpha^{3/2}\,A_z^{1/3}\,(\text{mg cm}^{-2}) \qquad (14.9)$$

A_z is the atomic weight of the absorber. Figure 14.5 shows the range of various charged particles in an aluminum absorber. The range of a 5 MeV α is 6 mg cm^{-2}; thus $\hat{R}_{Al} = 6 \times 10^{-3}/\rho_{Al}$ cm $= 0.002$ mm. Alpha-particles from radioactive decay are easily stopped even by the thickness of a sheet of paper.

The number of ion pairs formed per millimeter of range for α-particles, protons, and electrons are shown in Fig. 14.6a. The larger *specific ionization* of the α-particles

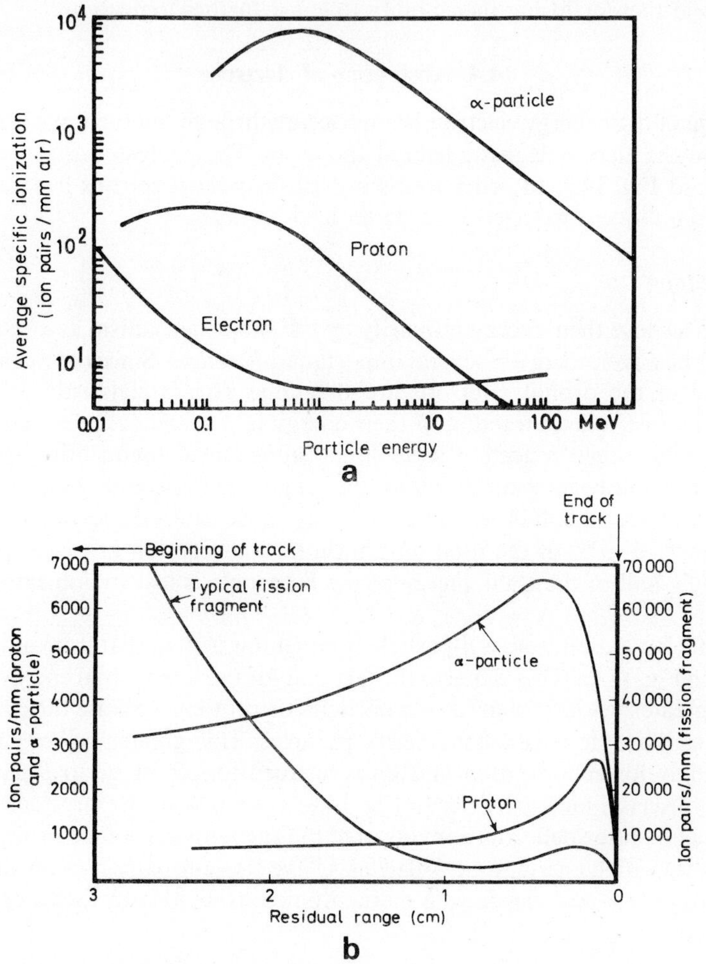

FIG. 14.6. Number of ion pairs formed per mm air at STP (specific ionization) as a function of the particle energy. (a) gives the average specific ionization for the maximum energy of the particle, (b) gives the specific ionization for the residual energy (given in residual range) of the particle. (According to H. A. C. McKay.

compared to the protons is related to the fact that the former are doubly charged. In general the specific ionization increases with the ionic charge of the particle for the same kinetic energy. Fission fragments that initially have very large energies also have very large ionic charges leading to quite high specific ionization in their absorption in matter; their range is 2–3 cm in air and 2–3 mg cm^{-2} in aluminum.

Charged particles decrease in velocity as they lose their energy in traversing an absorber. As a result they spend progressively longer times in the vicinity of any particular atom, which results in an increase in the probability of interaction with that atom. Consequently, there is a steady increase in the number of ion pairs formed along the path of the particle rather than a constant density of ion pairs. Near the end of the range for heavy charged particles a maximum is observed for the number of ion pairs formed per unit path length (the *Bragg peak*) (Fig. 14.6b). At a distance just beyond the Bragg peak maximum, the kinetic energy of the particles is comparable to those of the orbital electrons of the absorber. As a result, the particle can acquire electrons, becoming uncharged and thereby losing their ability to cause further ionization.

14.4. Absorption of electrons

Absorption of high energy electron beams occurs through interaction with the orbital electrons and the electromagnetic field at the atom. The processes are summarized in Table 14.1 and Fig. 14.7. In order to distinguish between electrons from accelerators and those from β-decay we refer to the latter as *β-particles.*

14.4.1. *Ionization*

Beta-particles lose their energy primarily by the same mechanism as α-particles (Fig. 14.7a and b); however, there are several important differences. Since the masses of the β-particles and of the orbital electrons are the same at nonrelativistic velocities, the β-particles can lose a large fraction of their energy in a single collision. The β-particle undergoes a wide angle deflection in such collisions and consequently β-particles are scattered out of the beam path all along the length. The secondary electrons ionized from the atom have such high energies that they cause extensive secondary ionization which provides 70–80% of the total ionization in β-absorption processes (Fig. 14.4b). Approximately half of the total energy of the β-particle is lost by ionization and half by excitation.

The specific ionization from a β-particle is much lower than that from a heavy ion as can be seen in Fig. 14.6a. This is due to the fact that for the same initial energy β-particles have much greater velocity than have α-particles or protons, because their mass is very much smaller than the mass of the heavy particles. This greater velocity results in a correspondingly lower ionization and gives a much longer range to β-particles. The erratic path observed for β-particles in Fig. 14.4c is a result of the large energy transfer and consequently large deflections involved in the encounters with the orbital electrons. However, at very high energies the β-particles have straight paths as a result of the fact that very energetic β-particles have a momentum considerably in excess of that of the orbital electron.

14.4.2. *Bremsstrahlung*

As a β-particle approaches an atomic nucleus, it is attracted by the positive field of the nucleus and deflected from its path. The deflection results in an acceleration

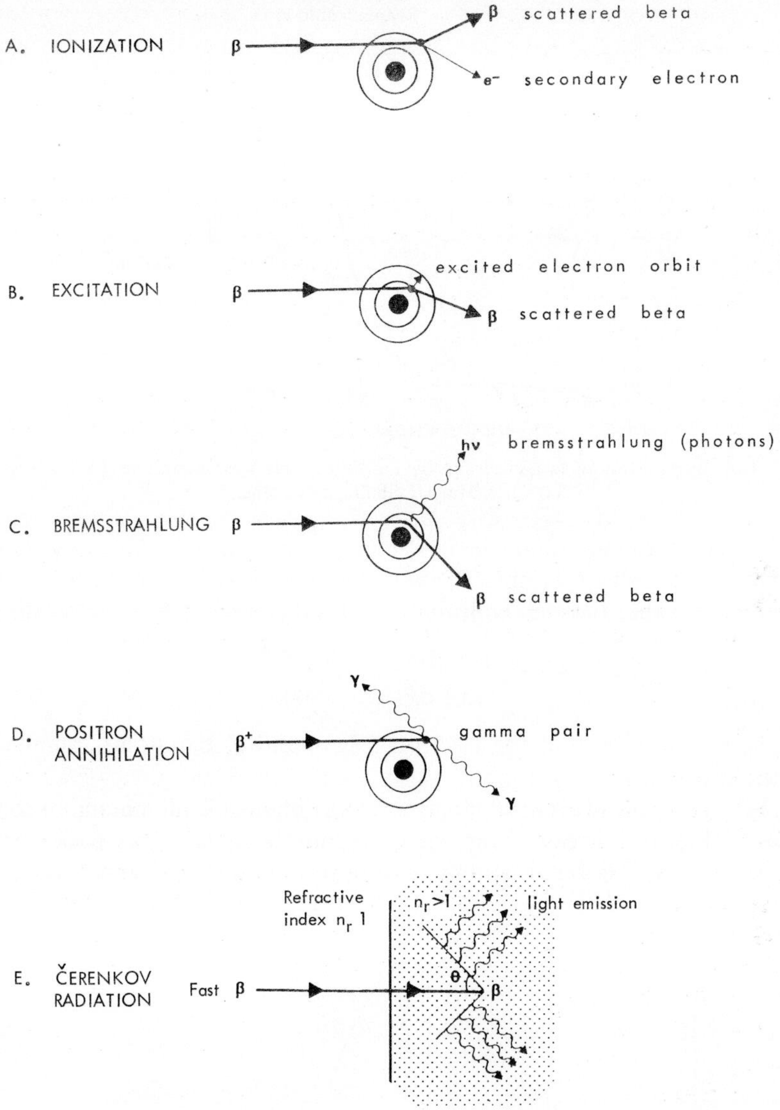

FIG. 14.7. Schematic description of the five processes accounting for β-particle absorption.

that, according to classical electrodynamics, leads to emission of electromagnetic radiation (Fig. 14.7c). Therefore the encounter with the positive charge of the nuclear field decreases the energy of the β-particle by an amount exactly equal to the amount of electromagnetic radiation emitted. This radiation is known as bremsstrahlung (braking radiation). The loss of energy by emission of bremsstrahlung radiation increases with the β energy and with the atomic number of the absorber material (Fig. 14.8). In aluminum approximately 1% of the energy of a 1 MeV electron is lost by bremsstrahlung radiation and 99% by ionization whereas in lead the loss by radiation is about 10%. For electrons of greater than 10 MeV energy, bremsstrahlung emission is the predominant mode of energy loss in lead. However, for the energies in radioactive

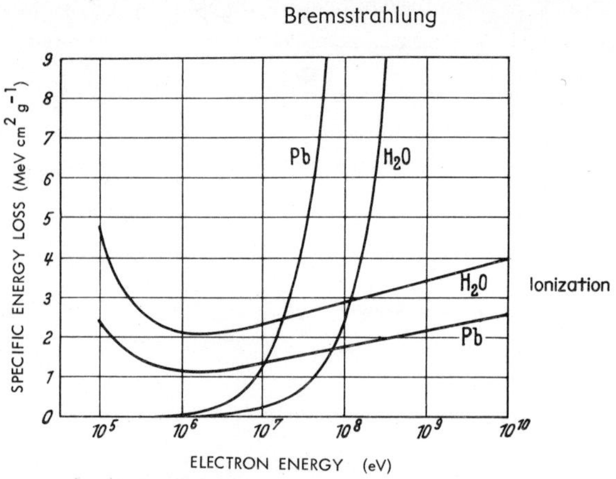

FIG. 14.8. Energy loss of fast electrons by ionization and bremsstrahlung. (According to Gentner, Maier-Leibnitz, and Bothe.)

decay, bremsstrahlung can usually be neglected—particularly for absorption processes in material of low atomic weight. The ratio of specific energy loss (dE/dx) through bremsstrahlung to that through collision (i.e. all other processes) is approximately:

$$\frac{(dE/dx)_{\text{brems}}}{(dE/dx)_{\text{coll}}} \approx \frac{E_e Z}{800} \tag{14.10}$$

where E_e is kinetic energy of the electron (MeV) and Z the atomic number of the absorber atoms.

Figure 14.9 shows the bremsstrahlung spectrum obtained in aluminum for β-particles emitted by ^{147}Pm. In this case a very small fraction of the β-energy is converted into radiation (Fig. 14.8). The bremsstrahlung spectrum is always of much lower energy

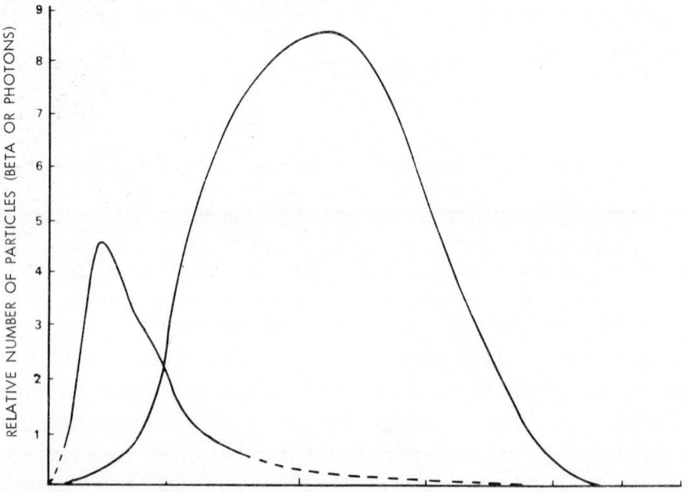

FIG. 14.9. Beta-spectrum (right curve) and bremsstrahlung spectrum in aluminum for ^{147}Pm. The ordinate of the bremsstrahlung spectrum is exaggerated by a factor about 100.

than that of the β-spectrum. Bremsstrahlung sources of a wide variety of energies are commercially available. They are used mainly for (analytical) X-ray excitation and for medical irradiation purposes.

14.4.3. *Čerenkov radiation*

The velocity of light in matter c depends on the refractive index n_r

$$c = \mathbf{c}\, n_r^{-1} \tag{14.11}$$

In water $n_r = 1.33$, in plexiglass 1.5. Figure 4.2 shows that β-particles with energies > 0.6 MeV move faster than light in water. When the particle velocity $(v_p) > c$, electromagnetic radiation is emitted coherently in a cone whose axis is the direction of the moving particle (Fig. 14.7e). The angle of the cone θ is obtained from

$$\sin\theta = c/v_p \tag{14.12}$$

This Čerenkov radiation is the source of the bluish light observed in highly radioactive solutions (Fig. 14.10) and around reactor fuel elements submerged in water. The radiation can be used for detecting β-particles and for measuring high particle energies (from θ). For a fast electron the energy loss through Čerenkov radiation is $\lesssim 0.1\%$ of the energy loss through other processes. Čerenkov detectors are described in §17.7.

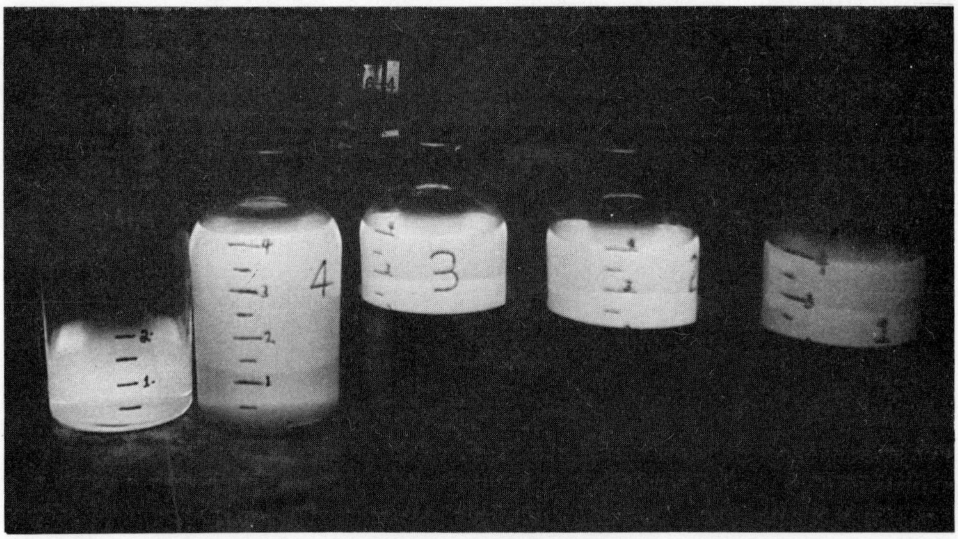

FIG. 14.10. Bottles containing highly radioactive ^{90}Sr solutions glow in the dark due to Čerenkov radiation from daughter ^{90}Y (E_{max} 2.3 MeV).

14.4.4. *Positron annihilation*

Positrons interact with matter through ionization, excitation, emission of bremsstrahlung, and Čerenkov radiation in the same manner as negative electrons. As the kinetic energy of the positron decreases in the absorber, there is an increase in probability of direct interaction between the positron and an electron (Fig. 14.7d) in which both the positron and electron are annihilated. The energy of the two electron masses is converted into electromagnetic radiation. This process, known as *positron annihilation*, is a characteristic means of identification of positron emission. Since an electron mass is equivalent to 0.51 MeV, and the kinetic energy of the particles of annihilation is essentially zero, the total energy for the annihilation process is 1.02 MeV. In order to conserve momentum at least two photons must be emitted with equal energy. These photons of 0.51 MeV each are referred to as *annihilation radiation*. The presence of γ-rays at 0.51 MeV in the electromagnetic spectrum of a radionuclide is strong evidence for the presence of positron emission by that nuclide.

14.4.5. *Absorption curves and scattering of β-particles*

An absorption curve for β-particles has a quite different shape than it has for α-particles (cf. Fig. 14.2). The continuous spectrum of energies in radioactive β-decay plus the extensive wide angle scattering of the β-particles by the absorber atoms account for the fact that range curves for β-particles continuously decrease. Even for a beam of initially monoenergetic electrons, the continuous removal of electrons from the beam path by wide angle deflection results in a plot showing a continuous decrease in the numbers of electrons with distance, with approximately 95% of the original β-particles stopped in the first half of the range. It is more common to speak of the absorber thickness necessary to stop 50% of the particles than to speak of the range itself. This *half-thickness value* is much easier to ascertain experimentally than is an apparent range. It should be remembered that the energy deposited at complete β-absorption is $E_{abs} \approx \frac{1}{3} E_{max}$ (Chapter 4).

The absorption curve for β-particles formed in radioactive decay can be described with fair approximation by the relationship (14.6). This is due to the continuous energy spectrum resulting in an exponential relationship for the range curve. In the E_{max} range 0.7–3 Mev the range in aluminum closely follows the relation (*Feather's rule*)

$$\hat{R}(\text{g Al/cm}^2) = 0.543\, E_{max}(\text{MeV}) - 0.160 \qquad (14.13)$$

This is the range C_3 in Fig. 14.3.

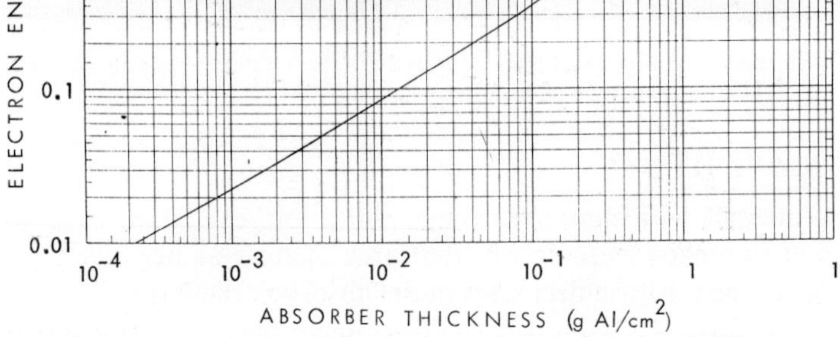

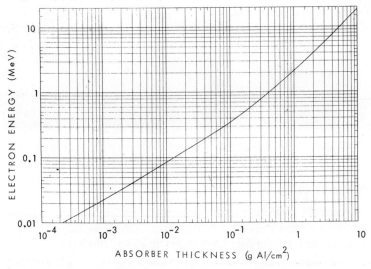

FIG. 14.11. Empirical relation for the maximum range of β-particles in aluminum.

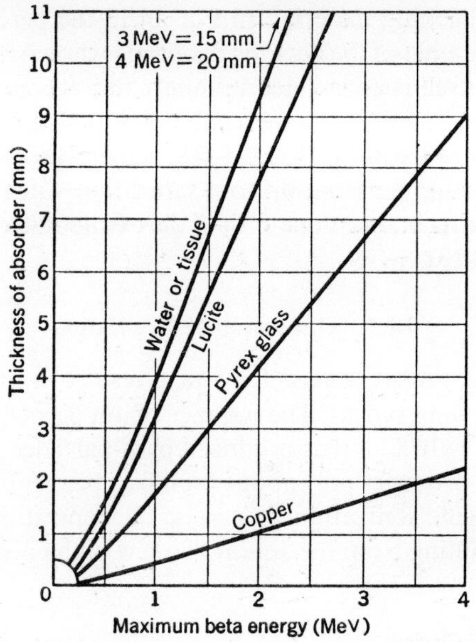

FIG. 14.12. Thickness of various materials needed to completely stop β-particles.

Figure 14.11 shows an empirical relationship between the maximum energy of β-particles and the extrapolated range in aluminum. Compared to α-radiation, β-radiation has a much longer range. For example, the range of an α-particle of 5 MeV is 3.6 cm in air while that of a β-particle of 5 MeV is over 17 m. A comparison of the range in air and water for electrons and heavy particles is given in Table 14.2. Figure 14.12 is useful for a rapid estimate of the absorber thickness needed to protect against β-particles.

An additional complication in the experimental measurements of absorption curves for β-particles is found in the fact that a certain fraction of β-particles which are not originally emitted in the direction of the detector may be deflected to the detector by the large angle scattering. This process is known as *backscattering*, since the *backing* (or *support*) for radioactive samples may cause scattering of a certain fraction of the particles through as much as 180°. The fraction of backscattered β-radiation depends on the geometry of the measuring system, the energy of the β-particles, and the thickness and

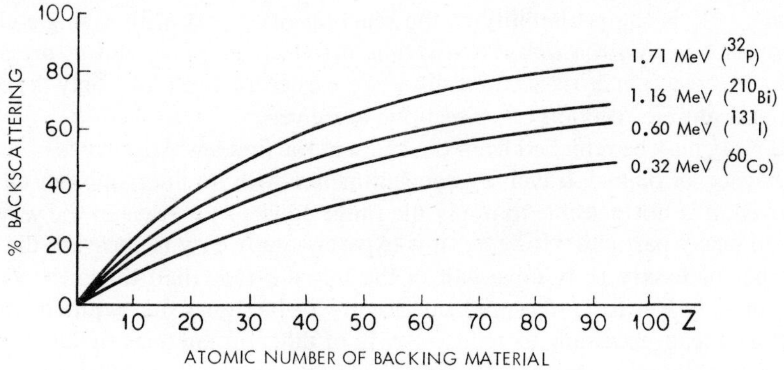

FIG. 14.13. Backscattering of β-particles of different energy as a function of atomic number of thick backing materials.

electron density of the backing material. In Fig. 14.13 the percent backscattering as a function of the atomic number of the backing material is shown for four β-energies (E_{max}); the radioactive sample itself is considered infinitely thin (i.e. no self-absorption). From the curve for ^{32}P on platinum ($Z = 78$) we see that about 40% of the measured radiation is due to backscattered radiation ($0.8/(1.0 + 0.8) = 0.4$). Backscattering increases with the thickness of the backing material up to a saturation value which is reached when the thickness of the backing is about one-fifth of the extrapolated range of the β-particles in that material (Fig. 14.13).

14.5. Absorption of γ-radiation

The absence of charge and rest mass for γ-rays results in little interaction with the absorbing atoms and in long ranges. The number of ion pairs produced in a given path length by γ-rays is only 1–10% of that produced by β-particles of the same energy (Fig. 14.4); e.g. a 1 MeV γ-ray produces only about 1 ion pair per centimeter of air. As a consequence of this low specific ionization of γ-rays, the ionization is almost completely secondary in nature resulting from the action of a few high energy primary ion pairs.

14.5.1. *Attenuation coefficient*

Unlike heavy particles and electrons which lose their energy as a result of many collisions, γ-rays are completely stopped in one or at most a few interactions. For thin absorbers the attenuation of γ-rays follows relation (14.6), where ϕ is the number of photons $m^{-2}\,s^{-1}$. The proportionality factor μ is called the (*total*) *attenuation coefficient*. When it has the dimension of m^{-1} and the thickness x is expressed in meters, μ is referred to as the *linear* attenuation coefficient. The attenuation coefficient can be expressed in other ways:

$$\mu_m = \mu/\rho = \sigma_a N_A/M = \sigma_e Z N_A/M \tag{14.14}$$

where ρ is the density, M the average atomic weight, and Z the average atomic number of the absorber. $N_A\rho/M$ can be replaced by N_v according to (9.5). By analogy with (9.7) we can define a macroscopic absorption cross-section Σ. Σ^{-1} is the mean free path or *relaxation length* of the radiation in the absorbing material. μ_m (in $cm^2\,g^{-1}$ when x is in centimeters) is the *mass attenuation coefficient*; σ_a is the probability of reaction between a γ-ray and the electron cloud of the absorber atom (*atomic reaction cross-section*, $m^2\,atom^{-1}$); σ_e is the probability of the reaction of a γ-ray with a single electron of the absorber (*electron reaction cross-section*, $m^2\,electron^{-1}$). σ_a and σ_e are analogous to the nuclear reaction cross-sections discussed earlier. In Table 14.1 only the equivalent nuclear and atomic reaction cross-sections are given.

Since a γ-ray may be removed from the beam in the first few Ångstroms of its entrance into the absorber or may travel several centimeters with no interaction at all and then be removed, it is not possible to apply the range concept to γ-rays in the way that it is applied to heavy particles. However, it is experimentally easy to measure the thickness of absorber necessary to remove half of the initial γ-rays (half thickness value) from a beam, or reduce it to 1/10, 1/100, etc. Figure 14.14 shows the required thickness of concrete and lead necessary to reduce γ-rays of different energies by factors of 10 (cf. §14.10 on shielding thicknesses). The half-thickness value is

$$x_{1/2} = \ln 2/\mu \tag{14.15a}$$

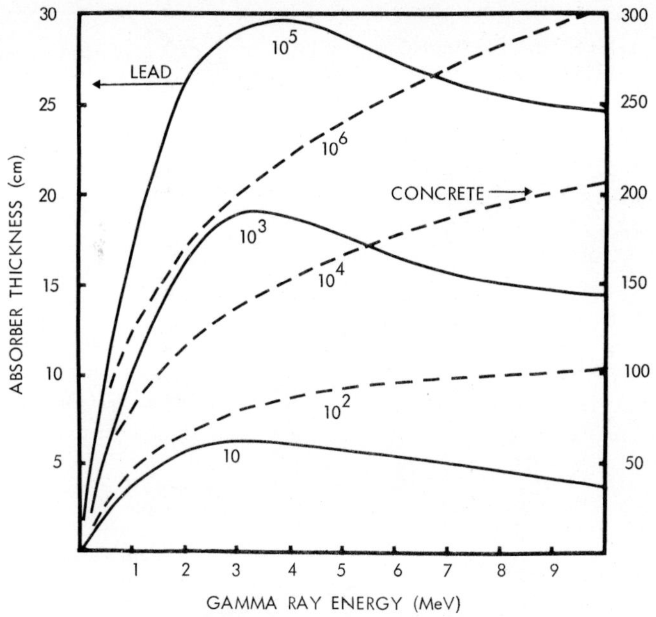

FIG. 14.14. Thickness (cm) of lead and concrete absorbers necessary to reduce γ-ray fluxes by different factors of 10, as a function of the γ-ray energy. These curves are for thick absorbers and include build-up factors (§14.10).

and the 1/10 value

$$x_{1/10} = \ln 10/\mu \tag{14.15b}$$

14.5.2. *Partial absorption processes*

Gamma-ray absorption occurs as illustrated in Fig. 14.15 by four different processes: coherent scattering, photoelectric effect; Compton effect, and pair production. For each of these processes, a partial coefficient can be expressed:

$$\mu = \mu_{coh} + \mu_{phot} + \mu_{Comp} + \mu_{pair} \tag{14.16}$$

Comparing with (14.7b), μ_{phot} and μ_{pair} are absorption processes, while μ_{coh} is all scattering; μ_{Comp} contributes to both the μ_s and the μ_a terms. In Fig. 14.16 the total attenuation, absorption, and the partial coefficients are given for water, aluminum, and lead as a function of the γ-ray energy. The corresponding linear coefficients are obtained by multiplying with ρ (for aluminum 2.7, for lead 11.3). It should be noted that the aluminum curves also can be used for absorption in concrete.

In *coherent scattering* (also called Bragg or Rayleigh scattering, denoted σ_r in Fig. 14.16) the γ-ray is absorbed and immediately re-emitted from the atom with unchanged energy but in a different direction. Coherently scattered radiation can give interference patterns, so the process is used for structural analysis of absorbing material in the same way as X-rays are. The probability for coherent scattering increases with the square of atomic number of the absorber and decreases with γ-ray energy. In lead, coherent scattering amounts to about 20% of the total attenuation for γ-energies of 0.1 MeV but decreases in importance for higher energy γ-rays.

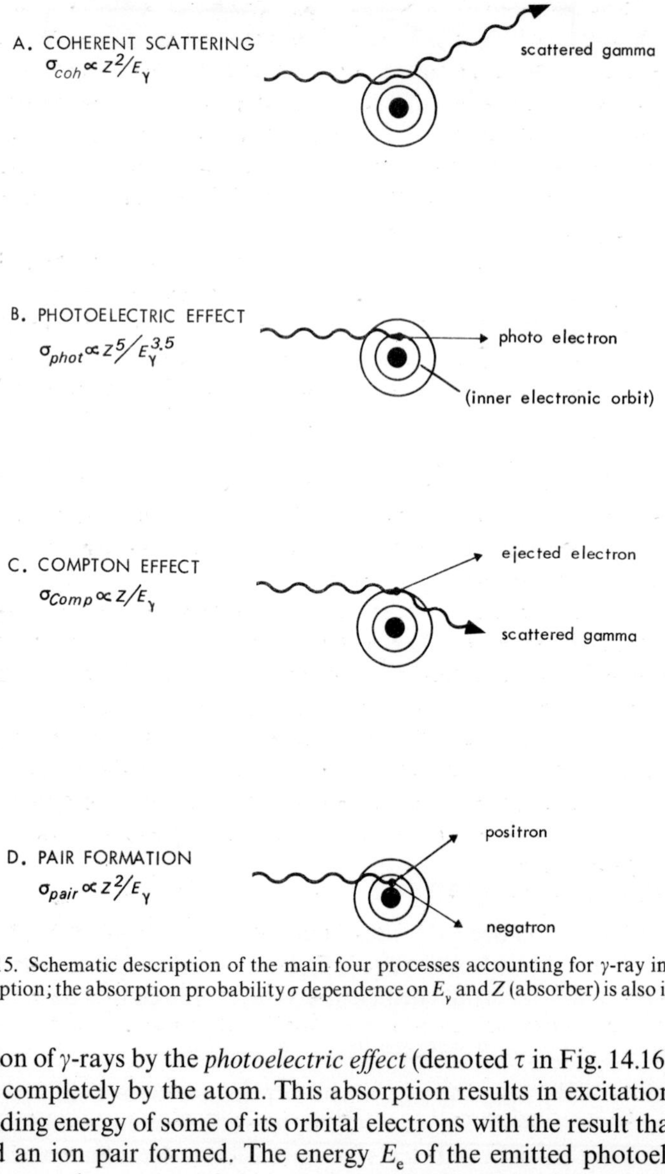

FIG. 14.15. Schematic description of the main four processes accounting for γ-ray interaction and absorption; the absorption probability σ dependence on E_γ and Z (absorber) is also indicated.

In absorption of γ-rays by the *photoelectric effect* (denoted τ in Fig. 14.16) the photons are absorbed completely by the atom. This absorption results in excitation of the atom above the binding energy of some of its orbital electrons with the result that an electron is ejected and an ion pair formed. The energy E_e of the emitted photoelectron is the difference between the energy of the γ-ray and the binding energy for that electron in the atom

$$E_e = E_\gamma - E_{be} \qquad (14.17a)$$

If the photoelectron originates from an inner electronic orbital, an electron from a higher orbital moves to fill the vacancy. The difference in binding energy of the higher and the lower energy orbital causes emission of X-rays and of low energy Auger electrons. The process of electron cascade, accompanied by X-ray and Auger electron emission, continues until the atom is reduced to its ground state energy. The photoelectron as well as the Auger electrons and the X-rays cause extensive secondary ionization by interacting with the absorber atoms.

The probability for the photoelectric effect decreases with increasing γ-ray energy. It is largest for the most tightly bound electrons and thus the absorption coefficient for the photoelectric effect decreases in the order of electron shells K > L > M >, etc. In Fig. 14.16 we see that in lead it is the dominating mode of absorption up to about 0.7 MeV. Discontinuities observed in the graph of μ vs E_γ are related to the differences in binding energies of the electrons in the different shells as the increasing γ-ray energy allows more tightly bound electrons to be emitted. These discontinuities coincide with the K, L, etc., edges observed in X-ray absorption.

Gamma-rays of higher energy, rather than interacting with the field of the whole atom as in the photoelectric effect, interact with the field of one electron directly. This mode of interaction is called the *Compton effect* after its discoverer, A. H. Compton. In the Compton effect an electron is ejected from an atom while the γ-ray is deflected with a lower energy. The energy of the scattered γ-ray, E'_γ, is expressed by the equation

$$E'_\gamma = E_\gamma - E_e \tag{14.17b}$$

where E_e is the kinetic energy of the Compton electron. The probability for Compton scattering increases with target Z and decreases with E_γ. Since the Compton interaction occurs only with the most weakly bound electrons and high energy γ-rays, the binding energy of the electron is negligible compared to E_γ. The Compton electrons and scattered γ-rays have angles and energies which can be calculated from the relationships between the conservation of energy and momentum, correcting for the relativistic mass of the electrons at these kinetic energies. The scattered γ-ray may still have sufficient energy to interact further by the Compton effect, the photoelectric effect or pair production. Again, emission of X-rays and Auger electrons usually accompanies Compton interaction and extensive secondary ionization follows. Since the Compton electron can have a spread of values, the scattered γ-rays exhibit a broad spectrum. The Compton electrons, as in the case of photoelectrons, are eventually stopped by the processes described for β-particles.

Figure 14.16 shows the division of energy between the scattered Compton γ and the Compton electron as a function of γ-ray energy. Only the energy of the electron is deposited in the absorber as the scattered γ-ray has a high probability of escape. Thus Compton electrons contribute to the *(energy) absorption coefficient* μ_a while the Compton γ contributes to the *total attenuation coefficient* μ through the scattering coefficient μ_s in (14.7).

The fourth mode of interaction for γ-rays with an absorber involves conversion in the Coulomb field of the nucleus of a γ-ray into an electron and a positron (Fig. 14.17). This process is termed *pair production* since a pair of electrons, one positive and one negative, is produced. The process can be considered as the inverse phenomenon of positron annihilation. Since the rest mass of an electron corresponds to 0.51 MeV, the γ-ray must have a minimum value of 1.02 MeV to interact by pair production. As the energy of the γ-ray is increased beyond this value, the probability of pair production increases (see Fig. 14.16, where μ_{pair} is denoted κ). The excess energy (above the 1.02 MeV) appears as the kinetic energy of the electron pair.

$$E_\gamma = 1.02 + E_{e-} + E_{e+} \tag{14.18}$$

The pair of electrons are absorbed as described in § 14.4. The annihilation of the positron produces a 0.51 MeV γ, which is absorbed by the processes described above.

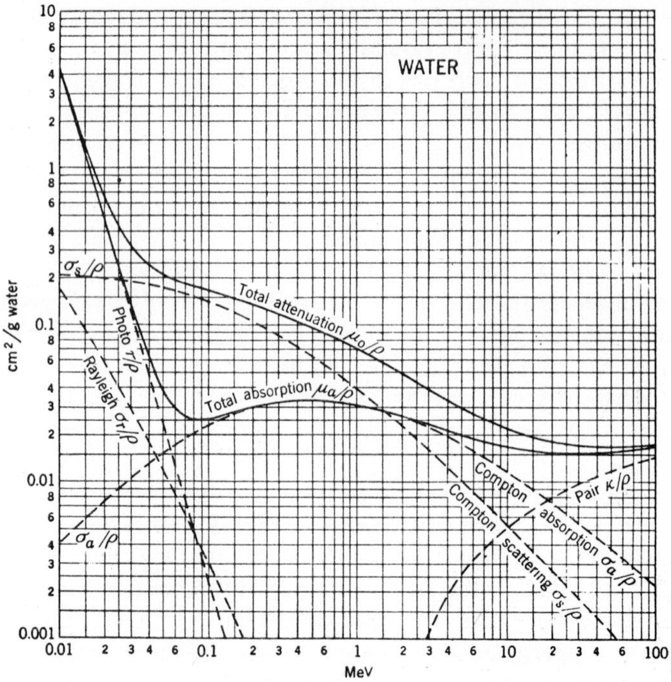

WATER
cm²/g water
MeV
Total attenuation μ_o/ρ
Total absorption μ_a/ρ
σ_s/ρ
Photo τ/ρ
Rayleigh σ_r/ρ
σ_a/ρ
Compton absorption σ_a/ρ
Compton scattering σ_s/ρ
Pair κ/ρ

(a)

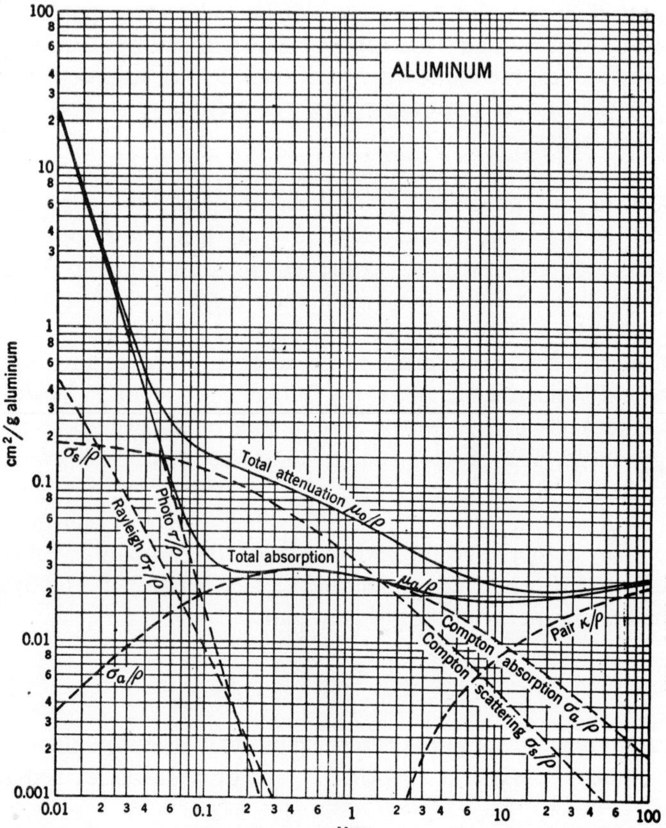

ALUMINUM
cm²/g aluminum
MeV
Total attenuation μ_o/ρ
Total absorption
μ_a/ρ
σ_s/ρ
Photo τ/ρ
Rayleigh σ_r/ρ
σ_a/ρ
Compton absorption σ_a/ρ
Compton scattering σ_s/ρ
Pair κ/ρ

(b)

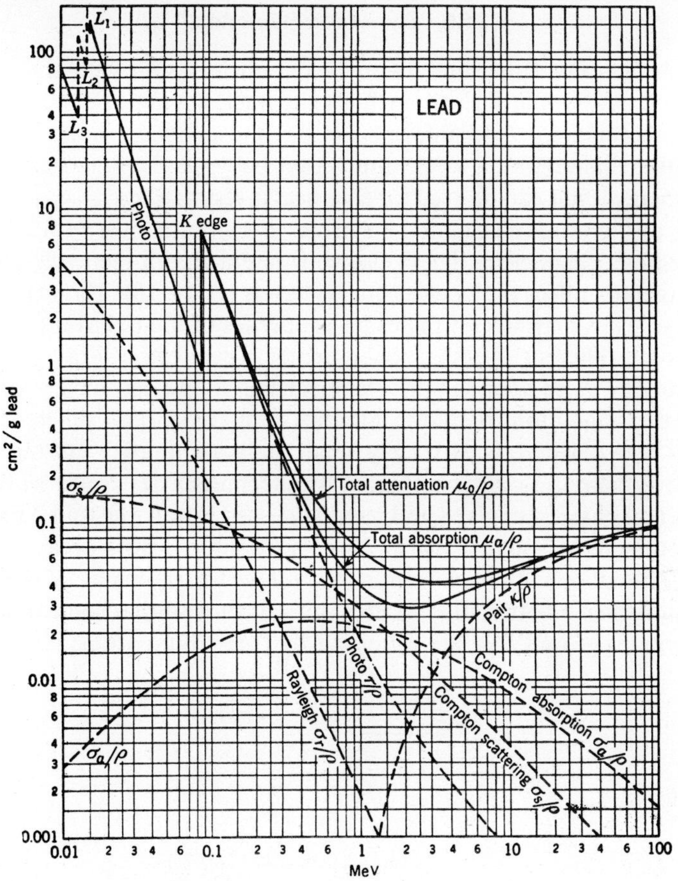

(c)

FIG. 14.16. Total and partial mass absorption and attenuation coefficients for γ-rays in water, aluminum and lead. (According to R. D. Evans.)

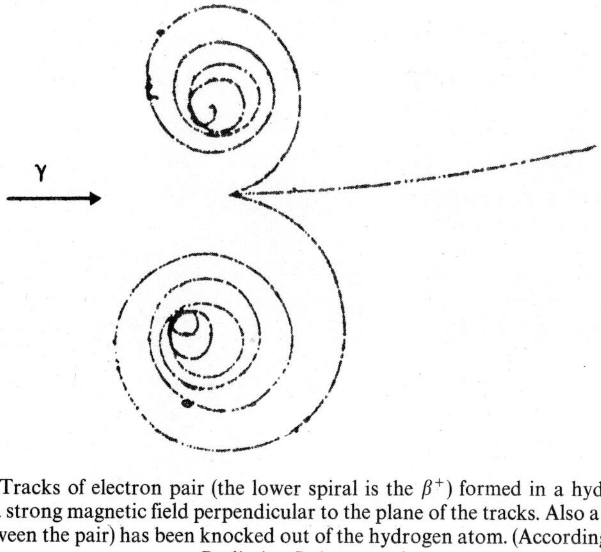

FIG. 14.17. Tracks of electron pair (the lower spiral is the β^+) formed in a hydrogen bubble chamber in a strong magnetic field perpendicular to the plane of the tracks. Also a third electron (emitted between the pair) has been knocked out of the hydrogen atom. (According to Lawrence Radiation Laboratory.)

14.6. The Mössbauer effect*

In §6.6.3 we described the interaction of the nuclear spin with that of the orbital electrons which allows that measurement of the nuclear magnetic moment in different chemical compounds to yield information on the chemical bonding (i.e. the nmr technique). In a similar way a weak interaction between the innermost electrons and the nuclear energy levels can be used for chemical investigations.

According to quantum mechanics, electrons in the innermost orbitals have a finite probability within the nucleus. These electrons interact with the nuclear charge distribution, and thereby affect the nuclear energy levels (cf. §6.3.3). The extent of the effect on the nuclear levels depends on the exact properties of the electron orbitals involved, which vary with different chemical compounds. Therefore a γ-ray emitted from an isomeric state of an atom bound in one chemical compound may have a slightly different energy than the same atom bound in another compound. This difference, referred to as the *isomer* (*energy*) *shift*, is extremely small, only about 10^{-10} of the energy of the emitted γ. Nevertheless, it can be measured by a technique developed by R. Mössbauer. The fundamental physics involved and technique used is well illustrated by Mössbauer's original experiment. Mössbauer placed an ^{191}Os source (whose decay scheme is given in Fig. 14.18) about a half-meter from a γ-ray detector shown as A in Fig. 14.19. An iridium foil absorber is between the source and detector so that some of the photons of 129 keV energy from the ^{191}Ir are absorbed by the iridium atoms in the foil, exciting

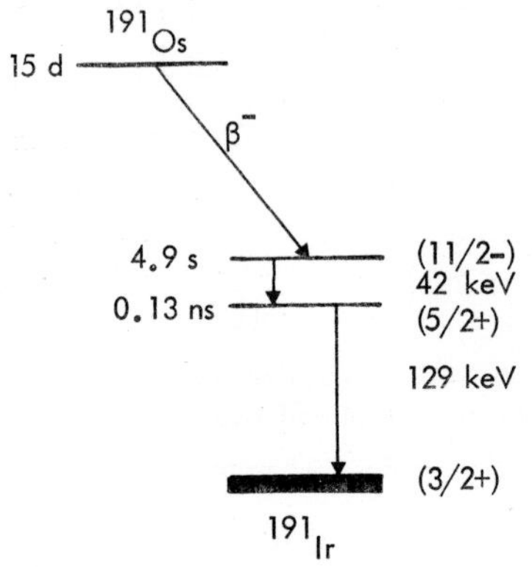

FIG. 14.18. Decay scheme of ^{191}Os.

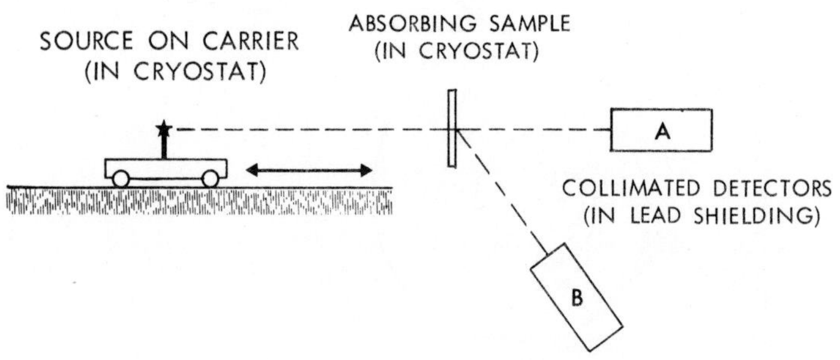

FIG. 14.19. Principle of a Mössbauer experiment.

these atoms from the ground state (3/2 +) to the 5/2 + state. Because of the short half-life of the latter state it immediately decays, re-emitting the γ-ray. The emission is isotropic, i.e. occurs in all directions. The result is a reduction in intensity measured by detector A but an increase in the count rate in detector B. The conditions for such a *nuclear resonance absorption* are very stringent. Using the Heisenberg relationship (5.8) we can estimate the half-value width of the 129 keV peak to be 5×10^{-6} eV. We can also use relation (4.34) to calculate the iridium atom recoil energy to be 46×10^{-3} eV. Thus the γ-ray leaves the source with an energy of (129×10^3 to 46×10^{-3} eV). Also, in order for the 129 keV γ-ray to be absorbed in ^{191}Ir, it must arrive with an excess energy of 46×10^{-3} eV to provide for the conservation of momentum of the absorbing atom. Thus there is a deficit of $2 \times 46 \times 10^{-3}$ eV, which is very different than the value of the very narrow energy width of the γ-ray. Consequently, no resonance absorption can take place.

The limitation posed by the recoil phenomenon can be circumvented. If the source and absorber atoms are fixed in a crystal, the recoil energy may be insufficient to cause bond breakage. The energy is absorbed as an atomic vibration in the crystal, provided the quantization of the vibrational states agree exactly with the recoil energy. If not, which is often the case, the absorber atom stays rigid in the lattice, and the recoil energy is taken up by the whole crystal. In this case it is necessary to use the mass of the crystal in (4.34) rather than the mass of a single atom. Under these circumstances the recoil energy becomes infinitesimally small for the emitting as well as the absorbing atom; this is called recoilless absorption. The probability for recoilless absorption is improved if the source and absorber are cooled to low temperatures.

The data of Fig. 14.20 were obtained by recoilless absorption in osmium metal containing ^{191}Os (source) and Ir (x_{191} 37.4%) metal, both cooled in cryostats. By slowly moving the source (with velocity v) towards or away from the absorber, some kinetic energy ΔE_γ is added or subtracted from the source energy E_γ as "detected" by the absorber (Doppler effect). The energy and velocity relationship is given by the Doppler equation

$$\frac{\Delta E_\gamma}{E_\gamma} = \frac{v}{\mathbf{c}} \tag{14.19}$$

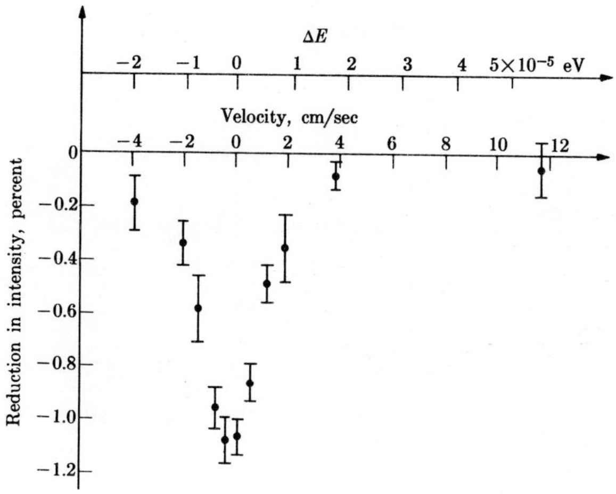

FIG. 14.20. Mössbauer spectrum of ^{191}Ir metal. (According to R. Mössbauer.)

The velocity is indicated in Fig. 14.20, where a value of v of 1 cm s^{-1} corresponds to 4.3×10^{-6} eV. The half-value of the γ-peak is found to be about 20×10^{-6} eV, i.e. a factor 4 times higher than calculated by the Heisenberg relationship. This is due to Doppler broadening of the peak as a consequence of some small atomic vibrations. Although the Mössbauer method can be used for measurements of γ-linewidths the results are subject to considerable errors.

One of the most striking uses of the extreme energy resolution obtainable by the Mössbauer effect was achieved by R. V. Pound and G. A. Rebka, who measured the emission of photons in the direction towards the earth's center, and in the opposite direction from the earth's center. They found that the photon increased its energy by one part in 10^{16} per meter when falling in the earth's gravitational field. This can be taken as a proof that the photon of $E_{h\nu} > 0$ does have a mass.

When, for a "Mössbauer pair" (like $^{191}Os/Ir$, or $^{57}Co/Fe$, $^{119m}Sn/Sn$, $^{169}Er/Tm$, etc.), source and absorber are in different chemical states, the nuclear energy levels differ for the two Mössbauer atoms by some amount ΔE_γ. By using the same technique as in Fig. 18.20, resonance absorption can be brought about by moving the source with a velocity corresponding to ΔE_γ. In this manner a *Mössbauer spectrum*, characteristic of the compound (relative to a reference compound) is obtained; the location of the peaks (i.e. the absorption maxima) with respect to a nonmoving source (the *isomer shift*) is usually given in mm s^{-1}.

Figure 14.21 shows the isomer shifts obtained for a number of neptunium compounds. The positions of the isomer shifts on the right show the effect of valence states due to different population of the 5f orbitals in neptunium. The different shifts for compounds of the same valence state is a measure of the variation in the covalency of the bonding.

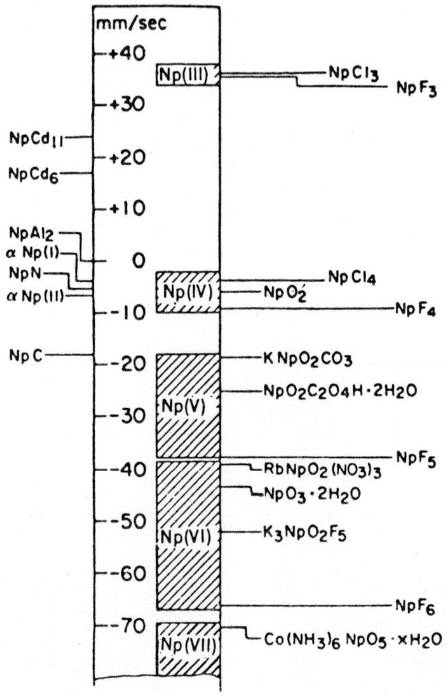

FIG. 14.21. Isomeric shifts observed for a number of neptunium compounds. The electron density increases towards the bottom. (According to R. L. Cohen.)

The compounds on the left are metallic. The shifts reflect the contributions of conduction electrons to the electron density at the nucleus of neptunium.

Mössbauer spectroscopy is limited to the availability of suitable sources. About 70 Mössbauer pairs are now available. The technique provides a useful method for studying chemical compounds in the solid state, especially compounds which are nontransparent to light and chemically or radioactively unstable.

14.7. Electron spectrometry for chemical analyses (ESCA)*

As first demonstrated by S. Hagström, C. Nordling, and K. Siegbahn, high resolution β-spectroscopy can be used to determine chemical properties. Figure 14.22 shows the experimental arrangement for electron scattering analysis (ESCA) or photoelectron spectroscopy. A sample is irradiated with monoenergetic photons of $E_{h\nu}$, leading to the emission from the sample surface of photoelectrons. The relevant equation is

$$E_{h\nu} = E_{be}(X, Y) + E_e \tag{14.20}$$

where E_e is the kinetic energy of the emitted electrons, which can be determined very accurately (presently to about 0.01 eV) in the magnetic spectrograph. This sensitivity is much greater than chemical binding energies, E_{be} (X, Y), where X refers to an atom in compound Y. The probability for ejection of photoelectrons increases with decreasing photon energy (Fig. 14.16) so low energy X-rays are used as a source.

Although it is the outermost (or most weakly bound) electrons which form the valency orbitals of a compound, this does not leave the inner orbitals unaffected. An outer electron (which we may refer to as e_L) of an atom X_1 which takes part in bond formation with another atom X_2 decreases its potential, which makes the inner electrons (which we may call e_K) more strongly bound to X_1. Thus $E_{be}(e_K)$ increases by an amount depending on $E_{be}(e_L)$. Although this is a somewhat simplified picture, it leads to the practical consequence that the binding energy of e_K, which may be in the 100–1000 eV range, depends on the chemical bond even if its orbital is not involved directly in the bond formation. Figure 14.23 shows spectra of two nitrogen compounds. The two lines NH_4NO_3 indicate the nonequivalence of the two nitrogen atoms; the one with formal oxidation number -3 and the other with $+5$. In the spectrum of sodium hyponitrite, $Na_2N_2O_2$, the two lines indicate two different nitrogen atoms; the charges of these nitrogen atoms are not known. It is observed that all peaks appear at different E_{be} values. The possibility of studying surface reactions is illustrated in Fig. 14.24,

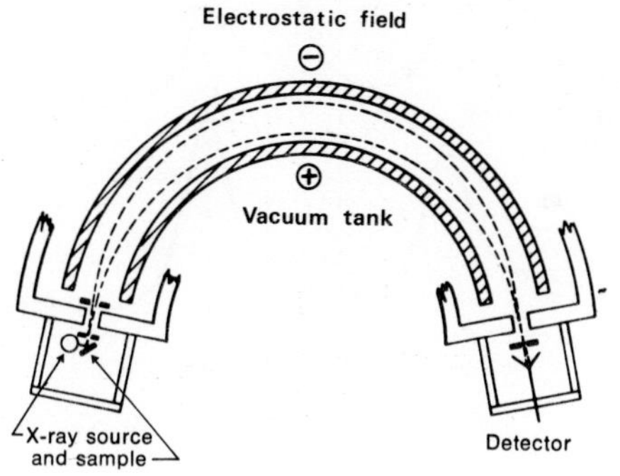

FIG. 14.22. Principle of the ESCA apparatus.

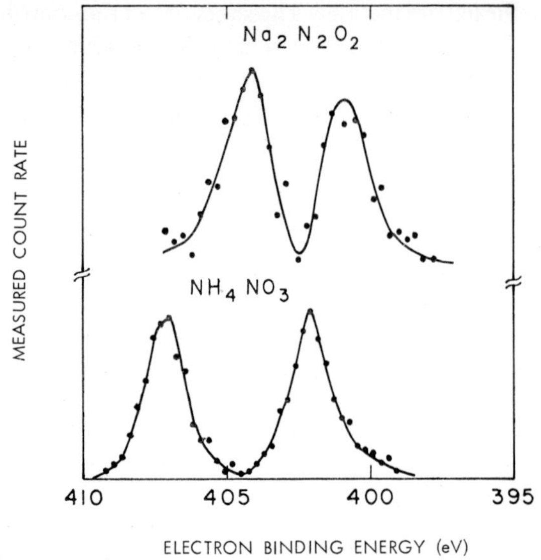

FIG. 14.23. ESCA spectrum for sodium hyponitrite ($Na_2N_2O_2$) and ammonium nitrate (NH_4NO_3).

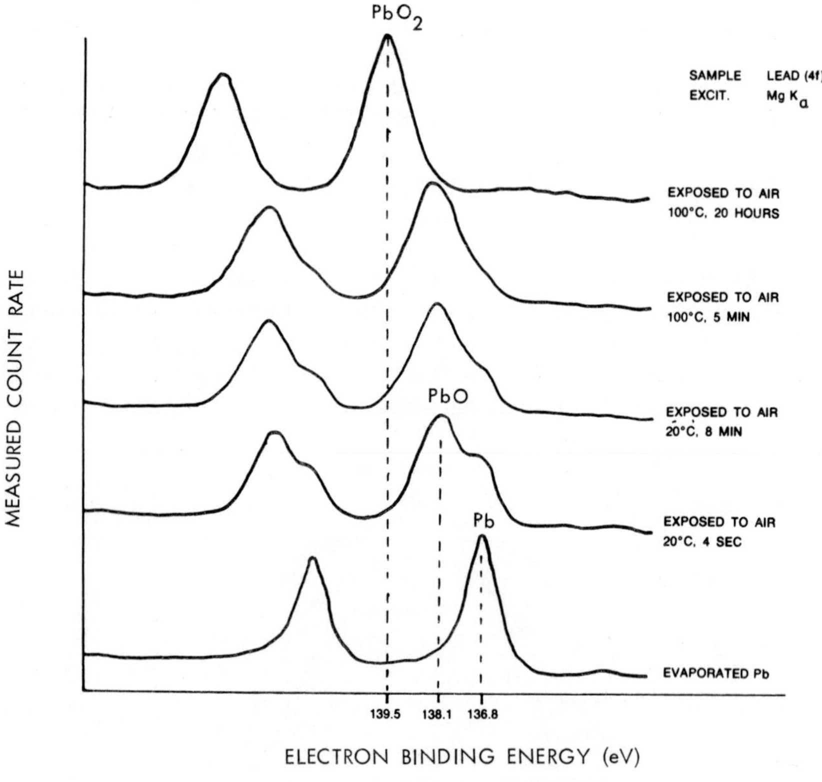

FIG. 14.24. ESCA spectrum measured on pure lead (bottom curve) while it is being oxidized in air to form (first) PbO and (then) PbO_2 (top curve). (From J. F. Rendina.)

showing the oxidation of Pb to PbO_2. The peaks at 136.8 and 141.8 eV are due to a pair of 4f electrons whose energies are shifted about 2.7 eV higher in energy due to the formation of PbO_2. The Pb–O bond involves the 6p orbitals in lead, so this shows the indirect effect of the bond on the energy of the 4f electrons.

14.8. X-ray fluorescence analysis*

If a sample containing atoms of a particular element (e.g. Ag) is irradiated with photons of energy high enough to excite an inner electron orbital (e.g. the K_α orbital in Ag at 22.1 keV), X-rays are emitted in the de-excitation. If the photon source is an X-ray tube with a target made of some element (Ag in our example), the probability is very high that the K_α X-ray emitted from the source would be absorbed by the sample atoms and re-emitted (fluorescence). (This is an "electron shell resonance absorption" corresponding to the Mössbauer effect, although the width of the X-ray line is so large that recoil effects can be neglected.) The spectrum of the scattered X-radiation (or, more correctly, photon radiation emitted from the sample) is referred to as the *X-ray fluorescence spectrum.* It contains lower energy radiation including K_α radiation emitted by atoms of lower atomic number. The height of these other peaks is lower because of a lower reaction cross-section (Fig. 14.25).

X-ray fluorescence analysis using vacuum tube sources have become a well-established analytical technique in the last decade. Nuclear interest stems partly from the possibility of using nuclear radiation as a source for stimulating X-ray fluorescence in a sample. These sources can be classified into several groups, depending on the mode of production of the X-rays:

(i) γ-ray sources: decay between closely located nuclear energy levels, e.g. a 59.5 keV

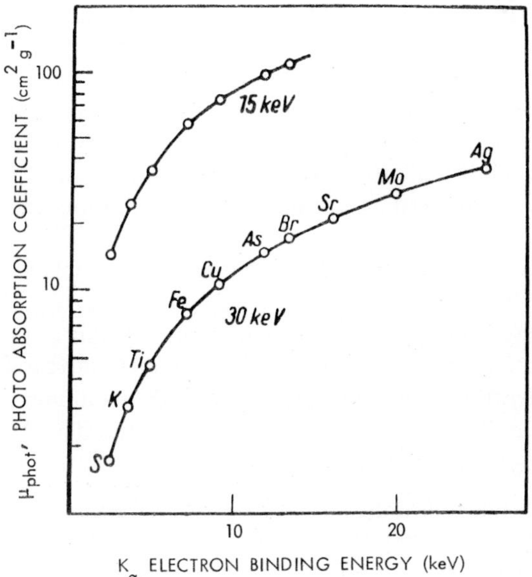

FIG. 14.25. Photoelectron absorption coefficients at K_α edges for 15 and 30 keV γ-rays as a function of absorber material. The photo effect is the dominating absorption mode.

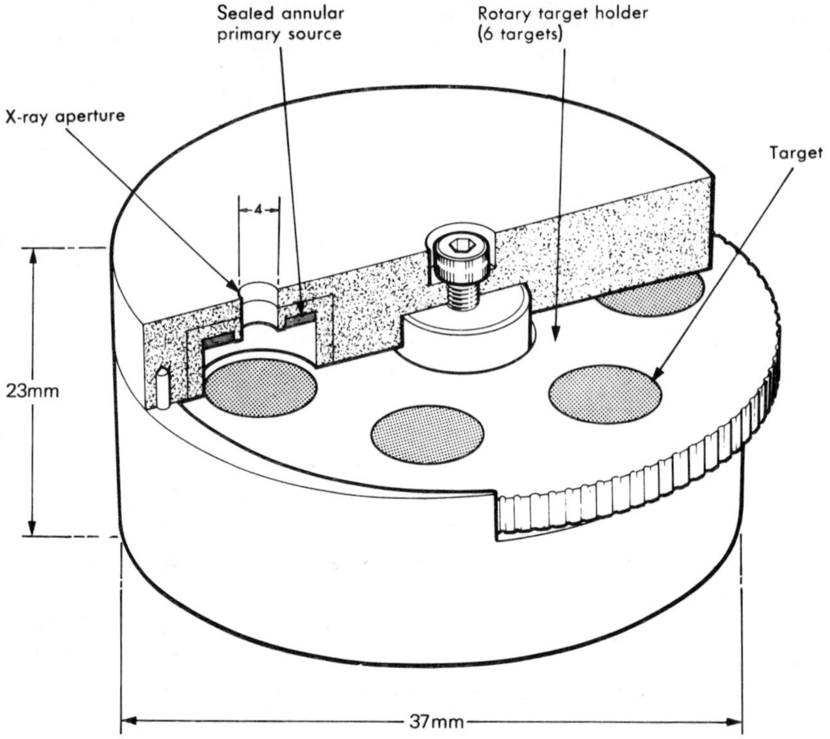

FIG. 14.26. Commercial secondary X-ray source of ^{241}Am (60 keV) with six target (Cu, Rb, Mo, Ag, Ba, Tb) arrangements. Dimensions are in mm. The X-rays emitted by the metal foils vary between 8 for copper and 50 keV for terbium. The yield is for copper 80 and for terbium 1750 photons s^{-1} steradian^{-1} mCi^{-1} ^{241}Am. (According to The Radiochemical Centre, Amersham.)

γ emitted in the α-decay of ^{241}Am: also "broad spectrum" γ-sources like ^{125}I are used.

(ii) X-ray sources: (a) radiation emitted in rearrangement of electron orbitals following α- or β-decay (primary X-rays), e.g. 11.6–21.7 keV uranium L-X-rays from ^{238}U formed in the α-decay of ^{242}Pu, or 41.3–47.3 keV europium K-X-rays from β-decay of ^{153}Gd; (b) irradiation of a target with α-, β-, or γ-radiation leading to ionization and excitation of the target atoms and its de-excitation by X-ray emission. Figure 14.26 shows a commercial arrangement in which an annular ^{241}Am source irradiates a number of target materials, which then emit characteristic X-rays.

(iii) Bremsstrahlung sources, e.g. T in titanium, or ^{147}Pm in aluminum (see Fig. 14.9).

(iv) Accelerator beams of charged particles (p, α, etc.) bombarding the source. In this case the acceleration energy may be far below the Coulomb threshold for nuclear transformation. Particularly the PIXE (proton induced X-ray emission) technique has become useful for studying very small samples.

An important advantage of radioactive X-ray fluorescence sources is that very small instruments requiring no (X-ray) high voltage or current can be designed for field applications. These instruments have been used in geological investigations (boreholes, mineral samples, etc.), in-line mineral analysis (Zn, Cu, and other ores in flotation cycles, Ca in cement raw-material, etc.), on-line analysis of surface depositions (Zn on

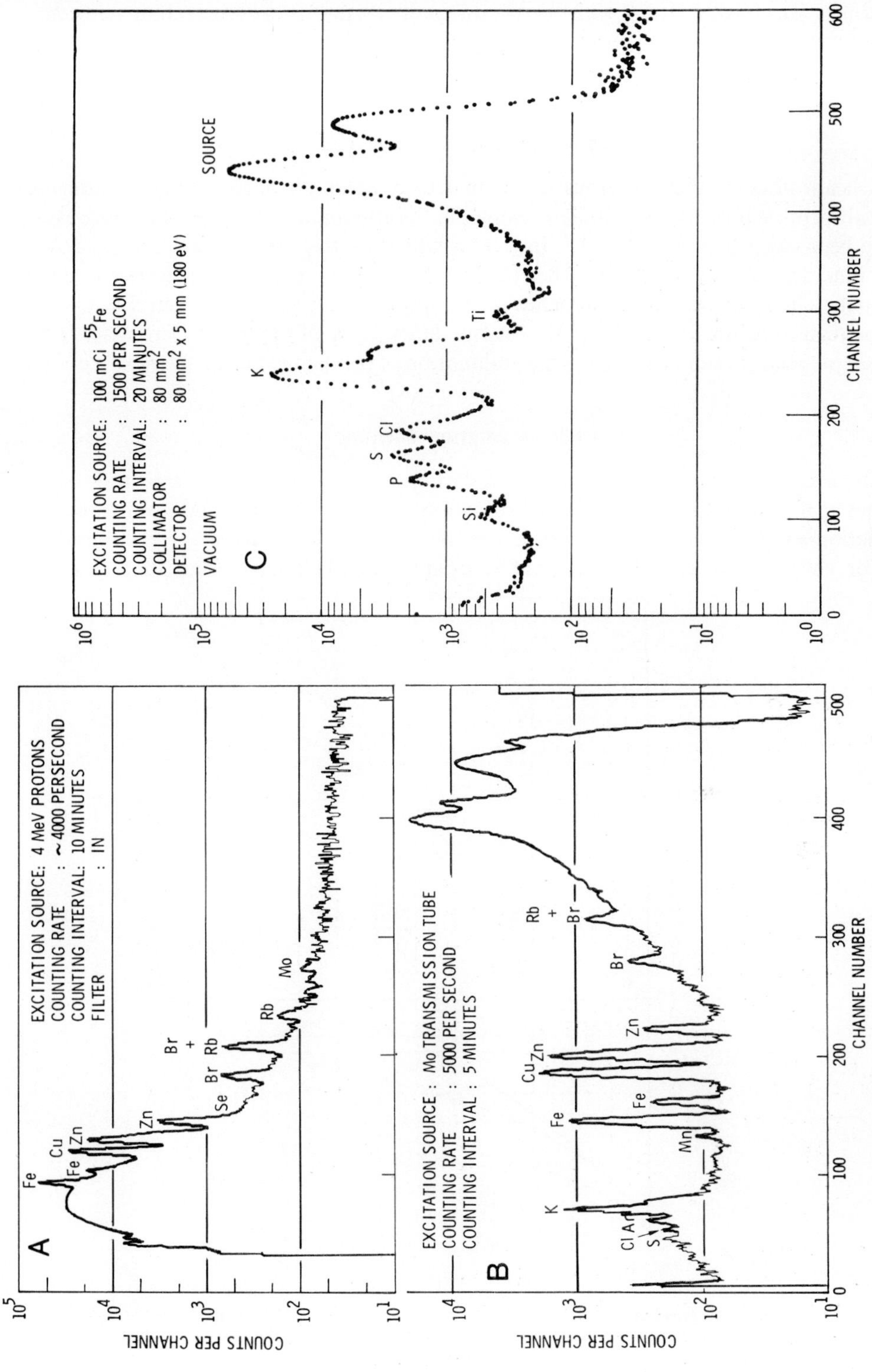

FIG. 14.27. X-ray spectra of the elements contained in bovine liver as measured with (a) 4 MeV protons, (b) a molybdenum transmission tube, and (c) an ^{55}Fe radioisotope source. (According to J. A. Cooper.)

iron sheets, Ag in photographic emulsions, etc.). Figure 14.27 shows fluorescent spectra obtained on bovine liver with three different excitation sources: (conventional) molybdenum transmission tube, radioactive ^{55}Fe and 4 MeV protons. The environmental and medical uses of sources of the latter kind are increasing.

14.9. Absorption of neutrons

A beam of collimated neutrons is attenuated in a thin absorber through scattering and absorption processes in a similar manner to the attenuation of γ-rays; these processes have been described in §§9.1–9.5. In a thick absorber the neutrons are slowed down from incident energy at the absorber face to thermal energies if the absorber is thick enough. The ultimate fate of the neutron is capture by an absorber atom. Because of the spread in neutron energy and the energy dependency of the capture cross-sections, no simple relation can be given for the attenuation of the neutron beam (cf. next section).

14.10. Radiation shielding*

The absorption properties of nuclear radiation in material must be known in order to design shielding to avoid unwanted radiation effects on the surroundings by nuclear radiation sources.

For charged particles the shielding is usually slightly thicker than that required

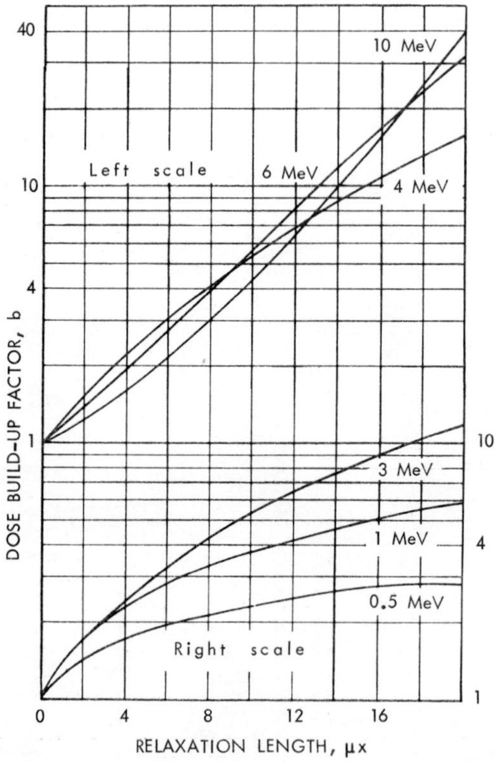

FIG. 14.28. Dose build-up factors in lead for a point isotropic source. The upper curves have the scale to the left, the lower curves to the right. (According to the *Radiological Health Handbook*.)

for the maximum range of projectiles in the material. Absorption thicknesses of 0.2 mm are adequate to completely absorb the particles from α-decay. By contrast 15 mm of materials of low Z such as water, plastic, etc., are required for absorption of β-radiation from radioactive decay. Radiation shielding constructed from materials of higher atomic number require correspondingly thinner thicknesses. The data in Table 14.2 and the curves in Figs. 14.5, 14.11, and 14.12 provide information on the thickness of absorber material required for the energy of various types of radiation.

Since γ-rays and neutrons have no definite range but exhibit a logarithmic relation between thickness and intensity, only a partial reduction of the radiation can be obtained. Combining (14.5) and (14.6)

$$\phi = \frac{nA}{4\pi r^2} e^{-\mu x} \tag{14.21}$$

it is seen that the intensity from a point source of radiation can be decreased by increasing either the distance from the source r or the thickness x of the absorber. Alternately, an absorber with a higher absorption coefficient can be chosen to reduce the thickness required.

Equation (14.21) is valid only for point sources with ideal geometry, i.e. no back or multiple scattering, etc. For γ-radiation the thicker the absorber the higher is the percentage of radiation which may be scattered backwards through secondary (mainly Compton) scattering. The effect of geometry and absorber thickness can be taken into account by including a constant b in the absorption equation:

$$\phi = b\phi_0 e^{-\mu x} \tag{14.22}$$

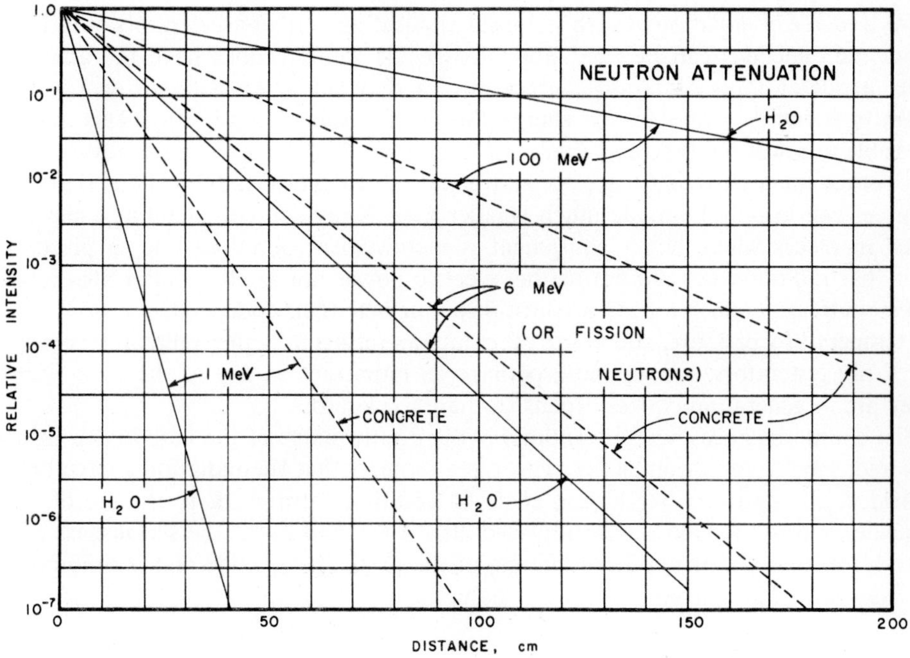

FIG. 14.29. Shielding thicknesses necessary to reduce neutron beams in water and in concrete. The 6 MeV lines can be used for fission neutrons. (According to 1960 *Nuclear Data Tables*.)

The "*dose build-up*" *factor b* not only takes into account multiple Compton and Rayleigh scattering but also includes correction for positron formation at high γ-energies and subsequent annihilation. Since for thick radiation shieldings and high γ-energies the factor b may reach several powers of 10, it is quite important to be considered in designing biological shielding for radiation. Calculation of b is difficult and empirical data are most commonly used. Figure 14.28 gives b-values for a lead shield. The thickness of the shielding is given in relaxation lengths μx. This value is obtained from diagrams like Fig. 14.16; e.g. for a 3 MeV γ, μ_m is found to be 0.046 $cm^2\ g^{-1}$. If the absorber is 0.1 m thick the linear density is $10\rho_{Pb} = 113\ g\ cm^{-2}$ and the relaxation length becomes 5.2. With Fig. 14.28 this gives (for the 3 MeV γ-line) a dose build-up factor b of 3. Thus the lead shield transmits three times more radiation than is expected by the simple relation (14.21). The flux reduction values for concrete and lead shielding in Fig. 14.14 have been adjusted to take the dose build-up into consideration.

Equation (14.22) is not directly applicable for neutrons. For an estimate of required shielding we can use diagrams like that in Fig. 14.29, which shows the attenuation of neutrons of three different energies in concrete and water, the most common neutron-shielding materials. It is necessary also to take into account the γ-rays emitted in neutron capture, which increases the shielding thickness required.

14.11. Technical applications of radiation sources*

Nuclear radiation absorption methods have many technical applications. These methods are not to be confused with radioisotope tracer methods, although radioisotopes may be used as radiation sources. In the tracer method the chemical properties of the radionuclide are important while in the applications discussed in this section only the type and energy of radiation are important.

As a *source* of radiation in such technical applications, either accelerators or radiation from radionuclides can be used. Interchangeable radionuclides have the advantage over accelerators as radiation sources in that they can cover a larger energy range from high energy γ-rays to low energy β-rays in a much simpler way. This makes it possible to select the type and energy of radiation which have the most advantageous properties for a particular use. An additional advantage of radionuclides is that the sources can usually be made much smaller than X-ray sources, enabling them to be used in places where larger equipment is inconvenient or impossible to place. The fact that radionuclides require neither electric power nor cooling water also renders them more suitable for field applications. Further, their independence from effects of temperature, pressure, and many other factors results in higher reliability compared to X-ray generators. Finally, and, perhaps as important as any of the other factors, they are in general much less expensive than accelerators.

To counterbalance these advantages is the disadvantage of the inability to turn off the radiation from radionuclides. This often requires that the radiation source be well shielded, adding to its weight and cost. An additional drawback to the use of radionuclides is that they have to be replaced after a few half-lives. The seriousness of this disadvantage depends upon the lifetime of the particular nuclide and is unimportant in cases where long-lived sources can be used.

14.11.1. *Radionuclide gauges*

Radionuclide gauges are a measurement system consisting of two parts, a radioactive source and a detector, fixed in some geometry to each other. They are used mainly for control in industrial processes but can also be applied for specific analyses. The gauges come in two types. In one type the radiation source and the detector are on opposite sides of the technical arrangement to be measured; these are known as transmission or absorption instruments. In the second type, known as reflection or backscattering instruments, the radiation source and the detector are on the same side. The instruments are also classified with respect to the kind of radiation involved. For example, γ-transmission, β-reflection, secondary X-ray instruments, etc. These radioisotope gauges are used for measurements of thicknesses, densities, etc.

In Fig. 14.30 a number of types of application for transmission instruments are illustrated. Illustrations A–C show measurements of level control. Type A can be used only for liquids while B and C can be used for all kinds of material. The latter instruments are uniquely suitable for application to large storage containers for grain, wood chips, oil, sand, cement, etc., and for material under extreme conditions such as molten glass and metal, explosives, etc. Level gauges are also used in the control of the filling of packages, cans, etc. in industry as illustrated in D.

Let us consider a somewhat unusual application of this type of instrument. In the manufacture of titanium it is important to keep liquid $TiCl_4$ at a particular high temperature and pressure in a vessel. $TiCl_4$ has a triple point (i.e. the pressure and temperature conditions at which all three phases—solid, liquid, and vapor—of a substance are at equilibrium) in the neighborhood of the particular technical conditions. By using a γ-density gauge it is possible to detect when the triple point is exceeded because of the disappearance of the vapor–liquid interface. This allows a simple method of control of the process conditions in the vessel.

The use of radioisotope gauges in density measurements is dependent upon (14.21) in which r and x are constant while the absorption coefficient is density dependent (i.e. dependent on the average atomic composition of the absorber). A practical arrangement is illustrated in Fig. 14.30, E, where the density of a medium in a pipeline is measured. This medium may be a mixture of gas and liquid such as water and water vapor, liquids with different composition and different amounts of dissolved substances as, for example, oil, salts or acids in water, process solutions in general or sludges, and emulsions such as fruit juices, latex emulsions, etc. From the variation in density the concentration and composition may be determined. Such density gauges are also used for control in filling of cigarettes and submerged in rivers and lakes for measuring the depth of the bottom silt, etc. Density gauges are used in the production and the fabrication of automobile tires and cigarette packages.

Thickness gauges are the most common type of instrument using the absorption technique. In this case x in (14.21) is varied. Measurements can be carried out on all kinds of materials with thicknesses of $\lesssim 100\ \mathrm{g\ cm^{-2}}$ and is independent of the temperature and of whether the material is stationary or in motion. Figure 14.30, F, illustrates the application of a thickness gauge in a rolling mill where material of constant thickness is produced by using the signal from the detector for control purposes. Thickness gauges are used in the fabrication of glass, metal, paper, plastic, rubber, candy bars, etc. They have been used for measuring the thickness of snow in polar regions, icing on airplane wings, and other applications in which it is necessary to use remote operation.

In order to measure very thin layers of material such as coatings of paint, wax, and plastic films on papers or other material, two thickness gauges are used with a differential coupling so that one detector measures the uncovered and the other the covered or treated portion of the material. Thickness gauges also are used in industry to measure the degree of wear in industrial machinery. For surface measurements of thicknesses $\lesssim 0.8\ \mathrm{g\,cm^{-2}}$ most thickness gauges use radiation sources with β-emitters while for thicknesses of 0.8–$5\ \mathrm{g\,cm^2}$ bremsstrahlung radiation sources are most suitable. For even thicker materials γ-emitters are used.

Use of reflection gauges depends on the fact that the intensity of the scattered radiation under conditions of constant geometry depends on the thickness of the scattering material

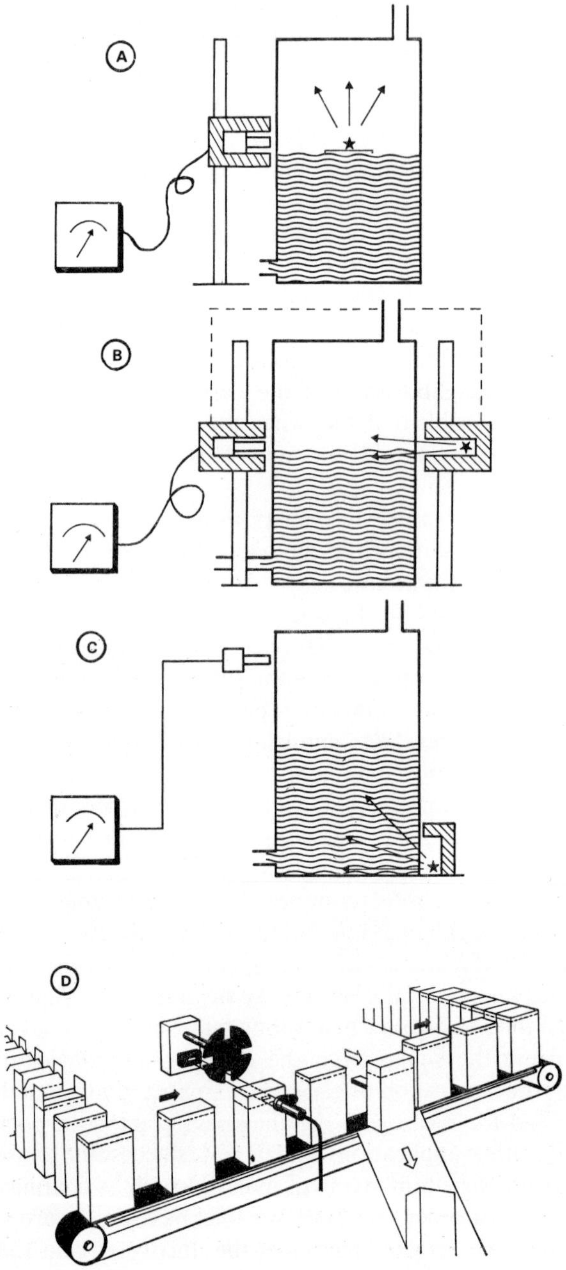

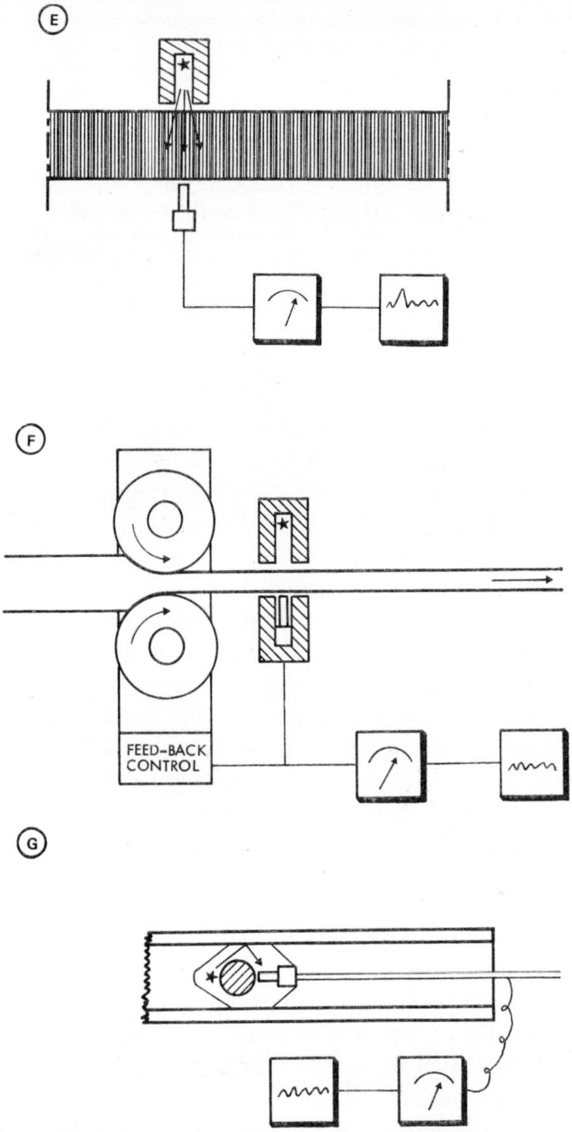

FIG. 14.30. Radionuclide gauges A–C for level measurements in tanks, D for control of package fillings, E for liquid flow density measurements, F for thickness control in rolling mills, and G for investigations of wall material in tubings and bore holes.

and its electron density (if β- or γ-sources are used). If neutrons are used the mass number of the scattering material is of prime importance. The electron density of the scattering material varies with the particular element and the chemical composition. Frequently, it is possible to determine the thickness and the nature of a surface layer by means of β-scattering. Reflection gauges have been applied to on-line analysis of tin-covered iron plates, metal coatings on plastics, paint layers, and to measuring the protective coating inside pipelines (Fig. 14.30, G). In some instances γ-radiation sources are preferred over β-emitters in measurements of material with greater wall thicknesses, particularly when transmission measurements are not feasible. Steel thickness from

TABLE 14.3. *Some commercially available radionuclide gauges and γ-sources for radiography (France)*

Radiation	Source	$t_{1/2}$	Application
α	U or Ra	Long	Thickness control in manufacturing paper, aluminum; $\leq 60\ g\ m^{-2}$
Soft β	^{147}Pm (0.2 MeV)	2.6 y	Thickness control: $\leq 400\ g\ m^{-2}$
β, soft γ	^{204}Tl (0.8 MeV)	3.8 y	Thickness: 1–10 mm steel, 3–50 mm glass; 8–100 kg m^{-2}
Hard β	^{144}Ce (3 MeV)	0.78 y	Thickness $\leq$ 1 mm steel; $\leq 10\ kg\ m^{-2}$
x	^{109}Cd (88 keV)	1.24 y	Detection of S-content in hydrocarbon
n, γ	RaBe, ^{137}Cs	30 y	Moisture–density meter for civil engineering and agriculture
γ	^{60}Co (1.3 MeV)	5.3 y	10^{-4} Ci source for backscatter on $\leq$ 20 mm steel
γ	^{60}Co (1.3 MeV)	5.3 y	0.01–1 Ci for remote level indication
Soft γ	^{192}Ir (0.3 MeV)	74 d	10 Ci, 26 kg: $\leq$ 40 mm steel radiography
Medium γ	^{137}Cs (0.7 MeV)	30 y	10 Ci, 45 kg: $\leq$ 70 mm steel pipeline inspection
Hard γ	^{60}Co (1.3 MeV)	5.3 y	250 Ci, 900 kg: $\leq$ 180 mm steel radiography

1 to 20 cm has been measured with 5% accuracy using backscattering from ^{60}Co or ^{137}Cs sources of 20 μCi intensity.

Scattering and reflection are dependent on the electron density of the absorber, which is approximately proportional to the value of Z/A. Backscattering of β-particles from organic compounds is therefore very dependent on the hydrogen concentration ($Z/A = 1$) but fairly independent of the concentration of C, N, and O ($Z/A = 0.5$). This has led to the development of sensitive instruments for hydrogen analysis for various organic and water-containing materials. In an instrument using 10 mCi ^{90}Sr, the hydrogen concentration in a 10 ml sample can be determined in 20 min with 0.03% accuracy, which is superior to other conventional analytical methods. It should be observed that this is a nondestructive technique.

Neutrons are slowed down most effectively by light elements (§7.6). As a consequence, neutron scattering can be used for the analyses of light elements, particularly hydrogen. In one type of instrument the radiation source consists of ^{252}Cf, which produces fast neutrons (from spontaneous fission), while the detector is sensitive only to slow neutrons. This system is used for studies of ground water and analysis of bore holes in wells (Fig. 14.30, G). These analyses are usually combined with density determinations using a γ-source, thereby making it possible to identify strata of water, oil, coal, etc.

A mode of analysis, which is growing rapidly in scope and utilization, involves the use of nuclear radiation to induce X-ray fluorescence; cf. §14.8. If the energy of the primary radiation is selected to be slightly above the X-ray fluorescence energy, a very selective analysis is possible. For example, using ^{133m}Xe, which emits 81 keV γ-rays, the K-level (80.7 keV) in gold can be excited; an instrument for use in the field has been developed in which a whistling sound is emitted if gold is found and the pitch of the sound is proportional to the amount of gold. A number of field instruments for selective analysis of other metals as, for example, copper, tungsten, silver, etc., have been developed. Nondestructive analysis involving excitation of X-ray fluorescence using solid state detectors is capable of extremely high resolution and accuracy. Some properties and uses of commercial radionuclide gauges are listed in Table 14.3.

14.11.2. *Radiography*

Radiography is a photographic technique in which nuclear radiation is used instead

FIG. 14.31. Gamma-radiography set up for obtaining a picture of the detonator of a bomb, and a picture obtained with the same arrangement on a World War II detonator.

of light: β-particles, X-rays, γ's, and neutrons. Medical examination and nondestructive industrial testing using X-rays generated by high-vacuum tubes are the most important areas. A number of suitable sources of radioactive nuclides for producing radiograms are given in Table 14.3.

Beta-radiography is only suitable for thin objects and not widely applied. On the other hand, γ-radiography is a common nondestructive test technique in which normally ^{137}Cs or ^{60}Co has been used. Figure 14.31 illustrates one application of γ-radiography and demonstrates two particular properties of this technique: the field use (albeit the picture is an indoor demonstration) and its utility for sensitive objects. The radiation source in Fig. 14.31 is 1.3 Ci of ^{60}Co, which is normally kept in a portable radiation shielding of 60 kg of lead. The radiation source is situated at the end of a rod so that the source can be pushed out of the shielding for exposition. The photographic film is located in a cassette surrounded by amplifying screens.

The exposure At (Curie hours) required for an optical density ($\hat{D}$ = log (incident light/transmitted light)) ~ 2 at an absorber (object) thickness x (cm) using a typical industrial X-ray film and a ^{60}Co γ-ray source positioned at a distance of 1 m from the film can be estimated by the expressions

$$At = -0.5 + 0.135\,x \qquad \text{for an iron absorber}$$
$$At = -0.5 + 0.040\,x \qquad \text{for a concrete absorber}$$

Exercise 14.14 is an example of the use of these expressions.

Gamma-radiography has been used for determining the number of reinforced iron bars in concrete construction, cavities in various kinds of castings (as explosives, plastics or metals), cracks or other defects in turbine blades in airplane parts (in the United States and the United Kingdom, regular γ-ray radiography of airplane wing construction is required), detonators in unexploded bombs, welded joints in pressure vessels, distillation towers and pipes (e.g. the Sahara oil pipeline), corrosion inside pipes and furnaces, and medical field X-rays, to mention only a few applications. Gamma-radiography is used throughout the world for product control leading to improved working safety and economy.

Because γ-absorption occurs through interaction with the electrons, objects of high atomic numbers show the strongest absorption. By using neutrons instead of γ-rays, the opposite effect is achieved, i.e. low Z objects are most effective in removing neutrons from a beam. This is used in neutron radiography in which both reactor neutrons from ^{252}Cf sources are applied. Because of the higher neutron flux from the reactor than from ^{252}Cf sources of normal size (i.e. $\lesssim 1$ mg) the exposure time at the reactor is much shorter. On the other hand, the small size of the ^{252}Cf source offers other conveniences.

In neutron radiography a collimated beam of neutrons must be produced. This is easily arranged by using thick-walled concentric cylinders, where the wall material is made of a combination of moderator and absorber, e.g. a mixture of paraffin and boric acid. Because this leads to γ-emission in the collimator, γ-shielding must also be added. In order to make a visible image in the film behind the object, the film is surrounded by gadolinium transfer screens; through (n, γ) reaction in ^{155}Gd (abundancy 15%; 61 000 b) and ^{157}Gd (abundancy 16%; 254 000 b) the film is blackened. In the neutron radiogram in Fig. 14.32, the light element material clearly shows up.

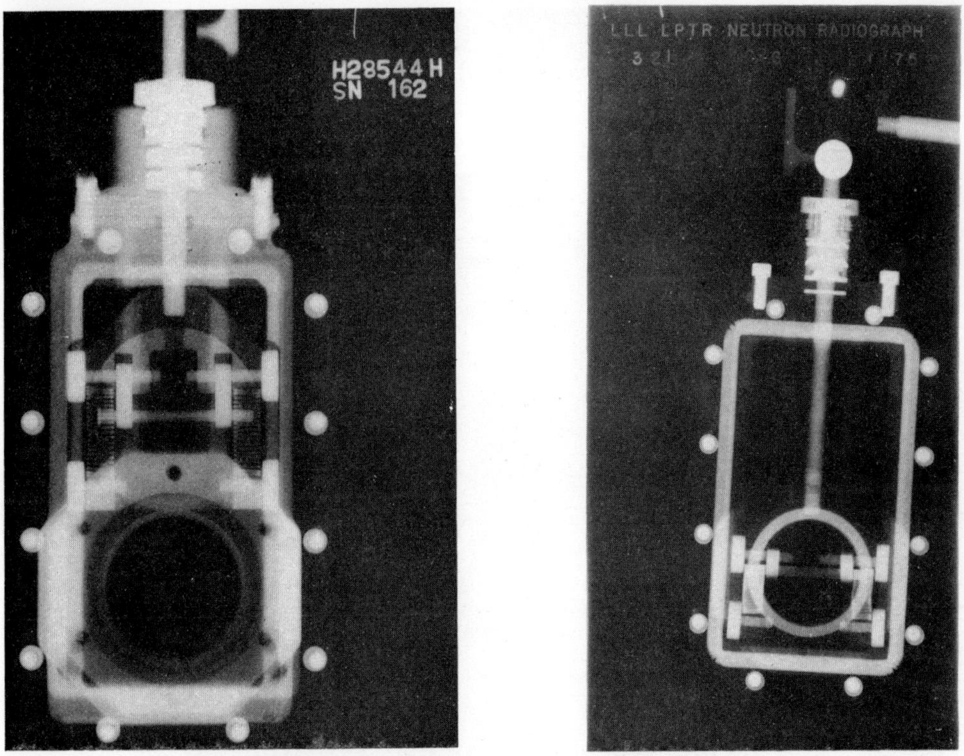

FIG. 14.32. X-Radiograph (left) and neutron radiograph (right) of a valve. Note the n-absorption of the gasket (slightly crackled at corners), o-rings, nylon wheels and cadmium-plated screws. (Courtesy W. Richards, R. Peterson and J. Prindle, Lawrence Livermore Laboratory.)

14.11.3. *Radionuclide power generators*

The absorption of radiation leads to an increase in the temperature of the absorber. An example of this is the absorption of the kinetic energy of fission products in nuclear reactor fuel elements which is a main source of the heat production in reactors. The absorption of decay energy of radioactive nuclides in appropriate absorbing material can be used in a similar—albeit more modest—way as an energy source.

Figure 14.33 shows the principle of a radioisotope power generator, which produces about 60 W. The radiation source consists of 15 rods (A) clad with hastalloy and containing approximately 7 kg of $SrTiO_3$ which has approximately 225,000 Ci of ^{90}Sr. This heat source is surrounded by 120 pairs of lead telluride thermoelements (B) and a radiation shield of 8 cm of depleted uranium (C). The whole arrangement is surrounded by a steel cover with cooling fins. The weight of the generator is 2.3 t with dimensions of 0.85 m in length and 0.55 m in diameter. It is estimated that the lifetime of such an energy source is at least 5 y, although the half-life of ^{90}Sr (30 y) promises a longer period.

Radionuclide generators in unmanned lighthouses, navigation buoys, automatic weather stations, etc., in sizes up to about 100 W have been in use for over 15 y in a number of countries, e.g. Japan, Sweden, the UK, the USA, etc. Since no moving parts are involved, these generators need a minimum of service. Their reliability makes them valuable in remote areas like the Arctic regions where several such generators have been installed.

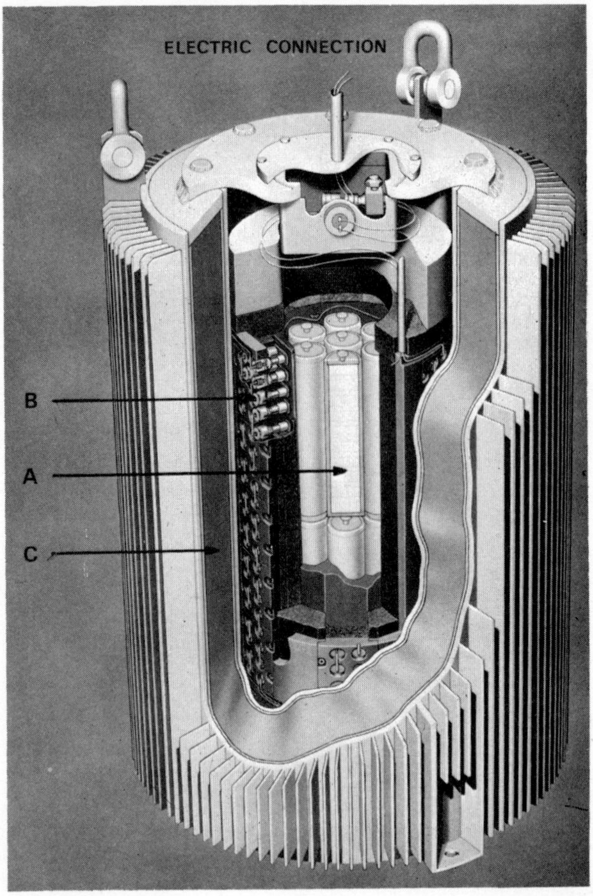

FIG. 14.33. A 60 W SNAP-7-type radionuclide power source. A shows rods of $^{90}SrTiO_3$; B, thermocouples; C, radiation shielding made of depleted uranium. The lower figure shows a light buoy near Baltimore powered by this generator.

^{90}Sr is the preferred radionuclide for terrestrial uses. Its power density is relatively high, 0.93 W g^{-1}, as compared to some other possible radionuclides as ^{137}Cs (0.26 W g^{-1}) and ^{238}Pu (0.55 W g^{-1}), but it is lower than ^{244}Cm (2.8 W g^{-1}) and ^{242}Cm (121 W g^{-1}). Recently ^{60}Co has come into use; the heating mainly comes from absorption of the γ's in a uranium shielding.

^{238}Pu has been used as an energy source in a number of satellites. For example, the satellite Transit-4A, which was put in space in 1961 and has operated flawlessly for many years, has a power source the size of a grapefruit, producing approximately 3 W of energy from ^{238}Pu. Later versions of these so-called SNAP generators have higher efficiency and are of even smaller size. Several satellites with radioisotope generators of 25 W have been placed in space, and the Apollo project employed a generator (SNAP-27) containing ^{238}Pu with a total weight of 14 kg and producing 50 W power. The Viking landers on the planet Mars uses ^{238}Pu as the main energy source. It was used also to power the Pioneer satellites which gave us the first close pictures of the planet Jupiter and in the Voyager missions to the outer planets where the ^{238}Pu power source produced 400 W.

Miniature generators have been developed for medical use in heart pacemakers.
landers on the planet Mars uses ^{238}Pu as the main energy source. It was used also to power the Pioneer satellites which gave us the first close pictures of the planet Jupiter and in the Voyager missions to the outer planets where the ^{238}Pu power source produced 400 W.

Miniature generators have been developed for medical use in heart pacemakers.
This is a device which is placed in the chest of a patient and connected to his heart muscles; the device generates regularly a small electric pulse ($\lesssim$ 10 mW), which causes the muscle to contract. The conventional pacemakers operate with a chemical battery, which presently has to be recharged every 24 months. With a nuclear-powered pacemaker the time between operations is extended at least by a factor of five. A standard nuclear pacemaker contains about 160 mg ^{238}Pu in a Ta–Ir–Pt encapsulation and weighs about 170 g (see Fig. 14.34). In 1976 almost 2000 pacemakers were in use in over 30 countries.

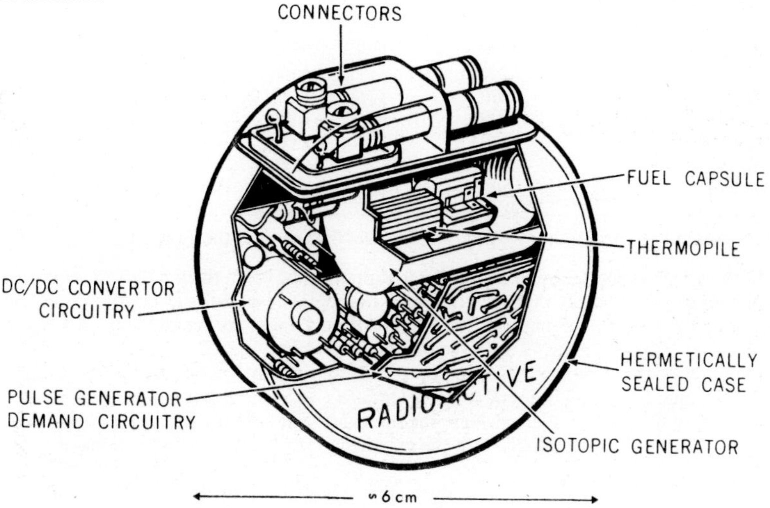

FIG. 14.34. Cardiac pacemaker containing ^{238}Pu. (According to Commissariat à l'Energie Atomique.)

14.12. Exercises

14.1. In § 14.3 two equations are given for calculating the range of α-particles in air and in other material from the particle energy as well as a curve for the range in aluminum. How different are the values from the equations and from the curve for a 5 MeV α?

14.2. What is the minimum energy that an α-particle must have to be detected by a GM tube having a mica (the density is approximately equal to aluminum) window of 1.5 mg cm^{-2}?

14.3. For an irradiation experiment it is necessary to extract a beam of deuterons from an accelerator. The projectile energy is 22 MeV D^+. For this purpose the beam is deflected and permitted to pass through a thin titanium foil (density 4.5 g cm^{-3}). Assuming that $\hat{R}_1 \rho_1 M_1^{-1/2} = \hat{R}_2 \rho_2 M_2^{-1/2}$ (Bragg–Kleman rule), what is the maximum thickness of the foil? Give the answer in millimeters.

14.4. Make a rough estimate of the range in air for a 1 MeV α, 1 MeV H^+, and 1 MeV e^- using the plot in Fig. 14.6. The energy to form an ion pair in air is 14.6 eV, but assume that twice as much energy is lost through excitation.

14.5. What is the range of a 6.3 MeV α-particle in (a) aluminum, (b) nickel, (c) platinum?

14.6. What is the γ-ray flux from a 100 mCi ^{60}Co source at a distance of 3 m? Assume $\psi_{sample} = 1$.

14.7. What is the maximum range in millimeters of β-particles from T, ^{14}C, ^{32}P, and ^{90}Sr in a photographic emulsion if its absorption efficiency is assumed to be the same as aluminum? The density of the emulsion is assumed to be 1.5 g cm^{-3}.

14.8. The E_{max} of ^{32}P β-particles is 1.71 MeV. To what electron velocity does this correspond?

14.9. In a laboratory an irradiation area must be designed for γ-radiography using a 10 Ci ^{60}Co source. For this purpose a cubic building is erected with an interior side length of 2 m. The desired flux reduction is 10^6. How thick must the wall be and how much will the shielding material cost (i.e. not including labor costs) if it is made of (a) concrete, (b) lead? Assume lead blocks cost $1.50 per kg and concrete $40 per m^3.

14.10. An experiment is done with ^{60m}Co which emits 0.058 60 MeV γ. The detector used is a NaI crystal. What photo peaks will be observed if the electron binding energies in sodium are K 1072 and L 63 eV, and in iodine K 33,170 and L 4800 eV?

14.11. A human body may be considered as consisting of water. Radiation from ^{137}Cs is absorbed by a 15 cm thick body. How much is the γ-ray flux reduced by the body, and how much of the beam energy (β plus γ) is absorbed?

14.12. For iron the mass attenuation coefficients are: at 0.5 MeV, γ 0.083 cm^2 g^{-1}; at 1.0 MeV, 0.059; at 1.5 MeV, 0.047 cm^2 g^{-1}. Calculate the corresponding one-tenth values.

14.13. An absorption curve of a sample emitting β- and γ-rays was taken with aluminum absorber using a gas-flow proportional counter. The data obtained were:

Absorber thickness (g cm^{-2})	Activity (counts min^{-1})	Absorber thickness (g cm^{-2})	Activity (counts min^{-1})
0	5800	0.700	101
0.070	3500	0.800	100
0.130	2200	1.00	98
0.200	1300	2.00	92
0.300	600	4.00	80
0.400	280	7.00	65
0.500	120	10.00	53
0.600	103	14.00	40

(a) Estimate the maximum energy of the β-spectrum. (b) Find the energy of the γ-ray.

14.14. A 40-story high modern business building is supported by 0.9 m thick pillars of reinforced concrete. The insurance company must check that the number of iron bars are as many as required, and therefore they want to investigate the pillars by γ-radiography. What exposure times are required for (a) a small 6 Ci ^{60}Co source, (b) for a large 4000 Ci source? Use the same film data as in § 14.11.2.

14.15. A swimming-pool reactor produces a flux of 3×10^{12} thermal neutrons cm^{-2} s^{-1} at 1 m from the reactor center. Assuming a parallel beam of neutrons diffusing up to the surface of the pool where the neutron flux is measured to be 10^4 n cm^{-2} s^{-1}, calculate the thickness (x m) of the water layer required. For thermal neutrons the flux is reduced exponentially with the exponent $x \cdot L^{-1}$, where L is the diffusion length (2.75 cm in H_2O).

14.16. In a sample of 280 Ci of old fission products, the average γ-ray energy is 0.5 MeV, and on the average 0.4 γ's are emitted per β-decay. (a) What is the lead shielding required to reduce the γ flux to 10^2 γ cm^{-2} s^{-1} at 1.5 m from the source assuming only exponential absorption? (b) What is the relaxation length? (c) What is the build-up factor?

14.13. Literature

See Appendix A, § A1.

IAEA, *Safety Series*, Vienna, 1960, and later.
B. T. Price, C. C. Horton, and K. T. Spinney, *Radiation Shielding*, Pergamon Press, Oxford, 1957.
S. Flügge (ed.), *Handbuch der Physik*, Band 34, 1958, and 38/2, 1959, Springer-Verlag.
R. L. Mössbauer, Recoilless nuclear resonance absorption, *Ann. Rev. Nucl. Sci.* **12** (1962) 1.
J. C. Rockley, *An Introduction to Industrial Radiology*, Butterworths, London, 1964.
C. S. Fadley, S. B. M. Hagström, J. M. Hollander, M. P. Klein, and D. A. Shirley, Chemical bonding information from photoelectron spectroscopy, *Science* **157** (1967) 1571.
D. A. Shirley, Chemical tools from nuclear physics, *Science* **161** (1968) 745.
IAEA, *Nuclear Well Logging in Hydrology*, Tech. Report 126, Vienna, 1971.
IAEA, *Commercial Portable Gauges for Radiometric Determination of the Density and Moisture Content of Building Materials*, Tech. Report 130, Vienna, 1971.
G. M. Bancroft, *Mössbauer Spectroscopy*, J. Wiley, 1973.
J. A. Cooper, Comparison of particle and photon excited X-ray fluorescence applied to trace element measurements on environmental samples, *Nucl. Instr. Methods* **106** (1973) 525.
H. W. Thümmel, Stand und Entwicklungstendenzen auf dem Gebiet der Isotopen und Strahlenanalytik. Physikalische Analysenverfahren mit Radionukliden, *Isotopenpraxis* **11** (1975) 1, 41, 87, 117, 172.

CHAPTER 15

Radiation Effects on Matter

Contents

Introduction

Soon after the discoveries of X-rays and radioactivity it was learned that radiation could cause changes in matter. In 1901 P. Curie found that when a radium source was placed on his skin, wounds were produced that were difficult to heal. In 1902 skin cancer was shown to be caused by the radioactivity from radium but 5 years later it was learned that radium therapy could be used to heal the disease. Large doses of radiation were found to kill fungi and microorganisms and produce mutations in plants.

Glass ampules containing milligrams of radium darkened within a few months and became severely cracked, allowing the leakage of radon gas. In the early years of the investigation of radioactivity in which emphasis was on radium, the contamination of laboratories by radon and its decay products was a common experience and a major problem in attempts to make accurate measurements of radioactivity. Among the early studies of radiation effects were the fluorescence induced in different salts and the changes in their crystallographic form. Metals were found to lose their elasticity and become brittle. Radiation was also found to have a profound effect on the chemical composition of solutions and gases. Compounds like water, ammonia, and simple organic substances were decomposed into more elementary constituents as well as combined into more complex polymeric products. Radiation decomposition (*radiolysis*) of water produced hydrogen and oxygen gas as well as hydrogen peroxide. Conversely, it was shown that water could be synthesized through irradiation of a mixture of H_2 and O_2.

In 1911 S. Lind found that 1 g of radium exposed to air resulted in the production of 0.7 g of ozone per hour. By relating this *radiation yield* to the number of ions produced by the amount of radiation, Lind initiated the quantitative treatment of radiation-induced changes. Radiation yields are measured by the *G-value*, which is the number of molecules decomposed, or the number of molecules, ions, or radicals formed per 100 eV of absorbed radiation energy. For example, the *G*-value for irradiated water is about 0.5 for production of H_2 and about 0.7 for production of H_2O_2.

15.1. Energy transfer and radiation dose

The effects of radiation depend on the composition of matter and the amount of energy that is imparted to it by the radiation. In this section we consider only the energy transfer. For this purpose it is practical to divide nuclear radiation into (1) *directly ionizing particles*, which include all charged particles (e^-, α, etc.), and (2) *indirectly ionizing particles*. The latter are uncharged radiations (n, γ, etc.), which can liberate directly ionizing particles or can initiate nuclear reactions.

The amount of energy imparted to matter in a given volume is

$$E_{imp} = E_{in} + \Sigma Q - E_{out} \tag{15.1}$$

where E_{in} is the sum of the energies (excluding mass energy) of all nuclear particles which have entered the volume, E_{out} is the sum of the energies of all particles which have left the volume, ΣQ is the sum of all Q-values for nuclear transformation that have occurred in the volume. For a beam of charged particles $E_{in} = E_{kin}$; for γ-rays it is E_γ. If no nuclear transformations occur, $\Sigma Q = 0$. For neutrons which are captured and for radionuclides which decay in the absorber, $\Sigma Q > 0$; in the latter case, $E_{in} = 0$.

The amount of absorbed radiation energy per unit mass is called the *absorbed dose* D. D is defined at a point by

$$D = \mathrm{d}E_{abs}/\mathrm{d}m \tag{15.2}$$

where m is the mass of the absorber.

In all practical cases we can put $E_{abs} = E_{imp}$. The dose is measured in J kg^{-1} and the unit is the gray (Gy):

$$1\ \mathrm{Gy} = 1\ \mathrm{J\ kg^{-1}} \tag{15.3a}$$

A special unit named the *rad* is commonly used, where

$$1\ \mathrm{rad} = 0.01\ \mathrm{J\ kg^{-1}} \tag{15.3b}$$

i.e. 1 Gy = 100 rad. Originally 1 rad was defined as the deposition of 100 erg g^{-1} matter. The *dose rate* $\mathrm{d}D/\mathrm{d}t$ is the dose produced per time interval, e.g. Gy s^{-1}, or rad h^{-1}.

15.1.1. *Directly ionizing particles*

We learned in the previous chapter (see also Table 14.2) that the energy of charged particles is absorbed mainly through ionization and atomic excitation. For positrons the annihilation process (at $E_{kin} \approx 0$) must be considered. For electrons of high kinetic energy bremsstrahlung must be taken into account. However, in the following we simplify

TABLE 15.1. *Ion pair formation energies for charged particles*
w is the mean energy expended by directly ionizing radiation per ion pair formed; j is the ionization potential of the gas

Absorber		w (eV)	j (eV)	$w-j$ (eV)
He(g)		43	24.5	18.5
H_2(g)		36	15.6	20.4
O_2(g)		31.5	12.5	19
Air		34	15	19
H_2O (g or l)		38	13	25
Ar(g) average		26	15.7	10.3
Ar(g)	5 MeV α	26.4		
Ar(g)	340 MeV p	25.5		
Ar(g)	10 keV e^-	26.4		
Ar(g)	1 MeV e^-	25.5		

by neglecting annihilation and bremsstrahlung processes. The bremsstrahlung correction can be made with the aid of Fig. 14.8 which gives the *average specific energy* loss of electrons through ionization and bremsstrahlung.

It has been found that the average energy w for the formation of an ion pair in gaseous material by charged particles is between 25 and 40 eV. For the same absorbing material it is fairly independent of the type of radiation and the energy. Table 15.1 lists some values of w in some gases. Because the ionization potentials j of the gases are lower than the w-values, the rest of the energy, $w-j$, must be used for excitation. Since the excitation energies per atom are $\lesssim 5$ eV, several excited atoms are formed for each ion pair formed. While it is easy to measure w in a gas, it is more difficult to obtain reliable values for liquids and solids. They also differ more widely; e.g. w is 1300 eV per ion pair in hexane (for high energy electrons) while it is close to 5 eV per ion pair in inorganic solids.

The *specific energy loss* of a particle in matter is called the *stopping power* $\hat{S}$,

$$\hat{S} = \mathrm{d}E_{\mathrm{loss}}/\mathrm{d}x \ (\mathrm{J\,m^{-1}}) \tag{15.4}$$

where x is the distance traversed by the particle. The stopping power is a function of w, which depends on the material but is relatively independent of the type of particle and its energy, and the *specific ionization*. The specific ionization J is the number of ion pairs per unit path length

$$J = \mathrm{d}N_{\mathrm{j}}/\mathrm{d}x \ (\text{ion pairs m}^{-1}) \tag{15.5}$$

The value of J depends on the projectile and its energy as seen from Fig. 14.6. The relation between $\hat{S}$ and J is

$$\hat{S} = wJ \ (\mathrm{J\,m^{-1}}) \tag{15.6}$$

The stopping power is expressed commonly in units of MeV $\mathrm{g^{-1}\,cm^{-2}}$.

Another important concept is the *linear energy transfer* (abbreviated as LET) of charged particles. It is defined as the energy absorbed in matter per unit path length of a charged particle

$$\mathrm{LET} = \mathrm{d}E_{\mathrm{abs}}/\mathrm{d}x \tag{15.7}$$

Values of LET in water are given for various particles and energies in Table 14.2. For the same energy and the same absorbing material, the LET values increase in the order:

Increasing LET ↓
- high energy electrons (also approximately γ-rays)
- β-particles (also approximately for soft X-ray)
- protons
- deuterons
- α-particles
- heavy ions (ions of N, O, etc.)
- fission fragments

The relationship between LET, which refers to the energy absorbed in matter, and the stopping power, which refers to the energy loss of the projectile, is

$$\mathrm{d}E_{\mathrm{loss}}/\mathrm{d}x = \mathrm{d}E_{\mathrm{abs}}/\mathrm{d}x + E_x \tag{15.8}$$

The difference in these two energy terms E_x is related to the energy loss by electromagnetic radiation, i.e. for charged particle absorption, bremsstrahlung radiation. For particles from radioactive decay, $E_x \approx 0$.

The dose received during time t for an internal radionuclide in the human body is

$$D\,(\mathrm{Gy}) = 1.44\dot{D}_0 t_{\mathrm{eff}}(1 - \mathrm{e}^{-0.693t/t_{\mathrm{eff}}}) \tag{15.9a}$$

where t_{eff} is the effective half-life (see (16.1) and text in §16.10.2) of the radioactive nuclide in the body, and $\dot{D}_0$ is the initial dose rate at $t = 0$. The dose rate at time t is:

$$\dot{D}\,(\mathrm{Gy\,s^{-1}}) = 6 \times 10^{-3}\,\mathrm{A}_0\,\mathrm{e}^{-0.693t/t_{\mathrm{eff}}} \times \bar{E}_{\mathrm{abs}} \times W^{-1} \tag{15.9b}$$

where A_0 is the radioactivity (in curies) in the body at time $t = 0$, $\bar{E}_{\mathrm{abs}}$ (MeV) is the average absorbed decay energy per disintegration and W(kg) is the mass of the body. For α-decay, $\bar{E}_{\mathrm{abs}} \approx E_\alpha$; for β-decay $\bar{E}_{\mathrm{abs}} \approx \frac{1}{3}E_{\mathrm{max}}$.

15.1.2. *Indirectly ionizing particles*

When indirectly ionizing particles having the incident particle energy E_{in} are absorbed, a certain fraction E_{tr} is transferred into kinetic energy of charged particles when traversing the distance dx. We can define an *energy transfer coefficient* as

$$\mu_{\mathrm{tr}} = \frac{1}{E_{\mathrm{in}}}\frac{\mathrm{d}E_{\mathrm{tr}}}{\mathrm{d}x} \quad (\mathrm{m^{-1}}) \tag{15.10a}$$

If we neglect the bremsstrahlung associated with the absorption of the secondary charged particles formed in the initial absorption processes, E_{tr} is the energy absorbed (designated as E_{abs}), and we can write (15.10a) as

$$\mu_{\mathrm{a}} = \frac{1}{E_{\mathrm{in}}}\frac{\mathrm{d}E_{\mathrm{abs}}}{\mathrm{d}x} \quad (\mathrm{m^{-1}}) \tag{15.10b}$$

Analogous to the definition of absorbed (radiation) dose for directly ionizing particles a quantity named *kerma* is defined for indirectly ionizing particles: "the kerma is the quotient of dE_{tr} by dm, where dE_{tr} is the sum of the initial kinetic energies of all the charged particles liberated by indirectly ionizing particles in a volume element of the specified material and dm is the mass of the matter in that volume element"

$$K = \frac{\mathrm{d}E_{\mathrm{tr}}}{\mathrm{d}m} \quad (\mathrm{J\,kg^{-1}}) \tag{15.11}$$

The kerma can be expressed in units of rad as well as $J\,kg^{-1}$ (15.3b). The kerma rate, dK/dt, may be expressed in units of rads per hour. For a given monoenergetic radiation

$$K = \theta_{in}\mu_{tr}t\rho^{-1} \tag{15.12}$$

where θ_{in} is the *energy flux density* ($J\,m^{-2}\,s^{-1}$) of all particles which enter the absorber. If the incident flux density is ϕ_0 of γ-rays of energy E_γ $m^{-2}\,s^{-1}$, then $\theta = \phi_0 E_\gamma$. Using (15.10b) we obtain

$$K = \phi_0 E_\gamma \mu_a t\rho^{-1} \tag{15.13}$$

where t is the time of exposure. When applying this expression care must be taken with respect to the thickness of the absorber the relation only holds for thin absorbers in which ϕ_0 is only slightly reduced. In radioactive decay, t is replaced by $(1 - e^{-\lambda t})\lambda^{-1}$.

Although the discussion has focused on γ-radiation, from the relation

$$\mu = \rho\sigma N_A M^{-1} \tag{15.14}$$

it is obvious that relations (15.10) to (15.12) also hold for neutrons.

15.2. Radiation dose units

The oldest radiation unit still in use is the *roentgen* (R). It is applicable only to photons and is defined as the *exposure* to photons which in air produces ion pairs of total charge dq (only the positive or negative ions are counted), the electrons of which are completely stopped in a volume element of air having mass dm. Thus the exposure is equal to dq/dm. The unit is defined as

$$1\,R = 2.58 \times 10^{-4}\,C\,kg^{-1} \tag{15.3c}$$

This corresponds to 1.61×10^{15} ion pairs per kg air, or an absorbed energy of $8.8 \times 10^{-3}\,J\,kg^{-1}$ (or 0.88 rad, or 88 erg g^{-1}) for an energy consumption of 34 eV per ion pair formed. The roentgen unit has been in use for over 50 years and is still common in medicine. An advantage is that it can easily be measured by air ionization chambers, but a disadvantage is that it only refers to air. However, in the range $E_\gamma = 0.2$–10 MeV the radiation which gives 1 R in air gives an absorbed dose (or kerma†) in water or body tissue corresponding to 0.93 rad. The common general unit of *absorbed dose* of radiation is the *rad*. While the roentgen expresses the intensity of the radiation field (the exposure) and not the dose absorbed by material in that field, by contrast the rad is a measure of the absorbed dose. The *specific γ-ray dose rate* is a practical measure for estimation of the hazards to man of γ-emitting nuclides. The unit used in Fig. 15.1 is roentgens per curie of radioactivity per hour at one meter from a radiation source (sometimes referred to as *specific γ-ray emission*). An approximate relation for calculating the dose rate obtained from a γ-emitting radioactive point source of A Ci is

$$D_\gamma\,(\text{rads per hour at 1 m}) = 0.6\,A\,\Sigma n E_{\gamma i} \tag{15.15}$$

where n_i is the fraction of disintegrations producing the γ-quantum of energy $E_{\gamma i}$. This equation is most nearly valid for E_γ between 0.1 and 3 MeV.

† Although energy absorbed from charged particles should be referred to as dose, while that absorbed from neutrons or γ-rays should be referred to as kerma, it is still common to refer to all radiation energy absorbed as the *radiation dose*, as we do in the subsequent text.

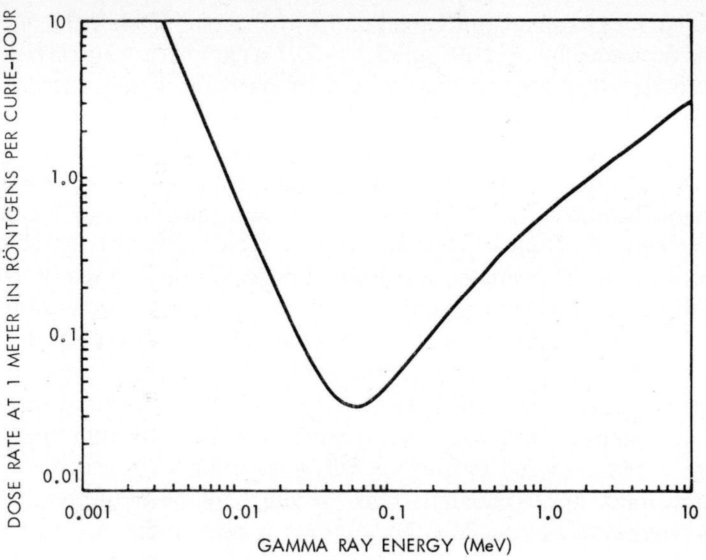

FIG. 15.1. Exposure in air at 1 m from a point isotopic γ-source of 1 Ci.

15.3. Radiation effects on metals

Metals consist of a solid lattice of atoms whose valence electrons cannot be considered to belong to any particular atom, but rather to a partially filled energy band (the conduction band) established by the total lattice network. Interaction of radiation with the metal can cause excitation of bound electrons in the atoms to the conduction band. De-excitation in which the electrons are returned to the original energy levels leads to only a minor change in the temperature of the metal.

While irradiation by gamma-rays and electrons has little influence on metallic properties, heavy projectiles cause serious damage through their collision with atoms of the metal lattice network. This results in displacements of the atoms from their lattice positions. The number of displacements (n_{disp}) depends on the amount of energy transferred in the collision event (E_{tr}) to the recoiling (target) atom, and the energy required for removing this atom from its lattice position. This so-called *displacement energy* (E_{disp}) is 10–30 eV for most metallic materials. According to the Kinchin–Pease rule,

$$n_{disp} \leq E_{tr}/2E_{disp}$$

The maximum energy transferred can be calculated assuming purely elastic collisions between hard spheres (§§4.1, 7.1). Thus for a 1.5 MeV fission neutron, E_{tr}(max) is 425 keV in C, 104 keV in Fe, and 25 keV in U. With $E_{disp} \approx 25$ eV, one finds that up to 8500, 2080 and 500 displacements, respectively, will occur in these metals due to the absorption of a fission neutron (neglecting nuclear reactions). In practice the numbers are somewhat smaller, especially at the higher E_{tr} (where it may be about one-third the calculated value).

Atomic displacements cause many changes in the properties of metals. Usually electrical resistance, volume, hardness, and tensile strength increase, while density and ductility decrease. Thus the resistivity of Cu increased by 9% after a fluence (i.e. the number of particles passing a unit area) of 6×10^{20} n cm^{-2}, while the ductile energy of carbon steel decreased by over 50% after 5×10^{19} n cm^{-2} at a neutron energy of 1 MeV.

The microcrystalline properties of metals are particularly influenced by irradiation. Although low-alloy steel in modern reactor tanks are rather radiation resistant (provided they are free of Cu, P and S impurities), stainless steel (e.g. of the 18% Cr, 8% Ni type) has been found to become brittle upon irradiation due to the formation of microscopic helium bubbles, probably due to n, α reactions in ^{54}Fe and impurities of light elements (N, B, etc.). This behavior is accentuated for metallic uranium in reactors because of the formation of fission products, some of which are gases. As a result of this radiation effect it is not possible to use uranium metal in modern power reactors, where high radiation doses are accumulated in a very short time. The fuel elements for power reactors are therefore made of nonmetallic uranium compounds.

The displaced atoms may return to their original lattice positions through diffusion if they are not trapped in energy wells requiring some activation energy for release. Such energy can be provided by heating or by irradiation with electrons or γ-rays (these do not cause new displacements). This "healing" of particle radiation damage is commonly referred to as *annealing*. The *thermal annealing* rate increases with temperature as does *radiation annealing* with radiation dose. Doses in the Mrad range are usually required for appreciable effect.

In breeder reactors the accumulated neutron fluence is higher than in water-cooled thermal reactors. Also the neutron energy spectrum is more energetic. The structural damage due to displacement is therefore more severe in breeders. This is considered to be one of the most serious technical problems in the development program of breeder reactors.

In fusion reactors the energy of the neutrons emitted in the T-D reaction is 14 MeV, and the neutron fluxes are expected to exceed those of breeder reactors. Extensive research therefore is required on the radiation resistance of possible construction materials.

15.4. Inorganic nonmetallic compounds

The time for a high energy nuclear particle to pass an atom is $\lesssim 10^{-16}$ s. In this time the atom may become excited and/or ionize, but it has had no time to move (the Franck–Condon principle) provided there is no direct collision. The excited atoms are de-excited through the emission of *fluorescence radiation*, usually within 10^{-8} s. The ionization can result in simple trapping of the electrons and production of "electron holes" in the lattice, especially at impurity sites. The local excess (or deficiency) of charge, which is produced in this way, leads to electronic states with absorption bands in the visible and ultraviolet regions of the spectrum. For example, irradiation of LiCl results in a change of the color of the compound from white to yellow. Similarly, LiF becomes black, KCl blue, etc. The irradiation of ionic crystals also leads to changes in other physical properties such as conductivity, hardness, strength, etc. Frequently, upon heating the properties and color return to the normal state (or close to it), accompanied by the emission of light; this forms the basis for a radiation dose measurement technique named "thermoluminescence dosimetry" (§ 15.9).

Following a collision between a heavy projectile (n, p. etc.) and an absorber atom in a crystalline material the recoiling ion produces lattice vacancies and upon stopping may occupy a nonequilibrium interstitial position (Fig. 15.2). The localized

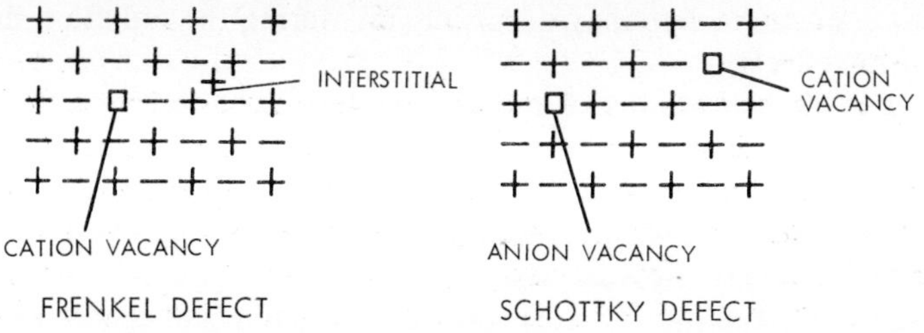

FIG. 15.2. Irradiated NaCl type crystal showing negative and positive ion vacancies.

dissipation of energy can result in lattice oscillations, terminating in some reorientation of the local regions in the crystal lattice. These crystal defects increase the energy content of the crystal. Semiconductors, where the concentration of charge carriers is very small, have their conductivity reduced by introduction of lattice defects during irradiation. The production of interstitial atoms makes the graphite moderator in nuclear reactors stronger, harder, and more brittle. Since these dislocated atoms are more energetic than the atoms in the lattice, the dislocations lead to an energy storage in the material (the Wigner effect), which can become quite significant. For reactor graphite at 30°C this frequently reaches values as high as 2000 kJ kg^{-1} for fluences of 2×10^{21} n cm^{-2}. At room temperature the interstitial atoms return to their normal positions very slowly (annealing), but this rate is quite temperature dependent. If the elimination of the interstitial atoms occurs too rapidly, the release of the Wigner energy can cause the material to heat to the ignition point. This was the origin of a fire which occurred in a graphite moderated reactor in England in 1957, in which considerable amounts of radioactive fission products were released into the environment.

Inorganic substances exposed to high fluences of neutron and γ-radiation in nuclear reactors are found to decompose. Thus:

$$KNO_3 \rightarrow KNO_2 + \tfrac{1}{2}O_2$$

At high fluences the oxygen pressure in the KNO_3 causes the crystal to shatter. However, some crystals are remarkably stable (although they may become colored), e.g. Li_2SO_4, K_2SO_4, $KCrO_4$, and $CaCO_3$.

Theory has not been developed sufficiently to allow quantitative calculation of the radiation sensitivity of compounds. Usually covalent binary compounds are highly radiation resistant. Two examples are UO_2 and UC, whose insensitivity to radiation has led to their use as reactor fuels. In the fuel elements of commercial water-cooled reactors UO_2 is used in the form of small, sintered pellets about 1 cm^3 in volume. Because of the build-up of pressure from fission gases, these pellets crack at high fluences ($\gtrsim 10^{22}$ n cm^{-2}). Another binary compound, CO_2, which is used as a coolant in some reactors, decomposes by irradiation to graphite and polymeric species.

In mixtures of inorganic compounds many unexpected and even undesirable reactions may occur. For example, radiolysis of liquid air (often used in radiation research) yields ozone, while radiation of humid air yields HNO_3.

Radiation damage or decomposition of inorganic solid compounds can often be completely restored by heating. Such thermal annealing may be considerable even at room temperature. Neutron irradiation of potassium permanganate at liquid air temperature yielded only 3% of isotopic ^{56}Mn in the parent compound (probably the rest was $^{56}Mn^{2+}$), while 22% formed when irradiated at room temperature. This indicates that some annealing may occur already during the radiation due to thermal and/or radiation effects.

One of the first observations of radiation-induced changes was the darkening of glass. Glass often contains iron, manganese, and other metals that can exist in several oxidation states with different colors. As a result of the irradiation, the oxidation state can change, resulting in change in color. Dislocations as well as trapped electrons also contribute to the color changes in glass.

In chemical and metallurgical work with highly active substances it is desirable to observe the experiment through a thick glass window, which provides protection from the radiation. In order to avoid the coloring of the glass, a small amount (1–2%) of an element which can act as an electron trap is added, e.g. CeO_2, which acts by the reaction $Ce^{4+} + e^- \rightarrow Ce^{3+}$. After an exposure of 10^6 rad the transmission to light of ordinary glass had been reduced to 44%, while for a CeO_2 protected glass it was still 89%.

Glass is very resistant to radiation damage because it is a noncrystalline solid liquid. Therefore there is nothing like a dislocation: the random structure of the glass allows it to absorb foreign species practically everywhere. This is the reason why glass has been selected as the "container" for high active waste (HAW) consisting of fission products and actinides, i.e. some 40 different elements. Glasses, especially of the $Li_2O{:}B_2O_3{:}SiO_2$ type, have been produced containing > 30% fission products, although a smaller waste content is recommended (§20.12). The radioactive decay of the actinides produces α-particles and recoiling atoms. Doses as large as 10^9 Gy (10^{11} rad), corresponding to ~ 10^{18} α-decays per g glass, only produce minor changes in the glass. The helium gas formed dissolves in the glass, without bubble formation. The stored energy may at these high doses amount to ~ 5 J g^{-1}; it is released slowly upon heating. The leach rate of glass in water, an important factor in HAW storage, seems not to be affected by these high α-doses.

15.5. Covalent compounds

It is tempting to seek to relate photochemistry and radiation chemistry by considering that ionization is the ultimate state of excitation. However, the large range in energy and in the type of projectile and the occurrence of approximately equal partition of energy between ionization and excitation make radiolysis much more complicated. Often the similarities in the results of photolysis and radiolysis are apt to be superficial.

15.5.1. *General consideration*

The overall effect of radiation on covalent compounds may be separated into several steps as depicted in Fig. 15.3. In passage through the material the radiation induces ionization along the projectile track. The electrons from the primary ionization often have sufficient kinetic energy to cause secondary ionization. Many times the amount of secondary ionization is larger than the primary ionization but the kinetic energies of the secondary electrons are lower on the average. The secondary electrons collide with molecules, causing excitation to higher electronic and vibrational states, followed by de-excitation by means of fluorescence, collisional transfer of energy, and various modes of molecular dissociation. Various molecular fragments existing as ions and *radicals* are the result of these processes. Because of an unpaired electron in one of

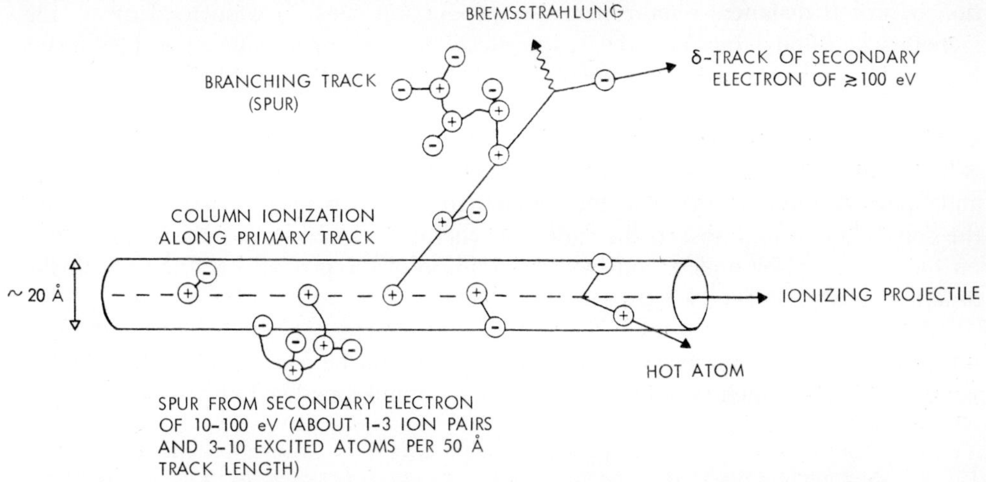

FIG. 15.3. Track formed by energetic ionizing particle in condensed matter. The distance between ion pairs formed along the track is about 10 000 A for γ's, 5000–1000 A for fast electrons, about 10 Å for slow electrons ($\sim$ 100 eV) and for α's. For each ion pair several excited atoms are also formed.

their outer orbits, these radicals are extremely reactive. The density of reaction products is naturally higher in condensed matter than in gas, and can be estimated from the LET values. The approximate time scale for the different events is given in Table 15.2.

TABLE 15.2. *Time scale for radiolysis of covalent compounds*

Event	Approximate time (s)
1 MeV electron traverses a molecule	10^{-18}
Excitation of a molecule	10^{-15}
One molecular vibration	10^{-14}
One molecular rotation	10^{-13}
Collision time in gas at 1 bar	10^{-10}
Collision time in liquid	10^{-9}
Lifetime for radiation of excited molecules	10^{-8}
Chemical equilibrium achieved in irradiated systems	10^{-6}

In the secondary stage of the process, these species interact to produce neutral molecules and new ions and radicals. Ion molecule reactions of the type

$$AB^+ + CD \rightarrow AC^+ + BD \tag{15.16a}$$

occur quite generally. The electrons from the ionizations recombine either with positive ions or with species of high electron affinity. The combination process is often sufficiently exothermic to cause bond rupture, producing new ionic and radical species. Since in a gas the ionization density is relatively low compared to a condensed system, no matter what the type and energy of the radiation, the ions and radicals diffuse, and recombination occurs at distances remote from the production sites. In condensed media the increased collisional rate and the "cage effect" of the surrounding molecules result in a much more rapid removal of energy, and enhance the probability of recombination within the local volume (the Franck–Rabinovitch effect). The concentration of ions and radicals in this local volume is a function of the ionization density and, therefore, is higher for α- than for γ-rays. The overall effects are more dependent on the energy and type of the radiation in condensed media than in gases. In the case of dilute solutions, the solvent molecules absorb the radiation energy to undergo radiolysis. The effect on the solute can be understood by consideration of the probable reactions with the radiation products of the solvent.

After about 10^{-7} s the track has dissolved and the reaction products have diffused into the surrounding matter. Chemical equilibrium can be considered to be achieved within 10^{-6} s, although the final products may contain long-lived metastable radicals.

15.5.2. *Experimental methods*

Radiation chemistry research follows two directions. Either the primary reaction steps, which take place before chemical equilibrium is reached, or the final reaction products are studied. While the latter are of practical importance (for reactor technology, biological radiation effects, etc.), the former approach provides the basis for understanding radiation processes.

Much insight into radiation processes has been gained from the method of *pulse* or *flash radiolysis* in which a sample is irradiated by pulsed emission of particles from an accelerator. With a pulse length of $\sim 1\ \mu$s and frequencies of 200 Hz, with 5–10 MeV electrons irradiation intensities of 200 000 rad per pulse are possible. Pulse lengths as short as 0.3×10^{-12} s can be achieved with recording of the emission spectrum from the irradiated sample beginning 10^{-9} s after the termination of the irradiation pulse. Commonly the absorption spectra of the products are recorded by illuminating the sample with an external flash or continuous monochromatic light source. The absorption spectra give information on the early products formed while the time dependency of the spectra provide data on the lifetimes of radiation products. A time resolution of 10^{-13} s is possible with this technique.

The products of radiolysis have been investigated by mass spectrometry often, combined with gas chromatography. The mass spectrometric technique can be used also for a direct investigation of the radiolytic products if the radiation process is allowed to take place in the ion source of the mass spectrometer.

The most commonly applied technique for investigating the production of free radicals in radiation chemistry uses electron spin resonance in which the number of unpaired electrons and their position in the molecule can be measured. For this purpose it is necessary to "trap" the free radicals before they react further. This is done by carrying out the irradiation either in a matrix containing molecules which react with radicals, or at very low sample temperatures which freeze the radicals so they can be studied by convenient techniques after the irradiation is completed. "Traps" for free radicals are NO, O_2, halogens, and some unsaturated organic compounds.

15.5.3. *Molecular excitation*

In Fig. 15.4 a potential energy diagram is given for a diatomic molecule in its electronic ground, first, and second excited states, A, A* and A** respectively. The ground state is usually a singlet state (antiparallel spin of valence electrons), while the excited states may be either singlet or triplet (parallel spin of the outer electrons) states. The horizontal lines in the diagram are associated with vibrational energy sublevels of the electronic state with the lowest sublevel for each electronic state being the zero point vibrational level. If the vibration energy gained by a molecule through a collision with another species, such as an energetic particle, has a value of G_n (Fig. 15.4) the resulting vibrations lead to dissociation of the molecules. This can occur within 10^{-13} s.

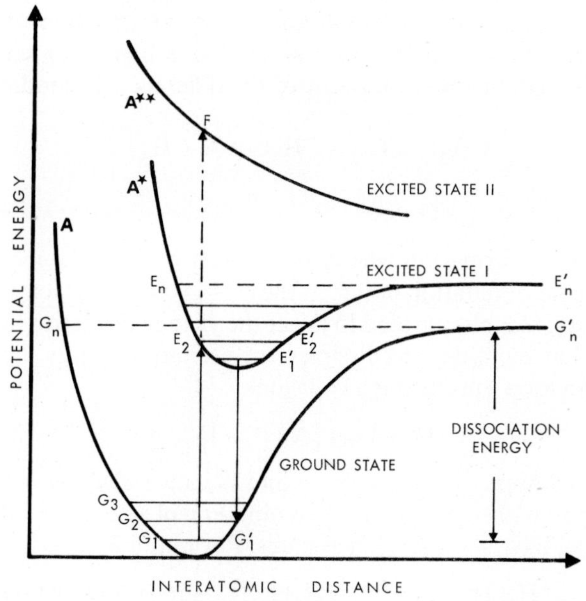

FIG. 15.4. Potential energy diagram for a diatomic molecule in different excited states.

The arrow from level G_1 to level E_2 indicates the excitation of the molecule from the ground state to a new state in which the molecule is both electronically and vibrationally excited. Such a dual state of excitation is designated as A_v^*. Through collision with surrounding atoms the vibrational energy may be lost:

$$A_v^* + B \rightarrow A^* + B_{kin} \tag{15.16b}$$

This process also occurs quite rapidly, within 10^{-11} s. Following this collision the molecule can emit a photon which causes it to change from the level E_1 to G_2. The time scale for this process depends on whether the molecule in the state A* is in the singlet or triplet state. The singlet state de-excites within 10^{-8} s by fluorescence whereas the triplet state requires much longer, 10^{-5}–10 s (phosphorescence). Subsequent to this emission the molecule can move from level G_2 to level G_1 by the loss of vibrational energy in an additional collision process.

$$A_v + C \rightarrow A + C_{kin} \tag{15.16c}$$

The symbols B_{kin} and C_{kin} indicate that the molecules (or atoms) B and C have received some form of kinetic energy in the collision.

In Fig. 15.4 the second electronic excited state, A**, is repulsive. If the molecule is excited to this state, which can easily occur by interaction with highly energetic projectiles, the molecule dissociates immediately (within 10^{-13} s).

15.5.4. *Reaction of the radiation products*

The primary products in radiation are excited molecules, radicals, and ions. These can de-excite through reactions with surrounding species or through emission of photons, dissociations, or internal transformations. Initially, a molecule is excited locally at a certain bond. However, the excitation energy is distributed rapidly over the whole molecule. If the excitation energy exceeds the energy for any bond in the molecule, that bond is broken. The probability for dissociation is higher for saturated than for unsaturated organic compounds as a result of the difference in bonding energies. The reaction

$$CH_3\underset{\underset{O}{\|}}{C}CH_3 \rightarrow CH_3CO\cdot + CH_3\cdot$$

is an example of radical formation where the weakest bond in the molecule is broken even though the primary excitation occurs at the C=O bond. All factors which increase the lifetime of the excited molecule tend to lower the yield of the decomposition products since the molecule has more time to dissipate the excitation energy through collisions.

Excited atoms can form ions through collisions:

$$A^* + D \rightarrow [A^+]^* + D^- \qquad (15.16d)$$

This process requires approximately 10^{-7} s and is called *electron transfer* (or *charge transfer*). It is a common phenomenon in the irradiation of aqueous solutions containing metal ions which can have several valence states. For example,

$$[Fe^{2+} \ldots H_2O]^* \rightarrow [Fe^{3+} \ldots H_2O^-]^* \rightarrow Fe^{3+} + H\cdot + OH^-$$

An electron is transferred in this reaction from Fe^{2+} to H_2O whereupon the H_2O dissociates into a hydroxide anion and a hydrogen radical.

Excited molecules and radicals often react in a similar manner. The excited molecules have a short lifetime and through collisions and dissociation produce free radicals which have much longer lifetimes.

15.6. Radiolysis of organic compounds*

Radicals which are formed in organic solvents can react rapidly with the surrounding molecules. Since the radical concentration is small, radical–radical reactions are uncommon. The product yield reflects radical–molecule reactions and is almost independent of the LET value.

The product yield in organic matter is normally less dependent than radiation yields in water on the presence of oxygen. However, the presence of small amounts of "foreign" substances may considerably affect the radiation yields of the main component; if the yields are increased it is called *sensitization*. Thus small amounts of CCl_4 in styrene increases the radiation polymerization yield. It has also been found that if hydrocarbons are adsorbed on silica gel, the $G(H_2)$ values (and other reactions) are increased by a factor of 10 (*radiation catalysis*).

Since organic molecules, in general, are complex structures, the radiolysis should be expected to give a variety of products. H_2, CO and CO_2 are formed in the radiolysis of H, C, and O systems, in addition to molecules larger and smaller than the original molecules. In hydrocarbons both C—C and C—H bonds are broken. The *G*-values for the major products in the radiolysis of methane (CH_4) are:

$G(H_2) = 5.7$	$G(\text{n-}C_4H_{10}) = 0.11$
$G(C_2H_6) = 2.2$	$G(\text{i-}C_4H_{10}) = 0.04$
$G(C_2H_4) = 0.7$	$G(C_5H_{12}) = 0.001$
$G(C_3H_8) = 0.36$	$G(C_6H_{14}) = 0.001$
$G(C_3H_6) = 0.04$	$G(\text{polymer }(CH_2)_n) = 2.1$

The formation of such simple species as C_3H_8 cannot be explained by simple radical combination; it probably involves ion–molecule reactions.

In the presence of air primary alcohols form aldehydes and secondary alcohols form ketones in small amounts. Carboxylic acids are de-carboxylated and the main products obtained are CO_2 and hydrocarbons. It is observed commonly that irradiation results in the breakage of the bond of the carbon chain at the functional group although this is not always true (cf. the example with acetone above). However, this may not occur in the primary process but may be a result of secondary reactions and also of the fact that the experimental detection methods are most sensitive to study of functional groups. Aliphatic halides are very sensitive to radiation since the carbon–halide bond is relatively weak. The products obtained are H_2, HX, X_2(X = halide), and higher halide hydrocarbons.

Polymerization increases with increasing unsaturated character and increasing length of the carbon chain. While it is rather small for simple low molecular weight aliphatics ($G \sim 2$), it is much larger for olefins and fairly high for acetylenes ($G \lesssim 70$). In some plastics such as polyethylene, radiation produces cross-linking of chains; degradation of the main chain or of side chains is the major effect with other polymers. The *G*-value for change in proteins (e.g. the side chain of amino acids of bovine serum albumin) is 5–10 for 2 MeV electrons, but lower for enzymes (e.g. 0.1 for inactivation of RNA).

Upon irradiation with γ-rays from a ^{60}Co source, the hydrogen yield from cyclohexane is 150 times greater than that from benzene. This has been interpreted to be a result of the greater stability of the excited states of aromatic systems. The presence of π-electrons diminishes the probability of a localization of excitation energy at a specific bond. As a result, the excitation energy is spread out over the whole carbon ring and de-excitation is more likely to occur through processes such as collisional transfer rather than by dissociation. The most radiation resistant compounds known contain aromatic rings (polyphenyls) and condensed ring systems (naphthalene, etc.). Their insensitivity to radiolysis has led to studies of such aromatic liquid hydrocarbons as cooling media in nuclear reactors. Radiation sensitivity of these compounds increases with increasing size of the aliphatic side chains but never reaches a *G*-value as high as that for a pure aliphatic compound. The primary radiolytic product of aromatic compounds is polymers. *G*-values for a large number of systems have been tabulated by M. Haissinsky.

In Table 15.3 effects of γ-radiation on some technical organic compounds are given. For oils the effects are mainly observed as changes in viscosity and acidity while for plastics they are associated with formation or rupture of the cross-linking. For elastomers like rubber there are changes in elasticity. For polyethylene, the following effects

TABLE 15.3. *Effect of γ-radiation on organic compounds of technical interest (all figures in Mrad)*

Compound	Observed change at	Useless[(a)] at	Compound	25% reduction[(b)] of desired property
			Teflon	0.01
Olefins	0.5	1	Cellulose acetate	0.2
Silicones	0.5	5	Polyethylene	0.9
Mineral oils	1	10	Polyvinylchloride	1
Alkyl aromates	10	50	Polystyrene	40
Polyphenyls	50	500	Neoprene, silicon rubber	0.06
			Natural rubber	0.25

[(a)]Useless means the radiation effects are sufficient to render the material of no value for its normal uses.
[(b)]"Desired property" is the one most important in the use of the material.

were observed: at 1–10 Mrad the tensile strength increased, between 10 and 200 Mrad the irradiated substance became rubber and jelly-like, at 200–500 Mrad it became hard, and at >500 Mrad it became glassy but with high elasticity.

The reprocessing of used reactor fuel elements involves solvent extraction processes with organic solvents. In these processes the solvents are subjected to high radiation fields with subsequent decomposition of the organic solvent. The design of chemical reprocessing systems must take into account any interference by the radiolytic products (§20.4).

Labeled compounds experience radiolysis induced by the radioactive decay. The extent of such radiation effects depends on the half-life, the decay energy, the specific activity of the sample, and the *G*-value for decomposition. The presence of other substances can considerably affect the amount of damage. Aromatic compounds such as benzene (as a solvent) can serve as a protective medium to minimize radiation self-decomposition, whereas water or oxygen enhance it.

Radiation doses of 10^7 rad can induce decomposition effects of the order of 1%. Samples whose specific activity exceed 1 Ci mol^{-1} for ^{14}C or about 10 Ci mol^{-1} for ^{3}H will receive a dose of this magnitude in a period of a year. Samples may be stored in benzene solution *in vacuo* or in deep freeze to minimize self-radiation effects and should be repurified before use if the decomposition products are likely to affect the experiment.

15.7. Radiolysis of water*

The radiolysis of water yields H_2, H_2O_2, H^+, OH^-, and as intermediate products $H\cdot$ and $HO\cdot$ in varying amounts depending on the LET value of the radiation, the pH and dissolved gases (especially O_2) (Fig. 15.5). The reactions are of considerable importance to the technology of reactors using water cooling and moderation.

The first step in the radiolysis of water is most probably

$$H_2O \rightsquigarrow H_2O^+ + e^- \tag{15.17}$$

where the symbol $\rightsquigarrow$ signifies action by radiation. In neutral and basic solutions, the electrons interact to become solvated:

$$e^- + n\,H_2O \rightsquigarrow e^-_{aq} \tag{15.18}$$

The time for this reaction is about 0.2 ps. (The solvation is much slower in organic solvents: 2 ps in methanol, 10 ps in ethanol.)

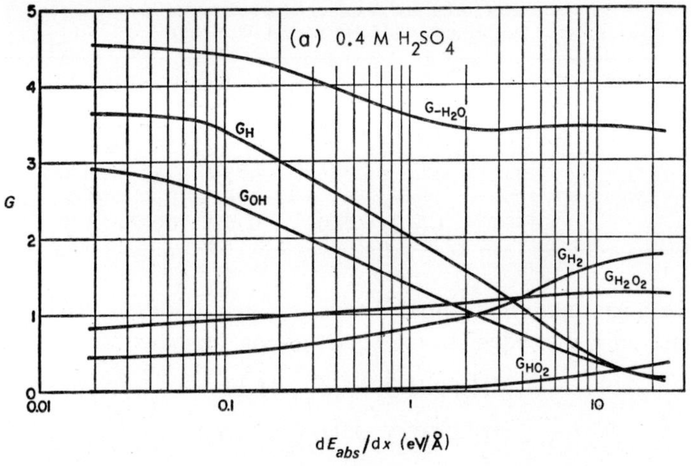

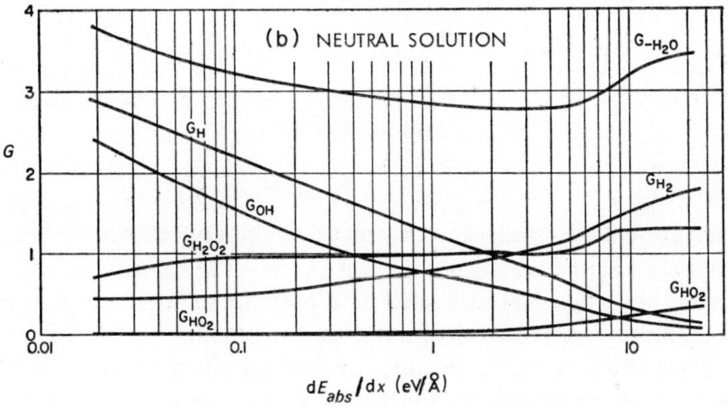

FIG. 15.5. *G*-values for radiolysis of water as a function of the LET value for the system; (a) 0.4 M H_2SO_4, and (b) neutral solution. (According to A. O. Allen.)

The solvated electron has been shown to be one of the principal reducing species in irradiated aqueous solutions

$$e^-_{aq} + H_2O \rightarrow OH^-_{aq} + H\cdot \quad k = 1\,b\,M^{-1}\,s^{-1} \tag{15.19}$$

where k is the reaction rate constant. In acid solution the electron reacts extremely rapidly with a proton to form a hydrogen radical:

$$e^-_{aq} + H_3O^+ \rightarrow H^{\cdot}_{aq} \quad k = 2 \times 10^{10}\,M^{-1}\,s^{-1} \tag{15.20a}$$

After 10^{-8} s the species present in irradiated water are $H_2, H_2O_2, H^{\cdot}_{aq}, HO^{\cdot}_{aq}, e^-_{aq}, OH^-$, and H_3O^+. The yields obtained in reactor cooling water are given in Table 15.4. Much research is required yet to understand how some of these species are formed initially.

In heavy particle irradiation of water the ionization density is high and the cage effect serves to cause the radicals to combine locally to form H_2 and H_2O_2. In electron and γ-irradiations the lower ionization density allows greater radical diffusion away from the local volume; in aerated solutions, $HO_2^{\cdot}$ (the hydroperoxy radical), H_2^+, and O^- are some of the species formed. In very pure water there is essentially no net de-

TABLE 15.4. *LET and G-values for reactor irradiations*
(according to E. J. Hart and J. W. Boag)

Energy source	Initial LET ($eV\ \text{Å}^{-1}$)	Yields of species (number per 100 eV) $-H_2O$	H_2	$H + e^-_{aq}$	H	e^-_{aq}	H_2O_2	OH	HO_2
β, γ	0.02	3.74	0.44	2.86	0.55	2.31	0.70	2.34	0.00
Neutrons	4.0	2.79	1.12	0.72	0.36	0.36	1.00	0.47	0.17
$^{10}B(n, \alpha)^7Li$	24	3.30	1.70	0.20	0.16[a]	0.04[a]	1.30	0.10	0.30

[a]Estimated.

composition with irradiation by γ- and X-rays as a steady state is reached in which the water dissociates at a rate equal to reassociation via the reactions:

$$H_2 + HO\cdot \rightarrow H\cdot + H_2O \tag{15.20b}$$

$$H\cdot + H_2O_2 \rightarrow HO\cdot + H_2O \tag{15.21}$$

$$H\cdot + HO\cdot \rightarrow H_2O \tag{15.22}$$

The presence of even small amounts of impurity, especially if this impurity has a high affinity for radicals, causes decomposition by competition with the above reactions as follows:

$$X^- + HO\cdot \rightarrow X + OH^- \tag{15.23}$$

$$X + H\cdot \rightarrow X^- + H^+ \tag{15.24}$$

Such impurities are called *scavengers* and increase the concentration of the radiolysis products H_2 and H_2O_2.

In nuclear reactors when water is used as a coolant or moderator, it should be as pure as possible to minimize dissociation during the time in the reactor. The formation of an explosive gas mixture of H_2 and O_2 must be carefully avoided in all reactors in order to prevent accidents. Moreover the decomposition products of water can increase the corrosion of fuel elements, structural material, etc. Many reactors use N_2 as a protective gas. In this case the radiolysis can lead to the formation of HNO_3 unless suppressed by an excess of H_2 which preferentially yields NH_3. The pH of the water may be regulated by the $H_2(g)$ pressure.

15.8. Radiolysis of aqueous solutions*

In dilute aqueous solutions the ratio of solute to solvent is so small that the energy absorption for all practical purposes takes place only with the water molecules. The chemical changes observed for the solutes are the result of reaction with the excited water molecules and with the decomposition products of the radiolysis of the water. The most abundant changes observed are those of oxidation resulting from reaction with the hydroxyl radicals and reduction induced by hydrogen radicals and the solvated electrons. In some systems both oxidation and reduction occur.

The radiation yields for the oxidation of Fe^{2+} and reduction of Ce^{4+} in a slightly acid solution are given in Fig. 15.6.

The LET value is the determining factor of the radiation effect, but the dose rate does have an effect on the curves in Fig. 15.6 such that at very high γ-ray dose rates the yields approach those of high LET radiation. This results from the formation of a

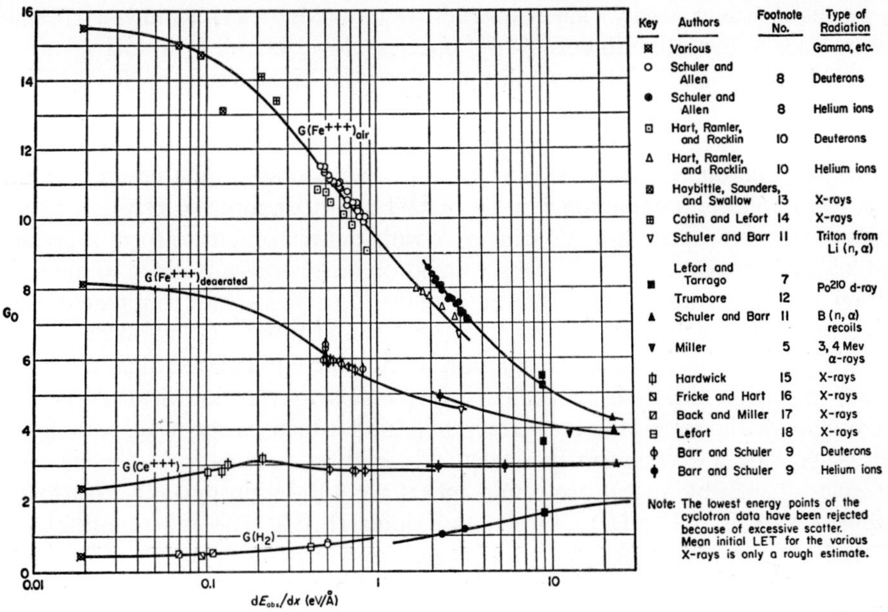

FIG. 15.6. Radiation yield for oxidation of Fe^{2+} and reduction of Ce^{4+} in slightly acid (H_2SO_4) solution as a function of the radiation LET value. (According to A. O. Allen.)

dense pattern of overlapping tracks before the tracks have time to dissolve. The presence of air has a strong influence on the yield of Fe^{3+}. It has been found that:

$$G(Fe^{3+})_{air} = 2G(H_2O_2) + 3G(H\cdot) + G(HO\cdot) \tag{15.25}$$

Reactions involving O_2, which interfere with H_2O_2 production and radical formation, influence the Fe^{3+} yield. Other substances may increase the $G(Fe^{3+})$ value even more than air, e.g. the addition of deaerated ethanol increases the $G(Fe^{3+})$ relative to air by a factor of 75, while formic acid increases it by a factor of 250.

The reduction of Ce^{4+} turns out to be independent of dissolved air, even though its overall G-value can be described by

$$G(Ce^{3+}) = 2G(H_2O_2) + G(H\cdot) - G(HO\cdot) \tag{15.26}$$

Both systems are used for dose determination.

When the concentration of the aqueous solutions is greater than ~0.1 M the solutes may undergo direct radiolysis. As a result the products of the radiolysis of the solute can react with water itself or with the radiolytic products of water.

Irradiation of solutions containing sulfate, sulfite, or sulfide ions with fast neutrons yields radioactive phosphorus, through the $^{32}S(n,p)^{32}P$ reaction, almost exclusively as orthophosphate. However, depending on the redox conditions of the solution, reduced species of phosphorus also appear in minor amounts. Somewhat in contrast to this, slow neutron irradiation of solutions of $NaHPO_4$ yields phosphorus in many different oxidation states; thus P(+1), P(+3), and P(+5) appears in species such as hypophosphite, phosphite, and *o*-phosphate. Since these species are all rather stable in aqueous solutions, they have been identified through paper electrophoretic analysis.

In general, it may be assumed that slow neutron irradiation of solutions of oxyanions changes the central atom to another (usually a more reduced) valence state through n, γ-reactions. For example, while manganese is in the Mn(+ 7) state in MnO_4^-, neutron capture leads to the formation of $^{56}Mn^{2+}$ species. However, if the product valence state is unstable at ambient solution conditions, it may be immediately oxidized to a more stable higher valence state.

The radiolysis of a solution can produce products which change the chemical equilibria of the solution components. For example, the α-decay of plutonium decomposes water; in a solution containing 1 mole of ^{239}Pu, $\sim$0.01 mole of H_2O_2 is produced per day. This hydrogen peroxide can react with the plutonium to form a precipitate of plutonium peroxide. To avoid the precipitation nitrite ions are added to the solution to react with the hydroxyl radicals formed by the radiolysis and to eliminate the H_2O_2.

The human body consists of about 65% water and from a radiation point of view can be considered as an aqueous solution. It is generally assumed that the hydroxyl radicals formed in radiolysis are primarily responsible for the biological effects. Certain substances which are biologically acceptable and which easily react with hydroxyl radicals provide a decrease in the extent of radiation damage in biological systems. Examples of such biological protecting agents are nitrate, propylamine, benzoic acid, cysteine ($HS\cdot CH_2\cdot CH(NH_2)\cdot CO_2H$), and thiols (RSH). The latter substances are oxidized by radiation of RSSR. It has been found that the lethal dose (LD-50) (see Chapter 16) increases by a factor of 5 upon administration of RSH.

15.9. Dose measurements

The amount of radiation energy absorbed in a substance is measured with *dose meters* (or *dosimeters*). These may react via a variety of processes involving (a) the number of ions formed in a gas, (b) the chemical changes in a liquid or in a photographic emulsion, (c) the excitation of atoms in a glass or crystal, and (d) the heat evolved in a calorimeter.

An important and accurate instrument for the measurement of ionizing radiation is the *condenser ion chamber*. This is a detector which has a small gas-filled volume between two charged electrodes. When radiation ionizes some of the gas between the two electrodes, the cations travel to the cathode and the electrons to the anode, thus preventing recombination of the ion pairs. Measurement of the amount of discharge provides a determination of the ionization and consequently of the dose delivered to the instrument. Such a dosimeter can be made very small with very thin walls so that secondary electrons formed through the interaction of radiation with the walls of the chamber are not likely to cause further ionization in the gas volume. The flexibility and accuracy of this dosimeter have led to it being widely employed for the exact measurement of γ-dose rates. The most common version of this type of instrument is the *pen dosimeter* (Fig. 15.7(b)), which either can be made to provide a direct reading of the absorbed dose or can be used with an auxiliary reading instrument. Ranges from 0.02 to 1000 rads (full scale) are available commercially. NRC regulations require direct reading pocket dosimeters to have a range from zero up to at least 200 mrem.

Photographic emulsions are sensitized by ionizing radiation resulting in darkening upon development. This is used in the *film dosimeter* (Fig. 15.7(a)) for measurement of β-, γ-, or n-doses. In order to discriminate between various types of radiation, the film is surrounded by filters or transfer screens. Although any type of film may be used, special nuclear emulsions have been designed. The dose received is directly proportional to the optical density of the exposed film. Film dosimeters are useful in the same range as the pen dosimeter, and both are used for personnel measurements. While

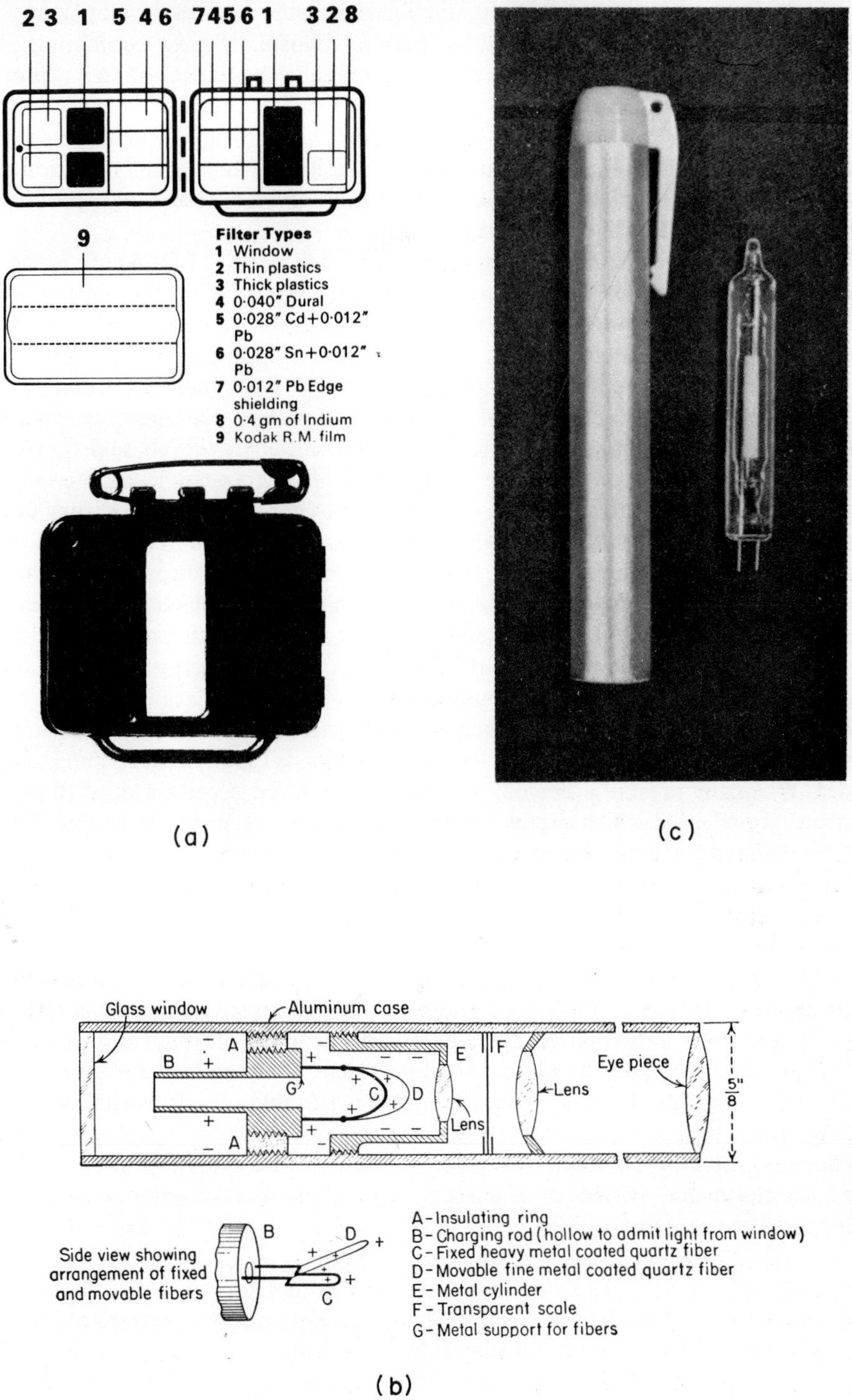

FIG. 15.7. (a) Film dosimeter. (b) Pen dosimeter with direct readout of the dose. (c) Pen-type thermoluminescence dosimeter showing arrangement of the LiF crystal rod and heating filament encasement.

the pen dosimeter can be directly read, the film dosimeter requires development.

The most common *glass dosimeter* is made of phosphate glass containing 5–10% of $(AgPO_3)_n$ polymer. The small piece of glass, a few cubic centimeters in size, is protected from light by means of a coating. When radiation strikes the glass, trapped electrons are produced which can be released by irradiation with ultraviolet light after removal of the protective coating. This results in the emission of fluorescent radiation which can be measured photometrically. The amount of fluorescent radiation is proportional to the dose received for dosages up to 1000 rad.

A recent development is the *thermoluminescent dosimeter* (TLD) which is the most useful in the range 0.0001–1000 rads. The detector consists of a crystalline powder of CaF_2, LiF, or similar compound, either pure or incorporated in a plastic material like teflon. The irradiation leads to ionization and trapping of the electrons in the crystal lattice. Upon heating, recombination occurs with light emission, which is measured photometrically. This is also the basis for thermoluminescence dating described in § 12.4.6. The electrons are trapped at different energy levels, and slow heating releases the electrons in order of increasing energy of the trapping levels. Consequently it is possible to take a dose reading on the lowest-energy trapped electrons and still retain a memory of the dose through the electrons left in more energetic traps. The dose can be read at a later time by releasing the remaining electrons at a higher temperature. The TLD can be designed like the pen dosimeter in which it is surrounded by screens to differentiate between different kinds of radiation; an example is shown in Fig. 15.7(c). By using a lead filter the dosimeter can be made energy independent in the range 0.02–20 MeV for X-rays and γ-radiation.

A recent modification is the *thermocurrent dosimeter*. In this case the detector may be a thin crystal of synthetic sapphire (ϕ 10 mm, thickness $\lesssim$ 1 mm) between thin metal electrodes. When the crystal is heated, after having received a certain dose the trapped electrons are released which causes a current to pass between the electrodes. The peak or the integrated current is a measure of absorbed dose. The effect is referred to as RITAC (radiation induced thermally activated current) and the technique is named TC (thermocurrent) dosimetry. Because it is instrumentally easier to measure an electric current than light, the TC dosimeter may replace the TLD in time.

Based on the radiolysis of chemical compounds, both organic and inorganic, numerous other *chemical dosimeters* have been suggested. An illustrative example is the $CHCl_3$ dosimeter. This is a two-phase aqueous-organic system. Through radiation HCl is produced, which changes the pH of the almost neutral aqueous phase. The color of a pH indicator reflects the dose. This dosimeter is suitable only for rather high doses, 10^4–10^7 rad.

The most common chemical dosimeter is based on the oxidation of Fe^{2+} to Fe^{3+}. The "iron dosimeter" (Fricke dosimeter) consists of an aqueous solution of approximately the following composition:

$$0.001\ \text{M}\ Fe(NH_4)_2(SO_4)_2,\ 0.001\ \text{M}\ NaCl,\ \text{and}\ 0.\ 4\text{M}\ H_2SO_4$$

The amount of Fe^{3+} formed through irradiation is determined spectrophotometrically and the dose absorbed in rads calculated by the equation:

$$D\,(\text{rad}) = \frac{1.60 \times 10^{-1} N_A(\hat{D} - \hat{D}_0)}{1000 \varepsilon x \rho G(Fe^{3+})} \qquad (15.27)$$

where $\hat{D}$ and $\hat{D}_0$ are the optical densities of irradiated and unirradiated samples at the molar extinction ε (= 2174 at 304 mμ), x is the length (in centimeters) of the cell, and ρ is the density of the solution (1.024 g cm^{-3} at 15–25°C). The G-value depends somewhat on the LET value of the radiation as is seen in Fig. 15.6. The Fricke dosimeter is reliable up to dose rates of about 10^8 rad s^{-1} and can be used in the range of 100–50,000 rad. In a common modification, the solution also contains NaSCN, leading to the formation of the intensely red complex ion $Fe(SCN)_6^{3+}$ upon irradiation.

In 1925 C. D. Ellis and W. A. Woorter, using RaE (^{210}Bi), obtained by calorimetric measurements the first proof that the maximum energy and the average energy of β-radiations were different. A precision of about 1% can be obtained in a calorimeter for an energy production rate of $\sim 10^{-6}$ J s^{-1} which corresponds to approximately 0.7 MeV average energy for a sample of 1 mCi. If the average energy of an α- or β-emitting nuclide is known, calorimetric measurement of the energy production rate can be used to calculate the specific activity.

Doses can be calculated from the product of the *dose rate* and the time of exposure. The most common dose rate meter is the ionization chamber. In this instrument the ions formed in a gas volume are continuously collected on two oppositely charged electrodes, leading to a current through the chamber which is proportional to the radiation intensity. Because of the close connection between this instrument and pulse type ionization counters, which measure individual nuclear particles entering the detector, the discussion of ionization chambers is deferred to Chapter 17, which deals in more detail with radiation measurement techniques.

If the number, energy, and type of nuclear particles being absorbed in a material can be measured or estimated, the absorbed dose can be calculated as described previously. Such calculations are very important, particularly in the field of radiation protection.

15.10. Radiation-induced synthesis*

Radiation effects have found practical use in the manufacture of some organic compounds. Since 1963 ethyl bromide (an anesthetic) has been commercially produced by irradiation of gas–liquid mixtures of ethylene and hydrogen bromide in ethyl bromide by γ-rays from ^{60}Co. The reaction steps are:

$$HBr \rightsquigarrow H\cdot + Br\cdot$$

$$Br\cdot + C_2H_4 \rightarrow BrC_2H_4\cdot$$

This product reacts with HBr

$$BrC_2H_4\cdot + HBr \rightarrow BrC_2H_5 + Br\cdot$$

The formation of a new bromide radical leads to a chain reaction and a high G-value of 10^5. With a source of 1800 Ci ^{60}Co, the production yield is 500 t y^{-1}.

Polymerization of organic compounds is a radical-induced process which can be promoted by nuclear radiation. The radiation can be used also to improve the properties of the polymers. Irradiation causes radical production and cross-linking of long carbon chains. In ordinary polyethylene only a few percent of the molecules are cross-linked; irradiation by a dose of 10^7 rad increases the cross-linking to 60%. The resulting polyethylene is more heat resistant, tougher, and has a lower tendency to crack. Ordinary polyethylene softens at about 90° whereas treatment with a dose of 2×10^6 rad increases the softening point to 150°. As a result of the improvement in properties by irradiation, it has become possible to produce very thin films of polyethlene. Using electron irradiation, several thousand tonnes per year of such film are now produced and used for coatings, insulating materials, and package film which shrinks with heat.

For mixtures of two or more monomers, the cross-linking leads to the formation of mixed polymers with properties different from those of the original material. New types of material have been obtained by copolymerization of synthetic and natural polymers such as wood, paper, cotton, etc., with vinyls, styrene, etc. An interesting process involves soaking a monomer into wood and then causing polymerization and copolymerization by γ-irradiation. The resulting material (graft-copolymer) is a harder, less elastic, but very stable woodlike material whose main use is for parquet flooring. A considerable improvement has been obtained by radiation cross-linkings of polyethylene foam plastics; the irradiated material has a smoother surface and the pores are uniformly closed even after a thirty-fold volume expansion.

A polymer may be irradiated with very low energy radiation leading to a change only in the properties of the surface of the material. As a result the surface properties can be different from those in the bulk so that it can be colored, made antistatic, oil resistant, etc.

Chemical synthesis by radiation has some advantages over photochemical or catalytic processes: (1) the energy deposition in the material is more homogeneous even in condensed systems; (2) the reaction can be carried out at ordinary pressure and temperatures; (3) the rate of production can be controlled and no expensive catalyst is required; (4) less energy is required for radiation treatment, especially by γ-irradiation, than by using heat and pressure; (5) an external source, which penetrates the reaction vessel, can be used; (6) reactions can be initiated evenly in solids. The major disadvantage of the radiation process is the relatively high radiation cost for materials with products of low G-values. In the examples cited, accelerator radiation or γ-radiation using ^{60}Co sources are used. The treatment of high level active waste from power reactors may make large amounts of ^{137}Cs available for irradiation use.

15.11. Special applications*

The ability of nuclear radiation to ionize gases is used in several connections. Static electricity can be eliminated through the installation of an α- or β-source in microbalances. Similarly, but on a much larger scale, elimination of static electricity is used in the paper, textile, and paint industries.

An ionization instrument for the analysis of gas has been developed in which the gas passes through a small chamber where it is irradiated by a small radioactive source. For a constant source of radiation, the ions produced in the gas depend on the flow velocity of the gas, and on its temperature, pressure, and atomic composition. The dependence of the ionization on the atomic composition is a consequence of the different ionization potentials of the different types of atoms of the gas and the different probabilities for electron capture and collision. The ion current is collected on an electrode and measured. This current is a function of the gas pressure and velocity since the higher the pressure, the more ions form, but the higher the velocity the fewer ions collect and the more are removed by the gas prior to collection. Such ionization instruments are used in gas chromatographs and other instruments as well as in smoke detection systems (the normal radiation source is ^{241}Am, usually $\lesssim 1\,\mu$Ci), where secondary electrons condense on smoke particles, leading to lower mobility for the electrons and a decreased ion current.

Both α- and β-emitters are used in luminescent paint. The fluorescent material is usually ZnS. T and ^{14}C are preferred sources since their β-energies are low, but ^{85}Kr, ^{90}Sr, and ^{147}Pm are also used. The amount of radioactivity varies, depending on the need (watches, aeroplane instruments, etc.) but it is usually <10 mCi, although larger light panels may require several Curies. For such high activities only ^{3}T or ^{85}Kr are acceptable because of their volatility and relatively low radiotoxicity.

15.12. Exercises

15.1. How many ion pairs are produced in 10 m of air of STP by one (a) 5 MeV α-particle, (b) 1 MeV β-particle, and (c) 1 MeV γ-quantum (μ_m (air) $\approx \mu_m$ (water))?

15.2. Estimate the fraction of energy lost through bremsstrahlung for a β-emission of $E_{max} = 2.3$ MeV, when absorbed in aluminum. The decrease of particle energy as well as the continuous β-spectrum must be taken into account.

15.3. A ^{60}Co source for γ-radiography has the strength of 120 Ci. If the flux decrease due to absorption in air and the body is neglected, (a) what is the exposure at 7 m from the point source reached after 8 h? (b) What is the dose in a human body for this distance and time?

15.4. An acidic aqueous solution is irradiated by α-particles from dissolved ^{239}Pu, which has a concentration of 0.03 M. The plutonium is originally in its hexavalent state, but is reduced to the tetravalent state by the reaction $Pu(VI) + 2H\cdot \rightarrow Pu(IV) + 2H^+$. How much of the plutonium can be reduced in one week?

15.5. An acid solution of fresh fission products contains 0.8 g l^{-1} cerium as Ce^{4+}. The γ-flux in the solution corresponds to 14 Ci l^{-1} of an average energy of 0.7 MeV. If half of the γ flux is absorbed in the solution, what fraction of cerium is reduced to Ce(III) in 24 h? Assume same *G*-values as in Fig. 15.6.

15.6. Estimate the LET value in water for β-particles from ^{90}Sr and ^{90}Y, and in aluminum for T.

15.7. A ^{60}Co irradiation source is calibrated by the Fricke dosimeter for which the *G*-value is assumed to be 15.6. Before the irradiation the optical density $\hat{D}$ of the solution at 305 nm was 0.049 in a 1 cm cuvette. After exactly 2 h the $\hat{D}$ had changed to 0.213. Calculate the dose rate when the molar extinction of Fe^{3+} is 2175 l M^{-1}.

15.8. A direct reading condenser chamber (pen dosimeter) is charged from a battery pack so that full scale (100) is obtained at 20 V. When completely discharged, the accumulated dose is 500 mrad; the gas volume is 4 cm^3 air (STP). What is the capacitance of the condenser chamber?

15.13. Literature

S. C. LIND, *Radiation Chemistry of Gases*, Reinhold, 1961.

A. O. ALLEN, *The Radiation Chemistry of Water and Aqueous Solutions*, D. van Nostrand, 1961.

M. HAISSINSKY and M. MAGAT, *Radiolytic Yields*, Pergamon Press, Oxford, 1963.

R. O. BOLT and J. G. CARROLL, *Radiation Effects on Organic Materials*, Academic Press, 1963.

B. T. KELLY, *Irradiation Damage to Solids*, Pergamon Press, Oxford, 1966.

J. L. DYE, The solvated electron, *Sci. Am.* Febr. 1967.

J. R. CAMERON, N. SUNTHARALINGHAM and G. N. KENNEY, *Thermoluminescent Dosimetry*, University of Wisconsin Press, 1968.

E. J. HART (ed.), *Radiation Chemistry*, Advances in Chemistry Series 81, Am. Chem. Soc., Washington DC, 1968.

E. J. HENLEY and E. R. JOHNSON, *The Chemistry and Physics of High Energy Reactions*, University Press, Washington DC, 1969.

W. SCHNABEL and J. WENDENBURG, *Einführung in die Strahlenchemie*, Verlag-Chemie GmbH, 1969.

N. E. HOLM and R. J. BERRY (eds.), *Manual on Radiation Dosimetry*, Marcel Dekker, 1970.

J. H. O'DONNELL and D. F. SANGSTER, *Principles of Radiation Chemistry*, Elsevier, 1970.

A. R. DENARO and G. G. JAYSON, *Fundamentals of Radiation Chemistry*, Butterworth, 1972.

G. E. ADAMS, E. M. FIELDEN, and B. D. MICHAEL (eds.), *Fast Processes in Radiation Chemistry and Biology*, J. Wiley, 1975.

J. W. T. SPINKS and R. I. WOODS, *An Introduction to Radiation Chemistry*, 2nd edn., J. Wiley, 1976.

J. SILVERMAN and A. R. VAN DYKEN, Radiation processing, *J. Radiation Phys. Chem.* **9** (1977) 1–886.

G. V. BUXTON and R. M. SELLERS, The radiation chemistry of metal ions in aqueous solutions, *Coord. Chem. Rev.* **22** (1977) 195.

CHAPTER 16

*Radiation Biology and Radiation Hazards**

Contents

Introduction

Although exposure to ionizing radiation from space and from naturally occurring radioactive elements in the soil and in the air has been a fact of life since the earliest days of man, it was not until the discovery of X-rays that the biological effects of ionizing radiation became known. Many serious accidents occurred as a result of the use of radiation before an adequate understanding of its biological effects led to formulation of rules for protection of workers. By 1922 approximately 100 radiologists (not patients) had died as a result of biological radiation damage. A rather famous example of the careless handling of radioactive substances resulting in many deaths was provided by workers whose occupation was painting radium dial watches. It was their practice to sharpen the tips of their brushes by licking them and, as a result, over a period of time they accumulated considerable amounts of radium in their bodies which led to an increased incidence of cancer.

Simply stated, radiation always represents a danger to biological systems. In the use of radiation the degree of the danger is directly related to the extent of the user's lack of adequate knowledge. The danger can be minimized by proper precautions and safety measures in much the same way as the danger attending the use of chemicals in the chemistry laboratory is minimized. In fact, since radioactivity can be detected in extremely small amounts, both the chemical and the radiological hazards are often much less of a problem in a properly operating radiochemical laboratory than in a normal chemical laboratory where safety rules are apt to be less rigorously followed.

16.1. Biomolecular effects of different types of radiation

No matter what the type of radiation, the amount of radiation energy necessary to produce observable biological effects is small. The quantity sufficient to cause death in mammals would result in the rise of body temperature of less than 0.01°C. Such a temperature rise would be the result of conversion of radiation energy to thermal energy and is only a symptom, not a cause, of the radiation injury. The energy of radiation is dissipated during the passage through an organism in the same manner as in any other material, by ionization and excitation of atoms or molecules of the material (Fig. 16.1). In a biological system this ionization causes damage directly by disruption of chemical bonds in the cell. The interaction of the radiation with water both inside and outside the cell produces free radicals which damage the cell by causing oxidation–reduction reactions.

The actual extent of the biological damage depends on many factors such as the type and energy of the radiation, its rate of administration, the organ of the body irradiated, the age and the state of health of the individual. Of particular importance is whether the radiation source is outside (*external radiation*) or inside (*internal radiation*) the body. For external radiation X- and γ-rays are most dangerous to man because of their penetrating power. Alpha-radiation is not dangerous externally as it can penetrate only the outer, rather insensitive skin layer, but it is the most dangerous radiation as an internal source. The many factors involved make it impossible to state that a definite amount of any radioactive material produces a certain definite amount of some biological effect.

In order to take into account the different extents of the biological effects of different radiations, the concepts of the *relative biological effectiveness* (the RBE value) and the *quality factor* Q have been introduced. This has led to the definition of two new quantities, the RBE dose and the *dose equivalent* H:

$$\text{RBE dose (rem)} = D\,(\text{rad}) \times \text{RBE} \tag{16.1}$$

and

$$H\,(\text{Sv}) = D \times Q \times N \tag{16.2}$$

1 Sv (Sievert) = 1 J kg^{-1} (= 100 rem). N includes all modifying factors, which must be considered in practical radiation protection work, e.g. weighing factors due to radiation sensitivity and physiological importance of different organs, etc. Presently N is set = 1 (ICRP, 1977); therefore, also, RBE = Q. Q-values and dose equivalent values for various radiation sources are given in Table 16.1.

From Fig. 16.1 it is obvious why α-radiation is more hazardous to the cell. The dense ionization column of α-radiation leads to disruption of a DNA molecule (Fig. 16.1(c)). On the other hand, a γ-ray photon may pass even a whole chromosome (which is made up of some 10^9 DNA molecules) without causing a chromosome disruption or aberration

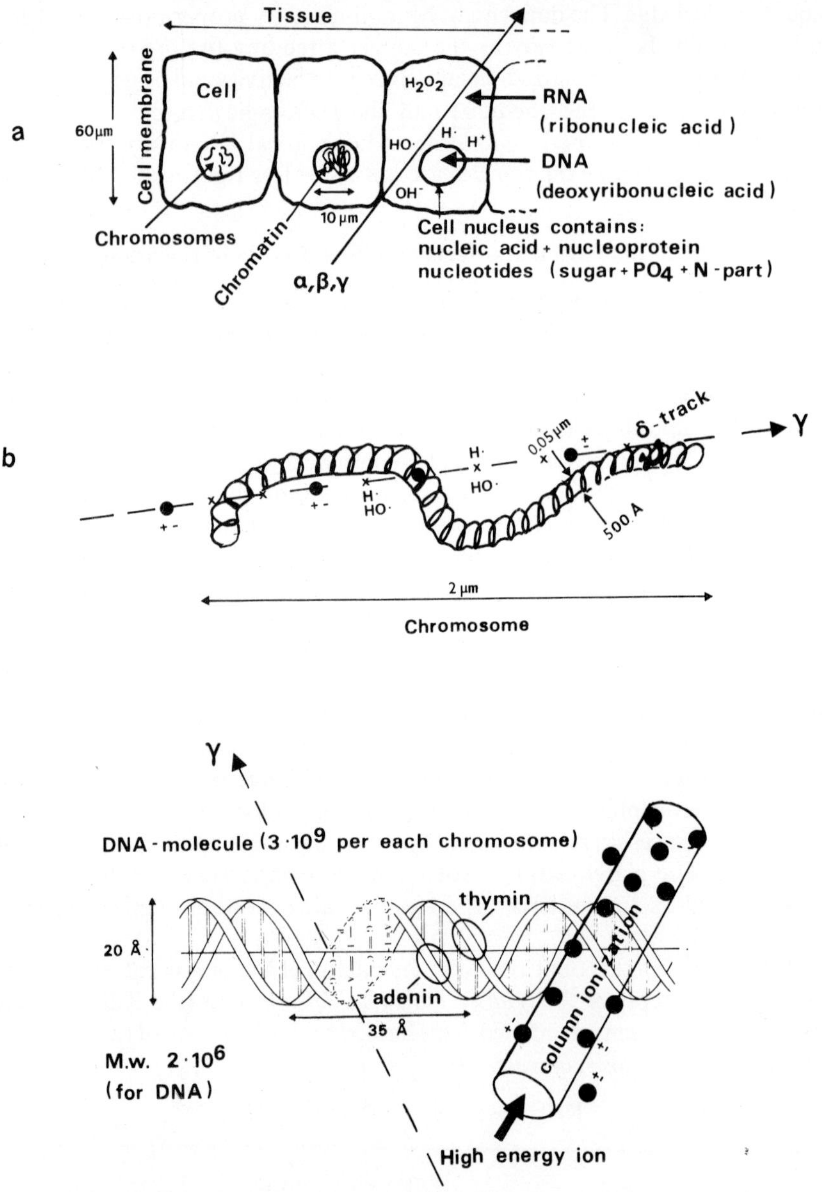

FIG. 16.1. Radiation effects on (a) cells, (b) chromosomes, and (c) DNA molecules.

(structural change). However, any radiation, whether α, β, or γ, causes radiolysis of the water in the cell (Fig. 16.1(a)), and the products may react with the RNA or DNA to such an extent that the cell dies. For an absorbed dose of 100 Gy (10^4 rad) the number of molecules destroyed is 0.06% in the case of a molecular weight of 10^4, but 1.6% for a molecular weight of 10^6 (which is about the weight of a DNA molecule). Since a human chromosome contains some 3×10^9 DNA molecules, such a high dose results in the death of the cell involved. This simple "hit theory" for radiation damage to the cell is illustrated by the dashed line in Fig. 16.2, which shows the "inactivation dose" to various particles and cells.

TABLE 16.1. *Q-values assigned to various kinds of radiation and dose equivalents for specified absorbed radiation doses*
Mainly from ICRP publ. 26, 1977

LET (keV/μm water)	Q (RBE)	Radiation absorbed dose or neutron fluency[(a)]	Dose equivalent (RBE dose)[(b)]
$\lesssim 3.5$	1	x, γ, β: 1 Gy (100 rad)	= 1 Sv (100 rem)
~7	2	High energy e^-: 1 Gy (100 rad)	= 2 Sv (200 rem)
~14	3	Thermal n: 10^{11} cm^{-2}	= 1 Sv (100 rem)
~53	10	Fast n (1 MeV) 3×10^9 cm^{-2}	= 1 Sv (100 rem)
~140	20	Natural α: 1 Gy (100 rad)	= 20 Sv (2000 rem)
$\gtrsim 175$	20	High energy ion: 1 Gy (100 rad)	= 20 Sv (2000 rem)

[(a)]ICRP publ. 4, 1964.
[(b)]For $N = 1$.

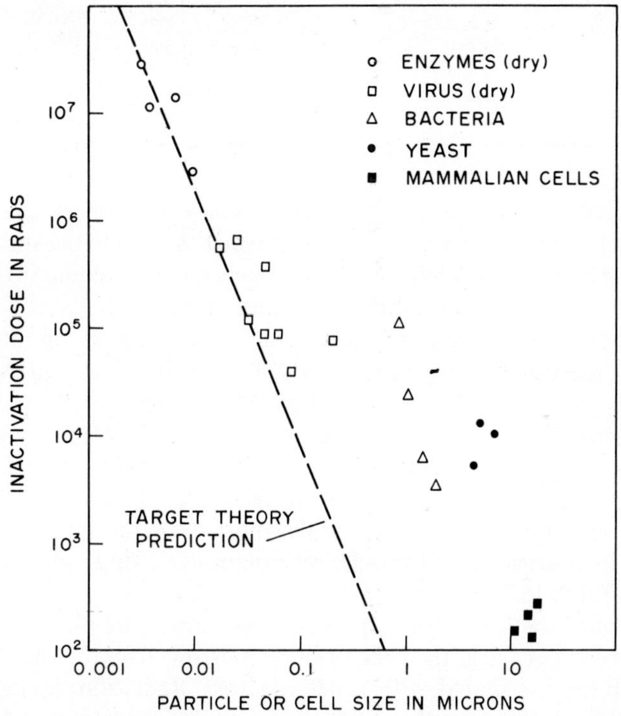

FIG. 16.2. Radiation dose required for inactivation of cells and particles.

The various types of radiation differ in the extent to which they produce biological effects, but not in the type of effects produced in a biological system. This is important since it means, for example, that the study of radiation effects induced by X-rays can be extrapolated to predict those induced by α-particles.

16.2. Radiation sensitivity of different organs and organisms

Radiation damage to biological cells is manifested in several ways. The rate of cell divisions is reduced, so a cell undergoes fewer divisions during its lifetime. Cell mutations

TABLE 16.2. *Effect of γ-radiation doses on different organisms*
$LD_{50(30)}$ refers to a dose which has been found lethal for 50% of the organisms within 30 d. For microorganisms about 10 times higher doses are required for killing them than for inactivation

Micro-organisms	Enzymes inactivated at	$> 2 \times 10^6$ rad	
	Virus (dry) inactivated at	30 000 – 500 000 rad	
	Bacteria inactivated at	2 000 – 100 000 rad	
	Human cells inactivated at	$\geqq$ 100 rad	
Plants	Flowers (Senecio) survive at	1 000 rad d^{-1}	during the growing season (normally the spring)
	Trees do not survive at	100 rad d^{-1}	
	Trees normally survive at	2 rad d^{-1}	
Animals	$LD_{50(30)}$ for: amoeba	100 000 rad	
	fruitfly (*Drosophila*)	$\gtrsim$ 60 000	
	shellfish	20 000	
	goldfish	2 300	
	tortoise	1 500	
	song sparrow	800	
	rabbit	800	
	monkey	600	
	man	250–450	
	dog	350	

may occur which can upset the normal biochemical evolutionary chain, resulting in the cells following an evolutionary development which is alien to the species (e.g. formation of cancers). Differentiated cells and cells which are undergoing division are much more sensitive to radiation. For example, the radiation hazard is about 100 times greater for a fetus during the third to seventh week than for the pregnant mother. Organs and tissues in which the cells are replaced slowly also exhibit high radiation sensitivity. The human tissues which exhibit a higher than normal sensitivity to radiation damage are the reproductive organs, the skin, the organs of the abdominal cavity, the eyes, the blood-forming organs in the spleen and bone marrow, and the tissues of the nervous system. In general the more differentiated the cells of an organ are, i.e. the higher the organ is on the biological evolutionary scale, the greater the sensitivity to radiation. This also holds to a large extent for different organisms (living individual) as reflected in Fig. 16.2 and Table 16.2.

We can distinguish between two cell types: those which are directly involved in the functioning of the organs (e.g. the cells of bone marrow, liver, or the nervous system) and those which are associated with genetic factors. Radiation damage gives rise in the former to *somatic effects* (i.e. limited to the organism irradiated) and in the latter to *genetic effects* (i.e. limited to future generations).

Neither for cells nor for higher organisms has any evidence been found for the development of true *radiation resistance*. Certain bacteria have been shown to develop a seeming resistance to radiation after receiving small radiation doses over a long period of time. This resistance is possibly due to the formation of mutated organisms with a different sensitivity to radiation than the original ones. It has been found that the radiation resistance of an organ can be increased if prior to irradiation the organ has had administered to it certain chemical compounds which act as *radiation protection agents*; cf. §15.8. Their effectiveness is evaluated by determination of the dose reduction factor (DRF), which is the ratio of $LD_{50(30)}$ for protected and unprotected animals. $LD_{50(30)}$ is the (lethal) dose required for killing 50% of the irradiated species within 30 days. These compounds are typically aminothiols, like the natural amino acid cysteine. The protective effect (DRF value) can amount to as much as a factor of two.

These agents probably function as scavengers for the products of water radiolysis, especially the hydroxyl radical. Unfortunately, because of their chemical toxicity many of these protective compounds can be administered only in small doses.

16.3. Somatic effects of large doses on man

Instantaneous whole body doses (i.e. those given within about a day) of $\gg$ 1000 rem led to death within 24 h through destruction of the neurological system. At about 750 rem, death through gastrointestinal bleeding occurs within several days to about a month, depending on medical care. Doses less than 150 rem are rarely lethal. For doses between these two levels intensive hospitalization is required for survival. At the higher end of this range, death usually occurs from 4 to 8 weeks after the accident through infection as a consequence of the destruction of the leukocyte ("white blood cells") forming organs. Those surviving this period usually recover completely. This demonstrates that the body has a *repair mechanism* for radiation damage, which also has been demonstrated in numerous animal experiments. For doses less than 50 rem the only proven effect is a decrease in the white blood count (leukopenia). The threshold value for early somatic damage appears, for short radiation times, to be about 25 rem.

Instantaneous doses of this order of magnitude occur in explosions of nuclear weapons, in accidents involving nuclear reactors, or from carelessness in working with accelerators, X-ray equipment of radioactive installations, such as large ^{60}Co sources used for technical and therapeutic purposes. Such doses are very unlikely to be received in radioactive work involving amounts of 1 Ci or less of radioactivity.

The most common type of overexposure in radioactive work involves high instantaneous doses to the hands. Fortunately the hands, where the skin is the most sensitive tissue, can stand fairly large doses (Table 16.3). If they do receive extremely high doses ($\gg$ 1000 rem β or γ) amputation is usually required although in some cases skin transplants have provided temporary relief.

The dose required to produce a certain damage is strongly dependent upon the time over which the dose is delivered to the organ. A dose which is lethal in humans within a short period of time may lead to very few if any symptoms if spread out over the normal lifetime of the individual.

As noted earlier, the body seems to have an ability to recover from radiation effects. However, some residual effects may not be evident for a considerable length of time (*late somatic effect*). The Japanese nuclear weapons victims, who barely survived after acute radiation sickness, are estimated to have a 2% probability of developing cancer;

TABLE 16.3. *Observed effects for radiation doses to the hands*

< 2 Gy	No proven effect
~ 4 Gy	Erythema, skin scaling, follicle deaths
6–7 Gy	Skin reddening after a few hours, which then decreases, later strongly increases after 12–14 d, and then finally disappears within a month; pigmentation
> 8.5 Gy	As above, but irreversal degeneration of the skin is visible to the naked eye (the skin becomes hard and cracks); with increasing doses degeneration of the binding tissue
50–80 Gy	Development of nonhealing skin cancer

the observed frequency of leukemia is 2–3 cases per 1000 individuals within 2–25 years after having received a dose of 100 rem. Similar late effects have been found for patients treated with large radiation doses ($\gtrsim$150 rem). In Japan the following latent periods have been observed: cataracts 5–10 y, leukemia 8–10 y, thyroid cancer 15–30 y. In the case of internal radiation (see §16.10.2) the latent periods for lung cancer to uranium miners (unvented mines) and bone cancer to the radium watch dial painters have been 10–20 y. The lung cancer frequency of the early Czechoslovakian mine workers was about 50% higher than for the population as a whole.

A third long term effect—acceleration of the aging process—has been estimated in a semiquantitative way as being rather small. If a person received the maximum permissible amount of radiation constantly for a 20 y period—an extremely unlikely situation—the predicted life shortening would be at the most a year. However, it has also been discovered that small doses up to about one-tenth $LD_{50(30)}$ slightly increase the life span of animals as well as growth of plants.

16.4. Biological and biotechnical applications of large radiation doses

Modern society has frequently converted destructive powers into useful servants. Thus the biological damage caused by radiation can be used for many beneficial purposes such as radiation therapy and food sterilization.

16.4.1. *Radiation therapy*

It was discovered early that radiation could heal as well as cause skin cancer. The healing effect has been related to the greater rate of division of most cancer cells. However, cell division may be much slower in some malignant (lethal) tumors. In such systems the cells may be highly differentiated, and it has been suggested that such cells seem to be more radiation sensitive than less differentiated cells.

Radiation therapy usually consists of the delivery of large instantaneous doses to tissues for which surgical operation is impossible or undesirable. In some cases local irradiation of some tissues is produced by the use of radioactive nuclides implanted in the tissue by means of needles or small capsules. For example, needles of ^{90}Sr–^{90}Y, pellets of ^{198}Au, etc., have been implanted in the pituitary gland (for acromegalia, Cushing's disease, and cancer), in the breast (for breast cancer), and in the nerves (to reduce pain), while γ-rays from ^{60}Co are used for irradiation of deeply located organs. The local dose may be several thousand rads.

A special form of radiation therapy uses high energy α-particles or protons from accelerators. The decreasing velocity of charged particles in matter results in a very high specific ionization near the end of the path (the Bragg peak, Fig. 14.6). The energy of the particles is varied with the depth and type of tissue to be penetrated, so that the particles have the proper range to provide a very high local dose in the proper volume of the diseased tissue. This technique has been particularly important in treating diseases of the pituitary gland, which is located deep inside the brain.

16.4.2. *Food sterilization*

Figure 16.2 and Table 16.2 illustrate that radiation has a more serious effect on the

FIG. 16.3. Comparison of irradiated onions and potatoes (left) with nonirradiated ones that have been stored for the same lengths of time. (From US Army Natick Research and Development Center.)

more developed organisms compared to simpler ones. This has a practical consequence, as radiation can be used for the conservation of food in a quite different manner than the conventional methods of heat, canning, and freezing. Radiation conservation attempts to achieve the complete destruction of all bacteria without changing the taste of the food; the change in the taste is caused by the formation of very small amounts of decomposition products, of the order of ppm (parts per million, 10^{-6}). Radiation pasteurization (i.e. partial sterilization with lower doses) and irradiation at low temperatures cause correspondingly smaller taste changes.

In 1976 about 40 countries had approved some 50 different irradiated food products for human consumption. Some types of foods were (country, maximum dose): potatoes (Canada, 10 krad), onions (Israel, 10 krad) (Fig. 16.3), shrimp (Netherlands, 100 krad), fried meat in plastic bags (Soviet Union, 800 krad), and wheat and wheat products (USA, 50 krad). At a joint meeting in 1976 between the FAO, IAEA, and WHO, the experts recommended the unconditional acceptance as wholesome of irradiated potatoes, wheat, chicken, papaya, and strawberries. They gave provisional approval (until 1980) for irradiated rice, fish, and onions.

The energy requirement for radiation pasteurization is 0.76 kWh t^{-1}, for radiation sterilization 6.3 kWh t^{-1}. Freezing requires 90 kWh t^{-1}, heat pasteurization 230 kWh t^{-1}, heat sterilization 300 kWh t^{-1}, and drying 700 kWh t^{-1}. Using the irradiation process, energy savings of 70–97% are possible. In addition, food irradiation does not require expensive packaging.

Although radiation sterilization has been investigated for almost 30 y, and at the present time there are 70 existing irradiation facilities capable of irradiating food either at pilot plants or on a commercial basis, opinion about its desirability remains divided. Advocates point out that no harmful effects have yet been established, and it is a technique which could considerably increase the amount of food available to the population of the earth. Particularly important is the sterilization of grain to destroy insects and fungi. It has been estimated that the quantity of food in developing countries could be increased as much as 50% by use of radiation.

Opponents of food preservation by irradiation argue that the potential risks are great. Toxic decomposition products are quite difficult to detect and it may require many generations to prove that radiation sterilization has not produced low levels of genetically harmful substances. However, recent extensive investigations have shown these genetic risks extremely low ("acceptably small", according to WHO). Moreover, in the light of other techniques for conservation of food (e.g. the addition of chemicals to bread for preservation, of nitrates to meat, etc.) the arguments against the preservation of food by irradiation may be somewhat exaggerated.

16.4.3. *Other uses*

Medical supplies, which must be sterile, can be manufactured and packed by conventional techniques, after which the packages may be exposed to high energy penetrating radiation, e.g. from ^{60}Co or ^{137}Cs. The radiation kills all bacteria with little damage to low molecular weight compounds (cf. Fig. 16.2). In this common technique, the irradiation source is 0.1–1 MCi. The packages slowly pass through the source, so the doses are of the order of 1–3 Mrad. The same facilities can be used also for food sterilization or other uses (cf. §15.10). The present problem of repeated replacement of expensive ^{60}Co ($t_{1/2}$ 5.3 y) may be solved in the future by using cheaper ^{137}Cs ($t_{1/2}$ 30.1 y) obtained in large amounts from reprocessing plants.

The organic waste produced by man in large population areas is a resource of high nutritional value, but it is infected with bacteria. Radiation alone can be used to destroy bacteria, but this is too expensive. Recently, sewage treatment plants have been built utilizing "thermoradiation", e.g. a combination of heat and γ-irradiation. Figure 16.4(a) shows the effect of each of these treatments alone, and the synergistic improvement when they are used together; (b) shows a projected thermoradiation sewage treatment plant in West Germany.

Radiation doses of $\lesssim$ 200 rem produces temporary sterility in man, while higher doses may give permanent sterility. Irradiation has been used to sterilize the males of certain insect species. When these sterilized insects are released to nature they mate with females and prevent further reproduction of the species. In the United States the screw worm flies were causing approximately \$20 million damage annually to the cattle industry. Approximately 1200 irradiated (8000 rem), sterilized male flies were released per week for each square meter in a total area of 50 000 m^2. The ratio between the sterilized and untreated males in this area was approximately 4 : 1. In 2 years at a cost of about \$8 million the screw worm flies were completely eliminated.

The sterile insect technique is now used on a practical basis against other insect species of agricultural importance, such as the Mexican fruit fly, the Mediterranean fruit fly, the codling moth, and the pink bollworm. The method is also being applied in large scale field trials for the control of tsetse flies, mosquitoes, *Heliothis* spp., the melon fly, oriental fruit fly, onion fly, and other important agricultural pests.

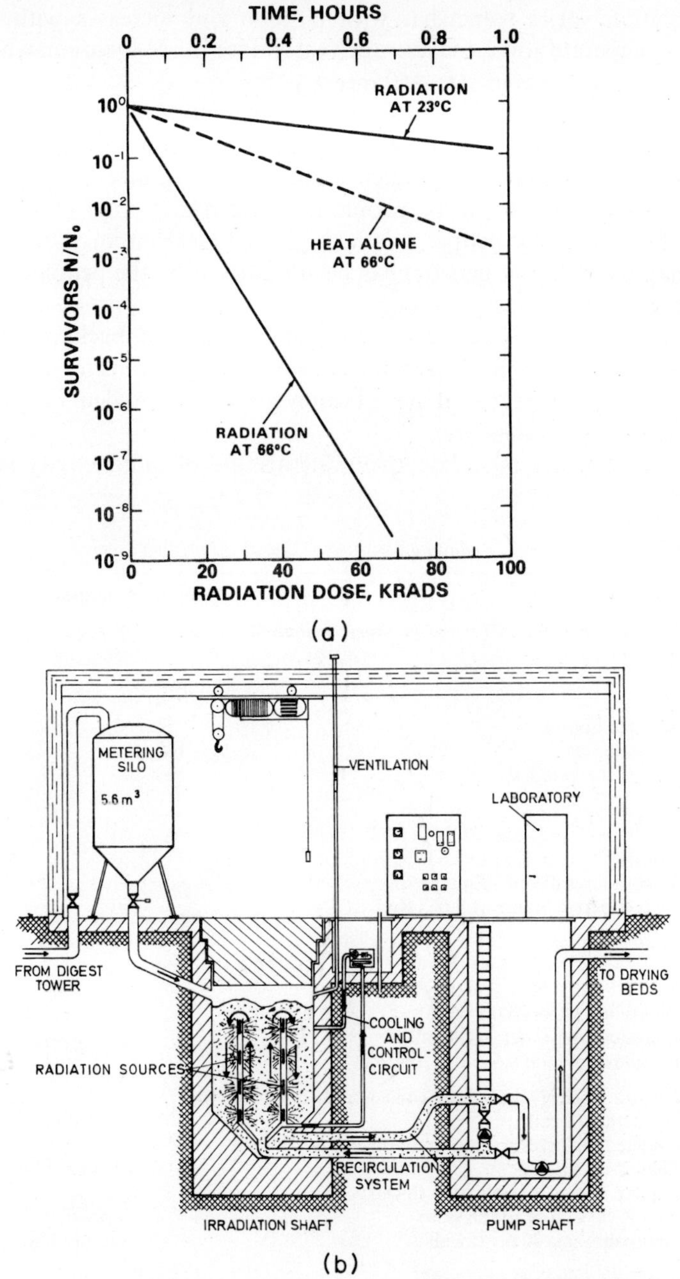

FIG. 16.4. (a) Comparison of the radiation treatment of T-4 bacteriophage at room temperature and at 66°C. (b) Scheme of plant for sewage sludge irradiation at Geiselbullach, West Germany.

16.5. Radiation background to man in present society

Through our lifetime we are constantly exposed to ionizing radiation from a variety of sources. In Table 16.4 typical values of the radiation doses occurring in society are given. The contribution of various sources to the average dose is given in Fig. 16.5.

Cosmic radiation varies somewhat with latitude and increases with altitude. At sea level in the temperate zones the average yearly dose is 30–40 mrem, while at a level of 3000 m the dose increases to 100–150 mrem y^{-1}.

The largest single "natural" radiation dose contribution comes from the ground upon which we live, because of its content of thorium and uranium minerals. A normal average value is 40 mrem y^{-1}, but much higher levels are measured in some areas; thus in Kerala (India) some 100 000 people live in a background of 1300 mrem y^{-1}, and in Brazil for about 100 people it is as high as 12 000 mrem y^{-1}. No particular radiation damage (somatic or genetic) has been observed by the people living in these locations of high natural background.

Houses made of wood give very little increase in the background radiation. Brick and concrete houses may contain considerable amounts of uranium and/or radium; the figures in Table 16.4, B, have been measured in Sweden, but similar values are found in many other countries.

Because of the existence of radioactive minerals and of radioactivity in the air, all

TABLE 16.4. *Radiation doses from natural and artificial sources*

A. *Cosmic radiation*	30–50 mrem y^{-1}
B. *Other external radiation (including cosmic radiation)*	
From clay soil	~ 50
Nongranite rocks	60–120
Granite	80–300
In wooden houses	50– 60
In brick houses	100–110
In houses made of light concrete	100–200
C. *Radiation from the air* (mainly Rn)	
Outdoors, average value (0.1 pCi Rn l^{-1})	20
In wooden houses (0.5 pCi Rn l^{-1})	70
In brick houses (0.9 pCi Rn l^{-1})	130
In light concrete houses (1.9 pCi Rn l^{-1})	260
D. *Internal radiation*	~ 20
(^{40}K (1.1×10^{-7} Ci) 15 mrem y^{-1}; ^{14}C (8×10^{-8} Ci) 1 mrem y^{-1}; Ra (1.2×10^{-10} Ci) 1 mrem y^{-1})	
Ordinary tap water $\lesssim$ 0.1 pCi l^{-1}	
Bad Gastein spa water ~ 100 pCi l^{-1}	
Daily intake of food 0.5–5.0 pCi ^{226}Ra	
E. *Diagnostic X-ray* (values differ considerably due to varying practices)[a]	
Lung picture (local)	0.04–0.2 rad per film
Complete lung investigation up to (local)	75 rad
Kidney X-rays (local)	1.5–3 rad per film
Complete kidney investigation (local)	15–30 rad
Dental X-rays (dose to the skin)	$\lesssim$ 1 rad per film
Dental panorama X-ray (local)	~ 60 rad per film
F. *Therapeutic radiation treatment*	locally $\leq$ 10 000 rad
G. *Other sources*	
Radium watch dial (~ 0.1 μCi Ra)	$\lesssim$ 5 mrem y^{-1}
Instrument table in airplane	1.300 mrem y^{-1}
Jet-flight (4 h)	~ 1 mrem
Arable fields covered with Florida phosphate	300–700 mrem y^{-1}
From nuclear tests	< 1 mrem y^{-1}
Average value	100–200 mrem y^{-1}

[a]The genetically significant doses are much less.

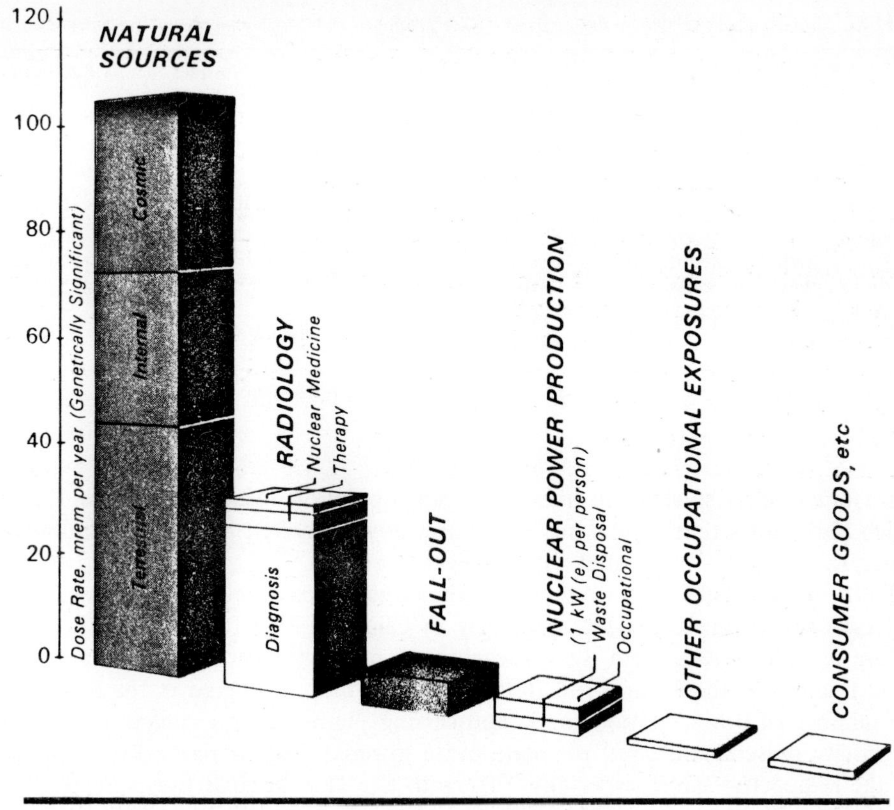

FIG. 16.5. Annual genetically significant dose rate as averaged through the OECD population. (From *OECD Observer*, 1976.)

foods contain some natural radioactivity. An approximate average value in northern Europe is 1 pCi ^{226}Ra per kg of food or beverage.

Testing of nuclear weapons in the atmosphere adds radioactivity to the environment. Measurement of 10 people in New York showed a level of 37 pCi or ^{137}Cs per kg of body weight in 1961. This rose to 110 pCi kg^{-1} in 1963. The test ban between the USA and USSR resulted in a steady reduction of environmental contamination via testing. The measured levels in the 10 people dropped to 20 pCi kg^{-1} by 1969.

Natural radon exhalation rate from the earth's surface is $\sim$0.5 pCi m^{-2} s^{-1}, which gives an average equilibrium air concentration of $\sim$0.07 pCi l^{-1}. Building materials like stone, brick, and concrete often contain small amounts of uranium and/or radium, which leads to a release of radon into the air of the building. Typical values range from 0.3 to 2 pCi l^{-1}, but they may increase manyfold if the room is poorly ventilated; a value of 0.5 will yield an annual lung dose of 30 mrad.

When the natural background from cosmic radiation, geological sources, food, and environmental sources such as housing, clothing, etc., is added together, the average values can be estimated to be approximately 100–200 mrem y^{-1} for each individual.

Radionuclides have a wide variety of uses, which exposes many people to ionizing radiation in excess of the natural background. Table 16.5 shows the different radionuclides and amounts used for various purposes in Sweden (pop. 8.2 million) and, for comparison, the number of X-ray exposures. These values are believed to be represen-

TABLE 16.5. *Radioactive sources in Sweden* (8 *million people*) *in* 1974, *excluding the nuclear power industry*

Use of radioactive source	Number	Nuclides	Radioactivity
Medical: diagnostic	~ 100 000 samples y^{-1}	^{99m}Tc, ^{131}I, ^{133}Xe	~ 500 Ci y^{-1}
radioimmunological tests	~ 500 000 samples y^{-1}	T, ^{14}C, ^{125}I	~ 20 Ci y^{-1}
therapeutic treatment	~ 350 000 treatments y^{-1}	^{60}Co, ^{137}Cs, ^{226}Ra	> 10^6 Ci
Research and teaching		All nuclides (α, β, γ)	~ 100 Ci y^{-1}
Industrially: level, density, etc., gauges	~ 130 instruments	^{60}Co, ^{137}Cs	~ 70 Ci
radiation sterilization	2 large installations	^{60}Co	~ 400 000 Ci
smoke detectors	~ 150 000	^{85}Kr, ^{226}Ra, ^{241}Am	~ 15 Ci
Wastes: medium activity	~ 1 m^3 y^{-1} collected	All long-lived nuclides	
low activity	~ 200 m^3 y^{-1} collected		

X-ray exposures: diagnostic 5.3 million, chest examinations (fluoroscopic picture) 0.9 million, dental 5.4 millions per year.

tative for a modern society and gives an overall population dose as indicated in Fig. 16.5, which also shows that at present the highest doses received by the population from unnatural sources is associated with medical treatment.

To the data in Table 16.5 we must add the large radioactivities in the nuclear power fuel cycle which are discussed separately in Chapters 19–21. However, the normal release of radioactivity from the nuclear power industry presently contributes a very small fraction to the natural radiation dose. This may not be so in the future when the number of power stations and reprocessing plants increase unless the amounts of activity presently released per station are lowered. This is particularly important for the radioactive gases, especially ^{85}Kr with $t_{1/2}$ 11 y, because they spread all over the globe.

16.6. Special dose concepts for human exposure

In order to relate the build up of radioactivity from nuclear power installations and the resulting dose to the population, the *dose equivalent commitment* concept has been introduced. The dose equivalent commitment H_c is the total dose contribution to the population over all future years of a specified release (for instance one year) of radioactivity from a nuclear installation. It is defined by

$$H_c = \int_0^{\infty} \bar{\dot{H}}(t)\mathrm{d}t \qquad (16.3)$$

where $\bar{\dot{H}}(t)$ is the per capita dose equivalent rate in a specified population ($\dot{H} = \mathrm{d}H/\mathrm{d}t$, i.e. the dose equivalent rate). The concept is best described by Fig. 16.6, where each rectangle is the dose delivered in one year from a one year release. A is the dose from the first year's (1) release; the next year (2) this gives a smaller dose B (because of radioactive decay); and the following year (3) an even smaller dose C; we assume that the following fourth year is the last year that the dose contribution (D) is significant. In this same fourth year we have a "first year release", A′, equal in amount to A, plus what is left from the previous years, B′ and C′. Thus the annual dose at release equilibrium is equal to the dose commitment for one year.

With this concept one can extrapolate the consequences of the large scale introduction

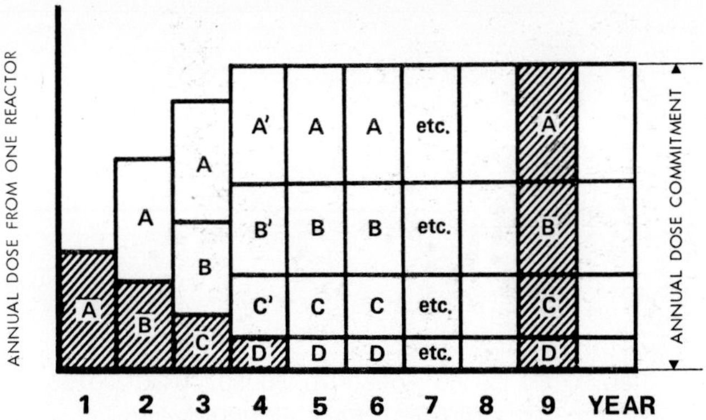

FIG. 16.6. Dose commitment for continuous radioactivity release.

of nuclear power, and take precautions that the total annual dose stays within agreed safe limits. Since it is not believed that we shall use fission power for more than about 500 y, the $\dot{H}$ dose commitment integral is commonly limited to 500 y (the "incomplete collective dose"). This also takes into account future medical improvements in cancer treatment.

Another important concept is the *collective dose equivalent*, i.e. the dose received to a population (N persons) by radiation exposure. The collective dose equivalent S is the sum of all individual equivalent doses

$$S = \sum H_i N_i \tag{16.4}$$

where $\sum N_i$ is the exposed population. The annual global collective dose equivalents from different sources have been estimated (D. J. Beninson):

Natural radiation	3×10^8 manrad
Air travel	3×10^5 manrad
Medical uses	1×10^8 manrad
Consumer goods	4×10^6 manrad
Nuclear power (1976)	4×10^5 manrad

Dividing the last value with the global population gives an average of 0.1 mrad y^{-1} from nuclear power in 1976. This figure is expected to rise, considering the increase in nuclear power plants, and that the dose commitment to the individual from a 1200 MWe light water reactor over 50 y is presently estimated to be $\lesssim 10$ mrem. The collective dose commitment from nuclear power production is estimated as 3–4 manrad per MWey for occupational exposure and 2–3 for exposure of the public.

16.7. Effects of low radiation doses

Cancer is the dominating late somatic effect of ionizing radiation. Figure 16.7 shows the malignant mortality rate of the US white population accumulated over 18 y in almost all states, as a function of the natural radiation background. Each point represents an average of about 100 000 deaths, and is thus very significant. The average background value for the whole of the US population is 130 mrem y^{-1} and the average mortality rate in cancer (horizontal line) corresponds to 0.16%. The only significant trend seems to be a decrease in cancer deaths with increasing background up to about 250 mrem y^{-1}.

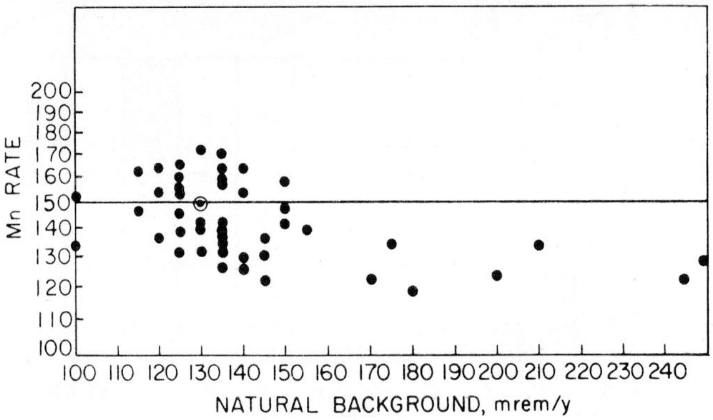

FIG. 16.7. Malignant mortality (Mn) rates per 100,000 of the US white population, 1950–1967, by state and natural background. The horizontal line and open circle indicate the rate and background for the United States as a whole. (From N. A. Frigerio, K. F. Eckerman and R. S. Stowe, 1973.)

Although attempts have been made to correlate the cancer decrease rate with other factors, such as living habits, no reliable explanation has been discovered.

This observance of a positive effect of low radiation doses is not unique. It has been found that the leukemia frequency is five times lower for the Japanese bomb victims who received an only slightly larger radiation dose than the average to the population. This may indicate that small radiation doses stimulate the cell repair mechanism. Although the Japanese group may not be representative (because of early elimination of the least resistant), it has been found that the health of cancer patients is improved by administering small radiation doses ("preirradiation") before radiotherapeutical treatment.

Radiation protection rules for workers in nuclear installations (e.g. medical X-ray personnel) and for the population as a whole must be such that the risk of radiation-induced cancer during an individual's lifetime and of genetic effects is extremely small. Because of the difficulty of basing such rules on adequate data for humans, it is necessary to rely on experiments with animals and on biological theory. A most important uncertainty is the existence of a *threshold dose* below which no somatic effects due to radiation are manifested during the normal life-span of the individual. Consider Fig. 16.8, which shows the (excess) tumor frequency for mice irradiated with doses up to 12000 rad. Below a value of about 500 rad (= rem for all except the ^{226}Ra curve) no excess cancer is observed, supporting the threshold theory.

From these and other recent investigations many radiobiologists hold probable that for low doses of low LET radiation there is a threshold value of about 100 rem below which no somatic effects occur. This threshold value probably differs between people. Other experiments indicate that for *certain cancer types* and animal groups no such threshold exists, but rather a linear relationship as indicated in Fig. 16.9. Even though radiobiologists disagree about the existence of a threshold dose, they agree that in order not to underestimate the risks at low dose values a *linear effect–dose rate relation* is assumed to exist. This assumption is used by the ICRP in estimating radiation risks and recommending dose limits. For low doses to the population (i.e. in contrast to the much larger doses to individuals in medical treatment) the probability of an effect (such as cancer) occurring is considered by the ICRP to be a function of the dose, without

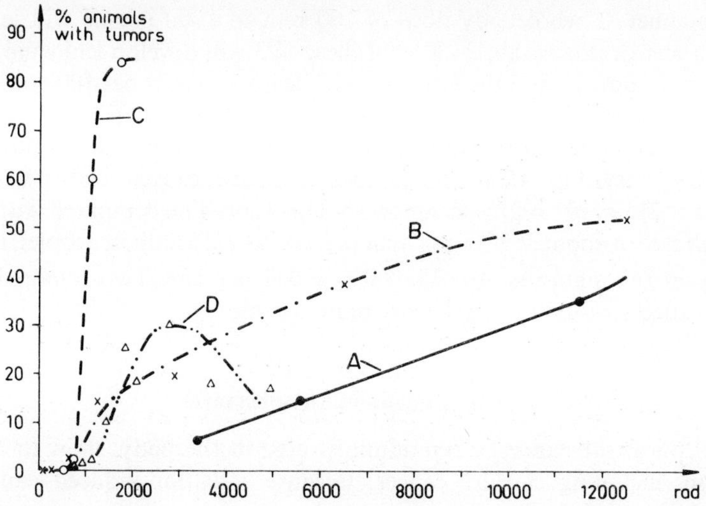

FIG. 16.8. Tumor frequency as a function of absorbed dose. A, ^{90}Sr-induced osteosarcomas in female CBA mice (after Nilsson, 1975). B, Bone tumors in man from incorporated ^{226}Ra (after Rowland, 1971). C, Kidney tumors in rats by X-rays (after Maldague, 1969). D, Skin tumors in rats by electrons (after Burns, 1968).

threshold. For such so-called *stochastic effects* the ICRP has recently (1977) published *risk factors*. For the purposes of radiation protection involving individuals, the ICRP concludes that the mortality (death) risk factor for radiation-induced cancers is about 10^{-2} Sv^{-1} (i.e. 1 in 100 individuals irradiated by 100 rem) as an average for both sexes and all ages; the morbidity (sickness) risk of cancer is about 2×10^{-2} Sv^{-1}. The average risk factor for hereditary effects, as expressed in the first two generations, would be substantially lower than this when account is taken of the proportion of exposures that is likely to be genetically significant, and can be taken as about 4×10^{-3} Sv^{-1}. For the eye lens the total dose equivalent of 15 Sv would be below the threshold for the production of any opacification that would interfere with vision. Cosmetically unacceptable changes in the skin may occur after irradiation with absorbed doses (to the skin only) of 20 Gy delivered in a short time (weeks to months).

The assumption of a risk of 10^{-4} cancer deaths per manrem can be restated in the

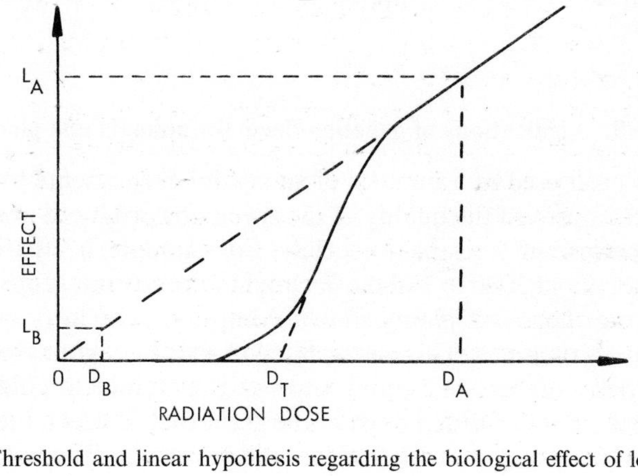

FIG. 16.9. Threshold and linear hypothesis regarding the biological effect of low level doses.

following manner: a wholebody dose of 100 rem to 1000 people will lead to about 10 excess cancer deaths within 2–30 y; of these $\lesssim 3$ will develop leukemia; the present leukemia rate in Europe and the USA is ~ 0.7 death annually per 1000 people. Normal death rate in industrialized countries is 9–11 in 1000 people. Thus if a population of 215 million accumulates an excess radiation dose (e.g. from the nuclear power complex) of 10 mrem y^{-1} (cf. Fig. 16.5) the number of excess cancer deaths would become $10^{-4} \times 0.01 \times 215 \times 10^6 = 215$ persons y^{-1}. This should be compared with the present cancer death rate of about 365 000 per year in the USA (215 million people); the statistical uncertainty in this figure is $\pm\sqrt{365000} = \pm 604$ persons. Two hundred and fifteen additional cancers would be statistically unnoticeable.

16.8. Radiomimetic substances

Many chemical substances, when administrated to the body, show the same effects as radiation, including causing cancer. Because radiation-induced cancer is more investigated than chemically induced, such substances are called radiomimetic (mimetic = imitative). In order to qualify as a radiomimetic agent, the substance must do all the following: stop cell division, stop tumor formation, give chromosome aberrations, cause mutations, kill lymphocytes, and be carcinogenic. Chemical substances which only meet a few but not all of these requirements are not radiomimetic. The effects depend on the concentration of the substance; e.g. cell division is interrupted by many radiomimetic substances at concentrations $\lesssim 10^{-5}$ mole l^{-1}.

Typical radiomimetic substances are organic peroxides (e.g. ethylene oxide) ethylene diimine, mustard gas and derivatives, aliphatic dichloro amines, etc. These compounds or chemical groups occur in many materials surrounding us, as in tobacco smoke. The effect of a certain amount of a radiomimetic substance can be calculated to correspond to a radiation dose. Thus smoking a pipe of tobacco corresponds to an average radiation dose of about 4 mrem; the extreme values fall between 1 and 100 mrem depending on individual sensitivity (cf. § 16.6).

It has been suggested to introduce a new unit named *radiation equivalent chemical* (rec), which is defined as the product of concentration of a chemical and time of exposure to it that produces the same mutagenic effect as 1 rem of radiation. It is estimated that an exposure to 5 ppm of ethylene oxide for 40 h corresponds to $\sim$4 rec, the present use of $NaNO_2$ in foods to $\sim$8 rec per 30 y, and exposure to all chemicals in modern society to 4–18 rec per 30 y.

16.9. Applications of genetic effects for animals and plants

Radiation has been used in a number of successful experiments to produce changes which have either improved the quality of the species or produced mutations disadvantageous to the species but desirable to society. For example, irradiation of plant seeds results in a ratio of about 1000 to 1 of the harmful to the advantageous mutations. However, by cultivating those few plants showing improvement in properties, new plant variations have been obtained. This has resulted in species of grains and legumes which have stronger stocks, higher yields, and improved resistance to cold and to fungi. In northern countries such as Sweden most of the grain that is grown today is a radiation-produced species possessing a much greater cold resistance. The following values give data on mutated rice produced in India compared with untreated rice:

1st year: mutated rice gave 5044; untreated gave 3492 kg ha^{-1}
2nd year: mutated rice gave 1804; untreated gave 1165 kg ha^{-1}
3rd year: mutated rice gave 4143; untreated gave 2621 kg ha^{-1}

The "green revolution", which has considerably lessened starvation in many areas of the world, has largely been achieved with the aid of mutated high-yielding pest-resistant corn (grain).

16.10. Genetic effects on man

In 1927 H. J. Müller showed how irradiation of the fruit fly (*Drosophila* M.) could produce new living species that were defective with regard to the parents (e.g. lacked wings), and that this property was carried on with the genes to later generations. This dramatic demonstration of the mutagenic effect of radiation, which later was extended to other primitive species, has caused great concern about the dangers of nuclear radiation. However, an investigation of 36 000 children born in Hiroshima to parents who had been exposed to $\leqq$300 rem (average exposure 20 rem) showed within statistical uncertainty no genetic differences from children born to nonirradiated parents. This is almost the only data on a large sample that we have for people; it is therefore questionable to assume that genetic effects have a threshold dose of 300 rem. Irradiation of the sexual organs of mice by K. G. Lüning with ~200 rem for 19 generations showed no genetic defects. However, because of the great genetic difference between the various living species (e.g. the much greater ease with which mutations occur for *Drosophila* as compared to man), one cannot directly extrapolate from animals to man.

Assuming a linear relationship between genetic effect and dose received over the sexual organs, the UNSCEAR and BEIR reports (1972) conclude that the number of genetic defects per live birth is 1.5×10^{-4} for each rem received by the father. Such genetic defects include sterility, poor health, color blindness, mongolism, etc. Thus in a population of 1 million people with a birth rate of 2%, if all the fathers have received an excess dose of 1 rem, the number of genetic defects would be 3. But of these 3 at least 1 would not be born (sterility, spontaneous abortion). Obviously such small numbers have a large uncertainty. The ICRP gives the genetic risk factor as 4×10^{-3} Sv^{-1}; cf. §16.6.

16.11. Radiation protection standards

All the uses of radioactive substances and other radiation sources expose people both to external radiation and to radiation from ingested or inhaled radioactive substances. In order to minimize and control these risks, national radiation protection agencies have issued rules with legal force on dose limitations and limits of intake of radioactivity as well as guidelines for working with radioactive substances. The ICRP and the International Atomic Energy Agency (IAEA) regularly issue recommendations for proper handling of radiation sources. In the discussion which follows we limit ourselves to the ICRP–IAEA joint recommendations, although some national restrictions set lower limits. Because the recommendations recently (1977) have been changed, both the older and newer rules are given.

TABLE 16.6. *Recommended radiation dose limits*

A. ICRP recommended MPD limits to individual adults (1956–1977)

Organ	Occupational workers: rem y^{-1}	Occupational workers: rem per 13 weeks	Members of the public (rem y^{-1})
Gonads, red bone marrow	5	3	0.5
Skin, bone, thyroid	30	15	3.0
Hands, forearms, feet and ankles	75	38	7.5
Other single organs	15	8	1.5

B. ICRP recommended annual dose equivalent limits, H_L (1977)

Nonstochastic effects (all tissues, except lens)[a]	500 mSv (50 rem) per year
Stochastic effects[a] (uniform irradiation of whole body)	50 mSv (5 rem) per year
Planned special (occupational) exposures for a single event (limited to a few workers)	100 mSv (10 rem) once in 2 years; 250 mSv (25 rem) once in a life-time
Critical groups of the public	5 mSv (0.5 rem) per year

C. Weighing factors for stochastic risks for tissues

Gonads ($w_T =$)	0.25
Breast	0.15
Red bone marrow	0.12
Lung	0.12
Thyroid	0.03
Bone surfaces	0.03
All other tissues	0.30

D. USA standards for protection against radiation (1977)

NRC: General Standards:

500 mrem y^{-1} for individuals, 170 mrem y^{-1} for population (whole body)[a].
1500 mrem y^{-1} (thyroid or bone).
"as low as reasonably achievable".

NRC: LWR Reactor Guides, per Reactor:

Individuals: 3 mrem y^{-1} to whole body and 10 mrem y^{-1} to any organ from water; 5 mrem y^{-1} to whole body, 15 mrem y^{-1} to skin, and 15 mrem y^{-1} to any organ from air.

Population: $1000 per whole body manrem and $1000 per thyroid manrem over 50 mile radius[a]

EPA rule 40 CFR (1977) regarding releases from whole nuclear fuel cycle:

Doses from all parts of nuclear fuel cycle: whole body < 25 mrem y^{-1}, thyroid < 75 mrem y^{-1}, all other organs < 25 mrem y^{-1}. Radionuclide emissions: $^{85}Kr < 50$ kCi, $^{129}I < 5$ mCi and transuranics < 0.5 mCi per GWey.

[a]See text.

16.11.1. *External radiation*

In Table 16.6, A, the older recommendations of ICRP regarding *maximum permissible dose* (MPD) are given. ICRP defines the MPD as follows:

"The permissible dose for an individual is that dose, accumulated over a long period of time or resulting from a single exposure, which in the light of present knowledge carries a negligible probability of severe somatic or genetic injuries; furthermore, it is such a dose that any effects that ensue more frequently are limited to those of a minor nature that would not be considered unacceptable by the exposed individual and by competent medical authorities. Any severe somatic injuries (e.g. leukemia) that might result from exposure of individuals to the permissible dose would be limited to an exceedingly small fraction of the exposed group."

The MPD values in Table 16.6, A, are based on the rule that the *maximum accumulated dose* received by anyone must not exceed 5 rem per year. The maximum life dose for any organ must be below 50 rem (for the eyes below 30 rem). The generally recommended maximum value per week is 0.1 rem, or approximately 2.5 mrem h^{-1} for a 40 h week. If it is necessary to work temporarily with high levels of radiation, the length of work time is based on the MPDs. No one below the age of 18 should be allowed to work with radioactivity. Such doses exceed by about a factor of 25–50 that due to the natural background described in the previous section and are based on the present evidence about radiation damage to biological systems. It is seen from the table that this dose is meant to apply only to those people working with radioactivity (occupational exposure) and is not for the population as a whole. The suggested limits for members of the public are in all cases a factor of one-tenth of the equivalent limits for adults exposed in the course of their work.

In the fall of 1977 the ICRP published revised recommendations for radiation protection:

The purposes of the recommended *system of dose limitations* are to ensure that no source of exposure is unjustified in relation to its benefits, that all necessary exposure is kept as low as is reasonably achievable, and that the dose equivalents received do not exceed the specified limits. The ICRP stresses that the individual *dose equivalent limits*, which are given in Table 16.6, must be respected. These values are based on total risk of all tissues irradiated. They are intended to limit somatic effects in individuals, hereditary effects in the immediate offspring of irradiated individuals, and hereditary and somatic effects in the population as a whole. However, the dose equivalent limits should not be regarded as a dividing line between safety and danger; when limits have been exceeded by a small amount it is generally more significant that there has been a failure of control than that one or more individuals have slightly exceeded a certain agreed dose.

The dose equivalent limit H_L refers to the sum of the dose equivalents from external sources and from radioactivity taken into the body. The dose equivalent limits do not include contribution from any medical procedure or from normal natural radiation. The recommended annual dose equivalent limits are summarized in Table 16.6, B.

Nonstochastic radiation effects are those for which a dose effect threshold may occur; they are usually specific to particular tissues (e.g. cataract of the lens, nonmalignant damage to the skin, gonadal damage leading to impairment of fertility, etc.). The dose equivalent limit of 500 mSv only refers to such known conditions; for the lens the limit is 300 mSv (30 rem). The limit is not a "per year" value, but rather a value that should not be exceeded in a single year.

All other limits refer to *stochastic* ("random") *effects*, of which carcinogenesis is considered to be the chief somatic risk of irradiation at low doses. For such effects

$$H_L \geqq \Sigma w_T H_T \tag{16.5}$$

where H_T is the annual dose equivalent in tissue T, and w_T is a weighting factor, tabulated for various tissues in Table 16.6, C. The weighting factor is mainly of use for calculating internal exposure from intake of radionuclides. This weighting is taken into account in the *annual limits of intake* (ALI) to be published by ICRP; the ALI values will not differ much from the *maximum permissible concentrations* (MPC) discussed in §16.10.2. The dose equivalent limit of 50 mSv (5 rem) is the fundamental rule for occupational work; it is the same value as the MPD limit used earlier.

The assigned radiation dose limits to the public of the ICRP compares with accidental risks from other causes. Although in the USA the accidental death rate from stochastic accidents (e.g. from using public transportation, climbing ladders, drowning, etc.) is $\sim 6 \times 10^{-4}$ per year (i.e. 6 deaths per year in a population of 10,000 people), it is assumed that the *level of acceptability for fatal risks* to the general public is of the order of $\lesssim 10^{-5}$. On the assumption of a total radiation risk of 10^{-2} Sv^{-1} this "acceptable risk" corresponds to a lifelong whole-body exposure of 1 mSv y^{-1}. By applying an annual dose limit of 5 mSv (500 mrem) to *critical groups* of the public (i.e. representative public groups likely to be most exposed such as people living close to nuclear installations) the average dose equivalent to *individual members of the public* is estimated to be below 0.5 mSv (50 mrem) per year. This is likely to ensure that the average dose equivalent to the whole *population* will not exceed 1 mSv y^{-1}.

Table 16.6.D summarizes recent radiation protection standards in the USA. They do not deviate to any important extent from the ICRP recommendations, but are more explicit with respect to dose contribution limits from nuclear power stations. Such are defined by the national radiation protection agencies. The table also contains cost–benefit values: if the collective dose of 1 manrem within a 50 mile (80 km) radius from the station can be reduced at a cost of less than $1000, the power company is required to take proper action.

It is generally suggested that the nuclear power industry shall not contribute more than 10 mrem y^{-1} to the dose values, and the releases must be adjusted to meet this requirement. Assuming a nuclear electric generating capacity of 10 kW per person in a highly industrialized society, the 10 mrem y^{-1} value corresponds to an "acceptable" maximum dose commitment of 1 manrem per MW and year from the nuclear industry; about half of this is allowed from the nuclear power stations and the rest from the fuel cycle. Experience shows lower values; thus for the CANDU reactors the collective dose commitment is 0.1 manrem per MW(e)y, with a value of < 1 mrem y^{-1} to individuals in the critical group (i.e. people living just outside the "fence").

16.11.2. *Internal radiation*

When radioactive material is taken into the body in sufficient amounts to provide a hazard, the procedure is to attempt to remove it as fast as possible so that it does not have time to become incorporated into tissues of relatively long biological lifetimes, such as the bones. A common practice in the attempts to remove these substances before they become permanently incorporated in the bones is to administer chemical complexing agents such as EDTA, which by forming very stable complexes with the radioactive substances provide a mechanism for their removal from the body. Radioactively contaminated wounds must be cleaned extremely well and be allowed to bleed heavily when possible. Such treatments must be carried out by medical experts (e.g. an overdose of EDTA will disturb the calcium balance, which may be fatal).

A substance which represents a hazard within the body due to its radioactivity is referred to as being *radiotoxic*. The radiotoxicity depends on the properties of the radiation and on a number of physical, chemical, and biological conditions such as mode of intake (via air, in water or food, through wounds, etc.), the size of the ingested or inhaled particles, their chemical properties (e.g. solubility), metabolic affinity, and ecological conditions. Most of these conditions are considered in the ALI, MPC, and MPBB concepts.

The maximum amount of a radioactive nuclide, which incorporated in the body

TABLE 16.7. *Maximum permissible amount of radionuclides (* = includes daughters) in body (MPBB) and concentration in air and water (MPC) for 168 h/week*

$t_{1/2}$ (eff) is the effective half-life of the radionuclide in the body. All values refer to the critical organ, which is considered to be that organ of the body *which suffers the greatest damage through radiation.* GI, gastrointestinal tract; LLI, lower large intestine; S, stomach; SI, small intestine

Nuclide	Critical organ	$t_{1/2}$(eff) (days)	MPBB (μCi)	MPC (Ci m^{-3})	
				Water	Air
^{3}H	Body tissue	12	10^3	0.03	2×10^{-6}
^{14}C	Fat	12	300	8×10^{-3}	10^{-6}
^{24}Na	GI (SI)	0.17	7	2×10^{-3}	4×10^{-7}
^{32}P	Bone	14	6	2×10^{-4}	2×10^{-8}
^{35}S	Testis	76	90	6×10^{-4}	9×10^{-8}
^{42}K	GI (S)	0.04	10	3×10^{-3}	7×10^{-7}
^{51}Cr	GI (LLI)	0.75	800	0.02	4×10^{-6}
^{55}Fe	Spleen	390	10^3	8×10^{-3}	3×10^{-7}
^{59}Fe	GI (LLI)	0.75	20	6×10^{-4}	10^{-7}
^{60}Co	GI (LLI)	0.75	10	5×10^{-4}	10^{-7}
^{64}Cu	GI (LLI)	0.75	10	3×10^{-3}	7×10^{-7}
^{65}Zn	Total	190	60	10^{-3}	4×10^{-8}
^{85}Kr	Total				3×10^{-6}
^{90}Sr	Bone	6.4×10^3	2	$4 \times 10^{-6(g)}$	$4 \times 10^{-10(a)}$
^{95}Zr	GI (LLI)	0.75	20	6×10^{-4}	10^{-7}
^{99}Tc	GI (LLI)	0.75	10	$3 \times 10^{-3(h)}$	$7 \times 10^{-7(b)}$
^{106}Ru	GI (LLI)	0.75	3	10^{-4}	3×10^{-8}
^{129}I	Thyroid	140	3	$4 \times 10^{-6(i)}$	$6 \times 10^{-10(c)}$
^{131}I	Thyroid	7.6	0.7	2×10^{-5}	3×10^{-9}
^{135}Xe	Total				10^{-6}
^{137}Cs	Total	70	30	2×10^{-4}	$2 \times 10^{-8(d)}$
^{140}Ba	GI (LLI)	0.75	4	3×10^{-4}	6×10^{-8}
^{144}Ce	GI (LLI)	0.75	5	10^{-4}	3×10^{-8}
^{198}Au	GI (LLI)	0.75	20	5×10^{-4}	10^{-7}
^{210}Po	Spleen	42	0.03	7×10^{-6}	2×10^{-10}
^{222}Rn*	Lung	(3.8)	(0.01)	(4×10^{-6})	10^{-7}
^{226}Ra*	Bone	1.6×10^4	0.1	$10^{-7(j)}$	$10^{-11(e)}$
^{230}Th	Bone	7.3×10^4	0.05	2×10^{-5}	8×10^{-13}
^{232}Th*	Bone	7.3×10^4	0.04	2×10^{-5}	7×10^{-13}
^{233}U	Bone	300	0.05	$4 \times 10^{-5(k)}$	$2 \times 10^{-10(f)}$
^{238}U	Kidneys	15	5×10^{-3}	6×10^{-6}	3×10^{-11}
^{238}Pu	Bone	2.3×10^4	0.04	5×10^{-5}	7×10^{-13}
^{239}Pu	Bone	7.2×10^4	0.04	5×10^{-5}	6×10^{-13}
^{241}Am	Kidneys	2.3×10^4	0.1	4×10^{-5}	2×10^{-12}

RCG_a values: (a) 3×10^{-11}; (b) 2×10^{-9}; (c) 2×10^{-11}; (d) 5×10^{-10}; (e) 2×10^{-12}; (f) 4×10^{-12}.
RCG_w values: (g) 3×10^{-7}; (h) 2×10^{-4}; (i) 6×10^{-8}; (j) 3×10^{-8}; (k) 3×10^{-5}. See text.

or a certain tissue produces an annual dose equivalent of 50 mSv (5 rem), is defined as *the maximum permissible body burden*, MPBB. Table 16.7 lists MPBB values of some radioactive substances as well as the *maximum permissible concentration* (MPC) values for water (including foodstuffs) and air. A lifetime exposure to the MPC values will produce the MPBB value. The MPC_W values correspond to the ADI values (*allowable daily intake*) given for chemical substances by WHO. The MPC values will be replaced by ALI values (*annual limit of intake*) in the coming years; the differences between present MPC and future ALI values will be minimal.

Although a daily consumption of water containing a radioactive substance of MPC_W value (e.g. 10^{-7} Ci ^{226}Ra m^{-3}) should constitute no hazard, local radiological authorities rarely permit such "high" levels simply because they can be avoided through technical arrangements (cleaning systems, etc.).

TABLE 16.8. *Classification of radionuclides according to their radiotoxicity*

I.	*Very high:*	^{90}Sr, Ra, Pu
II.	*High:*	^{45}Ca, ^{55}Fe, ^{91}Y, ^{95}Zr, ^{144}Ce, ^{147}Pm, ^{210}Bi, Po
III.	*Medium:*	^{3}H, ^{14}C, ^{22}Na, ^{32}P, ^{35}S, ^{36}Cl, ^{54}Mn, ^{59}Fe, ^{60}Co, ^{89}Sr, ^{95}Nb, ^{103}Ru, ^{106}Ru, ^{127}Te, ^{129}Te, ^{137}Cs, ^{140}Ba, ^{140}La, ^{141}Ce, ^{143}Pr, ^{147}Nd, ^{198}Au, ^{199}Au, ^{203}Hg, ^{205}Hg,
IV.	*Low:*	^{24}Na, ^{42}K, ^{64}Cu, ^{52}Mn, ^{76}As, ^{77}As, ^{85}Kr, ^{197}Hg

In Table 16.7 $t_{1/2}$(eff) is the effective half-life of the radionuclide in the body, defined by

$$t_{1/2}(\text{eff})^{-1} = t_{1/2}(\text{biol})^{-1} + t_{1/2}(\text{rad})^{-1} \qquad (16.6)$$

where $t_{1/2}$(biol) is the biological half-life. For example, $t_{1/2}$(biol) is 230 d for C, 19 d for Na, 38 d for K, 130 d for I, 190 d for Sr, and 20,000 d for Ra, taking the whole body into account. For strontium incorporated in bone $t_{1/2}$(biol) is 4000 d. In the case of radium all of it is assumed to enter the bone. In Table 16.7 the critical organ is selected by weighting of two factors; it is the organ for which the nuclide has the greatest metabolic affinity and that for which the damage by the radiation from the nuclide is greatest. Some of the commonly used nuclides are listed in Table 16.8 according to their relative radiotoxicities.

Heavy elements such as radium and plutonium are concentrated in the most sensitive areas of the bone where their α-emissions provide essentially lifetime irradiation since the rate of exchange is quite small. The energy is dissipated in the small volume where the element is concentrated, which considerably increases the local biological damage. Of the uranium miners in Erzgebirge from 1875 through 1912, 25–50% (the statistics are somewhat uncertain) died by lung cancers due to inhaled radon. However, it has been found that smokers among uranium miners in the United States have an incidence of lung cancer 10 times higher than nonsmokers. Such synergistic effects seem common to cancer.

It has been suggested that the MPC$_a$ value for plutonium should be much lower because if the plutonium is concentrated to one "hot particle", the local dose will be much higher than if it is evenly distributed. However, experiments have shown the opposite: an evenly distributed activity is more hazardous than radiation in a local spot because many more cells are exposed to radiation in the former case, while in the latter case the intense local radiation kills off the cells. Dead cells are not cancerogenous, and cancer tumors need living cells around to grow.

Beta-ray emitters dissipate their energy over a somewhat larger volume than that for α-emitters, but it is still localized sufficiently to be very damaging to the tissue with which they are in contact. Elements such as sodium and sulfur represent slight hazards as their body chemistry does not tend to localize them in any particular organ and their exchange rate is high, leading to rapid elimination. Strontium and iodine, on the other hand, are localized and retained, and therefore are more hazardous.

16.11.3. *Radiotoxicity values and risks*

A common practical question with regard to a particular amount of radioactivity is how "hazardous" a substance is. Considering both the intrinsic properties and the extensive (external) conditions, we may write this hazard (Ha) as a product

$$Ha = In \times Ex \qquad (16.7)$$

The intrinsic properties are the radioactivity amount and factors which give a measure of the risk for man, as, for example, the radiotoxic properties of the particular nuclide. Suitable factors are the MPC_w, MPC_a, and MPBB values. If we choose

$$In_w = A/MPC_w \qquad (16.8)$$

where A is the amount of radioactivity, the In_w value will have the dimension m^3 water. It means that if the radioactivity A is evenly dissolved in $In_w\,m^3$ of water, the water will be acceptable for drinking with regard to radiologic health aspects. As an example we will choose ^{90}Sr. Each tonne of uranium in light water reactor used fuel elements contain 60 kCi ^{90}Sr after 10 y cooling time (see Table 20.2). The MPC_w is 4×10^{-6} Ci m^{-3} (Table 16.7); hence In_w is 15×10^9 m^3. If this particular amount of fission strontium is homogeneously dissolved in 15×10^9 m^3 of water, the ^{90}Sr will not constitute any drinking (ingestion) risk. If several nuclides are compared, the one with the largest In_w value will be potentially most hazardous. We may call In_w the "water radiotoxicity value".

We can also define an "air radiotoxicity value" In_a and a "population toxicity value" $In_p(= A/\text{MPBB})$. For our example with ^{90}Sr one calculates In_a to be 15×10^{13} m^3; this air volume is needed to reduce the 60 kCi ^{90}Sr to a safe breathing level if it all could be evenly dispersed in air. When several nuclides are present, ΣIn must be used.

In Chapter 20 we see how the In_w and In_a values may be used to describe the risks from the high active waste (HAW) of the nuclear industry. Figures 20.23 and 20.24 show how the radiotoxicity values In_w and In_a decrease with time.

However, it is not possible to draw any definite conclusions about the true risk from a certain amount of a radiotoxic substance only from its radiotoxicity value. For that purpose it is also necessary to consider its chemical form and pathways to man. These may be considered to be extensive properties. The chemical form determines its degree of solubility in water, or vapor pressure, and the particular ways in which the substance may be released to the environment. The pathway of a radioactive substance from its point of release until it reaches man is the domain of *radioecology*; extensive knowledge in this field is essential to the evaluation of the hazards caused by nuclear power. Ex may be > 1 if concentration of the radioactive substance occurs (e.g. in milk or fish) or < 1 if the substance is highly insoluble (precipitates from the water). This is usually expressed through *transfer factors*. Thus the transfer factor for ^{90}Sr and ^{239}Pu between vegetables and soil is typically 0.2 and 0.0002 pCi kg^{-1} food per pCi kg^{-1} in reservoir, respectively, while values for the same nuclides in the fish/water system are typically 1 and 40, respectively. Thus plutonium is enriched in fish but not in vegetables.

By considering all possible transfer routes, one can estimate what amount of a radionuclide released to the environment may end up in plants, animals, or man. Return to the example of ^{90}Sr for which the weight of the strontium oxide is about 1 kg. Because it is very difficult to evenly disperse this amount in air and it is unlikely to occur through natural events, the dispersion and radioecological conditions must be carefully stated in order to evaluate Ha. Ex may also contain terms for different barriers set up against the spread of the substance.

If the amount and chemical composition of a radioactive substance taken in by man are known, tabulated dose factors may be used for calculating body doses (see ICRP publications). A few such values are given in the following table, where the first value refers to intake through food or water (w), and the second intake from breathing (a):

^{90}Sr	1.5×10^6 (w);	2.3×10^6 (a) rem Ci^{-1}
^{129}I	3.4×10^5	1.9×10^5
^{137}Cs	5.5×10^4	3.8×10^4
^{233}U	1.1×10^5	2.7×10^6
^{239}Pu	1.6×10^5	9.5×10^8

These values take into account the retention and metabolism of the radionuclide in the body. Thus if a person drinks 440 l of water a year containing 5×10^{-5} Ci ^{239}Pu m^{-3} (MPC_w value, Table 16.7) he will receive a dose of $1.6 \times 10^5 \times 5 \times 10^{-5} \times 0.44 = 3.52$ rem y^{-1}.

The *Ex* value depends on ecological conditions, living habits, etc., and is subject to large uncertainties. In some cases the risk may be a perpetual one. A natural such risk is due to the radium content of drinking water; an unnatural one to any traces of plutonium in the drinking water. In other cases the risk may be temporary. *Ha* may be calculated for a certain accident (e.g. a spill of radium in a hospital, release of fission products after a reactor excursion or fallout from nuclear weapons) and for different population groups. The *Ha* value may then include the probability *P* of occurrence. Many decisions between alternative nuclear technologies today are based on estimates of such values. This is discussed further in §21.2.

16.11.4. *Classifications, working rules, etc.*

Radioactive substances of various activities, concentrations (or specific activities), decay modes, etc., constitute quite different hazards, and must be handled accordingly. Various countries classify radioactive material differently and issue different working rules. We give a few classifications and rules, adhering primarily to the recommendations of the IAEA.

In Fig. 16.10 the international symbol for ionizing radiation and radiation sources is shown in several places. The symbol is either black or red against a yellow background. The symbol, which also comes on tape, should not be misused, e.g. for repairing students' textbooks.

The data in Table 16.7 refer to the concentrations and amount of radionuclide which must be avoided in terms of biological harm. If much smaller amounts than these levels are spilled in a scientific laboratory, the resulting increase in the radiation background may not be a hazard but can make the results of scientific investigation with tracer quantities uncertain. It is, therefore, important that radiochemical laboratories be maintained as free of radioactive contamination as possible.

In the US, *Radiation Concentration Guide* (RCG) values are applied instead of the ICRP values. Usually RCG_w and RCG_a are one-tenth the corresponding MPC_w and MPC_a values in Table 16.7; considerable deviations from this rule for given nuclides are indicated in the footnotes of the table.

Nuclear chemistry research laboratories are usually divided into low, medium, high, and α-active areas (Table 16.9). In low active areas, samples $\lesssim 100\,\mu$Ci β-γ emitters are handled. This is sufficient for tracer work and such samples do not involve any health risk, except for nuclides of radiotoxicity class I. Often such work can be carried out in regular chemistry laboratories without any restrictions. This is not the

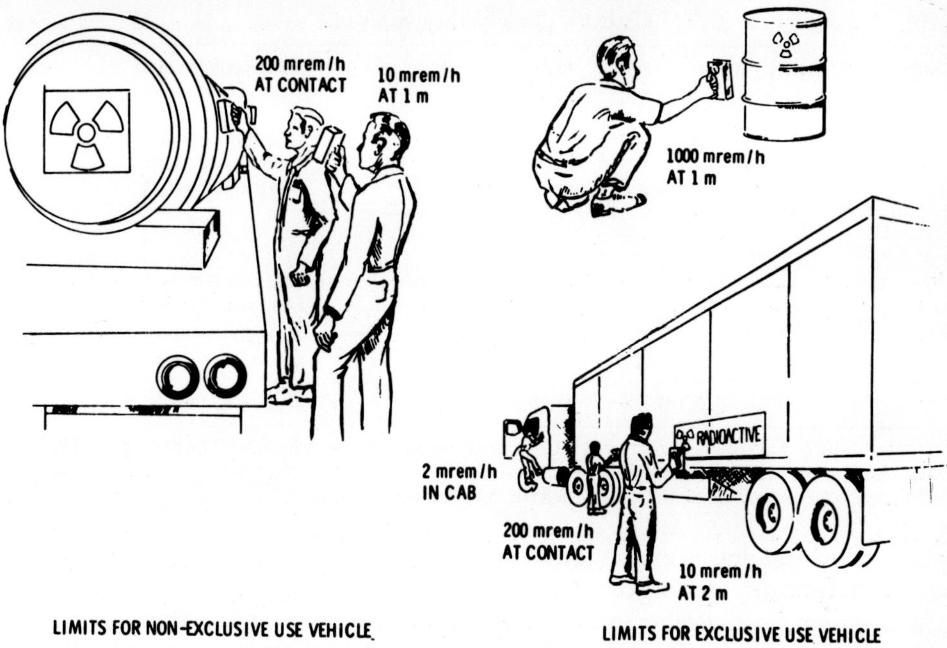

FIG. 16.10. Radiation dose limits for transportation in the United States.

case for medium level and high level ($\gtrsim 0.1$ Ci), or α-radioactive work, where special precautions must be taken.

In laboratories related to the nuclear power industry high activity usually refers to > 1000 Ci, and low activity to $\lesssim 0.1$ Ci. Samples of activities $\lesssim 1$ μCi are only handled in "ultrapure" laboratories.

In all work with radionuclides, radioactive waste is produced. Table 16.10 shows a suggested classification of such wastes according to IAEA, and the common sources of such wastes. It is common practice to let the MPBB value correspond to the amount of radioactivity which may be disposed of through common laboratory procedures. For example, if the substance is easily soluble, the MPBB value may be disposed of by normal flushing to the sewer with several liters of water. However, if the radionuclide has a half-life greater than 0.5 y, only 0.1 of the MPBB value may be disposed of through normal

TABLE 16.9. *Recommended working conditions for radionuclides of different hazards*
For inexperienced personnel, one-tenth of the given values should be applied. A refers to a low activity laboratory; B, medium active laboratory with good ventilation, and C, high activity laboratory (hot laboratory) with complete enclosure of the working area

Working space	Maximum amount of activity (mCi)		
	α	β	γ
A. Open laboratory bench (only if $t_{1/2} < 14$ d and the sample is dust free)	None	$\lesssim 0.1$	$\lesssim 0.1$
B. Fumehood with stock solutions behind lead shielding	0.001	5	5
Fumehood with frontal shield		300	100
C. Simple manipulator cell, ~ 10 cm Pb		~ 300	~ 1000
Advanced manipulator cell of the master–slave type		> 300	> 1000
Gloveboxes	> 0.1		

TABLE 16.10. *Classification of radioactive wastes*

Category	Activity ($Ci\ m^{-3}$)	Daily terminology	Radioactive source	
1	$\lesssim 10^{-6} (\lesssim 1\ \mu Ci\ m^{-3})$	Low active[a] up to $\sim 10^{-2}\ Ci\ m^{-3}$ (categories 1–2)	Uranium production Radioisotope uses[b]	
			Normal release	Nuclear power stations
2	10^{-6}–10^{-3}			
3	10^{-3}–10^{-1}	Medium active up to $\sim 10^{3}\ Ci\ m^{-3}$ (categories 3–4)	Cleaning circuits Solidified waste	Nuclear power stations
4	10^{-1}–10^{4}		Normal release (→ low active) Cleaning circuits	Reprocessing plants
5	$> 10^{4} (> 10\ kCi\ m^{-3})$	High active	High active waste	Reprocessing plants

[a] Substances with a specific activity $\lesssim 1\ n\,Ci\ g^{-1}$ may be classified as nonactive. The average radioactivity of the earth's crust is about 0.1 $pCi\ g^{-1}$.

[b] Under some circumstances the specific activity may be much higher.

sewage lines. Amounts in excess of those that may be disposed of in this fashion should be collected and disposed of according to the rules of the local radiation authorities.

Wastes from research laboratories are usually divided into α- and medium β, γ-emitters and packed in separate drums, which are transported to government-owned collection areas for storage or further treatment. Some procedures are discussed in Chapter 20 in connection with the handling of the large volumes of wastes from the nuclear energy industry. Figure 16.10 shows the radiation dose limit in the United States for transportation of radioactive materials.

16.12. Protective measures

Three basic principles are recommended for keeping radiation exposure to a minimum: shielding, control, and distance.

Radiation shielding has been discussed in Chapter 14. For α-emitters the main hazard is internal, not external, to the body. For solutions of nuclides which emit only β-particles, the glass walls of the container usually provide sufficient shielding. Sheets of glass or plastic (such as lucite) are commonly used to shield exposed solid samples. Because of the high penetrating power of n, γ-, and X-rays, thick layers of concrete, water, steel, or lead must be used.

Since the intensity falls off as the inverse square of the distance, isolation of unused nuclides and maintaining of maximum distance (by use, when necessary, of remote control apparatus such as tongs) in working with moderate levels of activity reduce the exposure appreciably.

To minimize the possibility of ingestion as well as the chances of ruining experiments through accidental contamination, limiting the radioactive work to a minimum area is essential. The ideal is no radioactivity except in the immediate working area, and even there, upon completion of the particular experiment, all activity should be removed and the area thoroughly cleaned (*decontaminated*) if necessary. For low levels this means working in a good hood with easily cleaned, nonporous surfaces.

One operation which commonly results in contamination involves evaporation of a solution to dryness either on a hot plate or under a heat lamp. Although the percentage of the sample carried away by the spray may be very small, it nevertheless may result in appreciable amounts of activity being spread around the area of the evaporation. Consequently, all evaporation should be performed in a hood and the vicinity should be protected from the active spray by a covering such as an asbestos pad or absorbent paper.

For moderate activity levels a glove box under slightly reduced pressure provides a simple and convenient closed chemical laboratory. Such a box of simple design can be constructed for a modest sum from ordinary supplies, and is an excellent means of confinement and containment of radioactive substances. Of course, it does not provide shielding from γ-rays, but if these have sufficient intensity to make shielding necessary, a modified glove box can be inserted in a lead shell. Operations must be carried out by remote control in this case.

If a radiochemical laboratory is designed properly and the work performed in such a manner that the general background contamination is sufficiently low enough to do valid low level tracer experiments, then the health aspects of radiation control is quite satisfactory. The regulations suggested in this chapter are designed to help ensure this and strict adherence is strongly recommended in all instances.

16.13. Radiochemical laboratory

The working surfaces and floors in a radiochemical laboratory should be even, nonporous, and with a minimum of seams. Fume hoods with porcelain, wooden benches, ceramic troughs, and pipes are unsuitable. Surfaces of plastic material, stainless steel, and artificial stone are acceptable bench materials. For flooring, linoleum has some advantages over vinyl, while sheets of flooring are preferred over squares. It is recommended that fume hood and benches be covered with an absorbing material such as absorbing paper, and that experiments be conducted when possible in trays of stainless steel or plastic. Such arrangements make sure that no radioactive material contaminates a larger area if a spill occurs. Moreover, it is much easier to recover expensive radioisotopes if they are contained in a spill. Wet and dry radioactive wastes are collected separately in closed vessels (Fig. 16.11).

In the design of the radiochemical laboratory it is important that airborne contamination be prevented from spreading to counting rooms and to offices. Therefore, a pressure difference between the laboratories and the other areas is desirable. The air velocity in the fume hoods should never fall below 0.25 m s^{-1}, and 0.5 m s^{-1} is recommended. With such a flow velocity, radioactive dust and fumes are retained in the hood and removed through the vents. The fume hoods should have filters for collecting radioactive particulates.

To reduce the danger of contamination of the laboratory, radionuclides with half-lives greater than 14 d should not be worked with in an open laboratory but only in a fume-hood. For half-lives of less than 14 d the amount of radioactive material used on an open bench should be limited to 100 μCi, an amount adequate in most cases for tracer experimentation but which eliminates any irradiation hazards. Workers in the laboratory should always wear surgical gloves and laboratory coats. If there is any danger of splashing, plastic facehoods are recommended. It is extremely important that oral contamination be avoided: thus pipetting by mouth is strictly forbidden. Furthermore, application of cosmetics, smoking, eating, etc., should not be allowed in the laboratory.

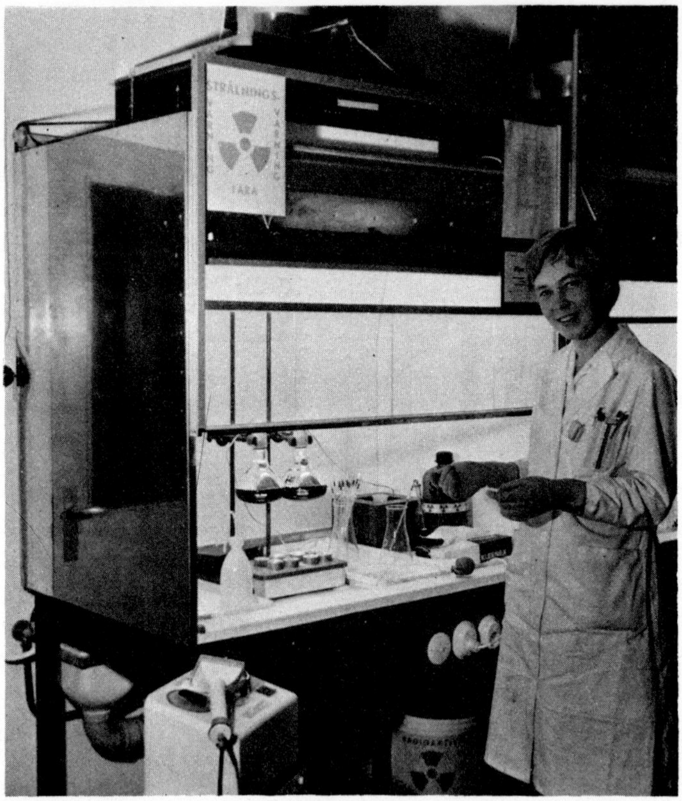

FIG. 16.11. Radiochemical laboratory. The fume hoods are made of stainless steel painted on the inside with removable plastic paint, and built-in filters (above) and plastic trays in the hood. The symbol of radiation hazards is mounted on the fume hood front and on the bottles for wet waste and on the can below the fume hood for dry waste. The worker wears surgical gloves and pen and film dosimeters.

16.13.1. *Medium level β-γ laboratory*

Beta-radiation from radioactive sources has ranges which rarely exceed 1 g cm^{-2}. Consequently, in a laboratory in which the level of β-emission is less than 300 mCi, protection from the radiation can be achieved with a 1 cm plexiglass shield. For γ-emitters heavier shielding must be used due to the greater penetrating power. Work with these samples can be conducted by the use of long tongs or manipulators which pass through lead shields (Fig. 16.12). The eyes can be protected in these cases by the use of lead glass windows of high density or by the use of mirrors.

16.13.2. *High level β-γ work*

In work with higher activities of β- and γ-emitters, special shielded cells must be used. These cells must be sealed from the atmosphere and kept at a pressure lower than that for the working personnel. The smallest cells usually have lead walls which sometimes reach all the way to the ceiling; through this arrangement no scattered radiation reaches the working personnel (Fig. 16.12). Viewing into the cells is facilitated with mirrors, periscopes, lead-glass windows, or windows filled with a heavy liquid (e.g. $ZnBr_2$ solution

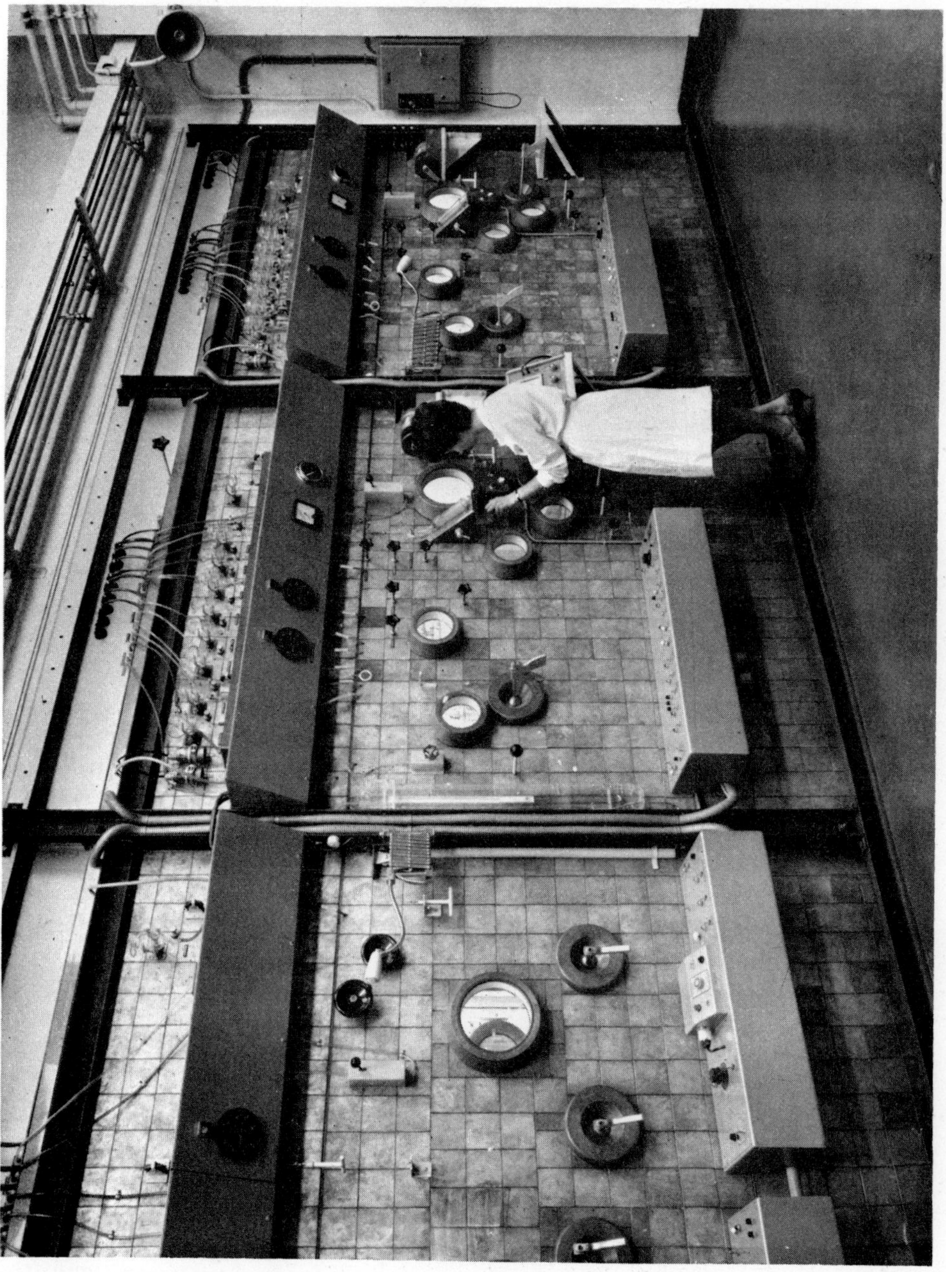

FIG. 16.12. Lead cell for medium and high activity $\beta-\gamma$ work. (According to Service Nucléaire, Saint-Gobain, France.)

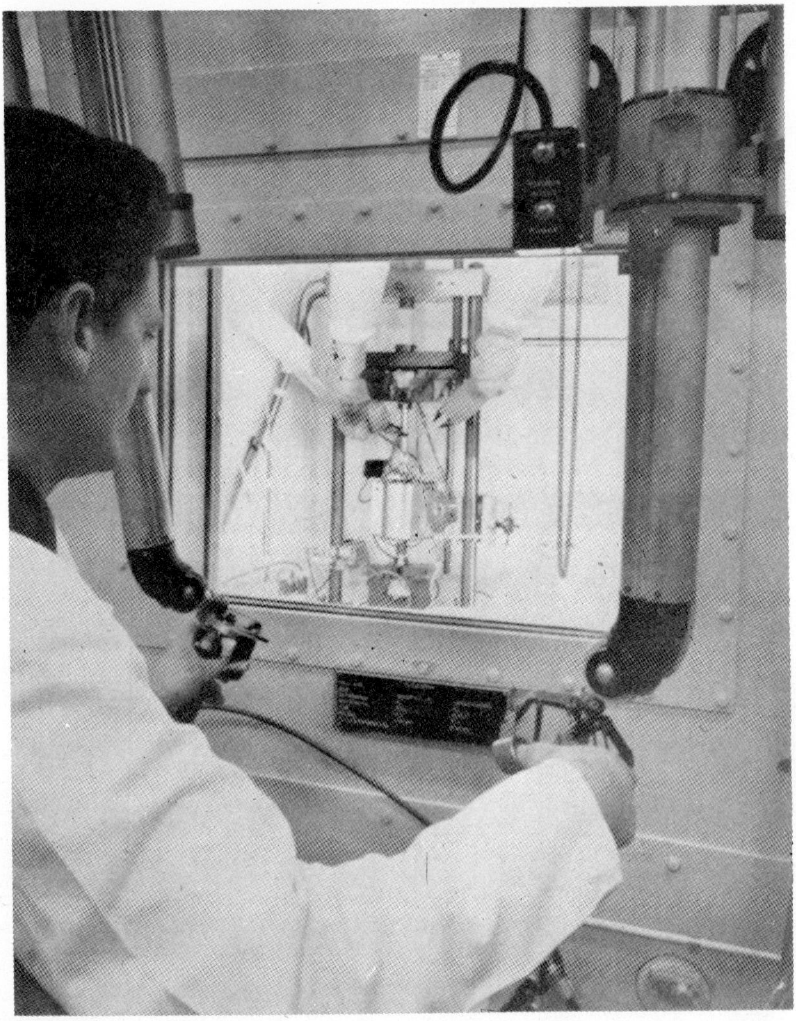

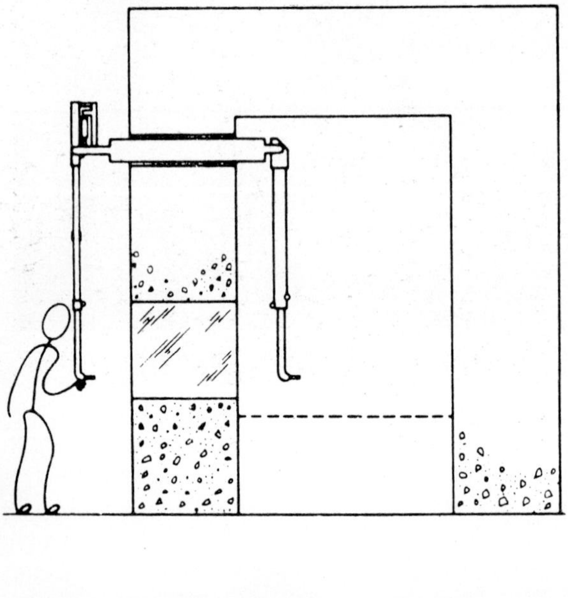

FIG. 16.13. High active work with master–slave manipulators. Wall thickness may exceed 1 m. Viewing is through lead-glass or—as in this picture—through a few centimeters thick radiation-resistant glass windows filled in between with a high density $ZnBr_2$ solution. The concrete cells are usually arranged in a row with five or more cells.

whose density is 2.5 g cm^{-3}). Experiments are carried out with the use of tongs passing through the lead walls or, for the thicker cells, with manipulators reaching above or through the walls. In cells for very high activities (> 1.000 Ci) these manipulators are either guided electrically or of the *master–slave* kind (Fig. 16.13). In the latter all movements of the operator are copied exactly by the slave hands inside the cell.

For very complicated work and extremely high radioactivity, robots have been constructed which can be guided to carry out repairs of such items as unshielded nuclear reactors, heavily contaminated radiochemical apparatus, radionuclide generators, etc. In smaller cells control of valves, of motors, etc., is always carried out remotely with the use of mechanical, electrical, pneumatic, or hydraulic couplings.

16.13.3. *Alpha-laboratory*

The actinide elements are of direct importance to nuclear technology. These elements are produced in high specific radioactivities. Since they are very harmful to the human body, great care must be exercised in working with them. Because these α-emitters usually are associated with weak γ-radiation or X-rays, chemical work with these elements is normally conducted in *gloveboxes* (Fig. 16.14).

The boxes are always kept at a pressure slightly below the surrounding atmosphere by circulation of pure air or inert gas through the box. The boxes may have an electrical

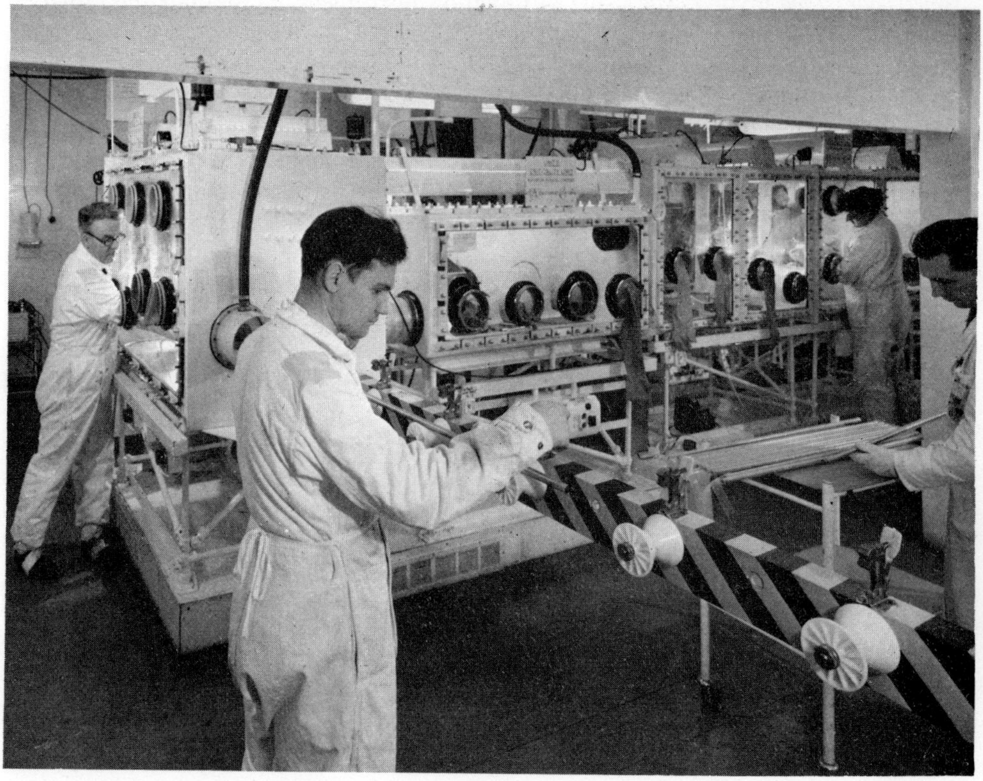

FIG. 16.14. Glovebox facility for making mixed oxide uranium–plutonium reactor fuel material. The fuel pins are seen outside the boxes, while filling is done from inside the boxes. (The Atomic Energy Authority, Windscale, UK.)

alarm system for monitoring hazards such as interrupted water circulation, electrical short circuits, oxygen in the protective gas, heat, etc. In Fig. 16.14 the control panel for these protective arrangements is shown above the box.

In such a laboratory the main hazard is radioactive dust. All room surfaces should, therefore, be made with as few seams and sharp corners as possible; particularly the floor must be of high quality. Electric power, water, waste, ventilation, etc., is connected from piping in the ceiling. The air into the laboratory passes through filters as does the air exiting from the laboratory. The exit air is monitored for α-activity. Entering and leaving the laboratory may occur via airlocks, and the hands and feet are always monitored for activity on exit. All these protective arrangments make an α-laboratory quite expensive, but smaller laboratories working with lower levels of α-activity can be constructed in simpler fashion for correspondingly less cost.

16.14. Control of radiation protection measures

In larger organizations the control of radiation hazards is the responsibility of specialists known as *health physicists* or *health chemists*, whose main duty is to ensure that work is carried out without hazard to the health of the people involved.

This protection follows three stages: *prevention*, *supervision*, and *after-control*. Preventive measures include use of radiation shielding, tongs, etc., as discussed above. The *supervision stage* involves the using of radiation instruments to monitor the radiation level (see Chapter 17). Small pocket pen and film dosimeters are used for individual monitoring (§15.9). For spills and contamination of hands, shoes, etc., *special contamination instruments* are used which are more sensitive than the monitoring instruments.

Contamination in the laboratory must be avoided. This is controlled by smearing tests, e.g. a filter paper is wiped over the surface and the paper is checked with a contamination instrument. In a so-called "clean area" the fixed contamination should not exceed 100 dpm for α and 200 dpm for β-γ on a surface of 100 cm^2. For an "active area" the rules are a maximum of 1000 dpm for α per 100 cm^2 and 1 mrad/h from β-γ at a distance of 2 cm from the surface. Radioactive aerosols are monitored by air sampling in which a certain amount of air is drawn through a fine filter paper after which the paper is counted.

The *after-control* usually consists of checking film dosimeters of the personnel and of medical examination. Depending on the kind of work, the film dosimeters are checked from twice a week to once a month. A medical examination is normally given once or twice a year. In the medical examination the fingertips are checked since if a high dose has been received it can make the skin smooth and free of fingerprints. At even higher doses the skin becomes harder and cracked. If danger of inhalation of α- or soft β-emitters exists, urine samples are analyzed. Such analyses are very sensitive and much less than a μCi in the body is easily detected. For workers who handle hazardous amounts of α-emitters (e.g. plutonium in more than milligram amounts), urine samples are taken regularly.

If necessary whole-body counts are also taken. *Whole-body counting* is carried out with the subject being surrounded by numerous scintillation or solid state detectors in a heavily shielded room. The natural body content of ^{40}K is easily detected by this technique.

16.15. Exercises

16.1. "Reference man" consists of 18% C, 66% H_2O, 0.2% of K per body weight. He may also have accumulated 10 pCi ^{226}Ra in the body. Assume 0.3 decay for each of the following 5 daughters and calculate for a body weight of 70 kg the total number of radioactive decays per unit time. Which radionuclide is the dominating one?

16.2. Using the information above, how many grams of the body's molecules (assume average mole weight of 10^5) will be destroyed in a year if the *G* (molecular destruction) value is 3?

16.3. Under the same assumption as above, what amount of destruction will be caused by cosmic radiation? Assume that the cosmic particles produce 3000 ion pairs s^{-1} cm^{-3} of the body.

16.4. With the information in exercise 16.1, calculate annual doses received from (a) ^{40}K, and (b) ^{226}Ra and daughters.

16.5. Ten mg ^{238}U has been collected in the kidneys. Considering the biological half-life of uranium and assuming only one α-emission in ^{238}U decay, calculate the dose (in rem) received by the organ if the uranium is evenly distributed. The weight of a kidney is 150 g.

16.6. A γ-dose rate of 100 rem is assumed to inactivate (kill) human cells. The body contains 6×10^{13} cells in a cell weight of 42 kg for a 70 kg man. (a) What average energy (in eV) has to be deposited in a cell to kill it? (b) Calculate the number of kidney cells destroyed for the dose received in exercise 16.5. For simplicity assume the cells to be cubic with a side length of about 11 μm.

16.7. A tumor has the weight of 80 g and we wish to destroy 20% of the cells by irradiating with 180 MeV protons with such penetration that 10% of the energy is deposited in the tumor. The particle beam is 5 μA. For what time must the irradiation be? A cell of weight 10^{-9} g is assumed to be killed on the absorption of 200 keV and no cell is assumed to be killed twice.

16.16. Literature

Recommendations of ICRP, esp. publ. 2, Report of Committee 11 on Permissible Dose for Internal Radiation, Pergamon Press, Oxford, 1959.

E. H. Schultz, *Vorkomnisse und Strahlenunfälle in kerntechnischen Anlagen*, Karl Thiemig, 1965.

Z. M. Bacq and P. Alexander, *Fundamentals of Radiobiology*, Pergamon Press, Oxford, 1967.

K. Z. Morgan and J. E. Turner (eds.), *Principles of Radiation Protection*, J. Wiley, 1967.

IAEA, *Radiosterilization of Medical Products*, Vienna, 1967.

A. P. Casarett, *Radiation Biology*, Prentice-Hall, 1968.

H. Dertinger and H. Jung, *Molekulare Strahlenbiologie*, Springer-Verlag, 1969.

K. I. Altman, G. B. Gerber, and S. Okada, *Radiation Biochemistry*, Academic Press, 1970.

BEIR Report, *The Effects on Populations of Exposure to Low Levels of Ionizing Radiation*, NAS-NRC, 1972.

UNSCEAR, *Ionizing Radiation: Levels and Effects*, UN, New York, 1972.

A. Martin and S. A. Harbison, *An Introduction to Radiation Protection*, Chapman & Hall, 1972.

H. Kiefer and R. Maushart, *Radiation Protection Measurement*, Pergamon Press, Oxford, 1972.

M. Eisenbud, *Environmental Radioactivity*, 2nd edn.. Academic Press, 1973.

IAEA Safety Series, especially No. 38, *Radiation Protection Procedures*, 1973.

C. W. Mays, Cancer induction in man from internal radioactivity, *Health Physics* **25** (1973) 585.

M. Oberhofer, *Safe Handling of Radiation Sources*, Karl Thiemig, 1974.

IAEA, *Population Dose Evaluation and Standards for Man and his Environment*, Vienna, 1974.

E. Schüfer (ed.), *Strahlung und Strahlungmesstechnik in Kernkraftwerken*, Elitera, Berlin, 1974.

E. Pochin, A. S. McLean, and L. D. G. Richings, *Radiological Protection Standards in the United Kingdom*, HMSO, 1976.

K. K. Manocha and R. K. Mohindra, Estimate of natural internal radiation dose to man, *Health Physics* **30** (1976) 485.

G. Walinder (ed.), *Tumorigenic and Genetic Effects of Radiation*, National Swedish Environmental Protection Board, 1976.

IAEA, *Nuclear Science and Technology in Food and Agriculture*, Vienna, 1977.

ICRP publication 26 (Radiation Protection), Recommendations of the International Commission on Radiological Protection, *Annals of the ICRP* **1** (3) (1977), Pergamon Press, Oxford.

UN, *Sources and Effects of Ionizing Radiation*, United Nations Publ. No. E. 77. IX.1, New York, 1977.

CHAPTER 17

Detection and Measurement of Nuclear Radiation

Contents

Introduction

Although animals have no known senses for detection of nuclear radiation, it has been found that sublethal but large radiation fields can affect animals in various ways such as disturbing the sleep of dogs or causing ants to follow a new pathway to avoid a hidden radiation source. Apollo astronauts observed scintillations in their eyes when their space ship crossed very intense showers of high energy cosmic rays. People who have been involved in criticality accidents experiencing high intensities of n and γ have noted a fluorescence in their eyes and felt a heat shock in their body.

However, we are not physiologically aware of the normal radiation fields of our environment. In such low fields we must entirely rely on instruments, some of which can be very simple. For example, a Geiger–Müller (GM) counter can be built by a handy student from easily available material (cost $\lesssim$\$100) in less than a day.

The ionization and/or excitation of atoms and molecules when the energies of nuclear particles are absorbed in matter is the basis for the detection of individual particles.

Macroscopic collective effects, as chemical changes and heat evolution, can also be used. The most important of the latter have been described before because of their importance for dose measurements (e.g. the blackening of photographic films and other chemical reactions, excitation of crystals (thermoluminescence), and heat evolved in calorimeters; §15.9).

In this chapter we consider only the common techniques used for detection and quantitative measurement of *individual* nuclear particles. We also discuss the problem of proper preparation of the sample to be measured as well as consideration of the statistics of the counting of nuclear particles necessary to ensure proper *accuracy* (i.e. agreement between measured and true value).

17.1. Track measurements*

The most striking evidence for the existence of atoms comes from the observation of tracks formed by nuclear particles in cloud chambers and in photographic emulsions. The tracks reveal individual nuclear reactions and radioactive decay processes. From a detailed study of such tracks, the mass, charge, and energy of the particle can be determined.

The tracks formed can be directly observed by the naked eye in cloud and bubble chambers, but the tracks remain only for a few seconds before they fade. For a permanent record we must use photography. On the other hand, in solid state nuclear track detectors (SSNTD), of which the photographic emulsion is the most common variant, the tracks have a much longer lifetime. Because of the much higher density of the absorber, the tracks are also much shorter and often therefore not directly visible. Thus the microscope is an essential tool for studying tracks in solid material.

17.1.1. *Cloud and bubble chamber*

The simple principle of a *cloud chamber* is shown in Fig. 17.1. A volume of saturated vapor contained in a vessel is made supersaturated through a sudden adiabatic expansion. When ionizing radiation passes through such a supersaturated vapor the

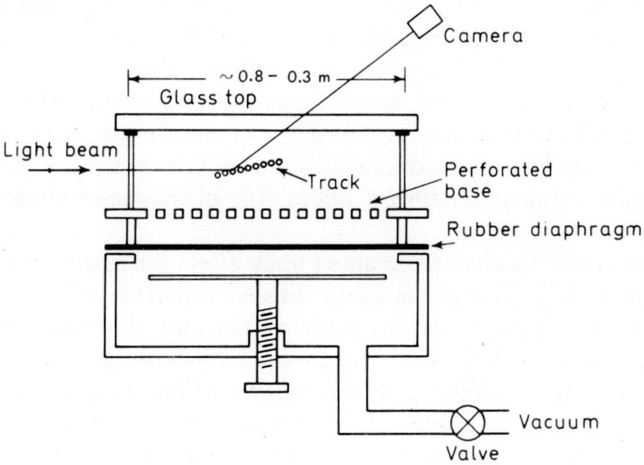

FIG. 17.1. Principle of a cloud chamber.

ionization produced in the gas of the vapor serves as condensation nuclei. As a result small droplets of liquid can be observed along the path of the radiation. These condensation tracks have a lifetime of a few seconds and can be photographed through the chamber window. The density of the condensation depends on the ionization power of the projectile as well as on the nature of the vapor, which is often alcohol or water. Cloud chamber photographs are shown in Fig. 14.4.

In a similar fashion *bubble chambers* operate with superheated liquids in which gas bubbles are produced upon the passage of ionizing radiation. The most commonly used liquid in bubble chambers is hydrogen, and, as a consequence, the chamber must be operated at quite low temperatures (23°K for H_2). Since the liquid medium in a bubble chamber is much denser than the vapor medium in a cloud chamber, the former are more suitable for studies of reactions of more energetic projectiles. The high energy p–p reaction shown in Fig. 5.6 has been recorded in the 0.8 m diameter bubble chamber at Saclay in France; see also Fig. 14.17.

17.1.2. *Solid state nuclear track detectors* (*SSNTD*)

The main type of SSNTD (or DTD, for dielectic track detector) are photographic-type emulsions, crystals, glasses, and plastics. Because the density of these materials is much higher than for the previous group (§17.1.1), nuclear particles can spend all their kinetic energy in these detectors, allowing the identification of the particle. Since the SSNTD retains the particle path, they can be used to record reactions over a long time period. These advantages have made SSNTD especially valuable in the fields of cosmic ray physics, radiochemistry, and earth sciences.

Nuclear emulsions are similar to optical photographic emulsions. They contain AgBr crystals embedded in gelatin to which small amounts of sensitizing agents have been added. The AgBr content is as much as four times (i.e. 80% AgBr) greater than in optical film. Also the crystals are much smaller (developed grain 0.1–0.6 μm) and well separated. The emulsions come in thicknesses from a few μm up to 1 mm. Nuclear radiation passing through the emulsion causes ionization and excitation which activates the AgBr crystals, producing a latent image of the particle path. Upon development the activated crystals serve as centers for further reduction of silver, leading to visible grains. It is assumed that at least 3 silver atoms must be activated to produce a visible grain, while about 30 atoms are needed for normal blackening. Each activated grain seems to require about 2.5 eV to be absorbed on the average.

While the memory effect of the developed film is almost infinite, this is not the case for the latent image which slowly fades, depending on the number of originally activated silver atoms, the film type and external conditions like temperature, humidity, etc. When stored under ambient conditions, about 80% of the latent image disappears in half a year.

The developed grains form an interrupted track along the original path of the energetic particle (Fig. 17.2(a)). The specific energy loss of the particle, dE/dx (i.e. the stopping power of the absorber), depends on the mass, charge, and velocity of the particle, and on the composition of the absorber. From the track length, grain density (i.e. grains per track length), and gap length between the grains, the particle and its energy can be determined (cf. §§14.3 and 15.1). For a given particle, the range $\hat{R}$ is proportional to the energy as shown in Table 17.1. The range decreases with increasing mass of the particle and density of the absorber. The grain density depends on the specific ionization of the

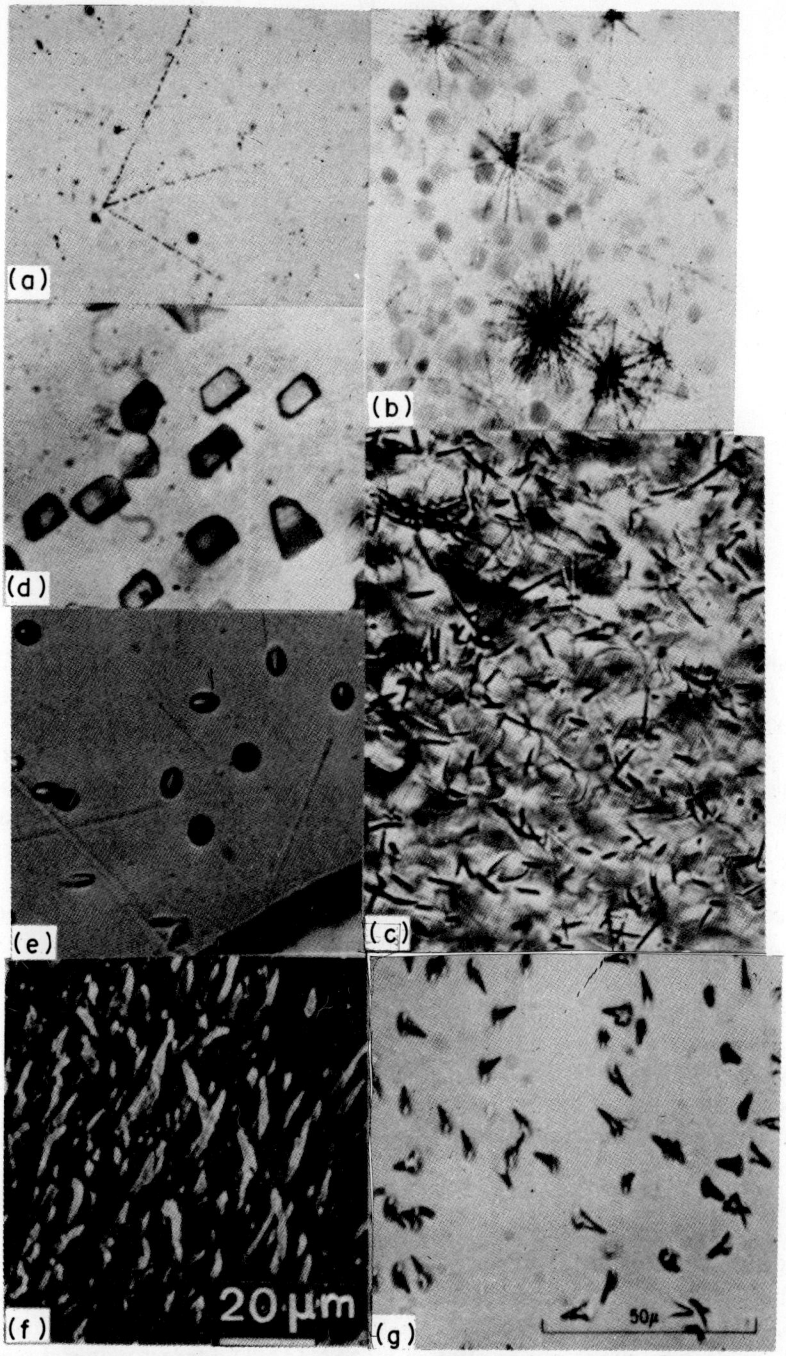

FIG. 17.2. Tracks of highly ionizing particles in solid absorbers. (a) Alpha-tracks originating in the same point in a nuclear emulsion. (According to P. Cuer.) (b) Autoradiography of a lung showing depositions of inhaled plutonium. (According to B. A. Muggenburg.) (c) Fission tracks in neutron-irradiated apatite containing some evenly distributed uranium. (According to E. I. Hamilton.) (d) Neutron-induced fission tracks in muscovite mineral. (According to Hamilton.) (e) Neutron-induced fission tracks in volcanic glass recovered from deep sea sediments. (According to J. D. Macdogall.) (f) Fission tracks in mineral from the Oklo mine. (According to J. C. Dran *et al.*) (g) Fission tracks from ^{252}Cf in Lexan polycarbonate. (From Fleischer, Price, and Walker.)

TABLE 17.1. *Range of energetic high-ionizing particles in various solid materials*

Particle	Energy (MeV)	Absorber (density)	Range (μm)
H	10	Ilford C2 (3.8)[a]	540
T	10	Ilford C2 (3.8)[a]	230
^{4}He	10	Ilford C2 (3.8)[a]	57
^{4}He(RaC′)	7.7	Eastman NTA (3.6)	38
		Mica (3.1)	36
		Glass (2.5)	41
		Water (1.0)	60
^{4}He(UI)	4.2	Mica (3.1)	13
^{4}He(U-series)		Pitchblende (7.0)	23[b]
		Carnotite (4.1)	32[b]
Light fiss. fragm. ^{235}U	} ~ 150	Eastman NTC (~ 3.4)	14 } ~ 25
Heavy fiss. fragm. ^{235}U		Eastman NTC (~ 3.4)	10.5
Both fiss. fragm. ^{238}U	~ 160	Leopoldite (~ 4)	~ 20

[a]Density of AgBr 6.47; of gelatin 1.31.
[b]Range of the predominating α-particles.

particle which does not vary linearly with the particle energy (or velocity), as seen from Fig. 14.6; thus the grain density changes along the track, being highest close to its end.

Other solid material may be used as SSNTD instead of AgBr emulsions: *plastics* (cellulose nitrate and polycarbonate films), *glass*, *crystals*, etc. In order to make the tracks visible in the microscope the surface of the SSNTD must be polished and etched, usually with alkali. The technique is very much of an art, each laboratory using its own recipe.

Because of the natural radiation background, every SSNTD has a memory of past nuclear events, which must be erased as far as possible before a new exposure. In nuclear emulsions an α-radiation background of 20–60 tracks cm^{-2} per day is normal. The technique of *background eradication* prior to exposure may consist of treating an emulsion with chromic acid or heating (annealing) a glass plate. Because this technique more easily removes weak images, it may also be used after exposure, e.g. to remove fainter α-tracks from heavier fission tracks.

Let us consider some examples of uses of SSNTD. Tracks obtained under various conditions are shown in Fig. 17.2.

(i) As mentioned, SSNTD has been used in cosmic ray experiments at high altitudes and in space journey where memory effect and simple construction make them especially suitable. Many elementary particles have been discovered by this technique, notably the π- and μ-mesons. Figure 5.3 shows tracks of high energy cosmic ray particles, probably iron atoms, which have been stopped in Apollo astronaut helmets.

(ii) Nuclear reactions can be studied by SSNTD. The target material may either be regular atoms of the detector (H, O, Ag, Si, etc.) or material introduced into the matrix, e.g. thin threads of target metals or uranium atoms. The former have been used in high energy physics for hadron-induced reactions, and the latter for studying fission processes. From experiments with uranium the frequency of spontaneous fission of ^{238}U has been determined, and also the rate of ternary fission and emission of long range α's.

(iii) When emulsions are dipped into solution, some of the dissolved material is soaked up or absorbed in the emulsion. For example, if the solution contained samarium, some α-tracks of its spontaneous decay (decay rate 127 dps g^{-1}) appear in the emulsion (cf. Fig. 17.2(a)). Since ^{147}Sm has an isotopic abundancy 15%, its half-life is calculated to be 1.1×10^{11} y. The lower limit of detection is about 500 tracks cm^{-2} d^{-1}, so quite low decay rates can be accurately measured, making this a valuable technique for determination of long half-lives.

(iv) ^{222}Rn is released through the earth's surface from uranium minerals. The amount released varies not only with the uranium content and mineral type, but also with the time of the day; variation from 30 pCi l^{-1} to 2000 has been registered during a 24 h period. To avoid this variation, cups containing a piece of plastic TD are placed upside down in shallow (0.5–1.0 m) holes for about 3 weeks, after which the SSNTD are etched and α-tracks from radon counted. Mineral bodies hundreds of meters underground can be mapped with this technique in great detail in a reasonably short time. The US Geological Survey uses the same technique to predict earthquakes; it has been observed that just before earthquakes the radon concentration first increases, then suddenly decreases, the minimum being observed about one week before the earthquake appears.

(v) Fission fragments make heavy tracks in all solid material. The tracks are short and thick; in a crystal material like zircon (a common mineral of composition $ZrSiO_4$) they may not be more than 10^{-2} μm (100 Å) in diameter, and 10–20 μm in length. They are therefore not visible even in the best optical microscopes. Using scanning electron microscopy, it has been found that the hole formed retains the crystal structure or regains it (Fig. 17.2(d)). On the other hand, if the track is formed in glass, a gas bubble appears instead of a track (Fig. 17.2(e)); these slightly elongated bubbles can be distinguished from other completely spherical bubbles formed by other processes. To make the tracks visible, the specimen is embedded in a resin, then one surface is ground and carefully polished after which it is dipped in an acid, e.g. HF. Because of defects in the crystal structure along the fission track, the track and its close surroundings are attacked by the acid, and the diameter of the track increases a hundredfold to a micron or so. The tracks are then visible under a microscope with a magnification of 500 to 1000.

This procedure has been used as an analytical tool for determination of uranium and plutonium in geological and environmental samples. In this technique, the sample (either a ground and polished surface of a mineral, or a dust sample on tape) is firmly pressed against a photographic film, and the package is irradiated by slow neutrons. From the fission track count of the developed film the uranium or plutonium content can be calculated. Thus a Swedish shale was found to contain 4 ± 1 ppm U, and a bottom sediment in a Nagasaki water reservoir 12 ± 1 pCi ^{239}Pu per kg sediment. In the latter case, the ratio between the number of fission tracks N_{ft} and α-tracks $N_{\alpha\mathrm{t}}$ from ^{239}Pu is

$$N_{\mathrm{ft}}/N_{\alpha\mathrm{t}} = \sigma_{\mathrm{f}}\phi/\lambda_{\alpha} \tag{17.1}$$

This technique is very useful for routine measurements of fissionable material in very low concentrations. Figure 17.2(c) shows fission tracks in uranium-containing mineral which has been exposed to neutrons.

In §12.4.5 we described fission track counting for dating geological samples.

17.2. Constant current ion chambers*

The ionization chamber is a gas-filled space between two electrodes. In Fig. 17.3(a) the electrodes are two parallel plates, but another common geometrical arrangement

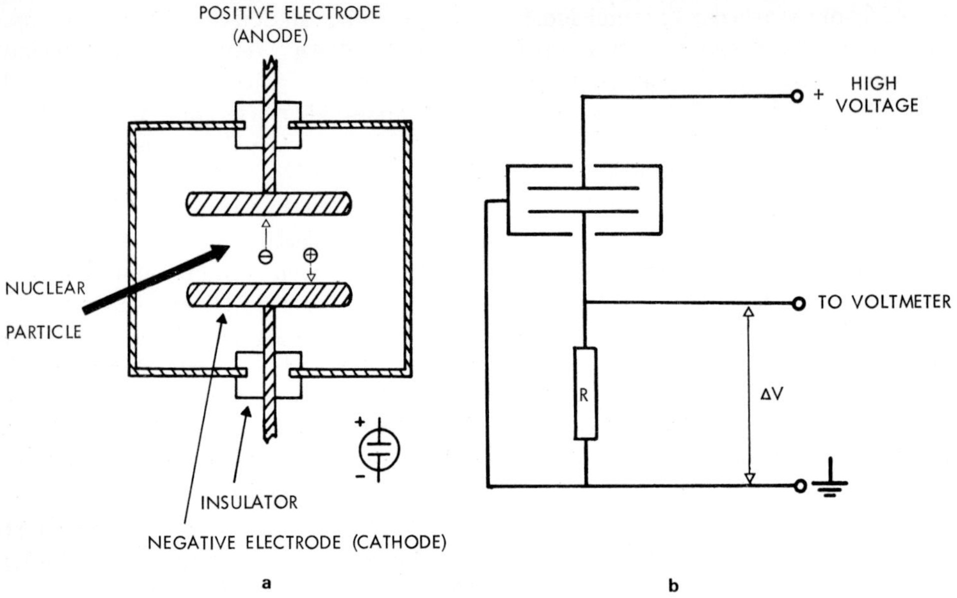

FIG. 17.3. (a) A parallel plate ionization chamber and (b) its measuring circuitry.

uses the cathode as a hollow cylinder and the anode as a thin wire in its center, e.g. the GM tube in Fig. 17.9(a). In other chambers the chamber walls serve as the cathode with a thin wire loop as anode as illustrated in Fig. 17.9(b). The chamber may be designed for recording radiation reaching it from the outside, or it may be used for measuring radioactive samples placed within it. Some chambers have additional electrodes, usually a thin central grid, to improve measuring conditions. The anode is kept at a positive potential 100–1000 V above the cathode.

Ions formed in the gas by nuclear radiation move towards the electrodes where they are discharged. If the gas is pure argon, only Ar^+ and e^- are formed. The electrons move rapidly towards the anode, and then through the electrical circuitry, as shown in Fig. 17.3(b), over the resistor R towards the cathode where they neutralize the argon ions: $Ar^+ + e^- \rightarrow Ar$. The gas is therefore not used up. The current i through the chamber and through the resistor R causes a voltage drop V,

$$V = Ri \tag{17.2}$$

which can be continuously recorded by a sensitive voltmeter. If the current or voltage drop, ΔV, is measured as a function of the voltage V applied over the electrodes, it is found that the current (or ΔV) increases with V up to a saturation value. The reason for this is that at low voltages some of the positive ions formed initially by the radiation recombine with the electrons, reducing the collected charge. With higher voltages the cations and electrons separate more rapidly with less recombination, and at saturation value essentially no recombination occurs. The ionization chamber will always be operated at saturation voltage.

Suppose a 2.7 μCi α-sample is placed within an argon-filled chamber of sufficient size that all the 5 MeV α's emitted are stopped in the gas volume. The saturation ion current will be found to be:

$$i = 1.60 \times 10^{-19} A E_{\text{loss}} \eta f_{\text{geom}} w^{-1} \tag{17.3}$$

where A is the radioactivity of the sample in dps ($2.7 \times 10^{-6} \times 3.7 \times 10^{10} = 1.0 \times 10^5$ dps), E_{loss} is the total energy lost per particle (5×10^6 eV) to the detector, η is the collection efficiency, f_{geom} is the geometric efficiency (for a thin solid sample we shall assume it to be half of a full sphere, i.e. 0.50), w is the energy required for the formation of an ion pair in the gas (for argon 26 eV), while the constant is the charge (Coulomb) of a single ion (the ion pairs must be regarded as single ions because only the electrons provide current through the resistor): thus $i = 1.54 \times 10^{-9}$ A. With a resistor of $10^9\ \Omega$, the voltage drop ΔV is 1.5 V.

Two types of ion chamber are common: (i) Simple, portable, rugged instruments with resistors $\lesssim 10^{13}\ \Omega$. With these, radiation intensities of $\gtrsim 10^3\ \beta\,s^{-1}$ and of $\gtrsim 10^5\ \gamma\,s^{-1}$ can be measured. They are usually calibrated in dose rate (e.g. roentgen per hour) and used for radiation protection measurements under field conditions (*dose rate meters*). (ii) Advanced, very sensitive instruments, with very high resistors ($\sim 10^{15}\ \Omega$) or special circuitry as in the vibrating reed electrometer. The chamber must always be designed with extreme care to avoid leaking currents from the anode over the chamber casing to the cathode. One way to minimize this is to ground the casing, as shown in Fig. 17.3(b). In the best of these instruments, currents as low as 10^{-18} A can be measured, corresponding to less than 1 $\alpha\,h^{-1}$, 10 $\beta\,min^{-1}$, or 10 $\gamma\,s^{-1}$. These instruments are best suited for measurement of radioactive gases, like tritium or radon in nature.

The beta-current neutron detector is a solid state ion chamber which has come into use in nuclear reactor technology. It consists of an emitter in which a nuclear reaction occurs, leading to the emission of primary β^- particles (e.g. through the reaction $^{103}Rh(n,\gamma)^{104}Rh \xrightarrow[4.2\,s]{\beta^-} {}^{104}Pd$) or secondary electrons (e.g. through absorption of the prompt γ's emitted in the neutron capture). These electrons represent a current and are collected by a collector. The radioactive decay type detectors have a response time depending on the product half-life, which the capture-γ detectors lack. These detectors have a limited lifetime; for the $^{59}Co(n,\gamma)\,^{60}Co$ it amounts to 0.1% per month at 10^{13} n cm^{-2} s^{-1}. The lifetime depends on the $\sigma_{n\gamma}$ value (37 b for ^{59}Co, 149 b for ^{103}Ru).

17.3. Pulse counting

A nuclear particle entering a detector produces excitation and ionization, both of which can be used for detection. When the excitation is followed by fluorescent de-excitation (§§15.4 and 15.5.3) the light emitted can be registered by light-sensitive devices, e.g. the photomultiplier vacuum tube (PMT) which transforms the light into an electric current. Scintillation and Čerenkov detectors are based on light emission. A similar current is produced when ionization takes place between the charged electrodes of an ion chamber. Detectors based on ionization are either gas-filled (proportional and Geiger–Müller tubes) or semiconductor crystals.

Because of the short duration of the absorption process for a single particle (10^{-4}–10^{-9} s) the current is referred to as a *pulse* of charge ΔQ

$$i = \Delta Q/\Delta t \tag{17.4}$$

If this current passes over a resistor R it will produce a voltage pulse (cf. (17.2))

$$\Delta V = R\Delta Q/\Delta t \tag{17.5}$$

The pulse is usually referred to as the *signal* (for the amplifier). While the signal from semiconductor detectors is a charge pulse, all other detectors produce a voltage pulse.

Pulse counting *per se* does not distinguish between different nuclear particles (α, β, γ, etc.) or particles of different energy. Such distinction is obtained by choosing detectors of unique (or exceptionally high) sensitivity to the particles of interest. With such detectors energy discrimination is achieved in the accompanying electronic circuitry because the pulse charge or size (ΔQ or ΔV) is proportional to the energy of the absorbed particle.

17.3.1. *Basic counting systems*

The block diagram of Fig. 17.4 indicates the most common components in the measuring circuit. The detector A is connected to a stabilized high voltage power supply B which furnishes the potential difference necessary for the detector to operate. The magnitude and type of the signal from the detector varies, depending on the type of detector, from a few hundred microvolts to several volts. For small signals it is often necessary to have a preamplifier C connected directly and physically close to the detector. The preamplifier increases the pulse size to the 1–10 V needed for the auxiliary system. This initial amplification also serves to overcome the pulse attenuation which results from passage through the connecting cable to the main amplifier D. The total gain in the system may vary from 10 for a Geiger–Müller counter to 10^4 for solid state detectors. The resolution must be better than 1 μs.

A limit is set on the voltage gain of an amplifier by the presence of electronic noise at the amplifier input due to stray fields, microphonics, and defective components which cause small, random voltage changes. This noise is amplified with the signal and can mask small input pulses. It is therefore of importance that the preamplifier gives pulses which exceed signals caused by external stray electromagnetic fields. Some of the noise may be eliminated by proper design and maintenance of the electronic circuitry, but a small inherent noise is always present (Fig. 17.5(a)). The *signal to noise ratio* is a ratio of output pulse (for a given input signal) to the noise level at the output. Noise pulses can be rejected by a discriminator E which serves as a filter to allow only

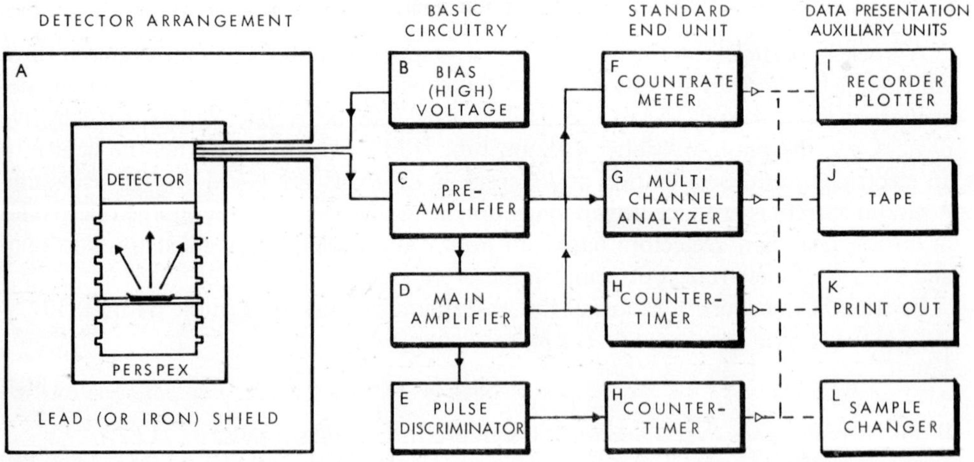

FIG. 17.4. The most common pulse-type measuring units and combinations.

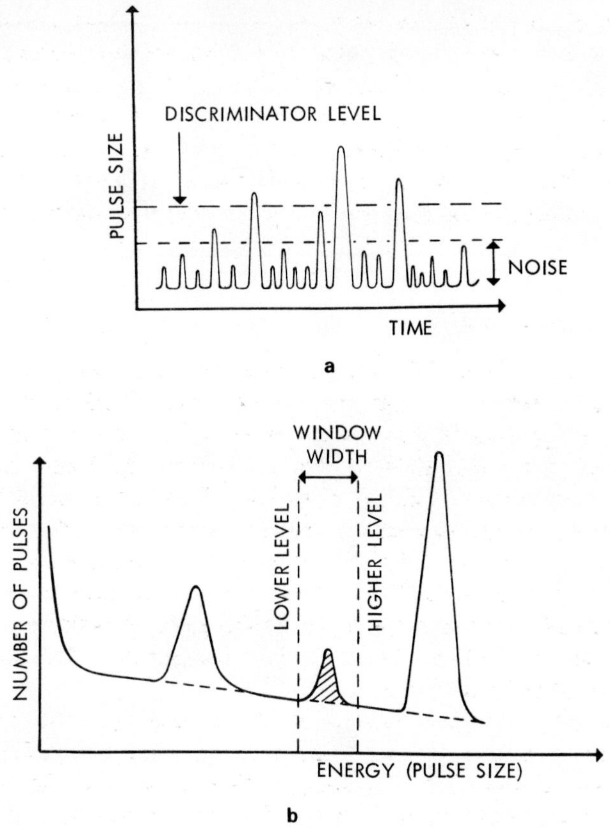

FIG. 17.5. Idealized pulse spectrum obtained (a) directly from a preamplifier, and (b) from a multichannel analyzer using a scintillation detector. The figure illustrates the effects of different discriminator settings.

pulses of a certain minimum size to be passed on to the rest of the system; in Fig. 17.5(a) the discriminator rejects all pulses below the dashed line. The signal pulse, after amplification, is sufficiently large to operate the electronic or mechanical counters H causing registration of the count. In modern counters the number of counts accumulated are directly displayed by light-emitting diodes (LED).

A wide variety of different counting systems has been developed for various purposes. Recently designed equipment is usually built in NIM (nuclear instrument modules) blocks, which fit into standard bins, as shown in Fig. 17.6. Connectors are also standardized (BNC type). Outputs of BCD (binary coded decimal) are usually available. This facilitates combinations of amplifiers, discriminators, counters, and other circuitry to fit any need (and most modern makes) as well as their connection to computers and various display units as plotters and cathode ray tubes (CRT). Many of these units have their own power supply and main amplifier. This is common for count rate meters F which are used to record, either linearly or logarithmically, the measured count rate.

For energy analysis the input pulses are sorted into channels depending on the size of the signal. In single-channel analyzers (SCA) only one channel exists which serves as a "window" to accept only pulses of a certain size corresponding to a limited range of energy, as indicated in Fig. 17.5(b). This window can be moved in steps through

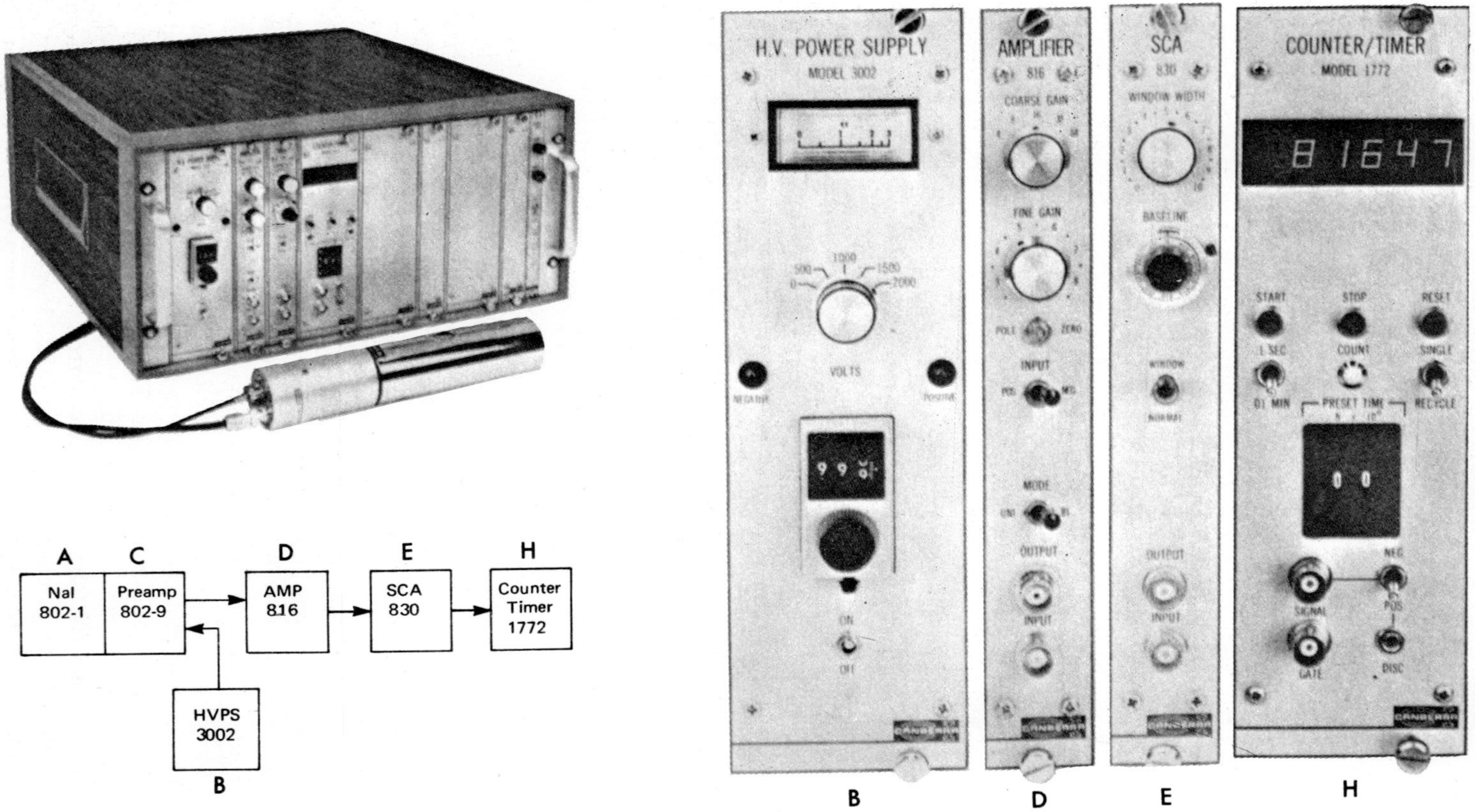

FIG. 17.6. Scintillation detector system. The letters for each unit are the same as in Fig. 17.4. (From Canberra Industries, New York.)

the entire energy range, thereby obtaining a measure of the count rate of the particles having different energies. In the figure the window position and width are set so as to cover only the middle peak. By narrowing the window, and moving it from zero to maximum pulse size, the whole particle spectrum is obtained. In Fig. 17.6 such a single-channel analyzer system is depicted. The final unit H has LED display of the number of counts for a preset time on the mechanical counter.

Multichannel analyzers G are in common use which have up to 16 000 channels (i.e. the energy scale is split up in that number of steps). In these cases the pulses are sorted immediately into the various channels which simultaneously record the counts rather than scan over an energy range in steps. In the most advanced systems the multichannel spectrometers are coupled on line to digital computers to provide analysis of the energy spectrum (see Fig. 18.4). Pulse height analyzers are probably the most versatile instruments for nuclear particle detection because of their usefulness both for qualitative identification and quantitative determination of radioactive nuclei. Practically all α- and γ-spectra reproduced in this book have been obtained through this technique.

17.3.2. *Pulse shape and dead time*

The purpose of a preamplifier is to shape the pulse so that the event can be recorded easily. We shall use Fig. 17.7 to describe the formation of a voltage pulse. A detector is connected between points A and B. The chamber has an *internal* resistance (because of the limited ion mobility in the detector) and capacitance (because of mechanical construction), indicated by R_i and C_i. Figure 17.7(a) does *not* show the physical design of the chamber, but only its electrical counterparts; this will make it easier to understand its function. When a particle enters the chamber it causes ionization (this is symbolized by the closing of the switch S), and the collection of ions at the electrodes gives a current flow and a voltage drop over R_i. R_e is the resistance between the chamber anode and the positive terminal of the high voltage supply (voltage $+V_0$); the other

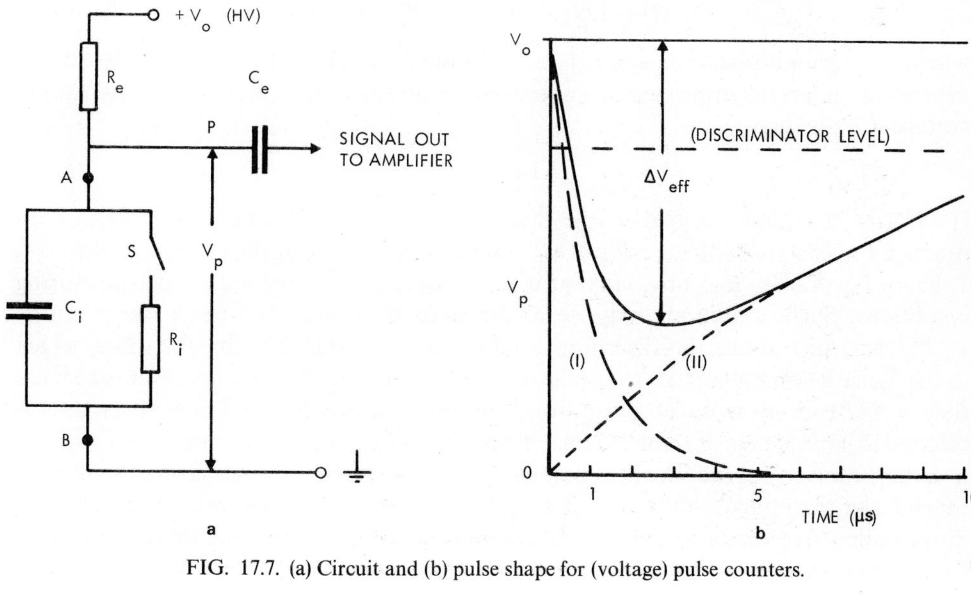

FIG. 17.7. (a) Circuit and (b) pulse shape for (voltage) pulse counters.

terminal is grounded. We shall concentrate our interest on the signal that, via the capacitance C_e, reflects the potential V_p at point P. In general $R_e \gg R_i$ (under conducting conditions), and $C_e \gg C_i$; typical values of R_e and C_e may be 10^8 Ω and 10 pF respectively.

When S is open (no ionization in the chamber), the potential at point P must be $V_p = V_0$, i.e. the potential of the high voltage. At time $t = 0$, S is closed (ionization has occurred in the chamber because of a nuclear particle), and the charge of $C_i + C_e$ leaks out over R_i. The potential in P decreases according to the relation

$$V_p = V_0 e^{-t/RC} \tag{17.6a}$$

where $R = R_i$ and $C = C_i + C_e \approx C_e$. For $R_i = 100$ kΩ and $C_e = 10$ pF, $RC = 10^{-6}$ s. RC is referred to as the *time constant* of the system (the decay time of the charge of the system); in the time RC the voltage has dropped to V_0/e. The long-dashed line in Fig. 17.7(b) indicates the discharge rate.

When all ions in the chamber have been collected, it is no longer conducting (S is "opened"). Now charge starts building up on the capacitors by current leaking from the high voltage through resistor R_e. The voltage in P begins to increase according to

$$V_p = V_0(1 - e^{-t/RC}) \tag{17.6b}$$

where $R = R_e$ and $C = C_e + C_i$. If we assume $R_e = 1$ MΩ the voltage build up will follow the short-dashed line in Fig. 17.7(b) provided $V_p = 0$ at $t = 0$.

In practice, as soon as V_p drops below V_0 current starts flowing through R_e, but because $R_e \gg R_i$, the capacitors discharge faster than they are charged. The overall voltage change in P follows the solid line in Fig. 17.7(b).

The voltage drop at P acts as a positive signal from the capacitance C_e. In the figure the size of the pulse is more than half the high voltage value; this is highly exaggerated, since in practice the pulse is only a very small fraction of the high voltage value. This is due to the fact that the current is carried only by the ions formed in the detector. If the number of primary ions formed is n_i, the charge transport ΔQ is

$$\Delta Q = 1.60 \times 10^{-19} n_i a \ \text{(Coulomb)} \tag{17.7}$$

where a is the multiplication factor (1 in solid state detectors, $\gg 1$ in gas-filled detectors). When a < 1 it has the same meaning as the collection efficiency η in (17.3). The maximum voltage drop is then

$$\Delta V = \Delta Q / C \tag{17.8}$$

If 10^5 ion pairs have formed in the detector and $C = 10$ pF, then $\Delta V = 1.6$ mV. In practice the effective voltage drop ΔV_{eff} is somewhat less than the calculated one, ΔV.

From Fig. 17.7(b) it is obvious that it takes some time to restore proper measuring conditions. Such conditions may be assumed to have occurred when the potential V_p has returned above the discriminator level as indicated. During this time, which in the figure is something like 15 μs, the detector cannot register any new events because they would overlap with the pulse from the previous event. This insensitivity time is referred to as *dead time* or *resolving time*. Thus the detector and measuring circuit requires a certain time to register each individual pulse. Normally the measuring circuitry is much faster than the detector and, therefore, the dead time is a function of the detector. Since radioactive decay is a statistical random process and not one evenly spaced in time, even for relatively low count rates a certain percentage of events can occur within the resolving time of the counting system. In order to obtain the true count rate it is necessary to know the correction that must be made for this coincidence loss.

The simplest technique for measuring the resolving time t_r of a counting system uses a method of matched samples. Two samples of similar but sufficiently low counting rates that the coincidence losses are almost negligible in both are counted separately and then together. From the difference between the measured count rate of the pair together and the sum of their individual rates, the resolving time is calculated using the equation:

$$t_r = \frac{R_a + R_b - R_{ab} - R_0}{R_{ab}^2 - R_a^2 - R_b^2} \tag{17.9}$$

where R_a, R_b, and R_{ab} are the measured count rate of samples a and b separately, and a and b together. R_0 is the background count rate for the system. The correction for the resolving time can be made according to the relation:

$$R_{corr} = \frac{R_{obs}}{1 - t_r R_{obs}} \tag{17.10}$$

where R_{corr} is the true activity for a measured value of R_{obs}.

17.4. Gas counters

All gas-filled counters are ion chambers (with the exception of the uncommon gas scintillation counters). In § 17.2 we learned that the ionization produced in an ion chamber by a single nuclear particle produced too low a current to be detectable except for α-particles. However, the ion chamber can be designed so that the number of ion pairs formed in each event is multiplied greatly.

Consider an ion chamber with a hollow cylindrical cathode and a thin central wire as an anode (see Fig. 17.9(a)). The ion pairs formed in the gas by the passage of the ionizing radiation are separated from each other by their attraction to the electrodes. The small, very mobile electrons are rapidly collected on the anode, which is maintained at a high positive potential above ground, $\gtrsim 1000$ V. Most of the voltage pulse which appears on the anode arises by induction by the positive ions as they move away from the immediate region of the anode. This step is responsible for the rise time of the pulse, curve I in Fig. 17.7(b). The potential decrease is only momentary as the anode is rapidly recharged by the power supply. The time necessary to restore the original potential is a function of the decrease in electric field near the anode due to the build-up of a layer of slow-moving positive ions, and of the time constant circuitry (curve II in Fig. 17.7(b)).

The electric field strength at a distance x from the anode is proportional to $1/x$. If the applied voltage of a cylindrical chamber of 1 cm radius is 1000 V, the potential in the immediate vicinity of a center wire of 0.0025 cm diameter is approximately 7×10^4 V cm^{-1}. As the primary electrons reach the vicinity of this high field and increase their kinetic energy, they cause secondary ionization which increases the pulse detected at the wire anode. The collected charge is given by (17.7), where a, *the gas multiplication factor*, is $\gg 1$.

This gas multiplication factor varies with the applied voltage as illustrated in Fig. 17.8. In region II of the curve a flat plateau over a relatively wide voltage range is observed. Prior to the attainment of the threshold voltage for the plateau, the ions would not have sufficient *drift velocity* to prevent elimination of some ion pairs by recom-

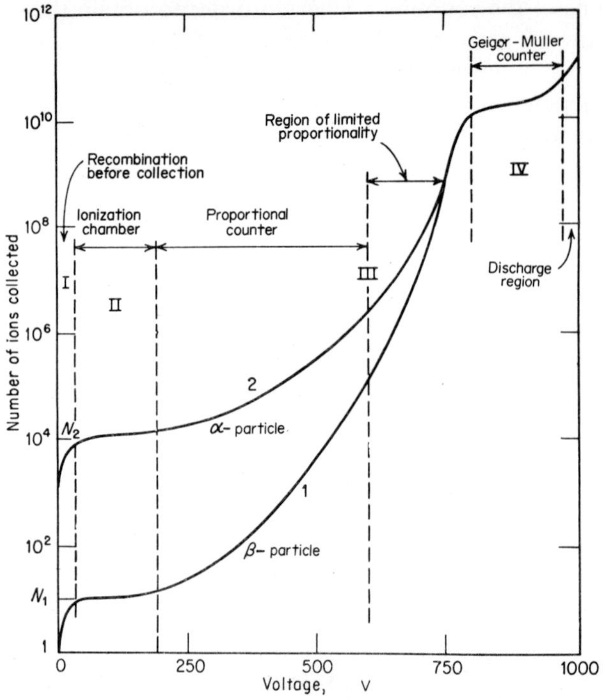

FIG. 17.8. Number of ion pairs formed in a gas-filled ionization chamber with a thin wire anode as a function of anode voltage. (According to C. G. Montgomery and D. D. Montgomery.)

bination. Throughout the plateau range the drift velocity is sufficient to make recombination negligible and since secondary ionization is not present, $a = 1$. Ionization chambers operate in this voltage region. In region III the electrons from the primary ion pairs receive sufficient acceleration to produce additional ionization and the process of gas multiplication ($a > 1$) increases the number of collected charges. This is the region of *proportional counter operation* as the pulses are proportional in size to the energy deposited in the detector by the passage of the initial radiation. Region IV is the one used for *Geiger–Müller counter* operation. In this region the gas multiplication is very high ($\gtrsim 10^6$) and the pulse size is completely independent of the initial ionization. Beyond region IV continuous discharge in the detector occurs.

17.4.1. *Proportional counters*

Values of a of 10^3–10^5 are commonly achieved in proportional counter operation. If $a = 10^3$, essentially all the gas multiplication occurs within 10 mean free path lengths from the wire for the electron in the gas ($2^{10} = 1024$). At 1 atm the mean free path length is approximately 10^{-4} cm, which means that the gas multiplication occurs within 0.01 mm of the wire.

The gas multiplication factor varies with the applied voltage, but for a given voltage a is constant so the detector pulse output is directly proportional to the primary ionization. As a result it is possible to use a proportional counter to distinguish between α- and β-particles and between identical particles of different energies inasmuch as different amounts of primary ionization are produced in these cases.

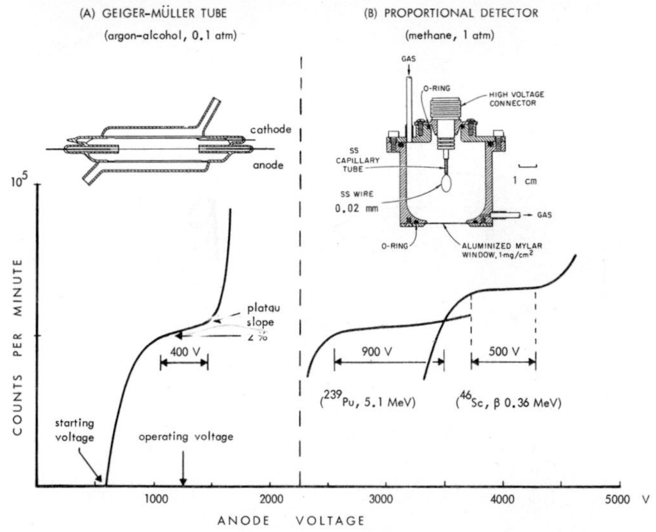

FIG. 17.9. Characteristics for GM and proportional counter tubes. The GM tube (a) is designed for flowing liquids. The proportional tube (b) uses a flowing counting gas.

The output pulse in proportional counter operation is not dependent on the collection of the positive ions by the cathode. Consequently the rate of detection depends on the time necessary for the primary electrons to drift into the region of high field strength near the anode wire. As a result, proportional counters have a much shorter resolving time than ion chambers which depend on the slow-moving positive ions. In fact the detector tube in a proportional counter can amplify a new pulse before the positive ion cloud of the previous pulse has moved very far if the new ionization occurs at a different location on the center wire. Time intervals necessary to enable the counter to measure two distinct pulses can be as low as 0.2–0.5 μs. Frequently the associated measuring equipment is a greater determinant of the resolving time than the detector itself. If, however, a proportional counter is being used for the measurement of particle energies, any residual positive ion cloud must have time to drift an appreciable distance before a new pulse is amplified. In this case the resolving time is closer to 100 μs.

Counting gases are usually one of the noble gases mixed with a small amount of polyatomic gas. The latter makes the gas multiplication factor less dependent on applied voltage, and increases the speed of electron collection. Typical counting gas mixtures are 90% Ar + 10% CH_4 ("P-gas") and 96% He + 4% i-C_4H_{10} ("Q-gas"). Many other combinations are possible, but molecules which readily attach electrons must be avoided. Figure 17.9(b) shows a proportional counter tube using pure CH_4 and the tube characteristics (i.e. count rate versus voltage) for ^{239}Pu α-particles and for ^{46}Sc β-particles entering through the thin aluminized mylar window.

The gas mixture and electrodes may be separated by a thin window from the radioactive sample, or the counter may be operated windowless. For windowless operation, after insertion of the sample the chamber must be flushed with gas to eliminate all the oxygen and water vapour as these molecules absorb electrons readily to form negative ions and, by so doing, reduce the pulse size. For α- and β-particles whose ranges do not exceed the dimensions of the chamber, windowless counters are often referred to as 2π counters since a solid angle of 2π is subtended above the sample. In these cases, with

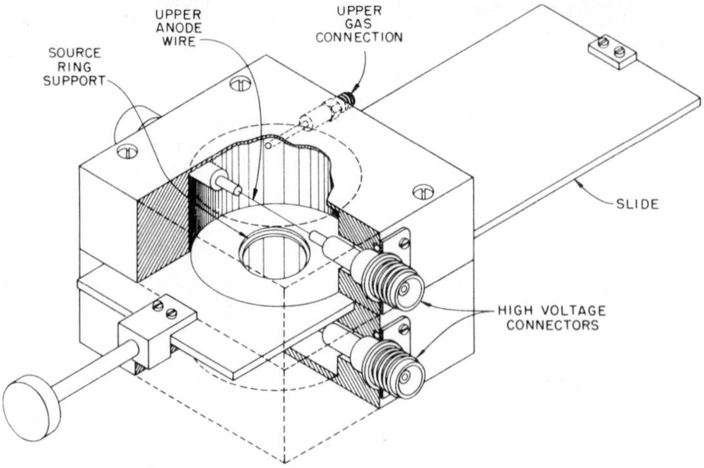

FIG. 17.10. A 4π proportional counter for measuring absolute decay rates. (According to G. D. O'Kelley.)

proper care, the measured count rate is very close to 50% of the true disintegration rate. Such windowless proportional counters are very useful for measuring low energy radiation such as the β-emissions of carbon-14 and tritium, and for absolute counting. Figure 17.10 shows a 4π proportional counter in which the sample is inserted in the middle between the two half-domes (cathodes).

Proportional counters can be used also for *neutron detection*. In this case the gas is BF_3, usually enriched in ^{10}B. With neutrons the reaction

$$^{10}B + n \longrightarrow {}^{7}Li + {}^{4}He \quad (Q = 2.78 \text{ MeV})$$

occurs (cf. Fig. 9.4). The ionization of the two products produces a heavy pulse, which is easy to discriminate against an intense γ-background. Some properties of a BF_3 neutron detector are given in Table 17.2. From Fig. 9.4 it is clear that the BF_3 counter has a higher efficiency for thermal than for fast neutrons.

Another technique for neutron detection uses a *fission chamber*. One design contains a stack of alternate anodes and cathodes, one of the electrodes being covered by a thin

TABLE 17.2. *Properties of some representative counter tubes*

	Geiger–Müller counters		Proportional counters	
Purpose	β, γ	α, β	n	α, β
Wall thickness	0.1 mm glass	1.5 mg mica cm^{-2}	1 mm steel	0.3 mg foil cm^{-2}
Filling gas	Ne + Ar + halogen	Ar + organic	BF_3, enriched in ^{10}B	Pure methane
Operating voltage	700 V	1250 V	2200 V	α 1900 V, β 2600 V
Plateau length	> 250 V	~ 300 V	> 300 V	α > 800 V, β ~ 400 V
Plateau slope	8% per 100 V	< 4% per 100 V	< 2% per 100 V	< 2% per 100 V
Lifetime	> 3 × 10^9 counts	3 × 10^8 counts		Infinite
Dead time	140 μs	300 μs		3 μs
Background count-rate	20–30 cpm	80 cpm	1–2 cpm	α 0.1 cpm; β 20–25 cpm
Background shielding	(50 cm Pb)	(10 cm Pb)		(50 cm Pb)
Special features	Insensitive to overvoltage		Insensitive to γ	

layer of uranium enriched in ^{235}U. The fission fragments produce large ionization even though the gas multiplication is quite low. This detector is more sensitive to fast neutrons than the BF_3 counter, and can be used for fast neutron fluxes up to $\sim 10^6\ n\ cm^{-2}\ s^{-1}$ with a background of a few cps.

17.4.2. *Geiger–Müller counters*

In region IV (Fig. 17.8), the proportionality between the primary ionization and the output pulse disappears and the latter becomes the same size for all initial ionization whether it be a 6 MeV α-particle or a 50 keV X-ray. Geiger–Müller counters which operate in this region have high sensitivity to all different kinds of radiation and the large size of the output pulse (from a tenth of a volt to one volt, compared to the several tenths of a millivolt output of ionization chambers) requires much less external amplification. This considerably reduces the complexity and the cost of the auxiliary electronic equipment. The detector tubes for GM counters are quite simple and allow a great deal of flexibility in design. Figure 17.9 shows a GM tube with a jacket for flowing liquids; Table 17.2 gives the properties of some other typical GM tubes. In general GM counters are limited to handling lower count rates than proportional counters.

As in the case of proportional counters, the primary electrons from the ionizing radiation cause secondary ionization near the center anode wire in GM detectors. This initial avalanche ends when these very mobile electrons are all collected by the anode. However, the neutralization of these electrons at the wire produces photons, which react with the gas leading to the emission of photoelectrons. These trigger further avalanches and an overall avalanche spreads along the complete length of the center wire and continues until the build-up of the positive ion sheath progresses to a point sufficient to reduce the field strength sufficiently to prevent further ionization. This build-up takes place because the heavy positive ions have such a slow rate of movement that they are essentially stationary during the time interval of the electron avalanches. The time required to reach this point in the process is of the order of a few microseconds. Since the resolving time is of the order of 100 μs due to the slow movement of the heavy cations to the cathode, it is necessary to make coincidence corrections in GM counting at much lower count rates than in the case of proportional counting. Typically, coincidence corrections are significant for count rates exceeding 10 000 cpm.

When the positive ion sheath reaches the cathode it may induce new avalanches as it collides with the cathode and causes emission of electrons. To avoid a recurring pulsing, which would render the counter useless, it is necessary to prevent further avalanches at this point by a process of quenching. This is usually accomplished by the addition of a small amount of organic compound such as ethyl alcohol or ethyl formate in the counting gas. Since the ionization potential of the organic molecule is lower than that of argon, the usual counting gas, when the positive argon sheath moves to the cathode and encounters organic molecules the following reaction occurs:

$$Ar^+ + C_2H_5OH \longrightarrow Ar + C_2H_5OH^+$$

In this reaction the charge of the argon ion is transferred to the organic molecule which gains an electron upon striking the cathode. The energy acquired in the neutralization causes dissociation into uncharged fragments rather than producing photon or electron emission. Inasmuch as the quenching gas is dissociated in the process of counting, such GM tubes have limited lifetimes which usually amount to approximately 10^9 discharges.

Halogen-filled GM tubes are now popular because of their infinite lifetime and lower operating voltage. Another advantage is that they will not be damaged by wrong polarity or excess voltage, as is the case with the organic quenched tube. Chlorine and bromine have strong absorption bands below about 2510 Å for the photons emitted, leading to dissociation. In the recombination the halogen molecule returns to its ground state via a series of low energy excited states.

Geiger–Müller tubes are available in a wide variety of shapes and sizes. Tubes have been used successfully which varied from approximately 1 mm to several centimeters in diameter and from 1 cm to almost a meter in length. The cathode can be made by coating the inside of a glass cylinder with a conducting material such as metal or graphite while the anode may be a tungsten wire mounted coaxially. The "end-window" GM tube has a thin mica window and the center wire is terminated with a glass bead. The cathode is always at ground potential while the anode is at a high positive potential.

When the radiation intensity is measured as a function of the electrode potential it is found for GM counters (as for proportional counters) that over a certain voltage interval there is little if any change in the measured count rate. This is known as the *plateau region* (Fig. 17.9). This is the voltage range in which the detectors are operated since they are relatively insensitive to small voltage changes in this region. For organic quenched GM tubes the plateau commonly occurs between 1200 and 1500 V and its slope should not exceed 5% per 100 V. Halogen-quenched tubes operate at a lower voltage but have higher plateau slopes (cf. Table 17.2).

17.5. Solid state detectors

This type of detector has become predominant for nuclear spectrometry (i.e. determination of the energy of nuclear radiation) but they are not commonly used for simple measurement of count rates. The solid state detector is similar to an ordinary semiconductor diode composed of p-type and n-type semiconductor material such as silicon.

Semiconductors are materials like silicon (resistivity $\sim 10^3\,\Omega$m) and germanium (0.6Ωm) with resistivities between metals (e.g. copper, $10^{-8}\,\Omega$m) and insulators (e.g. quartz, $10^{12}\,\Omega$m). A crystal of pure silicon placed between two electrodes is almost nonconducting. The electrons in the material are almost all *valence electrons*, bound to specific silicon atoms with an energy of 1.11 eV. If 1.11 eV is given to an electron, it moves to a band of overlapping energy levels which are not associated with specific atoms. The electron moves readily through the crystal in this "conduction band", i.e. the crystal conducts electricity. At a certain temperature some electrons, according to the Maxwell energy distribution, always have the necessary energy to be in the conduction band. These electrons provide a very small conductivity for pure silicon; this is referred to as the *intrinsic conductivity*. For diamond, the gap between the valency and conducting band is 7 eV, which is so large that essentially no electrons are found in the conduction band, and thus diamond is an insulator.

The energy needed for transferring valence electrons to the conduction band can be supplied by nuclear radiation. The average energy needed to produce an ion in silicon is not 1.1 eV but 3.5 eV because some energy is lost as crystal excitation. The electron removed from the valence band leaves a vacancy or "hole". The ionization is said to give an *electron–hole pair*. Just as the electrons move towards the anode, the hole moves towards the cathode.

Silicon has 4 valence electrons while phosphorus has 5 and indium 3. If we introduce

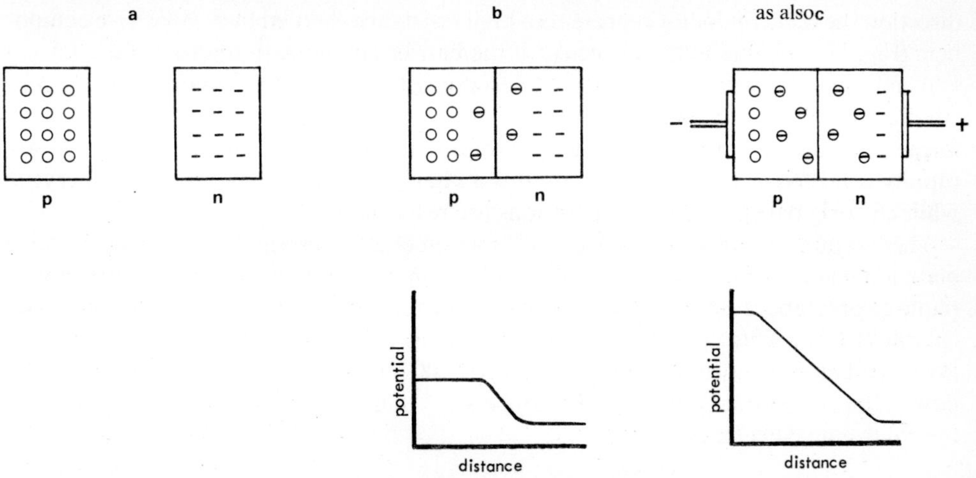

FIG. 17.11. Formation of a p–n junction. (a) Silicon crystal containing a trivalent impurity leading to the formation of electron–acceptor sites (or "positive holes", ○) to the left, and to the right the same with a pentavalent impurity, leading to the formation of an excess of electron–donor sites ("negative", –). (b) When combined, the two types of silicon form a p–n junction, where the acceptors and donors neutralize each other (⊖), thus forming a charge-depleted volume. The potential difference over the volume is about 0.6 eV. (c) When a reverse potential is applied over the crystal, the charges move so that the neutralized (depleted) volume increases, as also does the potential across the depleted layer.

a small amount of phosphorus into silicon, the phosphorus atoms substitute for silicon in the crystal lattice. Each such phosphorus has an excess of 1 electron. These electrons are not free but are very weakly bound, such that only 0.04 eV is needed to transfer them into the conduction bands. Because phophorus donates extra electrons to the system, it is referred to as a *donor material.* Silicon which contains small amounts of donor material (usually referred to as "impurity") is called *n-type silicon* since it has excess negative charge.

If, instead, indium is the impurity in the silicon crystal structure, the opposite effect is produced. Such material contains a number of energy levels only 0.06 eV above the valence bond; the result is holes in the valence bonds. Such material is referred to as *acceptor material.* Silicon with acceptor material is called *p-type silicon*, since the holes are considered to be positively charged. The addition of controlled amounts of impurity atoms provide *charge carriers* (as the electrons and holes collectively are called) and produces the desired properties in semiconductor materials.

The most interesting effect comes from the combination of two pieces of silicon, one n-type and another p-type (Fig. 17.11(b)). The contact surface of the pieces is referred to as a *p–n junction.* At such a junction some positive holes move to the n-type material, and vice versa. As a result at the junction a "depletion" layer a few microns thick is established where all the holes are filled with electrons and the layer is depleted of charge carriers. A p–n junction can be produced in a single piece of silicon by doping it with the proper impurity from either side of the crystal and by other techniques.

If a voltage is applied over the junction by connecting the negative terminal to the p-type region and the positive terminal to the n-type region, the junction is said to be *reversed biased.* With such reverse bias, the barrier height and depletion layer increase. As a result the crystal opposes any current as the resistivity is very high. In the reverse direction the semiconductor represents a high resistance shunted by a capacitive component (Fig. 17.7(a)) due to the dielectric of the barrier layer (the p–n junction diode).

A number of variations of this basic design exist. The point of importance for semiconductors as nuclear detectors is that a depleted layer with a high *space charge* is formed. A nuclear particle entering this volume forms electron–hole pairs, which are rapidly collected at the electrodes. By this a charge is transported through the crystal while the original space charge conditions are restored.

There is quite a similarity between the function of a solid state detector and a parallel plate ion chamber. In comparison with ionization chambers the solid state detector (i) requires only about 3.5 eV for an ion pair (as compared to about 35 in an ionization chamber), (ii) collects the charge much faster (no slow positive ions), (iii) requires a much lower voltage across the detector (usually < 100 V), (iv) has a much higher stopping power, but (v) but does not have the property of ion multiplication. The charge through the solid state detector is

$$\Delta Q = 1.60 \times 10^{-19} E_{loss} \eta w^{-1} \tag{17.11}$$

where $E_{loss} \leqq E_{kin}$ of the nuclear particles, η (the collection efficiency) is usually very close to 1, and w is 3.5 eV for silicon and about 2 eV for germanium. It is seen that the signal is directly proportional to the energy absorbed in the detector as long as η is constant. Because w is much smaller in solid state detectors than in gas and scintillation detectors, more primary electrons are released in each absorption event, which gives better "statistics" (see §17.11) and higher energy resolution. This makes solid state detectors useful for nuclear spectroscopy, although for α- and β-spectroscopy the precision is higher with magnetic deflecting devices (cf. §2.3).

17.5.1. *Surface barrier detectors*

These detectors are of p–n type silicon and are characterized by a rather narrow depletion layer (Fig. 17.12(a)). They are made of n-type silicon of which one surface has been exposed to air prior to coating with a thin layer of gold. This results in a p-layer which

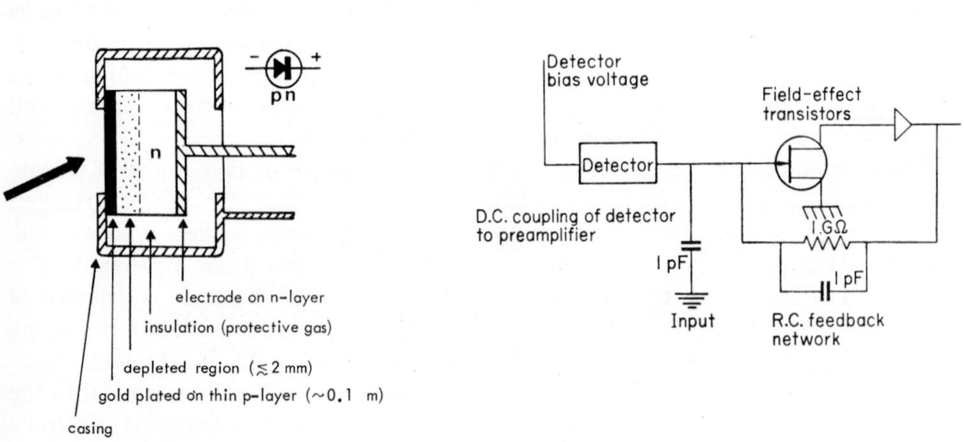

FIG. 17.12. (a) Surface barrier detector; the thickness of the gold and the p-layer is exaggerated. (b) Coupling of solid state detector to FET amplifier. (According to Nuclear Enterprises Ltd.)

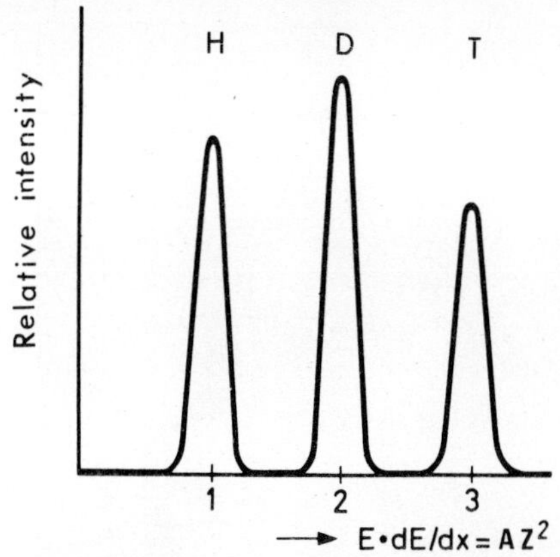

FIG. 17.13. Plot of relative particle intensities for H, D, and T versus $E\,dE/dx$ in a two-detector telescope, where the first detector records dE/dx and the second E.

including the gold is $\lesssim 0.1$ μm. Surface barrier detectors are used mainly for α- and β-spectroscopy and for dE/dx measurements for high energy particles, although the efficiency is limited by the sensitive surface diameter of $\lesssim 10$ mm.

The thickness of the radiation sensitive depleted layer is $\lesssim 2$ mm, enough to stop electrons of $\lesssim 1.5$ MeV, p of $\lesssim 20$ MeV, and α of $\lesssim 80$ MeV. A typical silicon surface barrier detector may operate at 100 V reverse bias and have a sensitivity of 40 Ωm and a capacitance of 100 pF. The resistance is $R = \rho d s^{-1}$, where ρ is the resistivity, d the distance between the connectors, and s the area of the conductor. There is a simple relation between dimensions, depletion layer, applied voltage, and the other electric properties, which usually is presented in nomograms by the manufacturer so that proper circuitry can be selected. The resolving time is about 10^{-8} s.

When used for α- or β-spectroscopy, a vacuum is applied between the detector and the radiation source. In the absence of a vacuum for α-radiation the energy loss is about 1 keV per 0.001 atm per cm distance between source and detector. The absorption in the detector window for a 6 MeV α is less than 6 keV. A resolution of about 30 keV FWHM (full width at half maximum) can be obtained for a 6 MeV α or a 0.6 MeV β.

In totally depleted silicon surface barrier detectors the sensitive region extends through the whole thickness of the silicon, which may be in the form of a very thin slice (e.g. 20 μm). A particle passing through such a detector loses a small fraction of its energy dE/dx and may then be completely stopped in a second (much thicker) detector to lose the remainder of its energy, which may essentially be its original total E_{kin}. Particles of mass A and charge Z are identified with the aid of the relation

$$E\,dE/dx \propto AZ^2 \qquad (17.12)$$

Figure 17.13 shows a hypothetical distribution when the recorded intensities of ^{1}H, ^{2}H, and ^{3}H are plotted against $E\,dE/dx$.

High energy particles not only cause ionization in the detector crystal but may displace some detector atoms from the crystal lattice. Such radiation damage to the crystals limits the lifetime of the detectors to about 10^6 rad.

17.5.2. *Lithium-drifted detectors*

The probability of γ-interaction is so small in the thin depletion depth of the surface

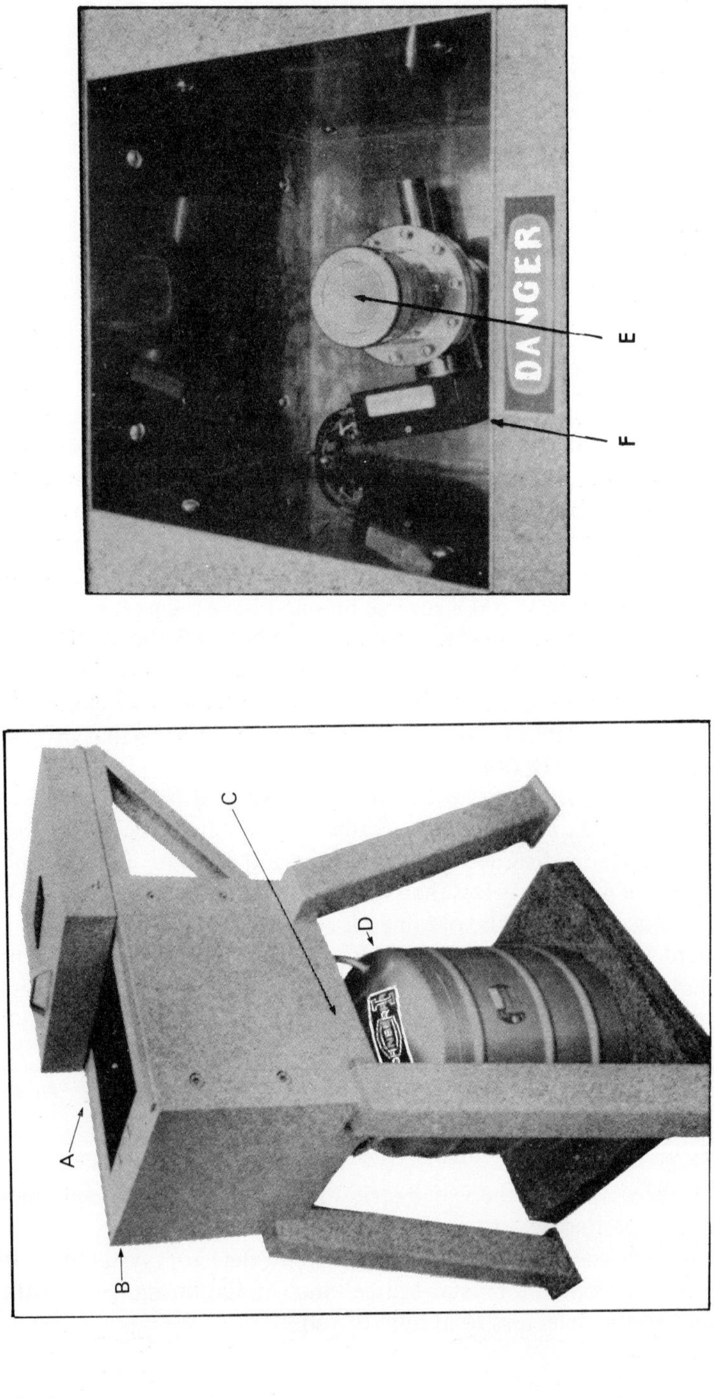

FIG. 17.14. Arrangement of shielding (ABC), Dewar vessel (D), solid state detector (E), and preamplifier (F). (According to Canberra Industries.)

barrier detectors that they are not useful for γ-spectroscopy. Thick sensitive volumes can be created by drifting lithium atoms into a silicon crystal. Lithium does not occupy a crystal site in the silicon, but is small enough to go into interstitial sites. The ease of ionization of Li to Li^+ makes it a donor impurity. The lithium is drifted from one side of the crystal so its concentration at the "entrance" side becomes high and then successively decreases towards the other end of the crystal. When a potential is applied over such a crystal, with the positive terminal at the high lithium side, three volumes are created, one of p-type, a middle "intrinsic" one, and an n-type one (p–i–n detectors). In the intrinsic volume the lithium donor electrons neutralize any original impurities, which are of acceptor p-type. The intrinsic volume is sensitive to nuclear radiation, and detectors with depths of up to 20 mm and surface areas of $\lesssim$ 50 mm are commercially available. Because of the high mobility of the lithium atoms at room temperature, these crystals must be kept at low temperatures (usually at 77 K by means of liquid nitrogen) even when not in use. Figure 17.14 shows the arrangement of the Dewar vessel with liquid N_2, detector, and preamplifier.

Lithium-drifted detectors are made either with silicon (Si(Li) detectors) or with germanium (Ge(Li) detectors). The latter has a higher atomic number and density than silicon and is therefore preferable for γ-spectrometry. For 60 keV X-rays, the efficiency of a Si(Li) detector may be 5%, while for a comparable Ge(Li) detector it may be 100%. At lower energy the Si(Li) detector is preferable, especially if the measurements are carried out in a high γ-background. Si(Li) detectors are of particular importance in X-ray fluorescent analysis (cf. § 14.8).

Both types give excellent resolution, about 1.71 eV FWHM at 1.33 MeV γ. This is far superior to the scintillation detector, as is seen from Fig. 17.15. On the other hand, the detection efficiency is less. For 1.33 MeV γ-rays an efficiency as high as 30% of that obtained with a 3″ × 3″ NaI (Tl) scintillation detector has been obtained in the best designs, but normal values are 10–20%.

17.5.3. *Intrinsic germanium detectors*

One reason for drifting lithium into silicon and germanium is the necessity to com-

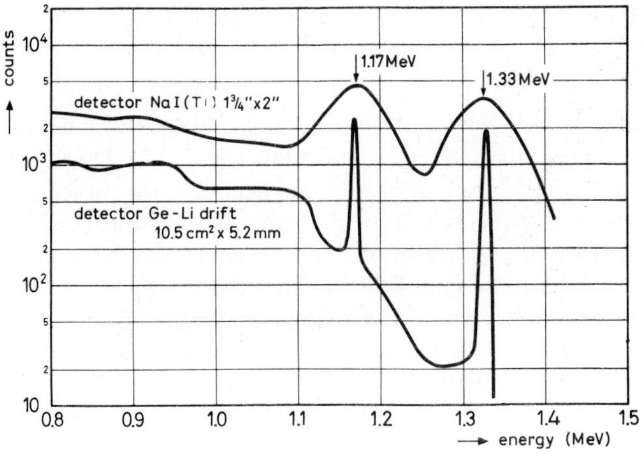

FIG. 17.15. Energy spectrum of ^{60}Co obtained with scintillation and Ge(Li) detectors. (According to J. A. Oosting.)

pensate some p-type (acceptor) impurities normally present in pure materials. It is now possible to increase the purity of germanium to $1:10^{13}$ (compared to earlier $1:10^{11}$) which makes lithium-drifting unnecessary. In addition, these intrinsic germanium detectors make cooling less important; it is not required when storing the crystals, but should be used when measuring in order to improve resolution. Intrinsic germanium is replacing the lithium-drifted crystals because of its greater simplicity at no higher cost or loss of resolution or sensitivity.

17.5.4. *Preamplifier*

The charge through the detector is given in (17.11). A 0.5 MeV γ completely absorbed in silicon will cause a charge $\Delta Q = 0.2 \times 10^{-13}$ C through the crystal. If the capacitance is 20 pF, this corresponds to a voltage pulse of 1 mV.

In proportional and scintillation detectors, which also are used for nuclear spectroscopy, the primary event is amplified in the detector system so that a much larger pulse is obtained. While the internal gain is about 10^3 for a proportional gas detector and about 10^6 for a scintillation detector, there is no gain in solid state detectors. The main purpose of the preamplifier for the first two types of detector is power amplification for simple noise-free driving of subsequent circuitry. The charge pulse from solid state detectors is both smaller and shorter. In this case the preamplifier must be capable of good signal (charge) amplification and have a very low noise. Preamplifiers with field effect transistors (FET) at the input stage satisfy these requirements and are used exclusively. Figure 17.12(b) shows such a coupling.

17.6. Scintillation detectors

In 1908 Rutherford and Geiger established the reliability of a method of counting α-particles by observing visually the flashes of luminescence† produced in a thin layer of ZnS by the α-particles. Since the development of reliable photomultiplier tubes in 1946, scintillating counting techniques have played an important role in nuclear science. A scintillation detector consists of a *scintillator* or *phosphor* optically coupled to a photomultiplier tube which produces a pulse of electric current when light is transmitted to the tube from the scintillator (see Figs. 17.6 and 17.16). The scintillating material can be an inorganic crystal or an organic solid, liquid, or gas.

17.6.1. *Scintillation counting*

In organic substances the absorption of energy raises the organic molecule to one of the vibrational levels of an excited electronic state (see § 15.5). Through lattice vibrations some of the excitation energy is dissipated as heat and the molecule decays to lower vibrational levels of the excited electronic state. After approximately 10^{-8} s, a time sufficient for many molecular vibrations, the molecule may return to the ground electronic state with emission of light photons. Since the energy which excites the molecule is in general larger than that emitted in any single step in the decay back to the ground state, reabsorption of these emitted photons is unlikely, and the crystal is consequently transparent to the emitted photon. This transparency is necessary if the scintillations are to escape the scintillator and reach the photomultiplier tube. Aromatic hydrocarbons such as anthracene and stilbene which have resonance structures are excellent

†*Luminescence* includes both fluorescence and phosphorescence (§ 15.5.3).

scintillators. Liquid and solid solutions of such organic substances as p-terphenyl are also used as scintillators. In these systems the energy absorbed through the interaction of radiation with the solvent molecules is transferred rapidly by the latter to the solute which undergoes excitation and fluorescence as described above. The exact mechanism of the transfer of energy from solvent to solute is not fully understood.

It is necessary to have small amounts of impurities in inorganic crystals to have luminescence. In ionic crystals in the ground state all the electrons lie in a lower valence band of energy. Excitation promotes the electrons into a higher conduction band of energy. If impurities are present they can create energy levels between the valence and conduction bands, as described in §17.5. Following excitation to the conduction band through absorption of energy an electron may move through the conduction band until it reaches an impurity site. At this point it can "decay" to one of the impurity electron levels. The de-excitation from this level back to the valence band may occur through phosphorescent photon emission. Again, since this photon would have an energy smaller than the difference between the valence and conduction bands, these crystals are transparent to their own radiation.

To be useful as a scintillator a substance must possess certain properties. First, there must be a reasonable probability of absorption of the incident energy. The high density in solid and liquid scintillators meets this condition. Following absorption, emission of luminescence radiation must occur with a high efficiency and—as mentioned—the scintillator must be transparent to its own radiations. Finally, these radiations must have a wavelength that falls within the spectral region to which the photomultiplier tube is sensitive. Since this is not always the case, particularly with liquid scintillators, "wavelength shifters" are added (e.g. diphenyloxazolbenzene (POPOP)) to solutions of p-terphenyl in xylene. Further, "quenching" substances which absorb the light emitted from the scintillator should be absent. This is a particular problem in liquid scintillation counting.

Table 17.3 lists the properties of some commonly used scintillators. The data indicate that the greater density of inorganic crystals makes them preferable for γ-ray counting. The resolving time is shorter for the organic systems whether liquid or solid. When large detector volumes are necessary a liquid solution system is the simplest and most economical. Liquid scintillation counting is used extensively for the low energy emissions of tritium and ^{14}C to obtain higher counting efficiency. ZnS(Ag) is a traditional phosphor for α-detection (though not much in use any more) while anthracene and stilbene can be used for β-particle detection. For γ-rays, sodium iodide with a small amount of thallium

TABLE 17.3. *Properties of some common phosphors*

Material	Density ($g\ cm^{-3}$)	Wavelength of maximum emission (Å)	Decay time for emission (s)	Relative pulse height (from β-particles)
Inorganic				
NaI(Tl)[a]	3.67	4100	2.5×10^{-7}	210
ZnS(Ag)[b]	4.10	4500	1×10^{-5}	200
Organic				
Anthracene	1.25	4400	3.2×10^{-8}	100
Stilbene	1.16	4100	6×10^{-9}	60
Plastic phosphors	1.06	3500–4500	$3\text{–}5 \times 10^{-9}$	28–48
Liquid phosphors	0.86	3550–4500	$2\text{–}8 \times 10^{-9}$	27–49

[a](Tl) and (Ag) indicate small amounts of these elements added as impurity activators.

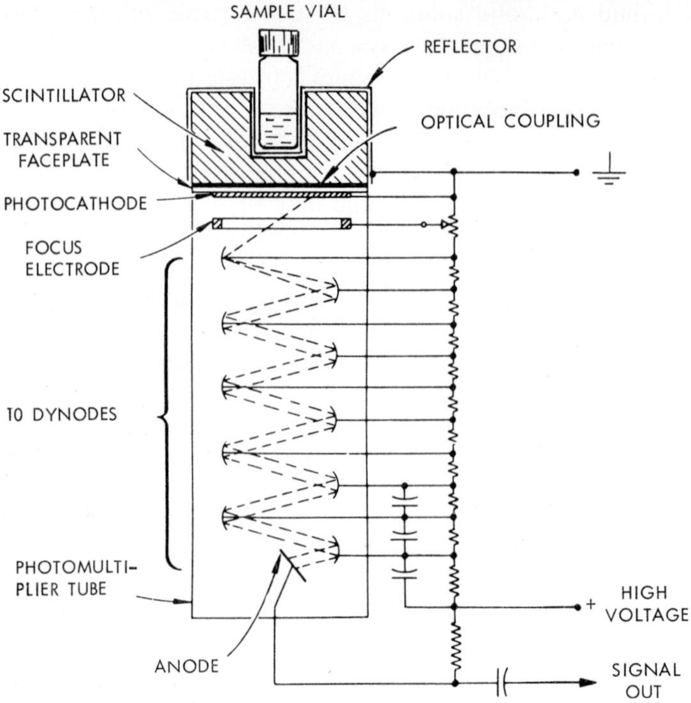

FIG. 17.16. Well-type scintillation detector with photomultiplier circuitry.

impurity, NaI(Tl), is the most common phosphor. Xenon and krypton are sometimes used as gas scintillators.

The scintillator must be coupled optically to the photomultiplier tube so that there is a high efficiency of transfer of the light photons to the tube. Since photomultiplier tubes are sensitive to light in the visible wavelength region, both scintillator and photomultiplier must be protected from visible light. Figure 17.16 shows a typical combination of a "well-type" crystal phosphor and photomultiplier tube. The light sensitive surface of the photomultiplier is a semitransparent layer of a material such as Cs_3Sb which emits electrons when struck by visible light. The emitted photoelectrons are accelerated through a series of 10–14 electrodes (*dynodes*) between which a constant voltage difference is maintained. When the photoelectrons strike the nearest dynode, secondary electrons are emitted as the dynodes are also covered with Cs_3Sb. Consequently, there is a multiplication of electrons at each dynode stage and at the last dynode the number of original electrons have been increased by about a factor of 10^6 over a total voltage drop in the photo tube of 1000–2000 V. The electrical pulse from the photomultiplier tube is passed to the rest of the detection circuit and recorded.

Scintillation detectors are used commonly for routine radioactivity measurements, particularly of γ- emitters, because of their reliability. As compared to GM tubes they have the advantage of shorter resolution time and higher γ-efficiency, although they require a more stable high voltage supply. Particularly the well-type crystal shown in Fig. 17.16 is popular because of the high counting efficiency for samples introduced into the well ($\psi \gtrsim 0.9$). For counting very large liquid volumes (e.g. environmental water samples) a specially designed sample vessel is used which fits over and around the cylindrical detector arrangement.

17.6.2. *Scintillation spectrometry*

The voltage pulse obtained from scintillation detectors is directly proportional to the energy absorbed in the scintillator. If all of the energy of the nuclear particle is absorbed in the scintillator, which is possible for all kinds of radioactive radiation, the energy of the particle can be determined from the size of the voltage pulse. Scintillation counting represents one of the principal techniques whereby the energy of radioactive emissions is measured. The single-channel spectrometer (Fig. 17.6) is a very economic and simple system for energy determinations, although multichannel systems are much faster. Although solid state detectors have superseded scintillation detectors for nuclear spectroscopy, the scintillation γ-spectrometer is still such a valuable tool that it warrants description in some detail.

As we learned in § 14.5, the capture of a γ-ray in an absorber such as a NaI(Tl) crystal can occur by any of three processes—photoelectric effect, Compton effect, and pair production. In energy analysis of γ-rays it is desirable to capture the total energy and to minimize the loss of energy by escape of the scattered γ-rays from Compton interaction. This increases the number of events contributing to the photopeak which corresponds to the total γ-ray energy. Also, if lower energy γ-rays are present their

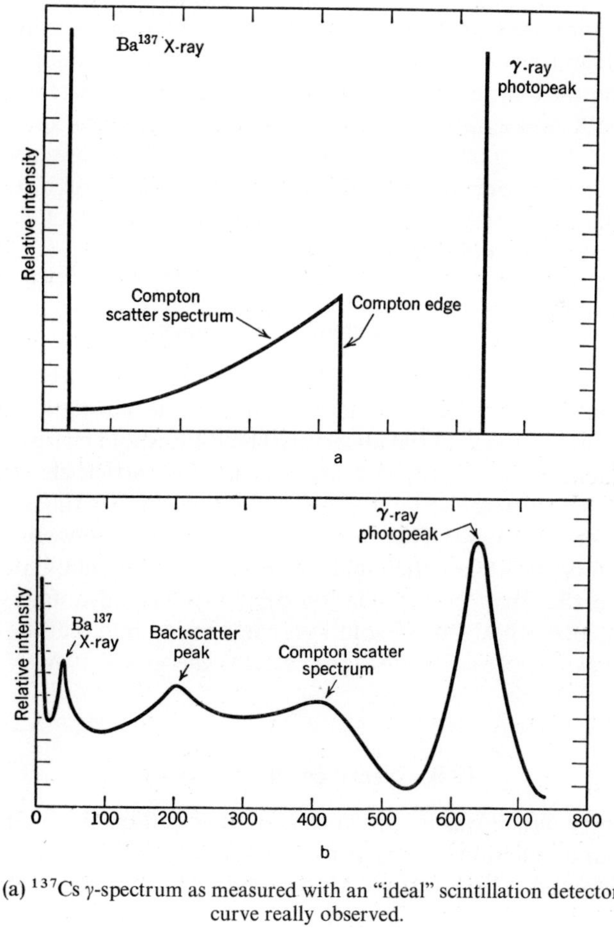

FIG. 17.17. (a) ^{137}Cs γ-spectrum as measured with an "ideal" scintillation detector, and (b) the curve really observed.

photopeaks may be obscured by the Compton distribution from higher energy γ-rays. With anthracene crystals little resolution is seen in γ-spectra as the low atomic number of the absorber makes the principal interaction the Compton effect. However, in NaI(Tl) crystals (and even more so in GeLi solid state detectors) the photoelectric effect is the more important. Increasing the size of the crystal increases the probability of photon capture and, therefore, the probability of capture of the scattered γ-ray in Compton events is increased. As a result, the increase in crystal size results in more capture of the total incident γ-ray energy which appears under the photopeaks.

For an ideal NaI(Tl) detector a γ-spectrum from ^{137}Cs would have the shape shown in Fig. 17.17(a), whereas the measured spectrum corresponds to that in Fig. 17.17(b). The broadening of the photopeak is due to many causes such as inhomogeneities in the crystals and variations in light reflection. However, the main cause is found in the phototube where nonuniformity in the photocathode, fluctuations in the high voltage imposed on each dynode, and statistical variations in the small number of photoelectrons formed at the photocathode are all contributing factors.

The resolution is the determining factor in the ability of the system to differentiate between photopeaks of γ-rays of similar energy. Other features of the spectrum in Fig. 17.17(b) are the backscatter peak, the X-ray peak, and the noise peak. The backscatter peak arises in the absorption in the crystal of scattered photons resulting from γ-ray absorption via Compton interactions in the material surrounding the crystal. Obviously the magnitude of this peak is dependent on the distance of this material from the crystal and on the nature and amount of the material. The X-ray peak is due to the absorption of the X-rays emitted in the electronic rearrangement following the nuclear disintegration or following internal conversion. Bursts of very low energy electrons are emitted spontaneously from the photocathode and are the cause of the noise peak at very low discriminator settings. This limits the photon energies that can be studied to a minimum of several thousand electron volts.

17.7. Čerenkov detectors*

The Čerenkov effect described in §14.4.3 can be used for detection of high energy β-radiation because the velocity of the nuclear particle must exceed the ratio $\mathbf{c}/n$, where n is the refractive index of the absorber. The β-threshold energy in lucite ($n = 1.5$) is 0.17 MeV, so lucite and similar plastics are often used as particle absorbers in Čerenkov detectors. In order to detect the light emitted, photomultiplier tubes are placed in the direction of the emitted light. There are many similarities between scintillation and Čerenkov detectors; however, the light pulse for the Čerenkov detector is faster, $\sim 10^{-10}$ s, but smaller than for scintillation detectors. The advantage of the Čerenkov detector is that aqueous (uncolored) solutions can be used, and that the angle of emitted light reveals the velocity of the absorbed particles. At energies < 10 MeV, only β-particles are detected.

17.8. Special counting systems*

For *low intensity measurements* in which a level of radioactivity comparable to the normal background radiation is to be measured, special electronic circuits incorporating two detectors are used. The detectors are coupled so that a signal registers only when both detectors are activated at the same time (*coincidence circuit*) or, alternatively, when only one but not the other is activated at the same time (*anticoincidence circuit*). With such arrangements the normal β, γ background of a detector may be decreased by more than a factor of 100. In the most advanced coincidence techniques both detectors are energy sensitive as well, providing information on the type of radiation being measured. For example, β–γ coincidence measurements are used for absolute determination of radioactivity for samples in which β-decay is immediately followed by γ-emission.

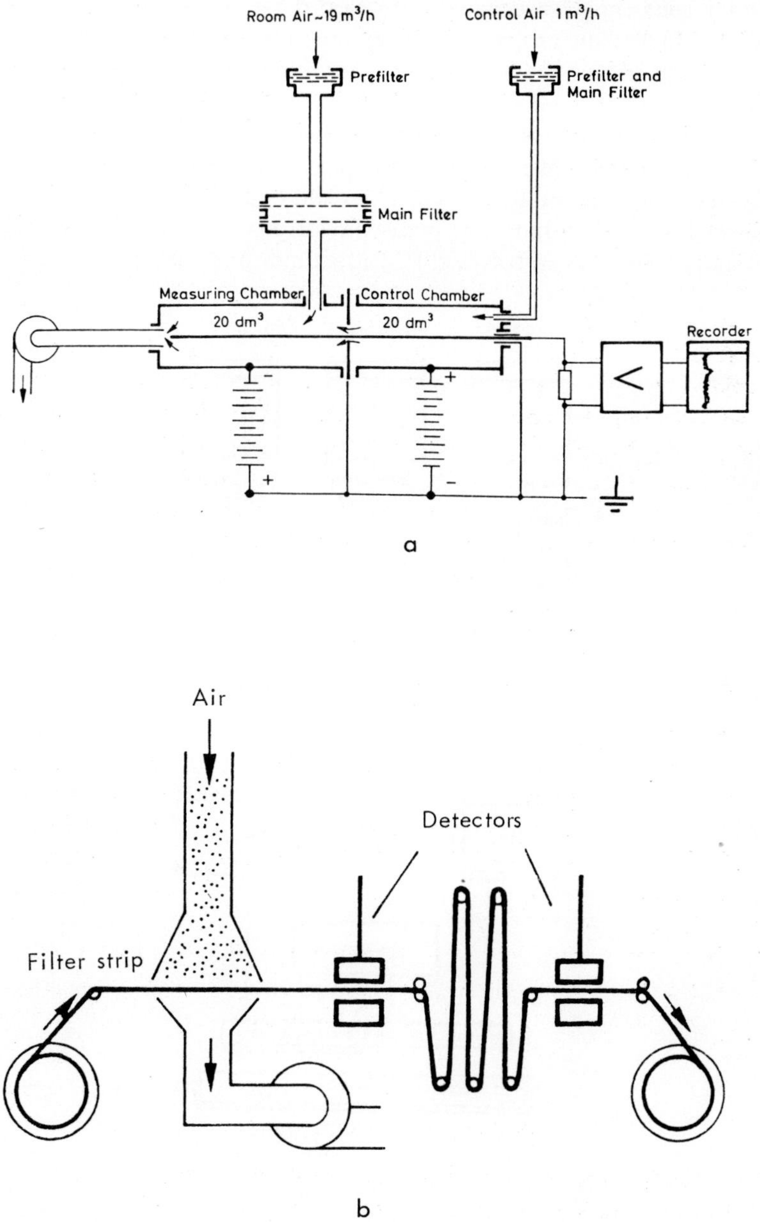

FIG. 17.18. Monitor (a) for radioactive gases, and (b) for radioactive aerosols, the latter with an arrangement for delayed time measurements. (From H. Kiefer and R. Maushart.)

Detectors which are *direction sensitive* have been developed primarily for use in medical diagnosis with radioactive isotopes. The simplest version involves a scintillation detector surrounded by a lead shield with a small hole through which radiation reaches the detector. Scanning instruments have been developed which permit the measurement of radioactivity as a function of two coordinates as illustrated in Fig. 18.12, which shows such a scintigram. Such instruments make it easy to detect the accumulation of a radioactive tracer in a particular organ of the body.

Whole-body counters were originally developed for investigation of poisoning by radioactive substances such as radium. They are now used diagnostically and consist of a large number of scintillation detectors surrounding the patient with the whole system, including the patient, placed in a heavily shielded room. The sensitivity is sufficient to detect a natural radioactivity in the human body from such nuclides as ^{40}K.

In nuclear installations like uranium mines, nuclear reactors, reprocessing plants, etc., it is necessary to continuously monitor gas and liquid effluents. Figure 17.18 shows two arrangements for air monitoring, in case (a) for pure gases (as, for example, $^{14}CO_2$, Ru, or Kr), and case (b) for aerosols (e.g. Tc, Ru, actinides). In the latter case two detectors

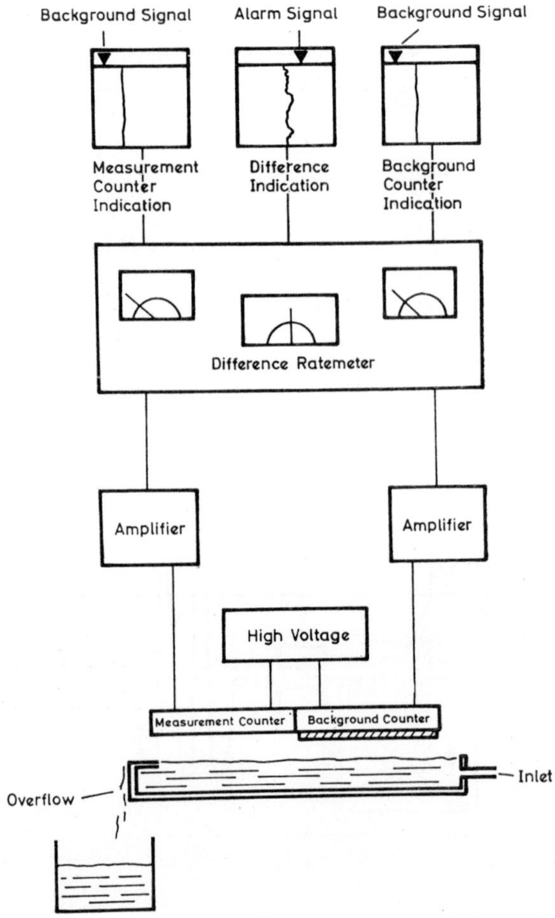

FIG. 17.19. Arrangement for continuous monitoring of radioactivity in water. (From Kiefer and Maushart.)

are used, so that some activity (e.g. mother or daughter activity) is allowed to decay between the two detectors. The delay time is adjusted by varying the length of the paper strip between the detector and the rate of movement of the paper strip.

Figure 17.19 shows a measuring arrangement for water monitoring. The detector may be energy sensitive or simple GM, proportional or scintillation devices. In this arrangement an alarm is engaged when the measured count rate exceeds the background and a certain preset value. In another arrangement the detectors (e.g. GM tubes) may dip into the streaming water, or the water may flow around the detector as in Fig. 17.9(a). The advantage of the arrangement in Fig. 17.19 is that no activity may adsorb on the surface of the detector.

17.9. Absolute disintegration rates

The determination of absolute disintegration rates is of great importance in all areas of nuclear chemistry, tracer work, age calculation, etc. Numerous methods have been employed, many using techniques described above, as track counting, liquid scintillation measurements, 4π proportional counters, etc. If the nuclei decay through β–γ emission, the absolute rate may be obtained by two solid state detectors placed close to each side of a thin sample, one detector β-sensitive and the other γ-sensitive.

When only a single detector in a conventional counting set-up is available (e.g. detector arrangement in Fig. 17.4), absolute counting rates can be obtained for unknown samples by comparison with known standards.

When standards are not available it is possible to obtain an approximate estimation of the absolute disintegration rate from a knowledge of the various factors that influence the counting efficiency. The detection efficiency ψ is defined as a ratio between the measured count rate and the absolute disintegration rate (e.g. (4.45)). This detection efficiency, which was discussed briefly in §14.2, is the product of all the factors which influence the measured count rate and may be expressed by

$$\psi = \psi_{det}\psi_{res}\psi_{geom}\psi_{back}\psi_{self}\psi_{abs} \tag{17.13}$$

where ψ_{det} = counting efficiency of detector,
ψ_{res} = resolving time correction (see §17.3),
ψ_{geom} = geometry factor (see §14.2),
ψ_{back} = backscattering factor,
ψ_{self} = self-absorption factor, and
(brace joining the two lines above:) $\psi_{back}\psi_{self} = \psi_{sample}$,
ψ_{abs} = absorption factor (see §14.2).

The efficiency of the detector is a measure of the number of counts registered compared to the number of particles that enter the sensitive volume of the detector. This efficiency is approximately 100% for α- and high-energy β-particles in most detectors, but often substantially lower for γ-rays. Inasmuch as it is quite difficult to apply simple geometric considerations to the solid angle subtended by a detector for a source which is not concentrated at a point, usually the factors ψ_{det} and ψ_{geom} are determined experimentally by using a very thin standard source of approximately the same area as the unknown. The factor ψ_{geom} can be calculated for circular samples and detector windows according to Appendix G.

It was noted in §14.4 that β-rays undergo large angle deflections. As a result, β-particles from the sample which may start in a direction away from the detector can be deflected by several scattering events back into the detector. Such backscattering is dependent upon the atomic number of the material upon which the sample is supported (cf. Fig. 14.1). ψ_{back} increases with backing material thickness up to a saturation thickness beyond which it is constant. Counting is done usually with either an essentially weightless backing ($\psi_{back} = 1$) or with a backing sufficiently thick as to have saturation. The saturation thickness corresponds to approximately 20% of the range of the β-particles in the backing material. Scattering can also occur from the walls of the sample holder, but this is usually less important than the backscattering from the sample backing material.

If the sample is thick, the count rate may be increased by internal backscattering, but the decrease of the count rate due to self-absorption within the sample is a greater factor. This sample self-absorption is inversely proportional to the β-ray energy and directly proportional to the thickness of the sample. For ^{32}P (1.72 MeV) sample thicknesses of 15 mg cm^{-2} show little self-absorption, while for ^{14}C (0.15 MeV), the absorption for sample thicknesses as small as 1 mg cm^{-2} is significant. The absorption factor can be determined by counting a series of samples of different thicknesses with the same total count rate, and then extrapolating that to zero sample weight.

The absorption factor ψ_{abs} is related to the absorption of the particles after they leave the sample by any covering over the sample, by the air between the sample and the detector, by the absorption in the detector window, etc. Again, this factor is usually determined by experimental comparison with a sample of known absolute count rate.

17.10. Sample preparation

From the discussion of the factors that enter into the counting efficiency it is obvious that the preparation of the counting sample must be done with care and must be reproducible if several samples are to be compared. Counting of α- and β-emitters in solution is best achieved by means of liquid scintillation counting. Since in this technique the emitters are included in the detection system itself, the efficiency is very high and reproducible. Alpha-emitters can only be counted efficiently from solid samples if the sample is very thin so that the self-absorption is eliminated. For α-spectrometry the surface density should preferably be < 0.1 mg cm^{-2}. Preferably, they should be counted by windowless proportional counters or internal ion chambers. Counting of solid samples of β-emitters may or may not be a problem depending on the energy of the β-emission. Again, care must be taken with uniform thickness of sample, backscattering, etc. The use of energy sensitive detectors makes possible a reliable measurement on one particular radioactive nuclide in the presence of other radiation of secondary importance.

Solid samples can be prepared by a variety of techniques such as precipitation, evaporation, and electrolysis (Fig. 17.20). When the precipitation technique is used the radioactive material must always be precipitated for comparative counting with the same amount of carrier and all samples must have the same surface density. The precipitate is filtered on a filter plate or filter paper of known reproducible geometry A. If filtration is not feasible the precipitate may be centrifuged in special vials C, or the precipitate, after centrifugation and decanting, may be slurried with ether or alcohol, and the slurry transferred by pipette to a counting disc of fixed geometry; when the organic liquid evaporates, it hopefully leaves a uniform deposit of the precipitate slurry D, F. Problems are plentiful in obtaining a deposit of uniform thickness by evaporation of a solution. However, an

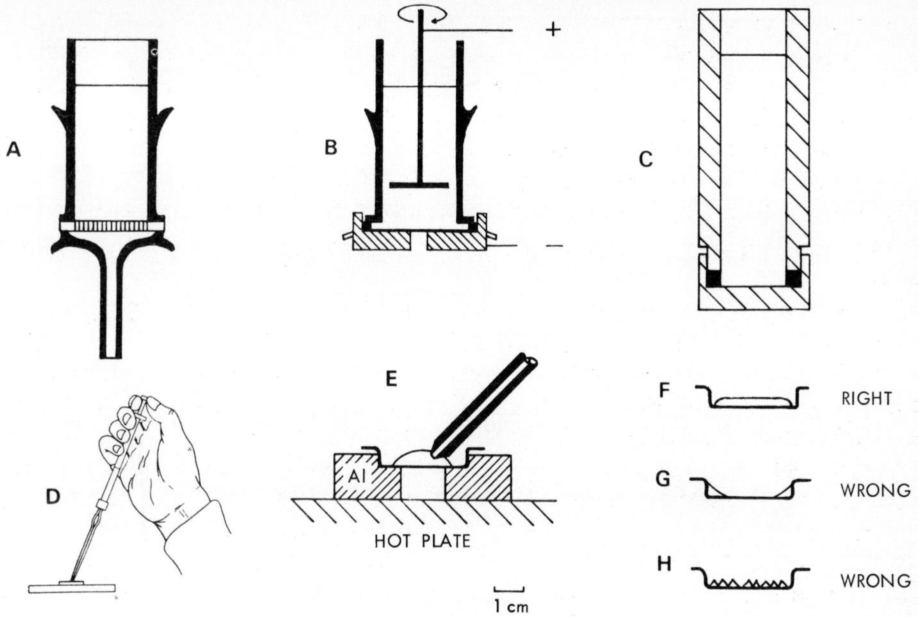

FIG. 17.20. The preparation of thin, even samples of known surface density. A, filtering arrangement; B, electrolytic plating system; C, centrifugation vessel; D–E, pipetting of a solution or slurry onto an evaporation tray (glass, stainless steel, platinum), yielding a correct even deposit F or uneven samples GH. The aluminum ring E should have a temperature only slightly exceeding the boiling point of the liquid.

arrangement such as that shown in Fig. 17.20 E has been found to be suitable: slow and even evaporation of 0.1–1.0 ml samples result in an even deposit if the amount of solid material is small. More even deposits can be obtained by electrodeposition of samples from solution B. This method can be used also for nonaqueous solutions provided that the organic solvents contain traces of water and a potential of several hundred volts per centimeter is used.

17.11. Statistics of counting and associated error

Even if the experimental design and execution are perfect so that the determinant error is eliminated in experiments involving radioactivity, there is always a random error due to the statistical nature of radioactive decay. Each radioactive atom has a certain probability of decay within any one time interval. Consequently, since this probability allows unlikely processes to occur occasionally and likely processes not to occur in any particular time interval, the number of decays may be more or less than the number in another similar interval for the same sample. It is necessary, when counting a sample, to be able to calculate the probability that the recorded count rate is within certain limits of the true (or average) count rate. The binomial distribution law correctly expresses this probability, but it is common practice to use either the Poisson distribution or the normal Gaussian distribution functions since both approximate the first but are much simpler to use. If the average number of counts is high (above 100) the Gaussian function may be used with no appreciable error. The probability for observing

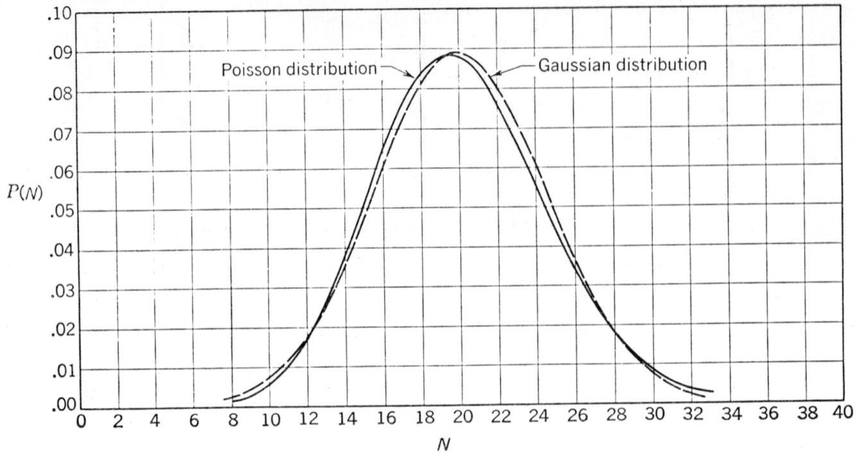

FIG. 17.21. Poisson (smooth) and Gaussian (dashed line) distributions for $\bar{N} = 20$.

a measured value of total count N is

$$P(N) = \frac{1}{\sqrt{2\pi N}} e^{-(\bar{N}-N)^2/2\bar{N}} \tag{17.14}$$

The standard deviation σ is given in such cases by

$$\sigma = \sqrt{N} \tag{17.15}$$

In these equations $P(N)$ is the probability of the occurrence of the value N while $\bar{N}$ is the arithmetic mean of all the measured values and N is the measured value. Figure 17.21 shows the Poisson and Gaussian distribution for $\bar{N} = 20$ counts. The standard deviation ("statistical error"), according to the theory of errors, indicates a 68% probability that the measured value is within $\pm\sigma$ of the average "true" value $\bar{N}$. For 100 measured counts the value 100 ± 10 indicates that there is a 68% probability that the "true" value will be in the interval between 90 and 110 counts. If the error limit is listed as 2σ the probability that the "true" count will be between these limits is 95.5%; for 3σ it is 99.7%.

Figure 17.22 shows the relationship between K, the number of standard deviations,

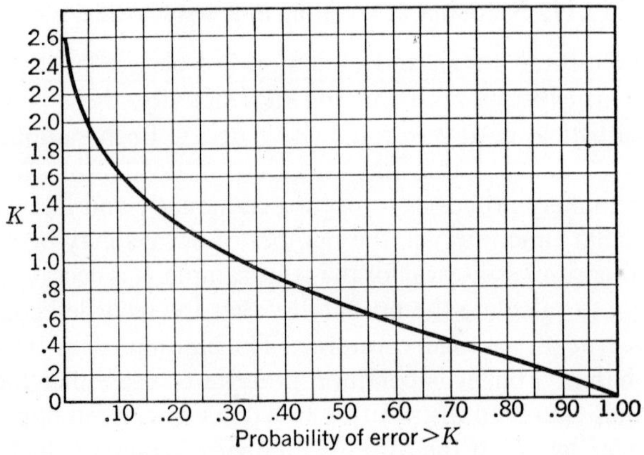

FIG. 17.22. The probability that an error will be greater than $K\sigma$ for different K-values.

and the probability that the true figure lies outside the limits expressed by K. For example, at $K = 1$ (i.e. the error is $\pm 1\sigma$) the figure indicates that the probability of the "true" value being outside $N \pm \sigma$ is 0.32 (or 32%). This agrees with the observation that the probability is 68%, i.e. the "true" value is within the limits of $\pm 1\sigma$. Figure 17.22 can be used to establish a "rule of thumb" for rejection of unlikely data. If any measurement differs from the average value by more than five times the probable error it may be rejected as the probability is less than one in a thousand that this is a true random error. The probable error is the 50% probability which corresponds to 0.67σ.

From the relationship of σ and N it follows that the greater the number of collected counts the smaller the uncertainty. For high accuracy it is obviously necessary to obtain a large number of counts either by using samples of high radioactivity or by using long counting times.

In order to obtain the value of the radioactivity of the sample, corrections must be made for background activity. If our measurements give $N \pm \sigma$ counts for the sample and $N_0 \pm \sigma_0$ for the background count, the correct value becomes

$$N_{\text{corr}} = (N - N_0) \pm \sqrt{\sigma^2 + \sigma_0^2} \tag{17.16}$$

If the sample was counted for a time Δt and the background for a time Δt_0, the measured rate of radioactive decay is

$$R = \frac{N}{\Delta t} - \frac{N_0}{\Delta t_0} \pm \sqrt{\left(\frac{\sigma}{\Delta t}\right)^2 + \left(\frac{\sigma_0}{\Delta t_0}\right)^2} \tag{17.17}$$

It is extremely important in dealing with radioactivity to keep in mind at all times the statistical nature of the count rate. Every measured count has an uncertainty and the agreement between two counts can only be assessed in terms of the probability reflected in terms of σ.

In experimental work with radionuclides many other errors occur in addition to statistical error in the count rate. Such errors may originate in the weighing or volumetric measurements, pH determination, etc. Such errors must also be considered in presenting the final results. For such composite errors, the *law of error propagation* must be applied:

$$\sigma_F = \sqrt{\left(\frac{dF}{dA}\sigma_A\right)^2 + \left(\frac{dF}{dB}\sigma_B\right)^2 + \ldots} \tag{17.18}$$

where σ_F is the (one standard deviation) error in F, which is a function of the uncorrected variables A, B, ..., with the standard errors σ_A, σ_B, etc. For the product $A \cdot B$ and ratio A/B, one obtains

$$F = (A \cdot B)(1 \pm \mathrm{s}) \tag{17.19}$$

$$F = (A/B)(1 \pm \mathrm{s}) \tag{17.20}$$

where

$$\mathrm{s} = \sqrt{\left(\frac{\sigma_A}{A}\right)^2 + \left(\frac{\sigma_B}{B}\right)^2} \tag{17.21}$$

For the function A^x and $\log A$ the following relations are valid:

$$F = A^x \pm x \cdot A^{x-1}\sigma_A \tag{17.22}$$

$$F = \log A \pm \frac{\sigma_A}{2.303A} \tag{17.23}$$

A useful technique for checking that the error in the measurements has a Gaussian distribution is the so-called "χ-square" test. The quantity χ^2 is calculated from

$$\chi^2 = \frac{\sum_{i=1}^{M} (\bar{F} - F_i)^2}{\bar{F}(k-1)} \tag{17.24}$$

where M is the number of measurements (e.g. points on a curve) for which the function F is (believed to be) valid. χ^2 would have a value 0.5–1.0 when the Gaussian fit exceeds 50%. k is the number of degrees of freedom, i.e. M plus the number of independent variables. This relation is generally valid; for simple counting systems F is replaced by N, the number of counts in the given time interval, and $k = M$.

17.12. Exercises

17.1. A detector has a 1 cm^2 efficient area perpendicular to a γ-particle flux produced by a source 7 m away. The sensitivity for the 0.73 MeV γ-radiation is 8.2%. (a) What must the source strength be for the detector to register 1000 cpm? (b) What fraction of the radiation is absorbed in the air space between source and detector?

17.2. A ~ 100 MeV fission fragment is stopped in a plastic plate with density ~ 1 and average atomic spacing of 2.5 A. Estimate (a) the range in the plate, and (b) the ionization density (ion pairs μ^{-1}). If the ionization along the track is spread out perpendicular from the track so that 1 in 10 atoms are ionized, (c) what would be the diameter of the track? From the track dimensions (d) calculate the average energy deposition to each atom within the "cylinder", and, using the relation $E = \frac{3}{2}kT$, (e) estimate the average temperature within the track volume. In lack of basic data for the plastic material, use data for water.

17.3. Plutonium in urine sample is soaked into a photographic emulsion so that the emulsion increases its volume by 20%. The 12 μ thick emulsion is dried to original thickness and left in darkness for 24 h. After development, α-tracks are counted for an average 2356 tracks cm^{-2}. If the plutonium consists of 67% ^{239}Pu and 33% ^{240}Pu, what was the plutonium concentration in the urine?

17.4. A ^{244}Cm sample is measured in an ion chamber (Fig. 17.3). The voltage drop over a $3 \times 10^{13}\ \Omega$ resistor is measured to be 0.47 V. What is the activity of the sample if all α's emitted in the chamber (2π geometry) are stopped in the gas?

17.5. In a proportional counter filled with methane of 1 atm the gas multiplication is 2×10^4. What is the maximum pulse size for a 5.4 MeV α, if the ion-pair formation energy is assumed to be 30 eV? The capacitance of the circuit is 100 pF.

17.6. In a GM counter, sample A gave 12 630, B 15 480, and A + B together 25 147 cpm. (a) What is the resolving time of the counter? (b) With the same counter, the distribution of radioactive samarium between an organic phase and water was measured according to $D_m = R_{org}/R_{aq}$. The measured R_{org} is 37 160 cpm, and that of R_{aq} is 2965. (b) What is the measured D_m? (c) Using corrections for resolving time, what is the true D-value?

17.7. Assume that 10^9 alcohol molecules are dissociated per discharge in a GM tube of 100 cm^3 filled with 90% Ar and 10% ethyl alcohol vapor at a pressure of 100 mmHg (25°C). What is the lifetime of the tube in terms of total counts assuming this coincides with the dissociation of 95% of the alcohol molecules?

17.8. A 1 mm thick surface barrier detector of 10 mm diameter has a resistivity of 7000 Ω cm and a capacitance of 50 pF at 300 V reverse bias. Calculate the resolving time (time constant).

17.9. A plastic scintillation detector was to be calibrated for absolute measurements of β-radiation. For this purpose a 2.13×10^{-5} M ^{204}TlCl$_3$ solution was available with a specific activity of 13.93 μCi ml^{-1}; ^{204}Tl emits β-particles with E_{max} 0.77 MeV. Of this solution 0.1 ml is evaporated over an area of exactly 0.1 cm^2 on a platinum foil. The sample is counted in an evacuated vessel at a distance of 15.3 cm from the detector, which has a sensitive area of 1.72 cm^2. The detector registers 2052 cpm with a background of 6 cpm. What is (a) the surface weight of the sample, (b) the backscattering factor, and (c) the detector efficiency for the particular β's?

17.10. A sample counted for 15 min gave 9000 total counts. A 30 min background measurement registered 1200 counts. Calculate (a) the count rate of the sample alone, with its standard deviation, and (b) with its probable error.

17.11. A certain sample has a true average counting rate of 20 cpm. What is the probability that 10 counts would be obtained in a 1 min recording?

17.13. Literature

See Also Appendix 1.

H. Yagoda, *Radioactive Measurements with Nuclear Emulsions*, Wiley, 1949.
G. B. Cook and J. F. Duncan, *Modern Radiochemical Practice*, Oxford University Press, 1952.
S. Flügge and E. Creutz (eds.), *Instrumentelle Hilfsmittel der Kernphysik II. Handbuch der Physik*, XLV, Springer-Verlag, 1958.
G. D. O'Kelley, *Detection and Measurement of Nuclear Radiation*, NAS-NS 3105, Washington DC, 1962.
E. Schram and R. Lombaert, *Organic Scintillation Detectors*, Elsevier, 1963.
W. H. Barkas, *Nuclear Research Emulsions*, Academic Press, 1963.
W. J. Price, *Nuclear Radiation Detection*, McGraw-Hill, 1964.
P. C. Stevenson, *Processing of Counting Data*, NAS-NS 3109, Washington DC, 1966.
W. B. Mann and S. B. Garfinkel, *Radioactivity and its Measurement*, van Nostrand, 1966.
Nuclear spectroscopy instrumentation, *Nucl. Instr. Meth.* **43** (1966) 1.
G. Bertolini and A. Coche (eds.), *Semiconductor Detectors*, North-Holland, 1968.
R. J. Brouns, *Absolute Measurement of Alpha Emissions and Spontaneous Fission*, NAS-NS 3112, Washington DC, 1968.
J. M. A. Lenihan and S. J. Thomson, *Advances in Activation Analysis*, Academic Press, 1969–.
K. Bächmann, *Messung Radioaktiver Nuclide*, Verlag-Chemie GmbH, 1970.
C. E. Crouthamel, F. Adams, and R. Dams, *Applied Gamma Ray Spectrometry*, Pergamon Press, Oxford, 1970.
O. C. Allkofer, *Teilchen-Detectoren*, Thiemig, 1971.
P. Quittner, *Gamma Ray Spectroscopy*, Adam Hilger, London, 1972.
H. Kiefer and R. Maushart, *Radiation Protection Measurement*, Pergamon Press, Oxford, 1972.
R. A. Faires and B. H. Parks, *Radioisotope Laboratory Techniques*, Butterworth, 1973.
J. A. Cooper, Comparison of particle and photon excited X-ray fluorescence applied to trace element measurements of environmental samples, *Nucl. Instr. Meth.* **106** (1973) 525.
J. D. Macdogall, Fission-track dating, *Sci. Am.* Dec. 1976.
J. Krugers, *Instrumentation in Applied Nuclear Chemistry*, Plenum Press, 1973.
Users' Guide for Radioactivity Standards, NAS-NS 3115, Washington DC, 1974.
D. L. Horrocks, *Application of Liquid Scintillation Counting*, Academic Press, 1974.
P. J. Ouseph, *Introduction to Nuclear Radiation Detection*, Plenum Press, 1975.
R. L. Fleischer, P. B. Price, and R. M. Walker, *Nuclear Tracks in Solids*, University of California Press, 1975.

CHAPTER 18

Applications of Radioactive Tracers

Contents

Introduction

For the most part so far we have been concerned with nuclear science. We have considered the principal interests of nuclear chemists, including nuclear spectroscopy and nuclear reactions. However, most scientists who use radioisotopes are not nuclear scientists, and to these former radioactivity is a tool to be used in research in somewhat the same fashion as optical spectroscopy. While nuclear chemists use chemical techniques to study nuclear phenomena, radiochemists use nuclear techniques to study chemical phenomena. In this chapter some of the ways in which radiochemistry has aided research in various areas of chemistry and related sciences are reviewed.

The first experiments with *radioactive tracers* were carried out in 1913 by G. de Hevesy (Fig. 18.1) and F. Paneth who determined the solubility of lead salts by using one of the naturally occurring radioactive isotopes of lead. Later, after discovery of induced radioactivity, de Hevesy and O. Chiewitz in 1935 synthesized ^{32}P and used this tracer in biological studies. In the same year de Hevesy and co-workers also carried out acti-

FIG. 18.1. (*Upper*) *George de Hevesy*, born 1885 in Budapest. Professor at the University of Copenhagen in 1920, later at Freiburg and Stockholm. Nobel laureate in chemistry in 1943 "for his work on the use of isotopes as tracers in the study of chemical processes". (Photograph around 1950.) (*Lower*) *Melvin Calvin*, born 1911 in St. Paul, Minnesota, professor of chemistry at the University of California, Berkeley. Nobel laureate in chemistry in 1961 "for his research on the carbon dioxide assimilation in plants".

vation analyses on rare earths. Despite the demonstration of the value of the tracer technique by these early studies the technique did not come into common use until after World War II when relatively large amounts of radionuclides became available through the use of nuclear reactors.

While it is not necessary to use radioactive isotopes for tracer studies, in general the use of radioactivity is simpler and less expensive than the use of stable isotopes. Research with the latter requires rather sophisticated and expensive measuring devices such as nuclear magnetic resonance and mass spectrometers, etc. The development of high field, high resolution, multinuclei nmr spectrometers which can use the natural isotopic abundance of ^{13}C, ^{14}N, ^{17}O, ^{19}F, ^{23}Na, ^{29}Si, ^{31}P, ^{33}S, ^{59}Co, ^{139}La, etc., has increased the value of this technique for studying a wide variety of systems. Such studies often provide a complement to radiotracer studies. We restrict our discussion to the use of radioactive tracers; for discussion of the use of stable nuclides as tracers, we refer the reader to the many textbooks on nmr, mass spectrometry, etc.

18.1. Basic assumptions for tracer use

In some experiments answers to the scientific questions can be obtained only through the use of a radioactive tracer, which indicates the presence and concentration of a specific element or compound at a certain place and at a certain time. For example, self-diffusion of metal ions in solutions of their salts cannot be studied by any other technique. However, in other cases the use of radioactive tracers is not necessary in principle but is justified by the greater convenience. In either type of investigation there are two assumptions implicit in such uses.

The primary assumption is that radioactive isotopes are chemically identical with stable isotopes of the same element, i.e. the substitution of ^{14}C for ^{12}C in a compound of carbon does not change the type or strength of the chemical bonds nor does it affect the physical properties of the compound. The validity of this assumption depends on the precision of measurement of the chemical and physical properties. The difference in mass between the various isotopes does cause some change in these properties (Chapter 2) but even in the case of ^{14}C and ^{12}C with a mass difference of approximately 15%, the isotope effect is rather small and difficult to detect, while for heavier elements it can be neglected in almost every situation. In fact, normally only for systems involving hydrogen–deuterium–tritium substitution must isotope effects be considered.

The second assumption in the use of tracer techniques is that the radioactive nature of the isotope does not change the chemical and physical properties. Until the moment of its disintegration the radioactive atom is indistinguishable from its stable isotope except for the isotopic mass difference. Upon radioactive disintegration the atom is counted; thereafter it is no longer the same element and its subsequent chemical behavior is usually of no interest. If the disintegration rate is very large, the energy released by the radioactive decay can cause observable secondary radiolytic effects (Chapter 15). However, in well-designed tracer experiments the level of radioactivity is high enough to provide accurate data but normally not high enough to produce noticeable radiation effects.

While the radioactivity of the tracers is assumed not to affect the chemical systems, the *parent–daughter relationship* of radioactive nuclides needs special consideration. For example, since strontium and yttrium are not chemically identical, a gross β-count of strontium samples including ^{90}Sr may include an unknown fraction of ^{90}Y activity present from ^{90}Sr decay because of the relationship

$$^{90}\mathrm{Sr}(\beta^- t_{1/2}\,29\,\mathrm{y}) \rightarrow {}^{90}\mathrm{Y}(\beta^- t_{1/2}\,64\,\mathrm{h}) \rightarrow {}^{90}\mathrm{Zr}\,(\mathrm{stable})$$

Beta-absorption measurements or γ-scintillation techniques which use energy discrimination are frequently useful in such parent–daughter cases. If equilibrium is rapidly established between the parent and daughter activities it is usually simpler to count the samples after sufficient time for this to occur. In the case of ^{90}Sr–^{90}Y, radioactive equilibrium is established in about 25 d; if ^{137}Cs is being used to study cesium chemistry it is necessary to wait only 15–20 min after separation to count a sample as the daughter ^{137m}Ba ($t_{1/2}$ 2.6 min) reaches an equilibrium level within that time. Since the ratio of the ^{137}Ba and the ^{137}Cs activity is the same in all samples at equilibrium, the total count rate before and after a chemical step is then a true measure of the behavior of cesium alone. If radioactive equilibrium is not re-established in a convenient time, it may be necessary to either discriminate against the activity not involved in the chemical system or to take into account its contributions to the net count rate (Chapter 4).

It may be necessary or expedient to use a radioactive nuclide which can undergo significant decay during the chemical investigation. In these cases, in order to compare results at different points in the process, it is necessary to correct all counts to the same point in time.

Among the advantages of using radiotracers we can list the following: (a) radiotracers are easy to detect and measure to sensitivities of 10^{-16} to 10^{-6} g with high precision; (b) the radioactivity is independent of pressure, temperature, and chemical or physical state; (c) radiotracers do not affect the system and can be used often in nondestructive techniques; (d) if the tracer is radiochemically pure, no interference from other elements can occur, as is common in chemical analyses.

18.2. Analytical chemistry

18.2.1. *Isotope dilution analysis*

In complex mixtures of compounds (for example, in organic synthesis or biochemical systems) it may be quite difficult to ascertain the exact amount of a specific component. A widely used technique of considerable value is *isotope dilution analysis.* A small radiochemically pure sample of the desired component is added to the mixture to be analyzed. This pure sample may either be an element or a labeled compound whose specific activity is known. The mixture is then separated by a process designed to isolate the desired element or compound in high purity but not necessarily in high yield (cf. §18.8). The separated sample is weighed and counted so that its specific activity can be calculated (Fig. 18.2). If the specific activity of the added pure sample is also known, the weight of the desired compound present in the original mixture w_u is calculated by (18.1):

$$w_u = \left(\frac{S_0}{S_m} - 1\right) w_0 \qquad (18.1)$$

S_0 and S_m are the specific activities respectively of the added radioactive sample and the final separated sample using an identical measurement technique, and w_0 is the weight of the added radioactive sample. The specific activity is defined by (4.49): $S = A/w$ dps g^{-1}; since only the ratio S_0/S_m is used in (18.1), A can be replaced by R because ψ is the same.

This technique is of particular advantage where quantitative separation of the desired

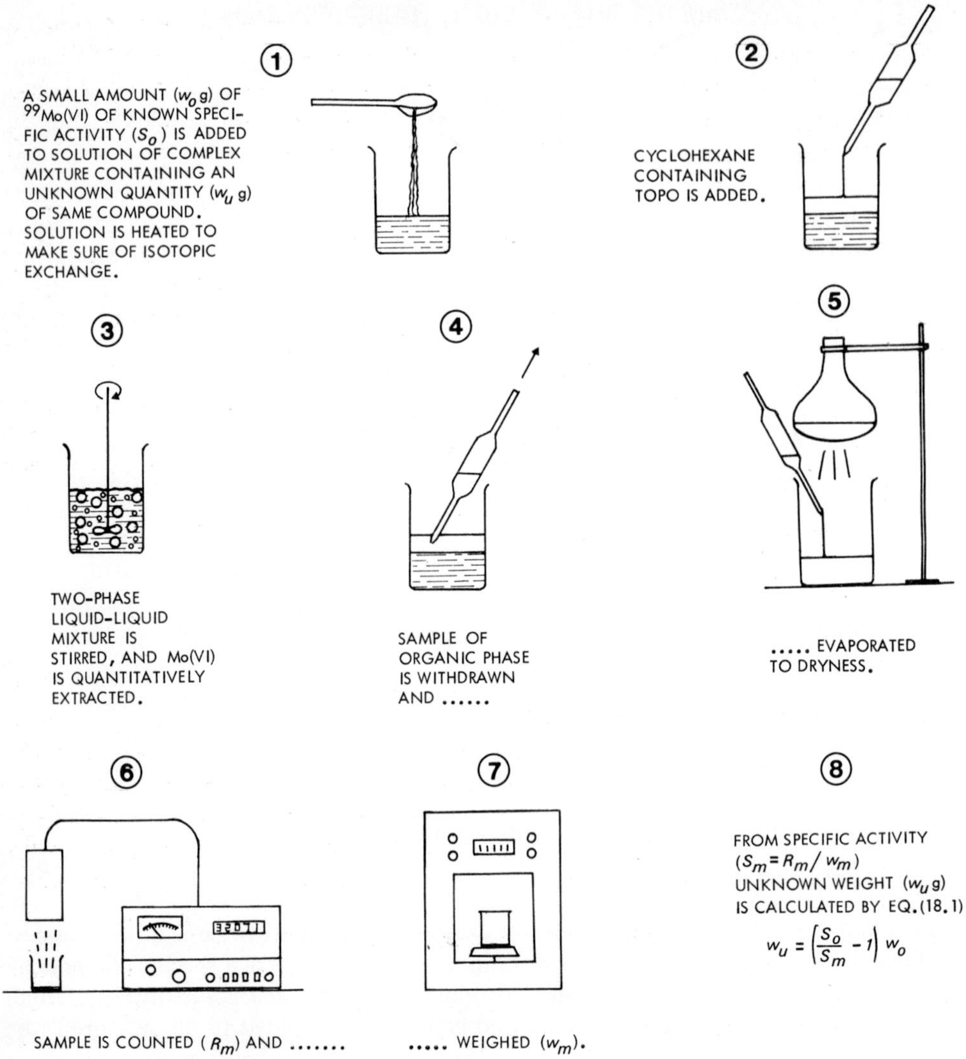
1
A SMALL AMOUNT (w_o g) OF ^{99}Mo(VI) OF KNOWN SPECIFIC ACTIVITY (S_o) IS ADDED TO SOLUTION OF COMPLEX MIXTURE CONTAINING AN UNKNOWN QUANTITY (w_u g) OF SAME COMPOUND. SOLUTION IS HEATED TO MAKE SURE OF ISOTOPIC EXCHANGE.
2
CYCLOHEXANE CONTAINING TOPO IS ADDED.
3
TWO-PHASE LIQUID-LIQUID MIXTURE IS STIRRED, AND Mo(VI) IS QUANTITATIVELY EXTRACTED.
4
SAMPLE OF ORGANIC PHASE IS WITHDRAWN AND ......
5
..... EVAPORATED TO DRYNESS.
6
7
8
FROM SPECIFIC ACTIVITY ($S_m = R_m / w_m$) UNKNOWN WEIGHT (w_u g) IS CALCULATED BY EQ.(18.1)
$w_u = \left(\frac{S_o}{S_m} - 1\right) w_o$
SAMPLE IS COUNTED (R_m) AND ....
..... WEIGHED (w_m).

FIG. 18.2. A sequence of steps for determining the amount of one compound in a complex mixture by the technique of isotope dilution. Mo (VI) is chosen as an example of an element to be determined, and solvent extraction as the separation technique.

compound is not feasible, as illustrated by de Hevesy in 1932. In determination of micro amounts of lead by anodic precipitation, quite varying results were obtained. By addition of a known amount of radio-lead and measuring the radioactivity of lead at the anode, the yield of the precipitation could be determined, and—although the electrolytic precipation was inefficient—an exact analysis was obtained.

In some cases the measurement of the final sample utilizes a technique other than weighing, but the principle remains the same. Isotope dilution is used, for example, in the determination of naphthalene in tar, of fatty acids in mixtures of natural fat, of amino acids in biological material, etc. For elements lacking a suitable radioactive isotope the technique is still useful if a highly enriched stable isotope is added and the analysis carried out by mass spectrometry.

J. Ruzicka and J. Stary have developed a radioanalytical technique referred to as *substoichiometric analysis* based on isotope dilution. Suppose the purpose is to determine small amounts of silver ions in an aqueous solution (w_u). A known amount (w_0, preferably of the same order of magnitude as w_u) of radioactive ^{110}Ag is added and silver is extracted by dithizone into chloroform. The specific activity of this solution is determined (S_m). A blank is run with exactly the same amount of dithizone on the w_0 amount of pure radiotracer (S_0). An important point is that the dithizone amount should be considerably less than required for extracting all the silver, i.e. substoichiometric. Again, (18.1) is to be used for calculating w_u.

Another application of this technique is immunoassay, developed by Rosalyn Yalow; it is now widely used for protein analysis in hospitals. A known mass w_0 of a labeled protein P* is allowed to react with a much smaller (substoichiometric) mass of an antibody A, so that a complex PA is formed. This is isolated and its radioactivity R_0 measured. At the same time and under identical conditions, an identical mass w_0 of the labeled protein is mixed with an unknown mass w_u of the protein to be determined. This sample is also allowed to react with the same amount of antibody A as before; the complex P*A is isolated (weight w_m) and its radioactivity R_m measured. The unknown weight w_u is calculated from (18.1).

18.2.2. *Radiometric analysis*

The term radiometric analysis is often used in a broad sense to include all methods of determination of concentrations using radioactive tracers. In a more restricted sense it refers to a specific analytical method which is based on a two-phase titration in the presence of a radioactive isotope. The endpoint of the titration is indicated by the disappearance of the radioisotope from one of the phases. A simple, rapid analysis of inorganic systems may be performed, for example, by adding a slight excess of radioactive reagent to a solution of the unknown which results in quantitative precipitation of the latter.

Suppose one wishes to determine the concentration of Ag^+ in a solution by titration with NaI solution. If radioactive iodine is included in the latter solution, the AgI precipitate will be radioactive but the solution registers no activity until all Ag^+ has been precipitated. If the radioactivity of the solution is measured and plotted against the volume of the titrant added, one obtains a curve A as in Fig. 18.3. Conversely, if the unknown to be determined is the concentration of iodide, the same titration technique may be used where trace amounts of radioactive iodine is added to the iodine solution and this solution titrated with Ag^+ ion. In this case curve B of Fig. 18.3 is obtained. In either situation the intercept with the abscissa indicates the endpoint of the titration and provides a direct measure of the concentration of the unknown solution. The technique is general and can be applied to any case of quantitative precipitation. If the precipitation is not quantitative, isotope dilution techniques must be used.

The radiometric titration technique can be extended to two-phase liquid–liquid systems. For example, to determine the concentration of Fe^{2+} in an aqueous solution, trace amounts of radioactive Fe^{2+} may be added to the solution and the solution adjusted to a pH of 2. A volume of chloroform is added to form a second liquid phase and the mixed system is titrated with a solution of oxine in chloroform. An iron–oxine complex forms which extracts into the organic phase, carrying with it the radioactivity. In this experiment the radioactivity of the aqueous phase follows curve (B), while that

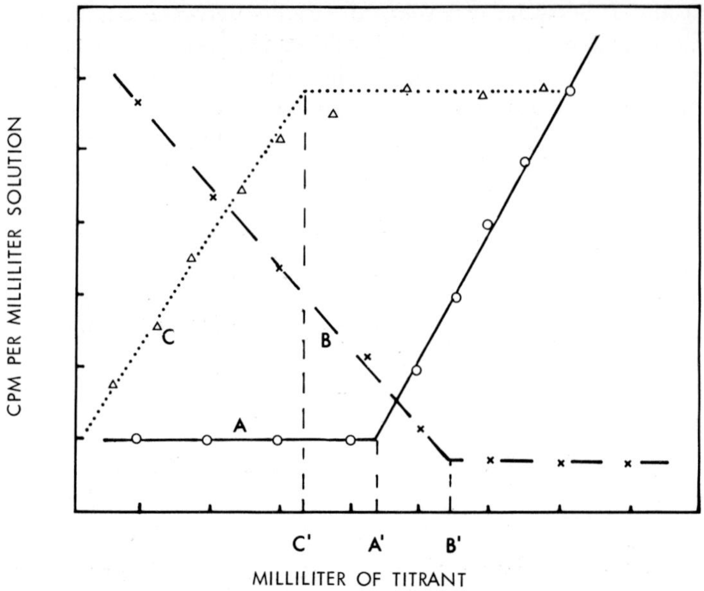

FIG. 18.3. Radiometric titration curves. A, Radioactive $^{129}I^-$ is added to solution of unknown Ag^+ content. B, Inactive Ag^+ is added to unknown iodide solution, to which trace $^{129}I^-$ was added initially. C, Two-phase titration: an aqueous phase with an unknown concentration of Fe^{2+} to which trace ^{55}Fe was added; the titration is carried out with oxine in $CHCl_3$. The measurements are of the organic phase.

of the organic phase follows curve C in Fig. 18.3. When all the Fe^{2+} has been extracted into the organic phase the radioactivity of this phase is constant upon the addition of more oxine to the system. Such a method permits the determination of very small amounts, as low as 10^{-8} M, with an error of only a few percent. Instead of using an organic solvent as the complex extractant phase, cation and anion ion exchange resins can be used.

Since radiometric analysis is so simple and rapid and the chemical purity of the counting sample relatively unimportant, it is very useful for checking yields of various steps in a complicated analytical or synthetic process. In the analysis of fission product yields it is often necessary to perform a series of chemical separations, isolating various elements in each step. The efficiency of separation as well as the level of decontamination can be checked for each step by the use of appropriate radioactive tracers.

Chromatography, distillation, electrodeposition, extraction, precipitation, and co-precipitation are some of the techniques employed in analytical chemistry that have been studied with radioactive tracers. The radiometric method is particularly useful when ordinary analytical techniques encounter difficulty due to the nature of the precipitate. Moreover, radiometric methods of analysis are relatively simple to automate, making them advantageous in routine work.

18.2.3. *Activation analysis*

Activation analysis is a highly sensitive nondestructive technique for qualitative and quantitative determination of atomic composition of a sample. It has been particularly

TABLE 18.1. *Limits of detection for 71 elements in a thermal–neutron flux of* 10^{13} n cm^{-2} s^{-1} (1 h irradiation)

Limit of detection (μg)	Elements
$1–3 \times 10^{-6}$	Dy
$4–9 \times 10^{-6}$	Mn
$1–3 \times 10^{-5}$	Kr, Rh, In, Eu, Ho, Lu
$4–9 \times 10^{-5}$	V, Ag, Cs, Sm, Hf, Ir, Au
$1–3 \times 10^{-4}$	Sc, Br, Y, Ba, W, Re, Os, U
$4–9 \times 10^{-4}$	Na, Al, Cu, Ga, As, Sr, Pd, I, La, Er
$1–3 \times 10^{-3}$	Co, Ge, Nb, Ru, Cd, Sb, Te, Xe, Nd, Yb, Pt, Hg
$4–9 \times 10^{-3}$	Ar, Mo, Pr, Gd
$1–3 \times 10^{-2}$	Mg, Cl, Ti, Zn, Se, Sn, Ce, Tm, Ta, Th
$4–9 \times 10^{-2}$	K, Ni, Rb
$1–3 \times 10^{-1}$	F, Ne, Ca, Cr, Zr, Tb
10–30	Si, S, Fe

useful for simultaneous determination of elements in complex samples (minerals, environmental samples, biological and archeological objects, etc.), because it provides a simple alternative to much more difficult, tedious, and destructive techniques. Its main limitation is the demand for a strong irradiation source.

In activation analysis advantage is taken of the fact that the decay properties such as the half-life and the mode and energy of radioactive decay of a particular nuclide serve to identify uniquely that nuclide. A sample is irradiated to form an amount R of radioactive nuclide according to the relationship (§10.2):

$$R = \psi\phi\sigma N(1 - e^{-\lambda t_{irr}})e^{-\lambda t_{cool}} \tag{18.2}$$

We assume that irradiation is carried out by a homogeneous particle flux ϕ, as is the case for neutrons in a reactor (neutron activation analysis, NAA), which is the most common technique.

The minimum amount of an element which can be detected in activation analysis increases with the sensitivity of the measuring apparatus ψ, the bombarding flux ϕ, the reaction cross-section σ, the irradiation time t_{irr} up to saturation activity), and the decay constant λ of the radioactive nuclide formed. Table 18.1 shows the limits of detection in NAA. Elements with very low sensitivity for thermal neutron bombardment (e.g. the lightest elements) can often be measured through irradiation with either fast neutrons (FNNA) or charged particles (CPAA). Thus oxygen can be analyzed by bombardment with 14 MeV neutrons ($\sigma = 37$ mb) yielding ^{16}N, which decays ($t_{1/2}$ 7.3 s) by emitting energetic β and γ (6–7 MeV). In FNAA and CPAA the flux may not be homogeneous, and (10.7)–(10.9) must be used.

Figure 18.4 shows a typical NAA spectrum obtained with a multichannel analyzer equipped with scintillation (upper curve) or solid state (lower curve) detectors. Each peak can be ascribed to a certain γ-energy, which in most cases identifies the nuclide. A number of nuclides can be identified simultaneously with solid state detectors (SSD), but with NaI(Tl) scintillation detectors the poor resolution limits simultaneous multi-nuclei analysis.

The area under the peak (shaded area in Fig. 17.5(b)) is proportional to the amount of

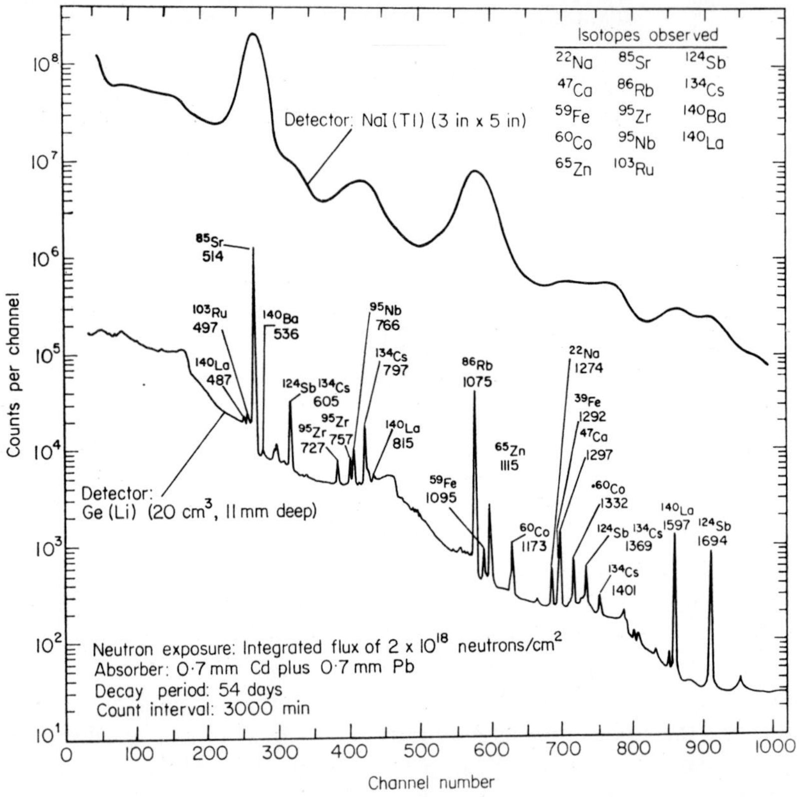

FIG. 18.4. Gamma-spectrum of neutron-activated sea water. (From J. A. Cooper, N. A. Wogman, H. E. Palmer and R. W. Perkins.)

the radioactive nuclide. If all other factors in (18.2) are known, the number of target nuclides N can be calculated.

When complex mixtures are irradiated, such as geological or biological samples, there may be some difficulties in peak assignment. The energy spectrum is scanned at repeated time intervals and from the decrease of the peak area with time the half-life of the peak may be established. This is a valuable additional aid in the assignment of the peak to a certain nuclide.

The magnitude of the information collected with this technique requires advanced instrumental systems (INAA, instrumental NAA). In INAA the information is stored on magnetic tapes, disks, etc., for later acquisition. Often a multichannel analyzer is directly connected to general purpose computers. Such instrumental technique is a requirement in prompt γ or photon analyses, i.e. when the nuclide of interest is detected through the immediate release of a γ or photon upon reaction with the bombarding particle. Spectra of this type are shown in Fig. 14.27.

While in principle it is possible to calculate the amount of the desired element through the use of the proper values for the cross-section, flux, irradiation time, and half-life in (18.2), a simpler approach has been developed that avoids errors implicit in the uncertainties of each of these values. The unknown and a known standard of similar composition are irradiated and counted in an identical fashion. A direct comparison can then be made according to the following relationship:

$$\frac{\text{weight of impurity in unknown}}{\text{weight of impurity in standard}} = \frac{\text{activity of impurity in unknown}}{\text{activity of impurity in standard}}$$

or

$$\frac{w_{\text{iu}}}{w_{\text{io}}} = \frac{R_{\text{iu}}}{R_{\text{io}}} \tag{18.3}$$

Sometimes such a large number of competing radioactivities are produced in the bombardment process that it may be necessary to conduct some chemical purification (RNAA, radiochemical NAA). This is particularly true if simple counting of β-activity or γ-ray spectrometry using NaI(Tl) counting is used. However, with the development of SSD and INAA the increased resolution in the spectrum allows a simultaneous determination of as many as 15 or 20 competing radioactivities usually without the necessity of chemical purification. For the technical details of activation analyses the reader is referred to the chapters on production of radionuclides (especially §10.2) and detection and measurements of radiation (especially §17.5 on solid state detectors and §17.6 on scintillation detectors). Using the technique indicated by (18.3), geometry factors are less important in sample counting although the considerations in §§17.9–17.11 about sample preparation and statistics are of importance for accurate activation analysis.

Analysis of trace constituents in air and water, in soil and geological samples, in marine and in biological systems are some of the interesting applications of the NAA technique. Examples of on-line NAA include sorting ore minerals and oilwell logging. In forensic science, by using NAA to measure the composition of the material adhering to a hand which has held a gun during firing, it is possible to determine the type of ammunition and even the number of shots fired. Trace metal analysis of plants can be used to determine the location in which that plant has been grown (used, for example, for identification of marijuana growers). The trace constituents of archeological and art objects play an important role in ascertaining their authenticity and the identification of place of origin; the use of nondestructive NAA has been extremely valuable in this field. Activation analysis of the mineral content of pigments has enabled scientists to determine the authenticity of paintings attributed to certain artists since, in times past, each artist prepared his own paints by distinctive and individual formulae. A painting entitled "Christ and Magdalen" done in the Old Dutch style was proved to be a twentieth century forgery when NAA showed < 7 ppm silver and < 1.3 ppm antimony in the white lead paint. The sixteenth and seventeenth century Dutch paintings have white lead of about 10–1000 ppm silver and 50–230 ppm antimony.

It has been found that hair contains trace metals (e.g., Cu, Au, Ce, Na) in ratios which are typical for a particular individual, and activation analysis can be used to identify hair from a particular person. This application achieved public notice some years ago when it was found that hair from Napoleon had a relatively large amount of arsenic, indicating that some time prior to his death he had received large doses of arsenic. Through analysis of the hair of the Swedish king Eric XIV (who died suddenly in 1577 after a meal of peasoup) it has been found that he must have received lethal amounts of arsenic as well as large amounts of mercury. The latter is assumed to have been taken into his body through the use of a mercury compound for treatment of an old wound.

TABLE 18.2. *Estimated minimum detectable concentrations of pollutant elements in seawater by INNA and by NAA with separations*

Trace element	Typical reported concentrations in open ocean ($\mu g\, l^{-1}$)	Minimum detectable concentrations ($\mu g\, l^{-1}$)	
		INAA[(a)]	NAA with separations[(b)]
Hg	0.02–0.2	0.05	0.001
Cd	0.06–0.7	16 000	0.001
Ag	0.002–0.05	1.0	0.003
As	2–3	Not possible	0.000 1
Cu	0.5–2	Not possible	0.002
Cr	0.02–0.6	0.3	0.003
Zn	0.5–10	0.2	0.01
Sn	0.02	Not possible	9
Se	0.08	0.2	0.02
Sb	0.2	0.02	0.000 03

[(a)]25 ml sea water; 24 h irradiation at 10^{13} n cm^{-2} s^{-1}; 40 d decay; 1000 min count on 20 cm^3 Ge(Li) detector; based on 3 × above background-Compton contribution in peak areas.

[(b)]500 ml sea water; elements chemically separated; 24 h irradiation at 10^{13} n cm^{-2} s^{-1}; 3 d decay; 500 min count on a 20 cm^3 Ge(Li) detector; based on twice background-contribution in peak areas.

The high sensitivity of activation analysis has made it very useful in environmental pollution studies. Table 18.2 lists the limits of detection for some elements in sea water under the conditions specified in the table.

18.2.4. *Autoradiography*

Autoradiography is a technique whereby the distribution of a radioactive element or compound in a composite matrix is made visible either to the naked eye or under a microscope. It is based on the blackening of photographic films when exposed to nuclear radiation (§§14.11.2 and 17.1). The technique is best illustrated by a few examples.

Lead is an unwanted impurity in stainless steel even in very small amounts. In order to investigate the mechanism of its incorporation, ^{212}Pb was added to a steel melt. After cooling, the ingot was cut by a saw and the flat surface machine-polished and etched in an electrolytic bath (electropolishing). This provides a very flat and "virgin" surface. A photographic film was placed firmly under even pressure against the metal surface, and the film was exposed to the radiation from ^{212}Pb in darkness in a cool room for about a week. After development of the film the darkened areas showed where on the metal surface lead was to be found. It turned out that it was rather evenly distributed in the last solidification section. By taking this into account in the production, the negative effect of lead in the raw material could be reduced. An autoradiograph produced by the same technique is shown in Fig. 17.2(b).

If it is difficult to incorporate a radioactive trace element in a matrix material, a pure surface can be irradiated by neutrons or charged particles. In this case not only the distribution of an element can be determined, but also the composition, by using both autoradiography and γ-spectrometry. This technique is only useful for trace concentrations in a simple matrix. A more elaborate technique for the same purpose is the electron scanning microscope (ESM) in which induced X-ray fluorescence is used for elemental identification.

In another technique, a polished surface of the specimen (metal, mineral, etc.) is

FIG. 18.5. (a) Photograph and (b) autoradiograph of mineral specimen treated with ^{35}S labeled xanthate solution. Zincblende has sorbed xanthate while pyrites are almost inactive. (According to L. G. Erwall.)

dipped into a solution containing a radioactive reagent, which selectively reacts with one of the constituents of the surface. Figure 18.5 shows an autoradiograph obtained after a mineral was dipped into potassium ethyl xanthate labeled with ^{35}S; the xanthate reacts selectively with zincblende, ZnS, in the sample. The low β-energy, E_{max} 0.2 MeV, is an advantage to the technique because the resolution of the autoradiogram will be better. This technique is similar to the staining of microscopic samples in order to make particular sections visible.

Autoradiography has become important in biology especially for the study of distribution and metabolism of compounds administered to the plant or animal. Some of these applications are described in §§18.3.1 and 18.6.

18.3. Inorganic, organic, and biochemistry

In no other area have radioactive tracers played such an important role as in the studies of chemical and biological reaction paths. This is, of course, due to that in each radioactive decay a single atom announces its position. Thus the detection sensitivity is at the ultimate limit. A radioactive nuclide, as for example ^{14}C, ^{32}P, or ^{198}Au, can be followed through any number of different chemical reaction steps and never lost, revealing details of metabolic or process reactions impossible to discover by other techniques. We can here only indicate some representative areas of application, and shall limit ourselves to more principal studies.

18.3.1. *Determination of chemical reaction paths*

The use of radioisotopes in the study of the steps in a chemical reaction system is well established. Let us consider a few examples to illustrate this technique.

If phenol is labeled with deuterium or tritium (T) in the hydroxyl group and then warmed to a temperature slightly below decomposition, one finds that the labeled hydrogen migrates to other hydrogen positions of the benzene ring either by intramolecular rearrangement or by intermolecular reactions. However, using C_6H_5OT and C_6H_4TOH, one obtains C_6H_4TOT, which can be formed only through intermolecular collisions, thus eliminating intramolecular rearrangement as the reaction mechanism.

The study of the reaction steps in the photosynthesis of carbohydrates from atmospheric CO_2 in the presence of light and chlorophyll is an outstanding example of the value of the tracer technique. The overall process (which involves many steps) can be written as

$$6CO_2 + 12H_2O \xrightarrow[\text{chlorophyll}]{\text{light}} C_6H_{12}O_6 + 6O_2 + 6H_2O$$

Using ^{14}C, ^{32}P, and T, M. Calvin (Fig. 18.1) and co-workers have been able to identify the intermediate steps involved. Plants were placed in atmospheres containing ^{14}C labeled CO_2 and irradiated with light. The plants were removed after different irradiation times and the molecular components separated using various types of partition chromatography techniques (§18.8). The presence of radioactive carbon in a compound was taken as proof that the compound was involved in the process of photosynthesis (Fig. 18.6). Similarly, the involvement of phosphorus and hydrogen was determined through the detection of radioactive phosphorus and tritium in the compounds.

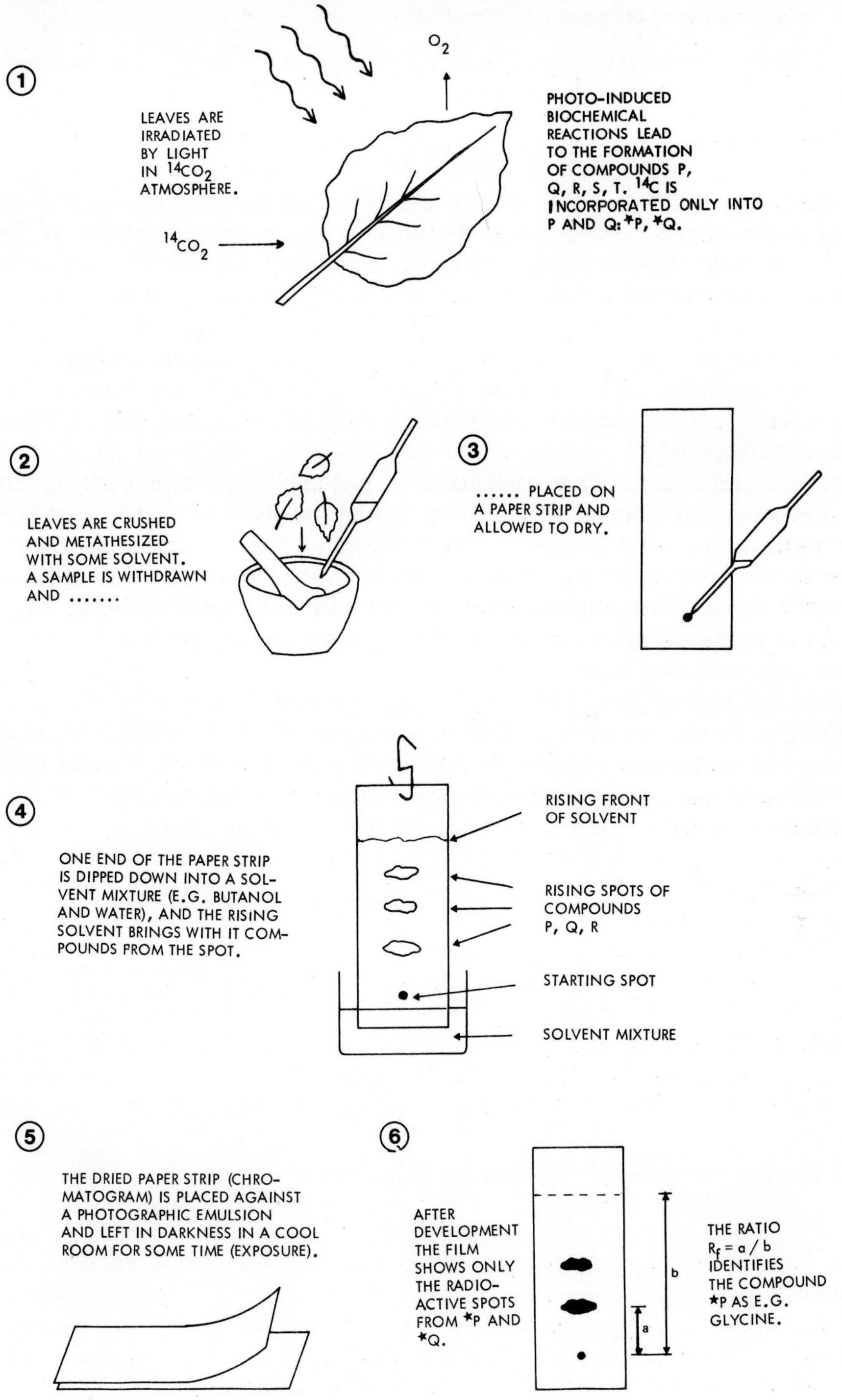

FIG. 18.6. Method used to determine which of the components P, Q, R, S, T in a leaf are involved in photosynthesis.

18.3.2. *Determination of chemical exchange rates*

If two different chemical species with some element in common are mixed in solution, exchange of this common component may occur. The chemical equation would have the form

$$AX + BX^* = AX^* + BX$$

Since the type and concentration of the chemical species remain unchanged, it is impossible to observe the exchange unless the atoms in one reactant are labeled (cf. §§2.6 and 2.7, and Appendixes B and C on isotope exchange). By using X*, a radioactive isotope of X, the reaction may be followed, and at equilibrium the activity should be uniformly distributed between the two chemical species, i.e. the specific activity of X* will be the same for both AX and BX. Of course, if AX and BX are both strong electrolytes, uniform distribution is essentially immediate upon mixing. If at least one of the reactants is an inorganic complex or an organic molecule, the exchange may be measurably slow if it occurs at all.

Since the chemical form of the reactants is not altered by the isotopic exchange, there is no change in heat content. However, the entropy of the total system is increased when uniformity in the distribution of the isotopes of X is achieved throughout the system. This entropy increase provides a decrease in the free energy, making isotopic exchange a spontaneous reaction. Despite this spontaneity, the exchange may be prevented or made very slow by a large energy of activation requirement in the formation of a necessary transition state.

For the exchange reaction represented above, the rate of increase of AX* is equal to the rate of formation minus the rate of destruction of AX*. The rate of formation is the product of the rate of reaction k_r, the fraction of reactions which occur with an active BX*, and the fraction of reactions which occur with an inactive AX. Using the following notation

$$a = [AX] + [AX^*] \tag{18.4a}$$

$$b = [BX] + [BX^*] \tag{18.4b}$$

$$x = [AX^*] \tag{18.4c}$$

$$y = [BX^*] \tag{18.4d}$$

the rate of formation k_f is equal to

$$k_r \frac{y}{b} \frac{(a - x)}{a} \tag{18.5a}$$

In a similar fashion, the rate of destruction k_d is equal to

$$k_r \frac{x}{a} \frac{(b - y)}{b} \tag{18.5b}$$

Therefore

$$\frac{dx}{dt} = k_f - k_d = \frac{k_r}{ab}(ay - bx) \tag{18.6}$$

The solution of this equation is

$$\ln(1 - F) = -\frac{(a + b)}{ab} k_r t \tag{18.7}$$

where $F = x_t/x_\infty$ (x_∞ is the value of x_t at $t = \infty$, i.e. equilibrium). The rate of exchange k_r is evaluated from the slope of a plot log $(1 - F)$ versus t. If more than one rate of exchange is present due to exchange with nonequivalent atoms in a reactant, it may be difficult to resolve this curve sufficiently to obtain values for the reaction rates. Isotopic exchange is a standard tool of the scientist studying the kinetics of chemical reactions whose half-lives are longer than a minute.

One example of isotope exchange can be used to illustrate the value of these studies. Consider the exchange between di- and trivalent chromium in $HClO_4$ solutions. If the total chromium ion concentration is 0.1 M it takes 14 days for the exchange to reach 50% completion at room temperature. Inasmuch as the di- and trivalent cations are both positively charged, it is unlikely that they can approach each other closely enough to exchange an electron directly to allow a reversal of oxidation state, and a more likely mechanism is that an anion is involved as a bridge between the two cations such that the intrusion of the anion reduces the repulsion between the two cations. If this model of the isotopic exchange mechanism is valid then it could be proposed that the reaction mechanism would be

$$\mathrm{Cr(III)^* + X^- + Cr(II) \rightarrow [Cr^*{-}X{-}Cr]^{4+} \rightarrow Cr(II)^* + X^- + Cr(III)}$$

Such a reaction mechanism would be fostered by the presence of anions that form complexes more readily than perchlorate ion. If HCl solutions are used rather than $HClO_4$ it is found that the exchange takes place more rapidly and the half time of exchange is only 2 min, which agrees with the proposed mechanism since chloride ions are known to be more favorable to complex formation than perchlorate ions. Without the use of radioactive chromium to label one of the original oxidation states there would be no means of identifying the exchange.

18.3.3. *Determination of equilibrium constants*

The sensitivity of tracer detection also makes measurement of solubilities of relatively insoluble compounds rather simple.

This may be illustrated by the first radioactive tracer experiment of de Hevesy and Paneth in 1913 in which the solubility of lead chromate was determined. Chromate ions were added to a $PbCl_2$ solution containing a known amount of ^{210}Pb (RaD), precipitating all lead as $PbCrO_4$. The precipitate contained 2030 "radioactive units", and had a weight of 11.35 mg. The specific activity was thus 2030/11.35 = 179 "units" mg^{-1}. By shaking the precipitate with water, some of it dissolved: 2.14 units per 1000 ml were measured. The solubility was calculated to be 2.14/179 = 0.012 mg l^{-1} or 3.7×10^{-8} M Pb^{2+}. If $[Pb^{2+}] = [CrO_4^{2-}]$ the solubility product would be $K_s = (3.7 \times 10^{-8})^2 \approx 1 \times 10^{-15}$. The modern value is 2×10^{-14}.

Since this early experiment, measurements of chemical equilibria using radioactive isotopes have become a well-established technique, particularly for determination of metal stability constants β_n. These are defined by the relation

$$\beta_n = [ML_n]/[M][L]^n \tag{18.8}$$

where the charges of the species have been omitted for the sake of simplicity (e.g. $M = Pu^{4+}$, $L = Cl^-$). Among the techniques for determination of β_n, radioactive tracers combined with solvent extraction and ion exchange have been particularly useful. Let us first consider solvent extraction.

Metals can be made soluble in an organic solvent (kerosene, chloroform, etc.) by formation of neutral complexes with organic ligands, e.g.

$$M^{2+} + zHA \rightleftharpoons MA_z + zH^+ \tag{18.9}$$

HA may be a weak organic acid like acetylacetone, oxine, dithizone, etc. The distribution of the metal D between the organic phase and water is governed by the relation

$$D^{-1} = \frac{[\text{metal}]_{aq}}{[\text{metal}]_{org}} = (\lambda_z \beta_z [A]^z)^{-1} \sum_{n=0}^{N} \beta_n [A]^n \tag{18.10}$$

The distribution D is easily measured for radioactive trace metals; z is constant (z = 2 for Sr^{2+}, 3 for La^{3+}, 2 for UO_2^{2+}, etc.), as are the β_n and the λ_z. The latter is defined for n = z by

$$\lambda_n = [MA_n^{z-n}]_{non\text{-}aq} / [MA_n^{z-n}]_{aq} \tag{18.11}$$

This equation is valid for all two-phase systems; in the present case the organic solvent is the non-aqueous phase. Therefore, D depends only on [A], the free ligand concentration in the aqueous phase, according to (18.10). Figure 18.7 shows measurements of the distribution of Pu(IV) between water and three different organic solvents as a function of the acetylacetonate ion concentration $[Aa^-]$ in the aqueous phase. From these curves the values of the β_n (i.e. the formation constants of $PuAa^{3+}$, $PuAa_2^{2+}$, $PuAa_3^+$ and $PuAa_4$) and that of λ_z (i.e. distribution constant for $PuAa_4$) were calculated.

If the system studied yields complexes MB_p which are not soluble in an organic solvent, the previous system (MA_n) is used in combination with the ligand B^-. For example, the complexation of Np^{4+} by HSO_4^- was studied in the two-phase system 0.1 M $NaClO_4$ and $CHCl_3$ containing the organic reagent ("extractant") thenoyltrifluoroacetone (HTTA). In this case the relation

$$D^{-1} = a + b \sum_{p=0}^{N} \beta_p [B]^p \tag{18.12}$$

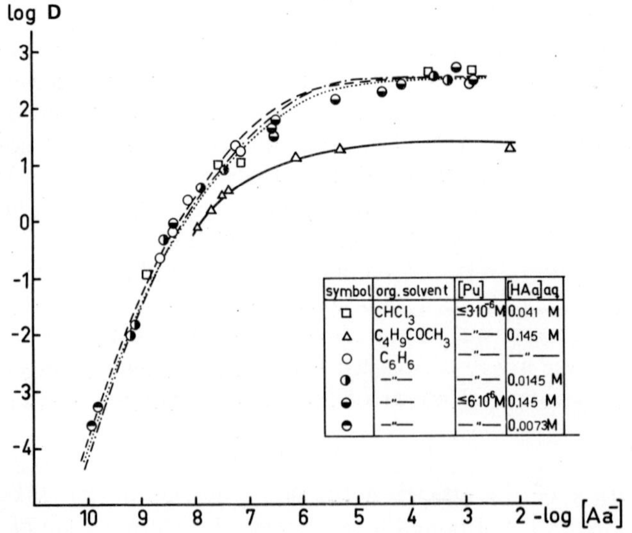

FIG. 18.7. Distribution ratio D of Pu(IV) between different organic solvents and 0.1 M $NaClO_4$ solution as a function of aqueous acetylacetonate (Aa^-) concentration.

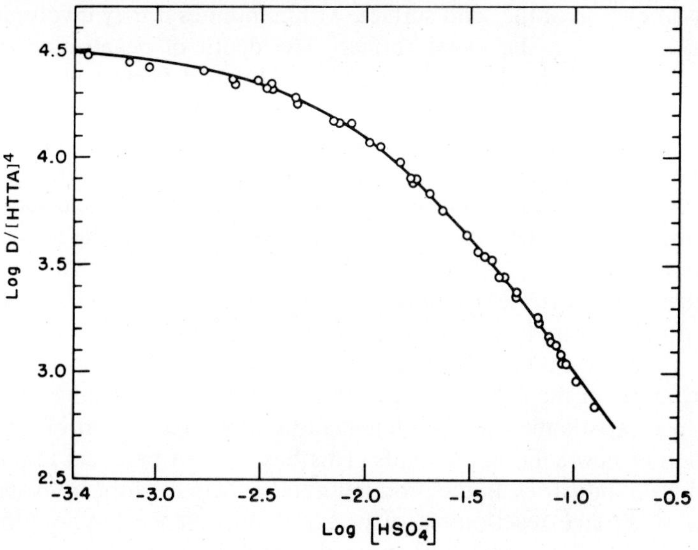

FIG. 18.8. Distribution of Np(IV)–TTA complex between benzene and water as a function of aqueous HSO_4^- concentration. (According to J. Sullivan and C. Hindman.)

is obtained, where *a* and *b* are constants for the Np–HTTA system (pH and [HTTA] must not change). From the data in Fig. 18.8 the formation constants for $Np(SO_4)^{2+}$ and $Np(SO_4)_2$ were determined.

These relations hold for metal concentrations down to $< 10^{-12}$ M. The technique has been used extensively for studying complexation for transuranium elements, for which it may not be possible or desirable (to avoid radiolysis) to use macroscopic amounts.

The ion exchange technique depends on the distribution of metal cations between organic cation exchange resins and water according to relation (18.11)

$$\lambda_p = [MB_p^{z-p}]_{resin}/[MB_p^{z-p}]_{aq} \tag{18.13}$$

where $z > p$, and λ_p is measured in amount of metal species per gram (air-dried) resin divided by amount of same metal species per milliliter solution. When several complexes are formed (MB^{z-1}, MB_2^{z-2}, etc.), the distribution Q of (radioactive) metal between the resin and the aqueous phase follows the relation

$$0 = \sum_0^N (Q - \lambda_p)\beta_p[B]^p \tag{18.14}$$

From measurement of Q as a function of [B], λ_p and β_p can be evaluated. For $p > 2$ this technique becomes more complicated than solvent extraction.

Because of the importance of ion exchange and solvent extraction for isolation of pure actinides and trace concentrations of metals, these techniques are described further in §18.8.

18.3.4. *Studies of surfaces and reactions in solids*

Investigation with radioactive tracers has shown that very rapid exchange takes place between atoms on a metal surface and the metallic ions in solution. While the exchange is a function of the solid surface, within minutes it may involve atoms several hundred layers deep in the metal surface. The depth of penetration of the sorbed radioactive isotopes from the solution can be obtained from a measurement of the absorption of the radiation. With the same technique the diffusion of atoms in their own solid matrix can be studied. For example, using single crystals of silver suspended in a solution containing silver nitrate labeled with ^{110m}Ag it has been possible to demonstrate different rates of diffusion into different faces of the crystal. The surface area of solids can be determined readily by measurement of the sorption of radiotracers on the surface.

If a radioactive gas is incorporated in a crystalline compound the amount of gas released can be measured as a function of the temperature (the *emanation ability*). It is found that the emanation increases considerably at certain temperatures, indicating structural changes in the solid matter at those temperatures. Studies of diffusion and emanation play a valuable role in understanding the mechanism of sintering and in the formation of new solid compounds. This has been of practical importance in the cement and glass industries, in the production of semiconductors, in the paint industry, etc. Studies of surface reactions is of practical importance for flotation, corrosion, metal plating and finishing, and detergent action to name only a few applications.

18.4. Studies of liquid flows

Radiotracers play an important role in *hydrology*. By using T_2O, $^{24}Na^+$, $^{82}Br^-$, ^{51}Cr-EDTA, and other tracer elements or compounds, the volume of natural water reserves (even underground) has been determined, and movement of ground and surface water and effluent pathways mapped. Also the consumption of water and water flow in industries are readily determined and leaking dams and pipes checked by radiotracers.

The simple technique of identifying *pipe leaks* is illustrated in Fig. 18.9. This technique

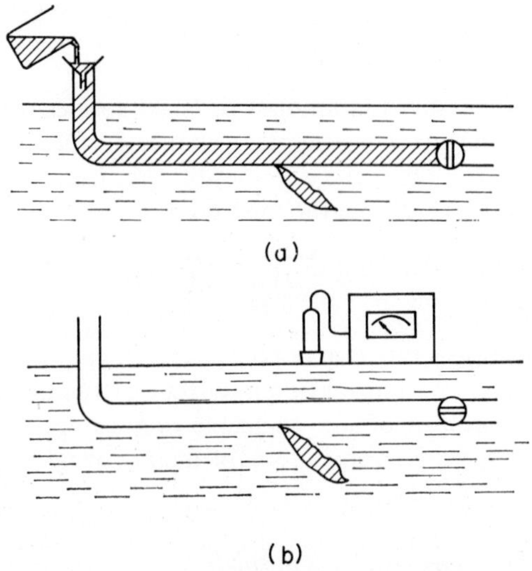

FIG. 18.9. Technique for identifying leaks in hidden pipe lines: (a) the radioactive nuclide is added to the liquid in the pipeline, (b) the radioactivity from the leak is identified when the radioactivity of the pipe has been flushed away.

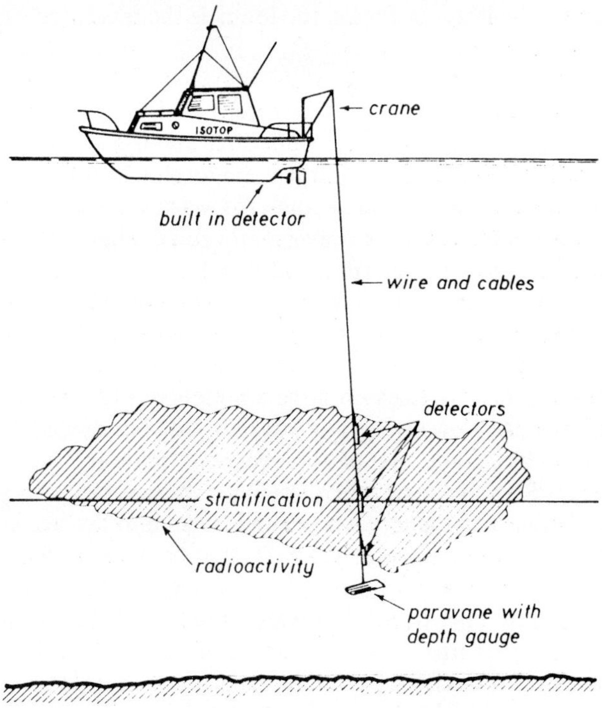

FIG. 18.10. Technique used for measuring stream lines in the Öresund strait. (According to Danish Isotope Centre, 1965.)

is used not only for liquids, but also for checking the seal of underground electric cable hoses in which case gas tracers like ^{85}Kr or ^{133}Xe are used.

The *liquid volume* of a closed system may be difficult to calculate from external container dimensions, particularly if there is a mixing action, either by external circulation or by internal stirring. For example, sulfuric acid volume is desired in an alkylation plant where extensive intermixing between acid and hydrocarbon prevents a well-defined level from forming. ^{134}Cs is added to the H_2SO_4 and from the dilution of the added tracer the total volume of H_2SO_4 is calculated.

Another method is applicable to tanks through which there is a known constant flow. A tracer batch is put into the incoming line. Mathematical treatment, assuming complete immediate mixing of the incoming stream with the vessel contents, predicts that the tracer concentration in the tank falls off exponentially with a rate determined by throughput F and volume V according to the equation

$$R = R_0 e^{-Ft/V} \tag{18.15}$$

where R_0 is the counting rate at zero time (break through of radioactivity at tank exit) and t is the time when the count rate R is obtained at the same point.

The *flow rate* F of rivers and streams can be measured by injection of a radionuclide and measurement of the time for its arrival at detectors placed downstream. Because of turbulence the radioactive "cloud" becomes quite diffuse. Therefore a more efficient technique called "total count" is used. A known amount (A_0) of the radionuclide is injected into the river and a downstream detector registers a total count (R_{tot}) as the radioactivity passes; the faster it passes, the lower is the measured radioactivity. Thus

$$F = \psi A_0 R_{\text{tot}}^{-1} \tag{18.16}$$

ψ is the calibration factor which has to be determined under known conditions. The technique takes into consideration both longitudinal and transversal mixing.

In order to avoid hazardous pollution it is important to discharge communal and industrial waste (silts, liquids, gases) so that the wastes are properly dispersed. The mapping of different dispersal sites is conveniently and commonly carried out by injecting a radioactive tracer at a testing spot, and then following its distribution at various depths, heights, and directions (Fig. 18.10). In such a test it was found that sludge emptied into the River Thames at one point traveled upstream, which led to a repositioning of the sludge pipe exit.

Nuclear weapons tests have released large amounts of radionuclides into the atmosphere, which through their own weight or by rain have been carried down to the earth's surface. Geophysics has made use of this weapon "fall-out". By measurements on T (as HTO water), ^{90}Sr, ^{137}Cs, and other fission products it has been possible to follow the movements of water from land via lakes and rivers into the sea, as well as to study the water streams of the oceans and the exchange between surface and deep water. As a result the circulation of water on earth has been mapped in quite detail. It has also been possible to analyze how tropical hurricanes are formed by measuring the water taken into the central part (the "eye") of the cyclone, since the HTO concentration in the normal atmosphere is different from that of surface ocean water.

18.5. Some industrial uses of radiotracers

Industry has applied radiotracers in a very large variety of ways. More than half of the 500 largest manufacturing concerns in the United States use radioisotopes in the production of metals, chemicals, plastics, pharmaceuticals, paper, rubber, clay and glass products, food, tobacco, textiles, and many other products. Radioisotopes are used to study mixing efficiency, effect of chamber geometry, residence time in reactors, flow rates and patterns in columns and towers for fractionation, absorption, racemization, etc. Some of the many uses are listed in Table 18.3 and a few are described below to reflect the scope and value of the industrial applications of radioisotopes. In general the radionuclide used is not isotopic with the system studied.

Mixing is an important mechanical operation in many industries. Poor mixing may

TABLE 18.3. *Some industrial uses of radiotracers*

The path of pellets in vanadium ore sintering kilns
Mixing study of bleaching tower pulp mill
Determination of degree of impregnation of wood with fungicides
The path of wood chips in a kraft pulp digester
Activation analysis of metals in pulp
Tracing the diffusion of sulfur in cable rubber
Efficiency studies of fabric washing techniques
Tracing detergents in sewage systems
Measurement of traffic paint abrasion
Studies of diffusion in semiconductor materials
Location of leaks in storage tanks, pipe lines, telephone cables, etc.
Leak testing of hermetically sealed components
Studies of intermetallic diffusion
Rate of carbon diffusion into iron

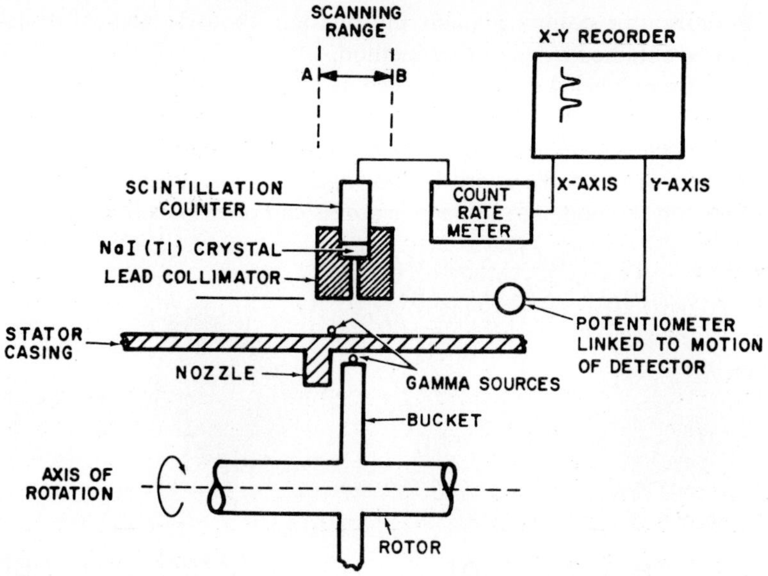

FIG. 18.11. Monitoring the axial position of a turbine. (According to W. W. Schultz, H. D. Briggs, R. A. Dewes, E. E. Godale, D. H. Morley, J. P. Neissel, R. S. Rochlin, and V. V. Verbinski.)

give an unsatisfactory product and low yield of the operation; unnecessary mixing is a waste of time and energy. By adding a radionuclide to the mixing vessel, or by labeling one of the components, the approach to mixing equilibrium can be followed either by external measurement or taking samples at different time intervals. Among examples of this technique are mixing of cement, gravel, sand, and water to concrete measured by using irradiated pebbles. The homogeneity of glass melts can be determined by adding $^{24}NaHCO_3$ to the melt; organic ^{95}Zr compounds have been used to follow homogenization of oil products. Other uses have been the addition of vitamins to flour, coal powder to rubber, etc.

Wear and materials transfer is easily followed if the material undergoing wear is made radioactive. This has been used for studying wear of parts in automobile engines, cutting tools, ball-bearings, furnace linings, paint abrasions, etc. In this case it is important that the surface undergoing wear has a high specific radioactivity. If the material cannot be tagged by adding (e.g. plating on) a radionuclide, the material has to be irradiated, e.g. by accelerator particles.

Figure 18.11 shows a somewhat uncommon but elegant use of radioactive tracers for studying how machine parts behave under *running conditions*. If this case it was possible to determine the bending of the turbine blades in addition to rotation speed and direction of rotation.

18.6. Radioactive tracers in biological sciences

The largest single user of radionuclides is medical science. It has been said that radioactive tracers have been of equal importance to medicine as the discovery of the microscope. In the field of diagnosis and medical research radioisotopes have provided completely new methods of great value. Presently, one in four patients admitted to hospitals in the United States is administered a radioactive tracer as part of the diagnostic process. Radioisotope scanning yields information about a medical disorder much before it can be obtained by any other method.

If a radioactively labeled compound such as an amino acid, a vitamin, or a drug is administered to a patient, the substance will be incorporated to varying degrees in different organs (*biological affinity*). The substance undergoes chemical exchange with other substances in the body, is broken down, and, finally, discharged from the body. The movement in the body of radioactive atoms can often be followed externally with

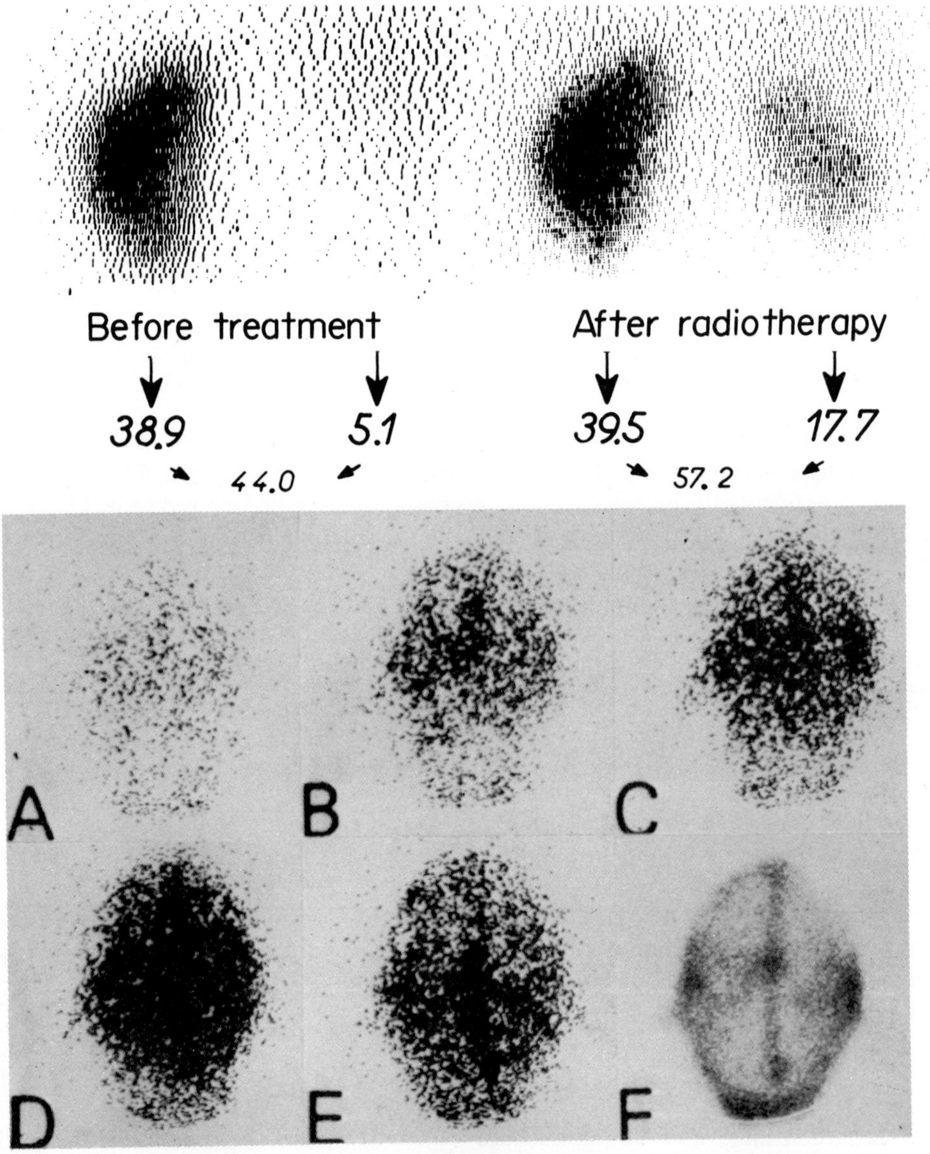

FIG. 18.12. Upper figure shows kidney function investigation by $^{197}HgCl_2$ using quantitative scintigraphy with a 16 000 channel analyzer. (According to C. Kellershohn, C. Raynaud, S. Richard, A. Desgrez, R. Riviere, P. de Vernejoul, L. Barritault, J. P. Morucci, and A. Lansiart.) Lower figure shows a cerebral CT scanning after injection of $Na^{99m}TcO_4$. (According to H. N. Wagner and A. F. Rupp.)

TABLE 18.4. *Diagnostic uses of radionuclides and absorbed dose in certain scanning procedures*
From J. R. Mallard, R. P. Parker and N. G. Trott.

Nuclide	Chemical form	Activity administered i.v. (mCi)	Organ examined	Absorbed dose (mrad)		"Efficiency"[(a)]		
				Organ examined	Whole body	Length of examination (h)	Assumed effective $t_{1/2}$ (h)	% dose during examination
^{99m}Tc	Pertechnetate	5	Brain	30	65	0.5	5.3	6
^{197}Hg	Chlormerodrin	0.5	Brain	—	250	0.5	24 (90%)	1.5
			Kidney	9300			65 (10%)	
^{18}F	Complex	1.5	Skeleton	300	110	1	1.8	30
^{87m}Sr	Chloride	2	Skeleton	200	20	1	2.8	20
^{85}Sr	Chloride	0.1	Skeleton	4400	2300	1	2.4 d (60%)	0.12
							65 d (40%)	
^{99m}Tc	MAA	1.5	Lung	290	2.6	0.5	6	6
^{131}I	MAA	0.3	Lung	1900	24	0.5	5.8 (90%)	3
							72 (10%)	
^{99m}Tc	Colloid	2	Liver	680	32	0.3	6	3
^{113m}In	Colloid	2	Liver	900	19	0.3	1.8	10

[(a)]These are rough estimates of the "efficiency" in terms of dose delivered during examination compared with the time integral of the dose resulting from the administration to complete decay.

scanning instruments of the type described in Chapter 17. Such studies give information about the relative concentration of the radioactive elements and the time dependency of its distribution in different organs. The rate of incorporation and discharge of the radioactive substances in the body provide a measure of the metabolism of healthy and of sick tissues. The upper part of Fig. 18.12 shows a kidney scan of a 38 year woman who has been administered $^{197}HgCl_2$. In the left figure, 38.9% of the compound has been fixed to the left kidney, and very little to the right. This was caused by a vaginal fibrous sarcoma blocking the urethra from the right kidney. After radiation therapy some improvement is seen (right figure).

With the increasing number of radiopharmaceuticals with specific biological affinities, visual scanning has become an important diagnostic tool with numerous clinical applications. The lower part of Fig. 18.12 shows serial vertex images of the transit of $Na^{99m}TcO_4$ through the brain after intravenous injection. The images were obtained at 2 s intervals beginning 5 s after injection. Because brain tumors have a very high affinity for and slow release of Tc, whereas its release from brain infarcts is fast and from healthy parts of the brain even faster, various constrictions to the cerebral blood are easily located. The amount of radioactivity administered varies depending on detection sensitivity ψ and permissible doses; for ^{99m}Tc, 20–40 mCi, for ^{131}I only 0.01 mCi. Table 18.4 summarizes the most common radionuclides used in medical diagnoses as well as information on radiation doses during the investigation.

Radioactive tracers can be used to measure the flow (Fig. 18.13) and amount of liquid volumes in other organs (exercise 18.1). For example, the amount of blood pumped by the heart per minute can be measured externally with scintillation detectors following the injection of ^{59}Fe into the body. In analogous fashion, the amount of extracellular water can be determined with T_2O, the circulation in the capillaries with ^{24}NaCl, etc.

Radiotracers are also used for therapy though to a much less extent than in diagnosis. The main application is the use of ^{131}I for treatment of the thyreotoxicos (Graves disease), an enlargement of the thyroid gland. Injection of ^{32}P compounds in the bloodstream is used to destroy excess red blood cells in polycytemia.

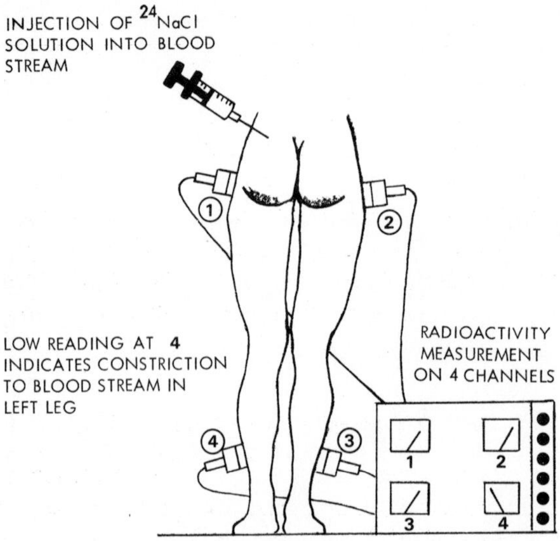

FIG. 18.13. Sequence of steps in using ^{24}Na in the bloodstream to diagnose the presence and location of constrictions to blood circulation.

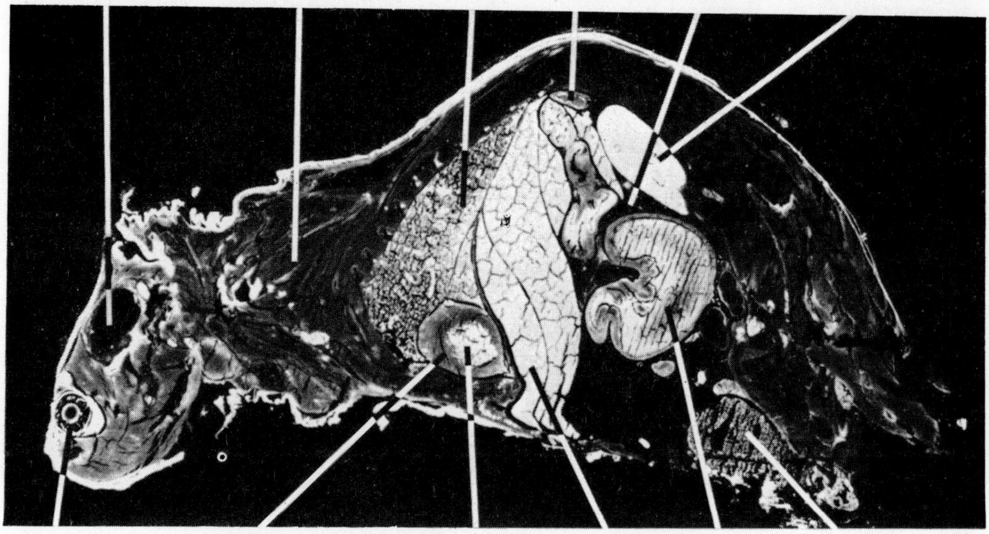

FIG. 18.14. Autoradiogram showing the distribution of ^{14}C PAS in a mouse 30 min after intravenous injection. (According to A. Hanngren.)

In addition to these routine medical applications, radiotracers are used extensively in medical research and other life science fields. An important research technique is to study the distribution of labeled substances in animals. Figure 18.14 shows the distribution of ^{14}C-labeled *p*-aminosalicylic acid, PAS, one of the first antiturbulostatic agents developed, in a 20 μ thick section of a mouse. The radioautogram was obtained after 20 μCi ^{14}C PAS had been injected, and the mice (weight 20 g) had been killed by immersion into a CO_2 acetone (– 80°C) mixture and sectioned. It is seen that PAS is mainly concentrated in the lung, where it is effective against lung tuberculosis, and the kidney and intestine as it is excreted through these organs. The technique is particularly useful in testing out new drugs.

In studying how living species interact with the environment (*ecology*) one can use radioactive tracers to follow the uptake of a trace metal (e.g. cobalt) from the soil by plants, and further by animals after having eaten the plant. In agriculture, this can be useful in studying the uptake of trace elements necessary for plant growth. For example, it has been found that sheep need plants containing selenium in order to combat white muscle disease. The turnover in nutrients fed to animals can be determined; it was found that 20% of the phosphorus in cow's milk comes directly from the feed, while 80% is taken from the cow's bone, i.e. 80% of the phosphorus passes via the bones.

18.7. Behavior of trace concentrations

Many applications of radiotracers involve combining the tracer atoms with a much larger amount of nonradioactive isotopic atoms prior to use. The nonradioactive component is called a *carrier* as it "carries" the radioactive and ensures normal chemical behavior. Consider a sample of 10^7 dpm, for a $t_{1/2}$ of 1 h the number of atoms is 1.7×10^9, while for a $t_{1/2}$ of 1 y it is 1.5×10^{13}. If such samples are dissolved in a liter of solution, the respective concentrations would be 2.8×10^{-15} M and 2.5×10^{-11} M. At such concentrations the chemical behavior may be quite different than it is for higher concentrations. Additions of a few grams of a nonradioactive carrier of the element results in concentrations of 10^{-3} to 10^{-1} M and "normal" behavior.

If a radionuclide is to follow the chemical properties of an isotopic carrier it is necessary that the radionuclide and the carrier undergo isotopic exchange. If it is not known *a priori* that such exchange takes place between two compounds with a common element this must be determined by experimentation before it can be assumed that the tracer and the carrier will act similarly in a chemical system. This consideration must be particularly borne in mind if the radioactive tracer and the inert carrier are in different oxidation states when mixed.

In the remainder of this chapter we discuss the behavior of trace level concentrations since it may be necessary in some applications to use pure radiotracers with no carrier.

18.7.1. *Adsorption*

Solutes in contact with surfaces have a tendency to be adsorbed on the surface. In order to cover the glass surface of a one liter vessel with a monomolecular layer of a hydrated cation only 10^{-7}–10^{-8} moles are required. As indicated in the previous paragraph, the amount of radionuclide in the solution is often considerably less than this and, in principle, all the radioactive atoms could be adsorbed on the walls of the vessel. The amount of radionuclide that is adsorbed on the walls of the container depends on the concentration, on the chemical state of the radionuclide and on the nature of the container material. Figure 18.15 shows the variation of the adsorption of thorium on the walls of glass and polyethylene containers as a function of concentration and pH. The variation of adsorption with pH reflects the adsorption of various hydrolytic species formed by thorium as the pH is increased.

In general adsorption increases with ionic charge in the order $M^+ < M^{2+} < M^{3+} < M^{4+}$. The importance of the nature of the surface can be seen in Fig. 18.15(b) and in the adsorption data for Pm(III) for which the order is platinum < silver < stainless

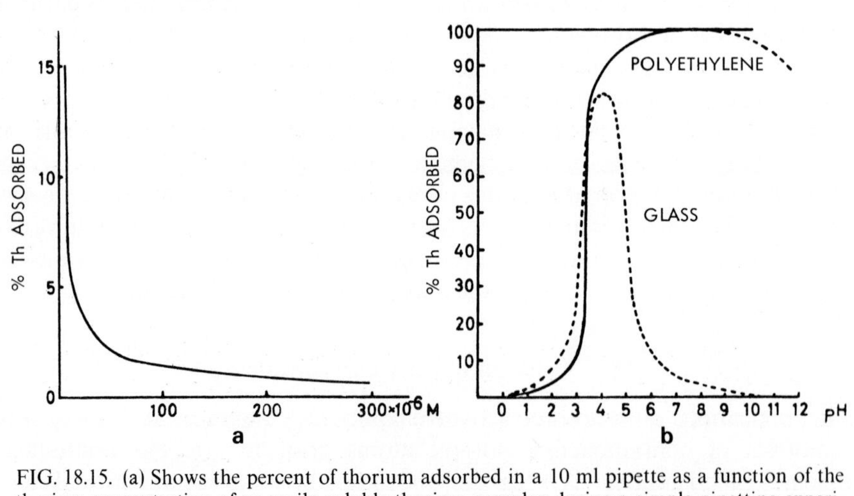

FIG. 18.15. (a) Shows the percent of thorium adsorbed in a 10 ml pipette as a function of the thorium concentration of an easily soluble thorium complex during a simple pipetting experiment. (b) Shows equilibrium adsorption of thorium (2×10^{-8} M thorium perchlorate) on walls of glass and polyethylene vessels as a function of pH.

steel < polyvinylchloride. Addition of isotopic carrier dilutes the radiotracer and a smaller fraction of tracer is adsorbed (Fig. 18.15(a)). Unfortunately, such isotopic dilution results in a decrease in the specific activity of the trace element, which can be disadvantageous in certain types of experiments. In some cases it is possible to avoid decreasing the specific activity by adding macro amounts of a nonisotopic element which is easily adsorbed and may block the available surfaces from adsorbing the tracer.

In addition to adsorption on the walls of the container, radioactive species frequently adsorb on precipitates present in the system. The nature of the precipitate as well as its mode of precipitation are major factors in the amount of adsorption. If silver iodide is precipitated in an excess of silver ion the precipitate has a positive surface layer due to the excess concentration of silver ions on the surface. By contrast if the precipitation occurs in excess iodide, there is a negative surface charge due to the excess iodide on the surface. When trace amounts of radioactive lead ions are added to a suspension of two such precipitates in water, the precipitate with the negative surface charge adsorbs greater than 70% of the tracer lead ions from the solution, while the precipitate with the positive surface charge adsorbs less than 5%. The amount of adsorption increases with the ionic charge of the radioactive tracer, e.g. it has been found that with a precipitate of Ag_2S about 7% of Ra^{2+}, 75% of Ac^{3+}, and 100% of Th^{4+} is adsorbed.

The Paneth and Fajans rule for tracer adsorption states that: "a microcomponent is adsorbed on a solid macrocomponent or precipitated together with it if it forms an insoluble compound with a counter ion of the macrocomponent."

The adsorption properties of trace elements have been used for the isolation of the trace elements as well as for the separation of different trace elements with different adsorption properties.

18.7.2. *Radiocolloids*

Radioactive tracers adsorb not only on solid container surfaces and precipitates but on any kind of solid material suspended or in contact with the solution. Dust, cellulose fibers, glass fragments, organic materials, etc., are examples of substances that readily adsorb radioactive tracers from solution. If the solution contains large molecules as, for example, polymeric metal hydrolysis products, these also tend to adsorb trace elements. The presence of such material in the solution leads to the phenomenon of *radiocolloid formation*, which is the grouping together of radionuclides into semicolloidal aggregates in solution. This phenomenon has been intensively investigated by techniques such as diffusion rates and autoradiography. If the solution is kept at sufficiently low pH and clean of large foreign particles, radiocolloid formation is usually avoided as a major problem.

18.7.3. *Equilibrium reactions*

The low concentration of radioactive tracers can lead to the formation of different amounts of solute species than are observed at equilibrium with macro amounts. For example, the hydrolysis of uranyl ions corresponds to the equilibrium

$$\mathrm{m\,UO_2^{2+} + p\,H_2O \rightleftharpoons (UO_2)_m(OH)_p^{2m-p} + p\,H^+}$$

With macro concentrations of uranium this equilibrium is shifted to the right with the formation of polymers with properties rather different than that of the uranyl UO_2^{2+}

ion. At a uranium concentration of approximately 0.001 M, more than 50% of the uranium is polymerized at a pH of 6, while for uranium concentrations less than 10^{-6} M the polymerization is negligible. The difference in the position of the equilibrium between macro amounts and trace amounts of material can be used to advantage if one wishes to study the properties of a metal ion at relatively high pH's without the disturbance of polymerization reactions.

An additional complication can arise in solution if the radioactive species in trace amounts react with trace concentrations of impurities. For example, in an investigation of the properties of pentavalent protactinium, Pa(V), it was found that the protactinium was extracted into pure xylene from 1 M $HClO_4$ solutions. Further experimentation showed that this extraction was due to the presence in the xylene of organic impurities at concentrations below the detectable limit of 0.01%. Support for this interpretation was provided when the solution was made 10^{-4} M in thorium, which was expected to form complexes with the probable impurity and thereby prevent its reaction with the protactinium, and, in fact, no protactinium was extracted into xylene from this solution.

18.7.4. *Precipitation and crystallization*

Due to the low concentration of radioactive tracers in solution the solubility product for a salt is rarely exceeded upon the addition of macro concentrations of a counter ion. For example, when ^{212}Pb is present with an activity of 5 Ci l^{-1} which corresponds to 1.7×10^{-14} M, even for sulfate ion concentrations of 10 M the solubility of lead sulfate is not exceeded. Even when the solubility product is exceeded, often precipitation is not observed. Consider the case of lead sulfide PbS for which the value of K_s is 3.4×10^{-28} Assuming $[^{212}Pb^{2+}] = 1.7 \times 10^{-14}$ M and $[S^{2-}] = 1$ M, it would be expected that PbS would precipitate; however, at these concentrations the precipitate is formed as a colloid in solution and even upon centrifugation the amount of precipitate is so small as to be unweighable by present techniques.

It is possible to remove ions at tracer level concentrations from solutions by precipitation using adsorption or coprecipitation. Coprecipitation occurs if the compound of the tracer and the oppositely charged ion of the precipitate is isomorphous with the precipitate. In these cases the active ion may be included in the crystal lattice of the precipitate at a lattice point, particularly if the tracer ion is close in size to the ion which it displaces. However, at trace level concentrations exceptions are found to this requirement of similarity in size as well as to the requirement of isomorphism. When the distribution of the tracer is found to be uniform throughout the precipitate it can be described by the homogeneous (according to M. Berthelot and W. Nernst) distribution law which is expressed as

$$\frac{x}{y} = D'\left(\frac{a-x}{b-y}\right) \tag{18.17}$$

where x and y are the amounts of A^+ and B^+ in the precipitate, a and b are the initial amounts of these ions, and D' is the "distribution coefficient". A more "true" distribution constant (D = concentration of tracer in solid/concentration of tracer in solution) can be obtained by using a conversion factor, e.g. C = g solute per ml of saturated carrier solution divided by the density of the solid

$$D' = DC \tag{18.18}$$

The entire precipitate is in equilibrium with solution in this system.

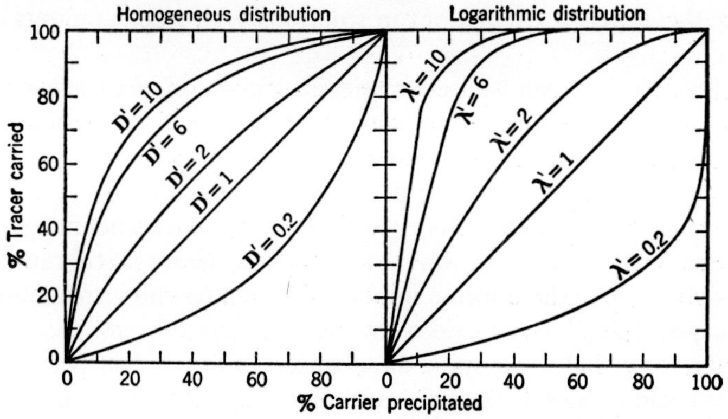

FIG. 18.16. Efficiency with which the tracer is carried for various values of the distribution coefficients D' and λ'. (According to A. C. Wahl and N. A. Bonner.)

If only the freshly forming surface of the growing crystal is in equilibrium with the solution phase, a nonuniform distribution is observed. In these cases the system is described by a logarithmic (according to H. Doerner and W. Hoskins) distribution law, which has the form

$$\ln\left(\frac{a}{a-x}\right) = \lambda' \ln\left(\frac{b}{b-y}\right) \tag{18.19}$$

where λ' is the logarithmic distribution coefficient, which is a constant characteristic of the system (Fig. 18.16).

The importance of isomorphism can be illustrated by the coprecipitation of Ra^{2+} in trace quantities with Sr^{2+} in strontium nitrate. If the precipitation is carried out at 34°C, the radium coprecipitates since at this temperature the strontium precipitates as $Sr(NO_3)_2$ with which radium nitrate is isomorphous. However, if the precipitation occurs at 4°C, the strontium crystallizes as $Sr(NO_3)_2 \cdot 4H_2O$, which is no longer isomorphous with the $Ra(NO_3)_2$. Due to the lack of isomorphism the radium is not coprecipitated at 4°C.

18.7.5. *Electrochemical properties*

For the redox equilibrium

$$M_{ox}^{z+} + n\,e^- \rightleftharpoons M_{red}^{z-n}$$

where M_{ox}^{z+} and M_{red}^{z-n} are the oxidized and reduced states of a chemical species, the *Nernst* equation is valid, i.e.

$$E = E^\circ + \frac{RT}{n_F} \ln \frac{[M_{ox}^{z+}]}{[M_{red}^{z-n}]} \tag{18.20}$$

In this equation E° is the potential for a standard state of 1 N concentration† and the species in brackets relates to the chemical activities in the particular solution phase. This relationship indicates that the redox potential E of a solution is independent of the total concentration of the species and depends only on the ratio of the oxidized and reduced forms. This has been confirmed since concentrations of trace amounts of ions show the same redox behavior as macro concentrations. Reduction and oxidation

† A1 N (N = normal) solution contains one equivalent weight per liter.

reactions can, therefore, be carried out in solutions with trace amounts of radioactive species.

Electrolysis of solutions can be used for electrodeposition of a trace metal on an electrode (cf. §17.10). The selectivity which would be present for electrolytic deposition of macro amounts of ions at a controlled potential is not present, however, for trace amounts. The activity of trace amounts of the species is an unknown quantity even if the concentration is known, since the activity coefficient will be dependent upon the behavior of the mixed electrolyte system for which the chemical theory is still rather poor. Moreover, the concentration of the tracer in solution may not be known accurately since there is always the possibility of some loss through adsorption, complex formation with impurities, etc. Nevertheless, despite these uncertainties it has been found that the Nernst equation can be used with some caution for calculating the conditions necessary to electrolytically deposit trace metals on electrodes.

It is also possible to precipitate insoluble species on electrodes. For example, if a fluorosilicate solution is electrolyzed thereby freeing a high concentration of fluoride ion at the electrode, a thin uniform layer of UF_4 can be deposited. Similarly, trace amounts of elements which form insoluble hydroxides can be deposited from solutions in which water is being electrolyzed as a region of extremely high pH is present at the cathode.

18.8. Tracer separation methods

All the analytical techniques used in conventional chemistry are used for the separation and isolation of radioactive elements and compounds. Normally they require the addition of a macro amount of isotopic carrier; however, in some cases analytical procedures are available for separation and isolation of carrier free radiotracer concentrations. Solvent extraction and various forms of partition chromatography have been found to be particularly advantageous in this connection since these methods are selective, simple, and fast.

18.8.1. *Liquid–liquid extraction*

Liquid–liquid (or solvent) extraction is a technique for selectively transferring a species from a mixture in an aqueous solution to an organic phase by equilibrating the aqueous phase with an organic solvent of suitable composition. Usually the organic phase contains a reagent A (extractant) which forms a neutral compound MA_N with the species M, which is to be removed from the aqueous phase. In §18.3.3 some conditions for the extraction of metals were described. The neutral compound is often combined with water or another neutral substance. If the substance is not water the formulation is shown as MA_NS_S in which S is called the adduct compound. Solvation and adduct formation play important roles in the extraction of metal compounds.

Organic solutions of tertiary amines are used to extract anions, while organic solutions of alkyl phosphates, phosphonates, and diketonates are the common cation extractants. The fraction extracted at equal phase volumes is

$$E = D(D + 1)^{-1} \tag{18.21}$$

where D, defined in (18.10), varies with the concentration of the reagent, the composition of the solvents, the pH and ionic strength of the aqueous phase, the temperature of the

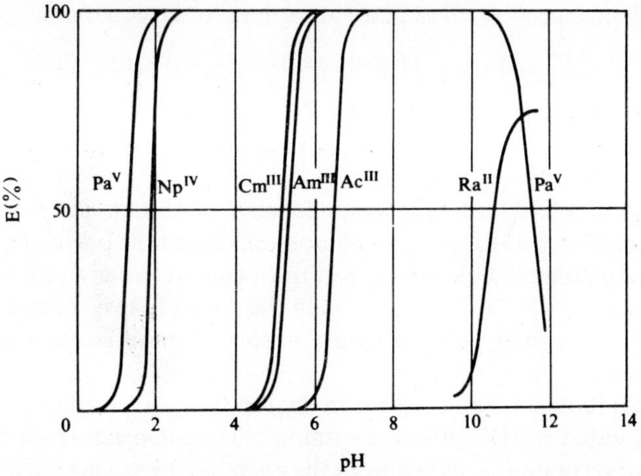

FIG. 18.17. The percentage ($E\%$) of tracer metals extracted from aqueous nitrate solutions of different pH by 0.1 M 8-hydroxyquinoline (oxine) into chloroform. $pH_{1/2}$ for V(IV) and Fe(III) is ~0.5, for Th(IV) 3.1, and for U(VI) 2.6 (not shown); $pH_{1/2}$ is the pH for 50% extraction ($E\% = 50$). (According to C. Keller and K. Mosdzelewski.)

system, etc. (cf. §18.3.3). Figure 18.17 shows the extraction of a number of metals into chloroform by 8-hydroxyquinoline. From such curves optimal separation conditions can be chosen.

18.8.2. *Ion exchange chromatography*

Solid organic resin ion exchangers consist of organic polymeric networks containing basic or acidic groups attached to the organic framework. In the acidic cation exchangers

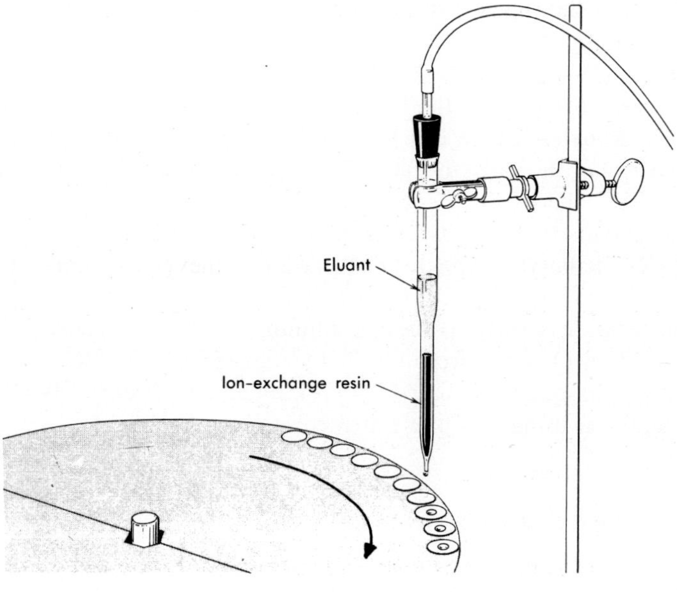

FIG. 18.18. Ion exchange column with simple fraction collector.

the overall exchange process takes place according to the equation

$$M^{n+}(aq) + n\,RH(resin) \rightleftharpoons MR_n(resin) + nH^+(aq)$$

$$K_{iex} = \frac{[MR_n]_{resin}[H^+]^n_{aq}}{[M^{n+}]_{aq}[RH]^n_{resin}} \quad (18.22)$$

The equilibrium constant for this reaction depends on the specific properties of the ion exchange material, such as the amount of cross-linking of the polymeric network as well as on such solution parameters as nature of the metal ion, ionic strength of the solution, temperature, etc. (cf. §18.3.3). The sorption in the resin phase increases with the valency of the cation so that multivalent ions are absorbed more strongly than divalent or monovalent ions.

In practice the radioactivity is sorbed in the top layer of a column containing the ion exchange resin either in the cationic or anionic form, depending on the type of resin used. Following sorption it is eluted from the resin bed by passage of a solution (*eluant*) through the column of resin (Fig. 18.18). The elution may be accomplished by use of another metal ion (e.g. M^{3+}), which shifts the equilibrium of the exchange reaction through competition with M^+ for positions in the resin. Alternately, it can be achieved by adding a complexing anion to the solution which, by reducing the concentration of free metal ions, also shifts the equilibrium of (18.22) to the left. Figure 13.3 shows the elution sequence for the separation of the rare earth and actinide cations from a column of cation exchange resin using a complexing agent.

The use of cation exchange resin has been extensive in the isolation and chemical identification of the transplutonium elements.

18.8.3. *Column partition chromatography*

In principle, partition chromatography is a liquid–liquid extraction where one of the liquid phases is stationary and attached to a supporting material, and the other liquid phase is mobile.

Liquid partition chromatography (LPC) can be carried out with either the aqueous or the organic phase stationary; in the latter case the technique is referred to as *reversed phase LPC*. The aqueous phase can be made stationary by adsorption on silica gel, cellulose powder, etc. In order to make the organic phase stationary, beads (usually 50–200 μm) of PVC, teflon, Kel-F, etc., are being used. Reversed phase LPC has been useful in radiochemistry for separating individual lanthanides or actinides with a stationary phase of tributylphosphate (TBP), di-2-ethylhexylphosphoric acid (HDEHP), etc.

The stationary phase is held at place in a column, just as for ion exchange. Figure 18.19 shows the separation of Am^{3+} from Cm^{3+} by reversed phase LPC. The capacity and efficiency of the technique is increased by working under pressure (30–400 atm) and at elevated temperatures using very finely divided support (5–10 μm).

18.8.4. *Sheet partition chromatography*

Instead of using columns in partition chromatography, a sheet of paper may be used to hold the stationary phase (*paper chromatography*) or an adsorbent coated on a glass

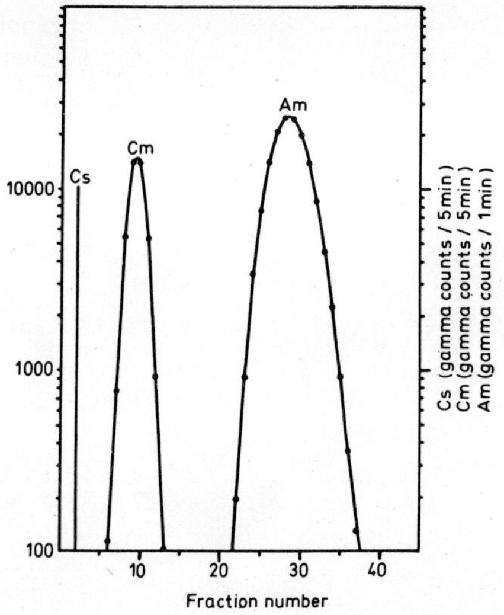

FIG. 18.19. Separation of americium and curium by reversed phase partition chromatography, using trilaury methyl amine nitrate ($TLMANO_3$) adsorbed on Kel–F. Eluant is 4 M $LiNO_3$ at pH 2.0. (According to J. van Ooyen.)

plate (*thin-layer chromatography*). This technique has an advantage over column separations because the positions of the radioactive species are easily identified on the sheet, either simply by autoradiography (§18.2.4) or by scanning instruments.

In paper chromatography the experimental technique is usually as follows (see Fig. 18.6). A few drops of the solution containing the substance to be separated (metal ion, organic molecule, etc.) is placed a few centimeters from the end of a paper strip. The paper strip is hung vertically and dipped into a solution so that the initial point of placement of the substance is near the bottom of the strip above the solution level. Capillary forces draw the solution upwards and bring it into contact with the adsorbed substances at the starting point. As this occurs the starting substance moves a certain distance up the paper from the starting point with the distance traveled dependent on the kind of paper, the solution used, and the chemical properties of the substance. For each substance a certain R_f value can be given as

$$R_f = \frac{\text{distance the substance of interest has traveled}}{\text{distance the liquid front has traveled}} \qquad (18.23)$$

A typical solvent system for metal ions may be a mixture of acetone, dilute HCl, etc., while for organic substances it is possible to use mixtures of phenol and water, acetylacetonate and water, etc. Figure 18.20 shows a two-dimensional paper chromatogram; in this case it has been run initially with a particular solvent mixture, then (after drying) turned 90° and run with a second solvent mixture, thus increasing the selectivity of the separation. The separated substance can be quantitatively recovered by cutting out the spot and leaching the compound from it. This technique has been further developed to make continuous separation possible.

Thin-layer partition chromatography (TLC) is run in the same manner as paper chromatography. It has, however, a broader usage, because it is not limited to cellulose

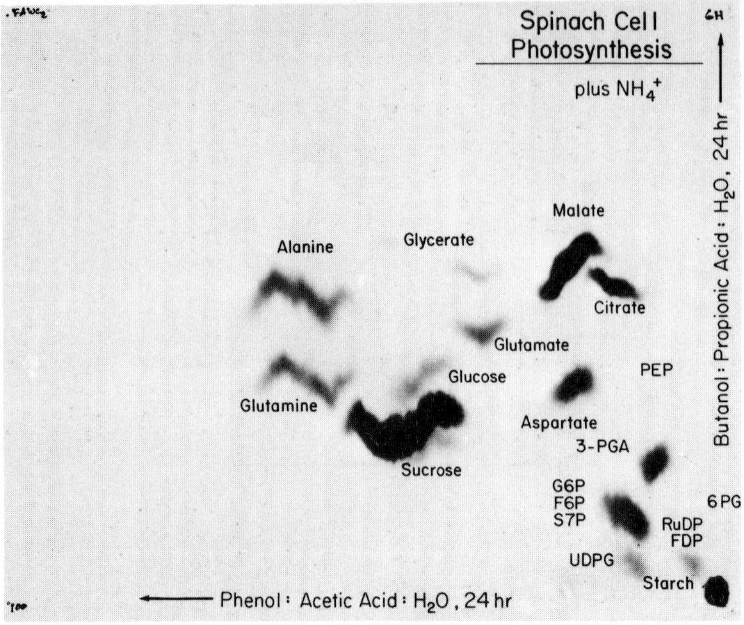

FIG. 18.20. Two-dimensional paper chromatogram (solutions used are indicated along the axes) of photosynthesis products (in $^{14}CO_2$—atmosphere) of spinach cells. (Courtesy M. Calvin and R. Lemmon, University of California, Lawrence Berkeley Laboratory.)

support. Any adsorbent which can be fixed in a thin (1–5 μm) layer on a glass plate is useful: silica gel, cellulose powder, alumina, etc. Sometimes an electric potential field is used to improve separation (*electrophoresis*). Continuous electrophoresis is used for selective separation of sugars, amino acids, etc.

18.9. Exercises

18.1. The blood volume of a patient is to be determined by means of ^{32}P. For this purpose 15.0 ml of blood is withdrawn from the patient and mixed with a very small volume of $Na_2H^{32}PO_4$ of high specific activity. In 1 h the erythrocytes (red blood cells) take up all ^{32}P; 1 ml is found to have an activity of 216 000 cpm in the detector system used. Exactly 5 ml of this tagged sample is reinjected into the patient, and 30 min later a new sample is withdrawn; 10 ml of this gives 2300 cpm. Calculate the blood volume.

18.2. A mixture of amino acids is to be assayed for cysteine. A 1.0 ml sample (density 1 g ml^{-1}) is withdrawn, and 2.61 mg of ^{35}S labeled cysteine of specific activity 0.862 μCi mg^{-1} is added. From this mixture pure cysteine is isolated by liquid–partition chromatography; 30.6 mg is isolated and measured to give 169 000 cpm in 27% detection efficiency. What is the percentage of cysteine in the original mixture?

18.3. In order to determine the lead content of a color pigment, 8.9871 g was dissolved in conc. HNO_3, and 5 ml ^{212}Pb solution added. After excess acid had been removed through evaporation, excess 1 M NaCl was added, the solution heated and filtered. After cooling and crystallization, the $PbCl_2$ was washed and recrystallized, 0.3276 g of the crystals was measured in a scintillation counter, giving 185,160 counts in 5 min. One milliliter of the original ^{212}Pb solution gave 57 000 cpm. The background was 362 cpm. Calculate the lead content (%) of the pigment.

18.4. What is the smallest amount of indium which can be determined in a 100 mg aluminum sample using NAA with a neutron flux of 10^{12} n s^{-1} cm^{-2}? Consider the neutron capture in both ^{27}Al and ^{115}In, as indicated in the nuclide chart. For $^{115}In(n,\gamma)^{116}In$, $\sigma = 45$ b. The lowest detectable activity for ^{116}In is assumed to be 10 dps, and the interference from ^{28}Al not more than 20%.

18.5. In order to determine the amount of gallium in meteorite iron, 373.5 mg meteorite iron (A) and 10.32 mg gallium oxinate (B) were irradiated in a reactor under similar conditions in 30 min. After a short cooling, A was dissolved in concentrated HCl and 4.53 mg inactive Ga^{3+} was added. After a number of chemical separation steps, which were not quantitative, a precipitate of 25.13 mg pure gallium oxinate was isolated (C). Sample B was also dissolved and diluted to 50 ml; 0.50 ml was removed, 4 mg inactive Ga^{3+} added, and gallium oxinate precipitated (D). The radioactive decay curves gave two straight lines: $\log R_c = 3.401 - 0.0213t$, and $\log R_d = 3.445 - 0.0213t$. What was the gallium content in sample A?

18.6. A 10 g sample of iodobenzene is shaken with 100 ml of 1 M KI solution containing 2500 cpm ^{131}I. The activity of the iodobenzene at the end of 2 h is 250 cpm. What percent of the iodine atoms in the iodobenzene have exchanged with the iodide solution?

18.7. A sodium iodide solution contains some radioactive ^{131}I. An ethanol solution was prepared containing 0.135 M of this sodium iodide and 0.91 M inactive C_2H_5I. In the exchange reaction

$$C_2H_5I + {}^{129}I^- \underset{k}{\overset{k}{\rightleftharpoons}} C_2H_5{}^{129}I + I^-$$

the reaction rate constant k is assumed to be the same in both directions. One part (A) of the solution was removed and heated to high temperature so that equilibrium was rapidly reached. Another part (B) was kept in a thermostated bath at 30°C. After 50 min ethyl iodide was separated from both solutions. The concentration of radioactive iodine in C_2H_5I in B was found to be only 64.7% of that in A. Calculate k ($k_r = k\,a\,b$ in §18.3.2).

18.8. Using the relation in §18.3.3, calculate the distribution constants for the uncharged plutonium acetylacetonate complex between (a) $CHCl_3$ and 0.1 M $NaClO_4$, (b) methyl isobutyl ketone (hexone) and 0.1 M $NaClO_4$. Data in Fig. 18.7.

18.9. With the equations in §18.3.3 the stability constants β_n for the formation of the $PuAa_n^{4-n}$ complexes can be calculated from Fig. 18.7. A simplified approach to estimate β_n is the use of the approximate relations $k_n = \beta_n/\beta_{n-1} = [A_a]^{-1}_{\bar{n}=n-0.5}$ and $\bar{n} = N - d(\log D)/d(\log[A_a])$; for Pu^{4+} N = 4. Make this estimation of β_n using the benzene curve in the figure.

18.10. One wants to determine the residual liquid volume of a closed sedimentation tank (nominal volume 80 m^3), which has been in use for many years, and in which $CaSO_4$ precipitates. 0.5 ml $^{24}Na_2SO_4$ (specific activity 3.2×10^8 cpm ml^{-1}) is added to the tank, and 10 ml withdrawn after 2 h of settling; measurements yield a net value (background subtracted) of 500 counts in 10 min. Calculate the free liquid volume in the tank.

18.11. Calculate the critical deposition potential ($E - E°$) for 10^{-12} M ^{210}Bi on a gold cathode (no overvoltage) from the Nernst equation (18.20), where the chemical activity of the reduced state (Bi°) is set at unity.

18.12. A mineral ore contains uranium and smaller amounts of vanadium. In order to determine the vanadium concentration it must be separated from uranium. Solvent extraction using 0.1 M 8-hydroxyquinoline in $CHCl_3$ is chosen. Which metal can be extracted from the other, and at what pH? Consider Fig. 18.17 and connected text.

18.13. In a solvent extraction system consisting of uranium and lanthanum in 1 M HNO_3 and 100% TBP, $D_U = 20$ and $D_{La} = 0.07$. If a phase ratio $\theta = V_{org}/V_{aq} = 0.5$ is chosen, how much uranium is removed from the aqueous phase in three repeated extractions? How much of the lanthanum is co-extracted? The fraction extracted with n fresh organic volumes (V_{org}) from one aqueous volume (V_{aq}) is:

$$E_n = 1 - (1 + D\theta)^{-n} \qquad (18.24)$$

18.10. Literature

A. C. Wahl and N. A. Bonner (eds.), *Radioactivity Applied to Chemistry*, Wiley, 1951 (still useful).
H. Piraux, *Radioisotopes and their Industrial Applications*, Philips Technical Library, Eindhoven, 1964.
PICPUAE, Vols. **19** (1958), **15** (1965), **13**, **14** (1972).
E. Broda and T. Schönfeld, *The Technical Applications of Radioactivity*, Pergamon, Oxford, 1966.
IAEA, *Radioisotope Tracers in Industry and Geophysics*, Vienna, 1967.
T. Braun and J. Tölgyessy, *Radiometric Titrations*, Pergamon Press, Oxford, 1967.
J. Ruzicka and I. Stary, *Substoichiometry in Radiochemical Analysis*, Pergamon Press, Oxford, 1968.
J. Molinari and J. Guizerix, Technical and economic aspects of the use of radioactive tracers for measuring flow in streams, *Kerntechn. Isotopentechn. und Chemie* **10** (1968) 264.
J. F. Duncan and G. B. Cook, *Isotopes in Chemistry*, Clarendon Press, 1968.
H. J. M. Bowen, *Chemical Applications of Radioisotopes*, Methuen, 1969.
J. M. A. Lenihan, S. J. Thomson, and V. P. Guinn (eds.), *Advances in Activation Analysis*, Academic Press, 1969 (Vol. I), 1972 (Vol. II).
J. A. Dean, *Chemical Separation Methods*, van Nostrand Reinhold, 1969.
A. K. De, S. M. Khopkar, and R. A. Chalmers, *Solvent Extraction of Metals*, van Nostrand Reinhold, 1970.
C. E. Crouthamel, F. Adams, and R. Dams, *Applied Gamma-ray Spectrometry*, Pergamon Press, Oxford, 1970.

D. L. MASSART, *Cation-exchange Techniques in Radiochemistry*, NAS-NS-3113, 1971.
E. K. HULET and D. D. BODE, Separation chemistry of the lanthanides and transplutonium elements, in *Lanthanides and Actinides* (ed. K. W. BAGNALL), MTP Int. Rev. of Science, Butterworth, 1972.
D. E. ROBERTSON and R. CARPENTER, *Neutron Activation Techniques for the Measurement of Trace Metals in Environmental Samples*, NAS-NS 3114, USAEC, 1974.

CHAPTER 19

Nuclear Chain Reactions

Contents

Introduction

After the discovery by Hahn, Strassman, and Meitner that neutrons induced fission in uranium, and that the number of neutrons released in fission was greater than one, many scientists realized that it should be possible to build a chain-reacting system in which large amounts of nuclear energy were released under controlled conditions. The first such system was constructed in Chicago in the early 1940s under the scientific leadership of E. Fermi. This first nuclear reactor became critical on December 2, 1942, as part of the Manhattan Project (§13.2.4). Since that time hundreds of nuclear reactors have been built throughout the world for research, for production of plutonium, and for power production. In early 1978, 218 nuclear power stations (103 GWe)† were in operation of which 69 (48 GWe total) were in the United States, 82 (30 GWe) in Western Europe, 27 (7 GWe) in the USSR. Presently about 380 more (~360 GWe) are ordered or under construction. In 1977 nuclear power contributed to electricity production with about 22% in Sweden, 17% in the United States and 7% in the United Kingdom. By 1978, the world's power reactors had operated for 1500 reactor years without any fatal accident. The nuclear power programs of the world are summarized in Appendix I.

Nuclear chemistry must play an essential part in achieving safety and reliability in this most rapidly expanding source of power. Nuclear chemists are also responsible for much of the nuclear fuel cycle from uranium ore processing to ultimate disposal of radioactive reactor waste.

19.1. The nuclear reactor

In the fission process

$$^{235}U + n \rightarrow FP + \nu n$$

For each neutron consumed on the average about 2.5 ($=\nu$) new neutrons are released. These new neutrons can be used to fission other ^{235}U nuclei leading to the release of even more neutrons. In a *nuclear reactor* the *nuclear chain reactions* are controlled so that an equilibrium state is reached, where for each fission only one of the new neutrons is used for further fission. Under these conditions the *neutron multiplication factor k* is said to be 1. If the factor is higher the number of neutrons, and consequently the fission rate, increase exponentially. Without any control mechanism the heat evolved would ultimately destroy the chain-reacting system. This is extremely improbable in conventional reactors because of various control mechanisms. However, in *nuclear weapons*, no control mechanisms are used, and, indeed, every effort is made to make the reaction as violent as possible.

There is quite a flexibility in choices of design for a controlled nuclear chain reaction. Because each concept has its advantages and drawbacks, more than a dozen different types of full scale or prototype nuclear power station have been developed and are in operation. Over 90% of these plants are of the *light water reactor* type (LWR); cf. Appendix I. We therefore begin our discussion of reactors by describing an LWR station with emphasis on the principles and main components. Figure 19.1 shows a schematic diagram and a picture of an operating LWR.

The principal component in any reactor is the *core* in which is located the fissionable *fuel material*. This is usually UO_2 enriched in ^{235}U to 2–3% contained in rods which are canned in zirconium alloy tubings (the *canning*). The *fuel rods* are connected into clusters forming *fuel elements* or *assemblies*.

When fission occurs in the nuclear fuel, the resulting fission fragments have high kinetic energies which they lose through collisions with atoms in surrounding material, resulting in the conversion of practically all of the kinetic energy into heat. In order to maintain a stable operating temperature in the reactor as this heat is evolved, it is necessary to provide a *coolant*. The coolant material, which is water in LWRs and enters the core from the bottom, is heated while passing along the fuel rods and leaves the core at the top as either superheated water (in the *pressurized water reactor* (PWR)

†GWe = gigawatt electric power installed.

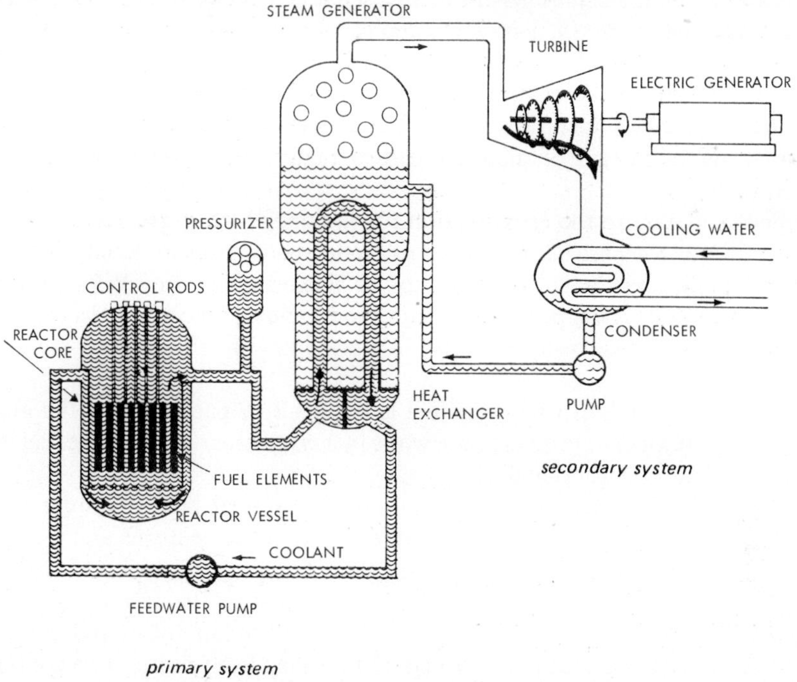

FIG. 19.1. Main components of a pressurized light water-colled and -moderated nuclear reactor (PWR) and the PWR plant (1130 MWe, 1975) at Trojan, USA, showing cooling tower to the left and reactor containment with auxiliary buildings to the right.

as indicated in Fig. 19.1) or high temperature steam (in the *boiling water reactor* (BWR) concept). In the PWR steam is generated in a separate heat exchanger. The steam drives turbines connected to electric generators and is condensed by an *external cooling system* on the backside of the turbine. Except for the external cooling system (a river, cooling towers, etc.), the steam–water flow systems are closed cycles.

The rate of the fission chain reaction in the reactor is controlled by the use of *control rods* which are fabricated of material with a high cross-section for the absorption of neutrons.

In order to maximize the cross-section for fission, which is greatest for low energy neutrons, the neutrons are slowed down or "moderated" by a material (the *moderator*) that elastically scatters but does not capture the neutrons. In LWRs ordinary (but very pure) water serves the purpose of both moderation and cooling. In other reactors the moderator may be a solid material like graphite and the cooling medium a gas like helium.

Reactors of this kind, in which the fuel is physically separated from the moderator, are said to be of the *heterogeneous types*, while in *homogeneous reactors* the fuel is directly dissolved in the moderator material.

The reactor core system is enclosed in a stainless steel tank. In order to provide safety to the operating personnel against hazards from the neutrons and γ-rays emitted in fission, the reactor tank is surrounded by a thick *biological shield*. The reactor building is completely enclosed in an *outer containment* so that in the case of an accident no radioactivity would leak to the surroundings. The small streams of air and water effluents from the reactor station are all monitored and purified from radioactive contaminants.

The fuel elements cannot be used to 100% consumption of the ^{235}U. Fission leads to the production of fission products, some of which have very high neutron capture cross-sections ($\gg 100$ b) and compete with the fission chain reaction for the neutrons. Before the reactor becomes "poisoned" by these fission products, the fuel elements have to be changed. This is done by means of a fuel *charging* (discharging) *machine*. Because of the large amounts of radioactive fission products, the used fuel elements are always allowed to "cool" (with respect to both radioactivity and heat) for several months in water-filled *storage basins* located in the reactor building (see Fig. 19.13).

The used fuel elements may later be *reprocessed* to recover the remaining amount of fissile material as well as any *fertile material*; fertile atoms are those which can be transformed into fissile ones, i.e. ^{232}Th and ^{238}U, which through neutron capture form fissile ^{233}U and ^{239}Pu, respectively. The chemical reprocessing removes the fission products, some of which are valuable enough to be isolated. However, for the most part the fission products are not individually isolated but are stored as *radioactive waste*. The recovered fissile material may be refabricated into new elements for reuse. This "back-end" of the nuclear fuel cycle is discussed in Chapter 20.

19.2. Energy release in fission

From Table 9.1 it is seen that in thermal neutron fission of ^{235}U the fission fragments are released with a kinetic energy of ~ 165 MeV (on the average), the 2.5 *prompt neutrons* have an average kinetic energy of ~ 5 MeV together, and the *prompt γ-rays* have an average of 7 MeV. This *prompt energy release* of ~ 177 MeV is absorbed in surrounding material.

The fission products are radioactive and decay through emission of β^-, γ, and X-rays;

TABLE 19.1. *Nuclear data for fissile and fertile nuclides (see also Figs. 19.2, 19.4, and 19.17)*

	^{232}Th	^{233}U	^{235}U	^{238}U	Nat. U	^{239}Pu	^{240}Pu	^{241}Pu
Radioactive decay	α	α	α	α		α	α, sf	α, sf
Half-life (years)	1.405×10^{10}	1.59×10^{5}	7.04×10^{8}	4.47×10^{9}	—	2.44×10^{4}	6.54×10^{3}	14.9
Specific rad. act. (dps μg^{-1})	4.06×10^{-3}	357	0.0800	0.01244		2270	8430	3.69×10^{6}
Thermal neutrons (0.025 eV)								
n, γ-capture (σ_γ barns)	7.40	48	99	2.70	3.42	269	290	368
fission (σ_f barns)	—	531	582	<0.5 mb	4.18	742	0.030	1009
neutron yield (ν)	—	2.49	2.42	—		2.87	2.90	3.00
fission factor (η)	—	2.29	2.07	—	1.33	2.11		2.15
Fast neutrons (~ 0.25; ~ 1.0 MeV)								
n, γ-capture (σ_γ barns)	0.18; 0.15	0.39; 0.08	0.24; 0.11	0.14; 0.17	0.16; 0.18	0.27; 0.09	0.20; 0.18	
fission (σ_f barns)		2.25; 1.95	1.35; 1.25	0.017	0.11	1.62; 1.65	0.11; 1.59	1.96; 1.65
neutron yield (ν)		2.52; 2.59	2.49; 2.58	2.85		2.93; 3.02		2.47; 3.3
fission factor (η)		2.29; 2.45	2.12; 2.39			2.53; 2.88		

their total amount of decay energy is ~23 MeV. About 10 MeV (the value is uncertain) escapes the reactor as radiation, and ~1 MeV of the decay energy is left in the fuel rods of the reactor; thus ~12 MeV β–γ decay energy (divided about equally between $\bar{E}_\beta$ and E_γ) is absorbed in the reactor. The prompt neutrons are captured in the reactor material with release of binding energy; it is estimated that this amounts to about 10 MeV. Thus the total amount of energy expended per fission in a shielded controlled reactor is about 177 + 12 + 10 = ~199 MeV. As a practical average value 200 MeV per fission is used:

$$E_f = 200 \text{ MeV per fission} = 3.20 \times 10^{-11} \text{ J} \tag{19.1}$$

Thus $(3.20 \times 10^{-11})^{-1} \approx 3.1 \times 10^{10}$ fissions s^{-1} correspond to the production of 1 W of reactor heat. The heat power of a reactor can be written

$$P = E_f \Delta N_f / \Delta t \tag{19.2}$$

where $\Delta N_f/\Delta t$ is the number of fissions per second. A reactor power station producing 3 Gigawatt heat (GWth; th for thermal) will have an electric output of 1 GWe at a 33% *efficiency* in converting the thermal energy into electric; this corresponds to 8.1×10^{24} fissions d^{-1}. Since the weight of a ^{235}U atom is $M/N_A = 3.90 \times 10^{-22}$ g, this would correspond to the fission of $8.1 \times 10^{24} \times 3.9 \times 10^{-22} = 3160$ g ^{235}U d^{-1}; the true consumption of 235U is slightly higher as discussed in §19.9.

19.3. Fission probability

When uranium is irradiated by neutrons, in addition to neutron capture followed by fission (n, f), several different processes occur: scattering, (n, γ), (n, 2n) reactions, etc. All these reactions are of importance for the reactor designer as well as the nuclear

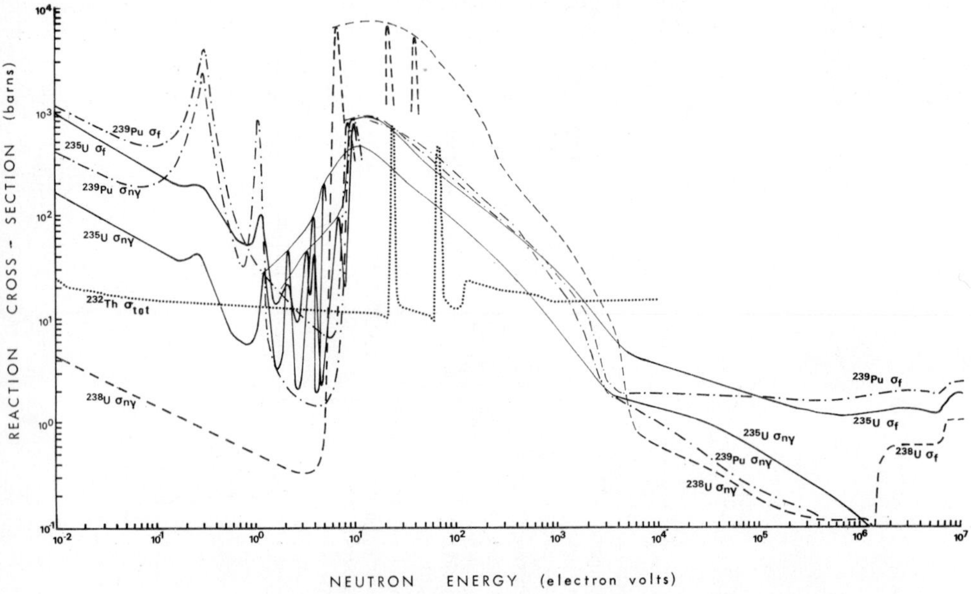

FIG. 19.2. Reaction cross-sections (barns) for scattering (σ_s), neutron capture ($\sigma_{n\gamma}$), fission (σ_f), and total (σ_{tot}) as a function of neutron energy (eV). In the energy region 1–5000 eV only the envelopes of the peaks of each curve are given.

chemists who have to take care of the used fuel elements. The probability for the various reactions depends on the neutron energy. As is seen in Fig. 19.2, three regions are clearly distinguishable: (1) For thermal neutrons with average kinetic energies ($\bar{E}_n$) $\lesssim 1$ eV fission of ^{235}U and ^{239}Pu dominates over neutron capture (i.e. $\sigma_f > \sigma_{n\gamma}$). Though $\sigma_{n\gamma}$ for ^{238}U is small considerable capture occurs because of the large fraction of ^{238}U present. (2) For epithermal neutrons ($1 \lesssim \bar{E}_n \lesssim 10^5$ eV) large radiative capture and fission resonances occur. In this region heavier isotopes are formed by (n, γ) reactions: ^{235}U(n, γ)^{236}U; ^{238}U(n, γ)^{239}U; ^{239}Pu(n, γ)^{240}Pu. (See Fig. 13.5.) (3) For fast neutrons ($\bar{E}_n \gtrsim 0.1$ MeV) the cross-sections are relatively small, $\lesssim 1$ b. Fission dominates over radiative capture. Of particular importance is that ^{238}U becomes fissionable at high neutron energies.

It is obvious that the neutron spectrum of a reactor plays an essential role. Figure 19.3 shows the prompt (unmoderated) fission neutron spectrum, with $\bar{E}_n \sim 2$ MeV. In a nuclear weapon almost all fission occurs with fast neutrons. Nuclear reactors can be constructed so that fission mainly occurs with fast neutrons or with slow neutrons (by moderating the neutrons to thermal energies before they induce fission). This leads to two different reactor concepts—the *fast reactor* and the *thermal reactor*. The approximate neutron spectra for both reactor types are shown in the figure. Because thermal reactors presently are of more importance, we shall first discuss this type.

In a thermal reactor we may assume that most of the neutrons are in thermal equilibrium with the moderator atoms, though in practice the neutron spectrum in power reactors is more energetic. From $E_n = \mathbf{k}T$ it follows that at about 300°C (a common temperature in an LWR) $\bar{E}_n$ is 0.05 eV; cf. Fig. 19.3. However, "thermal neutron energy" (E_{th}) cross-sections are standardized to monoenergetic neutrons with a velocity of 2200 m s^{-1}; this corresponds to E_n 0.025 eV according to (B.12). Cross-sections at this energy are given in Table 19.1. For most (but not all) nuclides the cross-section at low neutron energy outside the resonances may be estimated by the $1/v$ rule (§9.4). Because thermal reactors do not have pure thermal neutron energy spectrum, *effective cross-sections* must be used for calculating reactor product yields. Such cross-sections vary from reactor to reactor. In Fig. 19.4 typical LWR effective cross-sections are given.

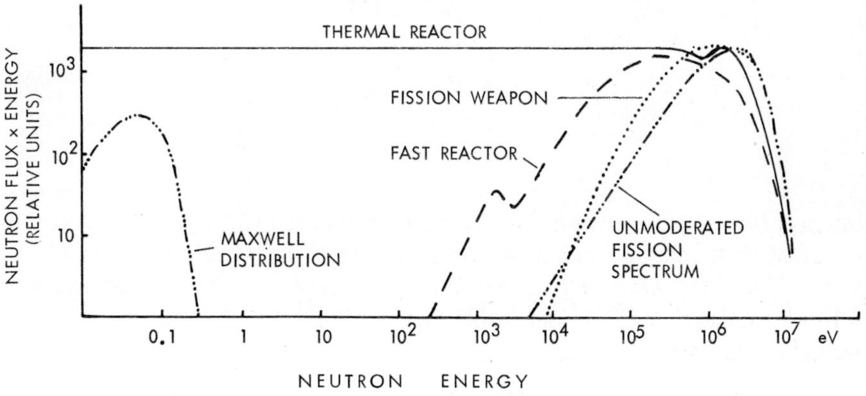

FIG. 19.3. Neutron spectra of some chain-reacting systems. The abscissa shows neutron flux energy of an arbitrary scale.

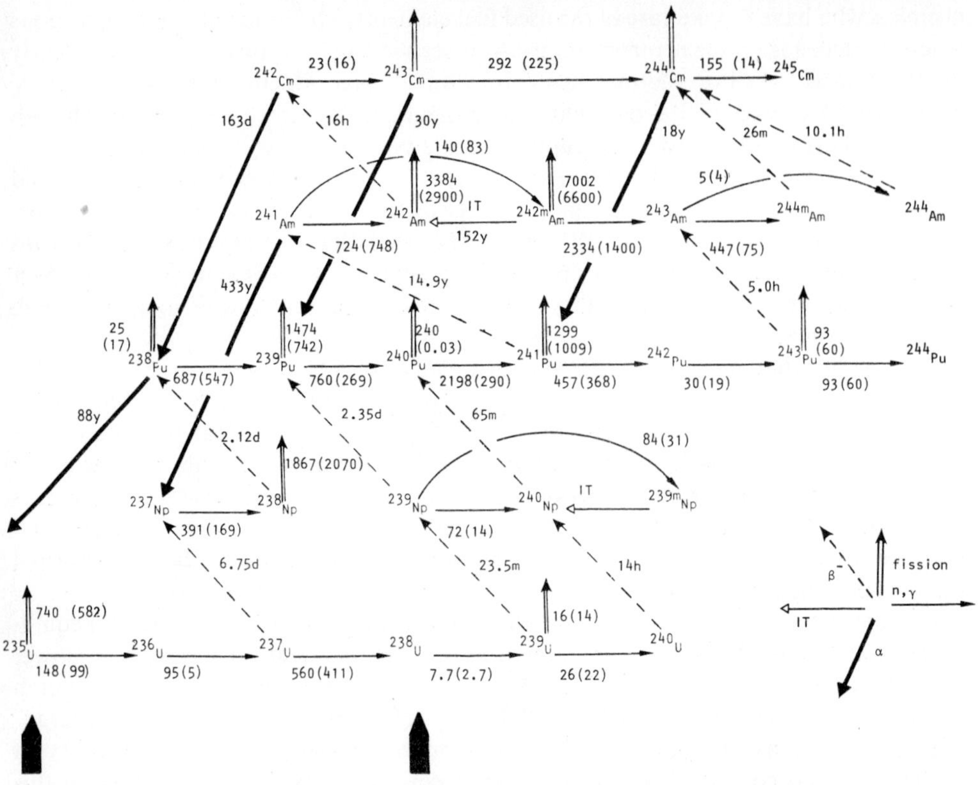

FIG. 19.4. Nuclear reactions in irradiation of uranium. Figures along the reaction lines are half-lives or effective reaction cross-sections (barns) for a standard power LWR, with thermal (0.025 eV) cross-sections within parentheses.

19.4. The fission factor

In thermal reactors ^{235}U is consumed mainly by fission and radiative capture

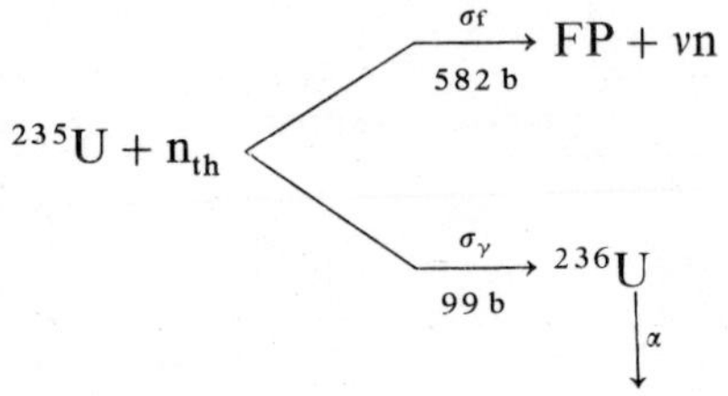

(19.3)

^{236}U decays through α-emission. Due to its long half-life (2.3×10^7 y) it accumulates in the reactor. ^{236}U may capture a neutron, forming ^{237}U, which over a few days' time decays to ^{237}Np. The *neutron yield per fission* ν depends on the neutron energy; it is 2.42 for thermal fission of ^{235}U (see Table 19.1).

If we define the ratio

$$\alpha = \sigma_\gamma/\sigma_f = \hat{\Sigma}_\gamma/\hat{\Sigma}_f \tag{19.4}$$

(cf. (9.6)) the probability that the captured neutron gives rise to fission is $\sigma_f/(\sigma_f + \sigma_\gamma) =$

$1/(1+\alpha)$. The number of neutrons produced for each neutron captured is

$$\eta = \nu/(1+\alpha) \tag{19.5}$$

where η is the *neutron yield per absorption*, also called the *fission factor*. A primary requirement for a chain reaction is that $\eta > 1$. Table 19.1 contains η-values for the most important nuclides. At thermal energies η is highest for ^{233}U, while for fast neutrons η is highest for ^{239}Pu.

In mixtures of isotopes the macroscopic cross-section $\hat{\Sigma}$ must be used in calculating α. Since for natural uranium $\hat{\Sigma}_f = 0.71 \times 582$ per 100 and $\hat{\Sigma}_\gamma = (0.71 \times 99 + 99.3 \times 2.70)$ per 100, one obtains $\alpha = 0.82$ and $\eta = 1.33$, which means a chain reaction is possible with thermal neutrons in natural uranium.

19.5. Neutron moderation*

The mode of moderation of the neutrons is one of the crucial design features of a thermal reactor. The fast fission neutrons lose their kinetic energy through elastic scattering with the atoms of the construction material. In §7.6 the equations are given for the energy change in such collision processes, provided only head-on collisions occur. Because most collisions involve angular scattering, the number of collisions required to reduce a fast neutron to thermal energy is larger. The *average logarithmic energy decrement* is given by

$$\xi = \overline{\ln (E_n/E'_n)} \tag{19.6}$$

where E'_n is the neutron energy after collision. From Table 19.2 it is seen that "light" water (i.e. H_2O) is most effective in reducing the neutron velocity. The average number of collisions n required to reduce the neutron energy from E_n^0 to E_n is given by

$$n = \frac{1}{\xi} \ln (E_n^0/E_n) \tag{19.7}$$

The *slowing down power* (SDP) of a moderator depends in addition on the neutron-scattering cross-section and number of scattering atoms per unit volume (N_0):

$$\mathrm{SDP} = \xi N_0 \sigma_s = \xi \hat{\Sigma}_s \tag{19.8}$$

The SDP is an average value over all the epithermal neutron energy region. A good neutron moderator should divert few neutrons from the fission process, i.e. the neutron absorption cross-section must be small. In this respect both heavy water (i.e. D_2O)

TABLE 19.2. *Physical properties of some moderator materials*

Property	H_2O	D_2O	Be	C
$N_0 \times 10^{-30}$ (atoms m^{-3})	0.033 4	0.033 4	0.123	0.080 3
ρ (t m^{-3})	1.0	1.10	1.84	1.62
σ_a (th) (barns)	0.66	0.000 92	0.009	0.004 5
$\hat{\Sigma}_a$ (th) (m^{-1})	1.7	0.008 0	0.13	0.036
σ_s (epith) (barns)	49	10.6	5.9	4.7
ξ	0.927	0.510	0.209	0.158
$\xi \times \hat{\Sigma}_s$ (epith) $\hat{\Sigma}_a^{-1}$ (th)	62	5860	138	166
L_m^2 (m^2)	7.62×10^{-4}	2.89	4.8×10^{-2}	0.287
τ (m^2)	0.002 8	0.011 5	0.010 0	0.038 0

and carbon are superior to H_2O, since for H_2O the reaction probability for $^1H(n,\gamma)^2H$ is relatively large (Table 19.2). In order to incorporate this property, the concept *moderating ratio* (MR) is used as a criterion of the moderator property in the thermal neutron energy region:

$$MR = SDP/\hat{\Sigma}_a = \xi\hat{\Sigma}_s/\hat{\Sigma}_a \tag{19.9}$$

The moderating properties are summarized in Table 19.2. The moderator qualities decrease in the order $D_2O > C > Be > H_2O$. In a commercial reactor, price and other properties must also be considered, so H_2O is favored over D_2O in many reactors.

19.6. The neutron cycle*

In order for a chain reaction to occur at least one of the neutrons released in fission must produce a new fission event. This condition is defined by the *multiplication factor* k:

$$k = \frac{\text{number of neutrons in generation 2}}{\text{number of neutrons in generation 1}} \tag{19.10}$$

When $k > 1$ the number of neutrons in the second generation exceeds the number of neutrons consumed. Under this condition the neutron flux, and thus the number of fission events, increases for each successive neutron generation with a resultant increase in the power production of the reactor (§19.8 discusses further the *reactor kinetics*). If $k = 1$ the number of fissions per unit time, and thus the energy production, is constant. For $k < 1$ the chain reaction cannot be maintained. A reactor operating with $k = 1$ is said to be *critical*, $k > 1$ supercritical, and $k < 1$ subcritical. k is regulated by means of the control rods.

In any neutron generation the neutrons experience a variety of fates (Fig. 19.5).

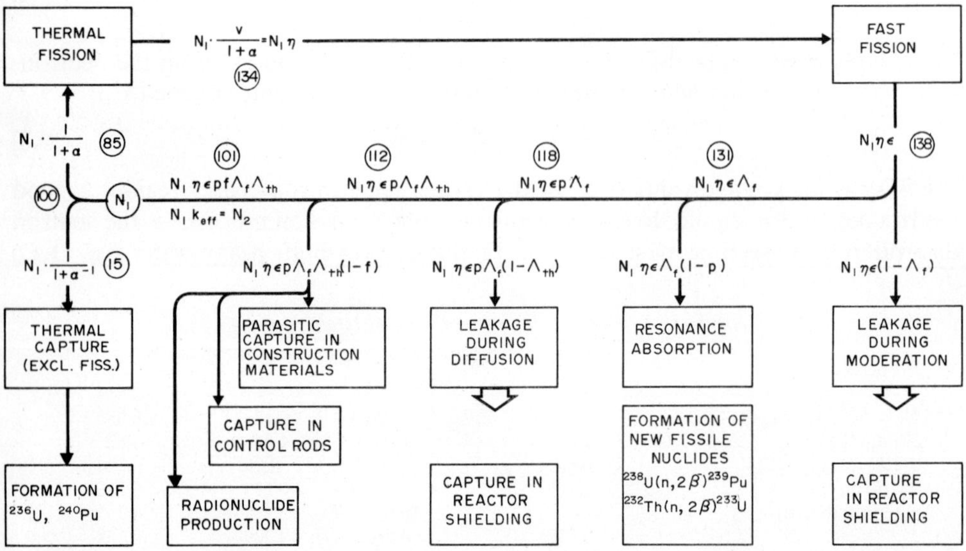

FIG. 19.5. The neutron cycle in a thermal reactor. The number of neutrons as a function of $N_1 = 100$ (first generation) are given in circles with values taken from the Belgian graphite-moderated air-cooled research reactor BR1.

Some neutrons escape the reactor or are absorbed in the reactor structural materials and shielding (i.e. control rods, moderator, coolant, etc.). This must also be taken into account. In practice two different multiplication factors are used: k_∞ refers to a reactor of infinite dimensions (i.e. no leakage) while k_{eff} refers to a reactor of practical finite size:

$$k_{\text{eff}} = k_\infty \Lambda \tag{19.11}$$

Λ is the fraction of neutrons which are *not* lost through leakage to the surroundings (*the nonleakage factor*). In order to minimize the neutron leakage, the reactor core is surrounded with a *neutron reflector* which for thermal neutrons is typically graphite, water, or beryllium, while for fast neutrons iron is frequently used.

Let us consider the neutron cycle in a reactor, i.e. the fate of the neutrons as they proceed from one generation to the next (Fig. 19.5, where, however, leakage, treated in §19.7, also is included). We shall start by assuming that we have a reactor containing natural uranium and a graphite moderator in a configuration which allows negligible leakage ($\Lambda = 1$). If 100 neutrons (N_1) are captured by the uranium fuel, the fission releases $N_1\eta$ or $100 \times 1.34 = 134$ new neutrons. The fast neutrons cause fission in uranium, thereby releasing more neutrons. This is measured by ε, *the fast fission factor*. Its value in a nuclear reactor depends to a large extent on the moderator. For natural uranium in graphite ε is about 1.03, so that the 134 neutrons released in fission by the original 100 thermal neutrons are increased to $N_1\eta\varepsilon$ (or $1.03 \times 134 = 138$) neutrons. In homogeneous reactors $\varepsilon = 1$.

These energetic neutrons are slowed down through collisions with surrounding atoms and decrease in energy steadily until thermal energies are reached. In the range 1–10^4 eV ^{235}U and ^{238}U have large resonance peaks for radiative capture, while fission in ^{235}U dominates at thermal energies (Fig. 19.2). In order to maximize the fission probability, the losses through radiative capture in the epithermal region must be minimized. This can be achieved by physically separating the moderator and the fuel elements. The fast neutrons, which diffuse out of the fuel elements, are slowed down mainly in the moderator. Because the moderator is of a material with light atoms, the number of collisions required to slow down these neutrons is minimal. Therefore it is unlikely that the neutrons diffuse into a fuel element (and are captured) until thermal energies have been reached. The probability that the neutrons pass through the energy region of the resonance peaks without capture is called the *resonance escape probability* and denoted by p. From the original N_1 neutrons at this point there would be $N_1\eta\varepsilon p$ neutrons. In a reactor of natural uranium and graphite, p is usually around 0.9. Therefore, from the original 100 neutrons 138×0.9 or 124 second generation neutrons reach the thermal energy range.

The cross-sections for neutron capture increase for all atoms for thermal energy neutrons. As a result, even though low cross-section materials are used some neutrons are captured by the structural and moderator materials. The probability for the non-capture of thermal neutrons in this fashion is signified by f, *the thermal utilization factor*, which in our case can be assumed to be ~ 0.9. Thus of the original N_1 neutrons 112 thermal neutrons remain in the second generation to cause fission in the nuclear fuel.

These 112 thermal neutrons constitute the second generation of neutrons which according to our definition of neutron multiplication factor is k_∞ so that

$$k_\infty = \eta\varepsilon p f \tag{19.12}$$

This relation is known as the *four-factor formula*. In our case $k_\infty = 1.12$.

TABLE 19.3. *Data for typical BWR power plant*

General: 640 MWe (1912 MWth) BWR plant at Würgassen, West Germany, in commercial operation since 1973.

Reactor physics: θ 4.5 × 10^{-5} (s), η 1.740, f 0.865, p 0.811, ε 1.041, k_∞ 1.270, k_{eff} 1.255, L^2 (cm²) 3.67, τ (cm²) 34.83, B^2 (cm^{-2}) 2.32 × 10^{-4}, thermal average flux 4.4 × 10^{13} ($n\ cm^{-2}\ s^{-1}$), d:o fast 1.9 × 10^{14}, Δk_{max} 25.5%, temp. coeff. − 2.2 × 10^{-3}% $\Delta k/k$°C.

Core: 86 600 kg U, 2.6% enrich. in ^{235}U; critical mass 38.5 kg ^{235}U; burn-up 27 500 MWd t^{-1}; refueling one-fifth of core annually; storage pool for 125 t fuel; core volume 38 m³.

Fuel element: UO_2 pellets, 1.43 cm diameter; 49 rods per assembly; 444 assemblies in core; enrichment 2.6% ^{235}U; Gd_2O_3 burnable poison; zircalloy cladding 0.8 mm.

Heat and coolant data: Max. fuel temp. 2580°C, max. cladding temp. 370°C; coolant inlet temp. 190°C, outlet 285°C, pressure 7 MPa (70 bar); 280 t coolant, 2 × 5200 $m^3\ h^{-1}$ internal pump capacity; condenser cooling water 95 000 t h^{-1} (River Weser).

The value of η depends on how much fissile material the fuel contains. The other three factors have a more complicated dependence of the reactor design including the fuel/moderator ratio, their amounts and form; some values are given for a practical case in Table 19.3.

For any homogeneous reactor

$$f = \Sigma_a\,(\text{fuel})[\hat{\Sigma}_a\,(\text{fuel}) + \hat{\Sigma}_a\,(\text{mod}) + \hat{\Sigma}_a\,(\text{other})]^{-1} \qquad (19.13)$$

where the denominator is the macroscopic cross-section of all absorption reactions in the assembly; the last term includes absorption by impurities in construction materials and by products formed in the course of the operation. When control rods are inserted $\hat{\Sigma}_a$ (others) increases. Effective cross-sections must be used for nuclides with large epithermal resonances.

In order to allow for a decrease in f during operation, commercial reactors are designed with k_∞ 1.2–1.3, rather than the 1.12 given in the example above. Even higher k_∞ values occur as is the case for reactors with highly enriched fuel. In reactors with large k_∞, which permits high burn-up values, a burnable neutron poison may be introduced at the start to decrease k_∞. Such a poison is gadolinium, especially ^{157}Gd, which occurs to 16% in natural gadolinium and has a thermal neutron cross-section of 254 000 b. Even a small amount of gadolinium considerably decreases f (see (19.13)). Although the gadolinium is continuously destroyed by the operation of the reactor, at the same time fission product poisons are formed, resulting in a compensating effect, which tends to maintain a satisfactory value of k_∞.

Inasmuch as some elements have very high neutron capture cross-sections for thermal neutrons, all materials in the reactor must be extremely pure in order to keep k_∞ as large as possible. The exact relations which define the dependency of the various factors in (19.12) on geometry, isotopic composition of the fuel elements, etc., are quite complicated. It is sufficient to note that the optimal geometric arrangement of fuel and moderator can be calculated.

An important factor is the temperature dependence of k. A temperature increase usually has little effect on $\eta\varepsilon$, but f usually increases (i.e. becomes close to 1) because of decreasing moderator density and average neutron energy increase; this causes more neutrons to be captured in ^{239}Pu, which has a low energy (0.3 eV) fission resonance. On the other hand, p decreases because of *Doppler broadening* of the capture resonances in the epithermal region. The Doppler broadening also increases capture in the control rods. These latter effects usually dominate, so that k_∞ slightly decreases with reactor temperature. This is referred to as a *negative temperature coefficient*. In a BWR a negative temperature coefficient is also caused by the bubbles of steam (or steam *voids*) occurring in the moderator, which reduces the thermalization of neutrons while increasing neutron leakage rate. In BWR the effect of the voids dominates over Doppler broadening in limiting the reactor power.

These are important safety features of a reactor. If k increases, so do the fission rate and temperature. If the temperature coefficient were positive, k would increase further, leading to further temperature increase, etc. Finally, the reactor would reach a temperature at which core destruction would result if controls to reduce k were not actuated. However, with a negative temperature coefficient the reactor controls itself: an increase in power (and thus temperature) decreases k, which tends to limit the power increase. The temperature coefficient is usually given in $\%\Delta k/k$ per °C.

19.7. Neutron leakage and critical size*

All practical reactors have some leakage of neutrons out of the reactor core. This leakage Λ is approximately described by the fast or thermal leakage factors Λ_f and Λ_{th} respectively:

$$\Lambda = \Lambda_f \Lambda_{th} \tag{19.14}$$

The effect of this leakage is included in Fig. 19.5, and results in the number of neutrons in the second generation being reduced from the 112 calculated previously for the k_∞ to $N_2 = 101$. Thus $k_{eff} = 1.01$ and $\Lambda = 0.9$, i.e. 10% of the neutrons are lost through leakage. The example in Fig. 19.5 refers to a small natural uranium research reactor. In nuclear power stations k_∞ is much larger due to the large η-values (because of enriched fuels), while the leakage is much smaller, $\Lambda \gtrsim 0.97$ (e.g. see Table 19.3).

The leakage is a function of the average distance the neutron travels after formation until it causes a new fission and of the geometrical arrangement of reactor core and reflector:

$$\Lambda_f = e^{-B^2\tau} \tag{19.15}$$

$$\Lambda_{th} = (1 + B^2L^2)^{-1} \tag{19.16}$$

where B^2 is the *geometrical buckling*, L the *thermal diffusion length*, and τ is referred to as the *neutron* (or *Fermi*) *age*.

Combining these equations gives the *critical equation*

$$k_{eff} = \frac{k_\infty e^{-B^2\tau}}{1 + B^2L^2} \approx \frac{k_\infty}{1 + B^2(L^2 + \tau)} \tag{19.17}$$

The quantity $L^2 + \tau$ is usually denoted by M^2 and is known as the *migration area* and M as the *migration length*:

$$M^2 = L^2 + \tau \tag{19.18}$$

Hence for a large thermal critical reactor one can write

$$k_{eff} = \frac{k_\infty}{1 + B^2M^2} = 1 \tag{19.19}$$

The thermal diffusion length L is calculated from

$$L^2 = L_m^2(1-f) \tag{19.20}$$

where L_m is the diffusion length in the pure moderator. L_m^2 and τ are given in Table 19.2 for various moderator materials.

The geometrical buckling B^2 depends on the neutron flux distribution in the reactor. This distribution in turn depends on the general geometry of the assembly including boundary conditions. The calculation of the buckling for heterogeneous reactors is quite complicated, but for homogeneous reactors the following approximate simple relations hold:

$$B^2 = \pi^2 r^{-2} \quad \text{(sphere)} \tag{19.21a}$$

$$= 3\pi^2 a^{-2} \quad \text{(cube)} \tag{19.21b}$$

$$= 33 b^{-2} \quad \text{(cylinder, height = diameter)} \tag{19.21c}$$

where r is the radius of the sphere, a the side of the cube, and b the cylinder height = cylinder diameter.

It is seen that k_{eff} in (19.17) increases with decreasing buckling, and because $B \propto 1/r$ (19.21) with increasing size of the reactor. This is a result of the neutron production k_∞ being a volume effect (proportional to r^3 for a sphere), whereas leakage is a surface effect (proportional to r^2). For each reactor there is a minimum *critical size* ($k_{eff} = 1$) below which the surface to volume ratio is so large that neutron leakage is sufficient to prevent the fission chain reaction.

The smallest critical sizes are obtained for *homogeneous systems* of pure fissile nuclides with maximum neutron reflection. For neutrons with the fission energy spectrum, the critical mass of a metallic sphere of pure ^{235}U is 22.8 kg, that of ^{233}U is 7.5 kg, and that of ^{239}Pu is 5.6 kg, assuming a 20 cm uranium metal neutron reflector. For fission by thermal neutrons the smallest critical size with no reflector of a *homogeneous aqueous solution* using a uranium sulfate solution requires 0.82 kg of ^{235}U in 6.3 l of solution. The corresponding figures for ^{233}U are 0.59 kg in 3.3 l, and of ^{239}Pu 0.51 kg in 4.5 l.

Homogeneous solutions of fissile nuclides are produced in the reprocessing of spent fuel elements, and care must be exercised that the critical size is not exceeded in any container in order to prevent an accidental chain reaction. Several such accidents have occurred in the past in reprocessing plants in which very high doses were received by personnel even though the duration of the chain reaction was very short and no violent explosion occurred.

For heterogeneous reactors it is more difficult to quote comparable simple values for critical size. These have to be calculated or determined empirically for each particular reactor configuration.

19.8. Reactor kinetics*

The mean lifetime θ for a neutron in a reactor is the time it takes on the average for the neutrons to complete one loop in the neutron cycle. In thermal reactors θ is 10^{-3}–10^{-4} s, while in fast reactors it is 10^{-6}–10^{-7}; in nuclear weapons (fast homogeneous reactors) it is even shorter, 10^{-8}–10^{-9} s. For each loop the number of neutrons is multiplied by a factor k_{eff}. Since one neutron is used for maintaining the chain reaction, the neutrons in the reactor change with time according to

$$\frac{dN}{dt} = \frac{N(k_{eff} - 1)}{\theta} + K \tag{19.22}$$

where N is the total number of neutrons in the reactor, and K is the contribution from any nonfission (e.g. constant) neutron source present. Solving this equation, we can distinguish three different cases:

(i) $k_{eff} < 1$; the reactor is subcritical,

$$N(t) = K\theta(1 - k_{eff})^{-1} \tag{19.23}$$

The number of neutrons is directly related to the constant neutron source. The reactor acts as a neutron amplifier, with amplification increasing with k_{eff}.

(ii) $k_{eff} > 1$; the reactor is supercritical,

$$N(t) = \left(N_0 - \frac{K\theta}{k_{eff} - 1}\right)e^{(k_{eff}-1)t/\theta} + K\theta(k_{eff} - 1)^{-1} \tag{19.24a}$$

where N_0 is the number of neutrons at $t = 0$. If $K = 0$ this reduces to

$$N(t) = N_0 e^{(k_{eff}-1)t/\theta} \tag{19.24b}$$

The neutron flux (and consequently also power) increases exponentially. In all reactors K is larger than 0, the reason being neutron production through spontaneous fission in ^{238}U or through other reactions. Usually, the resulting neutron source strength is not large enough to give a reliable indication on the control instruments of the power level of the reactor in the initial start-up procedure. Therefore, extra neutron sources are generally introduced in reactors to facilitate the starting. A common type is Sb–Be, where ^{124}Sb (half-life 60 d) has been produced through irradiation in a reactor using the (n, γ) reaction in ^{123}Sb (43% abundance). The neutrons are emitted from (γ, n) reactions in Be. Also ^{241}Am–Be sources are used. When the reactor has reached desired power, control rods are inserted, which decrease k_{eff} to 1.

(iii) $k_{eff} = 1$; the reaction is just critical,

$$N(t) = N_0 + Kt \tag{19.25}$$

This equation indicates that the number of neutrons would increase slightly with time, but the Kt term is usually negligible.

In reactor technology it is common to speak of the *reactivity* ρ and *excess reactivity* (Δk), which are defined by the expression

$$\Delta k \equiv k_{eff} - 1 = \rho k_{eff} \tag{19.26}$$

Since k_{eff} is close to 1 in a properly operating reactor, we have $\rho \approx \Delta k$. The *reactor time constant* (or *period*) is defined as

$$t_{per} = \theta/\rho \approx \theta/\Delta k = \frac{\theta}{k_{eff} - 1} \tag{19.27}$$

As the neutron flux increases in a supercritical reactor, the second term in (19.24a) becomes negligible compared to N_0 and the equation can be simplified to

$$N = N_0 e^{t/t_{per}} \tag{19.28}$$

The shorter t_{per} is the faster increases the neutron flux (and reactor power). With $\theta = 10^{-3}$

and $k_{eff} = 1.1$, Δk is 0.1 and t_{per} 10^{-2}. For a longer t_{per} of 1 s, the number of neutrons would increase with a factor of $e^{10} = 10^4$ every 10 s, which still is much too rapid for a safe and simple control of a reactor. As is described below, nuclear reactors are designed to avoid this problem.

In the equations above, θ is the mean lifetime for the prompt neutrons. Some fission products decay through neutron emission. For example ^{87}Kr decays by neutron emission; it is a daughter of ^{87}Br with a half-life of 56 s. A large number of neutron emitting fission products have been discovered, all with short half-lives. The delayed fission neutrons have lower kinetic energies ($\sim$ 0.5 MeV) than the prompt ones and amount to $<1\%$ of the total number of fission neutrons emitted: the fraction β is 0.28% for ^{233}U, 0.65% for ^{235}U, and 0.21% for ^{239}Pu. When the delayed neutrons are taken into account, the effective neutron generation time is approximately obtained by

$$\theta_{eff} = \theta + \Sigma\beta_i/\lambda_i \qquad (19.29)$$

where β_i is the fraction of fission neutrons which are delayed with a decay constant λ_i. The summation gives the mean lifetime of the delayed neutrons; for ^{235}U this value is 0.084 s. Thus $\theta_{eff} \approx 10^{-3} + 0.084$. Because the generation time determines the period of the reactor the delayed neutrons have lengthened the reactor period by almost a factor of 100, thereby making reactor control much more manageable.

t_{per} depends on Δk in such a way that if $\Delta k > \beta$ the delayed neutrons are not able to make their decisive influence on the reactor period, and the rate of neutron production is dependent only on the prompt neutrons. Such a reactor is said to be *prompt critical.* Nuclear weapons are designed to be prompt critical with a $\Delta k \gtrsim 1$ giving a $t_{per} \lesssim 10^{-8}$ s, while reactors for power production (thermal as well as fast reactors) are designed to be *delayed critical.* If in a thermal uranium reactor Δk is made $\geqq \beta$ ($k_{eff} \geqq 1.0065$), t_{per} becomes 13 s for $\theta_{eff} = 0.084$, and (19.24b) shows that the neutron amount (and power) will double in about 10 s. Usually k_{eff} is made smaller than 1.0065 and the power doubling time is correspondingly larger.

19.9. Fuel utilization*

Figure 19.6 shows the consumption of fissile ^{235}U while new fissile ^{239}Pu and ^{241}Pu (as well as some fission products and other actinides) are produced through radiative capture in fertile ^{238}U and ^{240}Pu. The figure relates to a particular reactor type, and different reactors give somewhat different curves. Fig. 19.4 shows the different capture and decay reactions. In § 19.2 it was concluded that in a reactor operating at a power of 3 GWth 3.16 kg $^{235}U\,d^{-1}$ would be fissioned. Because of the $^{235}U(n, \gamma)\,^{236}U$ reaction, some additional uranium is consumed (19.3). To account for this, we use $(\sigma_f + \sigma_\gamma)/\sigma_f = 1.17$; thus the ^{235}U consumption is $3.16 \times 1.17 = 3.70$ kg d^{-1}. This value holds for a fresh reactor core; however, after a core has operated for a time fission in ^{239}Pu formed from neutron capture in ^{238}U begins to contribute to the energy production, and even later, fission from ^{241}Pu also contributes. Therefore the ^{235}U consumption successively decreases for the same power production in an aging core. During the whole lifetime of a 2–3% enriched uranium core about 40% of the total energy production comes from fissioning of plutonium isotopes (Fig. 19.7). On the average a 1000 MWe (33% efficiency) LWR power station consumes 2.2 kg ^{235}U and 2.0 kg $^{238}U\,d^{-1}$. At the same time about 3.1 kg fission products are produced per day from the fissioning of uranium and plutonium. To calculate these yields effective cross-sections must be

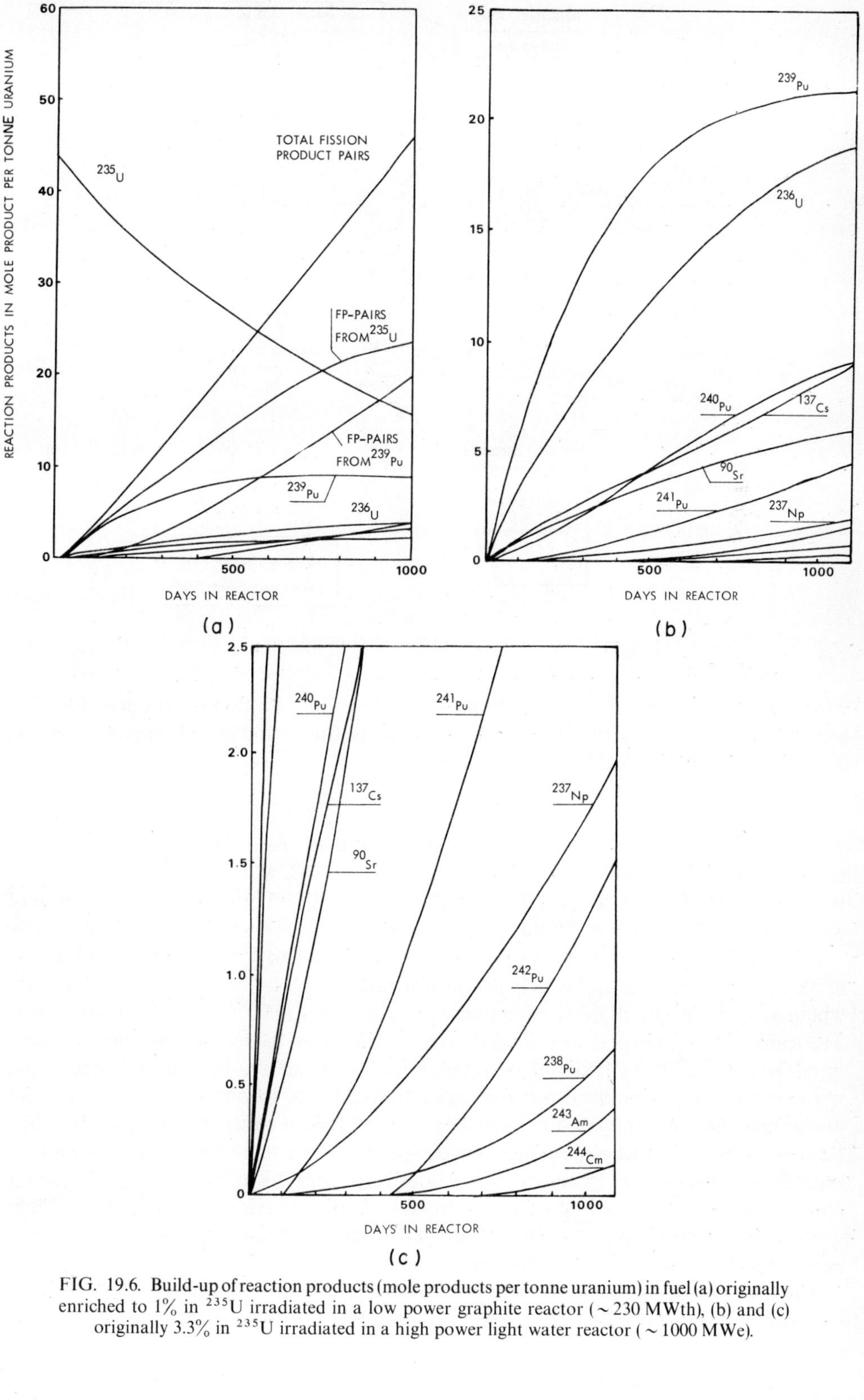

FIG. 19.6. Build-up of reaction products (mole products per tonne uranium) in fuel (a) originally enriched to 1% in ^{235}U irradiated in a low power graphite reactor ($\sim$ 230 MWth), (b) and (c) originally 3.3% in ^{235}U irradiated in a high power light water reactor ($\sim$ 1000 MWe).

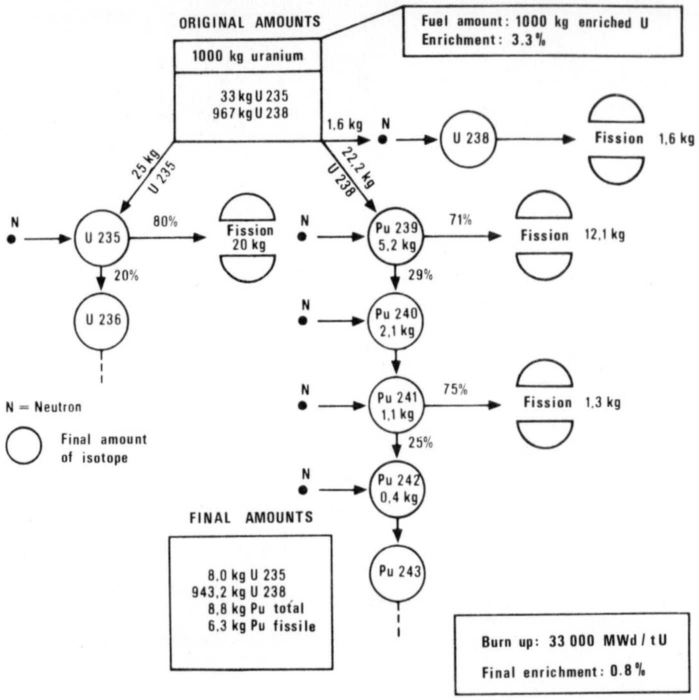

FIG. 19.7. Fission product balance in a light water power reactor. (From F. Hoop.)

used like those given in Fig. 19.4. The more ^{239}Pu and ^{241}Pu that is produced for each atom of ^{235}U consumed, the more efficient can be the utilization of fuel. This is expressed by the conversion factor C, which is given by

$$C = \eta - 1 - S \tag{19.30}$$

where S is the neutron losses by processes other than fission and capture reactions to produce fissile atoms in the fuel. If we assume $C = 0.8$, which is a valid figure for a heavy water moderated reactor, fission of 100 atoms of ^{235}U leads to the formation of 80 new atoms of ^{239}Pu and ^{241}Pu.

In the second generation one produces C^2 new ^{239}Pu and ^{241}Pu atoms. The fraction of fissile atoms in the fuel which is transformed is $x_i(1 + C + C^2 + \ldots) = x_i/(1 - C)$, where x_i is the original atomic fraction of fissile atoms (0.0071 for natural uranium). The ratio $x_i/(1 - C)$ expresses the maximum utilization of the fuel. If a reactor could be operated so that all the fissionable material, both original and produced in operation, were consumed, it would be possible, assuming $C = 0.8$, to obtain five times greater power production than would be provided by the original concentration of ^{235}U only. Because of the high neutron capture cross-section of the fission products, this unfortunately is not possible. Reactors with values of C close to unity are called *converter reactors*, while reactors with C considerably less than 1 are referred to as *burners*. Light water reactors may have $C \lesssim 0.6$ and are classified as burners. Figure 19.8 shows the possible fuel utilization in various reactor types.

The lowering of f (eqn. (19.13)) due to build-up of fission products and decrease in amount of fissile atoms are the main reasons for fuel replacement. It is obvious that if

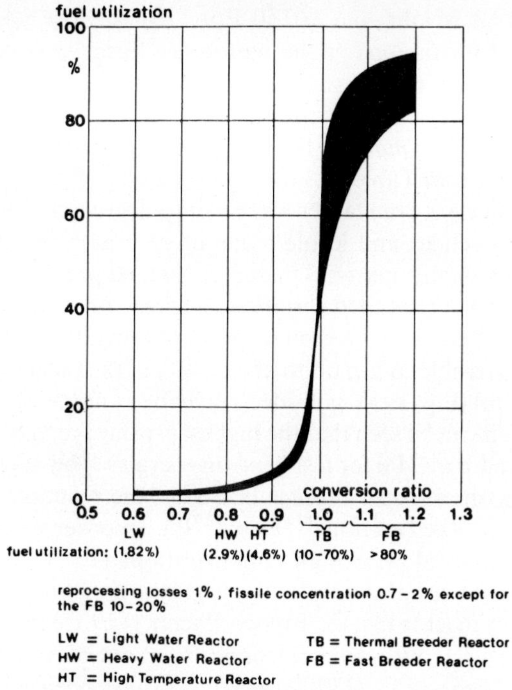

FIG. 19.8. Conversion ratios and fuel utilization efficiencies. (According to Thermal Breeder Consultants Group, Salzburg, 1977.)

$\hat{\Sigma}_a$ (fuel) is very large, as is the case for highly enriched fuel, higher amounts of fission products can be tolerated, i.e. more energy can be produced from the fuel before f becomes too small.

The fuel utilization is referred to as *burn-up*. The burn-up may be given as the percentage of fuel used before it must be replaced. For example, 1% burn-up means that for each tonne of fuel 10 kg of the fissile plus fertile atoms have been consumed (in fission *and* capture). However, usually the fuel burn-up is given in amount of energy taken out per tonne of fuel. The production of 33000 MWd of thermal energy (MWdth) from 1000 kg enriched uranium in a LWR consumes 25 kg ^{235}U and 23.8 kg ^{238}U (Fig. 19.7), or 4.88% of the original uranium amount. Then 1% burn-up corresponds to 33 000/4.88 = 6762 MWdth/t U. However, if credit is made for the 6.3 kg fissile plutonium formed (which "replaces" the ^{235}U), the total consumption is reduced to 48.8–6.3 = 42.5 kg (or 4.25%); a 1% burn-up is then equated with 7765 MWdth/t U.

Considering the amount of total consumed uranium ($\sim$7 t natural uranium for each tonne enriched to 3.3% in ^{235}U), the utilization of natural uranium is much less; however, if plutonium is recycled, utilization increases to the overall figures given in Fig. 19.8.

The maximum conversion factor at the start of a uranium-fueled reactor is given by

$$C = \eta\varepsilon(1-p)e^{-B^2\tau} + \hat{\Sigma}_a(238)/\hat{\Sigma}_a(235) \qquad (19.31)$$

where the two terms give the number of neutrons absorbed by ^{238}U in the resonance absorption region and the thermal region, respectively, per neutron absorbed in ^{235}U; the η-value refers to ^{235}U. The term $e^{-B^2\tau}$ is the fraction of fast neutrons which does not leak out of the reactor (19.15). Using this approximate expression with the reactor example in Table 19.3, we obtain $C = 0.50$. For a reactor core at equilibrium corrections must be introduced for fissions in the plutonium formed, as pointed out above; this gives a higher conversion ratio, usually 0.7 in LWR.

If $C > 1$ the possibility exists of producing more fissile material than is consumed, which is referred to as *breeding*. The breeding ability of a reactor is given by the *breeding gain*, $G = \eta - 2 - S$. From Table 19.1 one can see that ^{233}U, ^{235}U, ^{239}Pu, and ^{241}Pu all have $\eta > 2$; in practice η is even larger when using effective cross-sections (Fig. 19.4). In a reactor with these nuclides and fertile material such as ^{232}Th and ^{238}U it is possible to produce more fissionable material than is consumed provided that neutron losses are small; such reactors are called *breeder reactors* or simply *breeders*. While the utilization of the fuel can be increased to perhaps a maximum of 10% with converters, breeding makes it possible to use up to about 70% of the fuel material (natural uranium and natural thorium) (Fig. 19.8), provided recycling is utilized.

From Table 19.1 it can be seen that the highest η-value is exhibited by ^{233}U for thermal neutron energies and by ^{239}Pu for fast neutron energies. This suggests two different types of breeder reactor, the *thermal breeder* based on the reaction $^{232}Th \rightarrow {}^{233}U$ and the *fast breeder* based on the reaction $^{238}U \rightarrow {}^{239}Pu$. In order to operate efficiently, these breeders must be charged with high concentrations (15–30%) of ^{233}U or Pu. These materials can be produced at present only in conventional nonbreeder reactors.

For a fast breeder reactor like the French Phenix (§19.15) the fuel utilization is higher than in the LWRs; a 1% burn-up corresponds to about 8500 MWdth t^{-1} heavy atoms. Burn-up figures as high as 15% (130000 MWd t^{-1}) have been achieved. Breeder reactors are described in §19.15.

19.10. The Oklo phenomenon

The cover of this book is a picture of the Oklo mine in Gabon. Here uranium has been mined for many years and delivered to the Pierrelatte isotope enrichment plant of the French CEA. The ore body, which is estimated to contain 400 000 t of uranium, is quite inhomogeneous, and pockets very rich in uranium are found embedded in sandstone and granite. These pockets are often shaped like a lens, 10 m long and about 1 m in diameter, and contain on the average 10–20% pitchblende, although uranium concentrations up to 85% (pure pitchblende) are found in some spots. In 1972 it was discovered that the isotopic composition of some of the uranium received in France deviated from uranium from other sources, since the ^{235}U content was considerably lower than the natural 0.72%. Careful analysis showed some deliveries contained $< 0.5\%$ ^{235}U, a serious disadvantage in materials to be used for ^{235}U enrichment. Analysis at Oklo showed samples even lower in ^{235}U content, as well as other elements whose isotopic composition considerably deviated from the natural one. For example, natural neodymium contains 27% ^{142}Nd, while Oklo neodymium contained $< 5\%$. On the other hand, natural neodymium contains 12% ^{143}Nd, while in Oklo, samples containing 24% ^{143}Nd were found. Fission product neodymium contains about 29% ^{143}Nd, while ^{142}Nd cannot be produced by fission. This should be compared with the absence of ^{142}Nd and the excess of ^{143}Nd in the Oklo samples, a condition which was found to be directly related to a high total concentration of uranium but a deficiency of ^{235}U. The conclusion was obvious: in some ancient time the missing ^{235}U had fissioned, producing ^{143}Nd among other fission products. This conclusion was supported by similar investigations on the isotopic composition of other fission elements.

Because ^{235}U has a shorter half-life than ^{238}U, all uranium ores were richer in ^{235}U in the past. From ^{87}Rb–^{87}Sr analysis the age of the Oklo deposit is known to be 1.74×10^9 y old; at this time the ^{235}U content of natural uranium was 3.0%. Although the fission factor η rapidly increases with ^{235}U content (about 1.8 for 3% ^{235}U), conditions in the natural Oklo deposit were such ($\varepsilon \sim 1.0$, $p \sim 0.4$, $f \sim 1.0$) that $k_\infty < 1$. However, the deposit is sedimentary and was formed in the presence of water. Water greatly increases the resonance escape probability factor p; for an atomic ratio H_2O : U of 3 : 1, $p \sim 0.8$, and thus $k_\infty > 1$. Thus conditions existed in the past for a spontaneous, continuing chain reaction to occur in the Oklo deposit.

Further analysis and calculations have shown that these natural Oklo reactors (similar conditions occurred at several places) lasted for $\lesssim 10^6$ y. Possibly criticality occurred periodically as the heat from fission boiled away the water so that the chain reaction ceased after a while. After water returned (after heavy rains?) the chain reaction could resume. The neutron flux probably never exceeded 10^9 n cm^{-2} s^{-1}. The fluence is estimated to exceed 1.5×10^{21} n cm^{-2}. This would have consumed about 6 t ^{235}U, releasing a total energy of 2–3 GWy, at a power level probably $\lesssim 10$ kW. About 1.0–1.5 t of ^{239}Pu was formed from neutron capture in ^{238}U, but the relatively short (on a geological scale) half-life of ^{239}Pu allowed it to decay into ^{235}U. Since a few samples enriched in ^{235}U also have been found, it is believed that in some places in Oklo breeding conditions may temporarily have existed.

A serious problem for present reactors is the safe disposal of the very radioactive waste. What has then happened with the fission products and actinides produced in Oklo 1.7×10^9 y ago? Analyses show that the plutonium and rare earth elements have not moved from their place of origin. The same is true for most fission products, except for volatile and easily soluble ones; cesium and strontium seem to have been washed out completely.

It has been calculated that many uranium rich ore deposits 2–3 10^9 y ago must have been supercritical in the presence of moderating water. Therefore natural nuclear chain reactions may have had an important influence on the early geology of earth.

19.11. Nuclear fuels*

In all commercial reactors today the nuclear fuel is uranium, either with the natural 0.7% abundance of the fissile isotope ^{235}U, or—more commonly—enriched so that the ^{235}U content has been increased to a few percent. Some prototype power reactors also contain ^{232}Th, in which fissile ^{233}U is bred, or ^{239}Pu mixed with ^{238}U as a supplement to the ^{235}U ("mixed fuel").

Whether uranium, thorium, or plutonium, a fuel element must be capable of resisting temperatures considerably above 1000°C without either physical or chemical deterioration due to heat or to radiation. Metallic fuel elements have the high heat conduction necessary to minimize temperature effects, but, unfortunately, uranium melts at 1130°C and plutonium at 640°C. Moreover, metallic uranium has three and plutonium six allotropic forms between room temperature and their melting points. As a consequence, either the separate or combined effects of the radiation field and the high temperature can cause recrystallization into different allotropic forms with significantly different physical dimensions in the fuel rod. Such deformations within the fuel element reduce the mechanical stability and increase the problem of corrosion even if the elements are metal cladded.

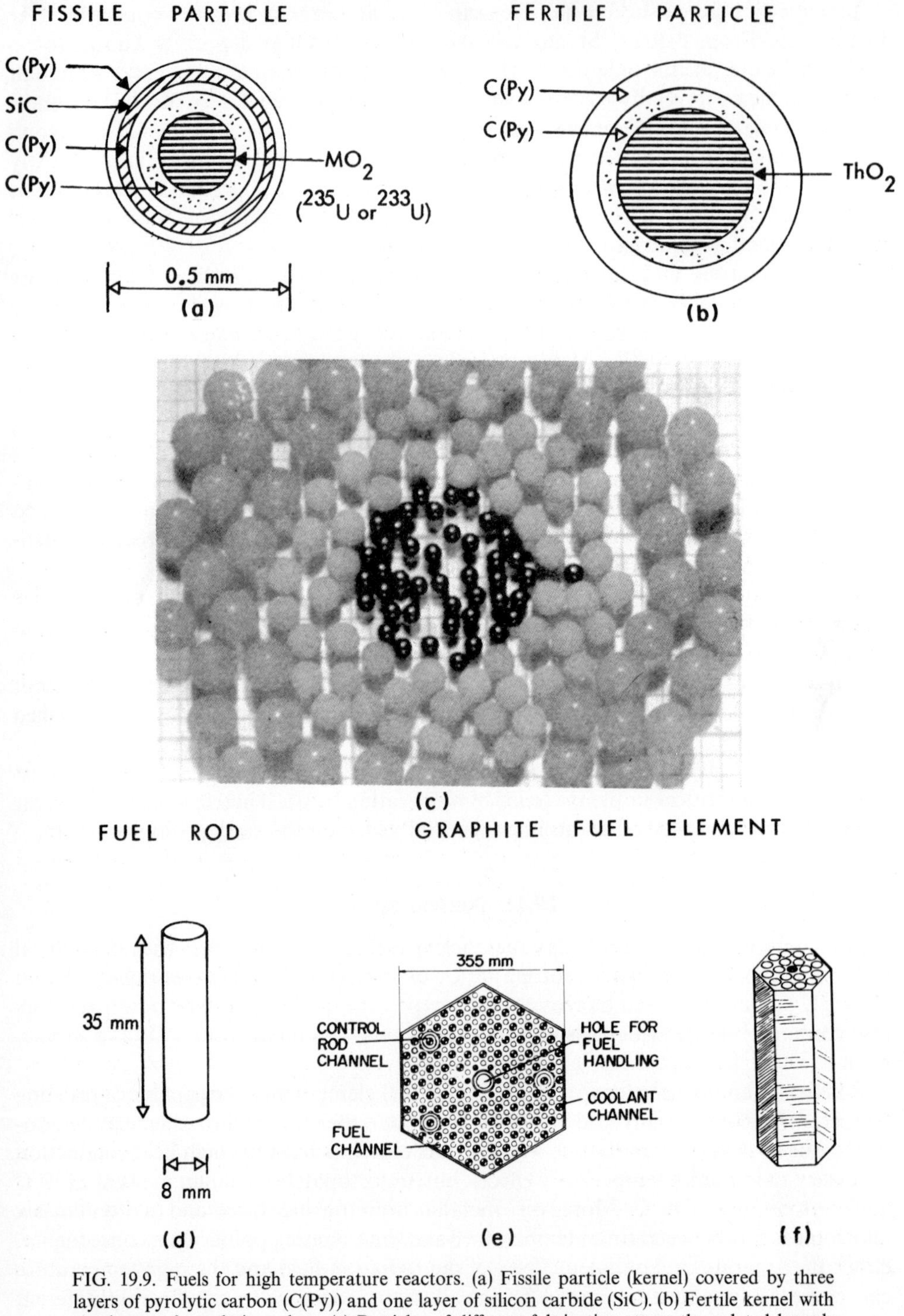

FIG. 19.9. Fuels for high temperature reactors. (a) Fissile particle (kernel) covered by three layers of pyrolytic carbon (C(Py)) and one layer of silicon carbide (SiC). (b) Fertile kernel with two layers of pyrolytic carbon. (c) Particles of different fabrication steps: the gelated kernels (largest), the dried kernels and the fired kernels (smallest). (d) Fuel rod of sintered particles. (e) Cross-section of graphite fuel element for Fort St. Vrain. (f) Isometric view of fuel elements showing a few of the channels.

Most power reactors today use ceramic encased pellets of UO_2, PuO_2, and ThO_2 as fuels. In some reactors UC is tested. The size of the cylindrical pellets is about 1 × 1 cm (diameter × height) (Fig. 19.11(a)). These ceramic fuel elements are quite temperature resistant, do not have the phase transformations of the metals, and have greater resistance to radiation effects. Unfortunately, the heat conduction is not as satisfactory as in the metallic fuel elements, and as a result rather high temperature gradients (up to 100°C mm^{-1}) often exist in the ceramic elements. As a result of these heat gradients it has been found that ceramic fuel elements may melt in the center (2700°C mp for UO_2) even though the surface temperature is well below the melting point. In order to provide maximum heat conduction in the ceramic fuel material, it is sintered and tightly compressed to densities of ~11 g cm^{-3} for UO_2.

When mixed uranium oxide (MOX) fuel elements are used as, for example, in plutonium recycling ($<5\%$ PuO_2) in LWRs or in fast breeders ($\lesssim 15\%$ PuO_2), the UO_2–PuO_2 mixture must be very intimate, preferably a solid solution. This is best achieved by coprecipitation of the tetravalent actinides, normally as oxalates, followed by calcination. MOX fuel elements have been regularly added to the cores of many European LWRs for many years without any technical difficulties. Also mixed uranium and thorium oxide fuels have been used in heavy water reactors.

Instead of pellets, spherical fuel particles can be used. This has advantages with respect to fabrication, reactor utilization, and fuel reprocessing. These oxides or carbide particles are very small, <1 mm in diameter (Fig. 19.9(c)). The particles are produced by the *sol–gel process*, which in principle consists of the following steps:

(i) An aqueous colloidal solution of the actinide or actinide mixture is prepared. The actinide(s) may be in the form of a hydrated complex of high concentration (3–4 M).

(ii) The solution is added to an inert solvent, which dehydrates the complex and causes the droplet to *gelate*. In one technique, hexamethylene tetramine, $(CH_2)_6N_4$, is added to the aqueous solution, which then is added dropwise to a hot (~95°C) organic solvent. The heat causes the amine to decompose, forming NH_3, which leads to hydroxide precipitation in the droplet. The droplet dehydrates and solidifies rapidly, forming a so-called "kernel".

(iii) The kernels are washed, air dried at 150–200°C, and—in the case of uranium—reduced by hydrogen gas at higher temperature to form UO_2.

(iv) The kernels are sintered at high temperature in an inert atmosphere.

Kernels of actinide carbides can be made in a similar manner. The kernels are placed in fuel rod casings, pressed into pellets (Fig. 19.9(d)), or further treated for use in high temperature reactors.

Metallic fuel is encased in a canning (cladding) of aluminum, magnesium, or alloys of these, while oxide fuel pellets are encased in alloys of zirconium or stainless steel. The purpose of casing the fuel is to protect it against corrosion and to protect the coolant from radioactive contamination by the fission products from the fuel element. Aluminum has been used in water-cooled reactors, but at temperatures >300°C zirconium alloys show superior strength. At steam temperatures >400°C zirconium absorbs hydrogen, which increases brittleness, so stainless steel becomes preferable. The most common alloys are zircalloy-2 (containing 1.5% Sn and 0.3% Cr, Ni, and Fe) and stainless steel type 302B (containing 18% Cr and 8% Ni). Stainless steel is not used at lower temperatures because of its larger neutron capture cross-section: σ 0.23 b for Al, 0.18 b for zircalloy-2, and about 3 b for 302B steel.

Satisfactory heat conduction between the fuel and the canning is obtained by using *bonding materials* such as molten sodium, graphite powder, etc. The canning material itself must not only be corrosion resistant to the coolant at all temperatures but should react with neither the fuel element nor the bonding material. The canning should be as thin as possible, consistent with satisfactory mechanical strength and corrosion resistance (Fig. 19.11(a)).

The fuel elements used in high temperature gas-cooled reactors consist of graphite rods or balls filled with oxide or carbide kernels produced by the sol–gel process. The kernels are covered by several layers of graphite and silicon carbide achieved by pyrolyzing methane or acetylene in a fluidized bed of the kernels (Fig. 19.9(a) and (b)).

The fuel is an important part of the economy of power reactors. Approximately 20% of the expense of the electrical production in a power reactor can be attributed to the cost of the fuel. This is due about equally to the expense of the consumption of fissile material and to the production and reprocessing costs. In the fast breeder reactors it is anticipated that fuel costs will be lowered substantially.

19.12. Reactor concepts

Nuclear reactors are designed for production of heat, mechanical and electric power, radioactive nuclides and weapons material, research in nuclear physics and chemistry, etc. The design depends on the purposes, e.g. in the case of electric power production the design is chosen to provide the cheapest electricity taking long term reliability in consideration. This may be modified by the availability and economy of national resources such as raw material, manpower and skill, safety reasons, etc. Also the risk for proliferation of reactor materials for weapons use may influence the choice of a national reactor system (cf. Chapter 21). Many dozens of varying reactor concepts have been formulated, so we must limit the discussion to a summary of the main variables and of the most common power and prototype power reactors.

We have already mentioned three basic principles for reactor design: according to neutron energy (thermal or fast reactors), according to core configuration (homogeneous or heterogeneous aggregation of fuel and coolant), and according to fuel utilization (burner, converter or breeder). In the homogeneous reactor the core can be molten

TABLE 19.4. *Abbreviations for reactor types*

AGR	Advanced gas-cooled graphite-moderated reactor
BHWR	Boiling heavy-water-cooled and -moderated reactor
BWR	Boiling light-water-cooled and -moderated reactor
FBR	Fast breeder reactor
GCFBR	Gas-cooled fast breeder reactor
GCR	Gas-cooled graphite-moderated reactor
HTGR	High temperature gas-cooled graphite-moderated reactor
HWGCR	Heavy-water-moderated gas-cooled reactor
HWLWR	Heavy-water-moderated light-water-cooled reactor
LWGR	Light-water-cooled graphite-moderated reactor
LMFBR	Liquid-metal-cooled fast breeder reactor
OMR	Organic-moderated and -cooled reactor
PHWR	Pressurized heavy-water-moderated and -cooled reactor
PWR	Pressurized light-water-moderated and -cooled reactor
SGR	Sodium-cooled graphite-moderated reactor

TABLE 19.5. *Characteristics of some power and prototype power reactors. All reactors are operative*

Type	Example	Power (MWe)	Fuel (enrichm. %)	Clad-ding	Mode-rator	Cool-ant	Coolant temp. (°C)	Coolant press. (MPa)	Net eff. (%)	Burn-up ($MWdt^{-1}$)	Temp. coeff.	Core size (h × d m^2)
PWR	Diablo Canyon, USA	1100	88 t UO_2 (2.5)	Za	H_2O	H_2O	317	16	32	33 000	Neg.	3.7 × 3.4
PWR	Novovoronezh, USSR	1000	75 t UO_2 (~4)	Za	H_2O	H_2O	322	16	32	26 500	Neg.	3.5 × 3.1
BWR	Forsmark, Sweden[a]	900	140 t UO_2 (2.6)	Za	H_2O	H_2O	290	7	34	30 000	Neg.	3.6 × 4.5
PHWR	Bruce 1, Canada	746	129 t UO_2 (nat.)	Za	D_2O	D_2O[b]	304	8.7	30	9 800	Pos.	5.9 × 7.1
GCR	Wylfa, UK	655	595 t U (nat.)	Mgn	C	CO_2	414	2.8	26	3 500	Pos.	9.2 × 17.4
AGR	Hinkley Point B, UK	620	129 t UO_2 (~2)	St	C	CO_2	645	4	42	18 000	Pos.	8.3 × 9.1
HTGR	Fort St Vrain, USA	330	1 t UC (93), 20 t ThC	C	C	He	770	5	39	100 000	Neg.	5.9 × 4.8
LMFBR	Phenix, F	233	0.8 PuO_2, 4 t UO_2 (0.3)	St	—	Na	510	0.1	41	70 000	Neg.	0.9 × 1.4

Za Zircalloy; Mgn, Magnox; St, Stainless.
[a] Commissioning planned for 1980.
[b] 338 tonnes.

metal, molten salt, an aqueous or an organic solution. In heterogeneous reactors the fuel is mostly rods filled with metal oxide. The fuel material can be almost any combination of fissile and fertile atoms in a mixture or separated as in the core (fissile) and blanket (fertile) concept (§ 19.15). The choice of moderator is large: H_2O, D_2O, Be, graphite, or organic solvent. The choice of coolant is even larger because it can be molten metal, molten salt, liquid H_2O, D_2O, or organic solvent, as well as gaseous CO_2, helium or steam.

Some of the more important combinations are summarized in Tables 19.4 and 19.5, while their use in power production is given in Appendix I.

19.13. Research and test reactors*

More than 400 research and test reactors have been built at university and government laboratories. While the smaller research reactors may have power $\lesssim 1$ kW, and thus do not need any forced cooling, the larger test reactors operate at $\lesssim 50$ MWth. Many of these reactors have facilities for commercial radionuclide production.

Research reactors are used in nuclear physics, in nuclear, analytical and structural chemistry, in radiobiology, and medicine, etc. They are usually easy to operate, inherently safe, and of low cost, $\lesssim$ \$ million for the smaller ones. Many of them are of the pool type (Fig. 19.10). The reactor core is located in the centre of a concrete vessel, 6–8 m deep and 2–5 m wide, containing purified water. The fuel assembly is attached to a steel framework

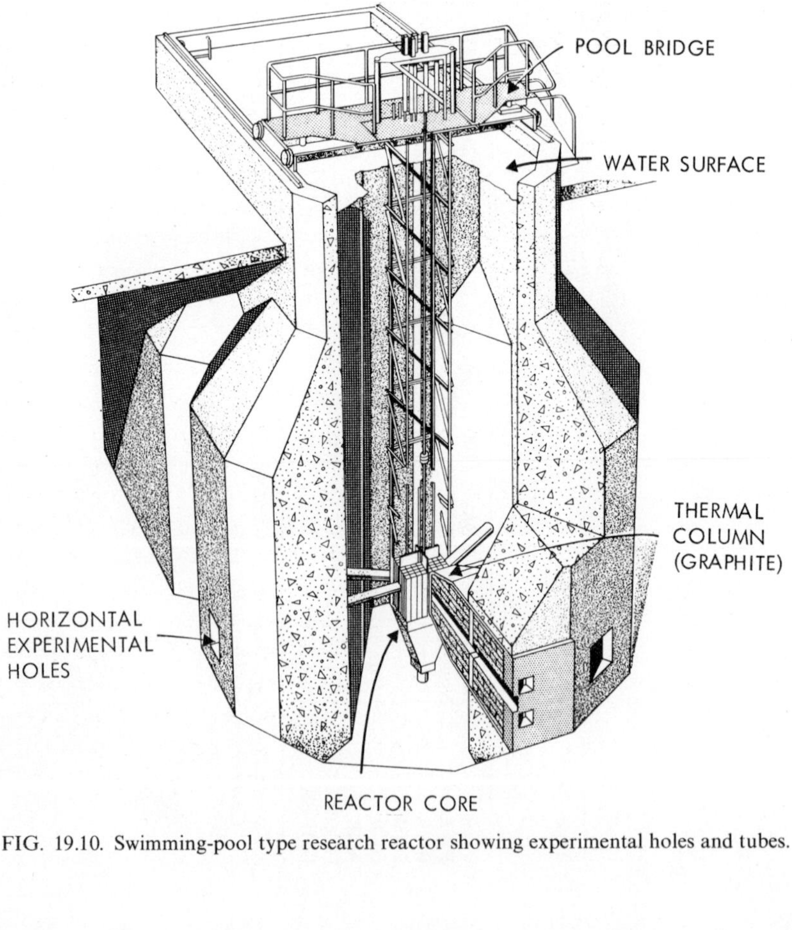

FIG. 19.10. Swimming-pool type research reactor showing experimental holes and tubes.

and suspended from a bridge, which spans the width of the pool. The reactor core is bare and can be seen from the water surface. The water provides the main radiation shielding, moderation, and cooling. The concrete walls are about 2 m thick when the reactor is located above ground; otherwise much thinner walls can be used. The main radiation protection demand comes from the $^{16}O(n,p)^{16}N \xrightarrow[7\,s]{\beta\gamma} {}^{16}O$ reaction, since the γ's emitted are very energetic (6.2 MeV). At power levels > 100 kW forced cooling may be required. The fuel is usually highly enriched uranium (up to 90%) rods or plates; the amount of ^{235}U is $\lesssim 3$ kg.

As example we choose the TRIGA reactor (Training, Research, and Isotope production reactor of General Atomics Div., USA). The core consists of aluminum clad fuel rods in a cylindrical array. Each fuel rod is 35 cm long and contains uranium–zirconium hydride; 8 wt% U enriched to 20% in ^{235}U. The total amount of ^{235}U is approximately 2 kg. The rods contain samarium as burnable poison. At 10 kW the neutron flux is 4×10^{11} cm^{-2} s^{-1} (thermal, maximum). Nominal life-time for the fuel is 10 y (2% burn-up). Reloading is carried out below water by transferring the fuel pins to a submerged lead-shielded fuel-handling cask. The temperature coefficient is $-1.3 \times 10^{-2}\% \Delta k/k$ °C. Control is achieved with rods containing B_4C powder. In the TRIGA-III version, the power can be pulsed, providing very intense neutron fluxes up to 10^{21} n cm^{-2} s^{-1} for ~ 100 ms.

In the CMEA countries (countries belonging to the Council for Mutual Economic Assistance, i.e. countries with centrally planned economies) the WWR is the dominating research reactor. This reactor is of the tank type, which is very similar to the pool type. The main difference is that the size of the pool is considerably reduced by using a steel tank around the core (in the pool). This makes it possible to operate at higher power, using forced cooling. Operating at 2–3 MWth the WWR provides 2×10^{13} thermal n cm^{-2} s^{-1}. The core consists of fuel pins with enriched uranium, 3–5 kg ^{235}U. The cover of the tank may be removed for fuel replacement. The MTR (Materials Testing Reactor) at Arco, USA, is of similar design. At a power of 30 MWth it provides an average neutron flux of 2×10^{14} cm^{-2} s^{-1}. At these high levels the reactors are classified as test reactors, their main purpose being testing materials for full-scale power reactors.

Some reactors are designed to produce very high neutron fluxes, either for testing materials (especially for fast reactors) or for isotope production. The high flux reactor at Grenoble, France, has fuel plates of highly enriched (93%) UAl_3 canned in aluminum. The construction is a swimming-pool tank type. Using 8.6 kg ^{235}U and 15 m^3 heavy water as moderator, it provides a maximum thermal neutron flux of 1.5×10^{15} at 60 MWth; 10 times higher fluxes can be achieved during shorter periods. At Oak Ridge, the HFIR mentioned in Chapter 13 has a flux of 3×10^{15}, the highest thermal neutron flux so far reported (cf. the Phenix reactor, §19.15.2). Some research reactors are built to test new reactor designs which may be candidates for future power reactors. The successful designs lead to prototype power reactors; some of these are described in the next section together with full-scale power reactors. Other designs never survive the first stage, although the design may be very interesting. An example of these short-lived products of imaginative reactor engineering is LOPO, the world's first "water boiler" reactor (only operated at 0.05 W) described in exercise 19.6, and the Clementine reactor operated at Los Alamos from 1946 to 1953. The latter had mild steel fuel pins, each containing 0.5 kg of δ-phase plutonium metal in a core cooled by liquid mercury. It operated at 25 kW maximal power and was the world's first fast reactor.

19.14. Thermal power reactors*

Electric power production reactors are presently built to 1300 MWe but all sizes in the range 500–1000 are common. With decreasing size the cost of the electric power usually increases. However, power reactors of ~200 MWe are competitive in some areas of high energy costs, such as in many developing countries. These smaller sized plants have some advantages: they are more reliable due to less stress of the components, they can more easily be built in remote areas because the smaller components are easier to transport, and the electric grid may be more capable of accepting the moderate power addition.

About half of the world's 190 nuclear power plants in operation in 1977 were of the PWR (pressurized water reactor) type, and about one-third of the BWR (boiling water reactor) type, indicating the dominance of light water reactors (LWR). The remaining 15% were of either older types of gas-cooled graphite-moderated reactors (GR) (being replaced by the advanced GR, AGR) and heavy-water-moderated reactors (HWR). These reactors are all heterogeneous thermal and usually (except possibly for the HWR) classified as burners. In the future high temperature reactors (HTR), which are similar to the AGR, and fast breeder reactors (FBR) will become of increasing importance. The HTR reactor may run as a converter or even as a thermal breeder. Future reactors are therefore expected to use the nuclear fuel more efficiently than present types. See Appendix Table I.2.B for a listing of reactors.

19.14.1. *Pressurized water reactors (PWR)*

The main components of a pressurized light-water-moderated and cooled reactor (PWR) station have been described in §19.1 and Fig. 19.1. The picture of the Trojan station, which started operation in 1975, is dominated by the large cooling tower; power stations located near large rivers or at the sea shore usually do not need such towers.

The important features of the PWR core are shown in Fig. 19.11, which, although taken from three different reactors, represents the typical Westinghouse design. A typical core contains about 40 000 fuel rods (a) in 193 assemblies, each with space for 208 fuel pins (b) and (c). The fuel is 88 t of UO_2 with an enrichment of 2.17% (inner region) to 2.67% (outer region). The reactor vessel (e) is made of low alloy steel, 13 m high and 4.4 m in diameter, with a wall thickness of 22 cm; (c) shows the distribution of control rods entering from the top of the vessel. The 1060 control rods are made of a silver alloy containing 15 wt% In and 5 wt% Cd; both these elements have high thermal neutron capture cross-sections. The main reactivity control is by boric acid, which is fed to the coolant through a special injection system. The boric acid circulates in the primary coolant loop, and acts as a burnable poison. At the start of a fresh core its concentration is ~1500 ppm (~0.025 M), but it is successively reduced to practically zero at time of fuel replacement. The concentration is adjusted in a side loop, containing either an anion exchange or an evaporator system. The ion exchange system is so designed that H_3BO_3 is fixed to the ion exchanger (the water is deborated) at low temperature (~30°C) while it is eluted at higher temperature (~80°C).

Some of the physical parameters are: k_∞ 1.29 (cold) and 1.18 (operational temperature), temperature coefficient $1 \times 10^{-3}\%\ \Delta k/k$ °C, average neutron flux 2.16×10^{13} (thermal), 3×10^{14} (fast). Data for heat production are given in Table 19.5.

The Westinghouse-type PWR (Figs. 19.1 and 19.11) reactor is presently the domi-

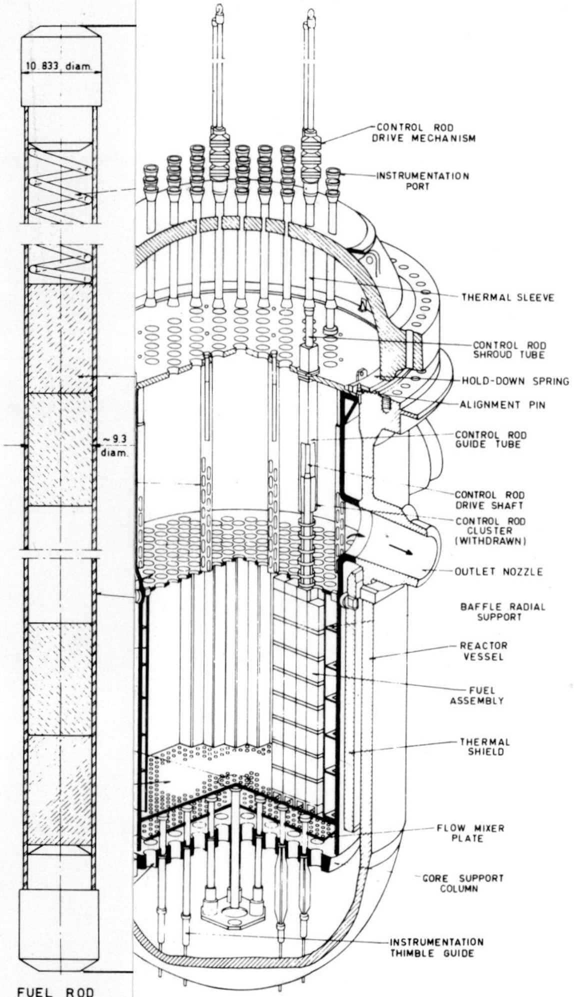

(a) (e)

ge, Belgium (370 MWe, 1975),

nating nuclear power reactor in the Western world. In the USSR and CMEA a similar reactor type is used, referred to as the VVER, in conjunction with the RMBK type; this latter is a graphite-moderated reactor with the fuel elements cooled by light water in pressurized tubes.

19.14.2. *Boiling water reactors (BWR)*

The reactor core and vessel of a typical BWR station is shown in Fig. 19.12 and supplemental data are given in Table 19.5. The fuel core and elements are rather similar

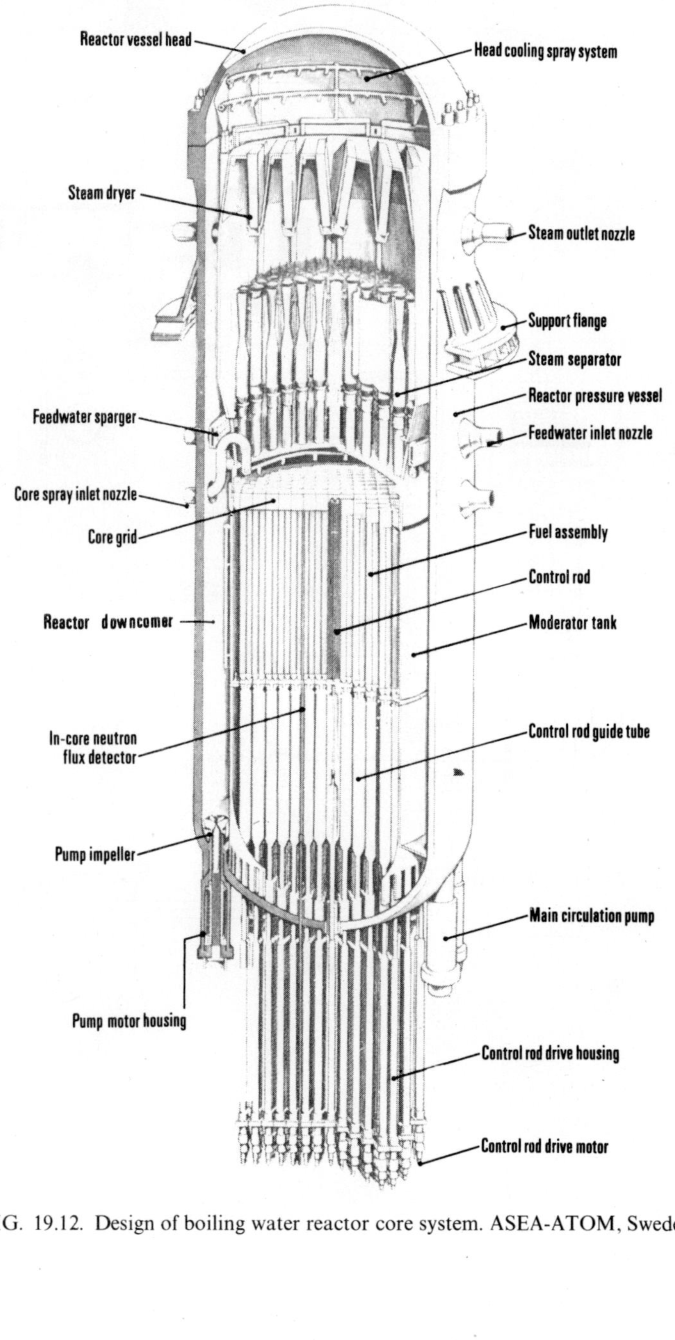

FIG. 19.12. Design of boiling water reactor core system. ASEA-ATOM, Sweden.

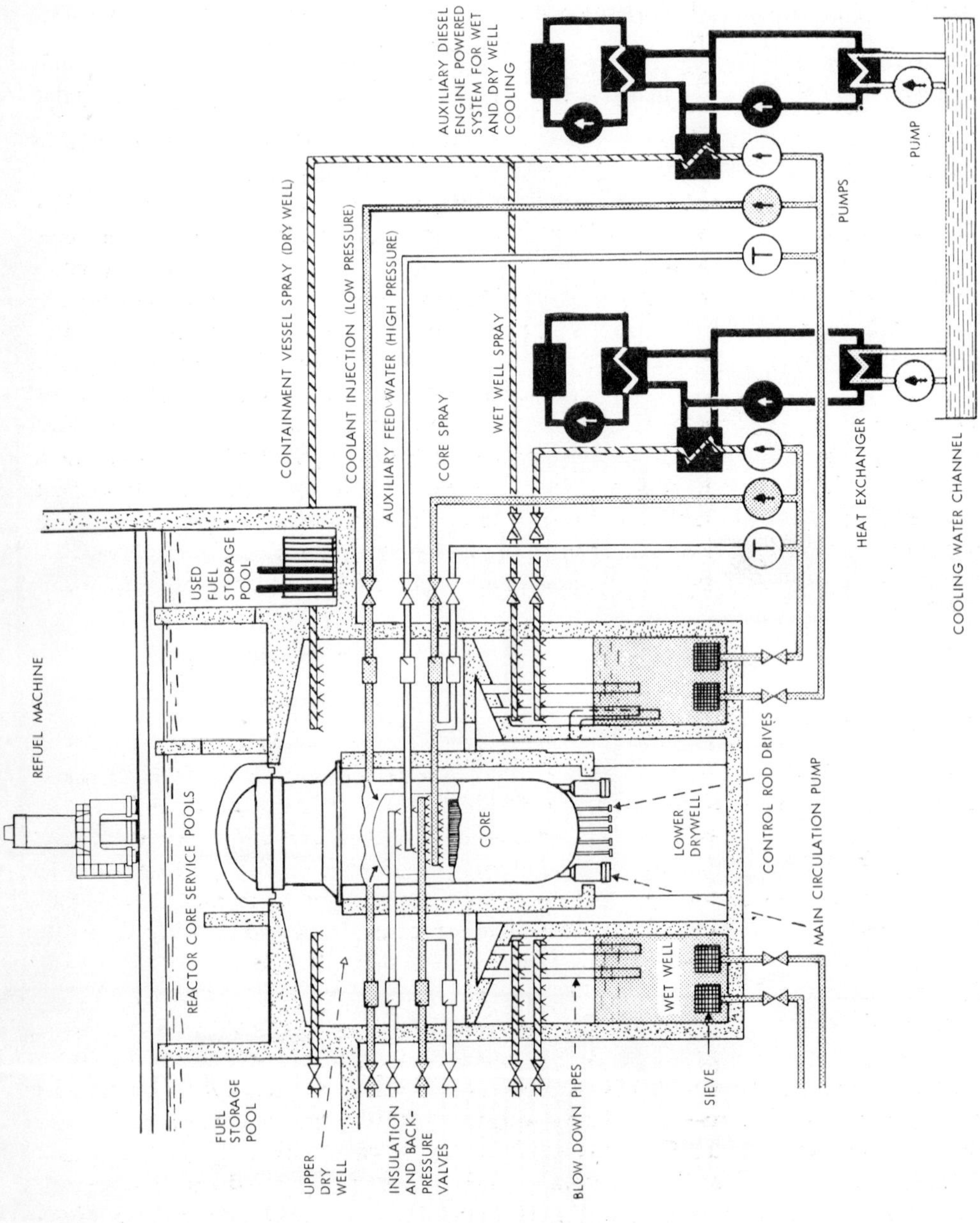

FIG. 19.13. Emergency cooling system of ASEA–ATOM boiling water reactor.

to those of the PWR. The main difference is that boiling occurs in the reactor vessel, so no external steam generator is required. To minimize the wear of the turbine it must be fed with steam that is as dry as possible. Therefore there is an extensive steam-drying system above the reactor core, inside the reactor vessel. Although the cores are about the same size for PWRs and BWRs, the reactor vessel is much taller for the BWR; typical values are 22 m in height and 6 m in diameter. Since the pressure in the vessel is lower for BWRs than for PWRs, the vessel wall can be thinner (a typical value is 16 cm). BWRs usually contain more uranium, but with lower enrichment than PWR fuel. While only about one-fifth of the core is replaced each year in a BWR, about one-third of the core is replaced annually for a PWR; in both cases about 30 t of used fuel is replaced every year.

The BWR is controlled by rods entering from the bottom of the reactor vessel. The rods are usually cross-shaped and fit in between the fuel assemblies. The fuel elements often contain burnable poison (e.g. Gd_2O_3). Because of the boiling in the tank boric acid cannot be used for reactivity control. However, it is provided for as a shutdown feature. The control rods contain boron carbide in stainless steel tubes. In addition to negative temperature coefficients both for fuel and moderator, BWRs have a negative *void coefficient*. This means that bubble formation (voids) along the fuel tubes reduces the reactivity. It is therefore common practice to control the reactor by the main circulation pumps: increased circulation reduces the voids thereby increasing power production.

In the BWRs the steam outlet from the reactor vessel is located in the so-called upper dry well (Fig. 19.13). In the case of excess pressure in the tank, safety valves at the steam outlet open, releasing steam into the wet well (which contains ~ 2000 m^3 water) where it condenses. Figure 19.13 shows some emergency systems for the ASEA-ATOM (Sweden) BWR reactor.

19.14.3. *Heavy water reactors* (*HWR*)

The advantage of using heavy water for neutron moderation is that its low thermal neutron capture cross-section ($\hat{\Sigma}_a$(th), Table 19.2) improves the neutron economy (i.e. $1 - p$ becomes much less than in LWRs) so that even unenriched natural uranium can be used as fuel. The good neutron economy and harder energy spectrum leads to about twice as high plutonium yield in heavy water reactors as compared to LWRs (see Appendix I). The first large heavy-water-moderated reactors were built in the United States around 1950 for production of materials for fission weapons. Presently the only commercial HWR is the CANDU (CANadian Deuterium Uranium) design, which is illustrated in Fig. 19.14. The reactor vessel is a horizontal cylindrical tank filled with D_2O, through which several hundred horizontal aluminum tubes pass. Within each of these *pressure tubes* are bundles of zircalloy clad natural UO_2 pins. The bundles are cooled by pressurized heavy water (~ 8 MPa), which in the heat exchangers produces light water steam for turbines. This pressure-tube-tank design is known as a *calandria*. In some designs the heavy water in the moderator can be dumped down into an empty tank below the reactor (5, Fig. 19.14) to allow an immediate shutdown, at the sacrifice of shielding in such an event. In other designs (Bruce, Table 19.5) the calandria is surrounded by a vessel full of ordinary water, which provides shutdown shielding; shutdown is then achieved by absorber rods and injection of gadolinium nitrate into the moderator. The calandria design permits on-line fueling.

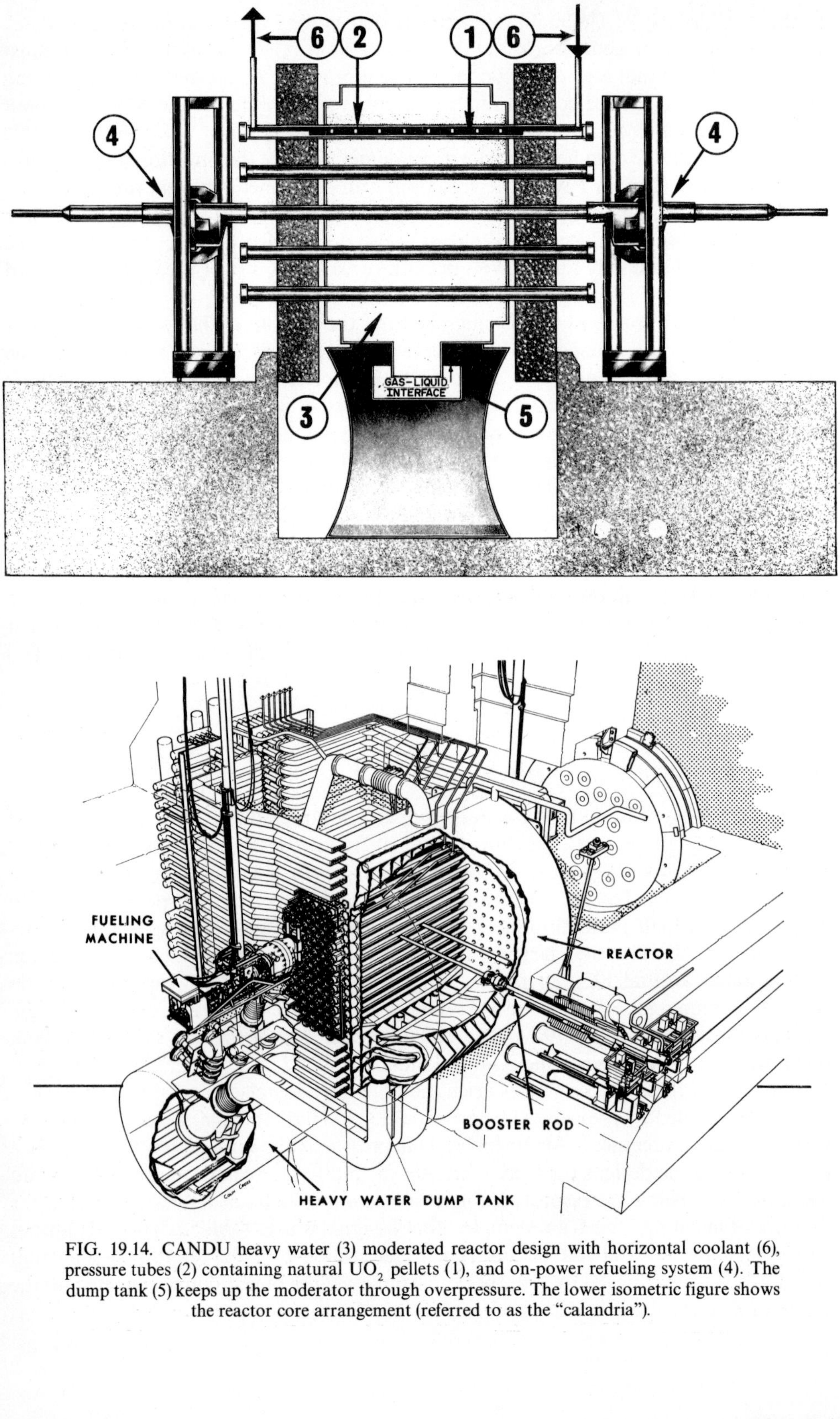

FIG. 19.14. CANDU heavy water (3) moderated reactor design with horizontal coolant (6), pressure tubes (2) containing natural UO_2 pellets (1), and on-power refueling system (4). The dump tank (5) keeps up the moderator through overpressure. The lower isometric figure shows the reactor core arrangement (referred to as the "calandria").

Early CANDU–PHW reactors were built in the 200 MWe size, but the design can be upgraded to 750 MWe (Table 19.5). Using natural uranium the burn-up is comparatively low, about 9000 MWd t.

The CANDU reactors have a higher plutonium yield than any other thermal reactor. The neutron economy makes it feasible to run HWRs on the thorium cycle with a conversion ratio of ~ 0.9. Such a cycle has to be started either with a mixture of enriched $^{235}UO_2$ and ThO_2, or—if isotope enrichment is not used—through the following three step fuel schedule: (i) running on natural uranium and extracting Pu; (ii) running on Th + Pu and extracting ^{233}U; (iii) running on Th + ^{233}U.

The investment cost for CANDU reactors is higher than for LWR's because of a more complicated design and the amount of D_2O required. The running costs are claimed to be lower; the standard loss of D_2O is ~ 1 kg d^{-1} (at about $100 per kg). The neutron capture by D forms tritium in considerable amounts, so the D_2O becomes contaminated by TDO. Also TD and DTO can leak into the surroundings, which may cause difficulties in biological safety. The CANDU reactors have been shown to have the highest reliability (availability cf. § 19.16) of all nuclear power reactors (> 90%) in terms of possible operating time.

The steam generating heavy water reactor (SGHWR) is a British 100 MWe prototype power reactor. The design has a calandria with vertical pressure tubes, cooled with light water, which is allowed to boil in the tubes. The fuel is slightly enriched, (1.2%) UO_2 in zircalloy. Refueling is carried out off-load (i.e. after power shut-off). Maximum burn-up is ~ 20 000 MWd t^{-1}. This design has both advantages and drawbacks compared to the CANDU.

19.14.4. *Chemistry of water-cooled reactors*

The conditions in water-cooled reactors are approximately: temperatures $\lesssim 350°C$, pressures $\lesssim 16$ MPa, and intense γ- and n-irradiation. This causes potentially severe corrosive conditions.

The radiation decomposition in water-cooled and/or moderated reactors is considerable. About 2% of the total energy of the γ- and n-radiation is deposited in the water. We have seen in §§15.7 and 15.8 that this produces H_2, O_2, and reactive radicals. In a 1 GWth BWR the oxygen production is about 1 l min^{-1}, but it is considerably less in PWR. Because of the explosion risk from $H_2 + O_2$, the two are recombined catalytically to H_2O in all water reactors.

Both the radicals formed and the O_2 increase the corrosion rate of reactor materials. In a BWR the corrosion rate of stainless steel (18Cr 8Ni) is about 10^{-4} mm y^{-1}, leading to the release of small amounts of Fe, Cr, Co, and Ni in the cooling circuit. Many of these are highly radioactive, e.g. ^{58}Co and ^{60}Co. The corrosion products form insoluble voluminous colloidlike products referred to as *crud* (originally said to be the acronym for Chalk River Unidentified Deposits). In addition to polymeric metal hydroxides the crud contains small amounts of other materials in contact with the wet circuit: zirconium from the canning material, copper from the condenser system, silicon and organic material from the water purification systems, boron from the boric acid control system, etc., and released fission products. Deposition of crud on the fuel element surfaces may block cooling canals, and, because of its poor thermal conductivity, may also lead to a burn-through of the canning and a subsequent release of fission products into the reactor water. The radioactive corrosion products carried with the cooling loop create a serious radiation problem for the reactor personnel. Considerable effort is therefore put into the development and selection of corrosion-resistant materials. Equally important in this respect is the selection of water conditions which minimize corrosion and the deposition of such products within the reactor core, and effective water-cleaning systems. In PWR the crud is transported through the whole primary cooling circuit, but removed in the purification circuit. In BWRs the crud accumulates in the reactor vessel; therefore BWRs have a special cleaning circuit attached to the reactor vessel. The cleaning circuits are discussed in §19.18.

The amount of crud deposited in the reactor core can be partly removed during shutdown by mechanical cleaning or washing with chemical decontamination solutions. These solutions contain mildly oxidizing agents as alkaline permanganate and/or organic complexing agents such as ammonium citrate (APAC treatment) or EDTA. Several companies offer decontamination services of reactor components on a commercial basis.

There are a number of ways to reduce corrosion. One is to increase the pH of the water to ~8 by adding alkali, e.g. LiOH or NH_3 ($\lesssim$ 10 ppm). While reactors of US type use ^{7}LiOH in order to reduce the formation of tritium from n,γ-capture in ^{6}Li, the CMEA country reactors use KOH. When ammonia is used, the radiolysis yields HNO_2 and HNO_3; it is then necessary to add H_2 gas to shift the equilibrium away from the acidic products. At an H_2 concentration of ~2 ppm, the concentration of dissolved O_2 is greatly reduced. Instead of NH_3, hydrazine or N_2 may be added to the water to increase the pH via the reactions

$$\tfrac{1}{2}N_2 + 3H \rightsquigarrow NH_3; \quad NH_3 + H_2O \rightarrow NH_4^+ + OH^-$$

The hydrogen content of the water must be kept reasonably low since it increases the brittleness of the zircalloy canning.

The purity of the reactor water is checked by measurement of pH, conductivity, turbidity, oxygen concentration, radioactivity, etc., either by sampling or—more usually—by on-line analyses.

19.14.5. *Graphite-moderated reactors*

The first large nuclear reactors were graphite moderated and fueled with natural metallic uranium. The first reactor, critical in Chicago in 1942, was (convection) air cooled, while the plutonium production reactors at Hanford, USA, were cooled by water taken directly from the Columbia river. Subsequent reactors at Windscale, United Kingdom and Marcoule, France were of similar design, but cooled with CO_2 gas. The world's first large civilian power station, which went into operation in 1956 at Calder Hall in the United Kingdom, was also of that type; the four reactors at this station produced 800 MWth, yielding 184 MW of electric power. These reactors were closed down some years ago because of obsolescence. In 1954, a 5 MWe (30 MWth) power demonstration reactor started operation in Obninsk, USSR.

The advantage of using graphite over water as a moderator is primarily that higher steam temperatures can be achieved, but it also has other advantages: as moderator it is superior to light water, it is less expensive than heavy water, and it has mechanical stability, leading to simpler core design.

The British Magnox and the French GR reactors are typical of the "first" generation gas-cooled graphite-moderated reactors (GCR); Wylfa in Table 19.5 is an example of this type. In the choice of reactor coolant the considerations are (i) nuclear properties,

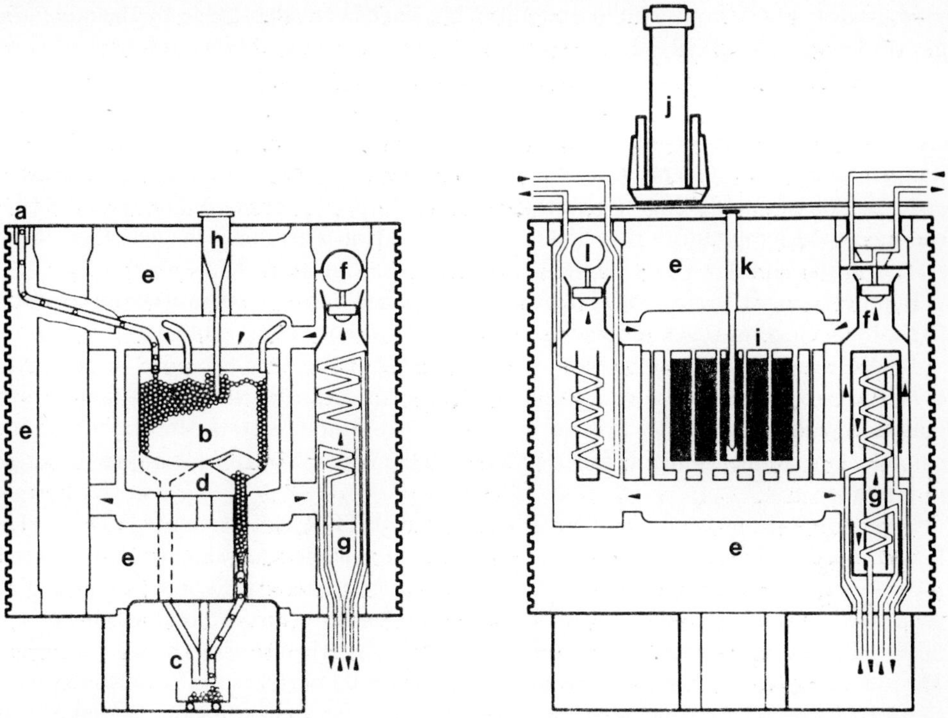

FIG. 19.15. Two high temperature gas-cooled reactor (HTGR) designs. *Left*: pebble bed thorium high temperature reactor (THTR), Uentrop, FRG. *Right*: gas-cooled Fort St. Vrain (FSV) design. a, fuel entrance; b, pebble bed core; c, used level exit; d, graphite reflector; e, prestressed concrete (reactor) vessel; f, main gas circulation pump; g, heat exchanger and steam generator ; h, control rod; i, core of hexagonal prisms; j, fueling machine; k, control rod; l, emergency cooling pump.

(ii) physical and heat transfer properties, (iii) chemical properties including stability and compatibility, and (iv) risk and health aspects (e.g. leakage). Many gases have been considered: air, steam, hydrogen, nitrogen, helium, carbon monoxide, and carbon dioxide. For the Magnox reactors CO_2 was chosen as the overall best compromise. Its heat transfer is less satisfactory than air, but it is compatible with (i.e. does not react with) graphite below 600°C; at higher temperatures $CO + O_2$ is formed. However, the graphite undergoes radiolysis: $C(s) + CO_2(g) \rightsquigarrow 2CO$. This is to some extent suppressed by adding methane, but through its radiolysis some H_2O is formed, which must be removed, because of its corrosive effect on the graphite.

The GCR have graphite cores made by multisided prisms stacked into a horizontal cylinder, about 10 m high and 15 m in diameter, surrounded by ~10 cm steel (forming the inner wall of the pressure vessel) and a 3 m thick prestressed concrete vessel ("PSC-vessel"). Holes for fuel and cooling pass through the graphite block. The fuel is natural uranium canned in a magnesium alloy (thus the name Magnox); the amount of uranium is about five times larger than for a water-moderated reactor of comparable power. The reactor can be refueled on-line (not always used) in a way similar to the HWR in Fig. 19.14. In order to achieve good heat transfer, the Magnox cans are provided with cooling fins. The coolant outlet temperature is about 100°C higher ($\gtrsim$400°C) than for water-cooled reactors. Steam is produced in gas–water heat exchangers (4 at Wylfa) within the pressure vessel.

Magnox type GCR are still in operation, but have been superseded by the advanced gas-cooled reactor (AGR). These reactors are similar to the Magnox type in general design. However, it has been possible to increase plant efficiency, fuel burn-up, and power without increase in plant size and capital costs. In the AGR the Magnox fuel elements are replaced by slightly enriched UO_2 clad in stainless steel, permitting surface temperatures $\lesssim 760°C$; enrichment is necessary to compensate for the higher neutron cross-section of the canning material. In the AGR the fuel operates red hot and the CO_2 coolant outlet temperature is $\sim 650°C$ at about 4 MPa. This raises the net efficiency to $\sim 42\%$ in the Hinkley Point, United Kingdom, AGR station (Table 19.5).

In order to increase efficiency further, higher coolant outlet temperatures are required. In the high temperature gas-cooled (graphite-moderated) reactors (HTGR) temperatures $\lesssim 1000°C$ are now contemplated; such temperatures have been exceeded in prototype HTGRs. At this temperature CO_2 is unsuitable because of thermal instability and reactions with the graphite. Instead helium is used. Stainless steel canning is also unsuitable. Unclad fuel elements are used with the fissile oxide or carbide atoms dispersed in graphite.

Figure 19.9 shows steps in the manufacture of HTGR fuel elements (§ 19.11). The kernels produced are shown in (c), and the covered particles in (a) and (b). The internal layer of pyrolytic carbon, C(Py), is porous in order to absorb fission products (FP), especially the gaseous ones. The next layer is high density C(Py); it eliminates the escape of the FP out of the particle. The fissile fuel particles are then covered by silicon carbide in order to facilitate the separation of fissile from fertile particles in the reprocessing of the used fuels. All particles have an outer layer of high density C(Py) to make them compatible with graphite.

Two principal HTGR designs are illustrated in Fig. 19.15, both with a design capacity of ~ 300 MWe. In the German pebble bed reactor (THTR), the fuel elements are graphite balls, 60 mm in diameter, containing a mixture of fertile and fissile particles; each ball contains 1.1 g $^{235}UO_2$ and 11 g $^{232}ThO_2$. The core consists of about 700000 such balls in a random arrangement, the overall dimension being ~ 5.6 m in diameter and 5.1 m high. As is seen from the figure, fueling is on-load. The core is surrounded by a graphite reflector and a 5 m thick steel lined PSCV. Although the THTR and the Fort St.Vrain (FSV) reactors have no special containment building, commercial HTGRs will be provided with a conventional outer containment building. In the US HTGR design (FSV), the core consists of stacks of hexagonal prisms of graphite (Figs. 19.15 and 19.9 (e) and (f)). The prisms are pierced with cooling canals and fuel rods (d), made by pressing graphite powder and fuel particles together. Each rod contains a mixture of fissile (UC + ThC) and fertile (ThC) particles. The uranium is enriched to 93% in ^{235}U. A core loading contains initially 936 kg U and 19 500 kg Th. Fueling is off-load. Control is achieved by B_4C rods.

Both HTGR designs can be classified as large prototypes. They provide average outlet temperatures of the helium coolant of $\sim 750°C$ (max. 1040°C) at ~ 5 MPa and net efficiencies of $\sim 40\%$. The burn-up is higher in these reactors than in any previously mentioned, $\sim 100\,000$ MWd t^{-1}. The fissile particles can tolerate a 75% burn-up, and the fertile ones are designed for $\sim 10\%$ burn-up. Both reactors have conversion ratios 0.5–0.6.

The HTGRs offer some advantages over LWRs with respect to heat properties, fuel flexibility, and safety. They have higher thermal efficiency. The high gas temperature makes it possible to bypass steam-producing heat exchangers by using direct cycle helium gas turbines, which should lower plant cost and increase plant efficiency and safety (no hazards from water and steam entering the core). The high temperature makes the HTGR promising for direct use of nuclear heat for such processes as coal gasification and production of synthetic gas. In several countries steel production through iron ore reduction using HTGR-produced hydrogen is being studied.

Figure 19.17 shows the build-up and decay reactions for a thorium-fueled thermal reactor. The effective η-value for ^{233}U is 2.24. A reactor running on the ^{233}U ^{232}Th fuel cycle would theoretically be able to breed more fissile material. However, the margin is quite small; $<11\%$ (0.24/2.24) of the neutrons can be lost if a breeding gain ($G > 0$) is to be achieved. In HTGRs starting with ^{235}U, the corresponding figure is $< 3\%$ and breeding is practically impossible. In practice a conversion ratio of only 0.6 is reached. This value would be increased when ^{235}U is replaced by ^{233}U. The HTGRs are more suitable for ^{233}U production from ^{232}Th than LWRs and HWRs because the neutrons spend longer time in the epithermal neutron energy region where σ_γ (Th) is large (in the 20–70 eV region ^{232}Th has a number of capture resonances with high peak cross-sections (Fig. 19.2)). By using highly enriched ^{235}U, very little neutron capture occurs in ^{238}U, which otherwise would compete with the ^{233}U production. The FSV reactor has the following physical values (cold): $k_\infty = 1.19$, $\Lambda = 0.95$, $k_{eff} = 1.13$, $\eta = 2.06$ (^{235}U only fissile), $f = 0.71$, $p = 0.71$, $\varepsilon = 1.15$. The low p-value reflects the high neutron capture in ^{232}Th, while the large ε-value both reflects the higher average neutron energy and high concentration of ^{235}U (σ_f ^{235}U $> \sigma_f$ ^{238}U at high neutron energies (Fig. 19.2)). A disadvantage of the ^{232}Th–^{233}U fuel cycle is that the reprocessing technology is not fully developed. However, used HTGR fuel can be stored safely for longer periods (thus reducing the urgency for reprocessing) than used LWR fuel because of the excellent radiation resistance, inertness, and fission product retention of the graphite matrix.

Although no graphite-moderated water-cooled reactor is in commercial operation in the free market countries, there is a successful reactor type in the CMEA countries. The fuel elements are pins of slightly enriched U–Mo alloy canned in stainless steel. They are cooled by light water in stainless steel pressure tubes passing through the graphite moderator. The construction cost is said to be low and the reliability high.

19.14.6. *Special reactor systems*

So far we have described reactors designed for conventional power production and research or isotope production. We have purposely not included reactor designs which presently are considered obsolete (e.g. the sodium-cooled graphite-moderated heterogeneous thermal reactor such as Hallam USA, 75 MWe), or designs which have little promise for large scale power production (e.g. the organic-moderated and cooled thermal heterogeneous reactor Piqua, USA, 11 MWe).

Nuclear power reactors may be used for other purposes than large scale production of electricity (or heat) for community consumption. The large amount of energy stored in the small volume of fuel elements makes nuclear reactors suitable as energy sources when repeated refueling is a disadvantage. For example, a nuclear reactor has been used for a long time as a power source in the Arctic. Several nuclear-powered cargo ships have been built to determine their feasibility but presently they are considered uneconomic. The USSR icebreaker *Lenin* (44 000 shaft horse power) launched in 1959 had three reactors, of which one was a spare. The reactor was of the PWR type with 5% enriched UO_2 producing 100 MWth. The ship had a speed of 18 knots in open sea and 2 knots in

FIG. 19.16. The USSR nuclear-powered icebreaker *Lenin* in polar seas.

ice of 2.4 m thickness (Fig. 19.16). Later two more Soviet nuclear icebreakers have been launched. The N/S *Arktika* (75 000 hpr) was the first surface ship to reach the North Pole (1977). Nuclear power is used in many naval vessels in the USA, the USSR, UK, and France. The first nuclear-powered vessel was the US submarine *Nautilus*, launched in 1955; in 1958 it made a famous voyage below the ice north of the North American continent from the Pacific to the Atlantic, passing under the ice at the North Pole. It could travel 250 000 km ($\sim$ 6 times around the earth) without refueling. The present nuclear fleets in the USA and USSR are dominated by more advanced submarines which now number several hundred. Representative data are (US *Lafayette* class): displacement 7300 t, length 130 m, 15000 shp, submerged speed $\sim$ 30 knots, crew 140, 4 torpedo tubes and 16 missiles with nuclear warheads. Because of the difficulty of detecting and destroying these submarines they are considered to act as deterrents against a nuclear war between the superpowers. The world's largest naval vessel is the US nuclear-powered aircraft carrier *Enterprise*, which has 8 PWR reactors, producing 300 000 shp. With a displacement of $\sim$ 80000 t (and $\sim$ 100 airplanes) it can cruise at 35 knots for 60 days without refueling.

19.15. Breeder reactors*

The possibility to breed more fissile nuclei from consumption of fissile + fertile nuclei increases with increasing fission factor η according to § 19.9. Since η depends on the fission cross-section, which is highly energy dependent (Fig. 19.2), the neutron spectra in reactors must be considered. In a reactor with large amounts of highly moderating atoms, the original unmoderated spectrum is strongly shifted towards lower energies. However, even under optimal moderating conditions, a thermal Maxwell neutron distribution is never achieved in a reactor, but can be closely approached in the thermal column of research reactors (Fig. 19.10). Because breeding occurs through neutron capture in fertile ^{232}Th and ^{238}U (the *only* fertile atoms on earth), with maximum probability between 600 and 700 eV, it is desirable for breeding that the neutron flux not be too low in this region. Just as breeding cannot be achieved with a thermal Maxwell neutron distribution it cannot be achieved with a fission neutron energy spectrum. Even if attempts are made to produce very "hard" neutron spectra (i.e. high energy), some moderation always occurs. Thus in a fast reactor, in which as few as possible of light atoms are used, the neutron spectrum is still shifted strongly towards a lower energy region (Fig. 19.3) so that typically the neutron flux at 0.25 MeV is higher than at 1 MeV. The curves shown for thermal and fast reactors in the figure are only indicative, and depend strongly on the particular reactor design.

Because breeding is only possible for $\eta > 2$, ^{233}U is the favored material for breeding in thermal reactors, and ^{239}Pu for breeding in fast reactors. Considering the shift towards lower neutron energies in fast reactors it is seen that η for ^{239}Pu is always greater than for ^{233}U. Thus more ^{239}Pu can be obtained by breeding than ^{233}U. Consequently much more effort has been put into fast plutonium breeders than into thermal ^{233}U breeders.

19.15.1. *The thermal breeder*

In the thermal breeder reactor ^{233}U is produced from ^{232}Th. The reactor may be designed either with a core containing a mixture of ^{232}Th and ^{233}U, or with a central zone (core) of ^{233}U surrounded by an outer layer (blanket) of ^{232}Th. In the previous section we described the THTR, which represents the former alternative with an almost homogeneous solid mixture of UO_2 and ThO_2 dispersed in graphite. Breeding is not achieved in the THTR, but the conversion ratio is raised when $^{233}UO_2$ replaces the $^{235}UO_2$. In the FSV modification of the HTGR, it is easier to use a core and blanket system. This makes it possible to increase the conversion ratio very close to or even slightly above one. However, it is necessary to minimize parasitic neutron capture in structural materials including monitoring systems, control rods, etc. Calculations have shown that a conversion ratio of 1.06 should be possible.

Instead of using a solid matrix, both core and blanket can be liquid. In the HRE-2 (homogeneous reactor experiment), USA, the core consists of a ^{233}U sulfate solution in D_2O, while the blanket is a suspension of ThO_2 in D_2O. The project has been canceled because of severe corrosion of the spherical reactor tank, of difficulties in avoiding precipitation in the core, and of other technical difficulties.

The molten salt reactor (MSR) is a similar project. Here UF_4 and ThF_4 are kept in an eutectic melt of BeF, ZrF_4 and NaF (or ^7LiF). The reactor may be homogeneous with respect to ^{233}U and Th, or of the core and blanket type in which case graphite may be used as moderator. The molten salt system has several advantages: (i) radiation effects are less in molten salts than in solid fuels, (ii) the expense of fabricating complicated fuel elements is avoided, (iii) the fission products can be removed continuously, (iv) fuel recharging can be made continuous, (v) the core contains no absorbing structural materials, and (vi) the system has a high negative temperature coefficient and is therefore easy to control. Points (iii) and (v) makes it possible to obtain a conversion ratio of 1.07, i.e. a 7% breeding gain; the fissile doubling time has been estimated to be 20 y. In the MSR high salt temperatures ($\lesssim 700°C$) are used. Nevertheless, the system can be

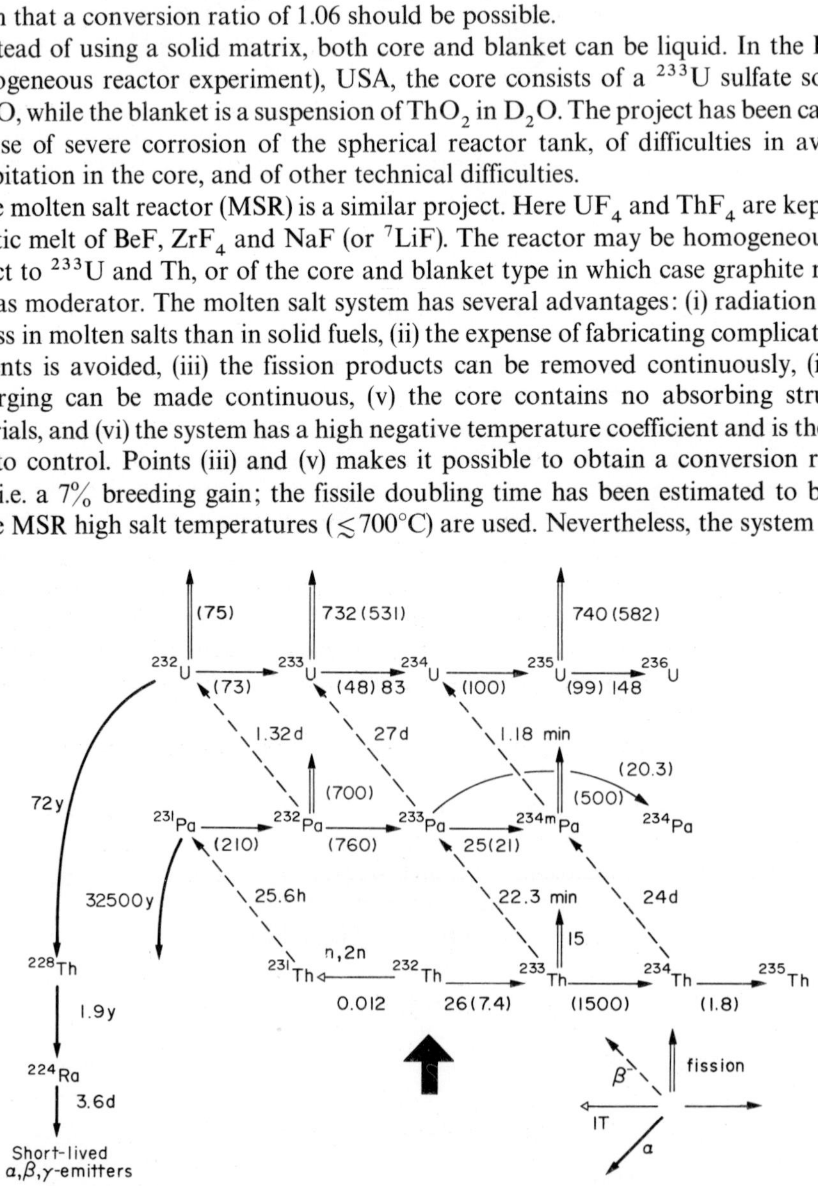

FIG. 19.17. Nuclear reactions in irradiation of thorium. Figures along reaction lines are half-lives or effective reaction cross-sections (barns) for a standard power LWR, with thermal (0.025 eV) cross-sections within parentheses.

operated near atmospheric pressure because of the low vapor pressure of the melt. The salt is also inert to air and water. By using a dump tank for the core, criticality accidents are minimized (cf. next section). The on-line fission product removal requires advanced chemistry, and these problems are not yet solved. The reactor concept has been tested at Oak Ridge, USA, for some years.

Figure 19.17 shows the production of heavy atoms in the ^{232}Th system. Because ^{233}Th is produced until the fuel is removed from the core (or blanket), the long half-life of ^{233}Pa (27 d) requires long cooling times before reprocessing, if a maximum yield of ^{233}U is to be obtained. The thorium would become contaminated by the isotope ^{234}Th and 240 d would be required to reduce this activity to 0.1%. The ^{233}U would be contaminated by ^{234}U and ^{232}U; the latter constitutes a problem because of its relatively rapid decay into $^{228}\mathrm{Th} \rightarrow {}^{224}\mathrm{Ra} \rightarrow$ chain. These isotopic contaminations of the fertile and fissile products in the ^{232}Th–^{233}U fuel cycle would cause handling difficulties, which are more severe than for the ^{238}U–^{239}Pu fuel cycle. At present no ^{233}U breeder exists.

19.15.2. *The fast breeder*

The principal design of a fast breeder reactor consists of a central core of plutonium in which fission occurs, surrounded by an outer blanket of ^{238}U in which neutrons are captured to form new ^{239}Pu. This blanket is surrounded by a reflector, usually of iron. The fission yield curve of ^{239}Pu is similar to that of ^{235}U (Fig. 9.8) but is shifted up by 4 atomic mass units. Some neutron capture occurs in ^{239}Pu yielding fertile ^{240}Pu, which through another neutron capture produces fissile ^{241}Pu; similarly, successively higher elements are formed in time. In a strong neutron flux their fate is destruction by fission. Thus, in the long run, by conversion of ^{238}U into ^{239}Pu, etc., fission energy is always released and in principle all the ^{238}U can be used for producing fission energy. In practice a value of ~70% is considered more realistic. Still this would mean 100 times more energy than is available from fission of ^{235}U only. Since the fast breeder reactor concept extends the fission energy resources by about a factor of 100, it makes uranium the largest energy resource presently available on earth.

The conversion efficiency of a reactor of this design can be estimated from

$$C = \eta + \varepsilon' - p' - \Lambda' - 1 \tag{19.32}$$

where ε' is the contribution to the neutron production from fertile nuclides, p' is the neutron loss due to parasitic absorption, and Λ' due to leakage. Some representative values are

	η	ε'	p'	Λ'	C
Core	2.56	0.27	0.049	1.15	0.73
Blanket	—	0.10	0.020	0.032	—
Total	2.56	0.37	0.069	0.032	1.82

It should be observed that the neutrons lost from the core are caught by the blanket. C is higher than in any other known chain-reacting system.

The greatest amount of experience with fast research reactors has been provided by the British reactor at Dounreay (DFR). It was operated at a power level of 200 MWth and at occasional temperatures $\gtrsim$950°C for 18 y until being closed down in 1977. Reprocessing of DFR core fuels has been carried out continuously most of this time. Larger

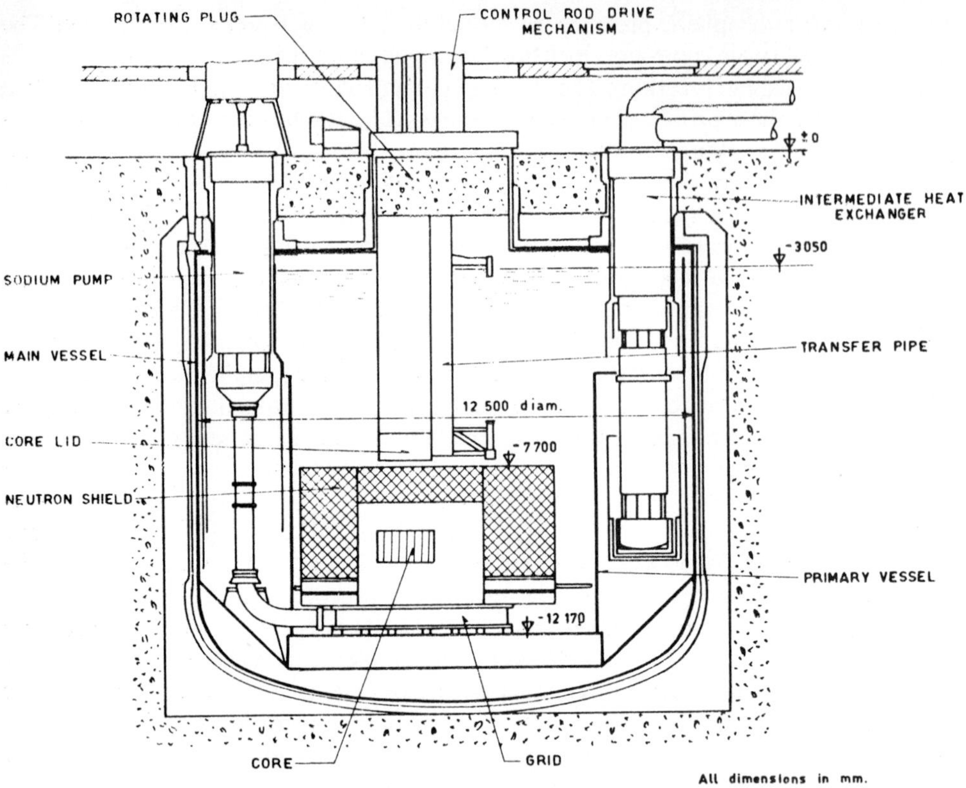

FIG. 19.18. Vertical section of French fast reactor Phenix. The vessel is filled with liquid sodium up to the level – 3050 mm.

(100–300 MWe) prototype power fast breeder reactors have been designed in the USSR, France, and the UK, and are under construction in other countries like Japan, Germany, and Italy. Presently the largest FBR being built is the Super-Phenix (France), which would produce 1200 MWe in the 1980s.

Figure 19.18 shows the French Phenix reactor, which went into operation in 1974. The thermal rating is 563 MW (gross 250 MWe); the neutron flux is very intense, $\lesssim 8 \times 10^{15}$ n cm^{-2} s^{-1}, which puts a severe strain on construction materials. The core consists of an inner and an outer zone surrounded by a blanket and a neutron shield, all immersed in a 35 cm thick steel tank filled with liquid sodium (850 t) at atmospheric pressure. The tank is surrounded by a concrete (biological) shield. The fuel is made of pins, the inner zone contains 20 vol. % PuO_2 (with 20% ^{240}Pu) and 80 vol. % UO_2, while the outer zone contains 27% PuO_2. The fuel pins are clad in 0.5 mm stainless steel. The total plutonium amount is 880 kg. The blanket is made of UO_2 pins, depleted in ^{235}U (0.2%). Outside of this is the neutron shield of borated graphite. Because the sodium becomes extremely radioactive due to ^{24}Na formation, the whole cooling system is contained within the radiation shielding. By means of a sodium/sodium intermediate heat exchanger within the reactor vessel (but outside the neutron flux) sodium of 550°C is transported to a sodium/water heat exchanger/boiler which produces 512°C steam of 17 MPa. Heat exchanger and piping are made of steel.

The prompt neutron lifetime in a fast reactor is about 1000 times smaller than in a thermal reactor. A reactor which can go critical on the prompt neutrons only would be exceedingly difficult to control. Therefore, fast reactors are designed to depend on the delayed neutrons (like thermal reactors). The time period (t_{per}) is large enough to allow reactor control through the use of neutron absorbing rods. Since the neutron spectrum is such that several percent of the flux is in the resonance region, control rods with boron can be used; in practice boron carbide (the carbon atom reduces the neutron energy further) and tantalum (which has large absorption peaks $\gtrsim 10^4$ b at 3–100 eV) are used.

Although the temperature coefficient of the Phenix reactor is negative, this is not so for fast reactors with harder neutron spectra, e.g. BN-350. If the fast reactor becomes overheated, the core may be deformed, leading it to become prompt critical. The power would then increase rapidly with a doubling time in the microsecond range, and a severe accident would be unavoidable. To avoid this the core is designed to achieve negative reactivity upon sudden power transients by using $^{238}UO_2$ in the core. When the temperature rises, Doppler broadening occurs in the ^{238}U resonance capture region, and consequently more neutrons are consumed by ^{238}U, which leads to a power decrease. The shorter the neutron lifetime, the less would be the power excursion. In addition the thermal expansion of the fuel contributes negatively to the reactivity.

The conversion ratio in the inner zone of Phenix is only 0.54, but the total breeding ratio is 1.16. In the USSR BN-350 a breeding ratio of 1.5 is expected because of less neutron absorbing material. High breeding gain may be achieved by some sacrifice of safety control of fast reactors.

19.16. Power station efficiency*

Not all of the heat produced in a nuclear reactor can be used for work, i.e. for turning the turbine blades connected to the rotor of an electric generator. According to the second law of thermodynamics the maximum *thermal efficiency* (η) is

$$\eta = \frac{Q_{in} - Q_{out}}{Q_{in}} \tag{19.33}$$

where Q_{in} is the heat input into a machine and Q_{out} is the heat discharged from that machine. Most reactors operate according to the *Rankin cycle* in which a liquid medium is heated up and vaporized at constant pressure, and work is carried out through expansion and decreasing temperature. While the steam entering the turbine should be as dry as possible to reduce turbine blade wear, the steam on the backside is "wet" due to some condensation. Because steam is not an ideal gas, the calculation of Q is not so easily made from heats of condensation and the ideal gas law.

The highest thermodynamic efficiency is achieved in the *Carnot cycle* in which energy input (heating up the coolant) and work both occur at different but constant temperatures, T_{in} and T_{out}. For a "Carnot engine"

$$\eta_{max} = \frac{T_{in} - T_{out}}{T_{in}} \tag{19.34}$$

A modern nuclear power plant may deliver steam of about 300°C to the turbine and have an outlet temperature from the condenser on the turbine backside of about 30°C. Using (19.34) $\eta_{max} = (300 + 273) - (30 + 273)/(300 + 273) = 0.47$. Due to less efficient energy cycle, friction, heat losses, etc., the net efficiency (η_{net}) is only about 0.30–0.35 (net electric output from generator divided by gross thermal output from reactor). In coal- and oil-fired power plants higher steam temperatures can be achieved, ~ 500°C with T_{in} 530°C and T_{out} 30°C, η_{max} 0.65 leading to an η_{net} ~ 0.50.

If we compare η_{net} for the two kinds of power plants we find that the fossile fuel power plant has about 50% higher efficiency, i.e. the amount at "*reject heat*" is 40% higher for the nuclear power plant ((0.50–0.30)/0.50). This lower efficiency is a consequence of the lower steam temperature of the nuclear reactor. However, in the last 10 years considerable development has taken place with respect to coolant, canning and fuel stability, and inertness, allowing considerably higher fuel temperatures.

The thermal efficiency of any reactor can be raised by using some of the heat directly for a beneficial purpose, such as for processing in the chemical industry (Fig. 19.10) or for district heating. Such schemes are contemplated, e.g. in the USSR, Czechoslovakia, and Sweden. Thus Barsebäck 3, Sweden, which was originally designed for 1000 MWe at 3000 MWth (η_{net} 33%) is now planned for 810 MWe *and* 950 MW heat, raising the thermal utilization η_{net} to ~60%. The heat is taken out as water at 165°C and returned at 70°C. In the USSR some nuclear power stations presently produce district heat.

The performance characteristics of power stations are commonly described by the availability and capacity factors, and by the forced outage factor. The *availability factor* is the time the station has been available for operation divided by the length of the desired time period. Thus if it has been desired to operate the station for 6000 h, but it has only been possible to run it for 5500 h (because of repair, etc.) the availability factor is 92%. Typical availability factors are 95–100% for water power, 75–85% for conventional fossile power, and 70% for nuclear power (usually the reactor has a higher availability, but with a lower value of 70% for the turbine). In the FRG, availability factors of 80% have been achieved for LWRs.

The *capacity factor* is the ratio between the energy produced by the station and the energy it would have produced running at maximum power (i.e. design power multiplied by the time). A 1000 MWe power station which has produced 5.80 TWh during the year has had a capacity factor of 66% ($= 5.80 \times 10^6$ MWh per 1000 MW $\times$ 8760 h y^{-1}). The capacity factor depends not so much on the power station performance but rather on the local energy demand conditions. Thus in Sweden in 1976 the capacity factor of water power was about 50%, conventional fossile power about 25%, and for a BWR nuclear power block 45%.

A better picture of the power station performance is given by the *forced outage rate* which is the ratio between the number of hours a power station has been shut down because of malfunction and the total number of hours during that period. A power station which has been shut off for one month in January for repair of the main feed water pump and three months (June, July, and August) because of no demand for electricity, during which time part of the fuel elements were changed, has an 8% forced outage rate (i.e. one month out of twelve). The forced outage rate is very similar for all power stations of similar size and age. Values for new 1000 MWe BWR or PWR stations vary from 15% up; a typical average value is 25%.

19.17. Reactor safety

Greater than 95% of all the operation shutdowns in nuclear power stations are due to failures common to conventional power stations. However, in this text we consider only safety with regard to the nuclear reactions in the reactor which account for the other 5%. We can distinguish between three safety (or risk) levels: (i) small deficiencies due to imperfect technique, which are accepted as inevitable in any human enterprise (e.g. small "chronic" releases of radioactivity to the environment), (ii) accidents, which normally should not occur with good equipment, but still are "probable" (e.g. pipe breaks, fire, etc.) and for which protective measures must be included in the reactor design (design basis accident, DBA), and (iii) maximum credible accident (MCA) which is the most serious reactor accident that can reasonably be imagined from any adverse combination of equipment malfunction, operating errors, and other foreseeable causes. The distinction between DBA and MCA is successively vanishing, as all possible accidents are being included in the DBA. The radioactivity releases of category(i) are discussed in the next section, the two other aspects here.

The broad functions of the safety systems are common to most reactors. In the event of an accident they shut down the reactor, ensure a supply of cooling water or gas for the fuel, and contain any fission products which escape from the fuel elements. The safety systems are designed for a high degree of redundance (duplication), so that if one safety systems fails another shall function. The several redundant safety systems are logically and physically separated from one another and from the reactor process systems. This confers immunity against common-mode events such as external explosions or control roomfires. Each safety system should be capable independently of protecting the reactor core and building from further damage.

The main threat to a reactor would involve the cooling capacity becoming insufficient either because of a sudden power excursion of the reactor or because of a blockage of the cooling circuit. In either case the core becomes overheated and a core meltdown may start.

The guarantee against a power excursion is a negative temperature coefficient. For example, if the design parameters of a reactor give a maximum available $k_{eff} = 1.041$, and the temperature coefficient is $-4.2 \times 10^{-3}\% \Delta k/k\,°C$, a temperature increase of 1000°C would be required to reduce k_{eff} to 0.999, i.e. to automatically stop the fission chain reaction without any of the safety systems operating. Such a reactor would also be prompt critical, as pointed out in § 19.8. This is not acceptable, so at starting conditions the maximum available k_{eff} is made < 1.0065 by introducing burnable poisons and control rods. After some time fission products with high neutron capture cross-sections are formed (Xe, Sm, etc.), reducing k_{eff}. With Δk 0.0065, a temperature increase of 150°C would reduce k_{eff} to <1 for the example given. This can be (and is) used for reactor self-regulation. The main contribution to the negative temperature coefficient comes from the decreased moderator density and increased voids in BWR with increasing temperature.

Three types of control rod are used: (*control*) *rods* for regulating small power fluctuations, *shim rods* for coarse adjustments of k_{eff}, and *scram rods* for suddenly shutting down the reactor. The scram rods can move very fast and are actuated by gravity, spring release, compressed air, etc.; a typical value may be 3 s for complete shutdown. Since the reactor time constant is much larger, a neutron excursion is effectively stopped. The number of control rods is large ($\gtrsim 100$) in order to facilitate an even power density in the reactor, but the number of scram rods may be <10. In light water reactors injection of boric acid into the core acts as an additional shutdown feature ("chemical scram").

The power of a reactor does not go to zero when it is shut off. Figure 19.19 shows the heat power due to fission product decay (logarithmic time scale) after a scram. For a

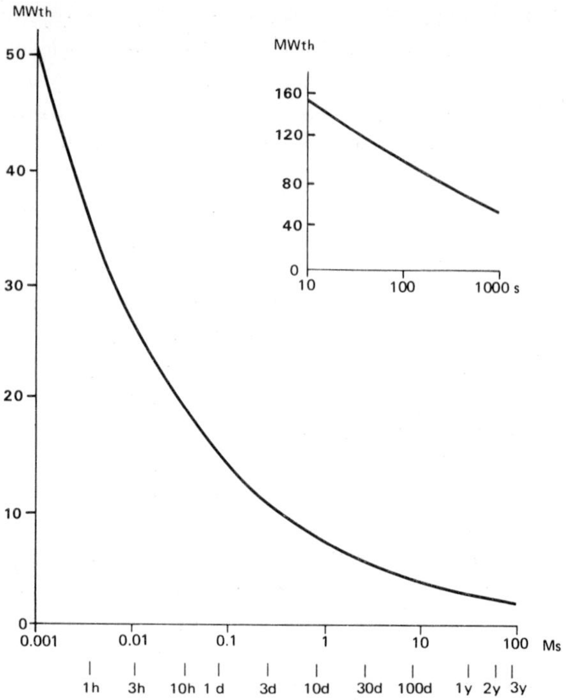

FIG. 19.19. Decay heat from a 3000 MWth LWR core, which has been in continuous operation for 3 years as a function of time (megaseconds) after shutdown. Immediately after shutdown the core heat output is reduced to ~ 200 MWth.

1000 MWe (3000 MWth) power reactor the rest (or after) power (after or decay heat) is 2% of the full power effect (i.e. ~ 60 MWth) after 1000 s. If this heat is not removed, the fuel elements would become overheated, leading to core meltdown.

The design basis accident is usually considered to start with a loss of coolant accident (LOCA), e.g. as a rupture on the main feedwater line. In LWRs this immediately leads to $k_{eff} \ll 1$ because of reduced neutron moderation, independent of the position of the shutoff rods. However, the rest power would continue to heat the core, so core cooling is required in order to avoid a meltdown. Due to lack of cooling a partial melt-down occurred in the Three Mile Island PWR reactor in 1979.

Figures 19.12 and 19.13 show the emergency cooling systems in BWRs. There are two such systems for the core—the core spray and the head spray cooling systems (Fig. 19.12). These two systems have a fourfold redundancy as Fig. 19.13 shows. The systems are located in four separate zones 90° apart around the reactor and at different levels above ground. The cooling water would be taken from the wet wells below the reactor. It is assumed that all these wells would not break simultaneously; such an accident could only be initiated by an earthquake of unanticipated magnitude. In addition to the high pressure core spray there is a low pressure coolant injection. Further protection, especially against fission products leaking out of a damaged reactor vessel, is provided by the containment vessel spray. Steam produced in the cooling is blown from the dry well into the wet well (the "blow-down" system). The water in the wet well is cooled by independent external cooling systems. The emergency cooling systems are much the same for BWRs and PWRs.

A special hazard, which is taken into account in the emergency systems, is that the emergency cooling would be actuated too late, i.e. after a core meltdown has begun. In practice the cooling system must start $\lesssim 1$ min after LOCA has occurred. If the core is very hot, a violent steam formation could occur when the cooling water contacts the core. In this case the blow-down system, the outer containment vessel, and the spray systems are expected to contain the energy excursion. Under very unfavorable conditions, hydrogen might be released from the cooling water; the risk for a fire or an explosion then depends on available oxygen (almost none in the reactor tank).

In HWR of the CANDU design, a LOCA would probably have little effect, as long as only a few pressure tubes were affected. If all cooling were lost, k_{eff} would increase because of less neutron absorption. Some CANDU reactors therefore have a dump-tank for the moderator. Without moderator the danger for core meltdown is less than for the LWRs.

The gas-cooled reactors would be less affected by LOCA for several reasons. The coolant has little effect on k_{eff}, and because the temperature coefficient is negative no large rapid power excursion can occur. The large graphite core has a large heat capacity and acts more or less as its own heat sink. The fuel would not be damaged even if it takes an hour before emergency cooling sets in (cf. the LWR requirement). Graphite melts above 3500°C, and such a high temperature is quite unlikely because of heat conduction to the surroundings and normal air convection. It is therefore very improbable that the core would deform even without any emergency cooling. However, to assist air cooling of the core several redundant pumps are installed (Fig. 19.15) in a manner similar to the BWR (Fig. 19.13).

A nuclear explosion is physically impossible because of the large t_{per}. Rupture of the reactor vessel due to overpressure is highly improbable because of a number of safety valves (not shown in any of the figures). The steam produced can be dumped directly into the condenser in case of emergency. All piping into the core has insulation valves, which can be closed if desired.

The reactor vessel is surrounded by several physical barriers. These barriers protect the surroundings from a core accident and protect the core from damage caused by external effects. In a PWR these barriers are the reactor tank (about 15–25 cm steel), the biological shield (1.5–3 m of concrete), and the outer containment. The whole primary circuit, including pressurizer dome, heat exchanger–steam generator, *and* connecting pipes, is surrounded by concrete. In gas-cooled reactors the biological shield is part of the reactor vessel. It is therefore very strong, 3–5 m thick, and made of prestressed concrete (PSC, i.e. concrete containing embedded stressed steel wires to assist against the internal pressure). This reduces the γ-radiation from the core by a factor of 10^{10} to 10^{12}.

All nuclear power reactors in the non-CMEA countries have a reactor building containment, the purpose of which is to contain steam and released radioactivities in case of severe core or pipe accident, and to protect the reactor from external damage. In CMEA countries this extra safety feature is considered unnecessary. The containment is calculated to withstand the internal pressure from a release of the entire primary cooling circuit (and in the case of PWRs of the additional loss of one of the secondary circuit boilers), corresponding to excess pressures of 0.4 MPa. The containment is provided with a spray, which cools and condenses the steam released and washes out radioactive contaminations. The steam produced can also be dumped directly into the condenser in case of emergency. All piping into the core has insulation valves, which can be closed if desired. Modern containment vessels are multilayered PSC and steel and can withstand the impact of a crash by a jumbo jet.

19.18. Radioactive releases and waste management

In the "ideal" nuclear reactor all fission products and actinides produced are contained in the fuel elements. In all practical reactors there are three processes through which radioactivity leaves the reactor vessel; in all cases the carrier of activity is the coolant:

(i) induced radioactivity in the cooling medium;
(ii) corrosion products containing induced activities from construction materials;
(iii) leakage of fission products and actinides from faulty fuel elements.

In all three respects, the HTGR approaches the "ideal" reactor whereas the radioactivity products of the BWR and PWR reactors present major concern.

19.18.1. *Gaseous wastes*

The intense flux of fast and thermal neutrons induces several radioactivities in H_2O: $^{2}H(n,\gamma)^{3}H$, $^{16}O(n,p)^{16}N$, $^{18}O(n,\gamma)^{19}O$, and $^{18}O(n,p)^{18}F$. Considerable activities of

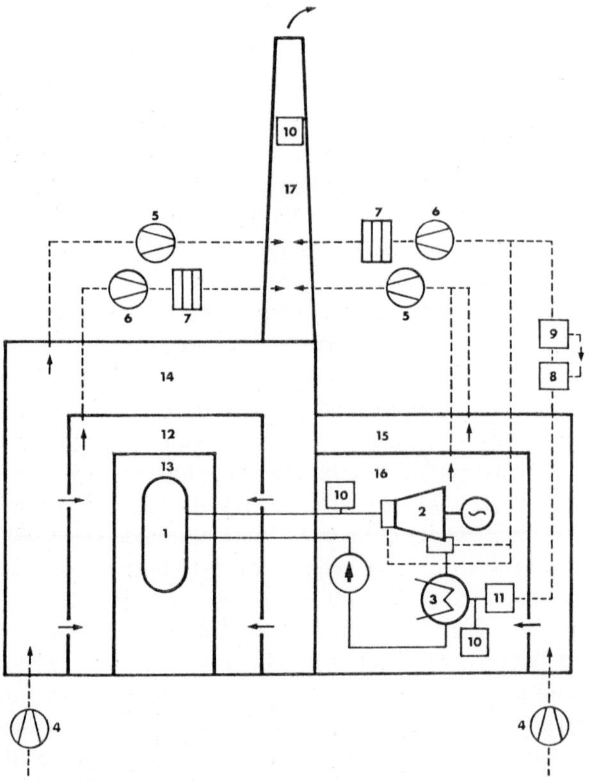

FIG. 19.20. BWR ventilation and gas purification system: 1, reactor; 2, turbine; 3, turbine condenser; 4, ventilation fans; 5, discharge fans for pure air; 6, discharge fans for contaminated air; 7, multiple filters; 8, delay tanks; 9, noble gas separator; 10, radioactivity meter; 11, vacuum ejector; 12, C-space in reactor building; 13, reactor vessel; 14, B-space in reactor building; 15, B-space in turbine building; 16, turbine containment; 17, stack.

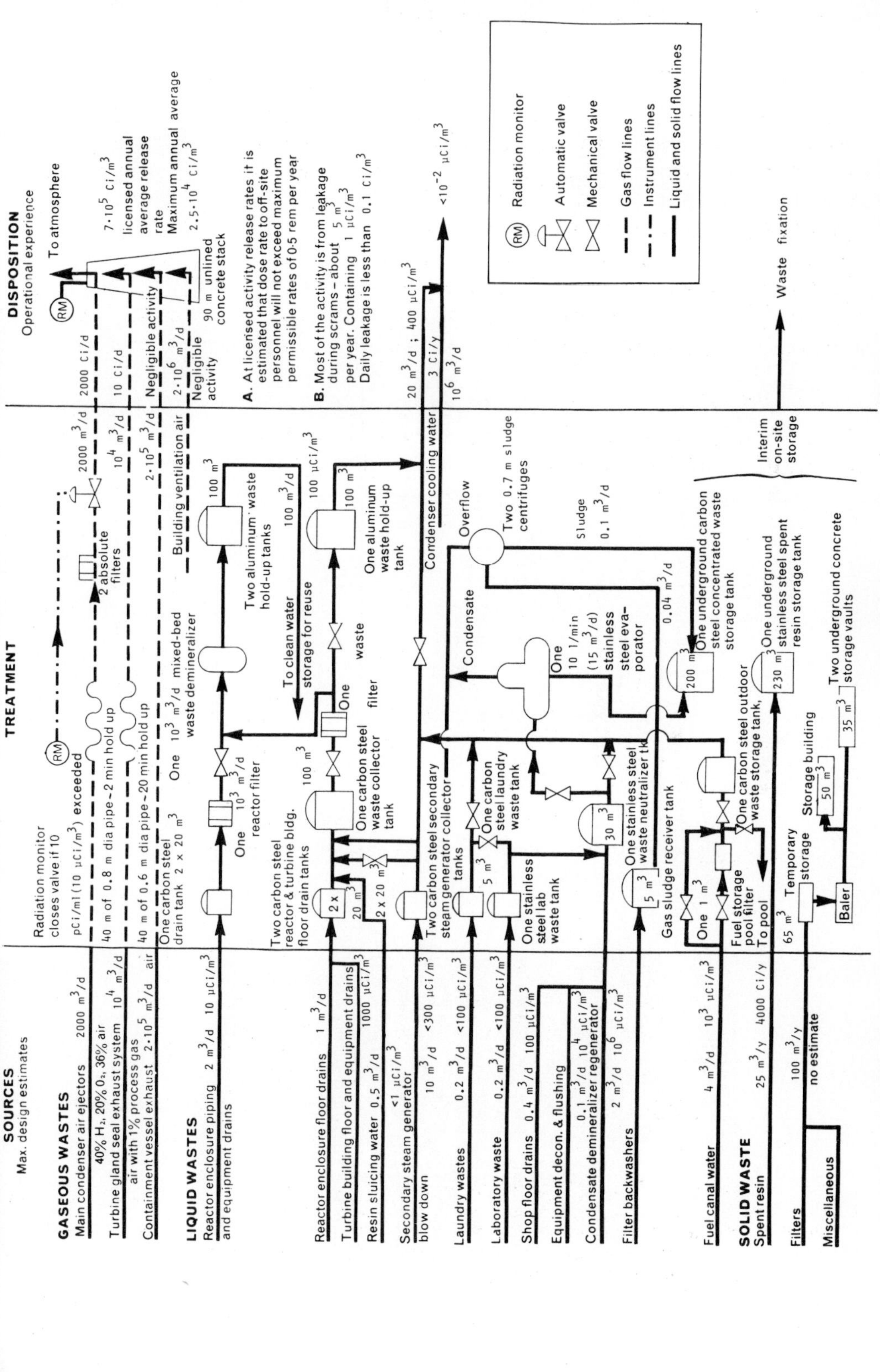

FIG. 19.21. Waste management system for a PWR. Flow sheet is only general and many variations occur. (Adopted from J. A. Richardson.)

TABLE 19.6. *Emission rates from German nuclear power plants 1975 in Ci y^{-1}*

Site of power plant	Type	First commercial operation	Maximum gross capacity (MWe)	Gaseous effluents: Noble gases	Gaseous effluents: Aerosols, Short lived	Gaseous effluents: Aerosols, Long lived[(a)]	Gaseous effluents: ^{131}I	Gaseous effluents: ^{14}C[(b)]	Gaseous effluents: ^{3}T	Liquid effluents: Fission and activation products (without ^{3}T)	Liquid effluents: ^{3}T
Gundremmingen	BWR	1966	250	7 400	9	0.008	0.25	2	~100	1.1	125
Lingen	BWR	1968	252	35 000	8.8	0.01	1.3	2	~ 30	0.045	16
Obrigheim	PWR	1968	345	8 000	0.003	0.026	0.012	3	27	1.8	150
Stade	PWR	1972	662	1 300	n.b.	0.03	0.01	5	15	0.13	114
Würgassen	BWR	1972	670	120	2	0.011	0.001	2	~2	0.89	3.6
Würgassen		Permitted releases		30 000	10.5		0.26			17	300
Biblis A	PWR	1974	1204	1 700	0.047	0.006	0.005	8	~13	0.38	106
Biblis A		Permitted releases		9×10^5	3.5		0.7			10	1600

(a) Half-time > 8 days, and except ^{131}I.

(b) Estimated values.

n.b. = not balanced.

^{13}N are produced through reactions $^{13}C(n, p)^{13}N$ and $^{14}N(n, 2n)^{13}N$ with carbon in the steel, and CO_2 and nitrogen in the water. ^{14}C is produced through reactions with ^{13}C, ^{14}N (mainly dissolved to 10–60 ppm in the fuel), and ^{16}O; 10–20 Ci ^{14}C is produced per GWe-year. Activation products from B, Li and Ar occur when the cooling water contains these atoms due to the reactions $^{10}B(n, 2\alpha)^3H$, $^6Li(n, \alpha)^3H$, and $^{40}Ar(n, \gamma)^{41}Ar$. In a PWR the tritium content may be 1–10 Ci m^{-3} because of the boric acid content, while 10^{-3} Ci m^{-3} is more typical for a BWR. Though most of these activities are short lived, they decay to stable products by emitting energetic γ-rays. In BWRs the active gases are transported with the steam to the turbines, where a considerable fraction of the decays occur, thus raising the background in this part of the power station to such a level that personnel cannot be allowed permanently in the turbine hall. Also some radioactivity may be released through leaking gland seals.

To minimize contamination from gaseous products transported with the steam, efficient drainage and suction (air-ejector) systems are installed at the condenser side of the turbines. Here most of the activity is swept out and allowed to pass a delay system (delay times of 20–30 min are used in BWRs) before the rest ($\gtrsim 99\%$) of the gases are caught in filters, consisting of several layers of absorbing material (e.g. charcoal). Because some of the activity may leak out into the reactor and turbine buildings, the building ventilation also contains filters before the air is vented through the stack. Figure 19.20 shows the principle of the gas purification system of a BWR.

The primary release of gaseous activities in a PWR is at the deaerator after the heat exchanger in the primary loop. The hydrogen–oxygen recombiner is also located here. The gases are compressed and stored for up to 2 weeks before they (via filters) are vented through the stack; these volumes are much smaller than for the BWRs. The PWR waste system is shown in Fig. 19.21.

If fission products are released by faulty fuel elements, the gaseous ones appear most rapidly in the cooling water: ^{133}Xe, ^{135}Xe, ^{85}Kr, ^{133}I, etc. The noble gases can be retained in charcoal filters and the iodine removed by various special techniques (§20.4.7). Some other activities are found in the steam: ^{24}Na, corrosion products, and fission products like ^{91}Sr, ^{99m}Tc, and ^{137}Cs. These activities appear partly as aerosols and are many powers of ten lower than for the gases. Table 19.6 shows releases of gaseous products from some commercial PWRs and BWRs in the FRG. These releases are $\lesssim 1\%$ of values permitted with respect to environmental contamination and doses.

19.18.2. *Liquid wastes*

Corrosion and fission products appear in dissolved ionic form and "precipitated" in the crud, depending on the chemistry and water conditions. The dominating corrosion products are ^{51}Cr, ^{54}Mn, ^{59}Fe, ^{58}Co, ^{60}Co, ^{65}Zn, and ^{124}Sb, and the dominating fission products are T, ^{131}I, ^{134}Cs, and ^{137}Cs. Other fission products and actinides are released in minor amounts depending on the kind and size of fuel element leak. These products are continually removed by the cooling circuit.

In PWRs a continuous liquid stream is withdrawn from the coolant in the primary circuit, on the back side of the main heat exchanger. After cooling in another heat exchanger, the water is allowed to pass a filter (e.g. of the precoat type, 25 μ particle size) and a demineralizer which removes ionic species as well as all particulates. The liquid waste streams are detailed in Fig. 19.21 with representative flow rates and activities. The system is designed to permit segregation of waste streams according to characteristics and type of treatment. The largest liquid waste source is the secondary steam generator blow-down, while the largest activities originate at the clean-up systems (condensate demineralizer and filter back-washes). The "demineralizer" may be an inorganic substance (e.g. kieselguhr) or an organic ion exchanger; among the most efficient ones is the Powdex system, which consists of very finely ground (~ 400 mesh) organic cation and anion exchangers. The boric acid adjustment system of some PWRs may be considered as part of the cleaning circuit; it is discussed in §19.14.1.

BWRs have two liquid purification circuits, one after the condenser and a second circuit with withdrawal of a small stream from the bottom of the reactor vessel. The condensate purification contains similar demineralizers as for the PWR. The reactor vessel purification system consists of a mixed bed ion exchange system; also other purification substances have been used (magnetite, kieselguhr, etc.). A typical capacity figure for the purification system is 50 $m^3 h^{-1}$.

The radioactivity becomes very high in the purification systems. The organic ion-exchange resins used are limited to absorbed doses of $\sim 10^8$ rad. The filters are back-washed and resins changed through remotely controlled systems, when measurements show that too much activity is leaking through. At present resins are not regenerated.

As an additional safety, when filters and demineralizers show insufficient purification of circulating water and of water for release to the environment, an evaporator system may be incorporated. Such a system is always a part of the PWR boric acid control circuit.

Liquid releases from PWR and BWR in the FRG are shown in Table 19.6. The releases are only a few percent of permitted values, and the increase in river activity due to the effluents minimal: 3×10^{-10} Ci m^{-3} of the River Weser (Würgassen) and 1×10^{-11} Ci m^{-3} of the River Rhine (Biblis A). Typical annually permitted releases to the environment are 10^4 Ci T, 10^2 Ci other $\beta\gamma$ activities, and 50 Ci α-activities. Permitted releases are presently calculated "individually" for each power station, taking its construction, location, and the surrounding population distribution into consideration.

19.18.3. *Solid wastes*

Solid radioactive wastes are produced from many components and purification circuits of the reactor station. Many contaminated items can be decontaminated by proper treatment, saving both money and waste storage space. Combustible solids may be concentrated through incineration. Several techniques are used for fixation of these wastes in such a way that they can be safely stored with a minimum of surveillance. These techniques all have in common an enclosure of the activity in an insoluble material and in blocks of sizes and activities small enough to be handled with a fork lift.

The waste amounts vary considerably between different stations depending on purification and concentration techniques used. Annual averages can be considered to have the following values for a 1000 MWe LWR: spent ion-exchange resins ~ 30 m^3 (~ 50% water), evaporator residues ~ 20 m^3 (~ 80% water), filter (condenser) concentrates ~ 60 m^3, other wastes from primary reactor loop ~ 5 t, general low activity wastes ~ 200 m^3 (~ 40 t). The activity in the resins is typically 200–500 Ci ^{60}Co, 300–3000 Ci ^{137}Cs, and < 200 Ci ^{90}Sr, and in evaporator and filter concentrates is < 1/10 of the resin amount.

Figure 19.22 shows waste solidification and fixation with cement, using prefabricated concrete boxes. This technique is useful both for contaminated or activated equipment

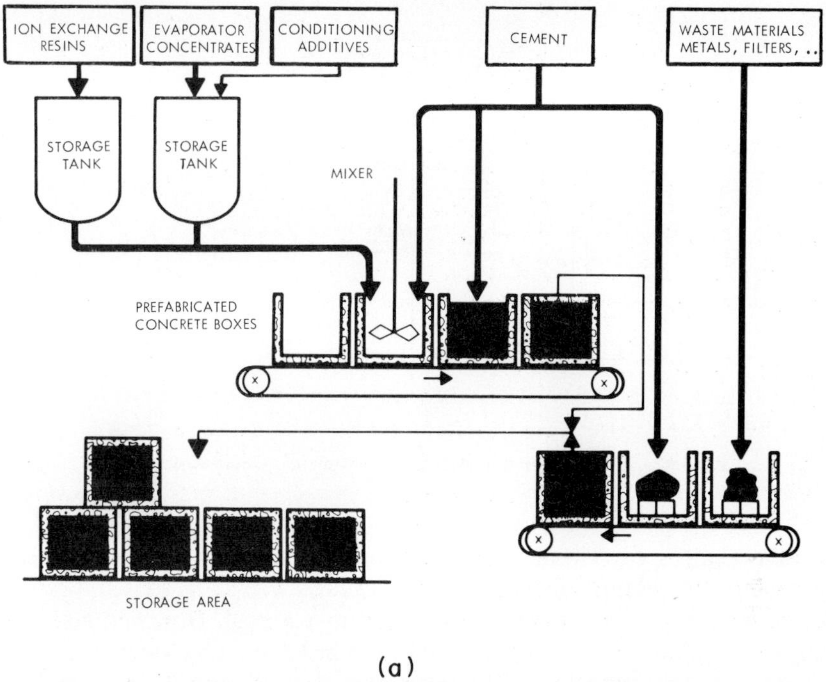

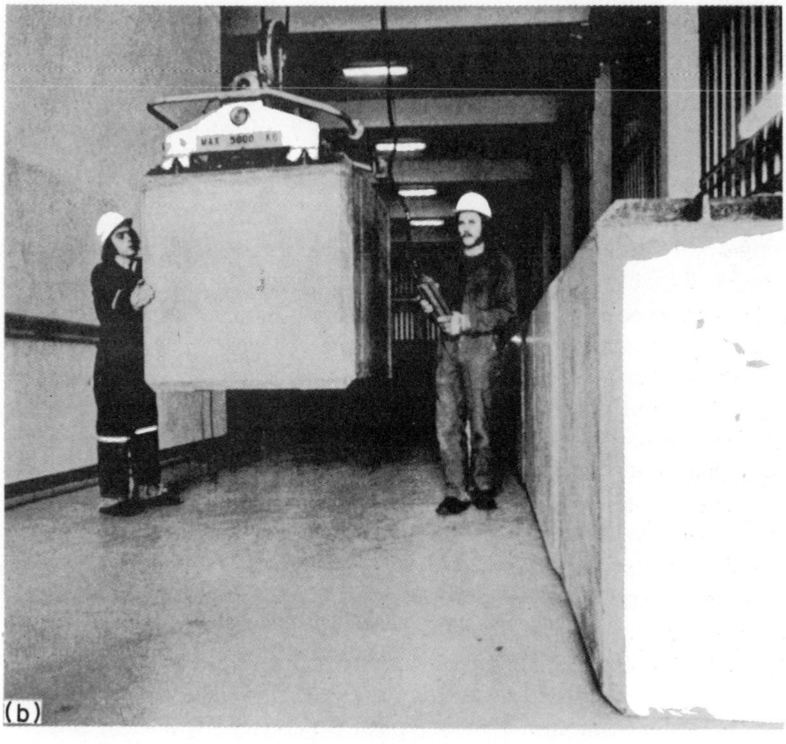

FIG. 19.22. Waste solidification in concrete blocks.

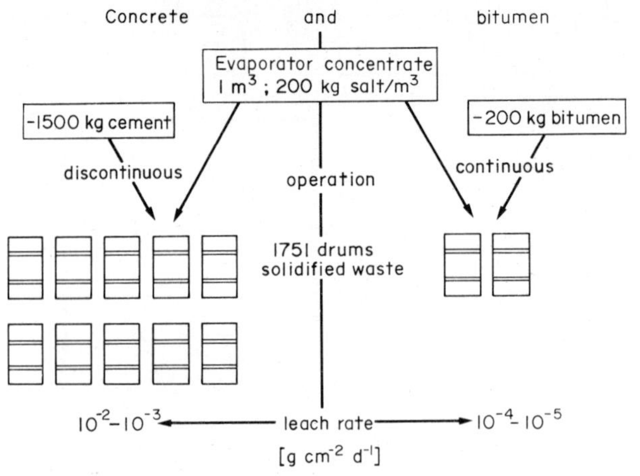

FIG. 19.23. Comparison of concrete and bitumen for solid waste storage.

and for resins, sludges, etc. The technique, however, increases the waste volume considerably, by a factor 4 to 40 times the unshielded waste.

Concrete is a cheap fire and corrosion resistant material. However, active species, especially easily soluble ions as cesium, can be leached from it by water. The addition of plastic binders to the concrete and bitumen have been suggested as alternate storage materials. In the continuous bitumen extruder process for semiliquid wastes all water is directly eliminated, considerably reducing the waste volume. The bitumen mixture is placed in steel barrels of standard size (150–200 l). When additional shielding is desired, the filled drums are placed into disposable or reusable sleeves of concrete, iron, or lead. Such a sleeve of 12 cm lead weighs 7 t, and reduces the surface dose rate by a factor of $\sim 10^3$. A typical unshielded bitumen drum may have a surface dose rate of 100 rem h^{-1}, necessitating remote handling.

In Fig. 19.23 the cement and bitumen processes are compared; with bitumen smaller waste volumes with much lower leach rates are produced. The disadvantage of bitumen is its inflammability and lower radiation resistance; a fire could release thousands of curies of long-lived nuclides. However, for low and medium active solid wastes, bitumen is presently the preferred fixation material. About 270 bitumen drums (200 l) with concentrated waste are produced annually per 1300 MWe reactor in West Germany; about 50 of these contain medium active waste and have outer concrete shielding (outer volume $\sim$400). Figure 19.24 shows the transport of medium active waste containers to a repository in the FRG.

If the solidified waste does not contain any fission products, a storage time of 50 y would be sufficient to allow all activity (mainly ^{60}Co) to decay to negligible amounts; storage at plant site would be feasible. If the amount of ^{90}Sr, ^{137}Cs, or actinides is high, much longer storage times are required, and a special storage site may be required. It is desirable to separate these products from other waste; the isolated ^{90}Sr, ^{137}Cs, and actinides could then be added to the high active waste of reprocessing plants.

The radioactive releases from the West German nuclear power stations in 1974 have been calculated to raise the total doses received by the population by <0.01 mrem, i.e. an unmeasurably small amount. This is typical experience for nuclear power stations worldwide.

FIG. 19.24. Transport of shielded waste containers.

19.18.4. *Gas- and sodium-cooled reactors*

In helium-cooled reactors small amounts of tritium are produced, and can be removed in a side stream by burning the hydrogen to water which is caught in special dryers, charcoal beds, or cold-traps. These traps will also catch the small amounts of CO, CO_2, and CH_4 formed as well as released fission products, including krypton and xenon. Filters (e.g. made of stainless steel fibers) will eliminate most particulates from the gas stream.

In CO_2-cooled reactors radiolysis produces CO and O_2 which have to be recombined. Induced activities and released fission products (mainly noble gases) are caught in cold traps and filters.

Considerable radioactivity is induced in the sodium in the primary cooling circuit of LMFBR: $^{23}Na(n,\gamma)^{24}Na$, $^{23}Na(n,p)^{23}Ne$, etc. ^{24}Na has a 15 h half-life and emits energetic γ's. The primary cooling loop must therefore be well shielded. Its activity is a nuisance only in case of repair demand, requiring considerable waiting time before the loop can be approached. The sodium dissolves many of the corrosion and fission products eventually released. To remove these the primary loop is provided with cold trap purification systems.

19.19. Nuclear explosives*

Nuclear explosives may be used for military purposes or for civilian applications. Assume that 50 kg of metallic ^{235}U are brought together to provide a critical configuration. Assuming a generation time of 3×10^{-9} s for neutrons, it can be calculated with (19.24) that it would take 0.2×10^{-6} s to increase the number of neutrons to that required for fissioning all ^{235}U atoms. However, long before this time, the energy released in the material would have resulted in blowing it apart, so 100% efficiency is impossible in a nuclear explosion. If 2% of the 50 kg of uranium has been fissioned, the amount of energy released corresponds to about 20,000 t of TNT (8×10^{13} J). The energy production is said to be 20 kt, which corresponds to a "conventional" atomic bomb. Only ^{233}U, ^{235}U, and ^{239}Pu of the more easily available nuclides have sufficiently high cross-sections for fission by fast neutrons to have reasonable critical sizes for use as nuclear explosives.

A rough estimate of the critical radius of a homogeneous unreflected reactor may be obtained simply by estimating the neutron mean free path according to (9.6). Assuming pure ^{235}U metal with a density of 19 g cm^{-3} and a fast fission cross-section of 2×10^{-24} cm^2, one obtains $\hat{\Sigma}_f^{-1} = 10$ cm. A sphere with this radius weighs 80 kg. For an unreflected metal sphere containing 93.5% ^{235}U the correct value is 52 kg. As for reflected systems (§19.7), ^{239}Pu has the smallest critical size; for ^{239}Pu (γ-phase, density 15.8) it is 15.7 kg, and for ^{233}U 16.2 kg.

Assume that one has two half-spheres of metallic ^{235}U, each of which is less than critical size but which together exceed criticality. As the distance between the two spheres is diminished the k_{eff} increases for the total system. At a certain distance k_{eff} will exceed 1 but can be maintained so close to 1 that control of the chain reaction is possible. If the distance between the two half-spheres is diminished rapidly to less than this critical distance, a single neutron can trigger a multiplying chain reaction. Since stray neutrons are always available, e.g. from spontaneous fission, a chain reaction would be started when the distance between the spheres is less than the critical distance.

19.19.1. *Fission devices*

Nuclear weapons contain "normally" either highly enriched ^{235}U (>93%) or fairly pure ^{239}Pu ($\gtrsim$95%). Some nuclear explosives operate by having two subcritical spheres blown together by the use of chemical explosives (shotgun type). To increase the explosive yield the fissile material is surrounded by a heavy material (*tamper*) which has the double function of reflecting the neutrons and of acting as a heavy mass (inertia) to increase the time in which the supercritical configuration is held together. If natural uranium is used as the tamper, secondary fission in this shell can increase the overall energy release.

A critical mass of material can be arranged as a subcritical spherical shell which, with conventional directive charges, can be compressed into the supercritical sphere. This process is called *implosion.* The bomb over Hiroshima used uranium and was of the shotgun type, while the Nagasaki bomb was of the implosion type using plutonium. In order for these devices to function efficiently, the neutron multiplication must not begin until the critical size has been well exceeded. Some microseconds after criticality a rapid neutron multiplication is initiated by a special neutron source. The amount of ^{240}Pu in a plutonium bomb must be maintained relatively low since this nuclide decays partly through spontaneous fission with the resultant release of neutrons. If the concentration of ^{240}Pu is too high, the neutron multiplication will start too early, and the energy release is decreased. To minimize the ^{240}Pu, plutonium for weapon usage is produced in special reactors by low burn-up (<2000 MWd t^{-1}), which is uneconomical to the nuclear power industry where high burn-up of the fuel (leading to higher ^{240}Pu concentrations; Fig. 19.4) is a necessity in order to obtain the lowest electrical production costs. Recently it has been stated that also power reactor grade plutonium even as oxide can be used as a weapon, though less effective and predictable. The critical mass for such plutonium containing 20% ^{240}Pu has been estimated to be 20 kg for metal and 37 kg for PuO_2.

Nuclear fission weapons have been developed in sizes ranging from 0.001 kt to about 500 kt yield. Because the critical amount can never be less than several kilograms of fissile material, a weapon of low yield such as 0.001 kt has very low efficiency ($< 5 \times 10^{-6}$). By 1977 six countries (USA, USSR, UK, France, China, and India) had carried out > 1000 nuclear weapons tests (fission and fusion devices). Since the Partial Test Ban Treaty in 1963, most explosions have been underground.

19.19.2. *Fusion devices*

In the explosion of fission weapons a temperature of approximately 10^8 K is obtained. This temperature is sufficient for producing fusion reactions between deuterium and tritium. Although no official information is available on how modern "hydrogen weapons" are constructed, it is generally assumed that they are based on the principle of using a fission bomb as the initiator of the fusion reaction. In the debris of hydrogen bombs, lithium has been discovered. It is therefore believed that in hydrogen bombs, deuterium is combined with lithium in the form of solid LiD. ^{6}Li will react with a neutron

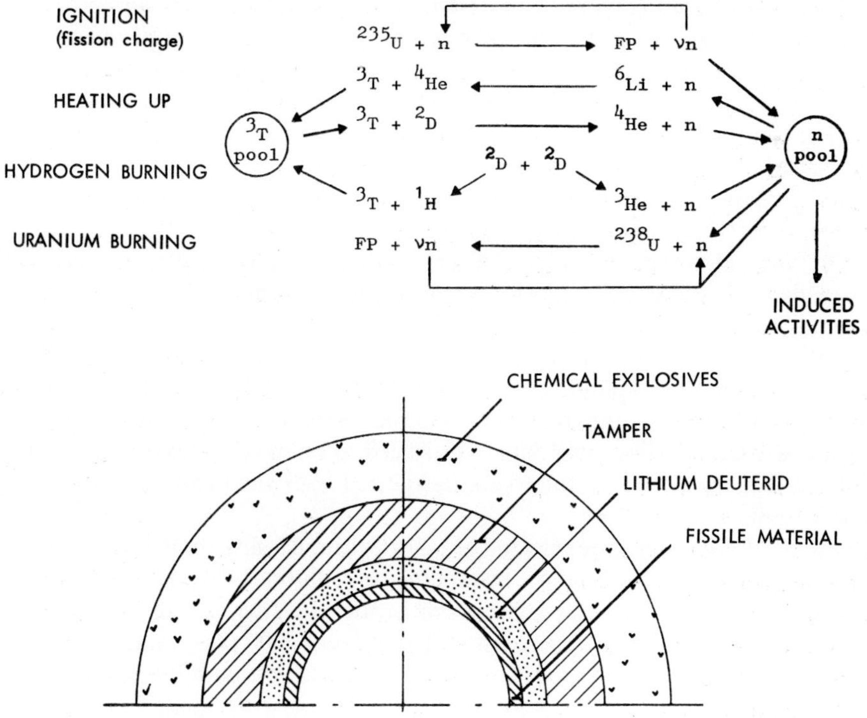

FIG. 19.25. Principle, design and reactions in a fission–fusion–fission charge (the "3-F-bomb").

to form tritium in the reaction

$$^{6}\text{Li} + \text{n} \rightarrow \text{T} + {}^{4}\text{He} \quad Q = 4\,\text{MeV}$$

with fast neutrons the reaction

$$^{7}\text{Li} + \text{n(fast)} \rightarrow \text{T} + {}^{4}\text{He} + \text{n}$$

also occurs.

The deuterium and the tritium can react as shown by equations (11.6)–(11.8). A reflector shell of natural uranium can be used to increase the neutron economy as well as to obtain secondary fission leading to an increased energy production in the weapon. A fission device may therefore be designed and react as indicated in Fig. 19.25.

The fission of ^{238}U produces large amounts of radioactive fission products, resulting in a "dirty" weapon. By contrast, if a shell of a nonfissionable heavy material is used the only fission products released are those obtained from the ignition process and the weapon is relatively "clean". Weapons of sizes greater than 50000 kt TNT have been tested. If it is assumed that half of the energy of a 50 Mt device comes from fusion and the other half from fission, this would require a weapon containing approximately 600 kg of ^{6}LiD and 1500 kg ^{238}U. If we further assume a 25% efficiency in the explosion, about 8 t of nuclear material is required.

19.19.3. *Effects of an explosion*

Very high temperatures are reached in the center of a nuclear explosion ($\sim 10^{8}$ K). The exploded material forms a fireball which rapidly expands and moves upward in the atmosphere. After the first wave has passed, a low pressure area is formed under the fireball which draws material from the ground to produce the typical mushroom-like cloud associated with nuclear explosions. This cloud contains fission products as well as induced radioactivity from the construction material of the device and from the immediate surrounding ground that has been irradiated. Some of the radioactive particles are carried to great heights and eventually fall from the atmosphere as *radioactive fallout*. Much of the fallout returns to the earth close to the explosion site, but some of the material reaches the stratosphere and returns to the earth only slowly over a long period of time and over great distances from the original explosion site. The radioactivity decays with a time t after the explosion according to the approximate relationship

$$A = A_0 e^{-1.2t} \tag{19.35}$$

Figure 19.26 shows the fallout from the US 15 Mt 3-F bomb on the Bikini atoll in 1954 in which test many inhabitants of the islands Rongelap and Rongerik received severe doses from the fallout, and Japanese fishermen on the ship *Lucky Dragon* received fatal doses. The accident was due to an unexpected high yield and a sudden change in wind direction.

The energy produced in the explosion of a 20 kt weapon is distributed approximately as follows: 50% pressure; 35% heat; 5% instantaneous radiation; 10% radiation from fission products. The radiation dose at 500 m from hypocentre (the vertical point on ground below the explosion center has been estimated to be $\sim$ 7000 rads; at about 1.1 km it is $\sim$ 400 rads. The pressure wave moves from the explosion site with the velocity of sound. This initial short pressure front is followed by a low pressure wave leading to a rapid change in wind direction. The pressure wave has an extended time period lasting several

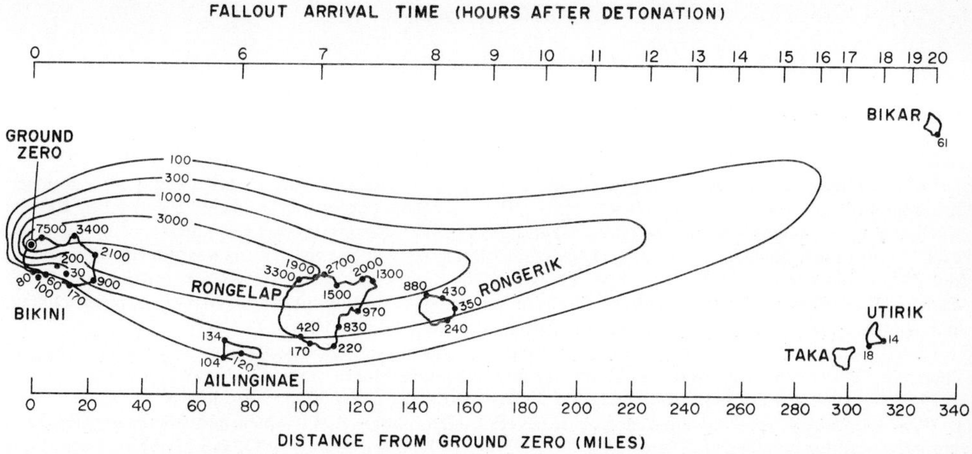

FIG. 19.26. Fallout from 15 Mt H-bomb test in 1954. The numbers give the total doses in rems received within 4 days after the explosion.

seconds, resulting in much greater damage than primary pressure waves from conventional explosions which last only milliseconds. The deaths in Hiroshima and Nagasaki resulted in 20–30% of the cases from primary burns, 50–60% from mechanical injuries and secondary burns, and approximately 15% from radiation injuries. The heat effect is even more dominating for a hydrogen weapon. For a 10 kt device the heat is 38 J cm^{-2} at a distance of 1.6 km, while for a 10 Mt device the heat effect is estimated to be 1250 J cm^{-2} at 8 km from the explosion center; as a comparison, paper, fabrics, etc., ignite at 30–60 J cm^{-2}.

19.19.4. *Peaceful uses of nuclear explosives*

A number of interesting ideas have been proposed for the peaceful use of nuclear explosives. It has been estimated that the explosive power of atomic weapons is sufficiently great that their use for large scale explosions would be much cheaper than the use of conventional explosives. For a 10 kt thermonuclear explosive the cost was estimated to be $350000 in 1966, while for a 2 Mt it was estimated to be $600000; for TNT a 10 kt device would cost $10 million.

So far only in the USSR have nuclear explosions been used to any extent for civil engineering projects. In at least 20 experiments the following projects have been studied: crater formation on land surface for water reservoirs, hole-blasting in rock for natural gas storage, underground blasts in oil shales for stimulation of oil and gas recovery, stopping a burning gas well, and canal building (Petjora–Kama, which would make a water way from the Polar Sea via the Volga to the Caspian Sea). The sizes have varied from < 1 kt to ~ 50 kt.

In the United States crater formation and gas stimulation experiments have been carried out. Suggested uses include the blasting of harbors on remote coasts to allow greater access to inland mineral deposits, crushing of ore bodies in order to obtain valuable minerals, crushing of underground rocks which may hinder the use of natural gas reservoirs, etc. One particular project that has been extensively studied is the construction of a new canal across the Isthmus of Panama to connect the Gulf of Mexico and the Pacific Ocean.

19.20. Exercises

19.1. If the energy developed by a 20 Mt fusion weapon could be used for producing electricity at a value of 2 cents kWh^{-1} what would the "electric value" of the device be? Such a weapon may be expected to cost $1 million. One tonne TNT releases 1 Gcal of energy in an explosion.

19.2. The bomb over Hiroshima contained ^{235}U. How many grams were fissioned to correspond to 20 kt of TNT?

19.3. Compare two 500 MWe electric power stations, one burning oil and the other using 3.0% enriched uranium. Both stations operate 6000 h y^{-1} at 35% efficiency (heat to electricity). The oil is carried by 100 000 tonne d.w. (dead weight, i.e. carrying capacity) oil tankers, and the uranium fuel by train cars of 20 t capacity each. Answer the following questions: (a) How many oil tankers will be needed every year for the oil-fired station? How many train cars will be needed every year for the nuclear power station for transporting (b) the enriched UO_2 reactor fuel, (c) the corresponding amount of natural uranium as U_3O_8 to the isotope enrichment plant, if the tail is 0.35% in ^{235}U? (See §19.9.) Reactor fuel rating 33 000 MWd/t U.

19.4. Calculate the number of collisions required to reduce a fast fission neutron ($\bar{E}_n^0 = 2$ MeV) to thermal energy ($\bar{E}_n$ 0.025 eV) in a light-water-moderated reactor, assuming that the data in Table 19.2 are valid.

19.5. Calculate the thermal fission factor for a mixture of 60% ^{239}Pu, 30% ^{240}Pu, and 10% ^{241}Pu.

19.6. The world's first water boiler reactor (LOPO, Los Alamos, 1944) was a homogeneous solution of enriched uranium sulfate as follows: 580 g ^{235}U, 3378 g ^{238}U, 534 g S, 14068 g O, and 1573 g H. From these values, and Tables 19.1 and 19.2, calculate η and f; neglect S. With $p = 0.957$, what will k_∞ be?

19.7. A large homogeneous thermal reactor contains only ^{235}U dispersed in beryllium in the atomic ratio 1 :30 000. The migration area is 230 cm^2. Assuming $p = \varepsilon = 1$, calculate the size of a cylindrical reactor with height equal to diameter.

19.8. A cubic unreflected graphite moderated natural uranium reactor contains 3% enriched uranium as UC homogeneously dispersed in the graphite matrix; the weight ratio C/U = 10. The resonance passage and thermal utilization factors are both assumed to be 0.9; $\varepsilon = 1.00$. Make an estimate of the critical size of the cube.

19.9. The LOPO reactor in exercise 19.6 has a neutron age $\tau = 31.4$ cm^2, and diffusion area L^2 1.87 cm^2. Calculate (a) the fast neutron leakage factor, and (b) the critical radius for the homogeneous sphere, if $k_\infty = 1.50$.

19.10. Our solar system is considered to be 4.5 billion years old. What was the ^{235}U percentage in natural uranium when the solar system was formed?

19.11. The radiometric sensitivities for discovering ^{59}Fe, ^{131}I, and ^{90}Sr are 2×10^{-6}, 7×10^{-7}, and 2×10^{-8} Ci m^{-3} of water. In the würgassen plant the total permitted aqueous annual release is 17 Ci β-emitters. Assume an activity ratio in the cooling water of 100: 10:1 for the three nuclides above and that none of these activities exceed 10% of the permissible release. How many times must a liquid sample taken each day be concentrated to meet these requirements?

19.12. What amount of tritium (curies) is produced in the Würgassen nuclear plant assuming that T is only produced through capture in the deuterons of the original cooling water, the amount of which is 50% of the core volume? Data of fluxes, cross-sections, and releases are given in Table 19.3. Neglect the tritium decay rate.

19.21. Literature

S. E. LIVERHANT, *Elementary Introduction to Nuclear Reactor Physics*, Wiley, 1960.
S. GLASSTONE, *The Effects of Nuclear Weapons*, USAEC, Washington, 1962.
R. G. HEWLETT and O. E. ANDERSON, *The New World*, Pennsylvania State University Press, 1962.
J. K. DAWSON and R. G. SOWDEN, *Chemical Aspects of Nuclear Reactors*, Butterworth, 1963.
S. PETERSON and R. G. WYMER, *Chemistry in Nuclear Technology*, Pergamon Press, Oxford, 1963.
ANL-5800, *Reactor Physics Constants*, USAEC, TID, 1963.
Reactor Handbook, 2nd edn., Interscience, 1960–1964.
J. R. LAMARSH, *Introduction to Nuclear Reactor Theory*, Addison-Wesley, 1966.
J. G. WILLS, *Nuclear Power Plant Technology*, Wiley, 1967.
C. O. SMITH, *Nuclear Reactor Materials*, Addison-Wesley, 1967.
E. TELLER, W. K. TALLEY, G. H. HIGGINS, and G. W. JOHNSON, *The Constructive Uses of Nuclear Explosives*, McGraw-Hill, 1968.
G. I. BELL and S. GLASSTONE, *Nuclear Reactor Theory*, van Nostrand, 1970.
W. FRATZSCHER and H. FELKE, *Einführung in die Kerntechnik*, VEB Deutscher Verlag für Grundstoffindustrie, Leipzig, 1971.
U.N. Conferences on peaceful uses of atomic energy, 1955, 1958, 1965, 1972.
J. R. LAMARSH, *Introduction to Nuclear Engineering*, Addison-Wesley, 1975.
IAEA, *The Oklo Phenomenon*, Vienna, 1975.
N. PICCINNI, Coated nuclear fuel particles, *Adv. Nucl. Sci. Techn.* **8** (1975) 256.
G. A. COWAN, A natural fission reactor, *Sci. Am.* July 1976.
IAEA, *Director of Nuclear Reactors*, Vienna, 1959–.
S. D. DRELL and F. VON HIPPEL, Limited nuclear war, *Sci. Am.* Nov. 1976, 235.

J. A. RICHARDSON, Radioactive waste quantities produced by light water reactors, *Nucl. Eng. Int.* Jan. (1974) 31.
IAEA, *Management of Radioactive Wastes from the Nuclear Fuel Cycle*, Vols. 1 and 2, Vienna, 1976.
Jahrbuch der Atomwirtschaft 1977, Handelsblatt GmbH, Verlag für Wirtschaftsinformation, Düsseldorf, Frankfurt, 1977.
Int. Conf. on Nuclear Power and its Fuel Cycle, Salzburg 1977, IAEA, Vienna, 1977.
H. EDELHÄUSER, H. BONKE, I. GANS, O. HUBER, K. J. VOGT, and R. WOLTER, *Radioactive Effluents and Present and Future Radiation Exposure to the Population from Nuclear Facilities in the Federal Republic of Germany*, ibid.

CHAPTER 20

Treatment of Spent Nuclear Fuel

Contents

Introduction

The nuclear fuel cycle comprises the handling of all fissile and fertile material necessary for nuclear power production and of the radioactive products formed in this process (Fig. 20.1). The fuel cycle is suitably divided into a front end and a back end part, where the nuclear power station is the dividing line. The front end comprises uranium exploration, mining, and refining (§12.6), isotope enrichment (§2.8), and fuel element

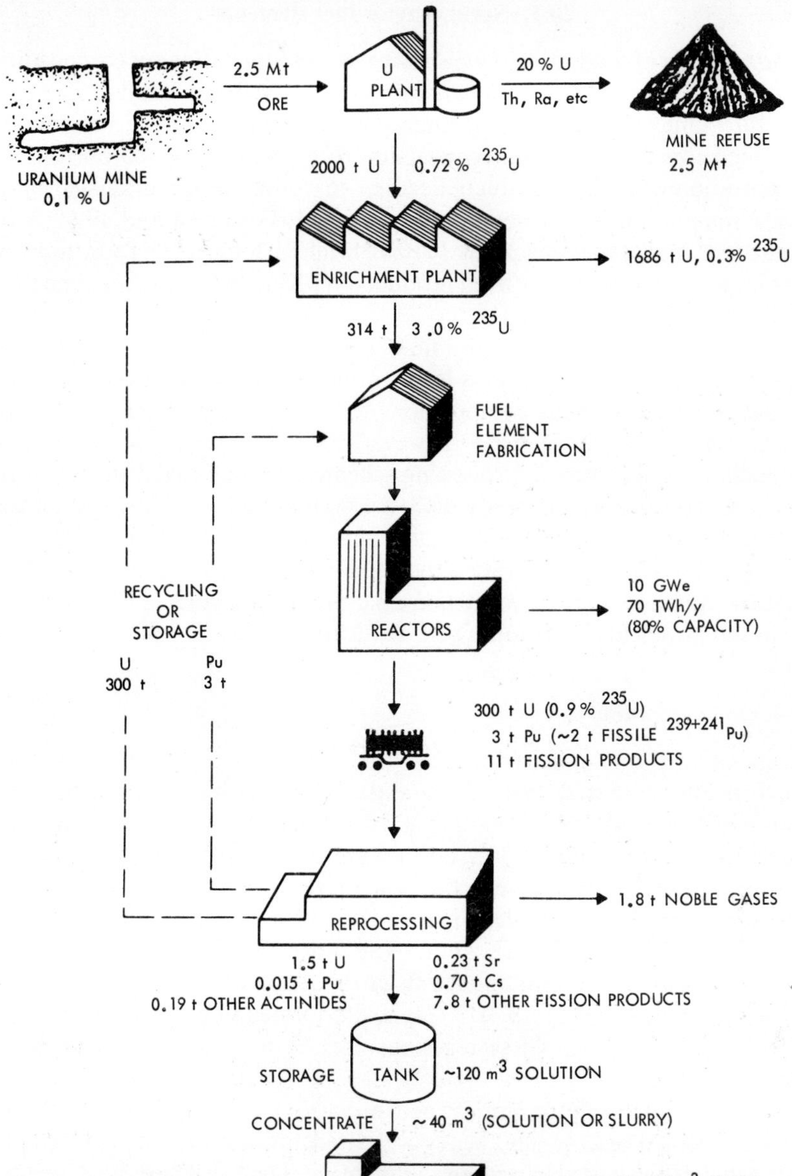

FIG. 20.1. Annual flow of materials in a 10 GWe LWR fuel cycle program.

fabrication (§19.11), while the back end involves reprocessing and radioactive waste ("radwaste") handling. Health and environmental aspects are important in all these steps, but being of a more general nature (see Chapters 16 and 21), they are not considered as "steps" in the nuclear fuel cycle. The long term world wide demand for nuclear fuel services is summarized in Appendix I.

In previous chapters we have discussed the front end of the nuclear fuel cycle. This chapter is devoted to the back end.

20.1. Spent reactor fuel elements

The composition of used reactor fuels varies as a function of input composition (kinds and amounts of fissile and fertile atoms), the neutron spectrum and fluency (or burn-up) and exposure time, and the cooling time after removal from the reactor. A harder spectrum increases fertile to fissile conversion (Fig. 19.2). Increased burn-up increases the concentration of fission products (leading to consequences discussed in §20.4 ff.) and larger amounts of higher actinides are formed (Figs. 13.4 and 19.6). A high flux results in more high order reactions (§10.3), while a long irradiation time produces relatively larger amounts of long-lived products. With increased cooling time the fraction of short-lived products is reduced.

Because of these effects, spent uranium fuel elements from PWR, BWR, HWR, and GCR differ in composition. (This leads to the confusing situation wherein different authors have quoted different compositions for spent fuel elements, even though the same type of reactor is considered.) However, the difference is not so large that very different fuel cycles (e.g. other reprocessing schemes) are required if the fuel is the same. In the next subsections we primarily discuss uranium fuel elements, and, more specifically, LWR elements. Fission product and actinide yields for typical PWR fuels are given in Appendix H. Figure 20.1 shows the annual materials flow in a mixed BWR–PWR conglomerate with a total power of 10 GWe (i.e. about ten plants of 1000 MWe capacity) running at full power for 7000 hours per year (capacity factor ~80%).

20.1.1. *Actinide formation*

Neutron capture and β-decay lead to the formation of higher actinides. This is illustrated in Figs. 13.5 and 19.4. ^{239}Pu and ^{241}Pu also fission, thereby contributing significantly to the energy production (Fig. 19.7). This also causes the plutonium in the spent fuel elements to be about 30% deficient in these fissile isotopes. Truly, all plutonium isotopes lead to fission, since the daughters ^{242}Am, ^{243}Cm, ^{245}Cm, and many other heavier actinides are fissile. Figures 13.5, 19.4, and 19.6 show the build-up of higher actinides through radiative capture and β-decay processes.

Fast neutrons in the reactor induce (n, 2n) reactions: $^{238}U(n,2n)\,^{237}U(\beta^-)\,^{237}Np(n,\gamma)\,^{238}Np(\beta^-)\,^{238}Pu$ (see Fig. 19.4). ^{237}Np is also formed through the reaction $^{235}U(n,\gamma)\,^{236}U(n,\gamma)\,^{237}U(\beta^-)\,^{237}Np$. These reactions are the main sources of neptunium and of ^{238}Pu.

The total amount of plutonium formed in various reactors is given in Table 20.1. The older gas-cooled reactors and heavy water reactors are the best thermal plutonium producers; they are also used in weapons fabrication. The LWR and AGR are the poorest plutonium producers.

The amounts and radioactivities of the main actinide elements formed in PWR fuels

TABLE 20.1. *Production of plutonium (kg/MWe y) in various reactor types*

Reactor type	Total plutonium	Fissile plutonium
Light water reactors	0.26	0.18
Heavy water reactors	0.51	0.25
Gas-cooled reactors	0.58	0.43
Advanced gas-cooled reactors	0.22	0.13
Liquid metal fast breeder reactor	1.35	0.7–1.0

TABLE 20.2. *Composition of 1 t spent uranium fuel, originally enriched to 3.3% in ^{235}U, after 33 000 MWd/t U burn-up at a flux of 3×10^{13} n cm^{-2} s^{-1} (30 MW/t U) at a cooling time of 10 y*
Straight figures are weight in kg, italics radioactivity in kCi. Total weight is 35 kg FP and 11 kg transuranium elements. Total radioactivity is 312 kCi from the FP and 2.35 kCi from the TU elements. Thermal power is 1.02 kW from the FP and 0.070 kW from the actinides. Most common corrosion products in HAW are shaded; P is a radiolysis product. Arrows indicate mother–daughter relations

H $4 \cdot 10^{-5}$ *0,39*																	He
Li	Be											B	C	N	O	F	Ne
Na	Mg											Al	Si	P (shaded)	S	Cl	Ar
K	Ca	Sc	Ti	V	Cr (shaded)	Mn	Fe (shaded)	Co	Ni (shaded)	Cu	Zn	Ga	Ge $4 \cdot 10^{-4}$ *0,007*	As $8 \cdot 10^{-5}$ *0,005*	Se 0,05 *$4 \cdot 10^{-4}$*	Br 0,015 *0*	Kr 0,36 *5,8*
Rb 0,35 *$2 \cdot 10^{-8}$*	Sr 0,78 *60* →	Y 0,47	Zr 3,78 *$2 \cdot 10^{-3}$* →	Nb $4 \cdot 10^{-6}$	Mo 3.47 *0*	Tc 0,84 *0,014*	Ru 2,15 *0,42* →	Rh 0,39	Pd 1.16 *$1 \cdot 10^{-4}$*	Ag 0,06 *$1 \cdot 10^{-5}$*	Cd 0,08 *$3 \cdot 10^{-5}$*	In 0,001 *0*	Sn 0,059 *$5 \cdot 10^{-4}$*	Sb 0,011 *0,61* →	Te 0,56	I 0,27 *$4 \cdot 10^{-5}$*	Xe 5,46 *0*
Cs 2,31 *92* →	Ba 1,81 *85*		Hf	Ta	W	Re	Os	Ir	Pt	Au	Hg	Tl	Pb	Bi	Po	At	Rn
Fr	Ra																

La 1,27 *0*	Ce 2,48 *0,10* →	Pr 1,2	Nd 4,1 *0*	Pm 0,007 *7,0*	Sm 0,86 *1,1*	Eu 0.16 *4,4*	Gd 0,12 *$1 \cdot 10^{-4}$* →	Tb 0,002	Dy 0,001 *0*	Ho *$7 \cdot 10^{-8}$*	Er	Tm	Yb		
Ac	Th *0,0003*	Pa *0,0003*	U 955 *0,0036*	Np 0,48 *0,018*	Pu 10,4 *86*	Am 0,14 *0,16*	Cm 0,023 *1,7*	Bk	Cf	Es	Fm	Md	No	Lr	

are given in Table 20.2 for cooling time 10 y. The amounts are about the same for shorter cooling times, except for ^{242}Cm, which decays into ^{238}Pu with a half-life of 163 d. Eventually all actinides decay into lighter elements and ultimately into stable lead and thallium.

20.1.2. *Fission product composition*

The fission product (FP) composition in spent fuel elements varies slightly between different reactor systems will the same parameters as the actinide concentration. In-

TABLE 20.3. *Long-lived fission products with activities practically independent of t_{cool} for thousands of years*
Activities at t_{cool} 1000 y; otherwise as in Table 20.2

Isotope	Half-life (y)	Radioactivity (Ci/t U)
^{79}Se	6.5×10^4	0.39
^{87}Rb	4.7×10^{10}	2×10^{-5}
^{93}Zr → ^{93m}Nb	1.5×10^6	2.0
^{99}Tc	2.1×10^5	14.3
^{107}Pd	6.5×10^6	0.11
^{126}Sn → ^{126m}Sb → ^{126}Sb; ^{126}Sn → ^{126}Sb	10^5	0.54
^{129}I	1.57×10^7	0.038
^{135}Cs	2.1×10^6	0.29

creased fission of heavier actinides displaces the yield curve (Fig. 9.8) towards higher mass numbers, and increased fast fission varies the yield of the products between the peaks. About 35 kg FP are formed in each tonne of uranium irradiated to 33,000 MWd.

From the yield values in the Isotope Chart the radioactivity of each fission product element has been calculated in Fig. 20.2 for a 2 y irradiation time at three different fluxes (10^{12}, 10^{13}, and 10^{14} thermal n cm^{-2} s^{-1}) taking decay and secondary reactions into consideration. Using effective cross-sections (Fig. 19.4) the amounts and radioactivities in Table 20.2 are calculated. It is seen that Xe, Zr, Mo, Nd, Cs, and Ru, which are the elements formed in largest amounts (both by mole percent and by weight), constitute about 70% of the fission product weight after a cooling time of 10 y.

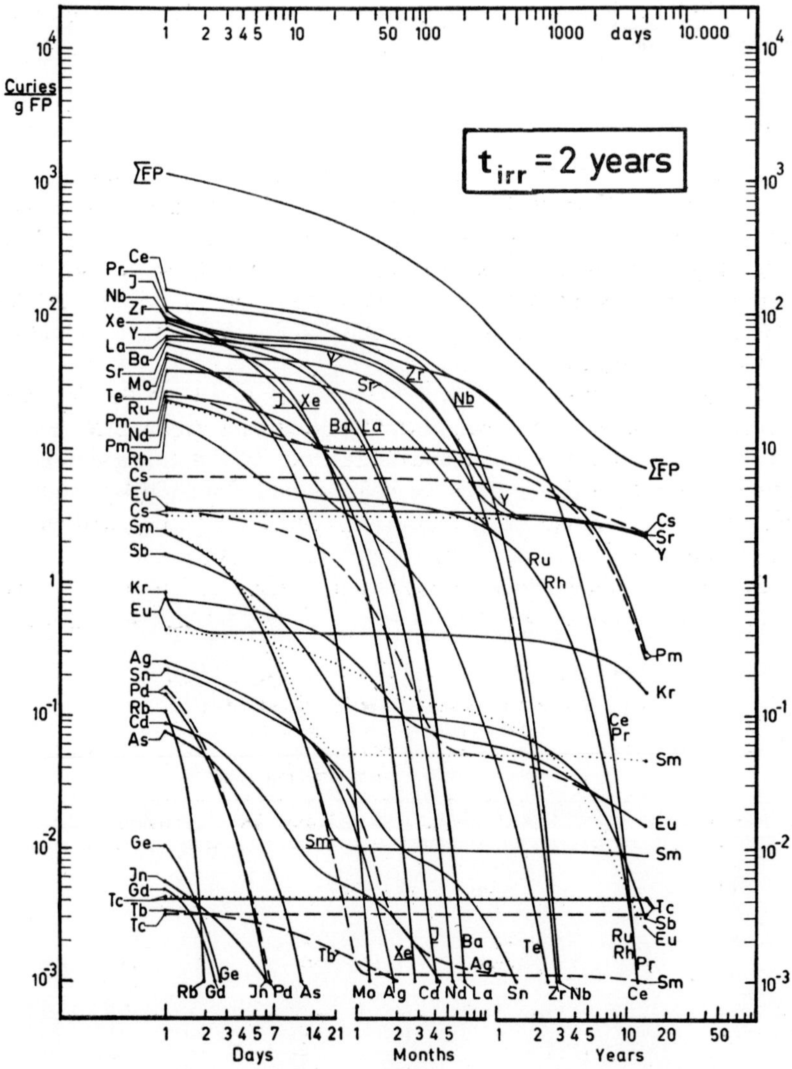

FIG. 20.2. Relative radioactivities from elements formed in thermal fission of ^{235}U at ...... 10^{12}, —— 10^{13}, and - - - 10^{14} n cm^{-2} s^{-1} after an irradiation time of 2 years. Inflexion points indicate the existence of several radioactive isotopes of same element. (According to J. Prawitz and J. Rydberg.)

The radioactivity of some fission products in Fig. 20.2 shows a strong flux dependency (e.g. Cs and Sm). At cooling times 10–1000 y the activities of ^{90}Sr and ^{137}Cs (with daughters) dominate (Fig. 20.23 b). From then on the FP activity comes from very long-lived nuclides of low activity (Table 20.3). ^{131}I, which for short cooling times is one of the most hazardous FP (because of its affinity to the thyroid gland), is harmless after a cooling time of 6 months. The $A = 95$ isobar chain, which is formed in highest yield (6.5%), leads to $^{95}Zr\,(t_{1/2}\ 64\,d) \rightarrow {}^{95}Nb\,(t_{1/2}\ 35\,d) \rightarrow {}^{95}Mo$ (stable). ^{95}Nb is the longest-lived niobium isotope formed in fission, which explains the disappearance of niobium in Fig. 20.2 at a cooling time of about 1 y.

From Table 20.2 it is seen that only stable isotopes remain for some fission elements at $t_{cool} \gtrsim 10$ y (Br, Mo, In, Xe, La, Nd), while some others are of very low activity (Se, Zr (see above), Pd, Ag, Cd, Sn, I, and Gd). Some of the more active ones after 10 y will have disappeared almost completely at 100 y (Sb, Ce, Eu, Pm, Rn, Kr, and T), leaving essentially only ^{90}Sr, ^{137}Cs, and ^{151}Sm.

20.1.3. *Cooling time*

The proper time to start reprocessing is a balance between interest loss on unused fissile and fertile material and storage costs of the unprocessed fuels, on the one hand, and, on the other, savings due to simplified reprocessing and waste handling. Originally a 180 d cooling time was considered appropriate (this time has been used in military programs). Presently the average cooling time for commercial fuel elements is 7–10 y because of lack of reprocessing capacity. This has led to some rethinking, and a cooling time of 2 y is now advocated.

As the radioactivity of the FP and actinides decreases by time, so also does the average energy per decay, since the longer-lived nuclides have lower decay energies. The radiation absorbed in the shielding material (in addition to self-absorption) produces *decay heat*, which diminishes slightly faster than the radioactivity. Table 20.4 gives data on the decay heat with contributions separately for the fission products and the actinides. For cooling times $> 10^3$ y the decay heat from the actinides dominates.

The decay heat is considerable at short cooling times (see Fig. 19.19) in the 30 t U annually discharged from a LWR. Before recharging a reactor, the used fuel elements

TABLE 20.4. *Decay heat from fission products and actinides in waste (only 0.5% U and Pu)*
Basic data as in Table 20.2

Cooling time	Decay heat (W/t U) from: Fission products	All actinides	Waste actinides
1 d	67 000		
90 d	24 000		
180 d	16 000		
1 y	8 800	840	240
5 y	1 600	810	80
10 y	1 000	780	70
100 y	100	410	9.6
10^3 y	0.023	65	2.1
10^4 y	0.021	45	0.52
10^5 y	0.013	4.5	0.065
10^6 y	0.001	1.8	0.076

are first allowed to cool in the reactor by forced circulation. Within a few weeks they are then transferred under water to the cooling basin at the reactor site for an additional cooling time of 6–12 months.

20.2. Transport of spent reactor fuel

The storage capacity of reactor pools is normally < 2 years' production, but can be increased somewhat by adding neutron absorbers to the pool. The fuel assemblies must then be transferred to special (interim) storage sites or sent for reprocessing. This is done in special transport flasks.

The properties of LWR used fuel assemblies are summarized in Table 20.5. The radioactivity requires shielding and remote handling, and the heat continual cooling. Because each transport is expensive the transport flasks are designed for carrying several assemblies. A 30 t (weight) flask may carry 4 PWR or 9 BWR assemblies (~ 1 t U), while 100 t flasks are designed for about 12 PWR or 30 BWR assemblies (~ 6 t U). The inner cavity of the transport flask is surrounded by $\gtrsim 0.2$ m of lead and (in most flasks) a neutron-absorbing shield. On the outside they have cooling fins and shock-absorbing material. From Table 20.5 it is seen that a filled 100 t flask with 1 year old fuel would contain ~ 15 MCi β, γ and develop ~ 60 kW heat; the design cooling capacity of the flask is ~ 100 kW. The cavity of such a flask is about 5 m in length and 1.2 m in diameter. In some flasks the cavity is filled with water (e.g. in the UK), for two reasons: the flasks are loaded and unloaded under water, the water functioning as heat conducting and neutron absorbing material (remember, some actinides decay by spontaneous fission; also α,n-reactions occur with light target atoms). Air-filled flasks have forced cooling, which is not necessary for the smaller water-filled flasks. Figure 20.3 shows a water-filled 70 t fuel cask for 5 PWR or 14 BWR assemblies (U pay load 2.7 t) designed for rail transport. Smaller flasks (up to ~ 50 t) may be transported by road using a special trailer. The flasks are designed for exceedingly severe treatment: free

TABLE 20.5. *Spent fuel data for 1000 MWe light water reactors*

	PWR (33 000 MWd t^{-1})	BWR (27 500 MWd t^{-1})
Assembly length (m)	5.1	4.5
Assembly diameter (mm^2)	214 × 214	140 × 140
Assembly number of pins	264	64
Assembly weight (kg)	780	250
UO_2 (kg/assembly)	600	220
Annual discharge, assemblies	52	170
Annual discharge, uranium (tonnes)	27	33
Annual discharge plutonium (kg)	258	260
Annual discharge, fission products (kg)	910	1150
Radioactivity at 150 d, kCi/assembly	2800	840
Radioactivity at 1 y, kCi/assembly	1400	420
Radioactivity at 2 y, kCi/assembly	800	370
Radioactivity at 5 y, kCi/assembly	300	190
Radioactivity at 10 y, kCi/assembly	190	64
Decay heat at 150 d, kW/assembly	13	3.7
Decay heat at 1 y, kW/assembly	6	1.7
Decay heat at 2 y, kW/assembly	2.8	0.8
Decay heat at 5 y, kW/assembly	0.7	0.3
Decay heat at 10 y, kW/assembly	0.6	0.12

FIG. 20.3. Flask of Excellox type for transport of spent LWR fuel elements.

fall from 9 m onto a concrete floor, 30 min gasoline fire ($\sim$ 800°C), submersion into 15 m of water, etc., without being damaged. Tests in which a 74t flask on a railway car has been run at 130 km h^{-1} into a concrete wall show the flask undamaged.

After loading a flask the outside of the flask must be decontaminated; often it is quite difficult to remove all activity, and a removable plastic covering is therefore used. Before unloading, water-filled flasks are flushed to remove any activity leaked from the fuel or suspended crud, which is present on the used fuel elements from BWRs and most PWRs. Decontamination is also carried out after the flask has been emptied.

Shielded transport is also required for solidified high level waste, hulls, plutonium containing material, and for some intermediate and α-active waste. For each type (and country) special containers are designed. As an example, in the United Kingdom plutonium containers made of wood and cadmium are limited to carrying 10 kg Pu; the container weighs 175 kg, is 1.3 m high, and 0.8 m in diameter.

20.3. Alternative fuel cycles

Fission energy can be obtained on the bases of uranium, forming the *uranium once-through* option and the *uranium–plutonium fuel cycle*, and of thorium, forming the *thorium–uranium fuel cycle*. Nuclear energy cannot be obtained with thorium alone because natural thorium contains no fissile isotopes; the thorium–uranium cycle must be started with uranium- or plutonium-fueled reactors.

Each fuel cycle offers a number of alternative routes with respect to reactor type, reprocessing, and waste handling. Although the uranium-based cycles are described with special reference to light water reactors, the cycles also apply to uranium-fueled gas-cooled reactors.

20.3.1. *The uranium once-through (UOT) option*

The heavy arrows in Fig. 20.4 indicate the steps in the nuclear fuel cycle presently used on a large commercial scale. It ends at the interim storage basin (block 7). From here two alternative routes are available, one (route 9) leaving to reprocessing and the uranium–plutonium fuel cycle described in the next section, and another (route 10) leading to final storage of the unreprocessed spent fuel elements; the latter is referred to as the once-through fuel cycle option. Its subsequent steps are waste conditioning (block 31) in which the used and cooled fuel assemblies are treated so that they can be stored for extensive periods without any risk, and deposition of the conditioned assemblies in permanent repositories (block 32). The technical aspects of the UOT option are further discussed in §20.4.

In this option the energy content of unused ^{235}U, fertile ^{238}U, and fissile and fertile plutonium is not saved for future needs, and a rather hazardous waste is produced

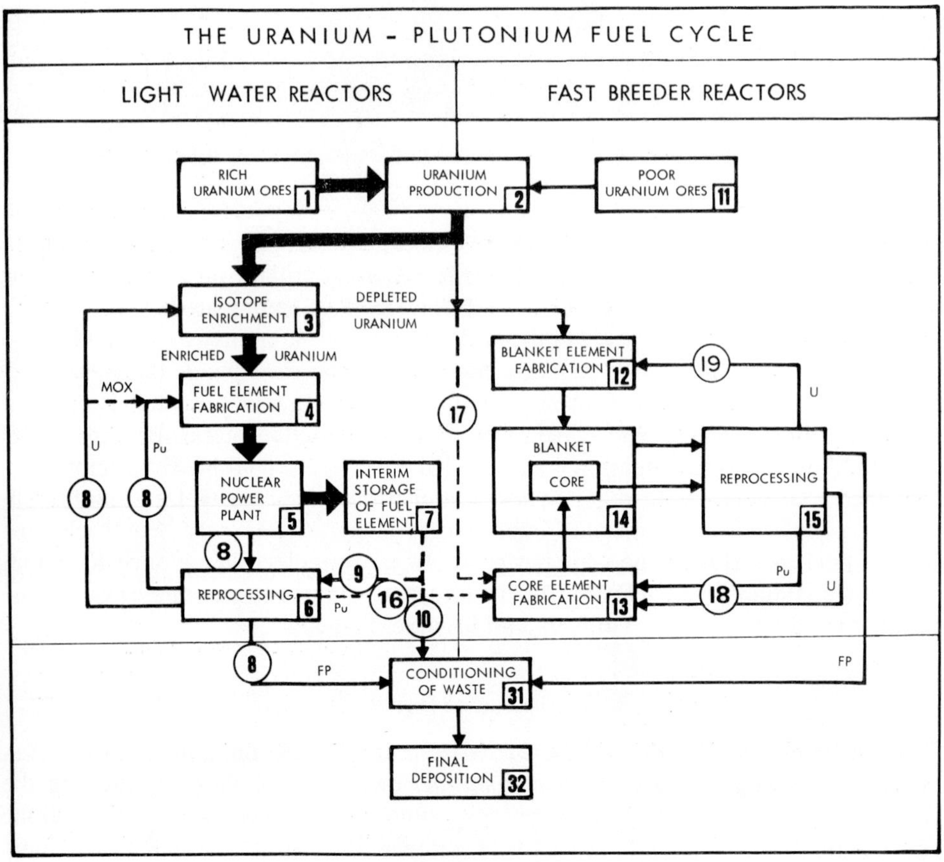

FIG. 20.4. The uranium–plutonium fuel cycle.

containing large amounts of fissile plutonium (see §21.6ff.) and toxic α-emitting nuclides. The advantage of the cycle is that it withdraws plutonium from possible weapons use.

20.3.2. *The uranium–plutonium (UPu) fuel cycle*

"Cycle" infers some kind of recirculation of material. The term "fuel cycle" was originally used for the steps in which fissile and fertile material was isolated from used fuel elements (*re*processing) at the back end part and returned to the front end part of the fuel cycle for use in new fuel elements. The flow scheme is marked 8 in the left part of Fig. 20.4. Instead of direct reprocessing after removal from the reactor pool, an interim storage (block 7) may be used (flow 9).

The fissile value of used LWR fuel elements amounts to $\sim 0.9\%$ ^{235}U and 0.5–0.7% $^{239+241}$Pu. By recovering these and returning them to the LWR fuel cycle the demand for new uranium and enrichment services is reduced by $\sim 30\%$. The uranium and plutonium are mixed into oxide (MOX) fuel elements either after their isolation in a reprocessing plant or—preferably—by coprecipitation in correct proportions at the plant. MOX fuel elements for LWR replacement contain about 3% $^{239+241}$Pu. Several tons of plutonium already have (on an experimental scale) been returned as MOX fuel elements into LWRs. In this cycle "old" plutonium from earlier production is successively exposed to increasing neutron fluency, which changes its isotopic composition. This is shown in Table 20.6 for both virgin plutonium and virgin + recycled plutonium. The fissile fraction (239 + 241) decreases requiring successively increased plutonium fractions in the MOX elements. The most toxic isotope, ^{238}Pu, substantially increases. Its short half-life (88 y) results in a high specific radioactivity which increases the heat evolution and radiolysis of the reprocessing solutions. ^{238}Pu, ^{240}Pu, and ^{242}Pu all decay partly through spontaneous fission, with the emission of neutrons ($\sim 2 \times 10^3$ n g^{-1} s^{-1}). This makes the MOX elements more difficult to handle. (All these changes also make the plutonium less suitable for weapons use.) ^{241}Pu has a critical mass of about half that of ^{239}Pu; as ^{241}Pu builds up, criticality risks therefore increase. The build-up of other actinides by recycling high A isotopes further increases reprocessing and waste problems.

In an alternative cycle, only uranium is recycled, while the plutonium is left with the waste. This would produce a waste with a very high concentration of plutonium; the weight ratio of plutonium to fission products will be 10:35 (kg/t U, Table 20.2). Such wastes have not yet been studied.

The reuse of plutonium in LWR will be only temporary if full-scale fast breeder

TABLE 20.6. *Composition of plutonium after each recycling in a LWR system with MOX fuel elements*

Isotope mass number	Original composition in spent uranium fuel (wt%)	Weight percent after recycling				
		1st	2nd	3rd	4th	5th
238	1	1.5	2	2	2.5	3
239	56	46	42	40	39	38
240	26	30.5	32	32.5	33	33.5
241	12	14	14.5	15	14.5	14
242	5	8	9.5	10.5	11	11.5

reactors are developed (Fig. 20.4), FBRs are designed with a core, containing ~15% fissile Pu and ~85% U in the form of mixed oxides or carbides (block 13, original feeds flows 16 and 17), surrounded by a blanket of depleted uranium. The actinide and fission product contents in discharged FBR and LWR fuel are roughly the same on a GWe y basis. Also the masses of total discharged fuel per GWe y are comparable. Breeding occurs in the blanket (cf. §19.15). The burn-up in the core elements is about 3 times higher than in LWRs, and the fraction of fission products in them is therefore also 3 times larger. Since only a small part of the plutonium is burnt, the used core fuel elements retain a high economic value, making it desirable to reprocess them after a short cooling time. The used FBR core elements may have a tenfold greater specific radioactivity than used LWR fuel elements at the time of reprocessing. More advanced reprocessing techniques than presently available would be required. The FBR blanket elements are simpler to handle and may be reprocessed in plants for LWR fuel elements. However, if the core elements are diluted with blanket elements in about a 1:1 mix they may be reprocessed together in a "conventional" plant for LWR fuels. The plutonium and uranium recovered are used for new fuel elements (flows 18 and 19).

In the FBR ^{238}U is consumed either through fission (i.e. energy production) or ^{239}Pu formation. Because the α-value of ^{239}Pu is 0.42, at least 70% $(100(1+\alpha)^{-1})$ of all ^{238}U is useful for energy production in the UPu cycle. This value should be compared with the $\lesssim 0.7\%$ of the natural uranium which is used in the UOT cycle (taking enrichment also into account), or $\lesssim 1\%$ in the LWR MOX fuel recycle. The FBRs not only increase the useful energy of natural uranium by a factor of ~ 100, they also make it economic to mine low grade uranium ore, vastly extending the uranium resources.

20.3.3. *The thorium–uranium (ThU) fuel cycle*

Fertile ^{232}Th can be transformed into fissile ^{233}U in any thermal reactor. The reactions in ^{232}Th irradiated by neutrons are given in Fig. 19.17. Of the thermally fissile atoms ^{233}U has the highest σ_f/σ_γ ratio, i.e. highest fission efficiency. The η-value is high enough to permit breeding in the thermal region. Capture of neutrons in ^{233}U is not a serious drawback, since a second capture (in ^{234}U) yields fissile ^{235}U. Since σ_{tot} increases with neutron energy (from ~ 7 b at 0.025 eV to ~ 26 b for reactor conditions; cf. also Fig. 19.2) the slightly harder neutron spectrum in HWR and GCR has made these reactors the prime candidates for a thorium–uranium fuel cycle together with the molten salt reactor (§ 19.15.1). The fuel may be arranged in a core (^{233}U) and blanket (^{232}Th) fashion, or mixed fissile and fertile material as, for example, in the HTGR graphite matrix fuels (§19.11), or as a metal fluoride melt. The initial ^{233}U must be produced in reactors fueled with ^{235}U (or ^{239}Pu). After sufficient amounts of ^{233}U have been produced, the ThU fuel cycle may become self-sustaining, i.e. thermal breeding is established (Fig. 20.5).

The advantage of the ThU fuel cycle is that it increases nuclear energy resources. In combination with the uranium fuel cycle it could more than double the lifetime of the uranium resources by running the reactors at a high conversion rate (~ 1.0) and recycling the fuel. Thorium is almost as widely distributed as uranium. Very rich thorium minerals are more common than rich uranium minerals. The presence of extensive thorium ores has motivated some countries, as, for example, India, to develop the ThU fuel cycle, although no full-scale ThU fuel cycle has yet been demonstrated. Reprocessing has only been demonstrated on an experimental scale. The fuel cycle has to

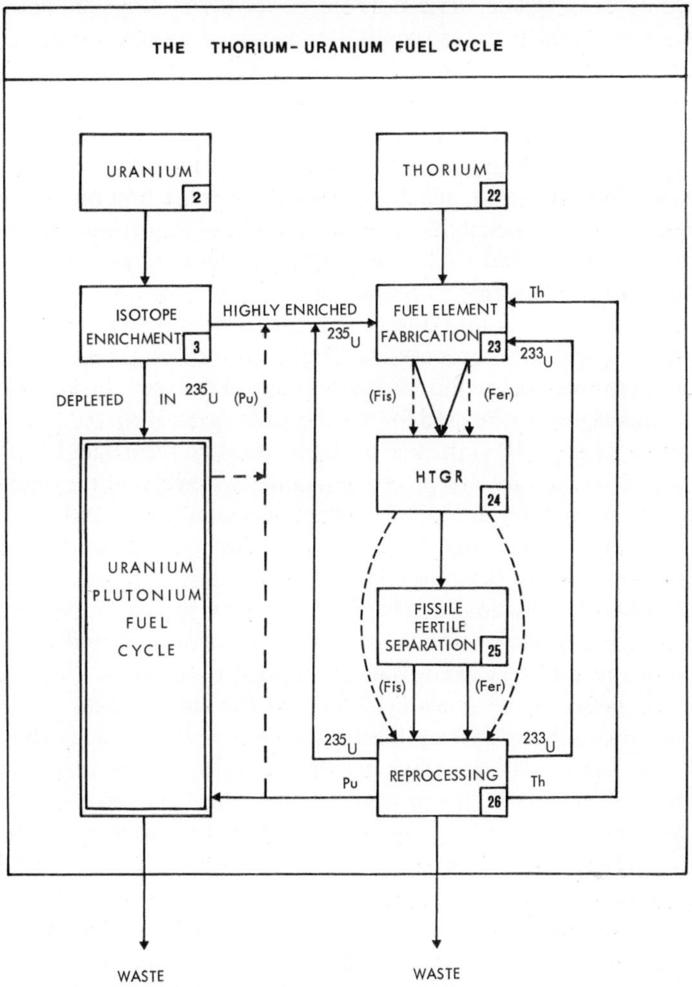

FIG. 20.5. The thorium–uranium fuel cycle.

overcome the high activity problems due to the presence of ^{228}Th formed in the thorium fraction and ^{232}U formed in the uranium fraction (Fig. 19.17). The ThU fuel cycle has a rather specific advantage over the UPu cycle in that its high active waste formed at reprocessing contains much less long-lived heavy actinides, and thus constitutes a smaller long term hazard.

20.4. Storing the spent fuel elements

Spent fuel elements can be stored in the reactor pools usually for a couple of years. During this time the radiation level and heat production decrease considerably (see Table 20.4). With limited capacity for reprocessing, separate spent fuel element storage facilities become compulsory. These facilities consist of large water-filled basins containing geometrically arranged stainless steel racks for the spent fuel elements. The water is circulated for cooling. The surrounding building is located either at ground level or underground at 30–50 m depth, according to present plans.

Zircalloy clad oxide fuel elements can be stored for decades in such pools with very little risk of leakage. Metal fuels, especially those canned in magnesium or aluminum alloys, are less resistant and should not be stored as such in this manner for a prolonged time.

Some of the stringent requirements on the storage pools are: $k_{eff} < 0.95$ (even if unused fuel elements are introduced), earthquake safety, no possible water loss, water level automatically kept constant, adequate leakage and radiation monitoring systems, water temperature $< 65°C$, acceptably low radiation level in working areas, etc.

Although some countries have indicated plans for pool storage up to 50 y for the spent fuel elements, it is desirable to reduce this time as much as possible for several reasons. In the case of the uranium–plutonium cycle, the economic value of the fertile and fissile atoms prompt early reprocessing. In the once-through option, the radiation hazards and proliferation risks require a safer storage of the spent fuel elements. Water-filled basins will therefore only be used for interim periods (5–50 y).

For long term storage ($\gg$ 10 y) the fuel elements must be recanned. For this purpose single fuel elements or bundles are placed in stainless steel cylinders, and the void is filled with some suitable material like lead, which has good heat conductivity and also provides some radiation protection. Depending on the external condition at the final storing place (humidity, temperature, etc.) the canisters are surrounded by an additional shielding to improve the lifetime of the fuel elements, which—preferably—should not be exposed to the biosphere until all radioactivity has disappeared.

In one concept the canisters are to be surrounded by concrete (see Fig. 20.32). A space may be left between the inner cylinder and the surrounding concrete wall to allow air convection cooling. These pillars are dispersed over a large dry area, e.g. a desert floor. As long as the climate stays dry the pillars will erode very slowly and last for tens of thousands of years. This procedure allows for easy retrievability. An alternative encapsulation is achieved by using hot ($\lesssim 1500°C$) isostatic (100–300 MPa) compression to surround the fuel bundles by a homogeneous, dense ceramic material, like corundum (microcrystalline Al_2O_3) or graphite. Since corundum and graphite are natural minerals, the long term resistance should be very high, even against water.

In another concept, the spent fuel elements would be stored irretrievably in deep underground geologic formations. If these formations are wet (e.g. percolated by groundwater), the steel canister must be surrounded by a more corrosion resistant material. Ceramics like corundum, zircon, graphite, etc., and metals like titanium, lead, copper, etc., have been suggested. In reducing groundwater (cf. §20.12.2) the lifetime of a 0.2 m thick copper layer is estimated to be $> 10^6$ y.

The various aspects of the long term (or "final") storage of high active waste is more thoroughly discussed in §20.12.

20.5. Reprocessing uranium fuel elements from thermal reactors

The main purpose of commercial reprocessing is:

(1) to increase the energy from fissile and fertile atoms;
(2) to reduce hazards and costs for handling the high level wastes.

Two other reasons are sometimes mentioned:

(3) to reduce the cost of the thermal reactor fuel cycle;
(4) to extract valuable byproducts from the high active waste.

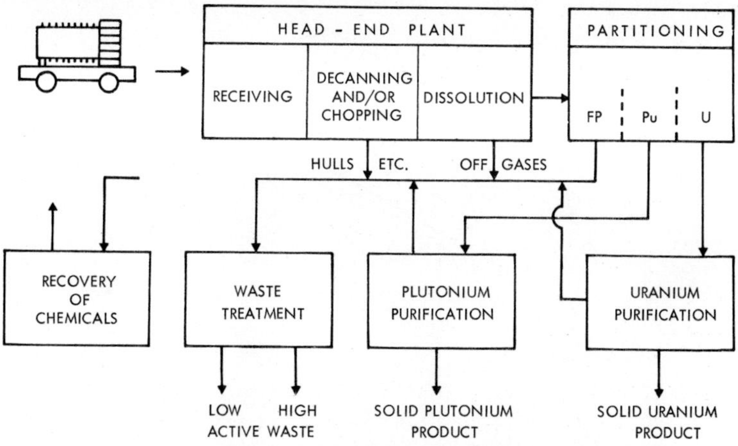

FIG. 20.6. Main steps in reprocessing spent LWR fuels.

The energy savings have already been discussed in the previous section: the 30% savings in LWR and similar reactors, and the hundredfold energy resource expansion by using FBRs.

The economic advantage of reprocessing depends on the cost of natural ("yellow cake") uranium, on enrichment and other front end activities, and on the prevailing energy price (mainly based on fossil fuels). In 1977 it was estimated that commercial reprocessing would amount to 10% of the fuel cycle costs. It is uncertain (in 1979) to what extent reprocessing reduces the high active waste handling problems and costs as compared to the once-through (or "direct deposition") philosophy (§20.3.1).

The reprocessing demand in the noncommunist world amounts to ~3000 t U/y, while the commercial capacity is ~1000 t y^{-1} (Appendix I–K). Military reprocessing is carried out in many countries (USA, USSR, UK, France, China, and perhaps others) on low burn-up metallic fuel elements. Though the technique is principally the same as for commercial high burn-up oxide fuel elements, the compositions differ to such an extent that a military type reprocessing plant cannot accept commercial fuel elements. Many countries have or have had pilot reprocessing plants with annual capacities from tens to hundreds of tonnes (see Appendix K). The first full-scale commercial plant to come on line is at La Hague, France, where the capacity will successively be stepped up from the present 400 t U/y.

The main steps in nuclear fuel reprocessing are shown in Fig. 20.6: (i) the head end section, in which the fuel is prepared for chemical separation; (ii) the main fractionation of U, Pu, and FP; (iii) purification of uranium; (iv) purification of plutonium; (v) waste treatment; and (vi) recovery of chemicals. Each step is described in the following sections.

20.5.1. *Head end plant*

Figure 20.7 is a simplified drawing of the French oxide fuel head end plant at La Hague. The flasks with the used fuel assemblies are lifted by a crane into water-filled stainless-steel-lined pits, where the flasks are unloaded and decontaminated. The assemblies are stored for a desired time and then transferred to a shielded dismantling and chopping section. Some BWR fuels have end parts, which can be mechanically

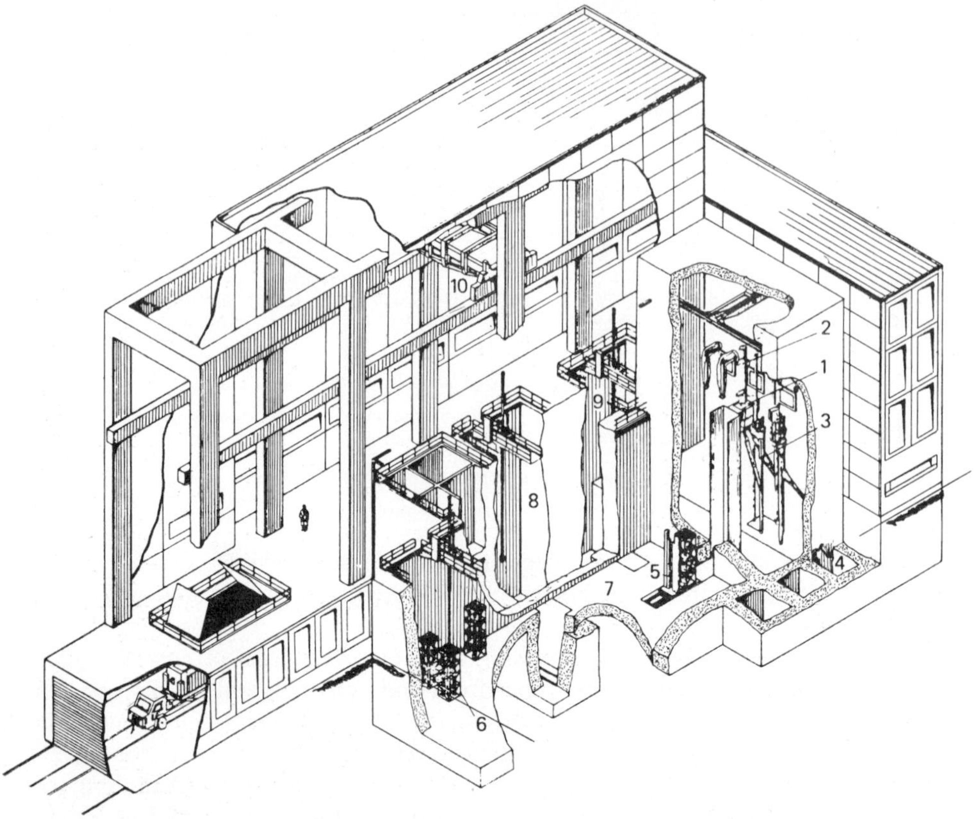

FIG. 20.7. Head end oxide fuel building at La Hague, France. 1, cutter; 2, supply store; 3, dissolution apparatus; 4, evacuation hulls; 5, store supply chariot; 6, storage baskets; 7, storage pool; 8, castle unloading pit; 9, special dismantling pit; 10, 1300 KN bridge.

dismantled; this is not the case for PWR fuels, in which the end parts have to be cut off. The fuel pins are then cut into pieces 3–5 cm long, either under water or in air, which fall into stainless steel baskets. At Windscale, UK, dry charging and chopping are used (Fig. 20.8).

The chopping is achieved with a shearing knife (cutter), but other techniques for removing or opening up the zircalloy (or stainless steel) cans have been tried, or are being developed, including electrolytic dissolution and laser cutting. Previously, chemical decanning was used (mainly for aluminum cans), but such techniques increase considerably the amounts of medium active waste.

The chopped pieces are transported to (or fall directly into) the dissolver unit, where the oxide fuel is leached out by boiling 6–11 M HNO_3 (the cladding hulls do not dissolve) in thick stainless steel vessels provided with recirculation tubes and condenser. The hulls are measured for residual uranium or plutonium, and, if sufficiently clean of fissile material, they are discharged to the waste treatment section of the plant. To improve the dissolution some fluoride ($\lesssim$0.05 M) may be added to the HNO_3. The F^- forms strong complexes with some metal ions (such as zirconium), while its corrosion of the stainless steel equipment has been found to be negligible. Soluble poisons, such as cadmium or gadolinium nitrate, are often added to the nitric acid to assure the criticality control of the dissolution operation.

FIG. 20.8. Shielded fuel-charging installation with transport flask in position at the head end of the Windscale reprocessing plant, UK.

When the fuel pins are cut the amount of volatile fission products contained in the gas space between the fuel oxide pellets and the canning is released (mainly ^{85}Kr, T, ^{129}I, and ^{131}I). These gases are ducted to the dissolver off-gas treatment system. Gases released during dissolution are Kr, Xe, I_2, T_2, THO, RuO_4, CO_2, minor amounts of fission product aerosols, and large amounts of H_2O, HNO_3, and nitrogen oxides. Oxygen or air is fed into the off-gas stream to allow recovery of part of the nitrogen oxides. The overall dissolution stoichiometry is

$$2U + 4HNO_3 + 3O_2 \rightarrow 2UO_2(NO_3)_2 + 2H_2O$$

The gas streams pass to a condenser which reclaims and returns some nitric acid to the dissolver; the noncondensibles are discharged to the off-gas treatment system.

When the dissolution is completed, the product solution is cooled and transferred by steam jet to the input measurement—clarification (filter and/or centrifuge) feed adjustment unit. At this point the uranium is in the hexavalent state, and plutonium in the tetravalent.

20.5.2. *Separation methods*

The specifications for purified uranium and plutonium to be recycled are summarized in Table 20.7. Comparing these data with those in Table 20.5 shows that at t_{cool} 1 y the fission product activity must be reduced by a factor of about 10^7 and the uranium content in plutonium by a factor of 2×10^4. Such large decontamination factors are not uncommon in analytical laboratories, but rarely required or achieved on an industrial scale. In addition to this, the large number of chemical elements involved (the FP and actinides in Table 20.2 *and* corrosion products during processing) and the lack of knowledge about the chemistry of some of the fission products make the separation a very difficult task. Additional complications arise from radiation decomposition and criticality risks and from the necessity to conduct all processes remotely in heavily shielded enclosures under extensive health protection measures. As a result reprocessing is the most complicated chemical process ever endeavored.

TABLE 20.7. *Specifications for reprocessed uranium and plutonium (according to IAEA* 1977)

Uranyl nitrate	
Uranium concentration 1–2 M	
Free $HNO_3 \gtrsim 1$ M	
Impurities: Fe, Cr, Ni $\lesssim$ 500 ppm	
	Boron equivalents $\leq$ 8 ppm[(a)]
Fission products $\leq 0.5\ \mu$Ci/g U	
Alpha-activity (excluding uranium) $\leq$ 15 000 dpm/g U	
Plutonium nitrate	
Plutonium concentration $\sim$ 1 M	
Free HNO_3 2–10 M	
Impurities: metallic $\leq$ 5000 ppm	
	uranium $\leq$ 5000 ppm
	boron equivalents $\lesssim$ 10 ppm
	sulfate $\leq$ 1000 ppm
Fission product activity ($t_{1/2} > 30$ d) $\leq 40\ \mu$Ci/g Pu[(b)]	
^{241}Am content (9 months after delivery to MOX-plant) $\leq$ 5000 ppm	

[(a)] The equivalent values are B 1.0, Cd 0.4, Gd 4.4, Fe 0.0007, etc. The amount measured for each of these elements multiplied by the factor indicated must not be more than 8 ppm.

[(b)] ^{95}Zr–Nb $\leq 5\ \mu$Ci/g Pu.

The primary problem encountered by the chemists of the Manhattan Project was the selection of satisfactory separation techniques. Advantage was taken of the relative stability of the oxidation state of uranium (+ 6) and most fission products, and the lability of plutonium (+ 3, + 4, and + 6). In the earliest process, only plutonium was isolated. This was done by precipitating plutonium in the reduced state by F^- as PuF_3 or PuF_4 together with all insoluble FP fluorides, and in a second stage dissolving the precipitate, oxidizing plutonium to the + 6 state, and carrying out a new fluoride precipitation, leaving relatively pure plutonium in the supernatant. In a final step plutonium was again reduced and precipitated as fluoride. The same principle was used for the first isolation of hundreds of kilograms of plutonium at the Hanford Engineering Works, but instead using phosphate: Pu(+ 3) and Pu(+ 4) form insoluble phosphates, but not Pu(+ 6). Since Bi^{3+} was used as a carrier for the precipitate, it is referred to as the "bismuth phosphate" process.

This principle of oxidizing and reducing plutonium at various stages of the purification scheme has been retained in all subsequent processes. No other element has the same set of redox *and* chemical properties as plutonium, though some fission products behave as Pu^{3+} (e.g. the lanthanides), some like Pu^{4+} (e.g. zirconium) and some like PuO_2^{2+} (e.g. uranium). The redox agents used have been $K_2Cr_2O_7$ (to PuO_2^{2+}), $NaNO_2$ (to Pu^{4+}), hydrazine, ferrous sulfamate, and U^{4+} (to Pu^{3+}), cf. §13.3.2.

The precipitation technique is not suitable for large-scale, continuous remote operations in which both uranium and plutonium have to be isolated in a very pure state from the fission products. It was replaced by solvent extraction processes in which the fuels were dissolved in nitric acid and contacted with an organic solvent which selectively extracted the desired elements.

The first solvent to be adopted was methyl isobutyl ketone ("MIBK" or "Hexone") at Hanford. This solvent forms adduct compounds with coordinatively unsaturated compounds like the actinide nitrates, e.g. $Pu(NO_3)_4S_2$, where S represents the adduct molecule

$$Pu^{4+} + 4NO_3 + 2S(org) \rightarrow Pu(NO_3)_4S_2(org)$$

The corresponding adduct compounds for 3- and 6-valent actinides are $An(NO_3)_3S_3$ and $AnO_2(NO_3)_2S_2$. These neutral compounds are soluble to different extents in organic solvents like kerosene, and—in the case of hexone—by hexone itself. The process using hexone is referred to as the *Redox process*.

The disadvantage of hexone is its tendency to decompose slowly by strong nitric acid. The high nitrate concentration necessary for extraction can be achieved by adding a salt ("salting-out" agent), such as $Al(NO_3)_3$. However, addition of salt increases waste volumes.

In the United Kingdom, hexone was replaced by β, β'-dibutoxy diethyl ether ("dibutyl carbitol" or "Butex"). It forms the same kind of adduct compounds as hexone. Though more expensive, it was more stable, less flammable, did not require a salting-out agent, and gave better separations.

Many other similar solvent systems, as well as chelating agents, have been tested. Thenoyltrifluoroacetone (HTTA) was found to form strong complexes with the actinides (e.g. $Pu(TTA)_4$, $UO_2(TTA)_2$), which show very high distribution ratios in favor of organic solvents. Though useful in the laboratory they were not found suitable for large scale commercial nuclear fuel reprocessing.

Presently, tributyl phosphate (TBP) is the favored extractant in all reprocessing plants. It acts as an adduct-former and is used normally as a 30% solution in kerosene.

It forms the bases for the *Purex process*. TBP is cheaper than Butex, more stable, less flammable, and gives better separations. Other extractants, especially tertiary amines, are being tested for some steps in reprocessing. The amines form organic soluble complexes with negatively charged metal complexes (§12.6.3). The switch in a Purex cycle to another type of extractive reagent increases the decontamination factor.

As alternatives to the aqueous separation processes, "dry" techniques have also been studied:

(a) *Halide volatility* Many FP and the high valency actinides have appreciable vapor pressures; this is particularly true for the fluorides. In *fluoride volatilization* the fuel elements are dissolved in a molten fluoride salt eutectic ($NaF + LiF + ZrF_4$, 450°C) in the presence of HF. The salt melt is then heated in F_2, leading to the formation of UF_6, which is distilled; it may be possible also to distill PuF_6, though it is much less stable. The process encounters several technical difficulties.

(b) *Molten salt extraction* The fuel may be dissolved as above or in another salt melt. With a heat resistant solvent of low volatility (e.g. 100% TBP), actinides and FP distribute themselves between the two phases analogous to solvent extraction. This technique is of interest for continuous reprocessing of the molten salt reactor. An advantage is the higher radiation resistance of the nonaqueous system.

(c) *Molten metal purification* Metallic fuel elements can be molten and/or dissolved in molten metals (e.g. a zinc alloy). In the presence of (deficient amounts of) oxygen, strongly electropositive fission elements form oxides, which float to the surface of the melt as slag and can thus be removed, while volatile FPs distill. The residual melt would mainly contain U, Pu, Zr, Nb, Mo, and Ru ("fissium alloy") and can be reused in new fuel elements. This melt refining technique has been tested on metallic breeder reactor fuel elements.

20.5.3. *Solvent extraction separations*

Solvent extraction is a technique which has been highly developed within the various national nuclear energy programs because of its suitability as a selective separation process for fission products, actinides, and other radioactive substances. The technique, its theory, and use is described in many sections of this book (particularly §§10.7, 18.3.3, and 18.8).

Solvent extraction is based on the formation of an organic metal compound which is preferentially soluble in an organic solvent. The three main types of such compounds are: (1) the ionic chelate complexes like plutonium–tetra-acetylacetonate, $PuAa_4$ (generally M_n, §18.3); (2) the inorganic metal complexes forming adducts with solvating organic compounds like TBP and hexone described in the previous section, e.g. $(Pu(NO_3)_4)$ $(Hexone)_2$, or generally ML_nS_s; and (3) ion pair complexes between organic cations (like alkylammonium ions) and negative inorganic complexes like $(R_3NH)_4UO_2(SO_4)_3$ (generally RB^+ML^-, §12.6). The distribution of such complexes between organic solvents and aqueous solutions is illustrated in Figs. 13.9, 18.7, 18.8, and 18.17. A single partitioning of such a complex between an organic solvent and water may not be sufficient for isolating the metal in acceptably pure form and good yield and various multiple extraction techniques may therefore be required. Such techniques have been described in §18.8.1 and their technical application for uranium production in §12.6.

Let us consider a solute (e.g. uranium) which is distributed between an organic and aqueous phase, independent of the kind of compounds it forms. After equilibrium the weight of solute in the organic phase (extract) is w_{org} and in the aqueous phase (raffinate) w_{aq}. Thus

$$w_{tot} = w_{org} + w_{aq} \tag{20.1}$$

the weight fraction in the organic phase is

$$\psi_1 = P/(P+1) \tag{20.2a}$$

where the index 1 refers to the conditions after one extraction. P is called the *extraction* (or partition) *factor* and defined by

$$P = D\theta \tag{20.3}$$

where D is the distribution factor (defined by (18.10)) and θ is the phase volume ratio v_{org}/v_{aq}. The percentage solute extracted, denoted $E(\%)$, is equal to $100\psi_1$. Because the fraction of nonextracted solute is φ_1, and $\psi_1 + \varphi_1 = 1$,

$$\varphi_1 = 1/(P+1) \tag{20.2b}$$

In the extraction of U(+6) with 100% TBP from 1 M HNO_3 $D_U = 20$, while for the fission product lanthanum $D_{La} = 0.07$. If we extract with a phase ratio of 0.5, according to (20.2a) and (20.3), $\psi_U = 0.909$ ($E_U = 100\psi_U = 90.9\%$) and $\psi_{La} = 0.034$ (E_{La} 3.4%). This may be unsatisfactory with respect to both uranium yield and purity.

The yield can be increased by repeated extractions of the same aqueous phase (multiple extraction with one stationary phase, or "crosscurrent extraction"). For n such extractions one finds that

$$\varphi_n = (P+1)^{-n} \tag{20.4}$$

Suppose $n = 3$ for our example, then $\varphi_{3,U} = 0.00075$, i.e. for the three organic phase volumes taken together $E_U = 100(1 - \varphi_{3,U}) = 99.92\%$. However, for lanthanum $\varphi_{3,La} = 0.902$, i.e. $E_{La} = 9.8\%$. Although the uranium yield is high, the lanthanum impurity may be intolerable.

A more elaborate technique must be employed in order to obtain both high yield and high purity under such conditions. Many such batch laboratory techniques have been described using alternatively fresh organic (extraction) and aqueous (washing) solutions. The extractions are carried out either with a battery of test tubes (or separatory funnels) or in special multistage equipment.

Continuous processes are preferred in industry, where the most common and simple solvent extraction equipment is the mixer–settler (Fig. 20.9(a)). This type of equipment is also becoming standard in laboratories engaged in process development. In the uranium industry a single mixer–settler may hold as much as 1000 m^3. The mixer–settlers, each closely corresponding to a single ideal extraction stage, are arranged in batteries containing any number of stages. In these batteries the aqueous and organic phases flow countercurrent to each other (see Fig. 20.10).

For countercurrent solvent extraction, either batch or continuously, one finds that for the stationary state and for n stages

$$\varphi_n = \frac{P-1}{P^{n+1}-1} \tag{20.5}$$

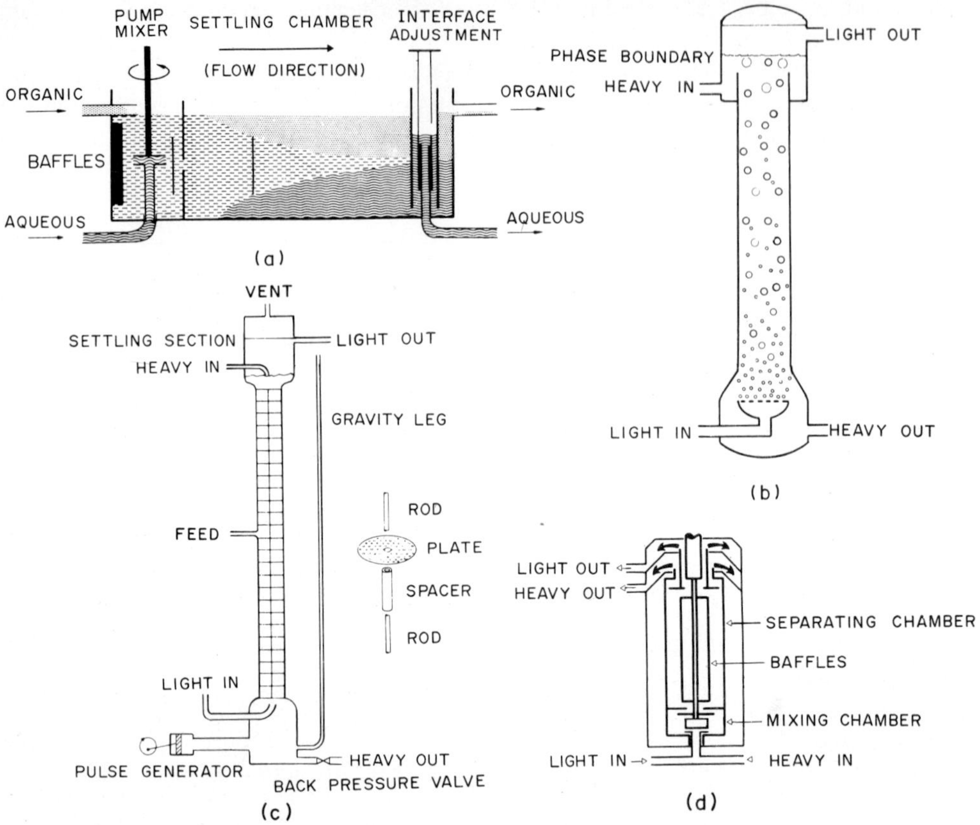

FIG. 20.9. Different types of solvent extraction equipment. (a) Mixer–settler unit for continuous extraction. The organic and aqueous phases flow by gravity and are pumped, respectively, into the mixing chamber. The mixture flows into the settling chamber, where the two phases separate by gravity and are withdrawn from top and bottom. A hydrostatic leg is used at the aqueous outflow to control the interface height in the settling chamber. (b) A spray column. The tower may be filled with packings, have baffles or sieve plates, etc., to improve solvent extraction efficiency. (c) Pulsed column with sieve plates. Phases separate in each section between the plates: suction draws the heavy phase down through the holes, while pressure from the pulse mechanism pushes the light phase up through the holes in the sieve plate. (d) A tubular liquid flow centrifugal extractor.

provided P is constant through all stages. In our example one obtains for $\theta = 0.5$ (ratio of flow organic : aqueous) and $n = 3$ that $E_U = 99.91\%$ and $E_{La} = 3.5\%$, thus about the same yield of uranium as before, but with a somewhat lower lanthanum impurity as compared to the crosscurrent extraction procedure. This impurity figure can be lowered by modification of the extraction process according to Fig. 20.11 so that the extraction battery contains n extraction stages with the extraction factor P_1, and $m - 1$ washing stages with the extraction factor P_2. Then

$$\varphi_{m,n} = \frac{(P_1 - 1)(P_2^m - 1)}{(P_1^{n+1} - 1)(P_2 - 1)P_2^{m-1} + (P_2^{m-1} - 1)(P_1 - 1)} \tag{20.6}$$

which in the case that $P_1 = P_2$ reduces to

$$\varphi_{m,n} = \frac{P^m - 1}{P^{m+n} - 1} \tag{20.7}$$

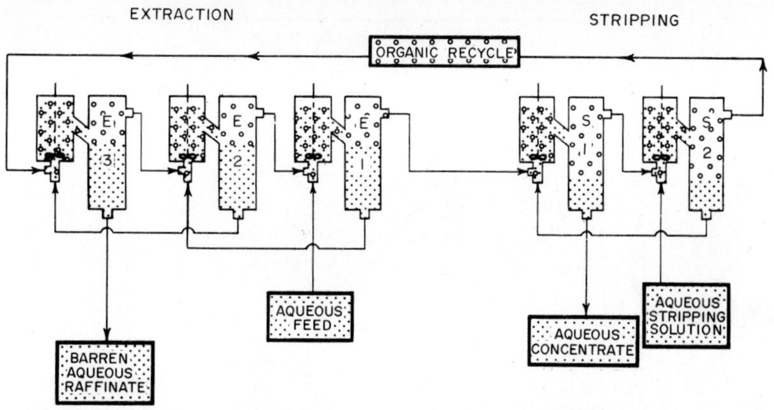

FIG. 20.10. Mixer–settler countercurrent solvent extraction battery. The value metal is extracted from the aqueous feed into the organic phase by the three extraction stages (E1–E3) and washed out into an aqueous stripping solution in two stripping stages (S1–S2).

In the latter case, using three extraction stages and one washing stage, we find for our example that E_U is 99.9% and E_{La} is 0.12%. Thus both high yield and high purity are achieved with the countercurrent, central or intermediate feed solvent extraction technique. This technique is extensively used in uranium production, nuclear fuel element reprocessing, and transuranium element separations. Because the conditions are selected so that a high ($\gg 1$) distribution factor is obtained for the desired product, and low ones ($\ll 1$) for the impurities (e.g. fission products), high purity and good yield are often obtained in relatively few stages ($m + n - 1 \lesssim 10$).

The purpose of the washing (or *scrubbing*) stages is to clean the desired metal from impurities, and should not be confused with the stripping, the purpose of which is to transfer the desired metal from the organic phase to a new aqueous phase. However, the stripping may be so arranged that further purification is achieved. If no additional purification is desired after extraction, the organic phase may be distilled, leaving the element as a pure solid compound. This technique can be used for hexone systems, but not for TBP, since the latter decomposes into phosphoric acid upon heating.

A further industrial requirement on solvent extraction processes is high capacity, which means high concentrations of solutes in the aqueous and organic solvents ("high loading"). The extraction factor P (or distribution factor D_M) may then vary from stage to stage. The calculation of the number of stages, flow ratios, etc., needed in order to obtain the desired product will then become very complicated. However, for one solute the problem is easily solved graphically with McCabe–Thiele diagrams (Fig. 20.12 and accompanying text).

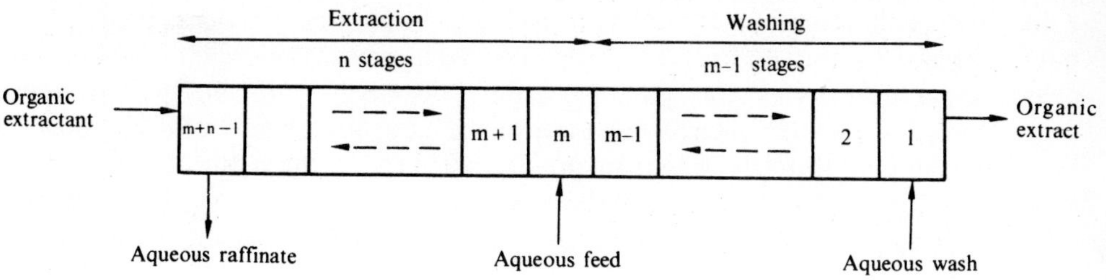

FIG. 20.11. Countercurrent solvent extraction with n extraction stages and m washing stages.

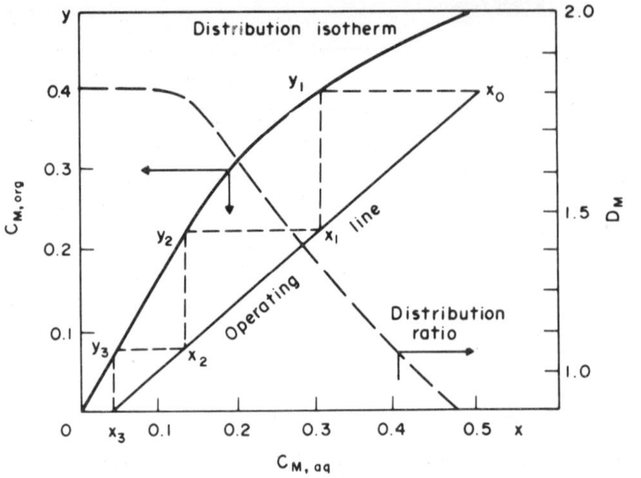

FIG. 20.12. Distribution isotherm and McCabe–Thiele diagram for a three-stage countercurrent extraction process. C_M in the feed is x_0. After equilibration and separation in the first mixer–settler stage, the outgoing organic extract contains $C_{M,org} = y_1$ and the raffinate $C_{M,aq} = x_1$. Because the slope of the operating line is equal to the flow rate ratio aqueous : organic (θ^{-1}), the extract and raffinate leaving the second stage will have the composition y_2 and x_2, respectively, and from the third stage y_3 and x_3. The extract yield is $(x_0 - x_3)/x_0 \approx 93\%$ in the example. D_M increases from 1.37 in the first stage to 1.80 in the third stage.

20.5.4. *Solvent extraction equipment*

The less conventional part—from a chemical engineering viewpoint—of a reprocessing plant is the solvent extraction equipment, even though the technique is becoming increasingly common in the chemical industry. The principle of all such equipment is illustrated by Fig. 20.9(a); it contains a mixing section for efficient transfer of materials between the phases and a settling section for efficient phase separation. The input and outputs provide for connecting stages in the countercurrent extraction scheme (Fig. 20.10). Mixer–settlers (Fig. 20.9(a)) provide good mixing and reasonably good phase separation performance but rather large hold-ups.

For reprocessing the equipment must be highly reliable, have high stage efficiency, short contact times, small liquid inventory (hold up), be easy to decontaminate and to service. High reliability usually means simple design and few (if any) moving parts. The *packed columns* meet this requirement. They are simply long columns (often 10–20 m with a diameter of 0.3–3 m) filled with small pieces of material obstructing a straight flow through the column, which is by gravity (Fig. 20.9(b)). However, high stage efficiency requires mechanical agitation of the two phases and clean phase separation, which cannot be met by packed columns.

In pulsed columns (Fig. 20.9(c)) the mechanical agitation provides good mixing but poor phase separation. Each plate is perforated (a sieve). The organic and aqueous phases separate between the plates ("settling chambers"). In the pulsed down movement ($< 1/4$ of the interplate distance) the aqueous phase is forced through the sieves, forming droplets, which by gravity fall through the lighter organic phase and coalesce when reaching the interface boundary. In the upward stroke, organic droplets form and rise through the aqueous phase until they meet the organic phase boundary.

The phase separation in a mixer–settler battery or in a column is usually not better than ~98%, i.e. each outgoing phase contains a few percent entrained droplets of the other phase. This separation efficiency can be improved to almost 100% by using centrifugal extractors (Fig. 20.9(d)). Centrifugal extractors effect good mixing, good phase separation, and have very small hold-ups. The organic–aqueous phase contact time in centrifugal extractors can be made much shorter than in mixer–settlers or columns, with the advantage that radiation decomposition is minimized.

Packed columns were used in the first Windscale plant (UK), pulsed columns in Hanford and NFS (USA), and Eurochemic plant at Mol (Belgium), while mixer–settlers are used in Savannah River (USA), the second Windscale plant, and at La Hague (France). Centrifugal extractors have been installed at Savannah River, La Hague, and are planned for newer plants.

20.5.5. *Purex separation scheme*

The distribution of uranium, plutonium, and some FPs between 30% TBP (in kerosene) and aqueous solutions of varying HNO_3 concentration is shown in Fig. 13.9. D_{Sr} and D_{Cs} are $\ll 0.01$ for all HNO_3 concentrations. At high HNO_3 concentration Pu(IV), Pu(VI), and U(VI) are extracted but very little of the FPs. At low HNO_3 concentration the D-value for actinides of all valency states is $\ll 1$, and consequently the tetra- and hexavalent actinides are stripped from the organic phase by dilute HNO_3.

A typical Purex flow-sheet is shown in Fig. 20.13. In the partitioning cycle > 99.5% U and Pu (in +4 state) are coextracted from ~6 M HNO_3 into the kerosene–TBP phase of the first column, leaving ~99% of the FP in the aqueous raffinate. In the partitioning column plutonium is reduced to the +3 state by a solution containing Fe(II) sulfamate; the plutonium is stripped to a new aqueous phase and transferred to the plutonium purification section. Uranium, which as U(VI) stays in the organic phase, is stripped by dilute HNO_3 in a third column. After concentration by evaporation it is sent to the uranium purification section.

The uranium purification contains two extraction-stripping column pairs. Plutonium tracers are removed by using a reducing agent, preferably U(+4):

$$U^{4+} + 2Pu^{4+} + 2H_2O \rightarrow UO_2^{2+} + 2Pu^{3+} + 4H^+$$

U(+4) is used as reducing agent instead of Fe(II) sulfamate in newer plants as it has the advantage that it introduces no foreign substances and can be used as a general reducing agent for oxidized plutonium; U(+4) is produced through electrolytic reduction of U(VI) in $HClO_4$ or H_2SO_4. More than 95% of the FP and plutonium entering this section is removed in the two extraction cycles. The final, concentrated uranium solution is percolated through a column filled with silica gel, which removes residual FP, particularly Zr–Nb.

The plutonium purification cycle contains a pair of extraction–stripping columns. Plutonium is oxidized by NO_2 and reduced by ferrous sulfamate, as indicated in the figure. The final plutonium purification may be achieved by additional TBP extraction cycles (which also remove the uranium), or by alternative processes. Solid cation and anion exchangers have been used (e.g. Fig. 20.13). The final uranium and plutonium products are nitrate solutions. Their transfer into oxides, fluorides, etc., have been presented in previous chapters.

Radiolysis of both of the aqueous and organic phases occurs, but the effects can be more easily controlled in the aqueous phase. Radiation decomposes TBP into lower

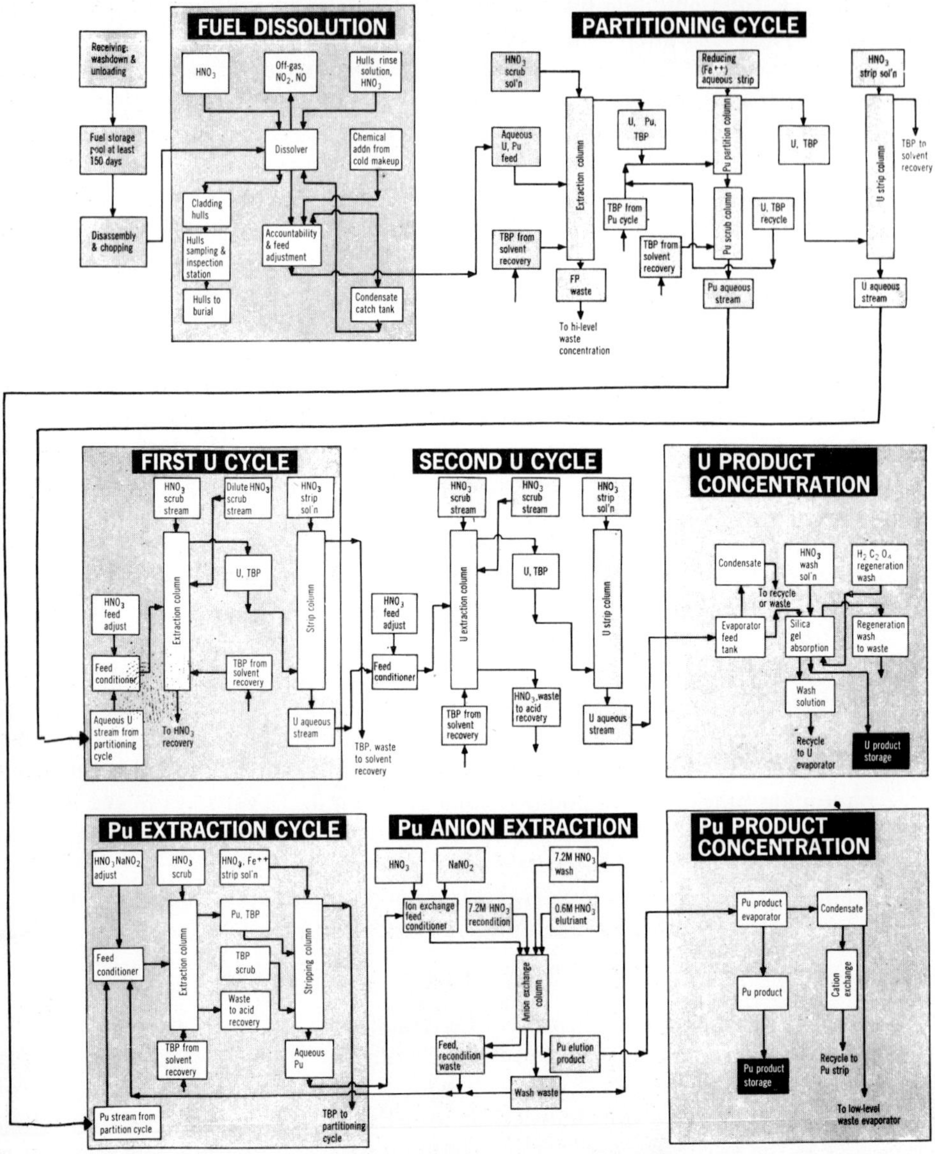

FIG. 20.13. Purex flowsheet used at Nuclear Fuel Services, West Valley, USA. Though the plant is closed down the flowsheet is very representative.

phosphates and butyl alcohol. The main products are dibutyl phosphate (DBP, $(BuO)_2POOH$) and monobutyl phosphate (MBP, $BuOPO(OH)_2$) which form strong complexes with many of the fission products as well as plutonium. As these radiolysis products are formed the decontamination efficiency decreases and losses of fissile material to the aqueous waste streams increase. The solvent is treated to remove the degradation products prior to recycle in the process, e.g. by washing the TBP solution successively with Na_2CO_3, NaOH, and dilute acid solutions.

During the decontamination steps, acid streams containing small amounts of actinides and fission products are produced. These streams are evaporated to concentrate the metal ions and recycle them. Nitric acid is recovered from the condensates and recycled. Excess HNO_3 may be destroyed by formaldehyde. Fission product concentrates are routed to the aqueous raffinate of the first extractor of the partitioning cycle which contains $> 99\%$ of the FP. This constitutes the high level liquid waste (HLLW, or alternatively called HAW, high active waste). All other liquid wastes can be subdivided into intermediate level waste (ILW or MLW) or low level waste (LLW). Waste treatment is discussed in §20.6.

20.5.6. *Engineering aspects and operation safety*

All operations in a reprocessing plant have to cope with the necessity of preventing nuclear criticality and of protecting operations personnel and the environment from exposure to or contamination by radioactivity. Thus all equipment has dimensions which are safe against criticality, as, for example, annular tanks for liquid storage instead of conventional tanks, or are provided with neutron absorbers. All equipment is made of stainless steel and is installed in concrete cells with wall thicknesses up to 2 m at the head end, partitioning, and waste treatment sections. All operations are carried out in airtight enclosures at reduced pressure relative to working areas.

In the event of failure of equipment within a radioactive area, three courses of action may be taken: (i) replacing equipment by remote methods, (ii) switching to duplicate equipment, or (iii) repairing by direct maintenance after decontamination.

Plants were originally constructed for either completely remote maintenance or completely direct maintenance. The plants at Hanford, USA, were designed for remote maintenance (Fig. 20.14). All equipment is installed in a large canyon with piping in a

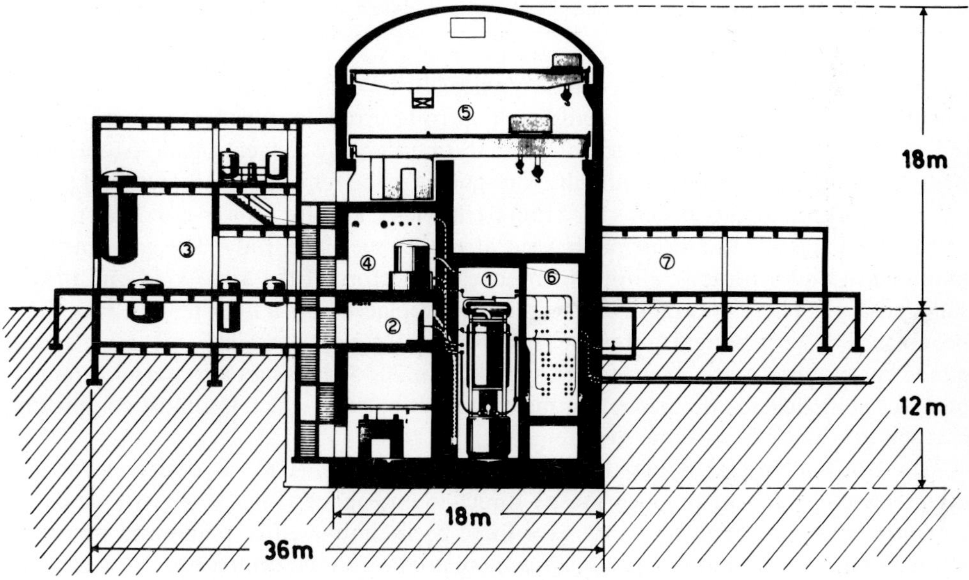

FIG. 20.14. Cross-section of the original reprocessing plant at Hanford, USA. 1, Canyon with process equipment; 2, control room; 3, feed preparations; 4, feed input; 5, cranes for remote servicing of canyon; 6, piping corridor; 7, analytical section.

FIG. 20.15. Large mixer–settler battery used in the Purex process for natural uranium fuels before installation in remotely maintained canyon at the Savannah River Plant, USA.

parallel corridor. The equipment can be replaced by an overhead crane operated from a shielded room; malfunctioning equipment is transferred to a decontamination and repair unit or to an equipment "grave". With this philosophy conventional type chemical plant equipment can be used, though redesigned for the remote replacement. Of particular importance are good joints for piping, electricity, etc., for remote connection.

A modification of this principle is used at the Eurex plant, Italy. The equipment is mounted in uniform racks, all of which have connectors of the same kind and in the same position. These unit racks can be remotely replaced: e.g. a filtering unit rack may be replaced by a centrifugal unit rack. This provides for great flexibility and easy maintenance but requires more engineering work. Figure 20.15 shows an alternative arrangement for complicated equipment of a mixer–settler battery, which is almost completely encased in a block of concrete shielding.

The British Nuclear Fuels Ltd. plant at Windscale, UK, is designed for no maintenance, which in practice means direct maintenance (Fig. 20.16). All kinds of common failure in chemical plants have there been eliminated or minimized by using welded joints and no moving parts. Thus there are no leaking fittings, frozen valves, or stuck pumps. All welding is carefully controlled by ultra-sound or radiography. Gravity flow is preferred in essential units. Active liquids are transported by gas (usually steam) jets or lifts, or by vacuum. Liquid levels and volumes are measured by differential

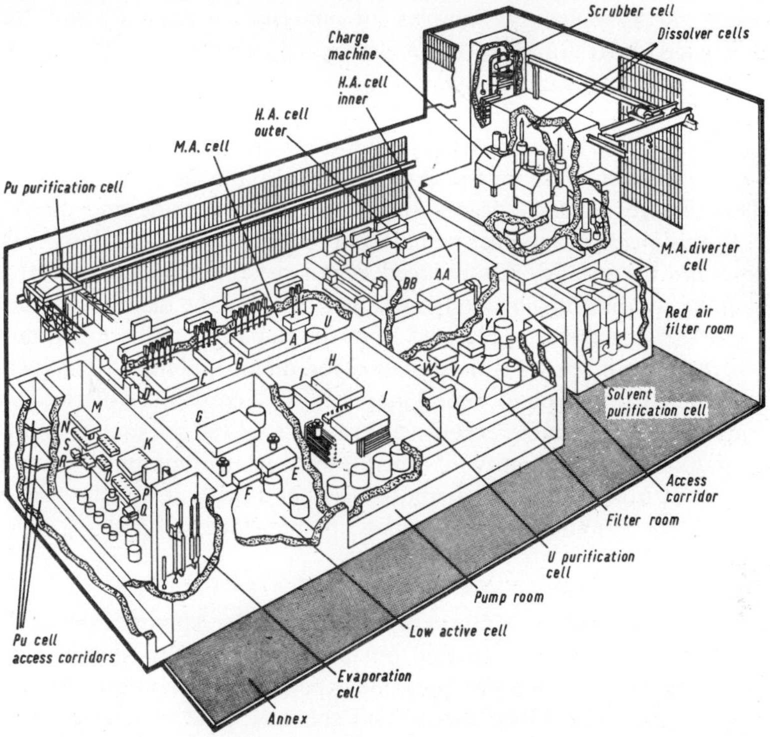

(a)

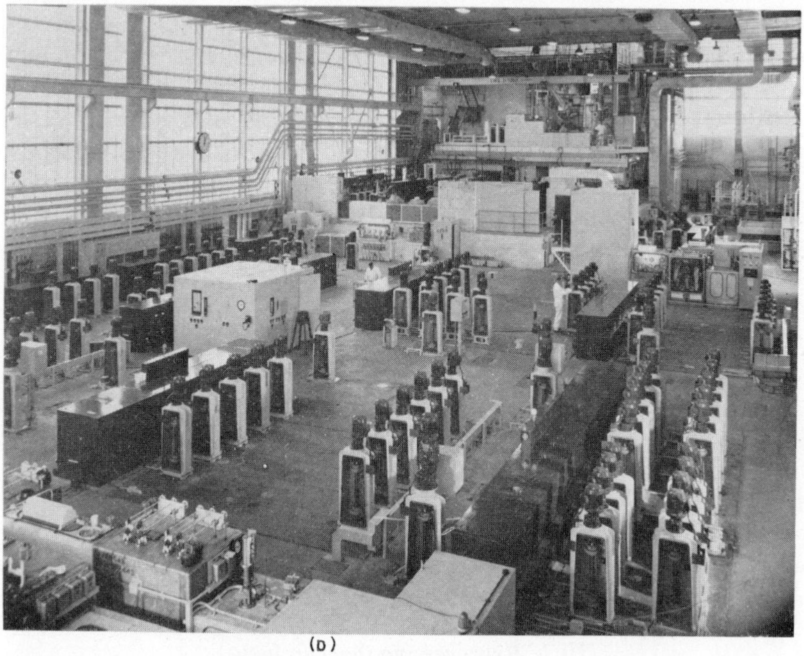

(b)

FIG. 20.16. (a) General layout of BNFL (British Nuclear Fuels Ltd.) reprocessing plant at Windscale, UK. (b) Cell top of primary separation plant showing stirrer motors through the floor from the mixer–settler batteries below the floor.

pressure gages and by weighing. Samples for analysis are remotely withdrawn and analyzed in shielded boxes or glove boxes depending on their activity outside the enclosure. Future, more advanced plants will probably use more on-line analysis techniques.

Even such systems may fail, and dual equipment is therefore installed in parallel cells. With the process running on the spare equipment, the failing equipment must be repaired. This requires efficient decontamination both on inside and outside of the equipment. Provisions for this must be incorporated in the original design. When the radiation level has been sufficiently reduced, repair personnel may start cutting down and replacing the equipment. This design requires dividing the plant in a large number of cells so that the workers are protected from the radiation of the functioning plant. However, there is a risk that plants designed for direct maintenance (Nuclear Fuel Services, USA, and Tarapur plant, India) may lead to overexposure of personnel.

Remote maintenance is more expensive, but safer to the personnel and desirable from the standpoint of continuity of operations because equipment replacements can be carried out quickly and interruption of operation is relatively brief. Modern robot technology successfully simplifies remote maintenance.

20.6. Reprocessing fuels from fast breeder reactors

No full-scale breeder fuel cycle reprocessing plant exists. Though extensive research and development work has been carried out in many countries (notably in the UK and the USA) during the last decades, the main experience in the near future is expected to come from France, the FRG, Japan, and the UK. The breeder reactor system yields a small excess of fissile atoms, in the range 10–30%. A growing breeder reactor power system is likely to face shortage in fissile material. There will therefore be a strong incentive to use the fuel in the reactor as efficiently as possible (i.e. high burn-up) and to reduce the fissile inventory in the rest of the fuel cycle, i.e. to use short cooling times, and rapid reprocessing and fabrication of new fuel elements. The high burn-up of fissile material leads to a high fission product content in the core fuel elements, resulting in the formation of seminoble metal fission product alloys, which are insoluble in boiling nitric acid. The insoluble material consists of 1–15 mm sized metal particles of Ru, Rh, Tc, Mo, and Pd. Presently these metal particles contain negligible amounts of uranium and plutonium and can therefore be filtered off as HLSW; should it later be found that they must be reprocessed, they can be dissolved in a mixture of HNO_3 and HF.

Presently a cooling time of about 6 months is considered. This will yield a heat emission of 1 MW/tonne spent fuel element, and a γ-radiation level of $\sim 2 \times 10^{17}$ γ-MeV $s^{-1} t^{-1}$ fuel (the corresponding figure for LWR spent fuel is 4×10^{15} γ-MeV $s^{-1} t^{-1}$ after t_{cool} 5 y). By mixing dissolutions of high burn-up core elements with low burn-up blanket elements the same Purex cycle as for LWR elements can be used. The high radiation levels will require improved reprocessing technique (e.g. shorter contact times) and additional precautions against accidental radiation exposures. The actinide content will be higher, which requires additional precautions against accidental criticality excursions. The same techniques can be used for waste treatment, and the releases to the environment will be about the same as for reprocessing LWR fuels.

20.7. Reprocessing thorium fuels

The thorium-containing fuels of present interest are those of the kind used in HTGR; in the future thorium-containing fuels from HWR or MSR may become important.

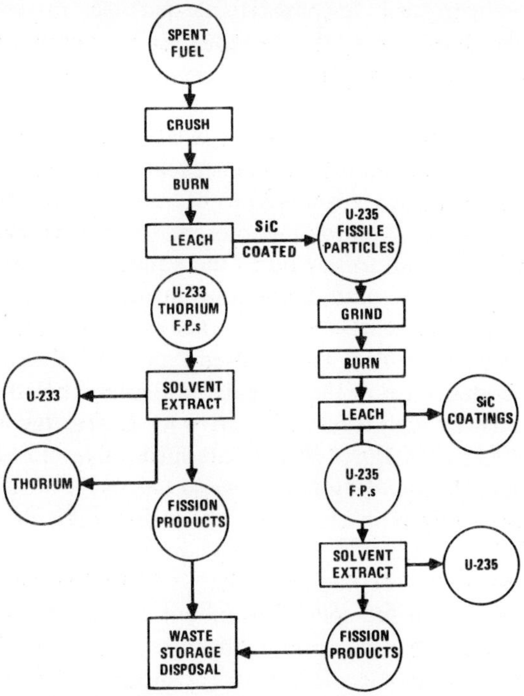

FIG. 20.17. Block diagram for reprocessing high temperature gas-cooled reactor (HTGR) fuel.

In the HTGR fuel elements the fertile ThO_2 and fissile $^{235}UO_2$ (or $^{233}UO_2$) particles are separately embedded in a graphite matrix (§§19.11 and 19.15.1). Figure 20.17 shows the various steps of HTGR fuel reprocessing.

The spent graphite fuel elements are mechanically crushed and burned to eliminate the graphite matrix and pyrolytic carbon coating from the fuel particles. Leaching permits separation of the fissile and the fertile particles because the fissile particles have a silicon carbide coating which remains intact during burning and leaching, while the all-pyrolytic carbon coatings on the fertile particles are burned away, allowing the oxide ash to be dissolved by a leach solution, consisting of HNO_3 and F^-. The solution is clarified and adjusted to proper acidity for solvent extraction (paragraph (i) below). The undissolved residue resulting from clarification is dried and classified for further treatment (paragraph (ii) below). The burner off-gas streams are passed through several stages of filtration, scrubbing, and chemical reaction to remove the entrained and volatile fission products prior to atmospheric discharge.

(i) The *acid Thorex solvent extraction process* is used to purify and to separate the U-233 and the thorium. Three solvent extraction cycles are used. In the first, the uranium and thorium are coextracted by 30% tributyl phosphate (TBP) from ~ 5 M HNO_3 (cf. Fig. 13.9) and then stripped into an aqueous phase. In the second cycle, the uranium and thorium are separated by controlling the extraction conditions using $\lesssim 1$ M HNO_3. The uranium is extracted and processed by an additional solvent extraction cycle for final purification, while the thorium remains in the aqueous raffinate stream. Following concentration and assay, the uranium is ready for fuel refabrication. The partially decontaminated thorium is concentrated and stored.

(ii) The separated silicon–carbide-coated fissile particles in the head end process are mechanically crushed to expose the fuel and are burned to remove carbon and oxidize the fuel material; the ash is leached to separate the fuel and fission products from the coating hulls. The ^{235}U is then separated from the fission products by solvent extraction using a Purex flow-sheet. A 3–5% TBP in an inert diluent is used to decontaminate and purify the uranium by two cycles of extraction and stripping. A reductant such as ferrous sulfamate is added to the extraction contactor scrub stream to reduce the small amount of plutonium present to the inextractable +3 valence state. This causes the plutonium to remain in the aqueous raffinate stream. After two cycles of solvent extraction, the uranium is concentrated by evaporation and passed through a silica gel column to remove the last traces of fission products.

Various waste streams arise from the head end and solvent extraction operations. Most of the fission products are in the aqueous raffinates from the first cycle solvent extraction contactor. These high level radioactive waste streams are concentrated by evaporation and treated as the Purex waste. Intermediate level radioactive wastes are reduced to small volume by evaporation and routed to storage tanks.

This process is plagued by both chemical and nuclear difficulties. The decay chain $^{233}Th \rightarrow {}^{233}U$ forms 27 d half-life ^{233}Pa. For a complete decay of all ^{233}Pa to ^{233}U, the spent fuel elements must be cooled for about a year. A still considerable amount of long-lived ^{231}Pa is present in the spent fuel (about 1/2000 of the amount of ^{233}U); protactinium complicates the reprocessing chemistry and constitutes an important waste hazard.

The ^{233}U isolated will contain some ^{232}U formed through reactions indicated in Fig. 19.17. Since the half-life of ^{232}U is rather short and its decay products even more short lived, a considerable γ-activity will grow in with the ^{233}U stock, considerably complicating its handling. Since the first decay product is ^{228}Th, some amounts of thorium will form in the fuel elements, also making the ^{232}Th contaminated by this isotope. Thus neither the ^{233}U nor the ^{232}Th produced will be free from γ-activity, however good the original separation from fission products is.

20.8. Wastes from reprocessing

Fuel reprocessing generates a large variety of wastes which can be classified in categories according to the diagram in Fig. 20.18. The amount of each category is given in Table 20.8 for a 1500 t U y^{-1} plant. These wastes are treated internally at the plant and not released to the environment. Because reprocessing plants vary considerably the amounts of wastes produced also differs, especially for the liquid medium and low level categories. In this section we shall briefly discuss the various waste streams appearing at the plant in the treatment methods.

20.8.1. *Gaseous wastes*

Gaseous wastes originate mainly from the chopping and dissolution operations. In current practice the volatile radionuclides are discharged to the stack after scrubbing with sodium hydroxide. The hydroxide scrubbing removes the acidic nitrous oxides which pass through the recombination unit above the dissolver.

The most hazardous volatile constituents are the iodine and ruthenium fission products. Though more than 95% of the iodine is volatilized in the dissolver (as I_2, HI and

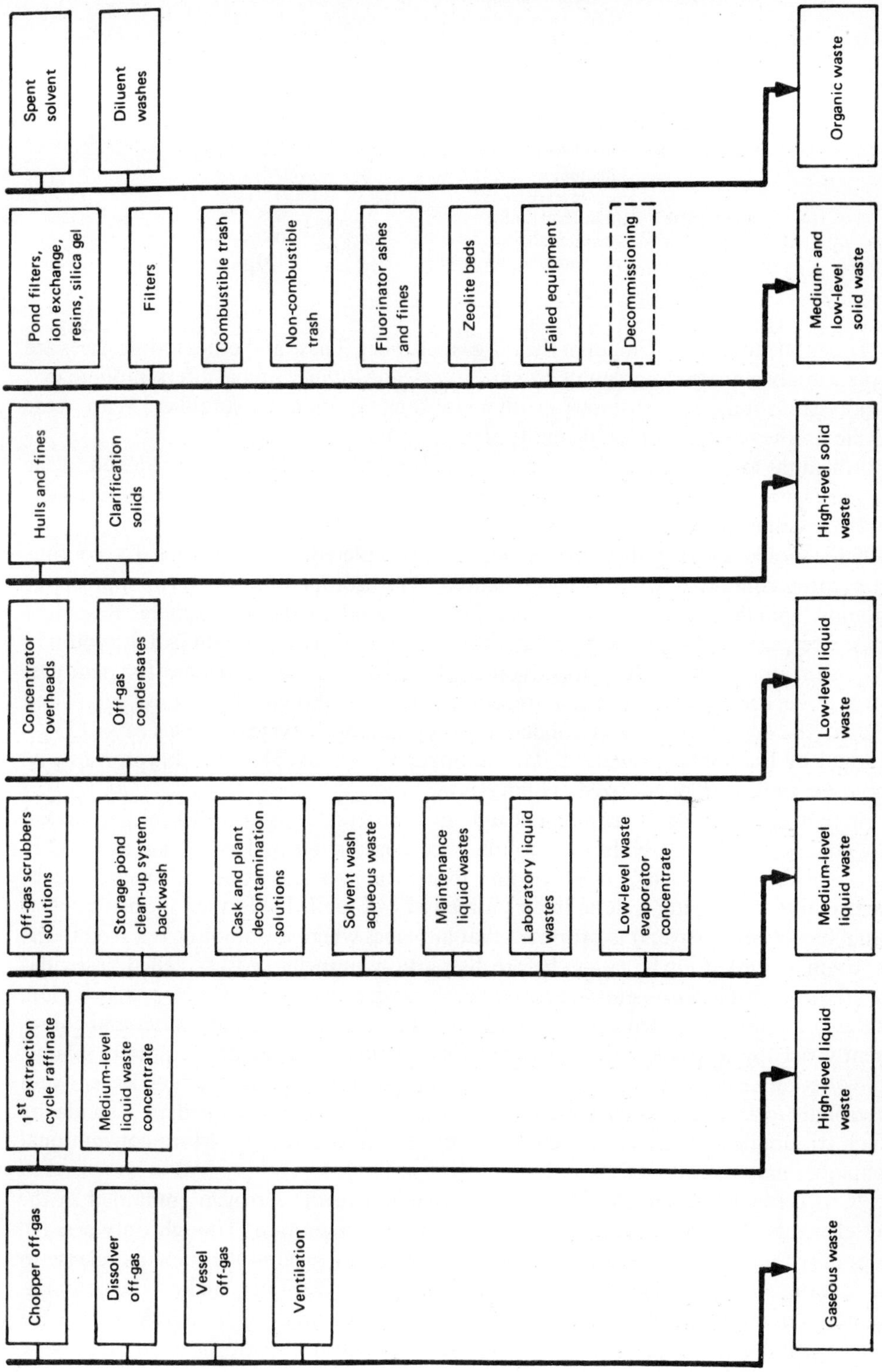

FIG. 20.18. Origins of various categories of waste at a reprocessing plant. (According to IAEA, 1977.)

TABLE 20.8. *Amounts of various wastes arising at a* 1500 t U/y *reprocessing plant; cooling time* ~ 3 y *(according to IAEA,* 1977)

Gaseous waste:	^{3}T	0.9 MCi
	^{14}C	1 000 Ci
	^{85}Kr	13 MCi
	^{129}I	60 Ci
Liquid waste:	high level (HLLW), conc.	700 $m^3 y^{-1}$
	medium level (MLLW)	$\lesssim$7 000 $m^3 y^{-1}$
	low level (LLLW)	150 000 $m^3 y^{-1}$
Solid waste:	high level (hulls)	600 $m^3 y^{-1}$
	medium and low level β, γ	3 000 $m^3 y^{-1}$
	medium and low level α	100 $m^3 y^{-1}$
Organic waste (liquid)		900 $m^3 y^{-1}$

HIO mainly) most of it is caught in the off-gas scrubber. These and other iodine gases are trapped in silver nitrate impregnated zeolite filters, and/or in charcoal filters impregnated by potassium iodide; both types of filter are located after the scrubber. With these techniques the retention of iodine in the plant is > 99.5%.

Ruthenium forms volatile RuO_4 in the dissolver. Almost all RuO_4 is retained in the gas purification system described. As an additional feature some plants use a steel wool filter (after the nitric acid recombination) to catche the RuO_4.

Of the noble gases, radioactive xenon has completely decayed after 1 y cooling, but krypton contains ^{85}Kr with 10.7 y half-life. This isotope is produced in appreciable amounts, and though commonly it had been released to the atmosphere, this is no longer acceptable. Many processes have been devised for krypton removal. Krypton in dry, clean air is effectively trapped on a charcoal filter at cryogenic temperature; however, because of explosion risk (reaction between radiolytically formed ozone and carbon), the favored process is condensation by liquid N_2 (krypton boils at − 153°C) followed by fractional distillation. This removes $\gtrsim$99% of ^{85}Kr. The krypton can be stored for cooling in pressurized cylinders.

Though the amount of tritium released in reprocessing is considerable, it caused little concern until recently. In the chopping section it is released as T_2, but as HTO in the dissolver, where > 90% of all tritium formed is present. While T_2 can be caught (particularly if chopping is done in air), a recovery of the HTO formed in the dissolver would be very expensive. It is proposed that at places where it cannot be released to the environment the tritium be caught before dissolution. In such a process, called Voloxidation, the chopped fuel elements are treated with oxygen at 450–700°C before dissolution. Tritiated water is generated which should be relatively free of ordinary water and consequently occupy a much smaller volume than tritiated wastes do in present plants. Voloxidation may collect ~ 99% of the tritium present in unprocessed fuel. Recently an electrolytic–catalytic isotope enrichment process has been developed in Canada by which tritium can be selectively removed from aqueous solutions. More conventional techniques may recover $\lesssim$90%.

^{14}C is formed through the $^{14}N(n, p)^{14}C$ reaction in the nitrogen contained in the fuel elements. It is released mainly as CO_2 at the dissolution. Though only a small amount is formed (~ 1 Ci per GWe y per each ppm nitrogen in the fuel), its release to the environment, estimated as ~ 0.1 MCi y^{-1} by the year 2000, may make it the dominating dose commitment of the fuel cycle back end. Presently $^{14}CO_2$ is released to the atmosphere, but techniques for its retention are available. It will probably finally be caught as $CaCO_3$, with a retention of ~ 80%.

To summarize, a 1500t U/y plant reprocessing LWR fuels is assumed to release the following approximate amounts of volatile gases to the environment: T ~ 10^5, ^{14}C ~ 10^2, ^{85}Kr ~ 10^6, and ^{131}I ~ 0.2 Ci y^{-1}. The retention factors cited are used in this estimate.

20.8.2. *Liquid wastes*

The high level liquid Purex waste (HLLW) contains > 99.5% of the FP in ~ 1 M HNO_3. Before it is pumped to the storage tanks it is concentrated to various final volumes depending on the cooling time and burn-up of the fuel. For 1 y cooling time the HLLW may be concentrated to only ~ 1 m^3/t U reprocessed (concentration ~ 35 g FP/l), while for a 10 y cooling time it may be ~ 0.2 m^3/t U (~ 180 g FP/l).

The medium level liquid waste (MLLW) results essentially from evaporating various streams from the chemical process, such as solvent clean-up, off-gas scrubbers, product concentration, etc. It may contain up to 0.2% of the uranium and plutonium processed. The radioactivity is usually < 1 Ci l^{-1} (an average value is 0.1 Ci l^{-1}). The solutions may also contain appreciable amounts of solids (e.g. $NaNO_3$, iron, etc.). The waste may be neutralized and is stored in steel tanks at the reprocessing site.

Liquid waste is generated in numerous places with activities < 0.01 Ci m^{-3}; such waste is classified as low level. Some of these liquids may be clean enough to be released directly into the environment; e.g. at Windscale reprocessing plant, ~ 180 000 Ci β (61% of permitted amount) and ~ 1600 Ci α (27% of permitted) were released into the Irish Sea in 1976. Others are cleaned by flocculation, ion exchange, sorption, and similar processes; the liquid then can be disposed of or recirculated in the plant. The general philosophy for liquid wastes is to concentrate all radioactivity to the next higher level because the waste volumes decrease in the order LLLW > MLLW > HLLW. Thus, in principle, the three kinds of wastes are reduced to two (HLLW and MLLW) and a cleaned aqueous effluent. The MLLW and residues from LLLW cleaning are treated as the wastes of the nuclear power stations, i.e. concentrated and put into concrete or bitumen (see §19.18). At coastal sites it has been the practice to release the LLLW to the sea. The nuclides of main concern are ^{90}Sr, ^{137}Cs, ^{106}Ru, and T, and the actinides.

At Windscale > 100kg Pu has been released into the Irish Sea. However, > 95% precipitates immediately as hydroxide, while the rest stays dissolved as organic and inorganic complexes. The liquid plutonium concentrations measured are < 10^{-9} M. No long-term build-up of plutonium occurs in animals. Marine species at the Bikini atoll (where large amounts of plutonium have been released in underwater detonations of nuclear weapons) receive much less radiation doses from plutonium than from natural polonium.

20.8.3. *Solid wastes*

High level solid waste (HLSW) originates at the dissolver. The hulls from the dissolution contain activation products and small amounts of undissolved fuel ($\lesssim$ 0.1%). Though not yet the practice, they may be compacted by a factor of ~ 5. The dissolver solution contains finely divided particles of undissolved seminoble metal alloys (Ru, Rh, Mo, Pd, etc.). This suspension is treated by filtering or centrifugation prior to the solvent extraction. Past practice in the USA and the USSR has been to put the HLSW in shielded containers which are transported to and stored at a dry disposal site. In the future the same disposal is expected to be used for the HLSW as for the solidified HLLW (§20.12).

Medium and low level solid wastes are produced at numerous places (Fig. 20.18).

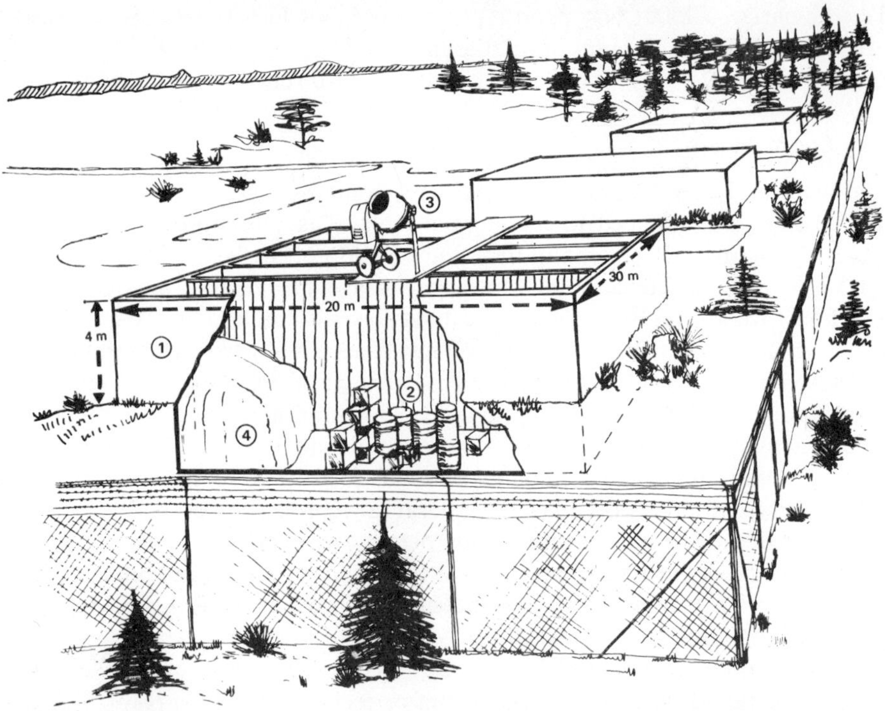

FIG. 20.19. Storage of low and medium solid waste at Zagorsk, USSR. 1, Concrete wall; 2, waste material; 3, concrete mixer; 4, concrete poured over waste material.

They are divided into combustible and noncombustible, and in α-bearing (in the USA "α-bearing wastes" is classified as such containing $> 10\,\text{nCi}\alpha\,\text{g}^{-1}$ material) and non-α-bearing fractions, and treated independently, when possible, to reduce volume. The wastes are then fixed in bitumen or concrete. In some places they are simply placed in drums and stored in a dry nonaccessible area (Fig. 20.19). This is further discussed in §20.12.

20.8.4. *Organic wastes*

The liquid organic waste consists of spent TBP diluent mixtures originating from the organic solvent clean-up circuits and from the diluent (kerosene) washings of the aqueous streams (to remove entrained solvent); in addition to degradation products of TBP and the diluent it contains small amounts of actinides (mainly U and Pu) and FP (mainly Ru, Zr, and Nb).

This type of waste is disposed of by incineration or decomposed by hydrolyses and pyrolyses leading to the formation of inactive hydrocarbons, which are distilled off, and active phosphoric acid, which is treated together with other aqueous wastes.

20.9. Long term storage of low and medium level wastes

It has become a common practice at reprocessing plants to store low level solid waste (and sometimes higher level wastes) in trenches dug from the soil. These trenches, which commonly are 5–8 m deep, are sometimes lined with concrete or simply have a gravel bottom. For this purpose dry areas are selected (deserts, when available) or

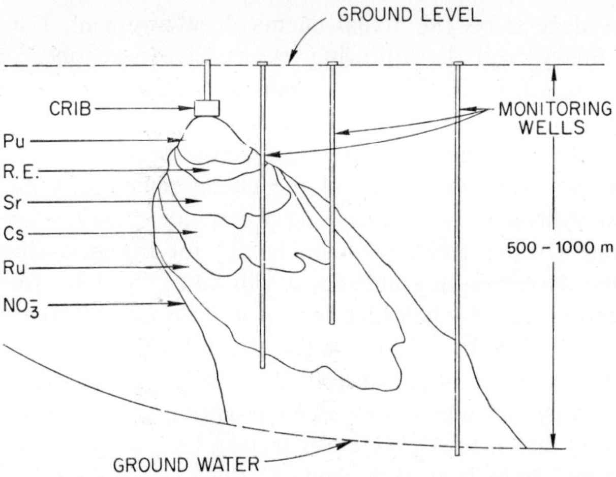

FIG. 20.20. Distribution of radioactive waste products leaking into the soil at Hanford, USA. The soil consists of ion exchanging clay covered by sandy gravel.

isolated areas with controlled groundwater conditions with respect to water table depth, flow rate, and direction. The disposed material should normally not exceed a dose rate reading of 100 mrem h^{-1} at 0.3 m distance, or contain a specific activity exceeding 10 μCi kg^{-1}. However, this varies, and may be a factor of 10 higher or lower in some places. When the trench is full, it is backfilled with ~ 3 m of earth, after which the surface dose rate usually is < 1 mrem h^{-1}. Trenches of this kind are used in the USA, the UK, France, etc., where tens of thousands of cubic meters are disposed of annually.

Since some of the waste nuclides are long lived, and the physical protection of the waste is poor, radioactivity will ultimately begin to leak into the groundwater before the radioactivity has died out. The ion exchange properties of the soil will retain the radionuclides to a varying degree, depending on the chemistry of the nuclide and composition of soil and groundwater (Fig. 20.20). The radionuclide retention factor (RF) is defined as

$$\mathrm{RF} = \frac{v_w}{v_n} = 1 + k_d \frac{\rho(1-\varepsilon)}{\varepsilon} \qquad (20.8)$$

where v_w is the groundwater velocity, v_n the nuclide transport velocity (which must be $\lesssim v_w$), ρ the soil density, ε the soil porosity (or void fraction), and k_d the nuclide distribution coefficient defined by

$$k_d = \frac{\text{conc. radionuclide per kg soil}}{\text{conc. radionuclide per m}^3\text{ water}} \qquad (20.9)$$

The groundwater velocity is calculated from Darcy's law

$$v_w = k_p\, i/\varepsilon \qquad (20.10)$$

where k_p is the soil permeability and i is the hydrostatic gradient (the height difference between the groundwater surface considered and the recipient surface water level, divided by the distance). Typical hydrologic soil values are $k_p = 10^{-5}\,\mathrm{m\,s^{-1}}$, $i = 0.05$ and $\varepsilon = 0.5$, which yields a groundwater velocity of $10^{-6}\,\mathrm{m\,s^{-1}}$, or ~ 30 m y^{-1}.

The k_d values depend on the actual chemical environment. For some important nuclides in "ordinary" soil the following values have been observed: ^{60}Co ~0.04, ^{90}Sr ~0.02, ^{137}Cs ~0.3, and ^{239}Pu ~1 $m^3\,kg^{-1}$; though these values are typical, they are nevertheless site specific. When these values are used in (20.8) with $\rho = 1500$ kg m^{-3} and $\varepsilon = 0.5$, the corresponding retention factors become ~60, ~30, ~450, and ~1500, respectively. Thus the order of increasing retention is Sr < Co < Cs < Pu. With the typical hydrologic soil values and a distance of 300 m between the leaking storage place and a water recipient, it takes 10 y for the groundwater to reach the recipient. For the corresponding nuclides it will take ^{90}Sr 300 y (most of it has then decayed, considering the 28 y half-life), 600 y for ^{60}Co (all decayed), 4500 y for ^{137}Cs (all decayed), and 15 000 y for ^{239}Pu (less than half of it decayed).

These examples show the importance of knowing ground conditions (geologic and hydrologic) at the disposal site, as well as the retention factors which can be expected. In some cases very much lower retention factors have been observed, but then the "normal conditions" have been disturbed. At the Maxey Flats facility in the USA, radionuclides were found to move from a storage basin much more rapidly through soil than would be expected from (20.8) and (20.9). This was caused by the addition of strong metal complex formers like chelates and nitrilotriaminopentaacetic acid (NTPA) to the basin liquid, leading to the formation of metal complexes of lower or no charge, which complexes have lower k_d values. Thus the chemistry of the waste storage is required information for safe storage.

20.10. Tank storage of high level liquid wastes

The main part of the HLLW is aqueous raffinate from the Purex "partioning cycle". It contains ~99.9% of the nonvolatile FPs and ≲0.5% of the uranium and plutonium. It also contains some corrosion products. For each tonne of uranium reprocessed about 5 m^3 of HLLW is produced. This is usually concentrated to 0.5–1 m^3 for interim storage; specific activity is in the range 10^5–10^6 Ci m^{-3}. The amounts of various elements in the waste and their concentration in 0.5 m^3 solution is shown in Table 20.9. The HNO_3 concentration may vary within a factor of 2 depending on the concentration procedure. The metal salt concentration is ~0.5 M; it is not possible to keep the salt in solution except at high acidity. The amounts of corrosion products, phosphate, and gadolinium (or other neutron poison added) also may vary considerably.

Wastes from the HTGR and FBR cycles are expected to be rather similar, though practical experience is quite limited. The HTGR waste volume is estimated to be about 4 m^3 HLLW/t heavy metal, containing ~5 kg fuel losses (~4 kg Th), 1 kg transuranium elements, and 80 kg FPs. The FBR waste volume is estimated as ~1.2 m^3 HLLW/t heavy metal (mixed core and blanket); because of some dissolution of the stainless steel cladding, the waste will contain considerable amounts of iron and chromium (~30 kg m^{-3}).

Presently most HLLW is stored in stainless steel tanks of 50–500 m^3 volume at the reprocessing plants. The tanks are rather elaborate (Fig. 20.21); they all contain arrangements for cooling (≲65°C to reduce corrosion) and stirring, removal of radiolytic gases, and for control of liquid level, pH, and radioactivity. The tanks are usually double-walled, have heavy concrete shielding, and are often placed underground. The tank farm is provided with spare tanks.

Storage in stainless steel tanks has been used in the last 20 y without failure. More

TABLE 20.9. *Composition of high level liquid waste (HLLW) from Purex plant reprocessing 1 ton of LWR fuel of 33 000 MWd/t U burn-up at a rating of 30 MW/t U; at t = 10y*
The original waste volume is ~ 5 m^3/t U reprocessed, which is converted to 0.5 m^3/tU

Component		Weight (g) in original waste volume	Approx. molar concentration in 0.5 m^3
H^+		1400	~ 1.0 M
NO_3^-		900 000	~ 2.4
Fission products:	Group I (Rb, Cs)	2 660	0.045
	Group II (Sr, Ba)	2 590	0.045
	Group III (Y, Ln)	10 660	0.140
	Zr	3 780	0.080
	Mo	3 470	0.070
	Tc	840	0.017
	Group VIII (Ru, Rh, Pd)	3 700	0.070
	Te	560	0.009
	Others	540	0.004
	Total fission products	28 800	0.484
Corrosion products:	Fe	1 100	0.040
	Cr	200	0.008
	Ni	100	0.003
	Total corrosion products	1 400	0.051
Phosphate (from TBP)		900	0.020
Actinides:	U (0.5%)	4 800	0.040
	Np (100%)	~ 500	0.004
	Pu (0.5%)	~ 50	0.000 4
	Am (100%)	140	0.001 2
	Cm (100%)	~ 40	0.000 3
	Total actinides	5 500	0.045
Neutron poison (e.g. Gd)		12 000	0.150

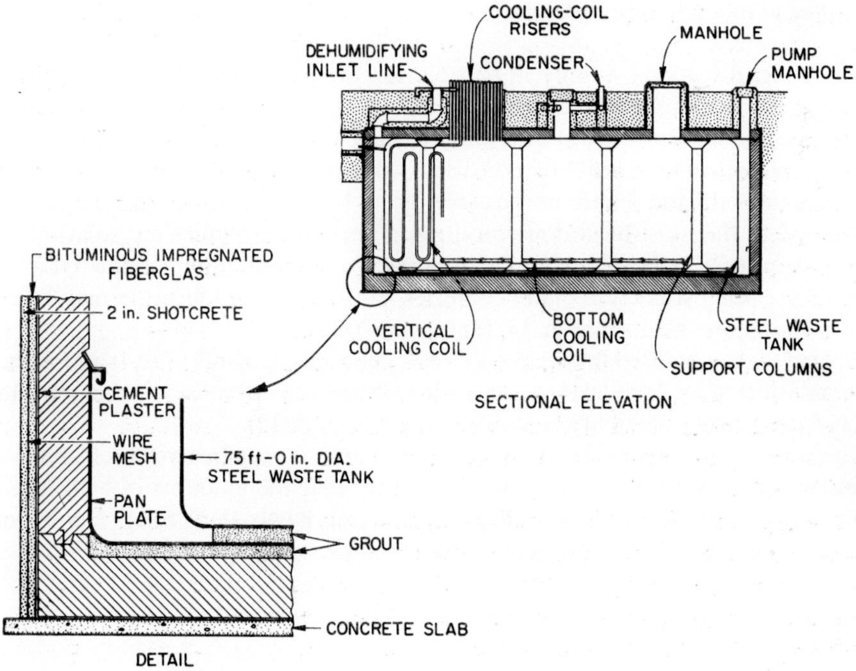

FIG. 20.21. Engineered tank storage for concentrated high level liquid waste at the Savannah River Plant, USA. (According to J. T. Long.)

than 300000 m^3 of acid and alkaline solutions and sludges have accumulated. Though the philosophy is that tank storage is only an interim procedure and will not last for more than about 1 y, present practice and planning call for at least 5 y of tank storage for the HLLW.

Some of the earliest storage tanks of ordinary mild steel built over 30 y ago at Hanford, USA, have failed, and radioactive waste has leaked out. At Hanford, radioactive waste has also been released directly into natural sites (in caverns, etc.). Figure 20.20 shows a leakage pattern. The least mobile metals are those of highest valency, i.e. Pu(+4). The low-charged metals (e.g. Cs^+) and those forming anions (e.g. Ru) move most easily.

Why this is so is clear from the considerations in §20.9. The concentration of plutonium in a relatively small volume leads to criticality risk; the minimum amount in an optimal configuration for criticality is 530 g for ^{239}Pu and 260 g for ^{241}Pu (cf. the Oklo natural reactor, §19.10). The risk increases with fissile content in the waste, and especially must be considered in the "once-through" uranium cycle in which all fissile material is left in the fuel elements stored.

20.11. Options for final treatment of high level wastes

Many concepts are being studied to treat the high level waste so that the living environment is protected against short and long term radiation damage (Fig. 20.22). The final treatment concepts are:

(a) dispersion to achieve environmentally acceptable concentrations;
(b) destruction through nuclear reactions (incineration);
(c) disposal into space;
(d) burial in a nonaccessible place.

In principle, dispersion is only applicable for the gaseous and liquid wastes, which would need negligible pretreatment. The limitations are practical (How efficient can the dispersion into air and sea be?) and radiological (What radioactive concentrations are acceptable in air and sea?). This is discussed in §20.11.1.

The options (b) and (c) are not feasible for *all* high level waste, and therefore has to be limited to the most hazardous products. They would require an isolation of these products, usually referred to as partitioning, or fractionation, of the HLW. These options are being studied in many countries, but practical application may be decades away. We discuss this further in §§20.11.2 and 20.11.3.

Option (d) can be used for unreprocessed spent fuel elements (§20.4) and is the main route considered for high level wastes from reprocessing. Because of its importance it is discussed in greater detail in a separate subsection (§20.12).

The options are obviously many and no definite choice has yet been made in any country. The reason for this indecision is partly that the amounts of high level waste are small and presently can be well controlled, and partly that "final" fuel cycles have not yet been selected because of economical, political, and social factors.

20.11.1. *Dispersion into sea and air*

As far as humans are concerned, the main danger in release of radioactive waste is the risk of ingestion or inhalation because it is rather easy to arrange physical pro-

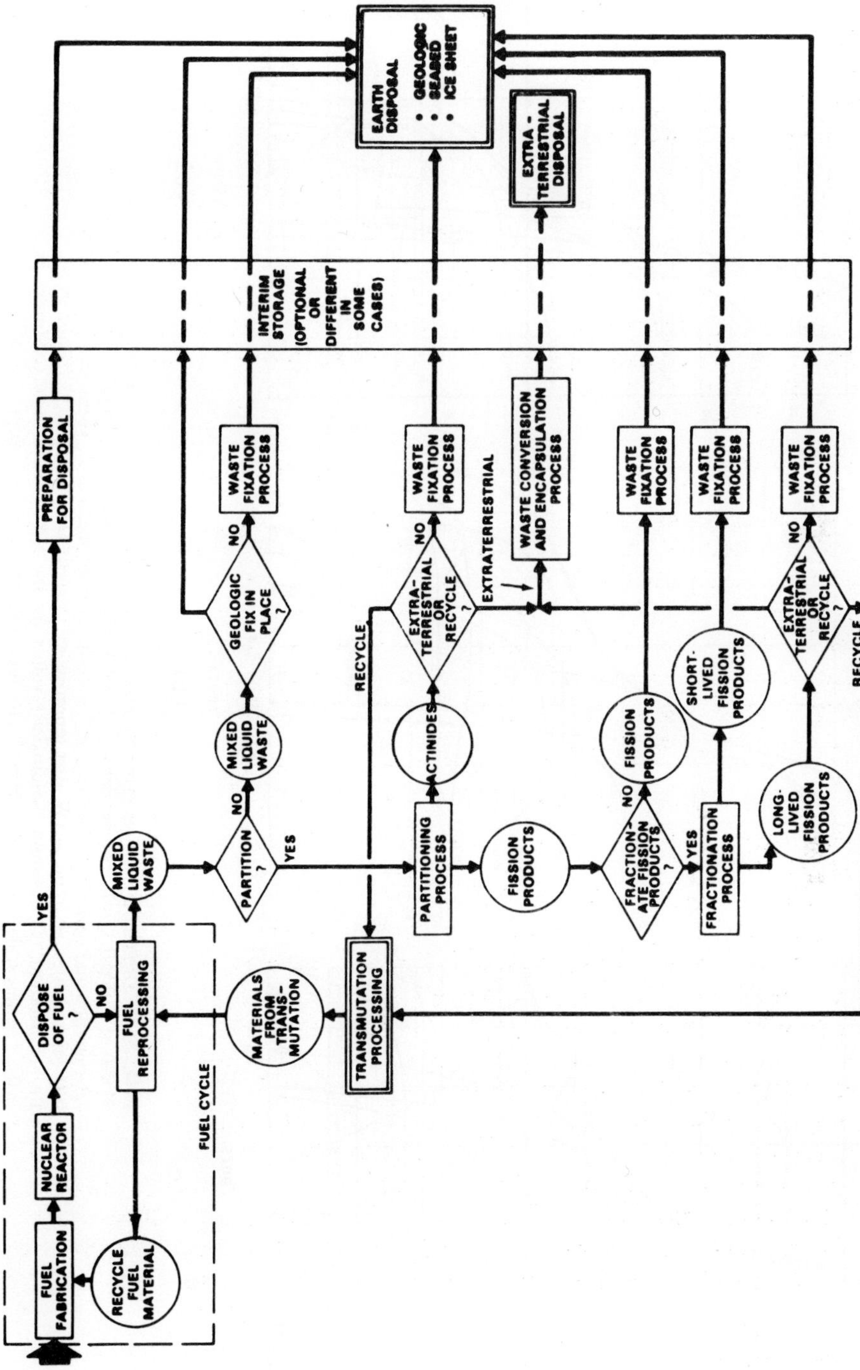

FIG. 20.22. General scheme of high level radioactive waste management options. From BNWL 1900.

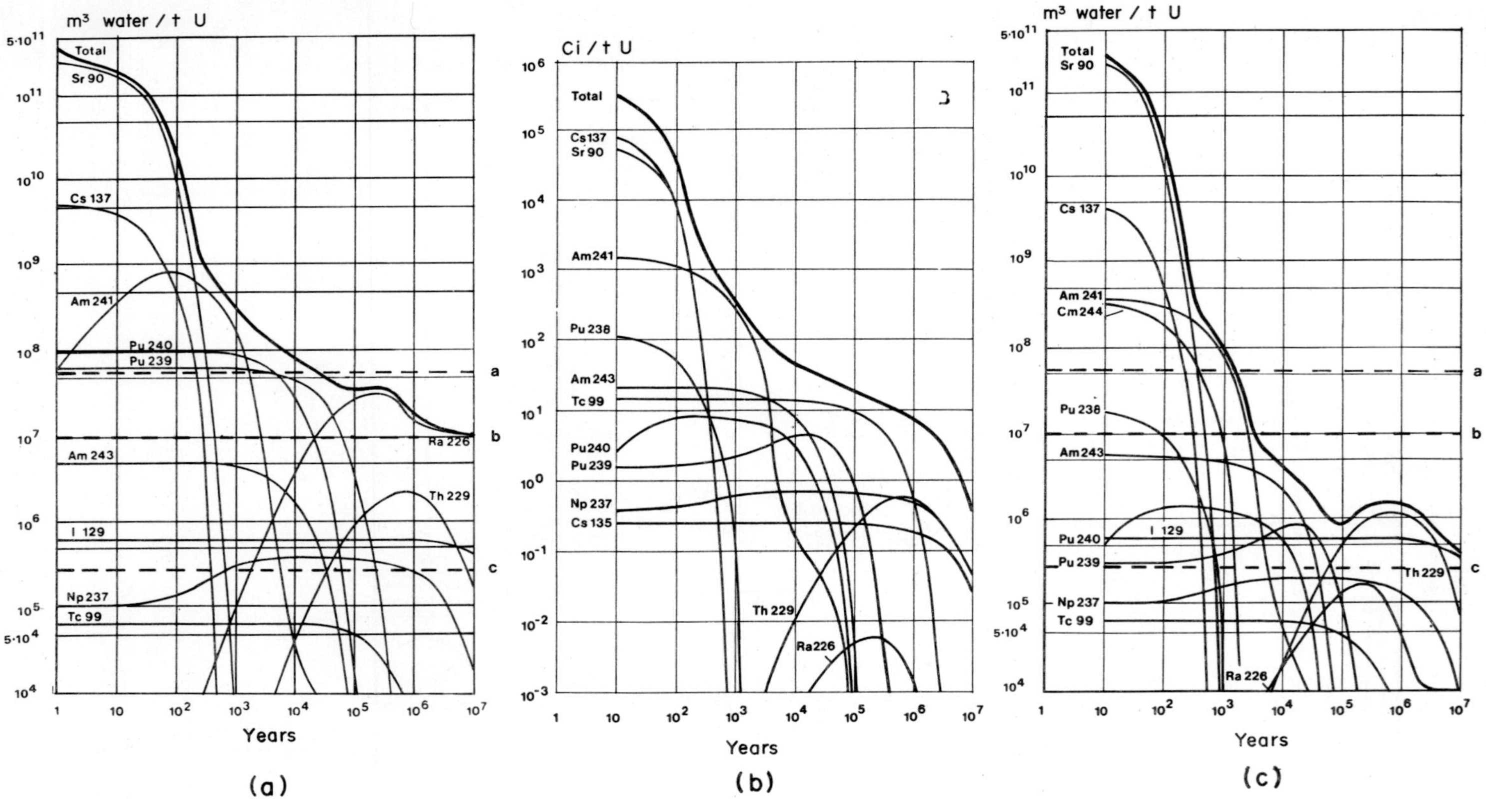

FIG. 20.23. Dominating fission products and actinides in used PWR fuel at a burn-up of 33 000 MW(th)d/t U and a power density of 34.4 MW(th)/t U; original enrichment 3.1% ^{235}U. (a) Ingestion radiotoxicity values (In_w m³ water/t U) from nuclides in spent unreprocessed fuel. (b) Activities left in the HAW after reprocessing 1 t spent U by which 99.9% U and 99.5% Pu are removed. (c) Ingestion radiotoxicity values for the HAW under same conditions as in (b). The time scale starts at the removal of the fuel from the reactor core. Reprocessing assumed at t_{cool} 10 y.

tection against the penetrating γ- and n-radiations. If the wastes can be contained for long periods (thousands to millions of years) so that there is no ingestion or inhalation risk, the additional requirement of a thick radiation shield is easily met.

The hazard differs considerably between the various waste products depending on their activity, half-life, and biochemical properties. These aspects are considered by calculation the radiotoxicity value (*In*). The radiotoxicity values in Figs. 20.23 and 20.24 are based on RCG values (*Radiation Concentration Guide*, NRC, USA) rather than MPC values according to the ICRP (Table 16.7). In Figs. 20.23 and 20.24 the In_w and n_a values have been plotted for the dominating fission products and actinides contained in unreprocessed uranium and in the HAW from reprocessing of 1 t used LWR fuel. The values for HTGR and FBR fuels are approximately the same. At time of reprocessing the toxicity values are about 2×10^{11} m^3 water (In_w) and about 2×10^{16} m^3 air (In_a). This should be compared with the global (free ocean) water volume, which is 1.4×10^{18} m^3, and the atmospheric volume (the troposphere up to 12 km) of 6×10^{18} m^3.

Multiplying the 1 t values by the global reprocessing demand of almost 10 000 t y^{-1} in 1985, we arrive at an annual aqueous volume of 2×10^{15} m^3 for total ocean dispersion, i.e. $\sim 0.2\%$ of the ocean capacity. Thus, in principle, the ocean capacity much exceeds the dispersion demand for quite a long time. However, considering that uniform dispersion is impossible, and that biological processes may enrich some radionuclides, local concentrations exceeding the MPC_w values would be expected from an ocean dispersion of all HAW. Therefore only limited amounts are allowed to be disposed of into the sea, within strict rules on the kind and amount of nuclide and the packaging prescribed by the London Convention.

The open sea contains more than 300 Ci $^{40}K/km^3$, or nearly 500000 MCi in total. Radium also accounts for more than 1000 MCi. The doses to marine organisms are usually of the order of 10–100 mrad y^{-1}. The London Convention on the prevention

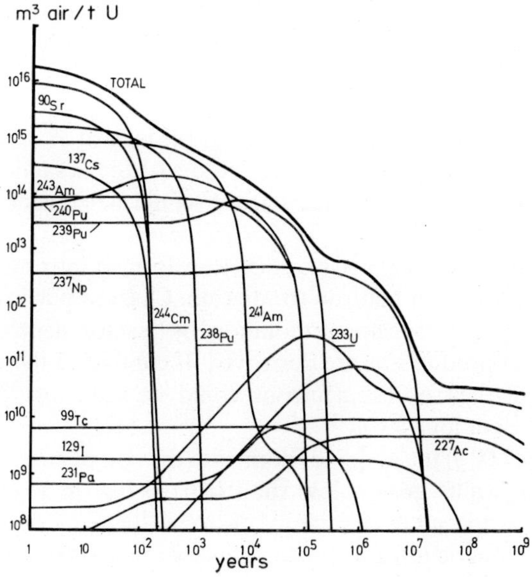

FIG. 20.24. Inhalation radiotoxicity values (In_a m^3 air/t U) of dominating fission products and actinides in HAW after reprocessing 1 ton U by which 99.5% U and Pu are removed. Fuels as in Fig. 20.23.

of marine pollution by dumping of wastes (1972) limits the amounts to 100 Ci y^{-1} for ^{226}Ra, 10 Ci t^{-1} for other α-active waste of $t_{1/2} > 50$ y, 10^6 Ci t^{-1} for tritium, 100 Ci t^{-1} for $^{90}Sr + ^{137}Cs$, and 1000 Ci t^{-1} for other β, γ-waste at any site. These figures are based on a dumping rate of not more than 100000 t y^{-1} at each site. Dumping is controlled by the IAEA.

A similar calculation of the In_a value for complete air dispersion of the 1985 HAW gives a value which exceeds the atmospheric volume by a factor 30. Such a dispersion is unacceptable and also practically impossible to achieve (i.e. $Ex_a \ll 1$, §16.11.3). Accidents, where plutonium has been spread out through fire or explosion, indicate that only a very small area (< 1 km²) becomes measurably contaminated. As a comparison, the In_a value for radon from 1 t U is 3×10^6 m³, i.e. below the diagram in Fig. 20.24.

In the diagrams Fig. 20.23(a) and (c), the horizontal dashed lines indicate the In_w values for the uranium involved in the fuel cycle. The three lines refer (a) to 7 t natural U (i.e. the amount that must be recovered from the ore to produce 1 t U enriched to 3.3% in ^{235}U), (b) 1 t enriched U, and (c) 0.050 t U (the amount $^{235+238}U$ consumed at 33 000 MWd in 1 t U); in the calculation of the In values all uranium daughters are taken into account. It may be argued that when the radiotoxicity value for the waste goes below such a line, the potential hazard is not greater than the natural material provided the external conditions (Ex, eqn. (16.7)) can be made the same. The In_w lines for HAW from reprocessing (Fig. 20.23(c)) and those from 7 and 1 t natural U cross each other at a decay time about 10^3 y. Thus if the HAW is evenly dispersed and equally well fixed as the original uranium and daughters in the mine, it does not constitute any greater hazard than the uranium ore itself. Consequently, it has been proposed to mix the HAW with cement and/or mine refuse and dispose of it in empty mines or natural underground cavities. In fact, this concept has been used for years in the USSR: low and medium active liquid wastes are injected at pressures $\lesssim 5$ MPa into permeable zones at 1000–1500 m below surface far from densely inhabited areas.

20.11.2. *Destruction through nuclear incineration*

The diagrams in Figs. 20.23 and 20.24 show that the hazard of the HAW is dominated by a few fission products (mainly ^{90}Sr and ^{137}Cs) and the actinides. If these could be destroyed through nuclear reactions, the hazard would be considerably reduced both in amount and time.

The fission products ^{90}Sr and ^{137}Cs can be transformed into shorter lived or stable products by charged particle or neutron irradiation. Charged particle irradiation would be energetically very expensive, and irradiation by reactor neutrons would produce almost as much fission products as are destroyed. If controlled thermonuclear reactors (CTR) are developed, their excess neutrons could be used for ^{90}Sr transformation, but would be less efficient for ^{137}Cs.

In the long term ($\gtrsim 600$ y) the actinides dominate the risk picture. It would take about 10^6 y before their In_w value goes below the 0.050 t U line in Fig. 20.23. Continuous neutron irradiation of the actinides finally destroys them by fission (cf. §13.2). The annual production of americium and curium is ~ 5 kg in a 1000 MWe LWR, but considerably less in an FBR. Thus if pins of these elements are inserted in an FBR, more americium and curium is destroyed than formed. It is estimated that 90% will have been transformed into fission products after 5–10 y in an FBR. In the future, CTRs could

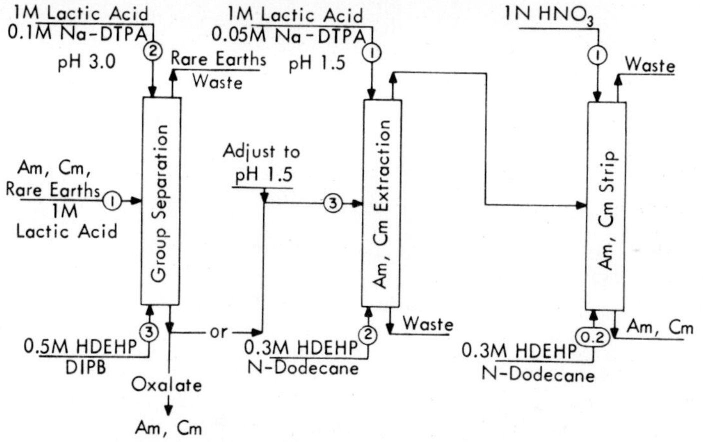

FIG. 20.25. The Talspeak process for separation of trivalent actinides and lantanides. (According to B. Weaver and F. A. Kappelmann.)

be used for the same purpose. As an alternative it has been suggested to leave the americium and curium in the uranium returned in the LWR cycle.

The transmutation concept (as well as the space disposal concept) requires that the actinides be relatively pure and free from lanthanides, since some of these (Sm, Eu, Gd) have very large neutron capture cross-sections which would take neutrons from the fission process. The annual amount of the HLLW from a 1000 MWe LWR contains about 100 kg U, 1 kg Pu, 30 kg Np, Am, and Cm, and about 1 t FP, of which ~ 300 kg are lanthanides. It is a difficult task to isolate the higher actinides and to separate them from the lanthanides, because these elements all are present in solution as trivalent ions of similar size and therefore have very similar chemical properties. The separation methods utilize their slightly different complex forming abilities in techniques such as solvent extraction, ion exchange, and reversed phase partition chromatography. Two processes have been run on a larger experimental scale:

(a) In the *Talspeak process* solvent extraction with di-2-ethylhexylphosphoric acid (HDEHP) dissolved in diisopropylbenzene (DIPB) is used; the flow scheme is illustrated in Fig. 20.25. By using lactic acid at pH 2.5–3.0 the actinide and lanthanides are co-extracted as HDEHP complexes. The actinides are stripped by an aqueous phase containing diethylenetriaminepentaacetic acid (DTPA) as complex former.

(b) In the *Tramex process* a tertiary amine (e.g. Alamine-336) dissolved in kerosene is used as the extracting agent (Fig. 20.26). The aqueous feed containing the actinide and the lanthanides is made 11 M in LiCl in dilute HCl. The actinide chloride complexes are extracted (leaving the lanthanides in the raffinate) and stripped by 8 M HCl.

20.11.3. *Disposal into space*

As discussed in the previous section, the actinides constitute the main long term hazard. In a successful partitioning cycle Np, Am, and Cm can be separated from the fission products, U and Pu, with the latter two being recycled. The amount of Np, Am, and Cm

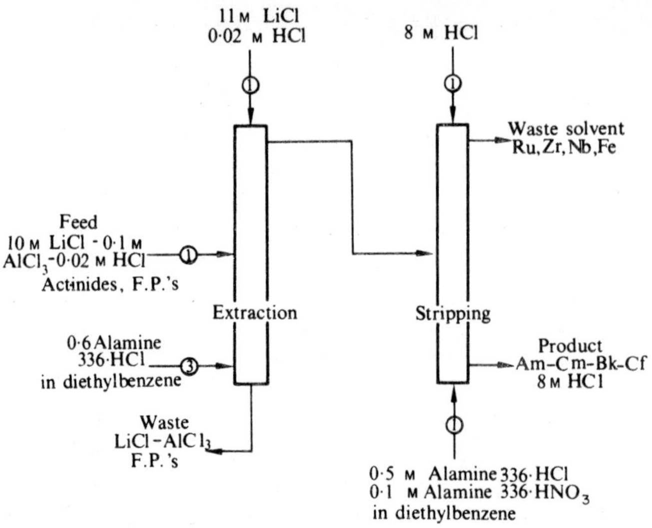

FIG. 20.26. The Tramex process for separation of trivalent actinides and lanthanides. (According to R. E. Leuze and M. H. Lloyd.)

is about 30 kg/1000 MWe LWR reactor year, or ~ 1 t y^{-1} for each of thirty large-scale nuclear power stations. This amount is small enough to make extraterrestrial disposal interesting, and it appears technically feasible for the "space countries" like the USA and the USSR.

The basic concept of extraterrestrial disposal includes packaging waste materials in a safe manner and transporting the material by rocket or other means to a location off the earth. Several different space trajectories have been considered. These include:

(i) a high earth orbit (of the order of 150 000 km); Δv 4.1 km s^{-1};
(ii) transport to the sun; Δv 24.1 km s^{-1};
(iii) a solar orbit; $\Delta v \sim 4$ km s^{-1};
(iv) solar system escape; $\Delta v \sim 8$ km s^{-1}.

Δv is the incremental velocity required to leave a low earth orbit and is a direct indication of the size and propulsion energy of the rockets required. Vehicles that could be used include existing rockets and the space shuttle.

A high earth orbit has the advantage of low Δv and possible later retrieval of the waste, but it requires a long term container integrity and very long orbit lifetime (not yet proven). Transport to the sun or solar escape has the advantages that the waste is permanently eliminated from earth, but very high Δv's are required. The solar orbit has low Δv but requires a very stable orbit, so that it does not return to earth.

Space disposal has an economic disadvantage since the cost of transportation is likely to exceed $2000 per kg of payload (1977 estimate). The weight of shielding could create significant economic penalty. There are also requirements for capsule integrity to provide a reasonable degree of assurance of survival in the case of an abort.

The actual quantities which might be considered for space disposal will vary with fuel cycle options. The earliest likely date for routine disposal in space, about 1990, is controlled mainly by the schedule for developing and achieving reliable operation of the space shuttle system.

20.12. Disposal into geologic repositories

A general consensus has developed that radioactive waste should be taken care of by the state in which the nuclear energy is produced. Thus, even if the spent fuel elements are sent from one state to another for reprocessing (like spent fuel from Japanese reactors being sent to the Windscale reprocessing plant in the United Kingdom), the producer state (Japan) must guarantee its readiness to accept the waste back. This condition has led to intensive national investigations of the safest way to dispose of the high level waste, either in the form of solidified HLLW, or spent fuel elements in the once-through option. In these studies geologic repositories have been the prime consideration, e.g. crystalline silicate rock, clay layers, salt domes, the sea bed, etc.

Before the final deposition, the waste must be conditioned, i.e. solidified and properly encapsulated to withstand environmental conditions. In one of the concepts for final storage (the Swedish Nuclear Fuel Safety Project, "KBS") it is proposed that (i) for *the uranium–plutonium cycle*, after 10 y of cooling time the spent fuel elements are reprocessed and the HLLW immediately solidified. The waste canisters produced are stored for a further 30 y in air-cooled underground vaults before they finally are recanned and deposited in granite bedrock. (ii) For the *once-through option* the spent fuel elements are stored in water-filled basins for 40 y, after which they are recanned and deposited in granite bedrock. In this section we shall briefly describe this and other high active waste projects suggested.

20.12.1. *Solidification of high level liquid wastes*

The solidification of HLLW is presently considered as the most reasonable and realistic technique to create conditions for a safe long term disposal of HAW. The objectives of solidification is to immobilize the radioactive elements and to reduce the volume to be stored. The solidified product must be nondispersable (i.e. not finely divided as a powder), insoluble, and chemically inert to the storage environment, be thermally stable, have good heat conductance (this determines the maximum radioactivity and volume of the final product), be stable against radiation (up to 10^{12} rad), and have mechanical and structural stability.

Figure 20.27 shows the options for solidification and encapsulation of HLLW. The first step in solidification is usually a calcination in which nitrates are destroyed and all metals converted to oxides. Thousands of cubic meters of HLLW have been solidified by fluidized bed calcination. However, calcine has low leach resistance, low heat conductivity, and can easily be dispersed in air (Table 20.10). It is, therefore, only considered as an interim product.

Most countries have focused development work on the fixation of the active waste in borosilicate or phosphate glass. Figure 20.28 shows a fluidized bed calciner combined with a continuous silicate glass melter. The additives consist mainly of silica ($\lesssim 50\%$) and borax ($\lesssim 30\%$ B_2O_3). In the melter the calcines and additives are heated to 1000–1200°C, leading to the formation of a homogeneous glass, which is poured into stainless steel cylinders. From a chemical standpoint (stability) the glass can contain up to 30 wt% fission product oxides, but 20% is a more normal value. Thus the waste nuclides in 1 t of LWR fuel can be contained in ~ 200 kg glass; the exact value depends on how long time the waste has cooled. For short cooling times a high waste content in the glass could cause it to melt and possibly crystallize (at ~ 700°C), which may

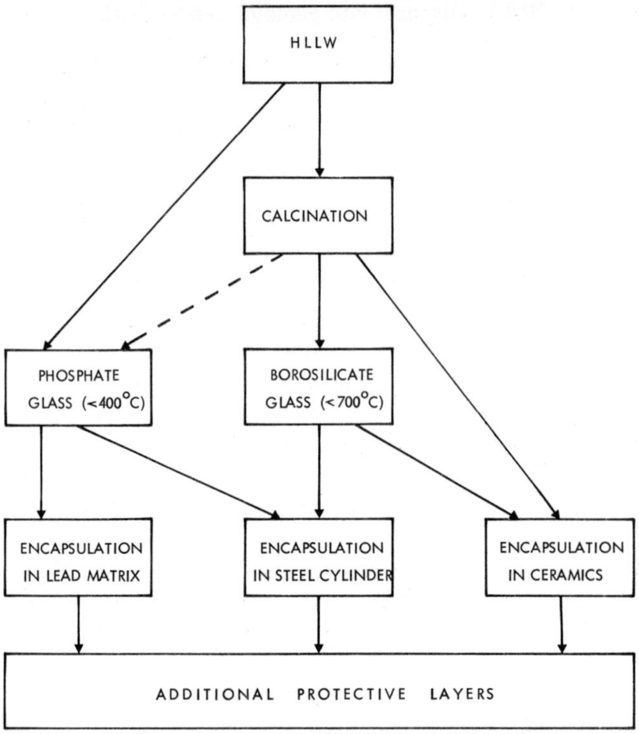

FIG. 20.27. Alternative flowsheets for high level liquid waste solidification and encapsulation.

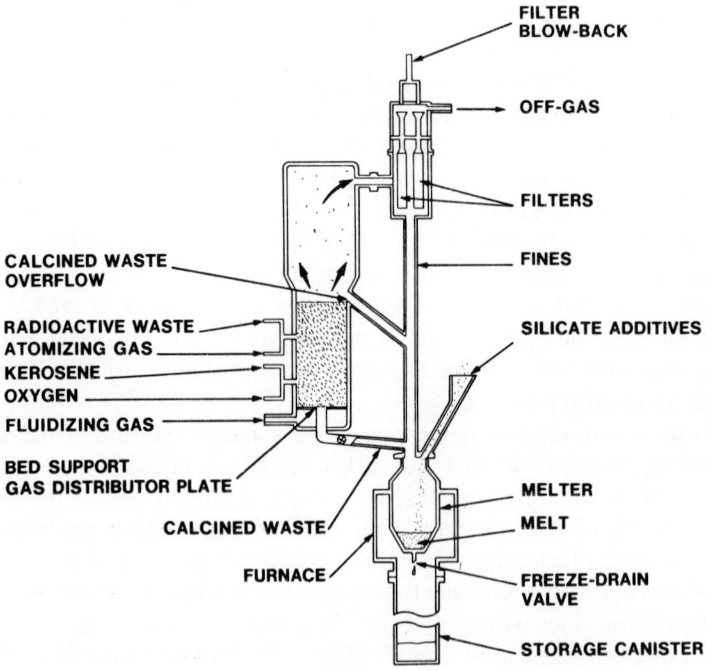

FIG. 20.28. Fluidized bed calciner and continuous silicate glass melter. (According to F. K. Pittman.)

TABLE 20.10. *Characteristics of solidified high level waste products*

Property		Fluidized bed calcine	Phosphate glass	Borosilicate glass
Physical form		Granular, 0.3–0.7 mm	Monolithic	Monolithic
Bulk density ($kg\,l^{-1}$)		1.0–1.7	2.7–3.0	3.0–3.5
Maximum weight FP		50%	35%	30%
Thermal conductivity ($W\,m^{-1}\,K^{-1}$)		0.2–0.4	1.0	1.2
Leachability (20°C) ($g\,cm^{-2}\,d^{-1}$)		1.0–0.1	10^{-4}–10^{-6}[(a)]	10^{-5}–10^{-7}[(f)]
Maximum center temperature		550°C[(b)]	400°C[(c)]	700°C[(c)]
Maximum allowable FP heat density ($W\,l^{-1}$)	forced water cooling[(d)]	45	53	127
	natural air convection[(e)]	38	36	92

[(a)]Devitrified (crystalline) glass has leach rates 10^{-2}–10^{-3} $g\,cm^{-2}\,d^{-1}$.
[(b)]Because of risk for FP volatilization.
[(c)]Devitrifies at higher temperature.
[(d)]For a cylinder of 0.3 m diameter and surface temperature $\lesssim$ 100°C.
[(e)]Cylinder, 0.3 m diameter.
[(f)]At 100°C the leach rate is ~ 10^{-3} $g\,cm^{-2}\,d^{-1}$.

TABLE 20.11. *Cans for final storage of highly radioactive waste products*
(According to the Swedish Nuclear Fuel Safety Project)

Canister condition	Reprocessed vitrified waste	Spent fuel elements
Waste equivalent amount[(a)]	From 1 t sUf[(b)]	~ 1.3 t sUf[(b)]
Central part composition	420 kg Na-B-Si-glass with 9% FP + 1% An[(c)]	~ 550 fuel pins plus lead in voids
Central part weight	470 kg including 3 mm stainless steel canister	2 t fuel material + 2.5 t lead
Outer canning	100 mm lead plus 6 mm titanium	200 mm copper
Total canister: weight	3.9 t	20 t
diameter	0.6 m	0.8 m
height	1.7 m	4.7 m
Can. power at t_{cool} 40 y	0.5 kW	0.8 kW
Can. surface temperature maximum[(d)]	65°C[(e)]	77°C
Can. surface temperature at 1000 y	35°C	50°C
Matrix dissolution rate	2×10^{-7} $g\,cm^{-2}\,d^{-1}$	10^{-6} – 10^{-7}
Bentonite layer	~ 0.2 m	~ 0.4 m
Rock hole diameter	1.0 m	1.5 m
Rock hole depth	5.0 m	7.7 m

[(a)]At ~ 30 000 MWd/t U burn-up.
[(b)]sUf = spent uranium fuel.
[(c)]55% SiO_2, 12% NaO, 20% B_2O_3, 2% Al_2O_3; 1% NiO + Fe_2O_3; 38 kg FP, $\lesssim$ 3 kg U, $\lesssim$ 0.05 kg Pu, ~ 1 kg other transuranium elements.
[(d)]Maximum temperature reached at about 15 y after deposition, which takes place 40 y after fuel discharge from reactor.
[(e)]At t_{cool} 10 y with air convection: surface temperature ~ 90°C, central temperature ~ 150°C.

reduce corrosion resistance because of the larger internal surface formed. The properties of some glasses are given in Tables 20.10 and 20.11.

A pilot plant vitrification facility (PIVER) has been in operation at Marcoule, France, since 1969, producing many m^3 glass. Larger continuous operating plants are under construction at Marcoule (AVM) and La Hague (AVH), the latter with a capacity

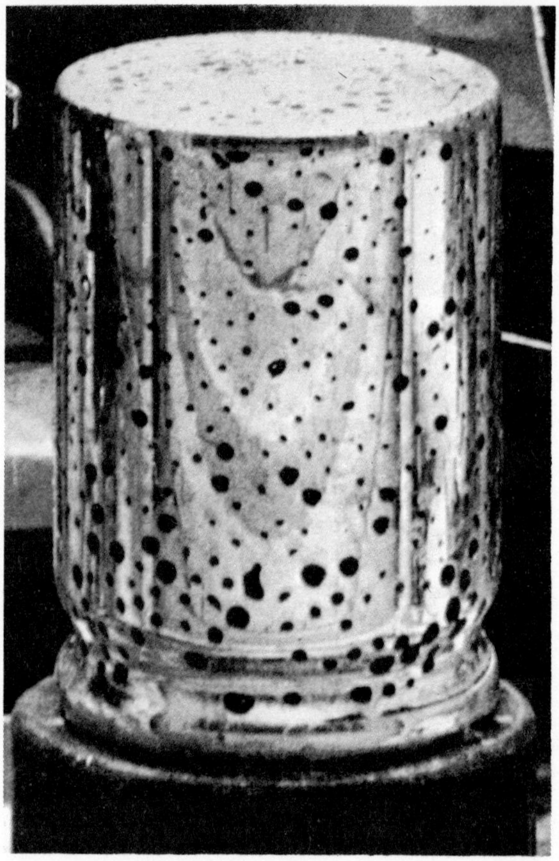

FIG. 20.29. HAW glass in lead matrix produced by the PAMELA process. (According to H. Esrich.)

of 800 m^3 HLLW/y. Similar development is underway in the UK (the HARVEST process) and other countries. The incorporation of the solidified waste into phosphate glass is an alternative to the use of borosilicate glass. In this case the calcination step can be bypassed. The HLLW, together with phosphoric acid, is evaporated and denitrated, and then fed to a continuous melter operating at 1000–1200°C, from where the molten glass flows into the storage pot. This process is more corrosive and produces a glass somewhat inferior to the borosilicate (Table 20.10). Further disadvantages are that the glass recrystallizes at relatively low temperature (~ 400°C), and that the devitrified (crystallized) phosphate glass exhibits a rather high leach rate, 10^{-2}–10^{-3} g cm^{-2} d^{-1}.

In the PAMELA process at Eurochemic (Belgium) granulated phosphate glass is incorporated into a metal alloy matrix in a steel cylinder (Fig. 20.29). This offers high chemical and mechanical stability, as well as good heat conductivity (~ 10 W m^{-1} K^{-1} for a lead matrix; cf. 1.2 W m^{-1} K^{-1} for borosilicate glass). This decreases the central temperature and allows the incorporation of larger amounts of radioactivity (up to 35% FPs) in a single cylinder. The high heat conductivity makes it feasible to use short fuel cooling times (down to 0.5 y) and diminishes the demand for interim fuel element storage basins or HLLW storage tanks.

The solidified waste described above is collected in stainless steel cylinders. As an

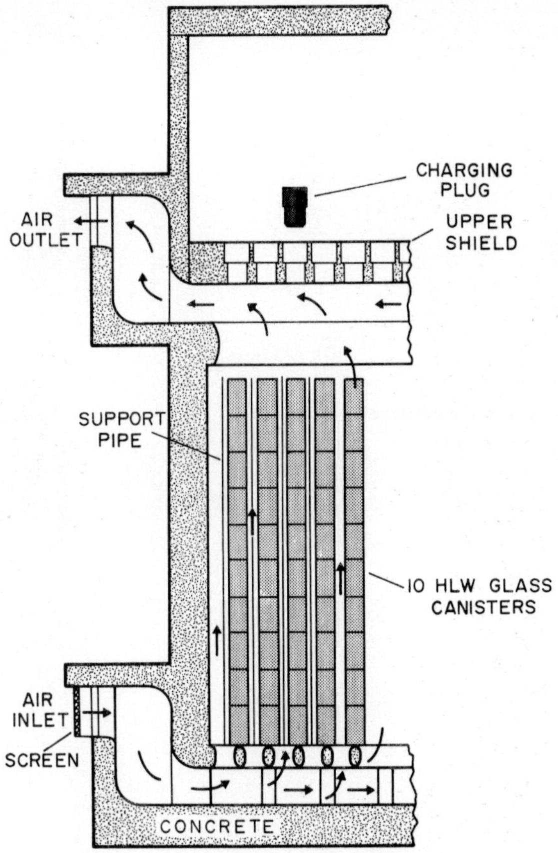

FIG. 20.30. Dry storage facility for solidified HAW cylinders using convection or forced air cooling.

alternative, ceramic vessels have been suggested. In the ASEA process (i) calcine or glass product is surrounded by a 0.2–0.4 m Al_2O_3 vessel; (ii) by applying heat and evenly distributed pressure, the Al_2O_3 is then sintered into corundum. Corundum is a natural mineral and has leachability $< 10^{-8}$ g cm^{-2} d^{-1}, very high mechanical strength (twice granite), and is heat resistant. Alternatively, graphite may be used instead of or combined with the Al_2O_3.

If waste of high fission content and short cooling time is placed at great depths in wet surroundings (e.g. at sea bottom), the high temperature and pressure condition will considerably reduce the lifetime of a glass canister. For such conditions the calcine may better be incorporated in a synthetic mineral produced through high temperature isostatic pressing (§20.4). Such a synthetic mineral will be about as stable as the natural one, i.e. almost indefinitely.

The reprocessed waste canisters (Table 20.11) contain ~9% FPs, emit ~10 kW at t_{cool} after 1 y, ~1 kW after 10 y, and ~0.5 kW after 40 y. At the higher power level they are stored in water-filled pools; after $\gtrsim$ 10 y air-cooled vaults may be satisfactory. Figure 20.30 shows the principal design of an air-cooled facility. Forced cooling is usually considered, but with precaution for sufficient convection cooling in case of ventilation failure.

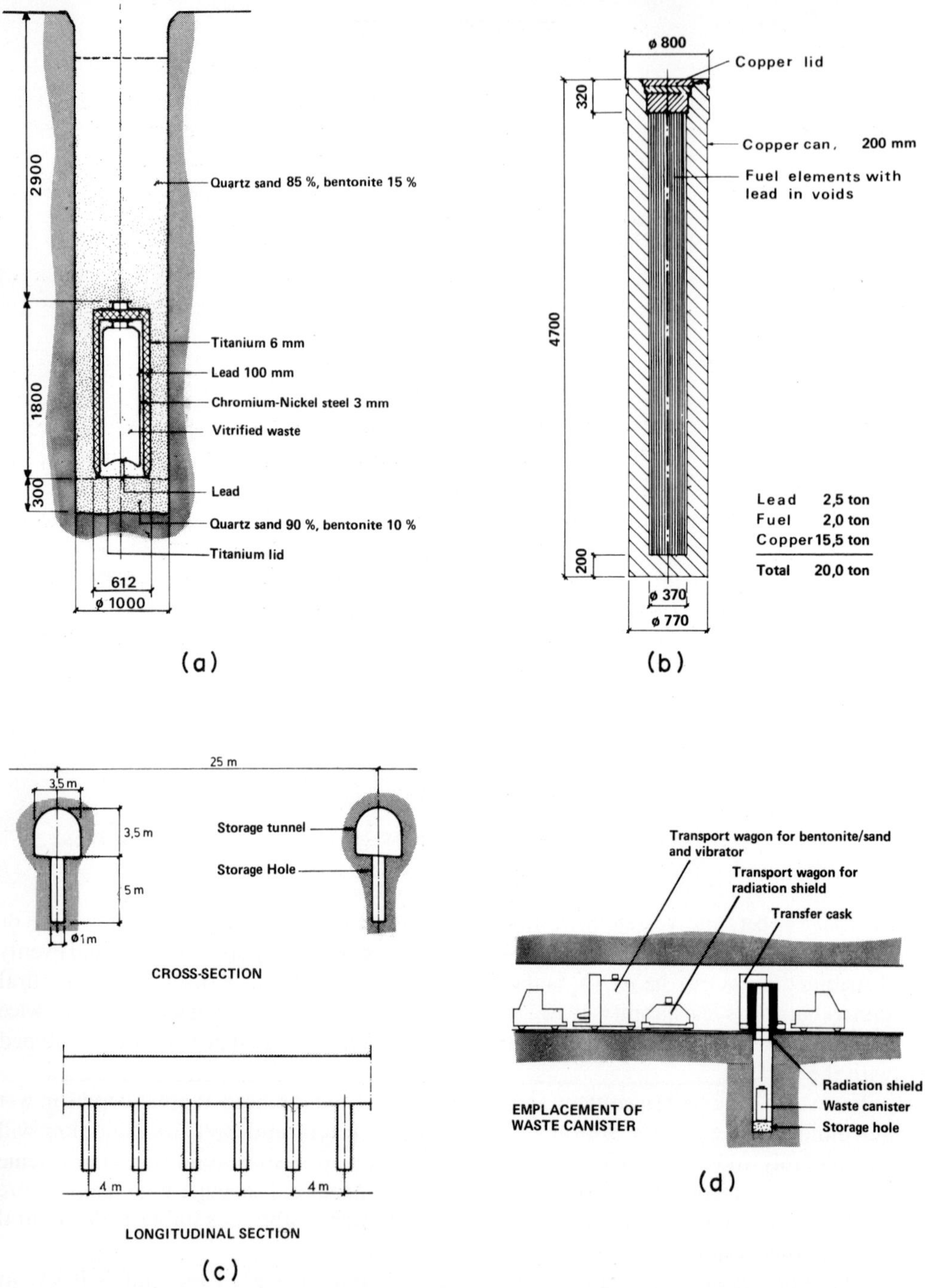

FIG. 20.31. Storage of high level waste. (a) Encapsulated canister of vitrified HAW deposited in clay-filled hole in crystalline rock. (b) Canister for spent (unreprocessed) fuel elements. (c) Storage tunnels for vitrified encapsulated HAW; for emplacement of spent fuel canisters, the hole distances are 6 m, diameter 1.5 m, and depth 7.7 m, also with slightly larger tunnels. (d) Emplacement of vitrified HAW canisters in storage holes. Holes and tunnels are back-filled with clay–bentonite mixture. (According to KBS.)

20.12.2. *Encapsulation before final storage*

In some countries the stainless steel cylinders containing the radioactive waste are not considered to provide sufficient protection against long term corrosion in the final repository. Additional encapsulation in various corrosion resistant materials is therefore suggested: lead, titanium, copper, gold, graphite, ceramics, etc. For example, in the KBS proposal canisters which contain 170 l vitrified waste in a 3 mm thick stainless steel cylinder would be surrounded by 100 mm lead in a casing of 6 mm titanium. The overall dimensions are length 1.6 m and diameter 0.6 m (Table 20.11 and Fig. 20.31(a)). One such cylinder is required to take care of solidified HLLW from reprocessing 1 t spent fuel from a 1000 MWe light water reactor; this corresponds to about 30 cylinders per year.

Although UO_2 dissolves more slowly than glass in groundwater, spent fuel elements must be recanned before entering the final storage facility. Table 20.11 gives the composition and Fig. 20.31 (b) the design of a spent fuel element canister where the fuel elements are embedded in lead and surrounded by 200 mm of copper metal. Alternatively, a ceramic outer casing may be used (Fig. 20.32) (cf. also §20.4).

The thick metal layer in these cases has two purposes: (i) to reduce the surface radiation to so low values that the radiolysis of groundwater will not contribute to canister corrosion; (ii) to protect, as long as possible, the waste-containing matrix from corrosion. The lifetime of this additional encapsulation is expected to greatly exceed 1000 y for oxygen containing groundwater, and 10^4 y for reducing groundwater (cf. next section, where groundwater composition is further discussed).

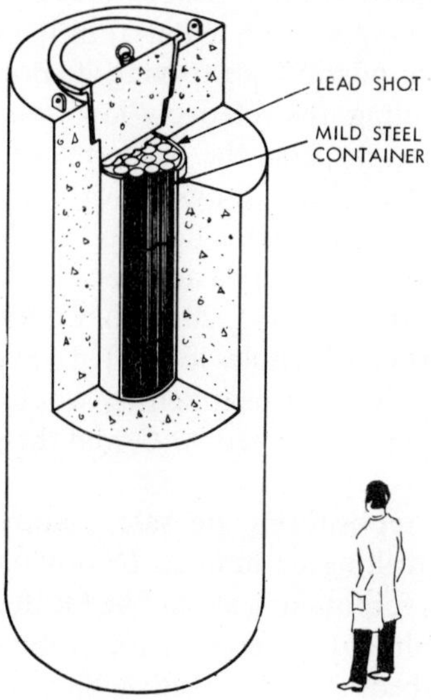

FIG. 20.32. Direct storage of spent fuel elements in lead with ceramic (e.g. concrete) casing.

20.12.3. *Properties of geologic repositories*

Many geologic formations are being studied for final storage of the waste canisters. The main requirements are

(i) geologic stability, i.e. regions of very low seismic, volcanic, or other geologic activity;
(ii) absence of cracks, holes, etc.;
(iii) impermeability to surface waters;
(iv) negligible groundwater circulation with no flow-lines leading to nearby potential intake sources;
(v) good heat conductivity;
(vi) of little interest to and as remote as possible from human activities.

Formations with many of these properties are now being evaluated: rock salt (France, FR Germany, Netherlands, USA), granite (Canada, France, Sweden, USA), clay (Belgium, Italy), etc. A cross-section of a geologic repository is shown in Fig. 20.31(c). Also disposal into polar ice sheets (USA) and the seabed (UK, USA) are being investigated. The most advanced projects are (i) at the abandoned Asse salt mine in FR Germany, where drums with low and intermediate level activities have been deposited since 1967, (ii) the waste isolation pilot plant (WIPP) at the Mendosa salt deposit in New Mexico, and (iii) the Swedish KBS project for disposal in granite (Stripa mine, etc.). These sites are all located in geologic formations, which have been unaltered for more than 100 million years. From the tectonic plate theory continual tectonic stability is expected for at least the next 10 million years. Although ice ages may alter surface conditions at northerly located sites, they are expected to have a negligible effect on repositories at great depth ($\gtrsim$500 m) in the granitic bedrock.

A limiting factor for geologic disposal is heat conductivity. If the thermal conductivity of the geological deposit is too poor, the material in the waste canisters will melt and react with the encapsulation, possibly destroying it. Because a safe disposal concept must rely on conduction cooling, the relatively low heat conductivity of the geologic deposit makes it necessary to disperse the waste material over a large volume. The thermal conductivity of $\sim 3\ W\,m^{-1}K^{-1}$ for granitic rock (compare rock salt ~ 9 and wet clay ~ 1) limits the waste heat density of the floor to $\lesssim 20\ W\,m^{-2}$. Therefore the Swedish granite storage repository would consist of a number of $\sim$ 1 km long tunnels at 25 m interspace and at a depth of $\sim$ 500 m; see Fig. 20.31(c). Canisters (Table 20.11) surrounded by clay would be inserted vertically in holes located at 4 m intervals. With a heat production of $\sim$ 500 W per canister the floor heat density would be 6–7 $W\,m^{-2}$. The internal glass temperature would then never exceed 90°C, and the surface temperature not more than 65°C.

Such deep (geologically) repositories are safe against surface activities (including nuclear explosions) except drilling or mining. In order to avoid such an occurrence after the repository has been forgotten, it should be located in a formation of no interest to man; e.g., the formation should not contain any valuable minerals. From this point of view, location in deep seabeds seems attractive. Such a location can be chosen either deep in the bottom silt or in a bedrock under the sea floor, or, possibly, in a tectonically active trench, where with time the waste will be pushed down into the interior of the earth.

The only important risk for a land-based geologic repository is that a combination of more or less unpredictable circumstances could lead to breaking the protective barriers around the waste matrix, followed by dissolution and transport of the most hazardous products by water to a place where it can enter into the food chain. One of the purposes of the clay layer around the canisters in the KBS project is to act as a mechanical buffer so that even considerable slippage caused by earthquakes will have little effect on the mechanical integrity of the canister.

As the metal encapsulation after a long time (see previous section) is dissolved, or if a break of the canister occurs, the radioactive products will release with the dissolution rate of the matrix. A reasonable figure for glass dissolution may be 10^{-6} g per cm^2 exposed glass surface per day, assuming no limit with regard to the solubility product (i.e. unlimited amount of water). This corresponds to a corrosion rate of the glass surface of $\sim 1.4 \times 10^{-6}$ m y^{-1}. Experience shows that this rate rapidly decreases with time for glass stored in ground; thus strontium release rates were reduced by a factor of 10^6 in 15 y. In order to avoid rapid dissolution, geologic repositories as dry as possible should be chosen.

The rocksalt deposits considered were formed > 100 million years ago. Often they push upwards in the form of a dome, the top of which may be at a few hundred meters below surface level, and with a depth up to more than 1000 m. The dome is usually protected from groundwater by a calcite ($CaCO_3$) cap. Salt has been mined in such formations for centuries. The formation contains only microcrystalline water and is extremely dry. Canisters emplaced in such formations are not believed to dissolve (as long as no water enters through the hole mined through the calcite layer). The good heat conductance will allow emplacement of waste of higher power density than for deposits in clay or granite. Since the salt is plastic, the holes and corridors, which are back-filled with crushed salt, will self-seal. However, the canisters will probably move in the plastic salt, and therefore could be irretrievable after a hundred years.

Granite and clay formations are percolated by groundwater except at depths which presently are considered impractical. The water flow rate is given by the equation

$$F_w = k_p i S \text{ m}^3 \text{ s}^{-1} \tag{20.11}$$

where k_p is the permeability (ms^{-1}), i the hydrostatic gradient (m m^{-1}), and S the flow cross-section (m^2) this is another version of Darcy's law, eqn. (20.10). A typical value for Swedish granite formations at 500 m depth is $F_w = 0.2$ l m^{-2} y^{-1}, which gives a groundwater velocity of 0.1 m y^{-1}, and a time of 5000y for the water to move 500 m. In this case the rock permeability is taken as 10^{-9} m s^{-1}; rock formations closer to the surface usually have higher permeabilities, but many formations with much lower values are also known ($\lesssim 10^{-13}$).

Clays consist of small particles, usually < 2 μm and with an average size of $\lesssim$0.1 μm, of various minerals like quartz, feldspar, montmorillonite (a hydrated aluminum silicate with high ion-exchange capacity), mica, etc. The overall chemical composition is mainly a mixture of silica, alumina, and water. The small particles in the clay give rise to a very large surface area (1 cm^3 of particles of 0.1 μm diameter have a total surface area of ~ 60 m^2) and correspondingly high sorption capacity. Although the clay can take up large amounts of water (up to 70%) without losing its plasticity, the water permeability is extremely low. For the sodium bentonite clay (~90% montmorillonite) buffer considered in the KBS project, the permeability is 2×10^{-14} m s^{-1} (10% water in clay compacted to a density of 2100 kg m^{-3}). Such clay can be considered impermeable to groundwater. Natural clays of this type occur in many places (the Netherlands, Italy, etc.).

When granite or clay is contacted with water containing dissolved cations, sorption or exchange of these ions with ions of the solid phase are observed. Montmorillonite has such a high exchange capacity that it is used as a natural ion exchanger, e.g. for water purification. The sorption capacity and distribution values (k_d, see eqn. (20.10)) for clay and crushed granite are about the same. Observed k_d values (and calculated retention factors, eqn. (20.9)) for Swedish granite are Cs 0.064 (400), Sr 0.016 (1500), Ln(III) 10 (200 000), and An(IV) 1.2 (23 000). Rather similar (about a factor of ten) values have been found for other geologic formations such as tuff from New Mexico, basalt from Idaho, and limestone from Illinois. The high retention values combined with a water transport time of a few years lead to lifelong retention of ^{90}Sr, ^{137}Cs, and most Pu, Am, and Cm isotopes. With the negligible water flow rate through clay, diffusion will be the dominating transport process. One can find it will take Cs 700 y, Sr 1800 y, Am 22000 y, and Pu 8000 y to penetrate 0.4 m bentonite. In the meantime most of the radioactivity of the waste nuclides have decreased to negligible values. Studies of the salt deposits in New Mexico indicate that of all waste nuclides released in the repository, only ^{14}C, ^{99}Tc, and ^{129}I can reach the atmosphere before they have decayed.

An important factor is the groundwater composition. If the water contains air, the following valency states are expected: Tc(VII), U(VI), Np(V or VI) and Pu(IV). Many groundwaters contain Fe(II), making it reducing (the redox potential E_h will be < 0). In such groundwater of $E_h \lesssim -0.1$ V, the following valency states appear: Tc(IV), U(IV), Np(IV), and Pu(III). These latter species are much more strongly retained than the corresponding ones in oxidizing groundwater.

Table 20.12 gives estimated amounts of released radioactivities carried by groundwater from a spent fuel repository to a recipient. The only important radioactivities come from uranium and daughter products, and from ^{129}I. All other important nuclides decay before they reach the repository. The high retention values for the tri- and tetravalent lanthanides and actinides are supported by studies of the Oklo natural reactor (§19.10), for which it has been shown that the lanthanides and actinides formed stayed

TABLE 20.12. *Calculated releases of radionuclides from a spent fuel waste repository in granite bedrock*
Water transport rate from repository to recipient is assumed to be 1000 y

Nuclide	Half-life (y)	1000 MWe reactor annual discharge (Ci y^{-1})	Sorption factor (k_d m^3 kg^{-1})	Breakthrough time (t_n y)	Annual release at t_n (Ci y^{-1})
^{90}Sr	28	1.8×10^6	0.016	7×10^5	0
^{93}Zr	1.5×10^6	57	3	1.3×10^8	0
^{99}Tc	2.1×10^5	430	0.05	2.2×10^6	0.3
^{129}I	1.6×10^7	1	0	1000	1.0
^{135}Cs	2.3×10^6	8.6	0.3	1.3×10^7	0.2
^{137}Cs	30	2.5×10^6	0.3	1.3×10^7	0
^{151}Sm	93	32 000	16	7×10^8	0
^{226}Ra	1600	~0	0.5	2.2×10^7	9
^{229}Th	7.3×10^3	~0	4	1.7×10^8	0
^{233}U	1.6×10^5	0.1	2	8.6×10^7	0
^{238}U	4.5×10^9	9	2	8.6×10^7	9
^{237}Np	2.1×10^6	10.2	2	8.6×10^7	0
^{239}Pu	2.4×10^4	972	1	4.3×10^7	0
^{242}Pu	3.9×10^5	41	1	4.3×10^7	0
^{241}Am	433	4700	30	1.3×10^9	0
^{243}Am	7.4×10^3	525	30	1.3×10^9	0

Assumed immediate complete dissolution, with no change in groundwater composition. $E_h \lesssim -0.1$ mV, pH ~ 8. $v_n = k_p i/\rho k_d$, where $k_p = 10^{-7}$, $i = 0.01$ (for $t_w = 1000$ y, this gives $\varepsilon = 0.06$), $\rho = 2700$ kg m^{-3}.

at the formation place until they had decayed, while strontium and cesium moved away during the last 1.7 billion years.

At the groundwater flow in a clay or granite repository the dissolution rate of canister encapsulation and waste matrix is limited by water solubility. In the granite repository the copper canister is expected to last $> 16^6$ y. The dissolution of the waste glass is assumed to take $> 10^7$ y, and the same holds for the UO_2 matrix. Although international agreements may exclude the ocean floor as a high level waste dump, calculations indicate that the release rate of radionuclides from glass blocks will be so low that the MPC_w values would never be exceeded in free-flowing water.

Both dissolution rate and transport rate of nuclides will depend on the existence of complex formers in the groundwater. Such complex formers are Cl^-, F^-, SO_4^{2-}, HPO_4^{2-}, CO_3^{2-}, and organic anions (e.g. humic acid). Complexes with these anions will in most cases increase solubility, and—through formation of less positively charged metal species—reduce the retention factors. The groundwater conditions therefore play a central role for evaluation of the risks of a waste repository. Since these conditions vary, they have to be evaluated at each place.

20.13. Beneficial utilization of nuclear wastes

The high level nuclear waste produced in the world 10 y from now will amount to $\gtrsim 10^4$ t annually with a radioactivity of $\gtrsim 10^4$ MCi. It contains large and potentially valuable sources of metals and radiation. Though today considered a liability it may in the future become a needed asset. Since the extraction and utilization of some of the fission products or actinides will probably not be economic after the waste has been vitrified and placed in permanent geologic storage, the nuclear fuel reprocessing scheme must therefore be designed for byproduct extraction.

Presently the most interesting products in the waste are the platinum group metals (due to their metal values) and ^{90}Sr, ^{137}Cs, ^{85}Kr, ^{238}Pu, and ^{241}Am (due to their radiation properties).

The waste contains considerable amounts of Ru, Rh, and Pd, all metals in scarce abundance on earth. These elements are used as catalysts in the chemical industry and as corrosion resistant materials. The United States demand exceeds the domestic production by about a factor of 100. Still, the United States would become independent of import in year 2000 if the radioactive waste amounts of these elements were recovered. This is particularly true if technetium is recovered, since it can replace platinum. The recovered elements would be radioactive, but the activities will be small enough to make the elements easy to handle.

Beta-irradiation from ^{85}Kr on phosphors causes visible light. Radiokrypton light sources have widespread applications where reliable lights are required as, for example, at airports, railroads, hospitals, etc., or where sources of electricity could cause dangerous explosions, as in coal mines, natural gas plants, etc. However, fission product krypton contains only 4% of ^{85}Kr, which makes it unsuitable for high intensity lighting applications. ^{85}Kr must therefore be enriched about a factor of 10, which presently can be done by thermal diffusion.

By the year 2000 over 100 MW heat will be produced by radiostrontium. Strontium-fueled thermoelectric generators are now used in several countries for powering unmanned weather data acquisition systems, lighthouses, and other navigation aids, etc. (§14.11.3). Their reliability surpasses any other remote power source. The current thermoelectric generators have a thermal-to-electrical efficiency of about 5%, while recently developed thermal-to-mechanical systems show efficiencies of 25–30%. The use of such systems could be expanded vastly by increased recovery of fission product strontium.

Food sterilization by radiation is potentially of global importance (§16.4.2). Though this presently is done by ^{60}Co or accelerator radiation, the advent of large quantities of radiocesium recovered from nuclear wastes may have a large positive impact on the economics and scale of food irradiation. An almost equally important use of ^{137}Cs would be for sewage sludge treatment (§16.4.3), which becomes increasingly important as the requirements for sterilization and secondary treatment of the sludge increases. The thermoradiation of sludge also makes it useful as a fertilizer and cattle feed product.

The low penetrating radiation, long half-life, and high power density of ^{238}Pu makes it ideal for special purpose power supplies (§13.3.4). The main present use is in space research, and ~ 30 kWth power sources have been launched into space, ^{238}Pu is also used in heart pacemakers and is a candidate as a power source for completely artificial hearts. Production of ^{238}Pu requires the isolation of ^{237}Np, which is then irradiated to produce pure ^{238}Pu.

The demand for ^{241}Am is much larger than present production capacity because of its use in logging oil wells, in smoke detectors, and for various gauging and metering devices. However, the potential source is large: the nuclear waste in the United States will contain 6.5 t in 1990.

If all the mentioned waste products are recovered, this will mean (a) that the waste is turned into an essential asset with benefits in food production, health, and safety, and (b) that the hazard of the remaining waste will be much lowered, considerably simplifying final waste storage.

20.14. Exercises

20.1. Using the data for the Würgassen reactor and the thermal neutron capture cross-section (Table 19.3) it can be calculated how many kg Pu should be formed per t U at a burn-up of 27 500 MWd. (a) Make this calculation assuming that plutonium disappears only through fission in ^{239}Pu. (b) According to Table 20.5 each t U from a BWR contains 6.8 kg Pu; why is your result much lower?

20.2. It is desired that 98% of all ^{233}Th formed by neutron capture in ^{232}Th decays to ^{233}U. How long a time must elapse between end of irradiation and start of reprocessing?

20.3. In a ^{233}U fueled reactor, some ^{233}U is converted into ^{235}U. Calculate the amount of ^{235}U formed in 1 t ^{233}U from neutron capture in ^{233}U and ^{234}U ($\sigma_{n\gamma}$ 97b) for a fluence of 10^{21} n cm^{-2} (a) assuming no consumption of ^{235}U formed, (b) taking ^{235}U fission and capture into account.

20.4. Explain why ^{238}Th and ^{232}U are a nuisance in the thorium–uranium fuel cycle.

20.5. A reactor starting with 3% ^{235}U produces 6000 M Wd energy/t U fuel each year. Neglecting fission in ^{238}U, (a) how much fission products have been produced after 5 years? (b) What is the ^{235}U concentration if plutonium fission also is taken into account?

20.6. In the example above, 1 t U fuel elements is removed from the reactor after 2 years. Using Fig. 20.2, (a) what is the total radioactivity from the fission products after 30 d cooling time? (b) Which FPs are the most radioactive at this time?

20.7. Calculate the decontamination factor required for (a) fission product activity, and (b) for gadolinium in commercial plutonium nitrade produced from PWR fuel (Tables 20.2 and 20.7) at t_{cool} 10 y.

20.8. In a solvent system the distribution ratio D_U is 2 for uranium and D_{Cs} is 0.003 for cesium. If 99.5% U is to be extracted in a repeated batch fashion (eqn. (20.4)), (a) how much cesium is coextracted? If instead a countercurrent process is used with 10 extraction and 2 washing stages, what percentage of (b) uranium and (c) cesium is extracted? In an extraction equipment D_{Cs} 0.003 cannot be maintained, because droplets are carried over; the practical value will be D'_{Cs} 0.02. (d) How much cesium is extracted in this latter case with the countercurrent equipment? Assume equal phase volumes.

20.9. 0.0015 Ci ^{239}Pu is released annually from a reprocessing plant. What will be the corresponding release of ^{238}Pu and ^{240}Pu for typical isotopic plutonium composition of LWR fuel?

20.10. Calculate the natural radiotoxicity value In_w of 1 km^3 of land (density 2.6 g cm^{-3}) containing 3 weight ppm ^{238}U with daughter products. Only ^{226}Ra has to be considered.

20.11. A tank contains 100 m^3 5 y old HLLW. (a) Calculate the heat production for a waste of composition in Table 20.9 left column. (b) How many 1000 MWe PWR reactor years does this waste correspond to?

20.12. Ru, Rh, and Pd are recovered from the waste from a 10 GWe program. What will the annual amounts and specific radioactivities be at $t_{cool} = 10$ y for each of them?

20.15. Literature

ICPUAE, Vol. **9** (1955), Vols. **17** and **18** (1958), Vols. **10** and **14** (1965), Vols. **8–11** (1972).

M. Benedict and T. H. Pigford, *Nuclear Chemical Engineering*, McGraw-Hill, 1957.

F. S. Martin and G. L. Miles, *Chemical Processing of Nuclear Fuels*, Butterworth, 1958.

J. Prawitz and J. Rydberg, Composition of products formed by thermal neutron fission of ^{235}U, *Acta Chem. Scand.* **12** (1958) 369, 377.

R. Stephenson, *Introduction to Nuclear Engineering*, McGraw-Hill, 1958.

C. B. Amphlett, *Treatment and Disposal of Radioactive Wastes*, Pergamon Press, Oxford, 1961.

E. Glückauf, *Atomic Energy Waste*, Butterworth, 1961.

J. Flagg (ed.), *Chemical Processing of Reactor Fuels*, Academic Press, 1961.

S. Peterson and R. G. Wymer, *Chemistry in Nuclear Technology*, Pergamon Press, Oxford, 1963.

IAEA, *Disposal of Radioactive Wastes into Seas, Oceans and Surface Waters*, Vienna, 1966.

IAEA, *Practices in the Treatment of Low- and Intermediate Level Radioactive Wastes*, Vienna, 1966.

IAEA, *Assessment of Airborne Radioactivity*, Vienna, 1967.

IAEA, *Disposal of Radioactive Wastes into the Ground*, Vienna, 1967.

D. Dyrssen, J. O. Liljenzin, and J. Rydberg, *Solvent Extraction Chemistry*, North-Holland, 1967.

J. T. Long, *Engineering for Nuclear Fuel Reprocessing*, Gordon & Breach, 1967.

Y. Marcus and A. S. Kertes, *Ion Exchange and Solvent Extraction of Metal Complexes*, Wiley Interscience, 1969.

A. K. De, S. M. Khopkar, and R. A. Chalmers, *Solvent Extraction of Metals*, van Nostrand–Reinhold, 1970.

S. Ahrland, J. O. Liljenzin, and J. Rydberg, (Actinide) Solution chemistry, in *Comprehensive Inorganic Chemistry*, Vol. 5, Pergamon Press, Oxford, 1975.

R. G. Post and K. Wirtz (eds.), Waste management symposium, *Nuclear Techn.* **24** (1974) 265–454; ibid., 1975, and later.

K. J. Schneider and A. M. Platt (eds.), *High-level Radioactive Waste Management Alternatives*, Battelle Pacific Northwest Laboratories, Richland, Wash. 99352, Report BNWL-1900, NTIS, Springfield, 1974.

J. P. Olivier, The management of fission products and long-lived alpha wastes, *Adv. Nucl. Sci. Techn.* **8** (1975) 141.

IAEA, *Transuranium Nuclides in the Environment*, Vienna, 1976.

IAEA, *Management of Radioactive Wastes from the Nuclear Fuel Cycle*, Vienna, 1976.

ERDA, *Alternatives for Managing Wastes from Reactors and Postfission Operations in the LWR Fuel Cycle.* ERDA-76-43, NTIS, Springfield, 1976.

AKA; National Council for Radioactive Waste Management (PRAV), Sweden; SOU 1976: 30, 31, 32, Stockholm, 1976.

IAEA, *Regional Nuclear Fuel Cycle Centres*, Vienna, 1977.

B. L. Cohen, The disposal of radioactive wastes from fission reactors, *Sci. Am.* **236** June (1977) 21.

Convention on the Prevention of Marine Pollution by Dumping of Wastes and Other Matter—"The Definition Required by Annex I, paragraph 6 to the Convention and the Recommendations Required by Annex II, Section D", IAEA, INFCIRC/205/Add. 1, 10 January 1975.

NEA-OECD, *Objectives, Concepts and Strategies for the Management of Radioactive Waste Arising from Nuclear Power Programmes.* OECD, 1977.

H. A. C. McKay and M. G. Sowerby, *The Separation and Recycling of Actinides*, EEC, EUR 5801e, 1977.

A. M. Aikin, J. M. Harrison, and F. K. Hare, *The Management of Canada's Nuclear Wastes*, Minister of Energy, Mines and Resources, Report EP 77-6, 1977.

KBS Project, Nuclear Fuel Cycle Back-End, 1978, Kärnbränslesäkerhet, Fack, 10240 Stockholm.

R. G. Wymer and B. L. Vondra, *Technology of the Light Water Reactor Fuel Cycle*, CRC-Press, 1980.

CHAPTER 21

*Nuclear Power: Problems and Promise**

Contents

Introduction

The success in the 1940s of nuclear scientists in designing and building nuclear reactors for the production of plutonium led to predictions of abundant, cheap energy from the nucleus. After more than three decades of development, nuclear power is now contributing to the energy needs of many nations. However, opposition to nuclear power exists in many countries and in some threatens to prevent its use. In this final chapter we discuss the basis of that opposition and attempt to evaluate the validity of those concerns. The reader should understand that our evaluation and final conclusions are somewhat subjective as they involve judgement of choice between the disadvantages (and there are some) and advantages (we believe there are many) of nuclear power.

By the mid-1950s, nuclear scientists realized that nuclear power reactors would be more complex and more expensive than reactors for plutonium production (for weapons). Natural gas and petroleum were abundant and inexpensive, so nuclear power development proceeded slowly. However, the drastic increases in the cost of petroleum of the last decade as well as the anticipated depletion of gas and oil resources in the next quarter century have accelerated active programs in nuclear power in many nations. Before discussing the problems of nuclear energy, let us review the world's energy situation.

21.1. World energy demand

The global energy consumption has steadily increased as civilization evolved. This demand accelerated sharply about 1900 as some nations moved into the age of modern technology. The energy demand is a function of two principal factors: the increase in population and the increase in the living standard of most individuals. Though there is no generally accepted way of defining living standard, we use the material living standard,

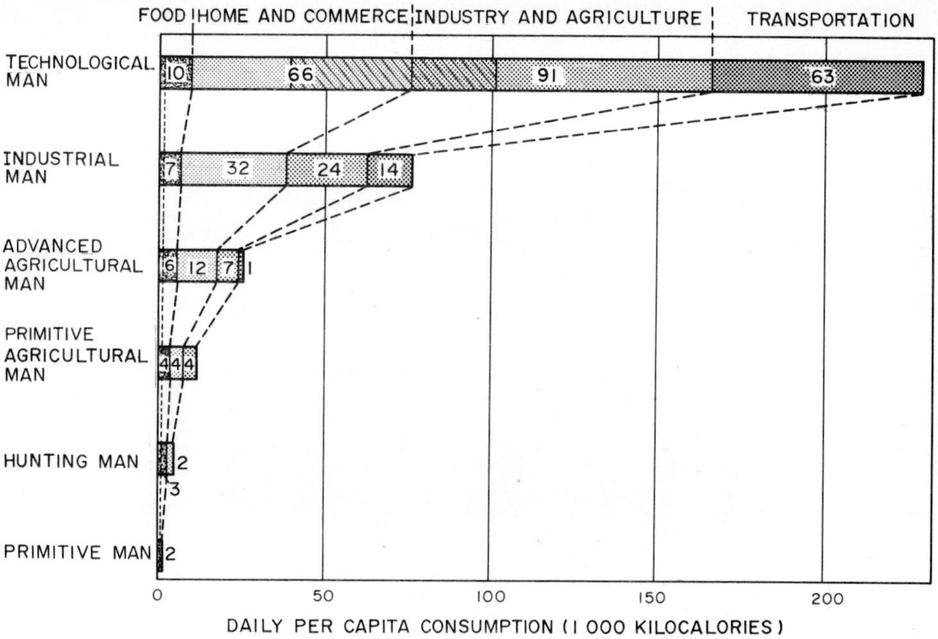

FIG. 21.1. Daily consumption of energy per capita is shown for six stages in human development. Primitive man (East Africa about 1 000 000 y ago) without the use of fire had only the energy of the food he ate. Hunting man (Europe about 100 000 y ago) had more food and also burned wood for heat and cooking. Primitive agricultural man (Fertile Crescent in 5000 BC) was growing crops and had gained animal energy. Advanced agricultural man (northwestern Europe in AD 1400) had some coal for heating, some water power and wind power, and animal transport. Industrial man (in England in 1875) had the steam engine. In 1970 technological man (in the United States) consumed 230 000 kcal (~ 270 kWh) per day, much of it in form of electricity (hatched area). Food is divided into plant foods (far left) and animal foods (or foods fed to animals). (According to E. Cook.)

which is closely related to the per capita energy consumption. Figure 21.1 shows the energy needs related to this living standard. The increased living standard of the people in much of the world since 1900 and the great increase in total population has resulted in a global energy consumption (1975) of 5.7×10^{19} cal y^{-1} or 7.5 TW (1 TW is equivalent to the burning of roughly 1 billion tonnes of coal per year).

A much-discussed question involves the values to be assigned as the limit for the world's population and for the per capita energy consumption. In year 2000 world population will exceed 6×10^9 people. It is generally expected that the population will level off but there is widely varying predictions of the "level" value. We shall use a medium estimate of 12–13 billion from the UN Population Conference (1974, Bucharest). The per capita consumption in the future is equally uncertain. At present, more than 70% of the world's people use less than the average (global) per capita value of 2 kW. This uneven per capita use of energy is related to the state of development of different countries. It is unlikely that such an uneven distribution will persist, and as the "under-privileged" 72% gain a better living standard, the per capita energy use will increase. In Fig. 21.2, total energy demand is presented based on per capita energy values of 2 (no increase from present), 3, and 5 kW years per year. Of course, higher values than 5 kW and 13 billion people are possible, so any prediction of future energy demands are

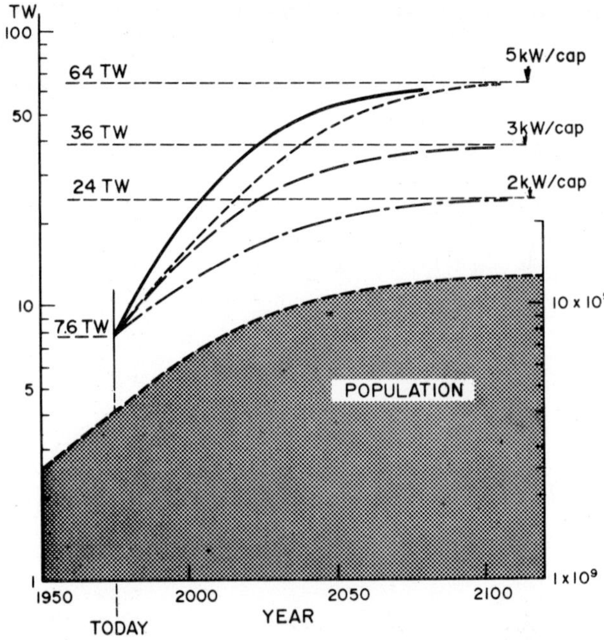

FIG. 21.2. Predictions of energy consumption based on different per capita values. Energy consumption is given in terawattyears per year, or simply TW. (According to W. Häfele.)

very uncertain. However, a value of 50 TW seems reasonable if, perhaps, conservative for evaluating the long range energy options open to us.

21.2. Energy reserves

Simple calculations show that "renewable" (or continuously flowing) energy sources like waterfalls, geothermal and solar heat, power from wind and waves, or from burning "biomass" (from rapidly growing "energy forrests"), etc., could make a considerable contribution to the world energy demand if these resources are utilized on a very large scale. However, a more thorough analysis of the technical, environmental, and social consequences of such a large scale deployment of renewable energy resources indicates numerous obstacles such as increased hazards to people, lack of material (for solar cells, batteries, etc.), or climatic changes.

Annually 15–20 PW y ($P \equiv 10^{15}$) of solar energy reach the land surface of the earth, so conversion of even a few percent would solve our energy problem. Unfortunately, except for home heating, conversion of solar energy is too expensive at present. Moreover, it is unlikely that solar energy will ever be used as the major industrial energy source since it is a very diffuse, low temperature source. Large land areas would be required to collect the sunlight: for a 1000 MWe plant, assuming a high conversion efficiency ($\sim 15\%$ for solar energy in the visible UV range) to electricity and an incidence of daily sunlight of temperate zones, a collector area of approximately 100 km^2 would be required. For the estimated 50 TW of future energy demand, a total of 5 million km^2 would be required for sunlight collection. In addition, the cost of the capital equipment involved, estimated to be \$15000 per capita based on present technology, would be quite prohibitive.

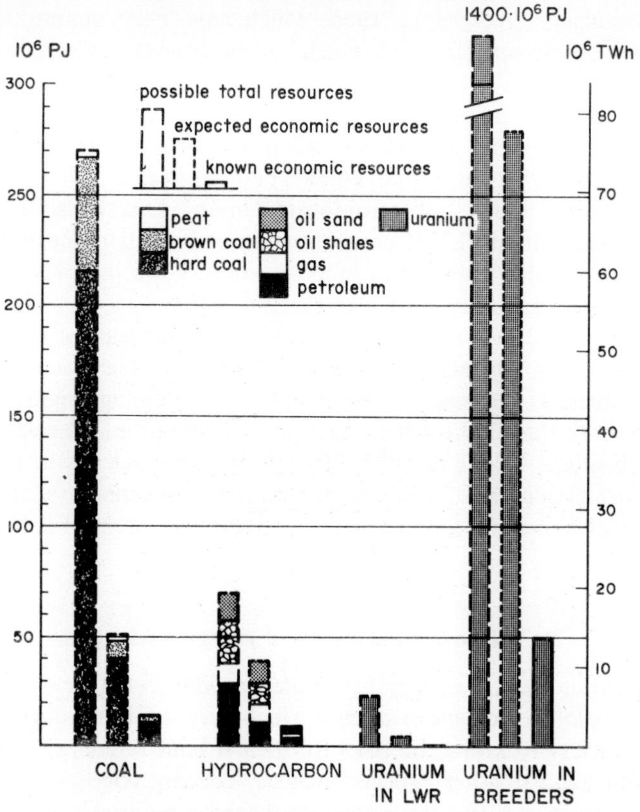

FIG. 21.3. World energy resources of coal, oil, fissile uranium (^{235}U), and fertile uranium (^{238}U, assuming 60% of it can be fissioned). The known hard coal reserves are one-third of the total coal reserves. One tonne natural uranium is assigned an energy value of 0.43 PJ (0.12 TWh; 30 000 MWd/6 tonnes natural uranium) in LWRs. (According to Bundesansfalt für Gewissenschaften und Rohstoffe, 1976.)

We can conclude that in the absence of a major and unforeseen technological development, solar power will contribute only 10–20% of the energy demand in the next 100 years (in the form of solar heating).

Independent studies by government organizations in different countries have concluded uniformly that renewable energy sources can contribute only a small fraction to the future world energy demand. This includes water power (hydroelectric), which in most projections will not contribute more than 2% to the world energy consumption.

Figure 21.3 shows the nonrenewable energy resources. The resources have been divided into:

(i) *Total resources*, i.e. the total amount in the earth which theoretically can be recovered. The estimates are based on general geologic considerations and includes undiscovered resources. For uranium it refers to resources of as low as 50 ppm U. For coal it refers to depths down to 2000 m in veins at least one foot thick.

(ii) *Recoverable reserves*, i.e. the amount of the total resources which are believed to be economically recoverable.

(iii) *Known economic reserves*, i.e. those which have been quantitatively estimated from geologic prospecting and which can be economically recovered.

Petroleum production (as a significant energy supplier) can be expected to cease within a century and, possibly, much earlier. Comparison of the data in Fig. 21.3 with the predicted energy demands show that known petroleum and LWR uranium reserves are insufficient even for man's short-term future demand. Even possible, "undiscovered" reserves of hydrocarbon and LWR uranium will not be able to sustain an increase in living standard. Coal and uranium in breeder reactors are the only energy resources which will last for at least a hundred years.

Presently we derive 92% of our energy from fossil and nuclear sources, 2% from water power, and 6% from biological sources (wood, dung, etc.). The fraction of energy from "renewable" sources, as discussed above, is unlikely to change significantly so it will account for no more than 10% of the total energy demand unless solar heating can be used on a much wider scale. Even if the average global efficiency in the use of energy could increase from the present $< 30\%$ to perhaps 50%, it seems we shall need to expand greatly our use of non-renewable energy sources unless or until solar power can be developed.

21.3. Detrimental effects of energy production

All production and use of energy for human needs—whether on a small or large scale—involves risks of accident or disease which may result in injury or death. It is of importance to society to know the risks from each kind of energy so that this factor can be balanced against other aspects such as working conditions, costs, land use, depletion of resources, etc., in selecting national energy policies.

For each energy source, the risks from the entire fuel cycle must be considered. The total evaluation must include the risks (i) in material acquisition (concrete, steel, etc.) and plant construction, (ii) in production of the fuel and its transport to the plant, (iii) in the operation and maintenance of the plant, (iv) to the public from emissions and waste

TABLE 21.1. *Public and occupational risks from different energy sources with a nominal output of* 1 *GWe y*

	Risks from entire fuel cycle[a]		Risks from plant operation[b]	
	Public lost mandays	Occupational lost mandays	Occupational disabilities	Occupational deaths
Natural gas	1000–2000	5 900	22	0.3
Nuclear	1400	8 700	35	0.07
Oil	1 920 000	18 000	42	0.5
Coal	2 010 000	73 000	875	4.1
Solar, thermal	510 000	101 000		
Solar, space heating	10 000	103 000		
Solar, electric	511 000	188 000		
Wind	539 000	282 000		
Water power			25	0.13
Wood			1380	3.5
Methanol	400	1 270 000		

[a] Adapted from H. Inhaber.
[b] Adapted from Swedish SOU 1977:56.

TABLE 21.2. *Emissions from 1 GWe power stations which annually produce six TWh electric energy at 35% thermal efficiency*
(Data from Swedish Energy Commission 1978 and USNRC)

Energy source	Oil[a]	Hard coal[b]	Wood	Natural gas	Nat. uranium[c]	
Heat of combustion (MWh t^{-1})	11.5	7.5	5.0	12.9	120 000[d]	
Fuel consumption (Mt y^{-1})	1.65	2.3	3.4	1.35	0.000 143	
Air emissions (t y^{-1})					Plant effluents (Ci y^{-1})[e]	
carbon dioxide (CO_2)	5 000 000	7 000 000	5 700 000	3 500 000	T	10 000[g]
carbon monoxide (CO)	700	1 200	15 000	200	^{14}C	20[g]
hydrocarbons	500	350	15 000	30	^{85}Kr	220 000[g]
nitrogen oxides (NO_x)	21 000	19 000	15 000	20 000	Pu(α)	0.001[h]
sulfurous oxides (SO_x)	10 000[j]	12 000[j]	1 700	15	Pu(β)	0.02[h]
smoke and dust	20	1 600	200	300	All others	9 000[h]
heavy metals	10	50	70	0	^{222}Rn	4000[f]
					^{226}Ra	0.002[f]
Solid waste; (t y^{-1})						
ash from filters[j]	2 000	230 000	17 000	0	100 solid LAW/MAW[i]	
desulfurization[j]	360 000	440 000	—	0	5–10 vitrified HAW	

[a]3% sulfur, no ash. [b]3% sulfur, 10% ash. [c]Data for a PWR–BWR mix as assumed for the 1975–2000 period in the Final Generic Environmental Statement on the use of recycle plutonium in MOX fuel in LWR reactors. US Nucl. Reg. Comm. –0002, 1976. [d]6 t nat. U per tonne enriched U at a rating of 30 000 MWd t^{-1} enriched U. [e]The effluent values are averaged over plant life and include all effluents at the reactor and associated fuel cycle. The total collective occupational dose is 650 manrem, and coll. population dose 720 manren. [f]Released at U mill. [g]No retention of these gases assumed. [h]Released in water. [i]Solidified LAW and MAW from reactor station only. [j]Smoke stack filters are assumed to have an efficiency of 99% for removal of particulates and 90% for desulfurization (lime scrubbers).

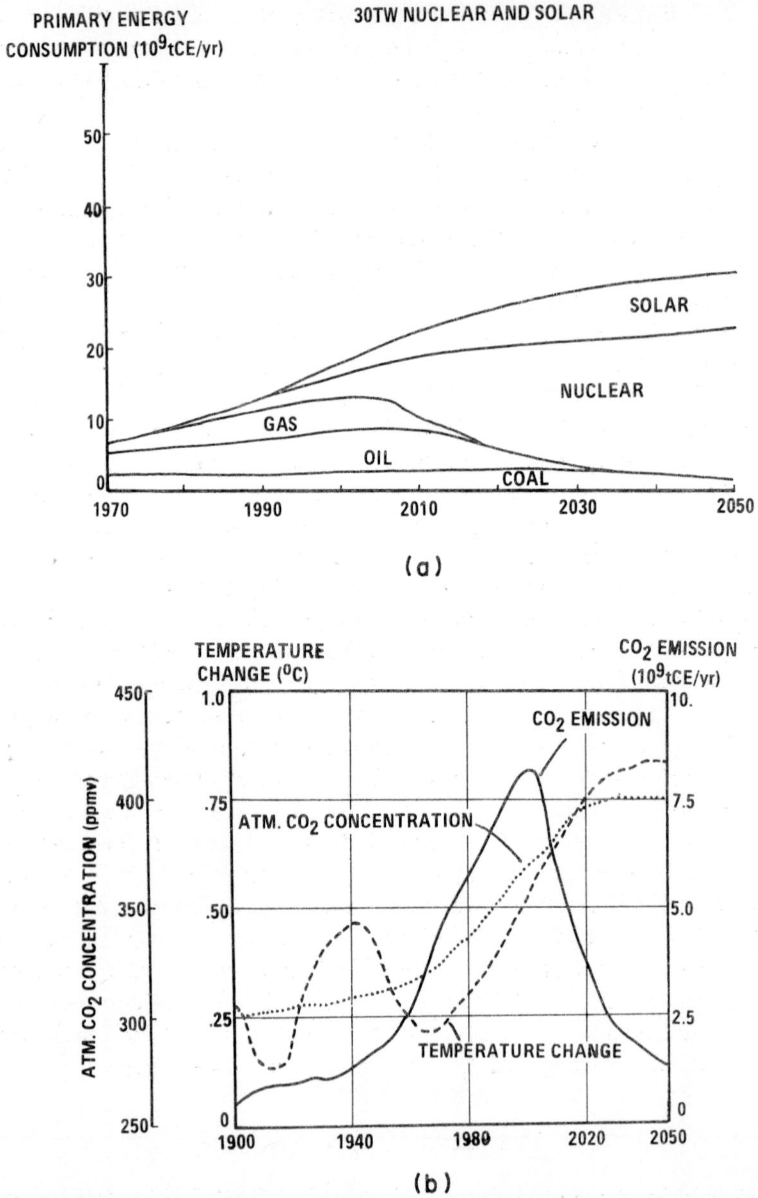

FIG. 21.4. Global energy scenario leading to a constant energy consumption of 30 TW y in year 2050 (a), and its consequences (b), due to CO_2 emission from combustion of gas, oil, and coal. 1 tCE (tonne coal equivalent) = 27 GJ. (According to J. Williams.)

disposition, and, finally, (v) associated with the demolition of the plant and restoring the land. The evaluation must be made over the entire lifetime of the plant and cover both normal operation and accidents. Points (ii) and (iii) are usually referred to as occupational risks, while point (iv) normally is the most hazardous to the public.

Points (i) and (v) are rarely included in risk analysis. However, all these points have been considered in a very extensive and thorough study by H. Inhaber. Some of his results are presented in Table 21.1, which also contains data on the occupational risks associated with the operation of power plants. The conclusion from this table is that the risks from natural gas, water power and nuclear energy are considerably smaller, both for the occupational worker and the public, than from oil and coal and from the free-flowing energy sources of wind and solar energy.

The major public risks in normal plant operation come from the emissions of waste products into air and water streams. Such normal emissions are listed in Table 21.2; it should be observed that the emissions from the fossil sources refer to plants with unusually high cleaning standards: 90% for SO_2 and 99% for particulate removal. Still, the large emissions of carbon, nitrogen, and sulfur oxides are a serious threat not only to the public health, but also to the global climate.

If we extrapolate the increasing rate of combustion of fossil fuels, the atmospheric concentration of CO_2 (presently $\sim$334 ppm) will have increased to 450 ppm in the year 2010, and to 600 ppm by 2030. Such CO_2 concentrations are expected to raise the global average temperature by 1°C (2010) and 2°C (2030). This may have several effects: reducing rainfall in the tropics, increasing precipitation in polar regions, melting polar ice with a rise in the sea level, etc. In fact, the CO_2 problem may curtail large scale future use of coal. Only with intense exploration of nuclear and solar energy and a limited growth energy policy (Fig. 21.4(a)) will it be possible to avoid considerable global climatic changes due to combustion of fossil fuels (Fig. 21.4(b)).

The SO_2 releases in western Europe have acidified all unbuffered lakes in this area, so that several now have pH < 4. It is estimated that 28 persons per day die in New York by excessive SO_2 emissions. Increased lethality can be observed at SO_2 concentrations as low as 0.1–1.0 ppm, if the exposure lasts for a day or more: chronic exposure to as little as 0.01 ppm SO_2 results in increased sickness absence of $\sim$ 1.5 days per person per year. The EPA air quality standard is 80 μg SO_2/m^3, or 0.023 ppm (1 ppm SO_2 = 2.86 mg SO_2/m^3); this value is much closer to observed health effects than the radiation health standards are (cf. Chapter 16). If the value of 80 $\mu g\, m^{-3}$ is used as a maximum limit, the coal- and oil-fired plants of Table 21.3 require about 1.4×10^{14} m^3 air/y for dilution to acceptable levels. If ^{85}Kr and ^{133}Xe from a nuclear plant are released in the same air volume, the MPC_a value of 3×10^{-7} Ci m^{-3} would mean an acceptable annual release of 42 MCi; however, for the Biblis A reactor (Table 19.6), the permitted release is 0.9 MCi y^{-1}, and the real release is $\sim$ 0.002 MCi y^{-1}, i.e. $\sim$ 25000 times less! Obviously, with regard to normal releases, the emissions from coal- and oil-fired plants are a much larger environmental hazard than the emissions from nuclear plants.

21.4. Nuclear reactor accidents

Opponents of nuclear power have provided dire accounts of the possible results of a catastrophic power excursion in a reactor. The safety aspects of reactors are discussed in Chapter 19, including the serious case of a core meltdown. In the more than 35 y since the first reactor was started, only one accident has occurred with release of significant amounts of radioactivity, the fire in the Windscale uranium–graphite reactor in 1957. Even this and the few other accidents (such as that of the Three Mile Island reactor) serious enough to cause extensive plant damage did not release enough radioactivity to be a danger to public health; in the Windscale case 25 persons received one-tenth of the maximum permissible dose.

In reactor safety studies (RSS) such as the Rasmussen Report (USA, 1976), an attempt has been made to assess the risk to the public of a major reactor accident. Risk is expressed as the product of the magnitude of the consequences that result when an event occurs and the frequency (probability) of that event:

$$\underset{\text{(consequences/y)}}{\text{Risk}} = \underset{\text{(consequences/event)}}{\text{Magnitude}} \times \underset{\text{(events/y)}}{\text{Frequency}}$$

In the calculation of this probability, Rasmussen uses an *event tree* (or fault tree) *methodology*. An event tree starts with an initial failure within the nuclear power plant. A "tree" is drawn starting with this initiating event to determine the course of events which results from the operation or nonoperation of various safety systems provided to prevent the core from melting with the possible release of radioactivity to the environment. All possible events and their combinations, whether fault or normal, are identified and their likelihood of occurrence determined. Many faults are hypothetical since they have never occurred; still, some probability must be assigned to their occurrence. The probability of each accident chain occurring is the product of the probability P_i of each item in the chain not working, or $P_1 \times P_2 \times P_3 \times$, etc.

An accident scenario for a BWR leads to a radioactivity release to the environment beginning about 2 h after the initial failure and lasting for 2 h. All the xenon and krypton, 40% of the iodine and cesium, 70% of the tellurium, 5% of the barium, and 50% of the ruthenium in the core are assumed to be released. The probability for this event is $\sim 10^{-6}$ per reactor year. The expected health effects are 450 latent cancers, 12000 thyroid illnesses, 450 genetic effects; normal incidence rates are 64000, 20000, and 100000, respectively.

Figure 21.5(a) and (b) show the result of the Rasmussen analysis; according to this,

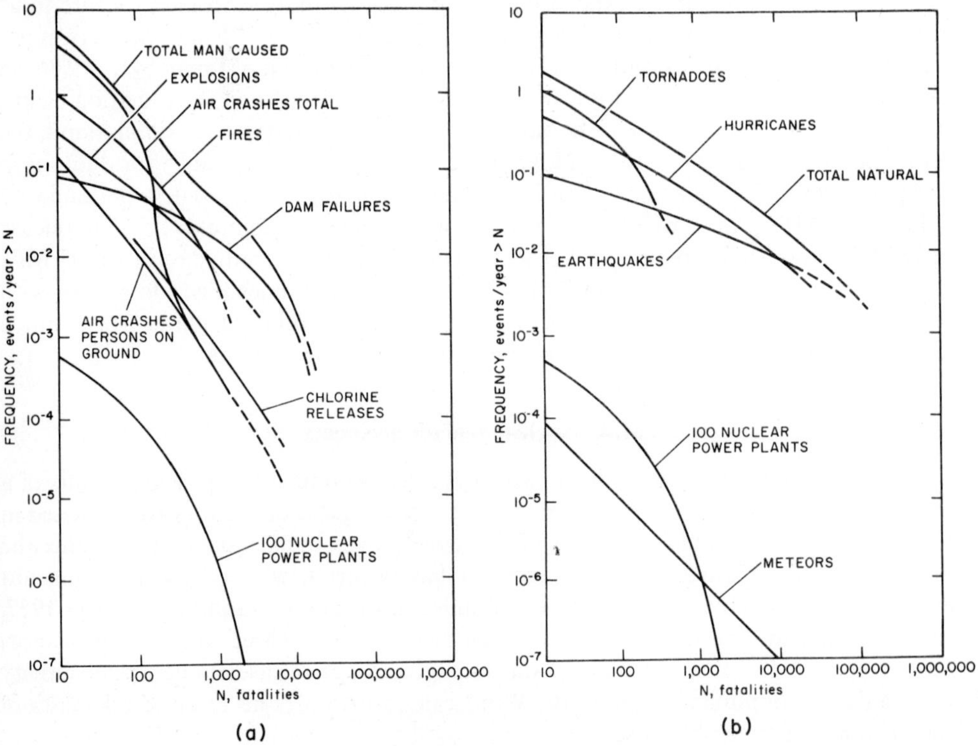

FIG. 21.5. Frequency of (a) man-caused events, (b) natural events with fatalities greater than *N*. (According to N. C. Rasmussen.)

the risk to the public of harm from a nuclear reactor accident is extremely small. The Rasmussen report has been much discussed and several critical reviews have suggested that the risk may be almost a factor 100 larger. However, even if the analysis is in error by a factor of 10^3, the risks of nuclear accidents are less than those of dam failures, hurricanes, etc. It is interesting that the Rasmussen analysis predicted an accident of the Three Mile Island type to have a probability of once in 8000 reactor years, whereas it actually happened after 3000 reactor years. Such an agreement would seem to add support to this analysis.

A reactor accident can also be caused by sabotage (not considered in the Rasmussen report). Nuclear reactors and other "sensitive" plants are designed to limit effects of sabotage by explosives so that only minor (if any) radioactivity would be released to the environment.

21.5. Normal reactor operation

A second source of potential danger to the public in nuclear power is the release of radioactivity during normal operation of a reactor. It is as impossible to have zero release of radioactivity as it is to have zero release of pollutants from other power stations. However, the small amounts of radioactivity released are constantly monitored to ensure they do not exceed the permissible levels. (Such monitoring of SO_2, nitrous oxides, etc, are almost never done at fossil-fueled plants.)

Xe, Kr, I_2, and $^3H_2$ can be released by reactors as gases. The amount released must be such that the exposure at the boundary of the plant area is less than 5 mrem y^{-1} (the particular value depends on local regulations). In practice, these releases have resulted in average exposures of less than 1 mrem y^{-1} or less than 1% of the normal background exposure everyone receives. Radioactivity escapes via liquid wastes discharged to the environment but, again, the amount is less than that present in normal sources. In 1970, for example, in the United States the nuclear power industry contributed only 0.007% by radioactive releases to the total annual radiation dose to the public (compared to 13.5% from medical X-rays).

Two kinds of biological effects due to radiation can be considered: somatic and genetic effects (Chapter 16). It has been estimated that 24 genetic mutations per year would

TABLE 21.3. *Estimated annual cancer mortality in the United States from radiation and other causes, 1970–2000*[a]

Cause of cancer	Cumulative cancer deaths to 2000	Average annual cancer deaths
Radiation		
Natural background	200 000	6 700
Medical X-rays	100 000	3 300
Jet airplane travel	7 000	230
Weapons fallout	7 000	230
Nuclear power industry[b]	90	3
Total from all radiation sources	314 090	10 500
All other causes	11 686 000	389 500
Total from all causes	12 000 000	400 000

[a]From *Nuclear Power and the Environment*, ANS, 1976.
[b]Assuming 300 power plant sites by the year 2000.

TABLE 21.4. *Life-shortening effects of various factors in human experience*[a]

Factors tending to decrease average lifetime	Decrease of average lifetime
Overweight by 25%	3.6 y
Male rather than female	3.0 y
Smoking 1 pack per day	7.0 y
City rather than country living	5.0 y
Actual radiation from nuclear power plants in 1970 ($\sim$ 20 GW)	Less than 1 min
Estimate for the year 2000 assuming hundredfold increase in nuclear power production	Less than 1 h

[a]From *Nuclear Power and Environment*, ANS, 1976.

occur in the total population of the United States (220 million) if all the present nuclear plants released the maximum permissible levels of radioactivity. However, the actual releases are much less and lead to an estimate of one mutation each 4 y. The normal genetic mutation rate is 800 000 per y, so the contribution by a greatly expanded nuclear industry would still remain an insignificant fraction of the normal rate. Table 21.3 gives estimates of cancer deaths while Table 21.4 shows the estimated life shortening. In both tables we see that the estimates attributable to the nuclear power industry in AD 2000 are quite small.

21.6. Reactor waste disposal

In 1952 James Conant, a prominent American scientist, stated that the safe, permanent disposal of radioactive waste was an insoluble problem and would prevent use of nuclear power. Among critics this view is still presented as one of the most compelling arguments against nuclear power. In principle, until 1972 it seemed Conant could be correct; since the waste would remain radioactive for generations, it could not be possible to prove that geologic disposal is safe over thousands of years by studies of a few years. However, the discovery of the remains of the Oklo natural reactors in 1972 provided definitive long term data at least for the Oklo type conditions.

At Oklo about 5 t of fission products and more than 2 t of plutonium were formed. The plutonium did not migrate at all during its lifetime. ^{137}Cs did migrate but ^{90}Sr seems to have remained fixed at least for a time much longer than its 30 y half-life. The rare gases escaped but the large bulk of the fission products as well as the transplutonium elements remained in place, decaying to stable nuclides, or to uranium, respectively.

While the Oklo data can be related directly only to the particular geology of that area, it is significant that the "waste" was not treated in any way to ensure its permanent retention.

Short term laboratory and field experiments on the behavior of plutonium and americium provide strong support for the generality of the Oklo data, at least for the actinides. These elements, which constitute the long term waste problem, show very little mobility in salt beds, granite, porous rocks, soils, and even water under natural conditions.

In Chapter 20 we have discussed waste disposal in some detail. Our conclusion is that this no longer represents an insoluble problem. Oklo data plus continuing studies in many laboratories indicate a number of acceptable modes of permanent disposal which pose a minimal threat to future generations.

21.7. Proliferation control of nuclear weapons

The risk of proliferation of nuclear weapons determines the nuclear energy policy of many countries and is used as a strong argument against widespread use of nuclear power. "Proliferation" refers to the spread of production capability for nuclear weapons. Often the word is used also to include diversion of fissile material to smaller nonnational groups; we shall refer to this as "nuclear theft" and discuss it in the next section.

Although heavy isotopes other than ^{235}U, ^{233}U, and ^{239}Pu are fissile, these latter nuclides can be most easily obtained in sufficient purity and amounts (10–50 kg, see §§19.7 and 19.19) for weapons use.

^{235}U is produced through isotope enrichment and most "first" nuclear weapons have been made with this isotope. The minimum requirements are some tens of tonnes of natural uranium and a working separations technique. While the amount of uranium is relatively easy to produce from low grade ores (§ 12.6), or possibly can be purchased illegally, the separations technique has always been a well-guarded secret. However, recent years have shown that it is possible for technically advanced countries to develop their own separations technique; this is particularly true for the relatively simple centrifugation method (§ 2.14).

^{233}U can be separated from neutron irradiated ^{232}Th. This requires not only a ^{232}Th-fueled reactor but also a separations technology for such spent fuel elements. Since only experimental thorium-fueled reactors are in operation (§ 19.15.1), and the reprocessing technology for thorium fuel elements is poorly developed (§ 20.7), ^{233}U is not considered as a proliferation sensitive material at present.

For ^{239}Pu the opposite is true. Hundreds of nuclear power stations are in operation in over 20 countries, annually producing several tonnes of plutonium. Even a small 50 MWth research reactor produces enough plutonium every year for one bomb. Reprocessing of spent uranium fuel elements has been carried out at the tonne level in 11 countries (Appendix K, plus the USSR and China). In addition, plutonium has been isolated from reactor irradiated uranium in at least 20 more national laboratories. Nuclear power and knowledge of reprocessing technology are thus available in almost every part of the world today. This condition can allow diversion of some plutonium for weapons fabrication.

As a result of the availability of knowledge of the technical processes for producing weapons grade ^{235}U and ^{239}Pu, the only certain way, at least at present, to avoid proliferation is through political agreements and international control. This can be achieved through the Treaty of the Proliferation of Nuclear Weapons (NPT) and the nuclear *Safeguards* system operated by the International Atomic Energy Agency.

The provisions of the NPT are as follows. Article I: the nuclear weapons states (NWS) undertake not to transfer nuclear explosive devices or control over them, or to help any nonnuclear weapons state (NNWS) to acquire such a capacity. Article II: the NNWS undertake not to acquire such devices. Article III: the states accept the control of this undertaking by the IAEA. Article IV: the rights of the NNWS to undertake research, production, and exploitation of nuclear energy for peaceful purposes is confirmed. The NPT covers the entire national nuclear program (including research) existing in the NNWS.

In August 1977 exactly 100 states had agreed to the NPT, including the USSR, the UK, and the USA. The NWS who have not signed the NPT are France and India.

Only four other states in the world are believed to have significant nuclear activities without being NPT parties or under Safeguards control: Egypt, Israel, South Africa, and Spain.

After signing and ratifying the NPT, a nation is required to negotiate a "safeguards agreement" with the IAEA. Detailed procedures are agreed upon for the control of each nuclear establishment: location, design, flow of nuclear material, location of equipment for producing or processing material, stock taking, verification procedures, etc. The Safeguards control is concentrated on "strategic points" since diversion of fissile material is only possible at a few key stages of the fuel cycle. For example, the top radiation shield of every Safeguarded reactor is not only sealed by IAEA but is also viewed by a locked automatic camera which takes pictures at irregular intervals. All containers transporting fissile material are sealed by IAEA. The nuclear plant operator must keep a precise account of all nuclear material within the plant, based on exact measurements of the material, which can be checked independently by IAEA inspection. It may be necessary to station inspectors permanently at fuel-reprocessing plants. However, this system does not include plants using classified technology (e.g. gas centrifugation) or of military importance.

Safeguards agreements can be signed for specific plants by a nonparty to NPT. Also parties to the NPT can exclude some plants from the Safeguards system; this is the case for the nuclear weapons facilities in the USSR, the UK, and the USA. In 1978 in the NNWS about 600 different plants (power reactors, reprocessing plants, etc.) with an amount of plutonium of over 20 tonnes were subject to Safeguards control.

21.8. Theft of fissile material

A nuclear weapon can be acquired by small, organized and very determined groups by (i) stealing the weapon from a military depot, (ii) stealing highly enriched uranium compound from a laboratory or during transportation (the HTGR and some research reactors use 90% enriched uranium), (iii) stealing a plutonium compound at a laboratory or during transportation, and (iv) stealing used reactor fuel elements. The theft of the uranium or plutonium may be spread over a period of time which could go unnoticed, or it could occur in a single action, which is likely to be detected. By careful materials accounting (e.g. under Safeguards supervision) and good physical protection all these actions may be hindered. The most unlikely action is (iv) because it would require transport of bulky stolen material and extensive long term work with highly radioactive material; the chance to survive capture seems highly improbable. The weakest point may be the plutonium transport (iii). This risk can be almost eliminated or reduced to almost the same level as that of spent reactor fuel elements by using "proliferation safe fuel cycles".

21.9. Proliferation safe fuel cycles

The present uranium–plutonium fuel cycle (§20.3.2) is considered to be especially vulnerable to diversion of fissile material since large amounts of pure plutonium compounds are produced and stored at the reprocessing plant and then shipped to fuel element fabrication centers. Even with Safeguards control, theft could occur along this line.

The establishment of *regional nuclear fuel cycle centers* under international control, instead of a larger number of national plants, could further reduce this risk. Spent fuel elements would be sent to such centers, where also new plutonium bearing fuel elements would be fabricated, whether (i) MOX fuel elements for LWRs containing $\leqq 5\%$ Pu, or (ii) MOX elements for FBRs containing $\leqq 20\%$ Pu. The "proliferation safety" is assured through the international status of the center. If the remixing is made from the outgoing plutonium and uranium streams from the Purex plant, pure plutonium would occur only within the processing line. In case of theft to make weapons out of this material, the uranium and plutonium would have to be separated chemically; considering the criticality problem of plutonium solution, this is not an easy task.

Several methods have been suggested in order to further improve the proliferation safety. (a) The new fuel elements would be "spiked" with a γ-emitting radionuclide (e.g. ^{137}Cs) so that remote handling would be required for weapons fabrication. (b) The new fuel elements would be slightly irradiated (up to 1000 MWd t^{-1} has been suggested), which would increase handling difficulty, and possibly also require some chemical processing before use. (c) The reprocessing procedure would be modified to leave some fission products with the plutonium, thus making it very radioactive and, hence, more difficult to handle. (d) All the actinides would be left with the FBR MOX fuel elements. Though the two latter suggestions would have the same effect as case (b), they have some further advantages.

According to the Civex process, which is, in principle, case (c), a reprocessing plant would consist of only a few cells. One cell would be used for chopping and dissolving. A second cell for chemical separation would have three streams: (i) almost all plutonium plus some uranium (to make the plutonium content $\leqq 25\%$) and some fission products (mostly Zr and Ru), (ii) the remaining uranium, (iii) the remaining fission products. Remaining cells would be used for transforming the Pu–U–FP mixture into new fuel elements and for uranium purification and fission product handling. In this case no pure plutonium appears in the plant.

Case (d) has been discussed partially in §§20.3.2 and 20.11.2. The same advantages as for case (c) can be achieved, but in addition the high active waste would be much more harmless and not constitute a very long term hazard.if the actinides are removed.

All four cases lead to new fuel elements with considerable radioactivity (for case (a)–(c) mainly γ, for case (d) mainly α- and n-radiation). Though this would create some difficulty in handling, the experience of the nuclear industry is sufficient to make this a minor problem, and it would have no major effect on the cost of nuclear power electricity.

The once-through fuel cycle (§ 20.3.1) has been considered highly proliferation proof by some. Depositing large amounts of spent uranium fuel elements containing tonnes of plutonium in the ground, however, constitutes a particular risk, sometimes referred to as a "plutonium time bomb". As the radioactivity of the fission products decays, it would become successively more easy to recover the fuel elements for purposes of using the plutonium content whether for energy production or for weapons. Since the need to protect the storage facility would increase with time, it would constitute a very complicated Safeguards situation.

The thorium–uranium fuel cycle (§20.3.3) is no more proliferation safe than the uranium–plutonium fuel cycle, since it is necessary to isolate pure fissile ^{233}U from the fertile ^{232}Th fuel in order to keep the cycle running (Fig. 20.5). The ^{233}U could be "denaturated" by addition of ^{238}U. However, in the next irradiation cycle this would lead to production of ^{239}Pu. The dilution of ^{233}U with ^{235}U would not denaturate the ^{233}U because ^{235}U is also fissile. Thus, problems similar to those for the uranium–plutonium cycle are present for the thorium–uranium cycle except in the use of the molten salt breeder reactor (§19.14.6 or §19.15) for which a closed reactor-reprocessing loop could be used.

All steps to secure the nuclear power fuel cycles against theft and proliferation would not protect against theft and proliferation through enrichment. Therefore, all uranium handling also must be protected. It has been suggested that all steps in the fuel cycle be Safeguarded, and that enrichment be carried out only in international centers. However, as long as some countries are not members of the NPT or withhold some plants from the Safeguards control, absolute protection against proliferation seems impossible.

21.10. Conclusions

Uranium and thorium used in LWRs constitute an energy source of the same magnitude as oil and natural gas. Fission energy may therefore have a time period in human civilization as short as we expect from these other energy sources. However, the success of the prototype breeder plants makes it probable that fission energy will last longer and reach a larger use than oil and gas have today. Using coal as a raw material, nuclear energy may be used to produce synthetic oil and gas when these natural resources have been depleted. It seems likely that both coal and nuclear power will be used to supply the energy demands of the world, until other—not yet developed—energy resources can take over.

Progress in research on fusion reactors is encouraging. This second source of nuclear power may provide the best answer to mankind's long-range need for cheap energy, without which no improvement of living conditions can be achieved for an increasing global population.

21.11. Literature

F. D. Sowby, Radiation and other risks, *Health Physics* **11** (1965) 879.

E. M. Mason, Nuclear reactors; transforming economics as well as energy, *Techn. Rev.*, March 1969, p. 27.

C. Starr, F. J. Dyson, M. K. Hubbert, D. M. Gates, W. B. Kemp, R. A. Rappaport, E. Cook, C. M. Summers, D. B. Luten, M. Tribus, E. C. McIrvine, and M. Katz, *Energy and Power*, Scientific American Book, Freeman, 1971.

J. W. Gofman and A. R. Tamplin, *Poisoned Power. The Case Against Nuclear Power Plants*, Chatto & Windus, London, 1973.

D. R. Inglis, *Nuclear Energy—Its Physics and its Social Challenge*, Addison-Wesley, 1973.

P. H. Pigford, Environmental Aspects of Nuclear Energy Production, *Ann. Rev. Nucl. Sci.* **24** (1974) 515.

Safeguards Against Nuclear Proliferation, SIPRI monograph, The MIT Press, Cambridge, USA, 1975.

N. C. Rasmussen, *Reactor Safety Study: An Assessment of Accident Risks in US Commercial Nuclear Power Plants*, WASH-1400 (NUREG 75/014), 1975.

P. Beckmann, *The Health Hazards of not Going Nuclear*, The Golem Press, Boulder, Colorado, 1976.

J. L. Weeks, *The Biological Costs of Industrial Activities*, AECL, 1975.

R. Avenhaus, W. Häfele, and P. E. McGrath, *Consideration on the Large Scale Deployment of the Nuclear Fuel Cycle*, Int. Inst. for Applied Systems Analysis, RR.-75-36, Schloss Laxenburg, 1975.

The Union of Concerned Scientists, *The Nuclear Fuel Cycle*, The MIT Press, Cambridge, Mass., 1975.

H. Ashley, R. L. Rudman, and C. Whipple (eds.), *Energy and the Environment: A Risk-Benefit Approach*, Pergamon Press, Oxford, 1976.

W. Häfele, Energy options open to mankind beyond the turn of the century, *Proc. Int. Conf. on Nuclear Power and its Fuel Cycle, Salzburg, 1977*, Vol. 1, p. 59, IAEA, Vienna, 1977.

A. B. Lovins, *Soft Energy Paths: Toward a Durable Peace*, Gallinger, 1977.

The Ford Foundation/The Mitre Coporation, *Nuclear Power, Issues and Choices*. Ballinger, Cambridge, Mass., 1977.

H. Inhaber, *Risk of Energy Production*, Atomic Energy Control Board, Report-1119, Canada, 1978.

CIVEX: the answer to blackmail, *Nucl. Eng. Int.* **23** July (1978) 59.

A. M. Weinberg, Reflections on the energy wars, *American Sci.* **66** (1978) 153.

R. H. Flowers, K. D. B. Johnson, J. H. Miles, and R. K. Webster, Long-term options for the FR fuel cycle, *Atom* **259** (1978) 138.

J. Williams (ed.), *Carbon Dioxide, Climate and Society*, Pergamon Press, Oxford, 1978.

APPENDIX A

Literature

In the last decades the literature in nuclear science and technology has grown to such proportions that it has become quite difficult for the nonspecialist to uncover more advanced but not too complicated reading, which he may need for a specific purpose. Therefore after most chapters we suggest a limited amount of further reading that goes moderately deeper into the subject just treated but that may also contain references to the more advanced literature.

In this part of the book we list some literature that may be used either as a general background for the chapters dealing with nuclear physics phenomena or as a "next step" for broadening the general knowledge in basic and applied nuclear chemistry. We have also included book series, which constitute the most recent survey of a specific field, as well as journals publishing current progress in areas of interest to the nuclear chemist. These listings are usually of older, more general texts, while the literature after each chapter is more specific and/or later.

A.1. Monographs and textbooks in atomic and nuclear physics

D. Blanc, *Physique Nucléaire*, Masson et Cie, 1975.
M. G. Bowler, *Nuclear Physics*, Pergamon Press, Oxford, 1973.
B. L. Cohen, *Concepts of Nuclear Physics*, McGraw-Hill, 1971.
H. Enge, *Introduction to Nuclear Physics*, Addison-Wesley, 1969.
R. D. Evans, *The Atomic Nucleus*, McGraw-Hill, 1955.
E. Fenyves and O. Haiman, *The Physical Principles of Nuclear Radiation Measurements*, Academic Press, 1969.
W. Gentner, H. Maier-Leibnitz, and H. Bothe, *An Atlas of Typical Expansion Chamber Photographs*, Pergamon Press, Oxford, 1953.
G. Hertz, *Lehrbuch der Kernphysik*, Bände I–III, Teubner Verlagsgesellschaft, Leipzig, 1958–1962.
P. E. Hodgson, *Nuclear Reactions and Nuclear Structure*, Clarendon Press, 1971.
P. E. Hodgson, *Nuclear Heavy Ion Reactions*, Clarendon Press, 1978.
I. Kaplan, *Nuclear Physics*, Addison-Wesley, 1958.
P. Marmier and E. Sheldon, *Physics of Nuclei and Particles*, Academic Press, 1969–1970.
E. B. Paul, *Nuclear and Particle Physics*, North-Holland–Wiley, 1969.
M. A. Preston, *Physics of the Nucleus*, Addison-Wesley, 1962.
M. A. Preston and R. K. Bhaduri, *Structure of the Nucleus*, Addison-Wesley, 1975.
K. Siegbahn (ed.), *Alpha-, Beta- and Gamma-ray Spectroscopy*, North-Holland, 1965 (3rd edn., 1974).
R. Vandenbosch and J. R. Huizinga, *Nuclear Fission*, Academic Press, 1973.
L. C. L. Yvan, *Elementary Particles, Science, Technology, and Society*, Academic Press, 1971.

A.2. Monographs and textbooks in nuclear chemistry

E. Broda and T. Schönfeld, *Die technischen Anwendungen der Radioaktivität*, VEB Deutscher Verlag, Leipzig, 1962.
G. D. Chase and J. L. Rabinowitz, *Radioisotope Methodology*, Burgess Co., 1962.
G. Choppin, *Experimental Nuclear Chemistry*, Prentice-Hall, 1961.
G. Choppin, *Nuclei and Radioactivity*, Benjamin, 1964.
D. Desoete, R. Gijbels, and J. Hoste, *Neutron Activation Analysis*, Wiley–Interscience, 1972.

J. F. DUNCAN and G. B. COOK, *Isotopes in Chemistry*, Clarendon Press, 1968.
R. A. FAIRES and B. H. PARKS, *Radioisotope Laboratory Techniques*, 3rd ed., Butterworth, 1973.
G. FRIEDLÄNDER, J. W. KENNEDY, and J. M. MILLER, *Nuclear and Radiochemistry*, Wiley, 1964.
M. HAISSINSKY, *La Chimie nucléaire et ses applications*, Masson et Cie, 1957. (Engl. edn., Addison-Wesley, 1964.)
B. G. HARVEY, *Nuclear Chemistry*, Prentice-Hall, 1965.
B. G. HARVEY, *Introduction to Nuclear Physics and Chemistry*, Prentice-Hall, 1969.
N. F. JOHNSON, E. EICHLER, and G. D. O'KELLEY, *Nuclear Chemistry*, Vol. II of *Tech. in Inorg. Chem.* (eds. H. B. JONASSEN and A. WEISSBERGER), Interscience, 1963.
M. LEFORT, *Chimie nucléaire*, Dunod, Paris, 1966. (Engl. edn., *Nuclear Chemistry*, van Nostrand, 1968.)
K. H. LIESER, *Einführung in die Kernchemie*, Verlag Chemie GMBH, Weinheim/Bergstrasse, 1969.
R. LINDNER, *Kern- und Radiochemie*, Springer-Verlag, 1961.
H. A. C. MCKAY, *Principles of Radiochemistry*, Butterworth, 1971.
A. N. NESMEYANOV, B. I. BARANOV, K. B. ZABORENKO, N. P. RUDENKO, and Y. A. PRISELKOV, *Handbook of Radiochemical Exercises*, Pergamon Press, Oxford, 1965.
R. T. OVERMAN and H. M. CLARK, *Radioisotope Techniques*, McGraw-Hill, 1960.
A. C. WAHL and N. A. BONNER, *Radioactivity Applied to Chemistry*, Wiley, 1951.
Y. WANG, *Handbook of Radioactive Nuclides*, The Chemical Rubber Co., 1969.
L. YAFFE (ed.), *Nuclear Chemistry*, Vols. I and II, Academic Press, 1968.
Source Material for Radiochemistry, Nucl. Sci. Ser. 42, Nat. Acad. Sci. Washington DC, 1970.

A.3. Textbooks in nuclear science and technology

M. BENEDICT and T. H. PIGFORD, *Nuclear Chemical Engineering*, McGraw-Hill, 1957.
J. K. DAWSON and G. LONG, *Chemistry of Nuclear Power*, George Newnes, 1959.
S. GLASSTONE, *Sourcebook on Atomic Energy*, 3rd edn., van Nostrand, 1967.
S. GLASSTONE and A. SESONSKE, *Nuclear Reactor Engineering*, van Nostrand, 1963.
S. PETERSON and R. G. WYMER, *Chemistry in Nuclear Technology*, Addison-Wesley, 1963.
J. G. WILLS, *Nuclear Power Plant Technology*, Wiley, 1967.

A.4. Book series in nuclear science and technology

Annual Review of Nuclear Science, Ann. Rev. Inc., Palo Alto, since 1951.
Annual Review of Physical Chemistry, Ann. Rev. Inc., Palo Alto, since 1950.
Proc. Int. Conf. on the Peaceful Uses of Atomic Energy (PICPUAE), United Nations, New York, 1st conf., Geneva, 1955, 2nd 1958, 3rd 1964, 4th 1971 (publ. 1972).
Advances in Inorganic Chemistry and Radiochemistry (ed. H. J. EMELEUS and A. G. SHARPE), Academic Press, since 1959.
Progress in Nuclear Energy, Pergamon Press, Oxford, since 1956. Condensed from PICPUAE; series in Process Chemistry, etc.
Progress in Nuclear Physics (ed. O. R. FRISCH **1** (1952) to **9** (1964); D. M. BRINK and J. M. MULVEY **10** (1969)).
Handbuch der Physik, volumes on nuclear particles, reactions, structure, etc., Springer-Verlag, since 1955.
Gmelins Handbuch der Anorganischen Chemie, volumes on all chemical elements, Verlag Chemie GMBH, Weinheim/Bergstrasse.
Advances in Nuclear Science and Technology, Academic Press, since 1962.

A.5. Journals in nuclear chemistry and connected fields

Applied Radiation and Isotopes, Int. J. of, Pergamon Press, Oxford, since 1956 (12)*
Health Physics official journal of Health Physics Society, Pergamon Press, Oxford, since 1958 (12).
Inorganic and Nuclear Chemistry, J. of, Pergamon Press, Oxford, since 1955 (12).
Inorganic and Nuclear Chemistry Letters, Pergamon Press, Oxford, since 1965 (12).
Isotopenpraxis Akademie Verlag, Berlin (DDR), since 1966 (12).
Labelled Compounds and Radiopharmaceuticals , J. of, Wiley, since 1965 (4).
Radioanalytical Chemistry, J. of, Elsevier Sequoia, Lausanne, since 1968 (14).
Radiochemical and Radioanalytical Letters, Elsevier Sequoia, Lausanne, since 1969 (6).
Radio Chimica Acta, Akademische Verlagsgesellschaft, Frankfurt a. M., since 1962 (2).
Radioisotopes official journal of Japan Radioisotope Soc., Tokyo, since 1952 (12).

*Volumes per year in parentheses.

Radiation Effects, Gordon & Breach, since 1973.
Radiation Physics and Chemistry, Int. J. of, Pergamon Press, Oxford, since 1969 (6).
Soviet Radiochemistry (Radiokhimiya), Consultants Bureau, New York, since 1962.

A.6. Journals in nuclear technology

American Nuclear Society, Transactions of, ANS, since 1958 (~2).
Annals of Nuclear Energy (formerly *J. of Nuclear Energy)*, Pergamon Press, Oxford, since 1959.
Atom, UK AEA, from 1956.
Atomic Energy Review, IAEA, from 1963 (~ 4).
Atomwirtschaft, Handelsblatt GmbH, Düsseldorf, since 1956 (6).
British Nuclear Energy Society, J. of, Inst. of Civil Engineers, London, since 1961 (12).
CEA, Notes d'Information, CEA, since 1962.
Energia Nucléare, Centro Informazioni Studi Esperienze, Milan, since 1951 (12).
Energie Nucléaire, Société de Productions Documentaires, Rueil Malmaison, France, since 1948 (6).
INIS Atomindex, IAEA, since 1971.
IAEA Bulletin, IAEA, since 1959 (6).
Nuclear Science and Technology, J. of, Atomic Energy Society of Japan, since 1964 (12).
Kerntechnik und Atompraxis, Karl Thiemig, since 1959 (8).
Meetings on Atomic Energy, IAEA, since 1969 (~4).
Nuclear India, Dept. of Atomic Energy, Bombay, (2).
Nuclear Instruments and Methods, North-Holland, since 1956 (24).
Nuclear News, official journal of ANS, since 1957 (12).
Nuclear Safety, USDOE, US GPO, since 1959 (6).
Nuclear Science Abstracts, USAEC, 1947–1975.
Nuclear Science and Engineering, ANS, since 1956 (12).
Nuclear Technology (formerly *Nuclear Applications*), ANS, since 1965 (12).
NRC News Releases, NRC, since 1976.
Nucleonics Week, McGraw-Hill, since 1960 (52).
Progress in Nuclear Energy, Series 1–12, ANS, since ~ 1965.
Soviet Atomic Energy (Atomnaya Energiya), Consultants Bureau, New York, since 1973 (~12).

A.7. Tables containing important nuclear (and other) data

A. I. Aliev, V. I. Drynkin, D. I. Leipunskaya, and V. A. Kasatkin, *Handbook of Nuclear Data for Neutron Activation Analysis*, Israel Program for Scientific Translation, Jerusalem, 1970.
W. H. Johnson (ed.), *Nuclide Masses*, Springer-Verlag, Heidelberg, 1965.
C. M. Lederer, J. M. Hollander, and I. Perlman, *Table of Isotopes*, Wiley, New York, 1967.
C. M. Lederer and V. S. Shirley (Eds.), *Table of Isotopes* (7th ed.), Wiley, New York, 1978.
A. E. Martell and R. M. Smith (Eds.), *Critical Stability Constants*, Vol. 1, Amino Acids (1974); Vol. 3, Other Organic Ligands (1977); Vol. 4, Inorganic Complexes (1976), Plenum Press.
J. H. E. Mattauch, W. Thiele, and A. H. Wapstra, *1964 Atomic Mass Table. Consistent Set of Q-values.* Adjustment of relative atomic masses, *Nucl. Phys.* **67** (1965) 1–120. Also to be found in appendix of B. G. Harvey, *Introd. to Nucl. Phys. and Chem.*, 1969.
W. Seelman-Eggebert, G. Pfennig, and H. Münzel, *Nuklidkarte*, Kernforschungszentrum Karlsruhe, 1974.
V. Viola and G. T. Seaborg, Nuclear systematics of the heavy elements, I and II, *J. Inorg. Nucl. Chem.* **28** (1966) 697, 741.
F. W. Walker, G. J. Kirouac and F. M. Rourke (Eds.), *Chart of the Nuclides*, 12th ed., General Electric Co., San Jose, Calif.
Y. Wang (ed.), *Handbook of Radioactive Nuclides.* (Emphasis on health physics and medical uses.) The Chemical Rubber Co., Ohio, 1969.
Atomic Data and Nuclear Data Tables/Nuclear Data Sheets, Academic Press, since 1965.
Handbook of Chemistry and Physics, The Chemical Rubber Co., Cleveland, Ohio.
Neutron Cross Sections, BNL-325, 2nd edn., 1958; suppl. no. 2, 1966; 3rd edn., *Resonance Parameters*, Superint. of Documents, U.S. Gov. Print. Office, Washington DC, 1973.
Stability Constants, 1964 (Spec. publ. no. 17), Suppl. I; 1971 (Spec. publ. no. 25), Chem. Soc., London.

APPENDIX B

Isotope Effects in Chemical Equilibrium

In a mixture of molecules AX and BX, with a common element X, an exchange of the atoms of the common element between the two molecules may occur. When the two compounds have different isotopes X and X*, we may have an isotope exchange according to

$$AX + BX^* \rightleftharpoons AX^* + BX \tag{B.1}$$

The equilibrium constant k in the reaction is given by

$$\Delta G^0 = -\mathbf{R}T \ln k = \mathbf{R}T \ln \frac{[AX^*][BX]}{[AX][BX^*]} \tag{B.2}$$

where ΔG^0 is the (Gibb's) free energy and $\mathbf{R}$ is the universal gas constant.

B.1. The partition function

It has been shown in §2.6 that k deviates slightly from 1, giving rise to the isotopic effects observed in nature for the light elements. This deviation can be not only explained but also calculated by methods of statistical thermodynamics. Only the main features will be given here; for a more detailed explanation the reader is referred to textbooks in physical chemistry and the literature on statistical thermodynamics.

According to this theory the equilibrium constant k can be written

$$k = F_{AX^*}\, F_{BX} / F_{AX}\, F_{BX^*} \tag{B.3}$$

where F is the *grand partition function*, which for each molecule includes all possible energy states of the molecule and the fraction of molecules in each energy state under given external condition. The grand partition function is defined by

$$F = f_{tr} f_{rot} f_{vib} f_{el} f_{nsp} \tag{B.4}$$

where each term f_j refers to a particular energy form: translation, rotation, vibration, electron movement, and nuclear spin. The two latter will have no influences on the chemical isotope effect, and can therefore be omitted. It can be shown that each separate partition function f_j can be described by the expression

$$f_j = \sum^{i} g_{j,i}\, e^{-E_{j,i}/\mathbf{k}T} \tag{B.5}$$

where E_i is the particular energy state i for the molecule's energy mode j; e.g. for $j =$ vibration, there may be 20 different vibrational states (i.e. the maximum i-value is 20) populated. $\mathbf{k}$ is the Boltzmann constant; (B.5) is closely related to the Boltzmann distribution law (see Chapter 11).

The term $g_{j,i}$ is called the *degeneracy* and corrects for the fact that the same energy state in the molecule may be reached in several different ways. The summation has to be made over all energy states i.

B.2. Kinetic energy and temperature

The total energy of one mole of an ideal gas at temperature T (K) is given by its translational energy, which according to the kinetic gas theory is

$$E_{tr} = 3/2\mathbf{R}T \quad (\mathrm{J\,K^{-1}\,mole^{-1}}) \tag{B.6}$$

Dividing by the Avogadro number **N** yields the average kinetic energy per molecule (or particle)

$$\bar{E}_{tr} = 3/2\,\mathbf{k}T \quad (\mathrm{J\,K^{-1}\,particle^{-1}}) \tag{B.7}$$

where **k** is $= \mathbf{R}/\mathbf{N}$. From mechanics we know that the kinetic energy of a single particle of mass m and velocity v is

$$E_{kin} = \tfrac{1}{2}mv^2 \tag{B.8}$$

Summing over a large number of particles, we can define an average kinetic energy

$$\bar{E}_{kin} = \tfrac{1}{2}m\bar{v}^2 \tag{B.9}$$

where $\bar{v}^2$ is the mean square velocity. Because (B.7) is the average kinetic energy at temperature T, it must equal (B.9):

$$\bar{E}_{tr} = \bar{E}_{kin} \tag{B.10a}$$

or

$$\tfrac{3}{2}\mathbf{k}T = \tfrac{1}{2}m\bar{v}^2 \tag{B.10b}$$

Thus for a given temperature there is a corresponding *average particle velocity*. However, the individual particles are found to move around with slightly different velocities. J. C. Maxwell has calculated the velocity distribution of the particles in a gas. For the simplest possible system (with $g = 1$) it is a Boltzmann distribution. In a system of n_0 particles the number of particles n_E that have kinetic energy $> E$ is given by

$$n_E = n_0 e^{-E/\mathbf{k}T} \tag{B.11}$$

In Fig. B.1 n_E/n_0 is plotted as a function of E for three different T's. For the line at 290 K we have marked the energies $\mathbf{k}T$ and $\tfrac{3}{2}\mathbf{k}T$. While $\tfrac{3}{2}\mathbf{k}T$ (or rather $\tfrac{3}{2}\mathbf{R}T$) corresponds to the thermodynamic average translational energy, $\mathbf{k}T$ corresponds to the *most probable kinetic energy*: the area under the curve is divided in two equal halves by the $\mathbf{k}T$ line.

In chemistry, the thermodynamic energy (B.7) must be used, while in nuclear reactions the most probably energy E' must be used, where

$$E' = \tfrac{1}{2}m(v')^2 = \mathbf{k}T \tag{B.12}$$

v' is the most probable velocity.

Although the difference between $\bar{E}$ and E' is not large (e.g. at 17°C $\bar{E} = 0.037$ eV and $E' = 0.025$ eV), it increases with temperature. The most probable velocity is the deciding factor whether a nuclear reaction takes place or not. Using (B.11) we can calculate the fraction of particles at temperature T having a higher energy than $\mathbf{k}T$: $E' > 2\mathbf{k}T$, 14%, $> 5\mathbf{k}T$, 0.67%, and $> 10\mathbf{k}T$, 0.0045%. This high-energy tail is of impor-

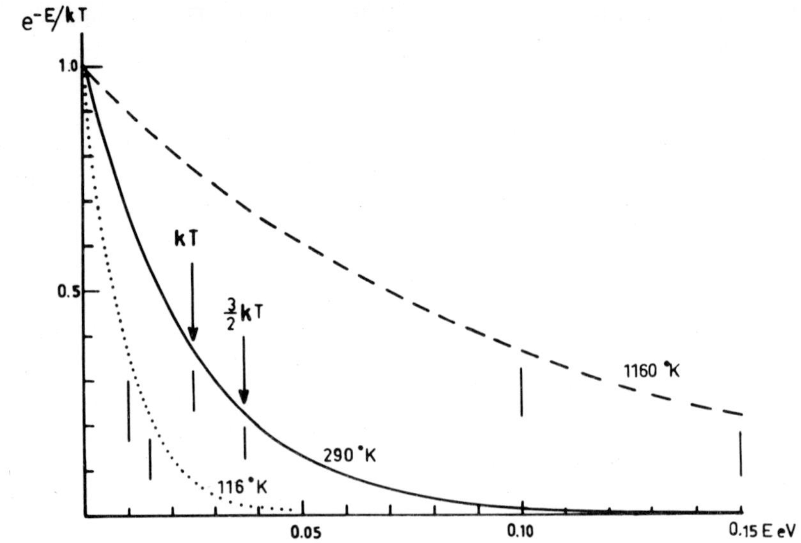

FIG. B.1.

tance for chemical reaction yields, because it supplies the molecules with energies in excess of the activation energy (see Appendix C). It is also of importance for nuclear reactions and in thermonuclear processes.

B.3. The partial partition functions

So far there has been no clue to why isotopic molecules such as H_2O and D_2O, or $H^{35}C$ and $H^{37}C$, behave chemically in slightly different manner. This explanation will come from a study of the energy term E_j, which will be shown to contain the atomic masses for all three molecular movements: translation, rotation, and vibration.

(a) *Translational energy.* The translational energy, as used in chemical thermodynamics, involves molecular movements in all directions of space. The energy is given by the expression (B.7). A more rigorous treatment leads to the expression

$$f_{\text{tr}} = (2\pi M\mathbf{k}T)^{3/2} V_{\text{M}} \mathbf{h}^{-3} \tag{B.13}$$

for the translational partition function, where V_{M} is the molar volume and **h** the Planck constant. It will be observed that no quantum numbers appear in (B.13); the reason is that they are not known because of the very small ΔE's of such jumps.

(*b*) *Rotational energy.* Taking the simplest case, a linear diatomic molecule with atomic masses m_1 and m_2 at a distance r apart, the rotational energy is given by

$$E_{\text{rot}} = I_{\text{rot}}\,\omega^2 \tag{B.14}$$

where I_{rot} is the rotational moment of inertia and ω the angular velocity of rotation (radians s^{-1}). I_{rot} is calculated from

$$I_{\text{rot}} = m_{\text{red}}\, r^2 \tag{B.15}$$

where the reduced mass $m_{\text{red}} = (m_1^{-1} + m_2^{-1})^{-1}$. Equation (B.14), derived from classical mechanics, has to be modified to take into account that only certain energy states are permitted:

$$E_{rot} = \frac{\mathbf{h}^2}{8\pi^2 I_{rot}} n_r(n_r + 1) \tag{B.16}$$

where n_r is the rotational quantum number. For example, for HCl one finds that $\Delta E_{rot} =$ 0.029 eV when n_r goes from 11 to 10. This corresponds to a wavelength of 42.8 μm, and a wave number $\tilde{\nu}$ of 233.4 cm^{-1}. For transformation of energy in eV to the wave number $\tilde{\nu}$ or wavelength λ of the corresponding photon energy, the relation

$$\Delta E \text{ (eV)} = 1.23980 \times 10^{-4} \tilde{\nu} \quad (cm^{-1}) \tag{B.17}$$

is used, where $\tilde{\nu} = 1/\lambda$. (For blue light of about 4800 Å the following relations are obtained: 4800 A $= 4.8 \times 10^{-7}$ m $= 0.48$ μm $= 20833$ cm^{-1} $= 2.58$ eV.) The rotational energies are normally in the range 0.001–0.1 eV, i.e. the wavelength region 10^{-3}–10^{-5}m. Figure 2.4 shows 12 rotational bands for HCl at about 1.7 μm.

The partion function for the rotational energy is obtained by introducing (B.16) into (B.5). More complicated expressions are obtained for polyatomic and nonlinear molecules.

(*c*) *Vibrational energy.* For a diatomic molecule the vibrational energy is given by

$$E_{vib} = \mathbf{hc}\omega_v(n_v + \tfrac{1}{2}) \tag{B.18}$$

where

$$\omega_v = \frac{1}{2\pi\mathbf{c}}\sqrt{\frac{k'}{m_{red}}} \tag{B.19}$$

is the zero point vibrational frequency (the molecule is still vibrating at absolute zero, when no other movements occur) and n_v the vibrational quantum number. The k' is the *force constant* for the particular molecule. For HCl k' is 4.8×10^5 dyne cm^{-1} and ω_v 8×10^{13} s^{-1}, leading to a ΔE of about 0.35 eV for a $\Delta n_v = 1$, corresponding to an IR absorption band at 3.5 μ ($\tilde{\nu}$ 2890 cm^{-1}). Figure 2.4 shows two higher vibrational transitions, one for $H^{35}Cl$ and one for $H^{37}Cl$, which are superimposed with a number of rotational levels each.

B.4. The isotopic ratio

It is seen from this digression that the mass of the molecular atoms enters into the partition functions for all three modes of molecular movement. The largest energy changes are associated with the vibrational mode, and the isotopic differences are also most pronounced here. Neglecting quantization of translational and rotational movements, one can chow that

$$F = f_{tr} f_{rot} f_{vib} = \left(\frac{2\pi M\mathbf{k}T}{\mathbf{h}^3}\right)^{3/2} V_M \frac{8\pi^2 I_r \mathbf{k}T}{\mathbf{h}^2 k_s} \frac{e^{-u/2}}{(1 - e^{-u})} \tag{B.20}$$

where k_s is a symmetry constant for rotation and $u = \mathbf{hc}\omega/\mathbf{k}T$. This expression holds for all molecules in (B.1). Thus for the ratio one gets

$$\frac{F_{AX}}{F_{AX^*}} = \left(\frac{M^*}{M}\right)^{3/2} \frac{k_s \, I_r^* \, (1 - e^{-u})e^{-u^*/2}}{k_s^* \, I_r \, (1 - e^{-u^*})e^{-u/2}} \tag{B.21}$$

where the asterisk refers to molecule AX*. This relation shows the mass dependency of the equilibrium constant in (B.2) (a similar relation holds for the BX–BX* combination). (B.21) contains factors all of which can be determined spectroscopically. It is therefore possible to calculate the equilibrium constant for an isotopic exchange reaction. This has been done for numerous cases, particularly by J. Biegeleisen. It should be observed that (B.21) contains the temperature in the exponent. Isotope exchange equilibria are thus temperature dependent. A practical use of this fact is the determination of paleo-temperatures (§2.6).

B.5. Exercises

B.1. When the rotational quantum number n_r goes from 0 to 1 in $H^{35}Cl$, it is accompanied by the absorption of light with a wave number of 20.6 cm^{-1}. From this it is possible to calculate the interatomic distance between hydrogen and chlorine in the molecules. What is this distance?

B.2. In one mole of a gas at STP (standard temperature and pressure, i.e. 0°C and 1 atm) a small fraction of the molecules have a kinetic energy $\geq 15\mathbf{k}T$. How many such molecules are there, and what would their temperature be if they could be isolated?

APPENDIX C

Isotope Effects in Chemical Kinetics

The rate constant for the reaction

$$A + BC \rightarrow AB + C \tag{C.1}$$

is given by the expression

$$\frac{d[A]}{dt} = k[A][BC] \tag{C.2}$$

The reaction is assumed to take place over an intermediate compound ABC, usually denoted $ABC^{\neq}$ where the $\neq$ indicates a short-lived transition state. According to the transition state theory, derived by H. Eyring, J. Biegeleisen, and others, it is assumed that the intermediate complex undergoes internal vibrations, with such an energy E_v that the bond is broken along the vertical line in the complex AB|C, leading to the fragments AB and C. The rate of reaction is the rate at which the complex $ABC^{\neq}$ decomposes into the products. It can therefore also be written

$$\frac{d[A]}{dt} = \nu[ABC^{\neq}] \tag{C.3}$$

where ν is the frequency at which the complex decomposes.

The reaction can be schematically depicted as in Figure C.1, where indices 1 and 2

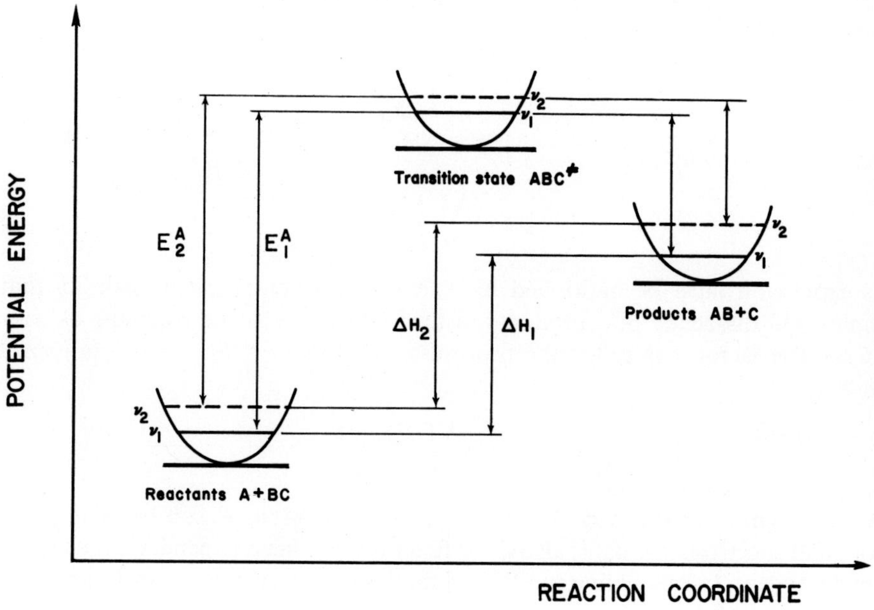

FIG. C.1.

refer to two isotopic reactant molecules (e.g. H_2O and HDO), which must have different zero point energies with frequencies ν_1 and ν_2, respectively. For simplicity only the vibrational ground state is indicated; thus the energy change when going from reactants to products corresponds to the heat of reaction at absolute zero, ΔH_1 (0°) and ΔH_2(0°), respectively. Because of the lower vibrational energy of the molecule indexed 1, this must contain the heavier isotope. In general, the difference in activation energy (E^{A}) is greater than the difference in heat of reaction ΔH for isotope molecules; thus, generally, isotope effects are larger in the kinetic than in the equilibrium effects.

When the molecule $ABC^{\neq}$ decomposes into AB and C, the vibrational energy, given by the Planck relation

$$E_{\mathrm{v}} = \mathbf{h}\nu \tag{C.4}$$

is changed into kinetic energy of the fragments, whose energy is (see §B.2)

$$E_{\mathrm{kin}} = \mathbf{k}\mathrm{T} \tag{C.5}$$

Because of the law of conservation of energy

$$E_{\mathrm{v}} = E_{\mathrm{kin}} \quad \text{and} \quad \mathbf{h}\nu = \mathbf{k}\mathrm{T} \tag{C.6}$$

This development assumes that the vibrational energy is completely converted to fragment translational energy. This assumption is not always valid for polyatomic fragments, in which internal excitation may occur. Introducing (C.6) into (C.3) and equaling (C.2) and (C.3) yields

$$k[\mathrm{A}][\mathrm{BC}] = \mathbf{k}\mathrm{T}\mathbf{h}^{-1}[\mathrm{ABC}^{\neq}] \tag{C.7}$$

It is assumed that $ABC^{\neq}$ is in dynamic equilibrium with the reactants A and BC. Thus

$$\frac{[\mathrm{ABC}^{\neq}]}{[\mathrm{A}][\mathrm{BC}]} = k^{\neq} \tag{C.8}$$

According to (B.3)

$$k^{\neq} = \frac{F_{\mathrm{ABC}^{\neq}}}{F_{\mathrm{A}} F_{\mathrm{BC}}} \tag{C.9}$$

which with (C.7) yields

$$k = \frac{\mathbf{k}\mathrm{T}}{\mathbf{h}} \frac{F_{\mathrm{ABC}^{\neq}}}{F_{\mathrm{A}} F_{\mathrm{BC}}} \tag{C.10}$$

This expression must be multiplied by a factor κ, which is the probability that the complex will dissociate into products instead of back into the reactants as assumed in (C.8). The factor κ is called the transmission coefficient. The final rate expression thus becomes:

$$k = \kappa \frac{\mathbf{k}\mathrm{T}}{\mathbf{h}} \frac{F_{\mathrm{ABC}^{\neq}}}{F_{\mathrm{A}} F_{\mathrm{BC}}} \tag{C.11}$$

As is shown in Appendix B, the grand partition functions F_i can be calculated from theory and spectroscopic data; also these functions are mass dependent, thus k must be mass dependent. In calculating the F_i's, all modes of energy must be included as well as the population of the different energy states.

APPENDIX D

Toll Enrichment and Separative Work

At the present, and probably for a considerable time in the future, only countries producing nuclear weapons have large scale uranium isotope enrichment plants which also offer commercial reactor grade enriched uranium. The specific enrichment task can be described in terms of quantity of material feed F and the quantity withdrawn from the cascade as product P and tails W, and the ^{235}U isotope content of each of these three streams x_F, x_p, x_w. The work required to carry out this separation of feed into product and tails is called separative work. The number of *separative work units* (SWU) which can be produced by a separation plant is directly related to the size of the plant, the power level, and the level of equipment, efficiency, and technology. Figure D.1 shows the operational scheme for the US gaseous diffusion plants†, which are the world's main producers of enriched uranium.

In order for the reactor operator to buy enriched uranium, the so-called *toll enrichment*

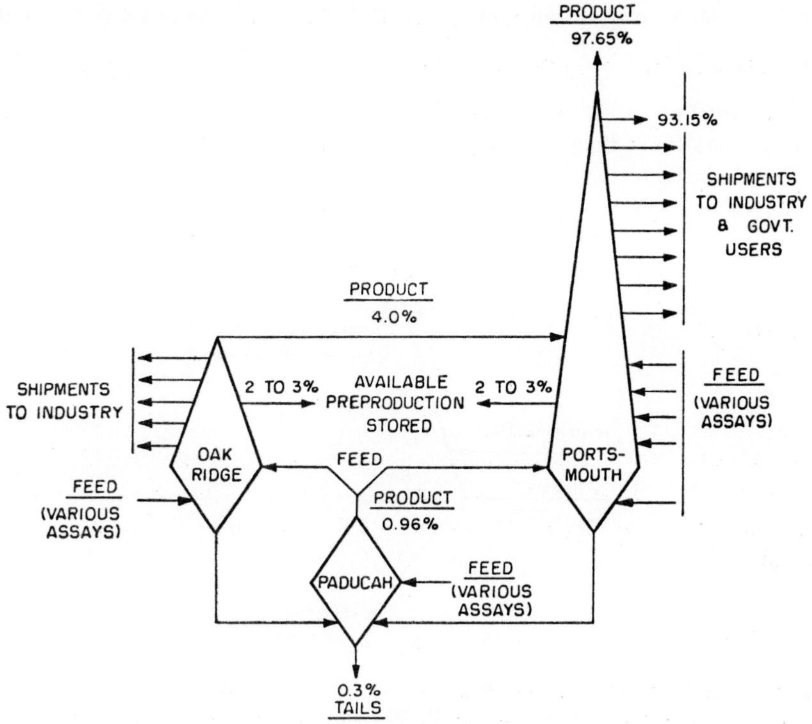

FIG. D.1. Mode of operation for gaseous diffusion plant complex (% values are wt% ^{235}U).

†Located at Oak Ridge, Tenn., Paducah, Ky, and Portsmouth, Ohio.

procedure has been developed. Under this system the reactor operator furnishes to the isotope separation plant either natural uranium, or uranium from a reprocessing plant, and receives in return uranium (normally in the form of UF_6) of the enrichment desired for reactor feed. Applying this to the US gaseous diffusion plants, the reactor operator supplies the additional uranium needed to maintain the normal output of highly enriched ^{235}U and pays for the incremental separative work required to prevent the tails concentration from increasing. There is a modification of this procedure in which the separative work is kept constant and the tails concentration is allowed to rise; it requires more uranium and is favorable for the reactor operator only at low uranium prices. This mode does not seem to be favored at present.

If a reactor operator under a toll enrichment plan desires to withdraw 1 kg of fresh reactor feed at concentration x_2 and returns 1 kg of reactor discharge at x_1, the net requirements of natural uranium (ΔF) and of separative work units (ΔSWU) are given by:

$$\Delta F = (x_2 - x_1)/(x_F - x_w) \quad \text{(kg feed/kg product)} \tag{D.1}$$

$$\Delta \text{SWU} = (\theta_2 - \theta_1) + \Delta F(\theta_W - \theta_F) \quad \text{(kg SWU/kg product)} \tag{D.2}$$

where θ_i (referred to as the separation potential)

$$\theta_i = (1 - 2x_i)\ln(x_i^{-1} - 1) \tag{D.3}$$

If, for instance, 1 kg of 3% enriched uranium is needed and the tails assay is to be 0.2%, then an input of 5.48 kg of natural uranium and 4300 SWUs are required. Figure D.2 illustrates a case with a normal feed of 100 t of natural uranium, to which the reactor operator furnishes 1 t uranium containing 2% ^{235}U, and receives 1 t enriched to 3%. ΔF is the amount of natural uranium which must be supplied to maintain constant product (x_p 0.93) and waste (x_w 0.002 53) streams. (The tails concentration of 0.253% ^{235}U is the 1962 optimum value, which, however, recently has been raised to 0.275%.) From (D.1) one finds that $\Delta F = (0.03–0.02)/(0.007\,113–0.002\,53) = 2.18$, i.e. for each tonne of enriched product 2.18 t of additional natural uranium and 1 t of discharge

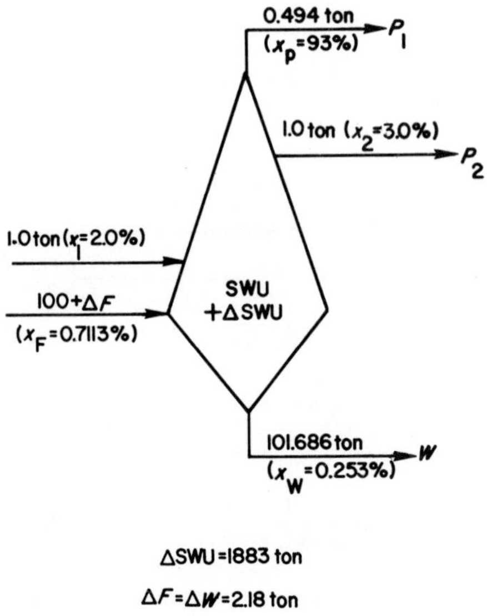

FIG. D.2.

uranium must be fed into the plant to maintain its steady operation conditions. The additional separative work required (ΔSWU) is (3.267 53–3.736 15) + 2.18 (5.946 76–4.868 06) = 1883 t of separative work.

In 1979 cost of separative work is set at about $100 per kg SWU by the US Department of Energy, which is the world's largest supplier of enriched uranium.

The capacity of a concurrent *gas centrifuge plant* is given by the relation

$$\Delta \text{SWU} = \frac{\pi \rho D \Sigma l}{2} \delta^2 \quad (\text{kg s}^{-1}) \tag{D.4}$$

where ρ is the gas density, D the gas diffusion constant, and Σl the total length of all centrifuges in series; δ is defined by eqn. (2.22). In the countercurrent gas centrifuge an internal thermal flow pattern is set up, which leads to an increase in the separative power of each centrifuge, i.e. larger α-values for the same Σl-values. Thus fewer stages are needed for a particular enrichment. The calculation of ΔSWU is more complicated than in (D. 4).

Exercises

D.1. In an isotope separation plant, weapons grade ^{235}U is produced from a natural uranium feed. Assuming the product ^{235}U concentration to be 93%, while the feed contains only 0.7113% ^{235}U, how many tonnes of natural uranium are needed for each tonne of weapons grade? The waste (tails) is assumed to contain 0.253% ^{235}U.

D.2. A 1000 MWe light water nuclear power station replaces annually one-third of its core of fuel elements. The discharged amount contains 0.9% ^{235}U in addition to ^{238}U, other actinides, and fission products. The used fuel elements are reprocessed with negligible uranium losses. The uranium is returned to an isotope separation plan from which the same amount of uranium, but now enriched to 3.2% in ^{235}U, is purchased under toll enrichment plan. (a) What number of SWU would annually be required to maintain the nuclear power station, if its core is assumed to contain 93 t of uranium, and x_w at the separation plant is assumed to be 0.002 75? (b) What would the annual and kWh enrichment cost be, using present US DOE prices and assuming that the overall load factor for the power station is 70%?

APPENDIX E

Conversion from the Laboratory to the Center of Mass (CM) Coordinate System

In many experiments it is necessary to convert the data from the laboratory system of coordinates to the center of mass coordinates. The following equations are useful for such a transformation. The mass of the target atoms is M_2 while the mass and energy of the projectile particles are M_1 and E_1. The heavy product nucleus has mass M_4 and the emitted light particles have mass M_3 and energy E_3. The reaction energy is Q.

$$E_{tot} = E_1 + Q \tag{E.1}$$

$$E_3\,(\text{in CM}) = E_{tot}\frac{M_2 M_4}{M_3 + M_4}(M_1 + M_2)\left[1 + \frac{M_1 Q}{M_2 E_{tot}}\right] \tag{E.2}$$

If the emitted particle is observed at laboratory angle of θ, the angle θ_3 in the CM system is:

$$\theta_3\,(\text{in CM}) = \sin^{-1}\left[\left(\frac{E_3}{E_{tot}A}\cdot\frac{M_1 + M_2}{M_3 + M_4}\right)^{1/2}\sin\theta\right] \tag{E.3}$$

where $A = M_2 M_4\left(1 + \dfrac{M_1 Q}{M_2 E_{tot}}\right)$

APPENDIX F

Suppliers of Enriched Stable and Radioactive Isotopes

The listing includes some major companies and organizations, many of whom in addition have sales agencies or subsidiaries in other countries. The services offered are listed within parentheses, with the following abbreviations: a, stable isotopes; b, radioactive nuclides; c, radioisotope labeled compounds; and d, radiochemical laboratory supplies. See also Buyers' Guide of common nuclear journals.

USA

Stohler Isotope Chemicals, 49 Jones Rd, Waltham, Mass. 02154 (a).
Amersham/Searle Corp., 2636 Clearbrook Dr, Arlington Heights, Ill. 60005 (b, c).
ICN Isotope and Nuclear Division (of Intern. Chem. Nucl. Corp.), 26201 Miles Rd, Cleveland, Ohio 44128 (a, b, c, d).
New England Nuclear, 575 Albany St, Boston, Mass. 02118 (a, b, c, d).
Schwartz/Mann, Orangeburg, New York 10962 (c).

Canada

Atomic Energy of Canada Ltd., Ottawa.

UK

The Radiochemical Centre, Amersham, Buckinghamshire, England HP79LL (b, c).

France

(CEA) Laboratoire de Metrologie des Rayonnements Ionisants, BP 2, 91190 Gif-sur-Yvette (b, c).

Belgium

CEN, Département des Radioisotopes, Mol-Donk (a, b, c).

Italy

SORIN, Uffico Radioisotopi, Saluggia (a, b, c).

Sweden

Isotopcentralen, Studevik Energitelinile AB, Fack, 61101 Nykoping (b, c).

Japan

JAERI, 1–1 Tamura-cho, Shiba, Rinato-ku, Tokyo.
Daiichi Pure Chemicals Co, No 1, 3-Chome, Edobaski, Nihonbashi, Chuo-ku, Tokyo (b).

Holland

N.V. Philips–Duphar, Cyclotron and Isotope Laboratories, Petten (a, b, c, d).

W. Germany

Amersham–Buchler, Harxbutteler Str. 3, D 3300, Braunschweig–Wenden.
NEN (New England Nuclear) Chemicals GmbH, Research Products, Siemensstrasse 1, D6072 Direichenhain (b, c).

Norway
Isotope Laboratories, Institutt for Atomenergi, Kjeller (b, c).
Switzerland
Stohler Isotope Chem., Im Baumgarten, CH3044 Innerberg.

APPENDIX G

Calculation of Geometric Efficiency for Circular Sample

If a detector with a circular window of radius r is at a distance h from a point source, the window extends over the fraction f_g^0 of full space, where

$$f_g^0 = 4\pi \tfrac{1}{2}[1 - (1 + r^2/h^2)^{-1/2}] \quad \text{(G.1)}$$

When the sample is larger than a point source and has a radius d, k_g can be calculated

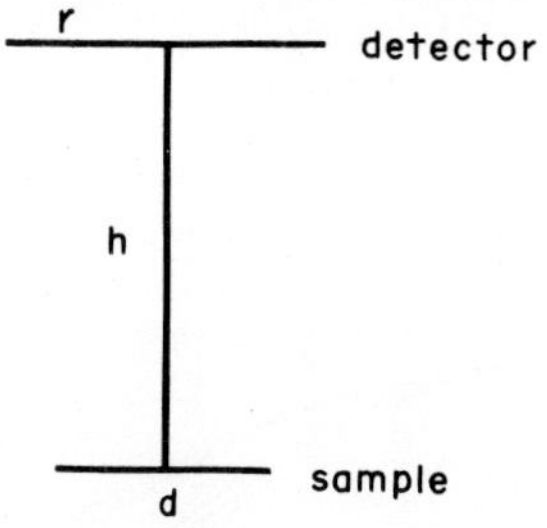

from the series of curves in Fig. G.1 (see overleaf), where k_g is the fraction in percent by which f_g^0 is to be multiplied to give the geometrical efficiency f_g:

$$f_g = k_g f_g^0 \quad \text{(G.2)}$$

Reference

G. B. COOK and J. F. DUNCAN, *Modern Radiochemical Praxis*, Oxford, 1952.

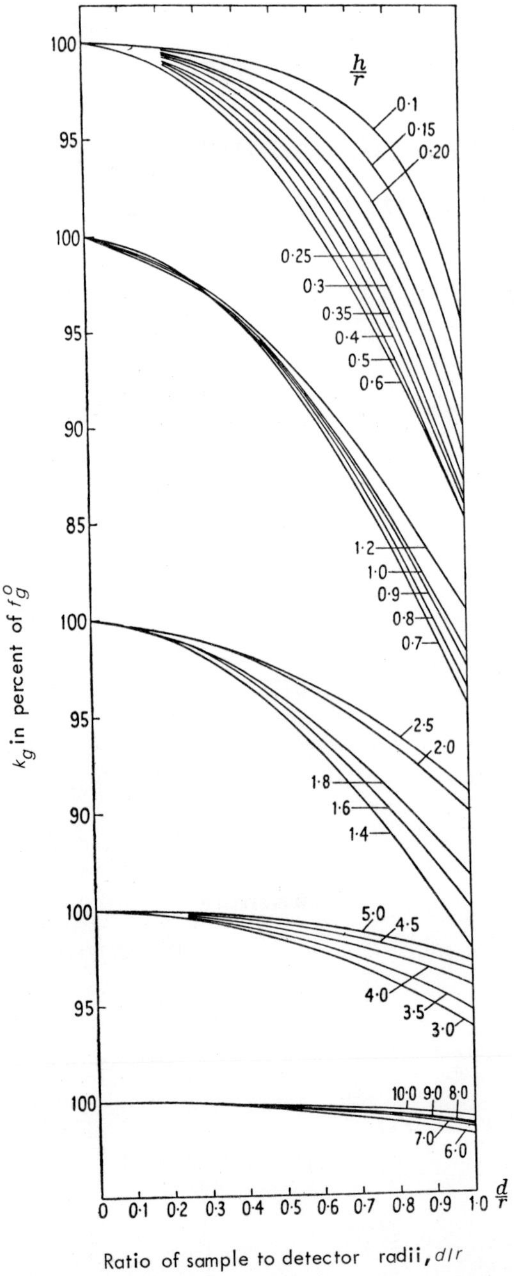

FIG. G.1. (From G. B. Cook and J. F. Duncan, 1952.)

APPENDIX H

Fission Products and Actinides in LWR Waste

The tables show the fission products and actinides in used reactor fuel elements from which 99.5% of uranium and plutonium (all isotopes) has been removed (but nothing else) through reprocessing. The amounts are given in grams of isotope per metric tonne of uranium (elementary amount, enriched to 3.3% in ^{235}U) charged to the reactor. A PWR reactor is considered with 33 000 MWd/t U burn-up at a power density of 30 MW/t U and a flux of 2.92×10^{13} n cm^{-2} s^{-1}. Data are taken from K. J. Schneider and A. M. Platt (Eds.), *High-level Radioactive Waste Management Alternatives*, Battelle Pacific Northwest Laboratories, BNWL–1900, NTIS, Springfield, USA, 1974.

REFERENCE PWR 3.3 PERCENT ENRICHED U PROPERTIES AFTER SEPARATION

POWER = 30.00 MW, BURNUP = 33000.MWD, FLUX= 2.92+13N/CM**2-SEC

Fission products

NUCLIDE CONCENTRATIONS, GRAMS
BASIS = MT OF U CHARGED TO REACTOR

	CHARGE	SEPARATION	1.0+00 YR	1.0+01 YR	1.0+02 YR	1.0+03 YR	1.0+04 YR	1.0+05 YR	1.0+06 YR	1.0+07 YR	1.0+08 YR	1.0+09 YR
H 3	0.00	7.14-02	6.75-02	4.07-02	2.55-04	2.43-26	0.00	0.00	0.00	0.00	0.00	0.00
GE 72	0.00	5.84-03	5.84-03	5.84-03	5.84-03	5.84-03	5.84-03	5.84-03	5.84-03	5.84-03	5.84-03	5.84-03
GE 73	0.00	1.51-02	1.51-02	1.51-02	1.51-02	1.51-02	1.51-02	1.51-02	1.51-02	1.51-02	1.51-02	1.51-02
GE 74	0.00	5.57-02	5.57-02	5.57-02	5.57-02	5.57-02	5.57-02	5.57-02	5.57-02	5.57-02	5.57-02	5.57-02
AS 75	0.00	8.46-02	8.46-02	8.46-02	8.46-02	8.46-02	8.46-02	8.46-02	8.46-02	8.46-02	8.46-02	8.46-02
GE 76	0.00	2.94-01	2.94-01	2.94-01	2.94-01	2.94-01	2.94-01	2.94-01	2.94-01	2.94-01	2.94-01	2.94-01
SE 76	0.00	1.85-03	1.85-03	1.85-03	1.85-03	1.85-03	1.85-03	1.85-03	1.85-03	1.85-03	1.85-03	1.85-03
SE 77	0.00	9.47-01	9.47-01	9.47-01	9.47-01	9.47-01	9.47-01	9.47-01	9.47-01	9.47-01	9.47-01	9.47-01
SE 78	0.00	2.54+00	2.54+00	2.54+00	2.54+00	2.54+00	2.54+00	2.54+00	2.54+00	2.54+00	2.54+00	2.54+00
SE 79	0.00	5.72+00	5.72+00	5.72+00	5.71+00	5.66+00	5.14+00	1.97+00	1.34-04	0.00	0.00	0.00
BR 79	0.00	5.85-04	6.46-04	1.19-03	6.68-03	6.13-02	5.79-01	3.75+00	5.72+00	5.72+00	5.72+00	5.72+00
SE 80	0.00	9.99+00	9.99+00	9.99+00	9.99+00	9.99+00	9.99+00	9.99+00	9.99+00	9.99+00	9.99+00	9.99+00
KR 80	0.00	2.82-01	2.82-01	2.82-01	2.82-01	2.82-01	2.82-01	2.82-01	2.82-01	2.82-01	2.82-01	2.82-01
BR 81	0.00	1.53+01	1.53+01	1.53+01	1.53+01	1.53+01	1.53+01	1.53+01	1.53+01	1.53+01	1.53+01	1.53+01
KR 81	0.00	1.52-01	1.52-01	1.52-01	1.52-01	1.52-01	1.52-01	1.52-01	1.52-01	1.52-01	1.52-01	1.52-01
SE 82	0.00	3.29+01	3.29+01	3.29+01	3.29+01	3.29+01	3.29+01	3.29+01	3.29+01	3.29+01	3.29+01	3.29+01
KR 82	0.00	4.83-01	4.83-01	4.83-01	4.83-01	4.83-01	4.83-01	4.83-01	4.83-01	4.83-01	4.83-01	4.83-01
KR 83	0.00	4.07+01	4.07+01	4.07+01	4.07+01	4.07+01	4.07+01	4.07+01	4.07+01	4.07+01	4.07+01	4.07+01
KR 84	0.00	1.12+02	1.12+02	1.12+02	1.12+02	1.12+02	1.12+02	1.12+02	1.12+02	1.12+02	1.12+02	1.12+02
KR 85	0.00	2.84+01	2.67+01	1.50+01	4.65-02	3.66-27	0.00	0.00	0.00	0.00	0.00	0.00
RB 85	0.00	9.51+01	9.69+01	1.09+02	1.23+02	1.24+02	1.24+02	1.24+02	1.24+02	1.24+02	1.24+02	1.24+02
KR 86	0.00	1.93+02	1.93+02	1.93+02	1.93+02	1.93+02	1.93+02	1.93+02	1.93+02	1.93+02	1.93+02	1.93+02
RB 86	0.00	2.34-05	3.09-11	0.00	0.00	0.00	0.00	0.00	0.00	0.00	0.00	0.00
SR 86	0.00	1.22-01	1.22-01	1.22-01	1.22-01	1.22-01	1.22-01	1.22-01	1.22-01	1.22-01	1.22-01	1.22-01
RB 87	0.00	2.39+02	2.39+02	2.39+02	2.39+02	2.39+02	2.39+02	2.39+02	2.39+02	2.39+02	2.39+02	2.36+02
SR 87	0.00	5.74-05	5.74-05	5.74-05	5.77-05	6.07-05	9.05-05	3.89-04	3.37-03	3.32-02	3.31-01	3.29+00
SR 88	0.00	3.50+02	3.50+02	3.50+02	3.50+02	3.50+02	3.50+02	3.50+02	3.50+02	3.50+02	3.50+02	3.50+02
SR 89	0.00	3.45+00	2.65-02	2.50-21	0.00	0.00	0.00	0.00	0.00	0.00	0.00	0.00
Y 89	0.00	4.63+02	4.67+02	4.67+02	4.67+02	4.67+02	4.67+02	4.67+02	4.67+02	4.67+02	4.67+02	4.67+02
SR 90	0.00	5.43+02	5.30+02	4.25+02	4.61+01	1.06-08	0.00	0.00	0.00	0.00	0.00	0.00
Y 90	0.00	1.41-01	1.38-01	1.10-01	1.20-02	2.75-12	0.00	0.00	0.00	0.00	0.00	0.00
ZR 90	0.00	2.86+01	4.18+01	1.47+02	5.26+02	5.72+02	5.72+02	5.72+02	5.72+02	5.72+02	5.72+02	5.72+02
Y 91	0.00	6.61+00	8.92-02	1.33-18	0.00	0.00	0.00	0.00	0.00	0.00	0.00	0.00
ZR 91	0.00	6.04+02	6.11+02	6.11+02	6.11+02	6.11+02	6.11+02	6.11+02	6.11+02	6.11+02	6.11+02	6.11+02
ZR 92	0.00	6.64+02	6.64+02	6.64+02	6.64+02	6.64+02	6.64+02	6.64+02	6.64+02	6.64+02	6.64+02	6.64+02
ZR 93	0.00	7.36+02	7.36+02	7.36+02	7.36+02	7.36+02	7.33+02	7.03+02	4.64+02	7.25+00	6.35-18	0.00

NB 93M	0.00	6.37-04	9.36-04	3.05-03	6.64-03	6.67-03	6.64-03	6.37-03	4.20-03	6.57-05	5.75-23	0.00
NB 93	0.00	3.92-05	7.94-05	1.03-03	2.80-02	3.34-01	3.39+00	3.32+01	2.72+02	7.29+02	7.36+02	7.36+02
ZR 94	0.00	7.91+02	7.91+02	7.91+02	7.91+02	7.91+02	7.91+02	7.91+02	7.91+02	7.91+02	7.91+02	7.91+02
ZR 95	0.00	1.31+01	2.67-01	1.61-16	0.00	0.00	0.00	0.00	0.00	0.00	0.00	0.00
NB 95M	0.00	1.60-02	3.27-04	1.97-19	0.00	0.00	0.00	0.00	0.00	0.00	0.00	0.00
NB 95	0.00	1.32+01	3.18-01	1.88-16	0.00	0.00	0.00	0.00	0.00	0.00	0.00	0.00
MO 95	0.00	7.45+02	7.71+02	7.72+02	7.72+02	7.72+02	7.72+02	7.72+02	7.72+02	7.72+02	7.72+02	7.72+02
ZR 96	0.00	8.32+02	8.32+02	8.32+02	8.32+02	8.32+02	8.32+02	8.32+02	8.32+02	8.32+02	8.32+02	8.32+02
MO 96	0.00	3.97+01	3.97+01	3.97+01	3.97+01	3.97+01	3.97+01	3.97+01	3.97+01	3.97+01	3.97+01	3.97+01
MO 97	0.00	8.33+02	8.33+02	8.33+02	8.33+02	8.33+02	8.33+02	8.33+02	8.33+02	8.33+02	8.33+02	8.33+02
MO 98	0.00	8.50+02	8.50+02	8.50+02	8.50+02	8.50+02	8.50+02	8.50+02	8.50+02	8.50+02	8.50+02	8.50+02
TC 99	0.00	8.41+02	8.41+02	8.41+02	8.41+02	8.39+02	8.14+02	6.07+02	3.20+01	5.35-12	0.00	0.00
RU 99	0.00	5.32-03	8.07-03	3.28-02	2.80-01	2.75+00	2.71+01	2.35+02	8.09+02	8.41+02	8.41+02	8.41+02
MO100	0.00	9.72+02	9.72+02	9.72+02	9.72+02	9.72+02	9.72+02	9.72+02	9.72+02	9.72+02	9.72+02	9.72+02
RU100	0.00	5.60+01	5.60+01	5.60+01	5.60+01	5.60+01	5.60+01	5.60+01	5.60+01	5.60+01	5.60+01	5.60+01
RU101	0.00	7.77+02	7.77+02	7.77+02	7.77+02	7.77+02	7.77+02	7.77+02	7.77+02	7.77+02	7.77+02	7.77+02
RU102	0.00	7.68+02	7.68+02	7.68+02	7.68+02	7.68+02	7.68+02	7.68+02	7.68+02	7.68+02	7.68+02	7.68+02
RU103	0.00	2.76+00	4.62-03	4.80-28	0.00	0.00	0.00	0.00	0.00	0.00	0.00	0.00
RH103M	0.00	2.76-03	4.62-06	4.81-31	0.00	0.00	0.00	0.00	0.00	0.00	0.00	0.00
RH103	0.00	3.85+02	3.88+02	3.88+02	3.88+02	3.88+02	3.88+02	3.88+02	3.88+02	3.88+02	3.88+02	3.88+02
RU104	0.00	5.45+02	5.45+02	5.45+02	5.45+02	5.45+02	5.45+02	5.45+02	5.45+02	5.45+02	5.45+02	5.45+02
PD104	0.00	2.45+02	2.45+02	2.45+02	2.45+02	2.45+02	2.45+02	2.45+02	2.45+02	2.45+02	2.45+02	2.45+02
PD105	0.00	2.95+02	2.95+02	2.95+02	2.95+02	2.95+02	2.95+02	2.95+02	2.95+02	2.95+02	2.95+02	2.95+02
RU106	0.00	1.22+02	6.13+01	1.24-01	1.36-28	0.00	0.00	0.00	0.00	0.00	0.00	0.00
RH106	0.00	1.16-04	5.80-05	1.17-07	1.29-34	0.00	0.00	0.00	0.00	0.00	0.00	0.00
PD106	0.00	3.27+02	3.88+02	4.49+02	4.49+02	4.49+02	4.49+02	4.49+02	4.49+02	4.49+02	4.49+02	4.49+02
PD107	0.00	2.31+02	2.31+02	2.31+02	2.31+02	2.31+02	2.31+02	2.29+02	2.09+02	8.58+01	1.16-02	0.00
AG107	0.00	3.60-05	5.88-05	2.65-04	2.32-03	2.29-02	2.29-01	2.28+00	2.18+01	1.45+02	2.31+02	2.31+02
PD108	0.00	1.56+02	1.56+02	1.56+02	1.56+02	1.56+02	1.56+02	1.56+02	1.56+02	1.56+02	1.56+02	1.56+02
CD108	0.00	5.01-07	5.01-07	5.01-07	5.01-07	5.01-07	5.01-07	5.01-07	5.01-07	5.01-07	5.01-07	5.01-07
AG109	0.00	5.97+01	5.97+01	5.97+01	5.97+01	5.97+01	5.97+01	5.97+01	5.97+01	5.97+01	5.97+01	5.97+01
CD109	0.00	1.29-09	7.36-10	4.82-12	6.98-34	0.00	0.00	0.00	0.00	0.00	0.00	0.00
PD110	0.00	3.34+01	3.34+01	3.34+01	3.34+01	3.34+01	3.34+01	3.34+01	3.34+01	3.34+01	3.34+01	3.34+01
AG110M	0.00	5.57-02	2.05-02	2.52-06	0.00	0.00	0.00	0.00	0.00	0.00	0.00	0.00
AG110	0.00	8.08-09	2.97-09	3.65-13	0.00	0.00	0.00	0.00	0.00	0.00	0.00	0.00
CD110	0.00	4.10+01	4.10+01	4.10+01	4.10+01	4.10+01	4.10+01	4.10+01	4.10+01	4.10+01	4.10+01	4.10+01
CD111	0.00	1.72+01	1.72+01	1.72+01	1.72+01	1.72+01	1.72+01	1.72+01	1.72+01	1.72+01	1.72+01	1.72+01
CD112	0.00	9.33+00	9.33+00	9.33+00	9.33+00	9.33+00	9.33+00	9.33+00	9.33+00	9.33+00	9.33+00	9.33+00
CD113M	0.00	1.98-04	1.89-04	1.21-04	1.41-06	6.31-26	0.00	0.00	0.00	0.00	0.00	0.00
CD113	0.00	2.25-01	2.25-01	2.25-01	2.25-01	2.25-01	2.25-01	2.25-01	2.25-01	2.25-01	2.25-01	2.25-01
IN113	0.00	1.06-05	2.02-05	8.80-05	2.07-04	2.09-04	2.09-04	2.09-04	2.09-09	2.09-04	2.09-09	2.09-09
CD114	0.00	1.23+01	1.23+01	1.23+01	1.23+01	1.23+01	1.23+01	1.23+01	1.23+01	1.23+01	1.23+01	1.23+01
IN114M	0.00	7.19-10	4.55-12	7.44-32	0.00	0.00	0.00	0.00	0.00	0.00	0.00	0.00
SN114	0.00	1.61-06	1.61-06	1.61-06	1.61-06	1.61-06	1.61-06	1.61-06	1.61-06	1.61-06	1.61-06	1.61-06

fission products (*cont.*)

	CHARGE	SEPARATION	1.0+00 YR	1.0+01 YR	1.0+02 YR	1.0+03 YR	1.0+04 YR	1.0+05 YR	1.0+06 YR	1.0+07 YR	1.0+08 YR	1.0+09 YR
CD115M	0.00	1.70-03	4.71-06	4.63-29	0.00	0.00	0.00	0.00	0.00	0.00	0.00	0.00
IN115	0.00	1.23+00	1.23+00	1.23+00	1.23+00	1.23+00	1.23+00	1.23+00	1.23+00	1.23+00	1.23+00	1.23+00
SN115	0.00	1.94-01	1.94-01	1.94-01	1.94-01	1.94-01	1.94-01	1.94-01	1.94-01	1.94-01	1.94-01	1.94-01
CD116	0.00	3.76+00	3.76+00	3.76+00	3.76+00	3.76+00	3.76+00	3.76+00	3.76+00	3.76+00	3.76+00	3.76+00
SN116	0.00	2.73+00	2.73+00	2.73+00	2.73+00	2.73+00	2.73+00	2.73+00	2.73+00	2.73+00	2.73+00	2.73+00
SN117	0.00	3.90+00	3.90+00	3.90+00	3.90+00	3.90+00	3.90+00	3.90+00	3.90+00	3.90+00	3.90+00	3.90+00
SN118	0.00	3.98+00	3.98+00	3.98+00	3.98+00	3.98+00	3.98+00	3.98+00	3.98+00	3.98+00	3.98+00	3.98+00
SN119M	0.00	2.46-03	6.95-04	9.88-08	0.00	0.00	0.00	0.00	0.00	0.00	0.00	0.00
SN119	0.00	4.12+00	4.12+00	4.12+00	4.12+00	4.12+00	4.12+00	4.12+00	4.12+00	4.12+00	4.12+00	4.12+00
SN120	0.00	4.30+00	4.30+00	4.30+00	4.30+00	4.30+00	4.30+00	4.30+00	4.30+00	4.30+00	4.30+00	4.30+00
SN121M	0.00	3.52-06	3.48-06	3.21-06	1.41-06	3.85-10	0.00	0.00	0.00	0.00	0.00	0.00
SB121	0.00	4.50+00	4.50+00	4.50+00	4.50+00	4.50+00	4.50+00	4.50+00	4.50+00	4.50+00	4.50+00	4.50+00
SN122	0.00	5.04+00	5.04+00	5.04+00	5.04+00	5.04+00	5.04+00	5.04+00	5.[illegible]+00	5.04+00	5.04+00	5.04+00
TE122	0.00	3.08-01	3.08-01	3.08-01	3.08-01	3.08-01	3.08-01	3.08-01	3.08-01	3.08-01	3.08-01	3.08-01
SN123M	0.00	5.56-02	7.33-03	8.93-11	0.00	0.00	0.00	0.00	0.00	0.00	0.00	0.00
SB123	0.00	5.41+00	5.46+00	5.46+00	5.46+00	5.46+00	5.46+00	5.46+00	5.46+00	5.46+00	5.46+00	5.46+00
TE123M	0.00	2.77-05	3.18-06	1.11-14	0.00	0.00	0.00	0.00	0.00	0.00	0.00	0.00
TE123	0.00	4.32-04	4.56-04	4.59-04	4.59-04	4.59-04	4.59-04	4.59-04	4.59-04	4.59-04	4.59-04	4.59-04
SN124	0.00	7.51+00	7.51+00	7.51+00	7.51+00	7.51+00	7.51+00	7.51+00	7.51+00	7.51+00	7.51+00	7.51+00
SB124	0.00	4.86-03	7.15-05	2.32-21	0.00	0.00	0.00	0.00	0.00	0.00	0.00	0.00
TE124	0.00	1.69-01	1.74-01	1.74-01	1.74-01	1.74-01	1.74-01	1.74-01	1.74-01	1.74-01	1.74-01	1.74-01
SB125	0.00	7.54+00	5.83+00	5.79-01	5.36-11	0.00	0.00	0.00	0.00	0.00	0.00	0.00
TE125M	0.00	1.79-01	1.42-01	1.41-02	1.31-12	0.00	0.00	0.00	0.00	0.00	0.00	0.00
TE125	0.00	3.55+00	5.29+00	1.07+01	1.13+01	1.13+01	1.13+01	1.13+01	1.13+01	1.13+01	1.13+01	1.13+01
SN126	0.00	1.92+01	1.92+01	1.92+01	1.92+01	1.91+01	1.80+01	9.62+00	1.88-02	1.54-29	0.00	0.00
SB126M	0.00	6.95-09	6.95-09	6.95-09	6.95-09	6.90-09	6.49-09	3.48-09	6.80-12	0.00	0.00	0.00
SB126	0.00	6.69-06	6.52-06	6.52-06	6.52-06	6.47-06	6.08-06	3.26-06	6.38-09	0.00	0.00	0.00
TE126	0.00	4.48-02	4.49-02	4.61-02	5.81-02	1.78-01	1.33+00	9.66+00	1.93+01	1.93+01	1.93+01	1.93+01
TE127M	0.00	6.51-01	6.39-02	5.36-11	0.00	0.00	0.00	0.00	0.00	0.00	0.00	0.00
TE127	0.00	2.31-03	2.27-04	1.90-13	0.00	0.00	0.00	0.00	0.00	0.00	0.00	0.00
I127	0.00	3.94+01	4.00+01	4.01+01	4.01+01	4.01+01	4.01+01	4.01+01	4.01+01	4.01+01	4.01+01	4.01+01
TE128	0.00	1.34+02	1.34+02	1.34+02	1.34+02	1.34+02	1.34+02	1.34+02	1.34+02	1.34+02	1.34+02	1.34+02
XE128	0.00	3.01+00	3.01+00	3.01+00	3.01+00	3.01+00	3.01+00	3.01+00	3.01+00	3.01+00	3.01+00	3.01+00
TE129M	0.00	2.21-01	1.29-04	1.03-3[illegible]	0.00	0.00	0.00	0.00	0.00	0.00	0.00	0.00
TE129	0.00	1.99-04	1.17-07	9.30-37	0.00	0.00	0.00	0.00	0.00	0.00	0.00	0.00
I129	0.00	2.29+02	2.30+02	2.30+02	2.30+02	2.30+02	2.30+02	2.29+02	2.21+02	1.53+02	3.90+00	9.59-16
XE129	0.00	9.06-02	9.06-02	9.07-02	9.15-02	1.00-01	1.84-01	1.03+00	9.27+00	7.70+01	2.26+02	2.30+02
TE130	0.00	4.25+02	4.25+02	4.25+02	4.25+02	4.25+02	4.25+02	4.25+02	4.25+02	4.25+02	4.25+02	4.25+02

XE130	0.00	1.06+01	1.06+01	1.06+01	1.06+01	1.06+01	1.06+01	1.06+01	1.06+01	1.06+01	1.06+01	1.06+01
XE131	0.00	4.41+02	4.41+02	4.41+02	4.41+02	4.41+02	4.41+02	4.41+02	4.41+02	4.41+02	4.41+02	4.41+02
XE132	0.00	1.15+03	1.15+03	1.15+03	1.15+03	1.15+03	1.15+03	1.15+03	1.15+03	1.15+03	1.15+03	1.15+03
CS133	0.00	1.01+03	1.01+03	1.01+03	1.01+03	1.01+03	1.01+03	1.01+03	1.01+03	1.01+03	1.01+03	1.01+03
XE134	0.00	1.54+03	1.54+03	1.54+03	1.54+03	1.54+03	1.54+03	1.54+03	1.54+03	1.54+03	1.54+03	1.54+03
CS134	0.00	1.65+02	1.18+02	5.61+00	3.43-13	0.00	0.00	0.00	0.00	0.00	0.00	0.00
BA134	0.00	9.36+01	1.41+02	2.53+02	2.58+02	2.58+02	2.58+02	2.58+02	2.58+02	2.58+02	2.58+02	2.58+02
CS135	0.00	3.24+02	3.24+02	3.24+02	3.24+02	3.24+02	3.23+02	3.17+02	2.57+02	3.22+01	3.01-08	0.00
BA135	0.00	3.54-02	3.55-02	3.61-02	4.29-02	1.10-01	7.83-01	7.43+00	6.68+01	2.92+02	3.24+02	3.24+02
XE136	0.00	2.31+03	2.31+03	2.31+03	2.31+03	2.31+03	2.31+03	2.31+03	2.31+03	2.31+03	2.31+03	2.31+03
BA136	0.00	2.50+01	2.50+01	2.50+01	2.50+01	2.50+01	2.50+01	2.50+01	2.50+01	2.50+01	2.50+01	2.50+01
CS137	0.00	1.23+03	1.20+03	9.74+02	1.22+02	1.14-07	0.00	0.00	0.00	0.00	0.00	0.00
BA137M	0.00	1.85-04	1.81-04	1.47-04	1.84-05	1.72-14	0.00	0.00	0.00	0.00	0.00	0.00
BA137	0.00	5.49+01	8.29+01	3.08+02	1.16+03	1.28+03	1.28+03	1.28+03	1.28+03	1.28+03	1.28+03	1.28+03
BA138	0.00	1.22+03	1.22+03	1.22+03	1.22+03	1.22+03	1.22+03	1.22+03	1.22+03	1.22+03	1.22+03	1.22+03
LA139	0.00	1.27+03	1.27+03	1.27+03	1.27+03	1.27+03	1.27+03	1.27+03	1.27+03	1.27+03	1.27+03	1.27+03
BA140	0.00	5.92-03	1.53-11	0.00	0.00	0.00	0.00	0.00	0.00	0.00	0.00	0.00
LA140	0.00	8.91-04	2.30-12	0.00	0.00	0.00	0.00	0.00	0.00	0.00	0.00	0.00
CE140	0.00	1.31+03	1.31+03	1.31+03	1.31+03	1.31+03	1.31+03	1.31+03	1.31+03	1.31+03	1.31+03	1.31+03
CE141	0.00	1.97+00	7.99-04	2.33-34	0.00	0.00	0.00	0.00	0.00	0.00	0.00	0.00
PR141	0.00	1.20+03	1.20+03	1.20+03	1.20+03	1.20+03	1.20+03	1.20+03	1.20+03	1.20+03	1.20+03	1.20+03
CE142	0.00	1.17+03	1.17+03	1.17+03	1.17+03	1.17+03	1.17+03	1.17+03	1.17+03	1.17+03	1.17+03	1.17+03
ND142	0.00	2.08+01	2.08+01	2.08+01	2.08+01	2.08+01	2.08+01	2.08+01	2.08+01	2.08+01	2.08+01	2.08+01
PR143	0.00	1.02-02	9.64-11	0.00	0.00	0.00	0.00	0.00	0.00	0.00	0.00	0.00
ND143	0.00	8.05+02	8.05+02	8.05+02	8.05+02	8.05+02	8.05+02	8.05+02	8.05+02	8.05+02	8.05+02	8.05+02
CE144	0.00	2.41+02	9.90+01	3.25-02	4.75-37	0.00	0.00	0.00	0.00	0.00	0.00	0.00
PR144	0.00	1.02-02	4.19-03	1.38-06	0.00	0.00	0.00	0.00	0.00	0.00	0.00	0.00
ND144	0.00	1.10+03	1.24+03	1.34+03	1.34+03	1.34+03	1.34+03	1.34+03	1.34+03	1.34+03	1.34+03	1.34+03
ND145	0.00	6.97+02	6.97+02	6.97+02	6.97+02	6.97+02	6.97+02	6.97+02	6.97+02	6.97+02	6.97+02	6.97+02
ND146	0.00	7.02+02	7.02+02	7.02+02	7.02+02	7.02+02	7.02+02	7.02+02	7.02+02	7.02+02	7.02+02	7.02+02
PM147	0.00	1.06+02	8.10+01	7.49+00	3.44-10	0.00	0.00	0.00	0.00	0.00	0.00	0.00
SM147	0.00	6.38+01	8.83+01	1.62+02	1.69+02	1.69+02	1.69+02	1.69+02	1.69+02	1.69+02	1.69+02	1.69+02
ND148	0.00	3.73+02	3.73+02	3.73+02	3.73+02	3.73+02	3.73+02	3.73+02	3.73+02	3.73+02	3.73+02	3.73+02
PM148M	0.00	1.89-02	4.57-05	1.27-28	0.00	0.00	0.00	0.00	0.00	0.00	0.00	0.00
PM148	0.00	1.95-04	4.72-07	1.31-30	0.00	0.00	0.00	0.00	0.00	0.00	0.00	0.00
SM148	0.00	2.45+02	2.45+02	2.45+02	2.45+02	2.45+02	2.45+02	2.45+02	2.45+02	2.45+02	2.45+02	2.45+02
SM149	0.00	6.23+00	6.23+00	6.23+00	6.23+00	6.23+00	6.23+00	6.23+00	6.23+00	6.23+00	6.23+00	6.23+00
ND150	0.00	1.77+02	1.77+02	1.77+02	1.77+02	1.77+02	1.77+02	1.77+02	1.77+02	1.77+02	1.77+02	1.77+02
SM150	0.00	3.19+02	3.19+02	3.19+02	3.19+02	3.19+02	3.19+02	3.19+02	3.19+02	3.19+02	3.19+02	3.19+02
SM151	0.00	4.21+01	4.17+01	3.89+01	1.90+01	1.46-02	1.07-33	0.00	0.00	0.00	0.00	0.00
EU151	0.00	1.92-01	5.26-01	3.41+00	2.33+01	4.23+01	4.23+01	4.23+01	4.23+01	4.23+01	4.23+01	4.23+01
SM152	0.00	9.11+01	9.12+01	9.12+01	9.12+01	9.12+01	9.12+01	9.12+01	9.12+01	9.12+01	9.12+01	9.12+01

Fission products (*cont.*)

	CHARGE	SEPARATION	1.0+00 YR	1.0+01 YR	1.0+02 YR	1.0+03 YR	1.0+04 YR	1.0+05 YR	1.0+06 YR	1.0+07 YR	1.0+08 YR	1.0+09 YR
EU152	0.00	5.80-02	5.48-02	3.26-02	1.80-04	4.82-27	0.00	0.00	0.00	0.00	0.00	0.00
GD152	0.00	1.44-01	1.45-01	1.51-01	1.60-01	1.60-01	1.60-01	1.60-01	1.60-01	1.60-01	1.60-01	1.60-01
EU153	0.00	1.28+02	1.28+02	1.26+02	1.28+02	1.28+02	1.28+02	1.28+02	1.28+02	1.28+02	1.28+02	1.28+02
GD153	0.00	6.32-03	2.22-03	1.81-07	0.00	0.00	0.00	0.00	0.00	0.00	0.00	0.00
SM154	0.00	3.69+01	3.69+01	3.69+01	3.69+01	3.69+01	3.69+01	3.69+01	3.69+01	3.69+01	3.69+01	3.69+01
EU154	0.00	4.66+01	4.46+01	3.02+01	6.12-01	7.21-18	0.00	0.00	0.00	0.00	0.00	0.00
GD154	0.00	2.53+00	4.50+00	1.89+01	4.85+01	4.91+01	4.91+01	4.91+01	4.91+01	4.91+01	4.91+01	4.91+01
EU155	0.00	4.96+00	3.38+00	1.08-01	1.17-16	0.00	0.00	0.00	0.00	0.00	0.00	0.00
GD155	0.00	9.13-01	2.49+00	5.76+00	5.87+00	5.87+00	5.87+00	5.87+00	5.87+00	5.87+00	5.87+00	5.87+00
EU156	0.00	3.93-03	1.84-10	0.00	0.00	0.00	0.00	0.00	0.00	0.00	0.00	0.00
GD156	0.00	8.36+01	8.36+01	8.36+01	8.36+01	8.36+01	8.36+01	8.36+01	8.36+01	8.36+01	8.36+01	8.36+01
GD157	0.00	4.15-02	4.15-02	4.15-02	4.15-02	4.15-02	4.15-02	4.15-02	4.15-02	4.15-02	4.15-02	4.15-02
GD158	0.00	1.34+01	1.34+01	1.34+01	1.34+01	1.34+01	1.34+01	1.34+01	1.34+01	1.34+01	1.34+01	1.34+01
TB159	0.00	1.78+00	1.78+00	1.78+00	1.78+00	1.78+00	1.78+00	1.78+00	1.78+00	1.78+00	1.78+00	1.78+00
GD160	0.00	9.51-01	9.51-01	9.51-01	9.51-01	9.51-01	9.51-01	9.51-01	9.51-01	9.51-01	9.51-01	9.51-01
TB160	0.00	2.68-02	8.01-04	1.52-17	0.00	0.00	0.00	0.00	0.00	0.00	0.00	0.00
DY160	0.00	3.89-01	4.15-01	4.16-01	4.16-01	4.16-01	4.16-01	4.16-01	4.16-01	4.16-01	4.16-01	4.16-01
DY161	0.00	2.31-01	2.31-01	2.31-01	2.31-01	2.31-01	2.31-01	2.31-01	2.31-01	2.31-01	2.31-01	2.31-01
GD162	0.00	7.23-02	3.62-02	7.08-05	5.79-32	0.00	0.00	0.00	0.00	0.00	0.00	0.00
TB162M	0.00	1.03-06	5.16-07	1.01-09	8.26-37	0.00	0.00	0.00	0.00	0.00	0.00	0.00
DY162	0.00	1.82-01	2.16-01	2.54-01	2.54-01	2.54-01	2.54-01	2.54-01	2.54-01	2.54-01	2.54-01	2.54-01
DY163	0.00	1.67-01	1.67-01	1.67-01	1.67-01	1.67-01	1.67-01	1.67-01	1.67-01	1.67-01	1.67-01	1.67-01
DY164	0.00	5.39-02	5.39-02	5.39-02	5.39-02	5.39-02	5.39-02	5.39-02	5.39-02	5.39-02	5.39-02	5.39-02
HO165	0.00	8.29-02	8.29-02	8.29-02	8.29-02	8.29-02	8.29-02	8.29-02	8.29-02	8.29-02	8.29-02	8.29-02
HO166M	0.00	4.15-05	4.15-05	4.13-05	3.92-05	2.33-05	1.29-07	3.45-30	0.00	0.00	0.00	0.00
ER166	0.00	2.28-02	2.28-02	2.28-02	2.28-02	2.28-02	2.28-02	2.28-02	2.28-02	2.28-02	2.28-02	2.28-02
ER167	0.00	4.07-03	4.07-03	4.07-03	4.07-03	4.07-03	4.07-03	4.07-03	4.07-03	4.07-03	4.07-03	4.07-03
SUBTOT	0.00	3.49+04	3.49+04	3.49+04	3.49+04	3.49+04	3.49+04	3.49+04	3.49+04	3.49+04	3.49+04	3.49+04
TOTAL	0.00	3.49+04	3.49+04	3.49+04	3.49+04	3.49+04	3.49+04	3.49+04	3.49+04	3.49+04	3.49+04	3.49+04

Actinides (0.5 % U, 0.5 % Pu)

	CHARGE	SEPARATION	1.0+00 YR	1.0+01 YR	1.0+02 YR	1.0+03 YR	1.0+04 YR	1.0+05 YR	1.0+06 YR	1.0+07 YR	1.0+08 YR	1.0+09 YR
HE 4	0.00	3.53-01	4.50-01	6.34-01	1.15+00	1.95+00	3.54+00	6.90+00	2.16+01	6.91+01	8.29+01	1.75+02
TL208	0.00	2.71-12	1.94-12	1.73-13	4.96-14	8.71-18	1.50-18	2.34-17	2.40-16	2.12-15	7.92-15	8.02-15
PB207	0.00	4.71-10	1.11-09	1.90-08	7.11-07	9.85-06	1.26-04	5.80-03	1.36-01	1.45+00	1.39+01	9.35+01
PB208	0.00	7.67-07	1.53-06	3.39-06	5.75-06	7.44-06	7.44-06	7.81-06	4.66-05	3.63-03	1.85-01	2.60+00
PB210	0.00	1.03-11	1.96-11	2.38-10	9.27-09	8.90-07	3.59-05	2.61-04	6.90-05	1.93-05	1.91-05	1.66-05
PB212	0.00	1.57-09	1.12-09	1.00-10	2.86-11	5.06-15	8.69-16	1.36-14	1.39-13	1.23-12	4.60-12	4.66-12

BI209	0.00	7.54-11	8.95-11	2.27-10	1.09-08	9.39-06	8.03-03	2.52+00	1.03+02	4.51+02	4.71+02	4.71+02
BI212	0.00	1.50-[illegible]	1.07-10	9.56-12	2.74-12	4.82-16	8.28-17	1.30-15	1.33-14	1.17-13	4.38-13	4.44-13
RN220	0.00	2.40-12	1.71-12	1.53-13	4.39-14	7.71-18	1.32-18	2.07-17	2.12-16	1.87-15	7.01-15	7.10-15
RA223	0.00	2.32-11	3.93-11	1.50-10	4.63-10	5.06-10	7.59-10	5.16-09	7.10-09	7.03-09	6.44-09	2.68-09
RA224	0.00	1.37-08	9.79-09	8.73-10	2.51-10	4.41-14	7.57-15	1.18-13	1.21-12	1.07-11	4.01-11	4.06-11
RA226	0.00	1.99-08	2.89-08	1.10-07	1.13-06	7.30-05	2.94-03	2.31-02	5.65-03	1.59-03	1.56-03	1.36-03
AC227	0.00	1.75-08	2.77-08	1.06-07	3.20-07	3.56-07	5.35-07	3.64-06	5.00-06	4.96-06	4.54-06	1.89-06
TH227	0.00	3.82-11	6.30-11	2.40-10	7.42-10	8.10-10	1.22-09	8.27-09	1.14-08	1.13-08	1.03-08	4.29-09
TH228	0.00	2.71-06	1.90-06	1.70-07	4.89-08	8.59-12	1.48-12	2.31-11	2.36-10	2.09-09	7.82-09	7.92-09
TH229	0.00	1.63-07	1.64-07	1.97-07	3.42-06	3.24-04	2.59-02	6.02-01	1.35+00	6.88-02	1.64-14	0.00
TH230	0.00	1.05-03	1.05-03	1.07-03	1.76-03	1.99-02	1.92-01	1.17+00	2.88-01	8.07-02	7.96-02	6.93-02
TH231	0.00	3.23-08	1.61-10	1.61-10	1.62-10	1.65-10	2.19-10	6.52-10	6.88-10	6.82-10	6.29-10	2.59-10
TH232	0.00	2.87-04	2.87-04	2.93-04	3.52-04	9.90-04	1.11-02	1.74-01	1.78+00	1.57+01	5.87+01	5.95+01
TH234	0.00	1.36-05	6.82-08	6.78-08	6.78-08	6.78-08	6.78-08	6.78-08	6.78-08	6.77-08	6.68-08	5.82-08
PA231	0.00	5.18-04	5.18-04	5.18-04	5.20-04	5.45-04	8.19-04	5.57-03	7.65-03	7.59-03	6.95-03	2.89-03
PA233	0.00	1.66-05	1.66-05	1.66-05	1.69-05	1.79-05	1.83-05	1.76-05	1.33-05	7.20-07	1.58-19	0.00
PA234M	0.00	4.57-10	2.30-12	2.29-12	2.29-12	2.29-12	2.29-12	2.29-12	2.29-12	2.28-12	2.25-12	1.96-12
U232	0.00	1.72-06	2.34-06	4.08-06	1.83-06	3.16-10	0.00	0.00	0.00	0.00	0.00	0.00
U233	0.00	2.43-05	1.94-04	1.58-03	1.55-02	1.61-01	1.65+00	1.35+01	3.04+01	1.55+00	3.70-13	0.00
U234	2.64+02	6.10-01	6.38-01	1.07+00	4.13+00	7.37+00	7.20+00	5.65+00	6.86-01	2.54-01	2.50-01	2.18-01
U235	3.30+04	3.99+01	3.99+01	3.99+01	4.00+01	4.07+01	5.41+01	1.61+02	1.70+02	1.69+02	1.54+02	6.42+01
U236	0.00	2.27+01	2.27+01	2.27+01	2.31+01	2.66+01	4.88+01	6.32+01	6.16+01	4.75+01	3.49+00	1.62-11
U237	0.00	1.66-07	1.46-07	9.70-08	1.52-09	7.65-11	4.02-11	4.99-14	0.00	0.00	0.00	0.00
U238	9.67+05	4.71+03	4.71+03	4.71+03	4.71+03	4.71+03	4.71+03	4.71+03	4.72+03	4.71+03	4.64+03	4.04+03
NP237	0.00	4.82+02	4.82+02	4.83+02	4.89+02	5.21+02	5.30+02	5.15+02	3.85+02	2.09+01	4.60-12	0.00
NP239	0.00	7.55-05	7.55-05	7.54-05	7.48-05	6.89-05	3.05-05	8.78-09	1.13-12	7.79-13	1.86-14	1.12-30
PU236	0.00	2.99-06	2.34-06	2.63-07	8.23-17	0.00	0.00	0.00	0.00	0.00	0.00	0.00
PU238	0.00	8.41-01	5.32+00	6.11+00	3.11+00	4.86-03	7.49-21	0.00	0.00	0.00	0.00	0.00
PU239	0.00	2.63+01	2.63+01	2.64+01	2.72+01	3.34+01	6.55+01	9.17+00	4.29-06	2.96-06	7.06-08	4.27-24
PU240	0.00	1.08+01	1.19+01	2.05+01	4.04+01	3.74+01	1.49+01	1.46-03	0.00	0.00	0.00	0.00
PU241	0.00	5.04+00	4.81+00	3.16+00	4.94-02	2.56-03	1.31-03	1.62-06	1.41-35	0.00	0.00	0.00
PU242	0.00	1.76+00	1.76+00	1.76+00	1.79+00	1.86+00	1.96+00	1.71+00	3.30-01	2.36-08	0.00	0.00
AM241	0.00	4.41+01	4.43+01	4.53+01	4.20+01	1.00+01	3.83-02	4.75-05	0.00	0.00	0.00	0.00
AM2-2M	0.00	4.14-01	4.12-01	3.96-01	2.62-01	4.34-03	6.56-21	0.00	0.00	0.00	0.00	0.00
AM242	0.00	4.97-06	4.95-06	4.75-06	3.15-06	5.21-08	7.88-26	0.00	0.00	0.00	0.00	0.00
AM243	0.00	9.12+01	9.12+01	9.12+01	9.04+01	8.33+01	3.69+01	1.06-02	1.37-06	9.42-07	2.25-08	1.36-24
CM242	0.00	5.62+00	1.23+00	9.54-04	6.32-04	[illegible]	1.58-23	0.00	0.00	0.00	0.00	0.00
CM243	0.00	1.19-01	1.16-01	9.56-02	1.36-02	[illegible]	0.00	0.00	0.00	0.00	0.00	0.00
CM244	0.00	3.11+01	3.00+01	2.12+01	6.77-01	7.33-16	0.00	0.00	0.00	0.00	0.00	0.00
CM245	0.00	1.76+00	1.76+00	1.76+00	1.75+00	1.03+00	8.37-01	1.04-03	8.98-33	0.00	0.00	0.00
CM246	0.00	2.24-01	2.24-01	2.23-01	2.21-01	1.97-01	6.32-02	7.20-07	0.00	0.00	0.00	0.00
CM247	0.00	3.17-03	3.17-03	3.17-03	3.17-03	3.17-03	3.16-03	3.15-03	3.04-03	2.09-03	4.99-05	3.01-21
SUBTOT	1.00+06	5.48+03	5.48+03	5.48+03	5.48+03	5.48+03	5.48+03	5.50+03	5.49+03	5.49+03	5.43+03	4.91+03
TOTAL	1.00+06	5.48+03	5.48+03	5.48+03	5.48+03	5.48+03	5.48+03	5.50+03	5.50+03	5.50+03	5.50+03	5.50+03

APPENDIX I

Nuclear Power Plants and Demand for Nuclear Fuel Cycle Services

Estimates of present uranium supply and future demands for uranium and nuclear fuel cycle services with respect to expected future nuclear power installations by R. Krymm, G. Woite, and M. Hansen of the International Atomic Agency (*IAEA Bulletin*, Vol. 18, No. 5/6, 1976).

I.1. Assumptions

In order to estimate the demand for uranium and nuclear fuel cycle services, it was assumed that the nuclear power plants will consist of light water reactors (LWR), heavy water reactors (HWR), and fast breeder reactors (FBR). It was further assumed that LWRs would contribute 93%, HWRs would contribute 5%, and FBRs and other reactor types 2% of the installed nuclear capacity in 1990. For the year 2000 it was assumed that the FBR portion would go up to 5% and the LWR portion down to 90%.

The nuclear fuel cycle characteristics of these reactor types are summarized in Table I.1.

TABLE I.1. *Power reactor characteristics*[a]

	PWR	BWR	HWR	FBR
1. Initial loading				
Uranium (t/GWe)	79	114	143	50
Average initial enrichment (w/o ^{235}U)	2.38	2.03	0.711	Depleted
Natural uranium required (t/GWe y)	372	444	145	—
Separative work required (1000 SWU/GWe y)	209	227	—	—
Fissile plutonium required (t Pu/GWe y)	—	—	—	2.5
2. Replacement loadings				
Uranium (t/GWe y)	33.8	39.4	168	20
Fresh fuel enrichment (w/o ^{235}U)	3.2	2.7	0.711	Depleted
Natural uranium required (t/GWe y)	221	211	170	—
Separative work required (1000 SWU/GWe y)	145	129	—	—
Fissile plutonium required (t Pu/GWe y)	—	—	—	1.2
3. Irradiated fuel				
Burn-up (MWd/kg)	32.5	27.5	7.5	2–66[b]
Uranium (t/GWe y)	32.8	38.4	166	18
Average enrichment (w/o ^{235}U)	0.90	0.83	Depleted	Depleted
Natural uranium equivalent (t/GWe y)	44.7	46.6	—	—
Separative work equivalent (1000 SWU/GWe y)	6.3	4.3	—	—
Fissile plutonium (t Pu/GWe y)	0.22	0.21	0.43	1.35

[a]Fuel amounts are in metric tonnes of heavy metal; tails assay = 0.25%; 1 GWe y = 8760 GWh.
[b]Depending on position in core or blanket.

TABLE I.2.A. *Regional breakdown of nuclear power estimates made in* 1976 *(installed capacity at year end in GWe)*

	1975	1980	1985	1990	2000
North America	43	88	150–170	230–310	650–750
Western Europe	18	68	150–170	220–290	600–700
Japan, Australia, New Zealand, South Africa	7	16	30–40	60–80	130–160
Developing countries	1	5	20–25	50–60	150–200
Subtotal (market economies)	69	178	350–400	550–750	1500–1800
Centrally planned economies	10	38		250	600–700
Total[a]	79	216		800–1000	2100–2500

[a]In view of the large margins of uncertainty, totals were rounded off.

TABLE I.2.B. *Distribution of reactor types*

		PWR	BWR	GCR/AGR	HWR	HTR	FBR	Total
1975	GWe	32.3	23.5	8.3/1.0	3.5	0.3	0.5	71
	%	45	33	9.6	4.4	0.4	0.7	100%
1980[a]	GWe	107	61	5.8/5.8	9.8	0.6	1.4	192
	%	56	32	6.0	5.1	0.3	0.7	100%
1990[b]	GWe	570	285	1.1/6.0	68	30	16.0	1000[c]
	%	57	29	0.7	6.8	3.0	1.6	100%

[a]Reactors in operation and under construction 1976.
[b]NEA/IAEA estimate in 1975.
[c]Includes also PTR 22 GWe.

A tails assay of 0.25% and a constant load factor of 70% were assumed. The delay times of the nuclear fuel cycle are taken into account: ~ 2 y from uranium mill to reactor start, and ~ 5 y for reprocessing until 1990, and a one year delay after 1990.

Based on the nuclear power forecasts in Tables I.2A and I.2B and the above assumptions, the requirements for uranium and nuclear fuel cycle services were computed. To show the influence of some key parameters, computations were carried out for four cases:

Case 1: High forecast, no recycling until 1990, uranium recycling after 1990;
Case 2: High forecast, uranium and plutonium recycling from 1981 onwards;
Case 3: Low forecast, no recycling until 1990, uranium recycling after 1990;
Case 4: Low forecast, uranium and plutonium recycling from 1981 onwards.

I.2. Estimated demand for uranium and nuclear fuel cycle services

All the following estimates only refer to countries with market economies.

Uranium requirements. It is estimated that uranium requirements will reach about 40 000 metric tonnes of heavy metal per year in 1980 (Fig. I.1). For 1990 the estimates for uranium requirements range from 90 000 t/a (Case 4: low forecast, uranium and plutonium recycling) to 140 000 t/a (Case 1: high forecast, no recycling). For the year 2000, the requirements are estimated to range from 200 000 to 300 000 t/a. Cumulative

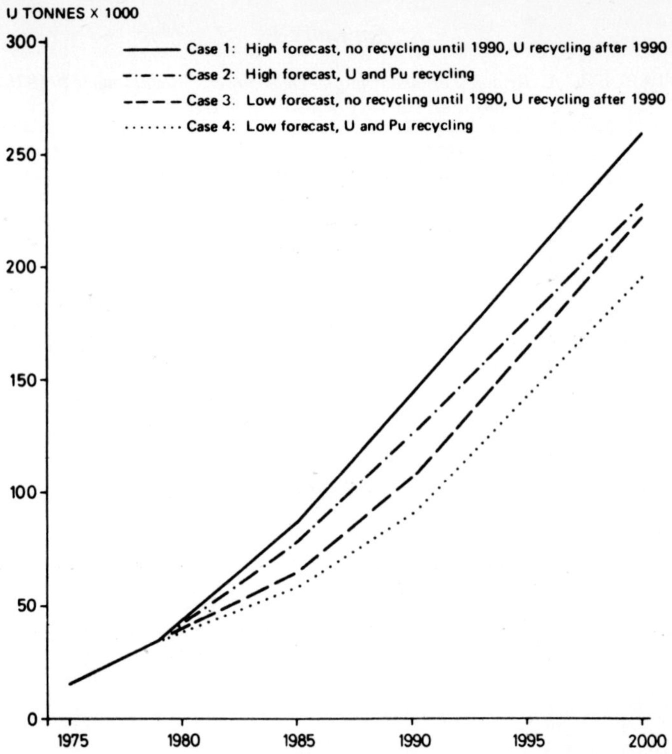

FIG. I.1. Annual world uranium requirements (not including countries with centrally planned economies).

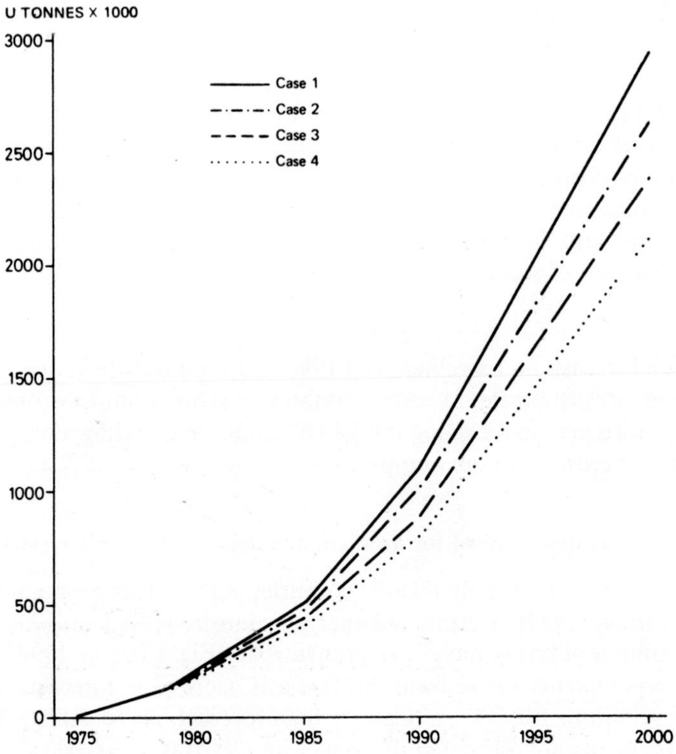

FIG. I.2. Cumulative world uranium requirements (not including countries with centrally planned economies).

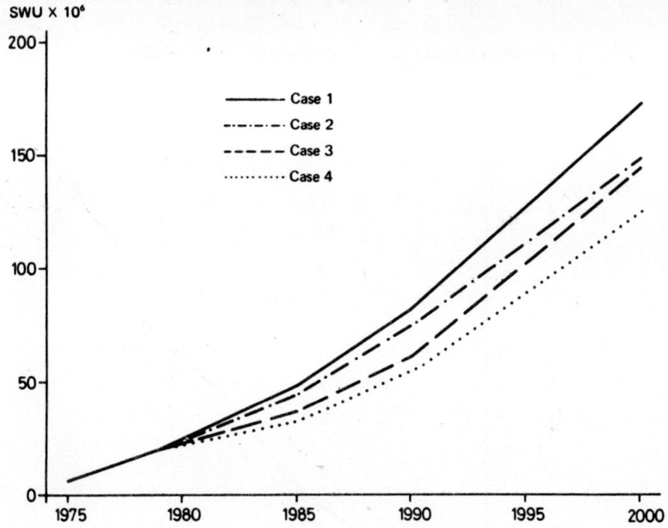

FIG. I.3. Annual world separative work unit requirements (not including countries with centrally planned economies).

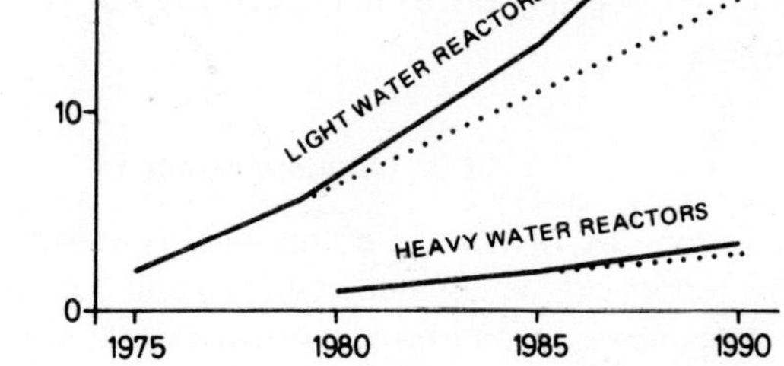

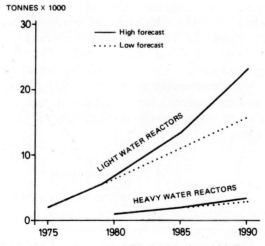

FIG. I.4. Annual fabrication requirements (not including countries with centrally planned economies).

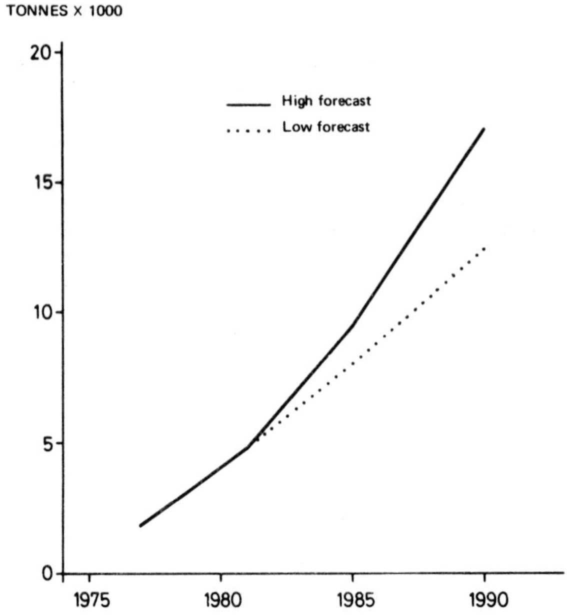

FIG. I.5. Annual reprocessing requirements for light water reactor fuel (not including countries with centrally planned economies).

uranium requirements are estimated to be about 0.8 to 1 Mt in 1990, and 2 to 3 Mt by the year 2000 (Fig. I.2). It is interesting to note that nearly the same amounts (1.8 to 2.6 Mt) would be required to meet the lifetime fuel requirements of the LWRs expected to be operating in the year 1990.

Separative work requirements. The annual separative work requirements are estimated at about 22×10^6 SWU/a in 1980, and 55 to 80×10^6 SWU/a in 1990 (Fig. I.3). For the year 2000, the estimates range from 120×10^6 (low forecast, uranium and plutonium recycling) to 180×10^6 SWU/a (high forecast, no recycling).

Fabrication requirements. The requirements for fabrication of LWR fuel are estimated at about 6000 t/a (heavy metal) in 1980, and 16 000–23 000 t/a in 1990 (Fig. I.4).

Reprocessing requirements. The processing requirements for LWR fuel are estimated to be about 4000 t/a (heavy metal) in 1980 and 12 000–17 000 t/a 1990 (Fig. I.5). The 1977 planned reprocessing capacities are inadequate to meet this demand. Further delays of the start-up of reprocessing plants may occur. It will be necessary to store irradiated fuel for many years.

I.3. Uranium resources

Uranium resources are divided into "reasonably assured resources", which for the purposes of this report can be equated reserves, and "estimated additional resources". Each of these categories is further broken down by cost of production: less than $15 per pound U_3O_8, and $15 to $30 per pound U_3O_8. (These values refer to the purchasing power of the US$ of January 1, 1975.)

TABLE 1.3. *Uranium resources in 1000 tonnes. Data available January 1, 1977*

	Reasonably assured		Estimated additional	
Cost range	< US \$80 per kg U (< US \$30 per lb U_3O_8)	US \$80 – 130 per kg U (US \$30 – 50 per lb U_3O_8)	< US \$80 per kg U (< US \$30 per lb U_3O_8)	US \$80 – 130 per kg U (US \$30 – 50 per lb U_3O_8)
Algeria	28	0	50	0
Argentina	17.8	24	0	0
Australia	289	7	44	5
Austria	1.8	0	0	0
Bolivia	0	0	0	0.5
Brazil	18.2	0	8.2	0
Canada[a]	167	15	392	264
Central African Empire[b]	8	0	8	0
Chile	0	0	5.1	0
Denmark (Greenland)	0	5.8	0	8.7
Finland	1.3	1.9	0	0
France	37	14.8	24.1	20.0
Gabon[b]	20	0	5	5
Germany, F.R.	1.5	0.5	3	0.5
India	29.8	0	23.7	0
Italy	1.2	0	1	0
Japan	7.7	0	0	0
Korea	0	3	0	0
Madagascar	0	0	0	2.0
Mexico[c]	4.7	0	2.4	0
Niger	160	0	53	0
Philippines	0.3	0	0	0
Portugal	6.8	1.5	0.9	0
Somalia[d]	0	6.2	0	3.4
South Africa	306	42	34	38
Spain	6.8	0	8.5	0
Sweden	1	300	3	0
Turkey	4.1	0	0	0
United Kingdom	0	0	0	7.4
United States	523	120	838	215
Yugoslavia	4.5	2.0	5.0	15.5
Zaire	1.8	0	1.7	0
Total (rounded)	1650	540	1510	590

[a]The material reported as "Reserves" is minable at prices up to \$104 per kg U and the other "Reasonably Assured Resources" are minable at prices between \$104 and \$156 per kg U.
[b]Source of data: *Uranium Resources, Production and Demand,* Paris, 1975.
[c]Data refer to resources *in situ*, rather than recoverable.
[d]Costs of recovery are not known so the resources are arbitrarily assigned to the higher cost category.
NB: A number of occurrences of uranium are not well enough defined to be included in these tables.

Reasonably assured resources

Table I.3 shows reasonably assured resources reported by the joint NEA/IAEA Working Party. Reasonably assured resources are in known ore deposits of such grade, quantity, and configuration, based on specific sample data and measurements and knowledge of ore body habit, that they could be recovered within a given production cost range with currently proven mining and processing technology.

Several countries have shown minor increases in reasonably assured resources during the last 11 years, but the most dramatic increase has occurred in Australia. Uranium resources in the United States have shown a fairly steady increase, while those in Canada and South Africa have remained more or less even. It should be noted that Algeria was added to the list of countries with uranium resources in 1975.

Estimated additional resources

Estimated additional resources are believed to occur in unexplored extensions of known deposits or in undiscovered deposits in known uranium districts. Such deposits can be reasonably expected to be discovered and mined in the given cost range. The quantities of estimated additional resources, however, are based primarily on knowledge of the characteristics of deposits within the same districts and are less accurate than estimates of reasonably assured resources. Table I.3 shows estimated additional resources reported. Only about one million tonnes uranium are predicted to be found within these parameters.

Resources at higher costs

Uranium resources were also reported to the NEA/IAEA Working Party at costs of \$15 to \$30. Within this increment, the reasonably assured resources totalling 730 000 t uranium are primarily estimated from a relatively small amount of data developed while exploring for lower cost resources. Estimated additional resources in this cost increment total 680 000 t uranium.

Resources estimated from geological information

In addition, several countries have started programs to determine the uranium potential of regions where the geological conditions may be favorable for uranium deposits. Because of the nature of the estimates, quantities estimated on the basis of the data available may be several orders of magnitude less accurate than either reasonably assured or estimated additional resources.

APPENDIX J

Nuclear Fuel Cycle Costs

Nuclear LWR fuel cycle costs in the FR Germany according to K. P. Messer, *Schweizerische Vereinigung für Atomenergie*, Tagung 13–14 Juni, 1977, in Zürich.

Overall costs in May 1977

Uranium	49% at \$40 per lb U_3O_8 (\$88 per kg U_3O_8)
Enrichment	31% at \$100 per kg SWU
Fuel element fabrication	10% at DM 375 per kg U
Back end and recycling	10% at DM 380 per kg U
This yields a total fuel cost of	DM 0.0192 per kWh

Back end expenses and recycling savings

The following costs all refer to 1 kg used uranium fuel. The price basis is January 1, 1976.

Reprocessing, including transportation (Back end)	DM	510 per kg U
Radioactive waste conditioning (Back end)	DM	300 per kg U
Final storage (Back end)	DM	90 per kg U
Savings in recycled uranium (at \$30 per lb U_3O_8)	DM	− 480 per kg U
Savings in enrichment (at \$90 per kg SWU)	DM	− 230 per kg U
Additional cost for MOX fuel fabrication (at \$600 per kg MOX)	DM	170 per kg U
Cost balance for recycling services	DM	360 per kg U

In May 1977 the back end alone was estimated at DM 1.000 per kg U, and cost balance increased to DM 380 per kg U.

Conversion value June 1977: DM 1 = US\$0.44.

APPENDIX K

Survey of Reprocessing Plants and Plans

Location (operating time)	Fuels	Capacity (t U/y)	Comment
Hanford, USA (1944–)[a]	Metallic	>10 000?	Military
Windscale I, UK (1952–64)	Metallic		Shutdown
Savannah River, USA (1954–)[b]	Metallic	>10 000?	Military
Marcoule, France (1958–); UP 1	Metallic	900	Military
Windscale II, UK (1964–)	Metallic + oxide	2 500 + 400	Military and civil
Trombay, India (1965–)	Metallic	75	
La Hague, France (1966–)[c]	Metallic + oxide	800 +	
West Valley, USA (1966–71)[d]	Oxide	300(+ 750)	Shutdown
Eurochemic, Mol, Belgium (1966–74)	Metallic, oxide	60	Standby[e]
Eurex I, Italy (1970–)	Oxide	15	Pilot plant
WAK, Karlsruhe, FR Germany (1971–)	Oxide	35	Pilot plant
Tarapur, India (1977–)		125	
Tokai Mura, Japan (1978–)	Oxide	210	
Barnwell, USA (1977)	Oxide	1 500	Cold standby[f]
Gorleben, FR Germany (1989)	Oxide	1 400	Firm plans
Thorp I, Windscale, UK (1987?)	Oxide	600–1000	Firm plans
EXXON, Oak Ridge, USA (1985?)	Oxide	1 500	Plans

[a]Three plants.
[b]Two plants.
[c]The oxide 400 t/y plant (UP 2), starting up in 1977, will be increased to 800 t/y in 1984. A new 800 t/y oxide plant (UP 3) will start around 1985.
[d]Shut-down, after 750 t/y enlargement, because of economy.
[e]This Euratom plant will be taken over by Belgium and restarted.
[f]Not started because of political decision.

APPENDIX L

Some Common Abbreviations of Nuclear Related Organizations†

ANS	American Nuclear Society
CEA	Commissariat à l'Energie Atomique
CEN	(Belgium) Le Centre de'Etude de l'Energie Nucléaire
CERN	Conseil Européen de Recherche Nucléaire (Geneva)
CFSTI	Clearinghouse for Federal Scientific and Technical Information, US Dept. of Commerce, Springfield, Va 22151, USA
CMEA	Council for Mutual Economic Assistance‡
CNEN	(Italy) Comitato Nazionale per l'Energia Nucleare
DOE	US Department of Energy (formerly ERDA), Washington DC 20545
EEC	European Economic Community§
ENS	European Nuclear Society
ERDA	US Energy Research and Development Agency (formerly AEC, now DOE)
Euratom	European Atomic Energy Community (part of EEC)
FAO	UN Food and Agricultural Organization
IAEA	International Atomic Energy Agency
ICRP	International Commission on Radiological Protection
IRPA	International Radiation Protection Association
JAERI	Japan Atomic Energy Research Institute
NAS	US National Academy of Sciences
NAS-NS	Nuclear Science Series, publ. by (US) NAS and NRC; available from CFSTI
NBS	US National Bureau of Standards
NEA	OECD Nuclear Energy Agency
NRC	US Nuclear Regulatory Committee (also NUREG)
NRC	US National Research Council
NTIS	US National Technical Information Services; same as CFSTI
OECD	Organization for Economic Cooperation and Development[P]
SORIN	(Italy) La Societa Richerche Impianti Nucleari
UKAEA	United Kingdom Atomic Energy Authority

†See also *World Nuclear Directory*, Harrap Research Publ., London, 1976.

‡Countries with centrally planned economies (East Europe).

§Belgium, Denmark, France, Federal Republic of Germany, Great Britain, Ireland, Italy, Luxemburg, Netherlands.

Nuclear research centers at Ispra (Italy), Geel (Belgium), Karlsruhe (W. Germany), Petten (Holland).

[P]All West European countries plus Australia, Japan, Turkey, and the USA.

UN	United Nations
UNICPUAE	United Nations International Conference on the Peaceful Uses of Atomic Energy 1955, 1958, 1964, 1971, 1977
UNSCEAR	United Nations Scientific Committee on the Effects of Atomic Radiation
USAEC	United States Atomic Energy Commission (replaced by ERDA and NRC in 1976)
USGPO	Superintendent of Documents, US Government Printing Office, Washington DC, USA
WHO	UN World Health Organization

APPENDIX M

Answers to Exercises

Chapter 2 (**1**) 1.52×10^{23} atoms ^{235}U. (**2**) 0.10. (**3a**) 2112 m s^{-1}. (**3b**) 0.046 V. (**4**) 472 V. (**5**) 11.009 34 u. (**6**) 1.040. (**7**) 4 stages. (**8**) 3.0 cm. (**9b**) $\alpha = 1.097$. (**9c**) 22 stages. (**9d**) 200 000.

Chapter 3 (**1**) 8.06 MeV nucleon^{-1}. (**2**) 1.5×10^6 times. (**3**) 136 000 km. (**4**) Fusion ~7 times more. (**5**) Neutron-binding energies are 1.10, 2.79, and 1.11 MeV below average value for ^{236}U, ^{239}U, and ^{240}Pu, respectively. (**6**) 7.54 MeV nucleon^{-1}. (**7**) $Z = 4.8(^{10}_{5}B)$, $12.6(^{27}_{13}Al)$, $26.4(^{59}_{27}Co$ stable), $92.3(^{239}_{24}Pu$ most stable).

Chapter 4 (**1**) 0.088 MeV, ^{235}U. (**2a**) 0.87 MeV. (**2b**) 0.46 MeV (measured 0.51 MeV). (**3**) 147 eV. (**4**) 2.53 keV. (**5**) 545 keV (measured 662 keV). (**6**) 1.31×10^9 y. (**7**) 6×10^9 y. (**8**) 2.5×10^{18} Jy^{-1}. (**9a**) 0.457 Ci. (**9b**) 3.0×10^{-10} m^3. (**11a**) 8.77×10^5 dpm. (**11b**) 8.05×10^5 atoms. (**12**) $t_{1/2}$ 108 h and 2.59 nCi, and $t_{1/2}$ 4.5 h and 25.9 nCi, respectively. (**13**) R_0 500 cpm for $t_{1/2}$ 106 min, R_0 380 cpm for $t_{1/2}$ 30 min (cf. ^{208}At 98 min, ^{206}At 31 min half-lives). (**14**) 2.3×10^5 y.

Chapter 6 (**1a**) 10. (**1b**) 10. (**2a**) 2.19×10^6 m s^{-1}. (**2b**) 1.000 026 7 $\mathbf{m}_e$. (**3**) 21 MHz. (**4**) $2.689 \times 2/3$. (**5**) li, 2g, 3d, 4s; 56 nucleons. (**6**) ~280 MeV (**7a**) 7/2, 1, 5/2. (**7b**) $3.81 \cdot \frac{1}{5}$, $4.50 \cdot \frac{2}{5}$. (**8a**) 0.152. (**8b**) 7/2 (also observed). (**9**) 1 p 1/2. (**10a**) 43.8 and 30.7 keV. (**10b**) 8.68 fm (meas. 5.48 fm). (**11**) 0.460 T. (**12**) 3.2×10^9 y.

Chapter 7 (**1a**) 37 MeV. (**1b**) 1304 m s^{-1}, 0.035 eV. (**2**) 46 fm. (**3a**) 32.8 b steradian^{-1}. (**3b**) 0.42 m. (**4**) 0.75 MeV. (**5a**) 38.0 fm. (**5b**) 7.56 fm. (**6a**) 8.03 MeV. (**6b**) −1.64 MeV. (**7**) 3.09×10^6 m s^{-1}. (**8**) 3.57×10^{-28} kg. (**9**) 76.7 MeV. (**10**) $Q > 0$, $^{5}_{2}He$ thus unstable. (**11a**) −5.85 MeV. (**11b**) 6.39 MeV. (**11c**) 2.75 MeV. (**11d**) 2.99 MeV.

Chapter 8 (**1a**) 0.104 m. (**1b**) 9 MeV. (**1c**) For $Z = 20$ (Ca) $E_{cb}(\min) = 9.1$ MeV. (**2a**) 10kW. (**2b**) 227 m s^{-1}. (**2c**) 2700 m s^{-1}. (**3**) 7.6×10^7 n cm^{-2} s^{-1}. (**4**) 4.1 m. (**5a**) $L_2 = 1.41$ cm, $L_3 = 1.73$ cm. (**5b**) 109 MHz. (**6**) H^+ 11.7 MeV, 0.80 T; D^+ 23.5 MeV, 1.59 T; He^{2+} 46.6 MeV, 1.56 T. (**7**) 0.9 mA. (**8**) 8.5×10^3 g.

Chapter 9 (**1**) 5.8×10^9 s^{-1}. (**2**) 6%. (**3a**) 2×10^{15} atoms Cu. (**3b**) 1 in 20 000. (**3c**) 29 kW. (**4**) 36 m^{-1}. (**5**) 1.95 MeV. (**6**) $E_\gamma \geqq 11.5$ MeV. (**7**) 10^{11} K.

Chapter 10 (**1**) 749 Ci/g Na (**2**) 0.102. (**3**) 5.2×10^{13} dpm. (**4**) 1.23×10^7 s. (**5**) 4870 cps. (**6**) 7.3 h.

Chapter 11 (**1a**) 1.3 keV. (**1b**) 2.2×10^{-34} (eq. (B.11)). (**2a**) 12.3 × Lawson crit. (**2b**) 49 MW. (**3**) 87 l. (**4**) 6.3×10^{11} kgs^{-1}. (**5**) 5.5×10^{18} atoms D/m^3. (**6**) Water power ~ 10^8 W, fusion ~ 2×10^{14} W.

Chapter 12 (**1**) 7.9 dpm m^{-3}. (**2**) 14×10^6 y. (**3**) 19 dpm/g C. (**4**) Probably true. (**5**) 1.5×10^9 y. (**6**) 5×10^8 y. (**7**) 3.7×10^9 y. (**8**) U 0.012, Th 0.011, K 0.0031 J m^{-2} s^{-1} (43% of total measured flow). (**9a**) 193 t. (**9b**) 447 000 t. (**10**) Japan 17.5 y, Argentine 52.7 y, France 210 y.

Chapter 13 (**1**) n, γ on ^{196}Hg, or n, 2n on ^{198}Hg. (**2**) 93%. (**3**) ^{239}Pu from n, γ in ^{238}U. (**4**) ^{248}Cm and FP; ^{253}Es and ^{249}Cm; ^{250}Cm and FP. (**5**) 2.87 n/fiss. (**8**) 0.085 W. (**9**) 5 f^{14} $7s^2$.

Chapter 14 (**1**) From (14.9) $\hat{R}$ 5.8 mg cm^{-2}; from Fig. 14.5 ~6 mg cm^{-2}. (**2**) E_α 2.0 MeV. (**3**) 0.6 mm. (**4**) e^- 3.8 m, H^+ 23 cm, α 3.3 mm. (**5**) Al 0.030, Ni 0.012, Pt 0.0074 mm. (**6**) 6500 γ cm^{-2} s^{-}. (**7**) T 5 μm, ^{14}C 270 μm, ^{32}P 4.7 mm, ^{90}Sr 1.1 mm. (**8**) 2.92×10^8 m s^{-1}. (**9**) Lead $129 000, concrete $3600. (**10**) Na_K 57.53, Na_L 58.54, I_K 25.43, I_L 53.80 keV. (**11a**) 72%. (**11b**) 52%. (**12**) 3.5, 5.0, and 6.2 cm. (**13**) E_{max} 0.62 MeV, E_γ 0.8 MeV. (**14a**) 210 h. (**14b**) 0.31 h. (**15**) 54 cm. (**16**) 7.0 cm. (**16b**) 11.9 cm. (**16c**) 2.5.

Chapter 15 (**1a**) 1.47×10^5. (**1b**) 2.94×10^4. (**1c**) 1.18×10^3. (**2**) ~1%. (**3a**) 25.9 R. (**3b**) 0.23 Gy. (**4**) ~0.3%. (**5**) ~11%. (**6**) ^{90}Sr 0.101, ^{90}Y 0.076, ^{3}T 3.05 keV μm^{-1}. (**7**) 2274 rad h^{-1}. (**8**) 38 pF.

Chapter 16 (**1**) Total 2.56×10^6 dpm; ^{40}K (2.62×10^5) and T (2.05×10^6 dpm) dominate over ^{14}C (2.52×10^5) and Ra plus daughters (50 dpm). (**2**) 5×10^{-7}%. (**3**) 1.8×10^{-6}%. (**4a**) 14 mrem y^{-1}. (**4b**) 0.58 mrem y^{-1}. (**5**) 1 rem. (**6a**) 4.4×10^6 eV. (**6b**) 690 million. (**7**) 5.7 μs.

Chapter 17 (**1a**) 34 mCi. (**1b**) 6%. (**2a**) 25 μm. (**2b**) 1.05×10^{-5} μm^{-1}. (**2c**) 46 Å, (**2d**) 3.8 eV atom^{-1}. (**2e**) 7000 K. (**3**) 0.14 μM. (**4**) 24 pCi. (**5**) 5.8 V. (**6a**) 760 μs. (**6b**) 12.53. (**6c**) 22.8. (**7**) 3×10^{10} counts. (**8**) 45 ns. (**9a**) 6.6 μg cm^{-2}. (**9b**) 1.60. (**9c**) 71%. (**10a**) 560 ± 6.4 cpm. (**10b**) ± 4.3 cpm. (**11**) 0.0054.

Chapter 18 (**1**) 4.7 l. (**2**) 24%. (**3**) 21%. (**4**) ~10^{-10} g. (**5**) 45 ppm Ga. (**6**) 20.4%. (**7**) 3.32×10^{-4} s^{-1} M^{-1}. (**8a**) 400. (**8b**) 25. (**9**) Calculated $\log \beta_n$ values are: 10.5, 19.7, 27.8, and 33.8. (**10**) 32 m^3. (**11**) −0.22 V. (**12**) $D = 1$ for V at pH 0.5 and for U at pH 2.6; optimal separation at pH 1.55. (**13**) 99.9% U with 9.8% La.

Chapter 19 (**1**) $465 million. (**2**) 1.02 kg ^{235}U. (**3a**) Eight oil tankers (7.37). (**3b**) One train car (12.3 t). (**3c**) Five train cars (91.4 t). (**4**) Twenty collisions (19.6). (**5**) 1.90. (**6**) $\eta = 2.022$, $f = 0.751$, $k_\infty = 1.45$. (**7**) 1.39 m. (**8**) 56.6 m^3. (**9**) $\Lambda_f = 0.62$, $r = 25.7$ cm. (**10**) 23%. (**11**) 1096 times. (**12**) 17.07 Ci y^{-1}.

Chapter 20 (**1a**) 3.44 kg. (**1b**) Because effective cross-sections (Fig. 19.4) are larger than the thermal ones.

(2) 152 d. **(3)** 2.2 kg ^{235}U have been formed, while 1.8 kg remain unfissioned. **(5a)** 31.7 kg FP. **(5b)** 0.78%. **(6a)** 5.3 MCi. **(6b)** Ce, Pr, Nb, Zr, Y, Sr, and Ru. **(7a)** 7.5×10^5. **(7b)** 5.1×10^3. **(8a)** 1.5%. **(8b)** 99.9%. **(8c)** 2.7×10^{-6}%. **(18d)** 8×10^{-4}%. **(9)** 0.0073 Ci ^{238}Pu, 0.0025 Ci ^{240}Pu. **(10)** 2.6×10^{10} m^3 (water). **(11a)** 265 kW. **(11b)** 5.4 reactor years. **(12)** Ru 714 kg, 139 kCi, 0.195 Ci g^{-1}; Rh 129 kg, none radioactive; Pd 385 kg, 33.2 Ci, 86 μCi g^{-1}

Appendix B **(1)** 1.29 Å. **(2)** 1.84×10^{17} molecules at average translational energy, corresponding to more than 6143 K.

Appendix D **(1)** 202 t. **(2a)** 118 000 kg SWU. **(2b)** \$7.1 million and 1.16 million kWh^{-1}.

APPENDIX N

Chart of the Nuclides

On the following pages the *Karlsruher Nuklidkarte* (by W. Seelmann-Eggebert, G. Pfennig and H. Münzel is reproduced by the kind permission of Prof. Seelmann-Eggebert. The chart shows Z on the ordinate and N on the abscissa. The axes are thus interchanged when compared with Fig. 3.1 (but same as for Fig. 3.8). In the original chart colors are used to make it easy to distinguish between different decay modes.

The following abbreviations and symbols are used:

μs; ms; s	Mikrosekunde, Millisekunde, Sekunde microsecond, millisecond, second
m; h; d; a	Minute, Stunde, Tag, Jahr minute, hour, day, year
Iγ	Isomerenübergang isomeric transition
sf	Spontanspaltung spontaneous fission
e−	Konversionselektronen conversion electrons
p	Zerfall durch Protonen-Emission decay by proton emission
(n); (p); (α)	n-, p- oder α-Emission nach einem β-Übergang n, p or α emission following β decay
m; g	Nuklid zerfällt in den metastabilen und/oder Grundzustand des Tochternuklids nuclide decays to the metastable and/or ground state of the daughter nuclide
σf; σ	Spalt- und (n, γ)-Querschnitt für thermische Neutronen (b) fission- and (n,γ)-cross section for thermal neutrons (b)

Element

Cd	symbol of the element
112,40	atomic weight
σ 2450	absorption cross section for thermal neutrons (b)

Unstable Nuclides

Ge 81	symbol of the element, number of nucleons
10,1 s	half-life
β⁻ ~ 5,3; ~ 5,6 γ 336; 198 ...	β⁻-energies in MeV, γ-energies in keV
0,21	isobaric yield for ^{235}U fission with thermal neutrons (%)

Stable Nuclide

Te 126	symbol of the element, number of nucleons
18,7	isotopic abundance (atom %)
σ 0,135+0,90	(n, γ) cross sections for the formation of ^{127m}Te and of ^{127g}Te (b)

Sr 85		symbol of the element, number of nucleons
67,7 m Iγ 232... ε γ 151	64,9 d ε no β⁺ γ 514...	left hand side: metastable state right hand side: ground state

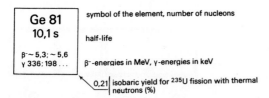

Nuclide with Metastable State

Se 77		left part: metastable state right part: ground state
17,5 s	7,5	
Iγ 162		energy of the isomeric transition (keV)
	σ 42	(n,γ) cross section (b)

Am 240		
sf	·50,8 h ε α 5,378... γ 988; 889... g	left hand side: spontaneous fission isomer T < 0,1 s right hand side: decay data of the ground state; "g" indicates that the daughter ^{240g}Pu is formed to at least 95% and ^{240m}Pu to at most 5%

Primordial Nuclide

Pt 190	symbol of the element, number of nucleons
0,013	isotopic abundance (atom %)
6,1 · 10¹¹ a α 3,17 σ 150	

KARLSRUHER NUKLIDKARTE 4. Auflage 1974

CHART OF THE NUCLIDES – TABLEAU DES NUCLEIDES – TABLA DE NUCLIDOS

W. Seelmann-Eggebert, G. Pfennig, H. Münzel

Kernforschungszentrum Karlsruhe
Institut für Radiochemie

Gesellschaft für Kernforschung mbH, Karlsruhe

Zu beziehen von: Gersbach u. Sohn Verlag
For sale by: 8 München 34, Barer Str. 32
En vente chez: ISBN 3 87253 084 4
Puede adquirirse en:

Gedruckt von: Ernst Klett Druckerei
Printed by: 7 Stuttgart 1
Imprimé par:
Impreso por:

									18	Ar 39,948 σ 0,678	Ar 33 173 ms β^+ 9,8; 10,6 ... (p 3,17 ...) γ 810
									17	Cl 35,453 σ 33,2	Cl 32 291 ms β^+ 9,5; 11,7 ... γ 2230; 4770 ... (α 2,10; 1,61 ...) (p 1,02; 0,76; 1,35)
					16	S 32,06 σ 0,520			S 29 187 ms β^+ (p 5,44; 3,72...)	S 30 1,22 s β^+ 4,4; 5,1 ... γ 678 ...	S 31 2,61 s β^+ 4,4 ... γ 1266 ...
					15	P 30,97376 σ 0,180			P 28 268 ms β^+ 11,5 ... γ 1779; 4497...	P 29 4,2 s β^+ 4,0 ... γ ...	P 30 2,50 m β^+ 3,2 ... γ ...
					14	Si 28,086 σ 0,16	Si 25 218 ms β^+ (p 4,08; 1,87; 3,33 ...)	Si 26 2,2 s β^+ 3,8 ... γ 830 m	Si 27 4,2 s β^+ 3,8 ... γ ...	Si 28 92,2 σ 0,17	Si 29 4,7 σ 0,28
				13	Al 26,98154 σ 0,230	Al 23 470 ms β^+ (p 0,83)	Al 24 129 ms | 2,05 s Iγ 439 | β^+ 4,4; 8,7 ... β^+ 13,3 ... | γ 1369; 2754; 7066 ... γ 1369 ... (α 1,40; 1,77) | (α 1,98; 1,57 ...)	Al 25 7,2 s β^+ 3,3 ... γ ...	Al 26 6,35 s | 7,16·10^6 a β^+ 3,2 | β^+ 1,2 γ 1809; 1130 ...	Al 27 100 σ 0,230	Al 28 2,246 m β^- 2,9 γ 1779
12	Mg 24,305 σ 0,063			Mg 20 ? 0,62 s β^+	Mg 21 122,5 ms β^+ (p 1,94; 1,77...)	Mg 22 3,86 s β^+ 3,2 ... γ 74; 583 ...	Mg 23 12,0 s β^+ 3,1 ... γ 439	Mg 24 78,99 σ 0,052	Mg 25 10,00 σ 0,180	Mg 26 11,01 σ 0,038	Mg 27 9,46 m β^- 1,8 ... γ 844; 1014...
11	Na 22,98977 σ 0,530				Na 20 446 ms β^+ 11,2 ... (α 2,15; 4,44...) γ 1634 ...	Na 21 22,8 s β^+ 2,5 ... γ 350	Na 22 2,60 a β^+ 0,5; 1,8 γ 1275 σ 29000	Na 23 100 σ 0,40 + 0,13	Na 24 20 ms | 15,03 h Iγ 472 | β^- 1,4 ... β^- ~ 6 | γ 2754; 1369 ...	Na 25 59,6 s β^- 3,8 ... γ 975; 390; 585; 1612 ...	Na 26 1,09 s β^- 7,4 ... γ 1809 ...
10	Ne 20,179 σ 0,038		Ne 17 109 ms β^+ 8,0; 13,5 ... (p 4,59; 3,77; 5,12 ...)	Ne 18 1,67 s β^+ 3,4 ... γ 1042	Ne 19 17,4 s β^+ 2,2 no γ	Ne 20 90,5 σ 0,037	Ne 21 0,27 σ 0,692	Ne 22 9,2 σ 0,048	Ne 23 38 s β^- 4,4 ... γ 439; 1639...	Ne 24 3,38 m β^- 2,0 ... γ 874 m	Ne 25 602 ms β^- 7,3; 7,4 ... γ 90; 980 ...
9	F 18,99840 σ 0,0095			F 17 64,5 s β^+ 1,7 no γ	F 18 109,7 m β^+ 0,6 no γ	F 19 100 σ 0,0095	F 20 11,0 s β^- 5,4 ... γ 1634 ...	F 21 4,4 s β^- 5,3; 5,7... γ 350; 1395...	F 22 4,23 s β^- 5,5 ... γ 1275; 2078; 2165 ...	F 23 2,23 s β^- γ 1701; 1822; 3431 ...	F 24

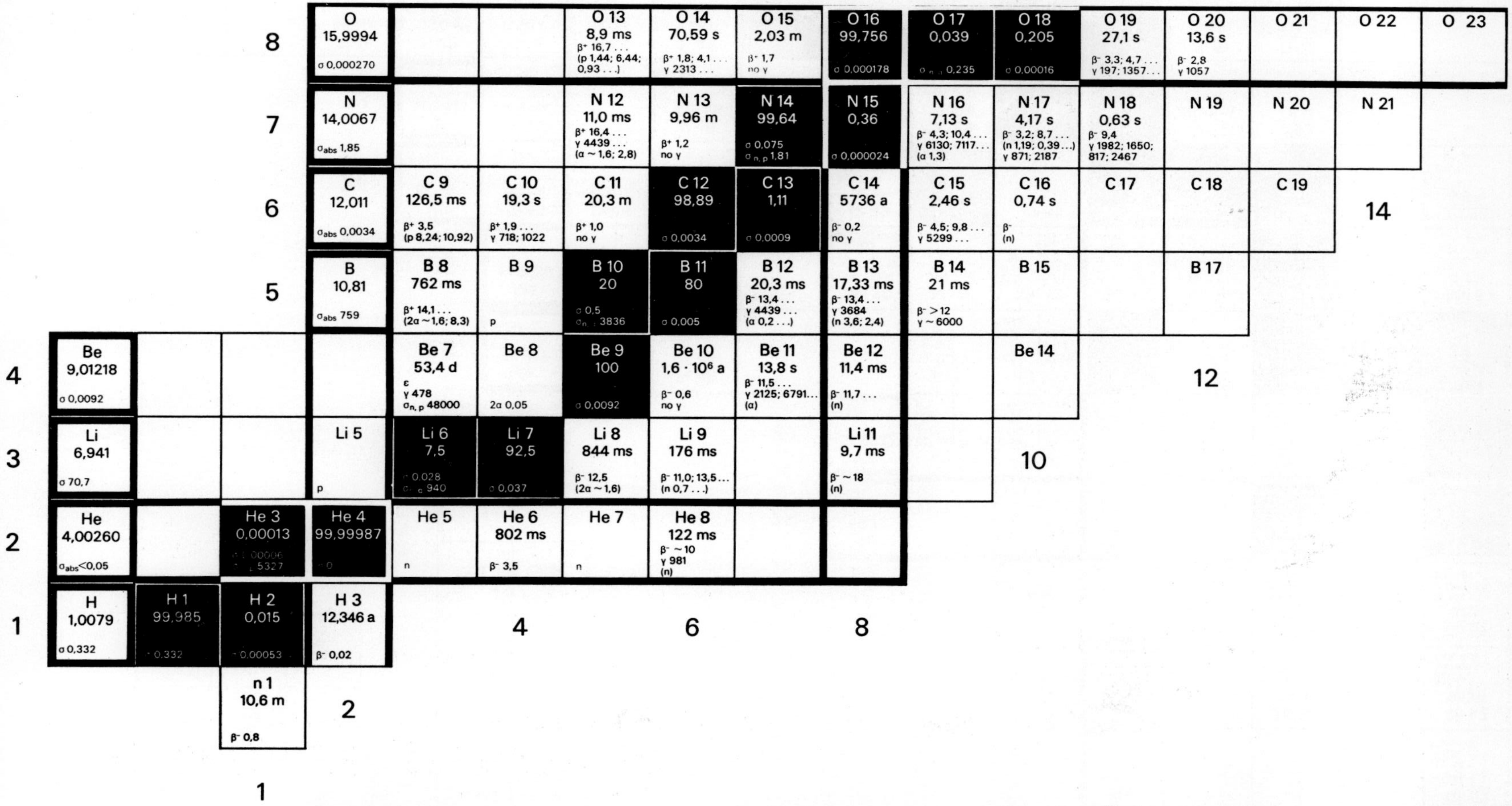
8
7
6
5
4
3
2
1
O 15,9994 σ 0,000270
O 13 8,9 ms β+ 16,7 . . . (p 1,44; 6,44; 0,93 . . .)
O 14 70,59 s β+ 1,8; 4,1 . . . γ 2313 . . .
O 15 2,03 m β+ 1,7 no γ
O 16 99,756 σ 0,000178
O 17 0,039 σ n,α 0,235
O 18 0,205 σ 0,00016
O 19 27,1 s β- 3,3; 4,7 . . . γ 197; 1357 . . .
O 20 13,6 s β- 2,8 γ 1057
O 21
O 22
O 23
N 14,0067 σabs 1,85
N 12 11,0 ms β+ 16,4 . . . γ 4439 . . . (α ~ 1,6; 2,8)
N 13 9,96 m β+ 1,2 no γ
N 14 99,64 σ 0,075 σ n,p 1,81
N 15 0,36 σ 0,000024
N 16 7,13 s β- 4,3; 10,4 . . . γ 6130; 7117 . . . (α 1,3)
N 17 4,17 s β- 3,2; 8,7 . . . (n 1,19; 0,39 . . .) γ 871; 2187
N 18 0,63 s β- 9,4 γ 1982; 1650; 817; 2467
N 19
N 20
N 21
C 12,011 σabs 0,0034
C 9 126,5 ms β+ 3,5 (p 8,24; 10,92)
C 10 19,3 s β+ 1,9 . . . γ 718; 1022
C 11 20,3 m β+ 1,0 no γ
C 12 98,89 σ 0,0034
C 13 1,11 σ 0,0009
C 14 5736 a β- 0,2 no γ
C 15 2,46 s β- 4,5; 9,8 . . . γ 5299 . . .
C 16 0,74 s β- (n)
C 17
C 18
C 19
14
B 10,81 σabs 759
B 8 762 ms β+ 14,1 . . . (2α ~ 1,6; 8,3)
B 9 p
B 10 20 σ 0,5 σ n,α 3836
B 11 80 σ 0,005
B 12 20,3 ms β- 13,4 . . . γ 4439 . . . (α 0,2 . . .)
B 13 17,33 ms β- 13,4 . . . γ 3684 (n 3,6; 2,4)
B 14 21 ms β- > 12 γ ~ 6000
B 15
B 17
Be 9,01218 σ 0,0092
Be 7 53,4 d ε γ 478 σn,p 48000
Be 8 2α 0,05
Be 9 100 σ 0,0092
Be 10 1,6 · 10⁶ a β- 0,6 no γ
Be 11 13,8 s β- 11,5 . . . γ 2125; 6791 . . . (α)
Be 12 11,4 ms β- 11,7 . . . (n)
Be 14
12
Li 6,941 σ 70,7
Li 5 p
Li 6 7,5 σ 0,028 σ n,α 940
Li 7 92,5 σ 0,037
Li 8 844 ms β- 12,5 (2α ~ 1,6)
Li 9 176 ms β- 11,0; 13,5 . . . (n 0,7 . . .)
Li 11 9,7 ms β- ~ 18 (n)
10
He 4,00260 σabs<0,05
He 3 0,00013
He 4 99,99987
He 5 n
He 6 802 ms β- 3,5
He 7 n
He 8 122 ms β- ~ 10 γ 981 (n)
H 1,0079 σ 0,332
H 1 99,985 σ 0,332
H 2 0,015 σ 0,00053
H 3 12,346 a β- 0,02
n 1 10,6 m β- 0,8
1
2
4
6
8

RADIOACTIVE DECAY

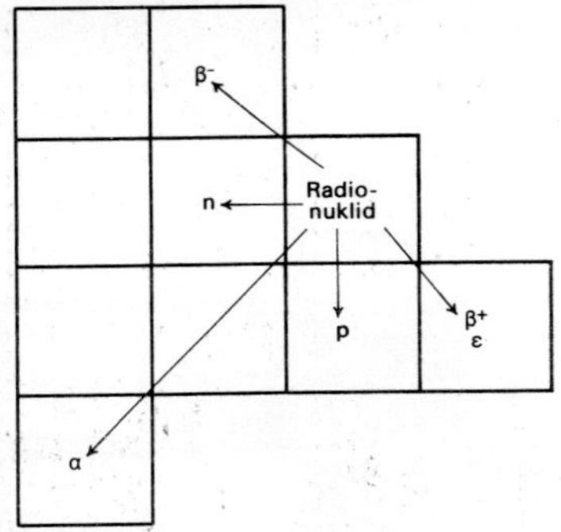

NUCLEAR REACTIONS

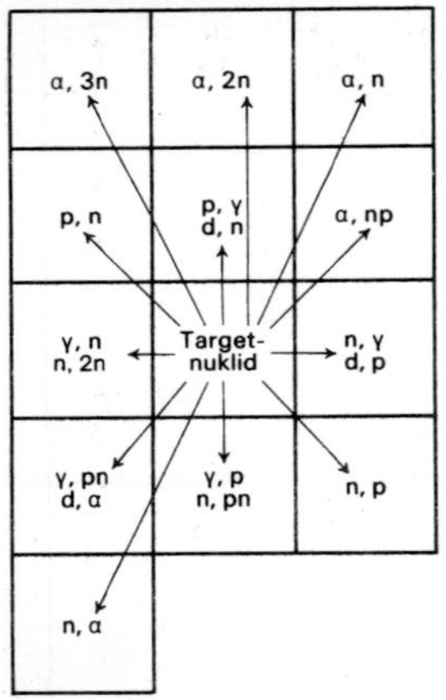

Z																	
32																Ge 72,59 σ 2,3	Ge 64 64 s β^+ 3,0; 3,3 ... γ 427; 667; 128 ...
31													Ga 69,72 σ 2,9			Ga 62 118 ms β^+ 7,8	Ga 63 31,4 s β^+ ~4,5 ... γ 637; 627; 193; 650 ...
30													Zn 65,38 σ 1,10		Zn 60 2,4 m β^+ 2,5; 3,1 ... γ 670; 61; 273; 334 ...	Zn 61 1,5 m β^+ 4,4 ... γ 475; 1660; 970 ...	Zn 62 9,13 h ε β^+ 0,7 γ 41; 597; 548; 508 ...
29													Cu 63,546 σ 3,79	Cu 58 3,20 s β^+ 7,5 ... γ 1454; 1448; 40 ...	Cu 59 81 s β^+ 3,8 ... γ 1302; 878; 339; 465 ...	Cu 60 23 m β^+ 2,0; 3,9 ... γ 1332; 1792; 826 ...	Cu 61 3,3 h β^+ 1,2 ... γ 283; 656; 67; 1186 ...
28										Ni 58,70 σ 4,43			Ni 56 6,1 d ε; no β^+ γ 158; 812; 750; 480; 270 ...	Ni 57 36,0 h ε; β^+ 0,8 ... γ 1378; 1920; 127 ...	Ni 58 67,76 σ 4,6	Ni 59 $7{,}5 \cdot 10^4$ a ε; no β^+ no γ σ 92 $\sigma_{n,\alpha}$ 12	Ni 60 26,42 σ 2,8
27										Co 58,9332 σ 37,2	Co 53 247 ms: β^+, p 1,56, m 262 ms: β^+, g	Co 54 1,5 m: β^+ 4,3, γ 411; 1130; 1407 194 ms: β^+ 7,3	Co 55 18 h β^+ 1,5 ... γ 931; 477; 1409 ...	Co 56 77,3 d ε; β^+ 1,5 γ 847; 1238; 2599; 1771; 1038 ...	Co 57 270 d ε γ 122; 136 ... e^-	Co 58 8,94 h: Iγ, e^-, 136000 70,78 d: ε, β^+ 0,5; 1,3, γ 811 ..., σ 1880	Co 59 100 σ 20 + 17
26						Fe 55,847 σ 2,55		Fe 49 75 ms β^+ (p 1,92)			Fe 52 8,2 h β^+ 0,8 γ 169 m	Fe 53 2,5 m: Iγ 701; 1328; 1011; 2340 ... 8,51 m: β^+ 2,8 ..., γ 378	Fe 54 5,8 σ 2,25	Fe 55 2,7 a ε no γ	Fe 56 91,7 σ 2,63	Fe 57 2,14 σ 2,48	Fe 58 0,31 σ 1,15
25						Mn 54,9380 σ 13,3				Mn 50 1,76 m: β^+ 3,5; 3,7 ..., γ 1098; 783; 1443 ... 286 ms: β^+ 6,6	Mn 51 46,2 m β^+ 2,2 ... γ ...	Mn 52 21 m: β^+ 2,6 ..., γ 1434 ..., Iγ 378 5,7 d: ε, β^+ 0,6 ..., γ 1434; 936; 744 ...	Mn 53 $3{,}7 \cdot 10^6$ a ε no γ σ 33	Mn 54 312,2 d ε γ 835	Mn 55 100 σ 13,3	Mn 56 2,58 h β^- 2,9 ... γ 847; 1811; 2113 ...	Mn 57 1,7 m β^- 2,6 ... γ 122; 692; 136; 352 ...
24					Cr 51, 996 σ 3,1	Cr 45 50 ms β^+ (p 2,05)	Cr 46 0,26 s β^+		Cr 48 23 h ε γ 308; 112	Cr 49 42 m β^+ 1,4; 1,5 ... γ 91; 153; 62 ...	Cr 50 4,35 σ 15,9	Cr 51 27,70 d ε γ 320	Cr 52 83,79 σ 0,76	Cr 53 9,50 σ 18,2	Cr 54 2,36 σ 0,36	Cr 55 3,6 m β^- 2,5 ... γ ...	Cr 56 5,9 m β^- 1,5 γ 26; 84
23					V 50,9414 σ 5,04	V 44 90 ms β^+ (α 2,78)		V 46 426 ms β^+ 6,0	V 47 32,6 m β^+ 1,9 ... γ ...	V 48 15,97 d ε; β^+ 0,7 γ 983; 1312; 944 ...	V 49 330 d ε no γ	V 50 0,25 σ 70	V 51 99,75 σ 4,88	V 52 3,75 m β^- 2,5 ... γ 1434 ...	V 53 1,6 m β^- 2,5 ... γ 1006; 1289 ...	V 54 43 s β^- 3,0; 5,2 ... γ 835; 986; 2255 ...	
	Ti 47,90 σ 6,1			Ti 41 88 ms β^+ (p 4,75; 3,09; 1,00 ...)	Ti 42 0,20 s β^+ 5,4; 6,0 ... γ 611 ... m; g	Ti 43 0,50 s β^+ 5,8 no γ	Ti 44 47,3 a ε γ 78; 68 ... g	Ti 45 3,08 h β^+ 1,0 ... γ ...	Ti 46 8,0 σ 0,6	Ti 47 7,5 σ 1,7	Ti 48 73,7 σ 7,8	Ti 49 5,5 σ 2,2	Ti 50 5,3 σ 0,179	Ti 51 5,8 m β^- 2,1 ... γ 320; 928 ...	Ti 52 1,7 m: β^- 1,8, γ 17; 124 49 m: β^-		

Sc 44,9559 σ 26,5			Sc 40 183 ms β+ 5,7; 9,6 ... γ 3737; 755... (p 1,05; 1,24...)	Sc 41 596 ms β+ 5,5 no γ	Sc 42 61 s / 0,68 s β+ 2,8; γ 438; 1525; 1227 / β+ 5,4...; γ ...	Sc 43 3,89 h β+ 1,2 ... γ 373	Sc 44 2,44 d / 3,92 h Iγ 271; ε; γ ... / β+ 1,5 ...; γ 1157 ...	Sc 45 100 σ 9,6+16,9	Sc 46 18,7 s / 84 d Iγ 142 / β- 0,4 ...; γ 889; 1121; σ 8,0	Sc 47 3,42 d β- 0,4; 0,6 γ 159	Sc 48 43,67 h β- 0,7 ... γ 983; 1312; 1038 ...	Sc 49 57,2 m β- 2,0 ... γ ...	Sc 50 1,7 m β- 3,7; 4,2 γ 1554; 1121; 524	Sc 51 12 s β- 5,0 ... γ 1429; 2136	
Ca 40,08 σ 0,43	Ca 37 175 ms β+ (p 3,10; 1,72...)	Ca 38 439 ms β+ 5,2 ... γ 1568 ... m	Ca 39 860 ms β+ 5,5 no γ	Ca 40 96,94 σn,α 0,0025 σ 0,40	Ca 41 $1,3 \cdot 10^5$ a ε no γ	Ca 42 0,65 σ 0,65	Ca 43 0,14 σ 6,2	Ca 44 2,08 σ 1,0	Ca 45 163 d β- 0,3 ... γ ... e−	Ca 46 0,003 σ 0,7	Ca 47 4,54 d β- 0,7; 2,0 ... γ 1297; 808; 489 ...	Ca 48 0,19 σ 1,1	Ca 49 8,72 m β- 2,2; 2,9 ... γ 3084; 4072 ...	Ca 50 13,9 s β- 3,1 γ 257; 1519; 72; 1591	
K 39,098 σ 2,10	K 36 341 ms β+ 9,9 ... γ 1970; 2434; 2208 ...	K 37 1,22 s β+ 5,1 ... γ 2794 ...	K 38 929 ms / 7,7 m β+ 5,0 / β+ 2,7...; γ 2167...	K 39 93,3 σn,α 0,0043 σ 1,96	K 40 0,012 $1,28 \cdot 10^9$ a β- 1,3; ε; β+ ... γ 1461; σn,p 4,4 σ 30; σn,α 0,39	K 41 6,7 σ 1,46	K 42 12,36 h β- 3,5 ... γ 1525 ...	K 43 22,2 h β- 0,8; 1,8 ... γ 373; 618 ...	K 44 22,2 m β- 5,7 ... γ 1157; 2151 ...	K 45 16,3 m β- 2,1; 4,0 ... γ 174; 1706 ...	K 46 115 s β- 6,4 ... γ 1347; 3700 ...	K 47 17,5 s β- 4,1 ... γ 2013; 586; 565	K 48 ~9 s β- γ 3833	K 49	K 50
Ar 34 839 ms β+ 5,0 ... γ 665 ... g	Ar 35 1,78 s β+ 4,9 ... γ 1219; 1763 ...	Ar 36 0,34 σ 5 σn,α 0,0055	Ar 37 35,1 d ε no γ σ 600	Ar 38 0,07 σ 0,8	Ar 39 269 a β- 0,6 no γ	Ar 40 99,59 σ 0,660	Ar 41 1,83 h β- 1,2; 2,5 ... γ 1294 ... σ 0,5	Ar 42 33 a β- ~0,6 no γ	Ar 43 5,4 m β- γ 975; 738; 1440 ...	Ar 44 11,9 m β- γ 1705; 182; 1887 ...		28		30	
Cl 33 2,47 s β+ 4,5 ... γ 1966; 2867; 841	Cl 34 32,0 m / 1,53 s β+ 2,5 ...; γ 2127; 1177; 3305...; Iγ 146 / β+ 4,5; no γ	Cl 35 75,77 σ 43 σn,p 0,489 σn,α 0,00008	Cl 36 $3,0 \cdot 10^5$ a β- 0,7 ε; β+ ... no γ σ < 10	Cl 37 24,23 σ 0,433	Cl 38 37,18 m β- 4,9 ... γ 2167; 1642 ...	Cl 39 56 m β- 1,9; 3,4 ... γ 1267; 250; 1517 ...	Cl 40 1,35 m β- 3,2; 7,5 ... γ 1461; 2840; 2622 ...	Cl 41 34 s β- 3,8 ... γ 167 – 1354	Cl 42	26					
S 32 95,0 σ 0,53 σn,α 0,004	S 33 0,75 σn,p 0,002 σn,α 0,140 σ 0,015	S 34 4,2 σ 0,240	S 35 87,5 d β- 0,2 no γ	S 36 0,015 σ 0,15	S 37 5,1 m β- 1,8; 4,9 ... γ 3103 ...	S 38 2,83 h β- 1,0; 2,9 ... γ 1942 ...	S 39	S 40							
P 31 100 σ 0,180	P 32 14,3 d β- 1,7 no γ	P 33 25,3 d β- 0,2 no γ	P 34 12,4 s β- 5,4 ... γ 2127 ...	P 35 47,4 s β- 2,3 γ 1572	P 36	P 37	P 38	24							
Si 30 3,1 σ 0,107	Si 31 2,62 h β- 1,5 ... γ ... σ 0,48	Si 32 280 a β- 0,1 no γ	Si 33 6,18 s β- 3,9; 5,8 ... γ 1848 ...	Si 34	Si 35	Si 36									
Al 29 6,6 m β- 2,5 ... γ 1273; 2426; 2028 ...	Al 30 3,3 s β- 5,1; 6,3 ... γ 2235; 3498; 1263 ...	Al 31 644 ms β- 5,6; 7,9 ... γ 2317; 1695 ...	Al 32	Al 33											
Mg 28 21,1 h β- 0,5 ... γ 31; 1342; 401; 942 ...	Mg 29 1,49 s β- γ 2224; 1398; 960 ...	Mg 30 0,85 s β-	Mg 31 (n)												
Na 27 295 ms β- 7,9 ... γ 985; 1699 (n)	Na 28 35,7 ms β- 12,3 ... γ 1474; 2389 (n)	Na 29 48,6 ms β- 11,0 γ 2570; 1470; 2100; 3150 (n)	Na 30 55 ms β- (n)	Na 31 17,7 ms β- (n)	Na 32 14,5 ms β-	Na 33 20 ms β-									
16		18		20		22									

32

													45	**Rh** 102,9055 σ 145	Rh 94 ~25 s ~80 s
											44	**Ru** 101,07 σ 2,56		Ru 92 3,2 m ε γ 259; 214; 135	Ru 93 10,8 s Iγ 734; β^+ γ 1396; 1111; 2039 … 58 s β^+ γ 680; 1435 …
										43	**Tc**			Tc 91 3,2 m β^+ γ 503; 2451; 1639 … g; m	Tc 92 4,4 m β^+ 4,2 γ 1510; 773; 329; 148 …
										42	**Mo** 95,94 σ 2,65	Mo 88 8,2 m β^+ γ 171; 80; 131 m		Mo 90 5,7 h ε β^+ 1,1 γ 257 … m; g	Mo 91 65 s Iγ 653 β^+ 2,5; 4,0 … γ 1508; 1208 …; m 15,5 m ε β^+ 3,4 … γ … g
									41	**Nb** 92,9064 σ 1,15	Nb 86 1,4 m γ 751; 1003	Nb 87 3,9 m β^+ 3,8 m 2,6 m β^+ 4,2 … γ 201; 471 … g	Nb 88 7,8 m β^+ 3,4; 4,0 … γ 1057; 1083; 340 … 14,3 m β^+ 3,1; 3,6 γ 1057; 1083; 503; 671 …	Nb 89 66 m β^+ 2,4; 2,9 … γ 588; 508 …; m 1,9 h β^+ 3,3 … γ 1628; 1834; 2574 …; g	Nb 90 18,8 s Iγ 122 … e^- 14,6 h β^+ 1,5 … γ 1129; 2319; 141 …
					40	**Zr** 91,22 σ 0,185	Zr 81 ~10 m	Zr 82 9,5 m	Zr 83 5–10 m	Zr 84 5,0 m g	Zr 85 8 m γ 454; 416; 1199 … m 1,4 h g	Zr 86 16,5 h ε; no β^+ γ 243; 28; 612 g	Zr 87 14,0 s Iγ 201; 135 1,6 h β^+ 2,3 γ 1227; 1210; 1024 …; m	Zr 88 83,4 d ε γ 393	Zr 89 4,16 m Iγ 588 ε β^+ 0,9; 2,4 γ 1507; g 78,4 h ε β^+ 0,9 γ 1713 … m
					39	**Y** 88,9059 σ 1,28				Y 83 2,85 m β^+ γ 421 m 7,1 m β^+ 3,5 … γ 36; 490; 882 … g	Y 84 28,5 m? Iγ 113 38,5 m β^+ 1,6; 2,1 … γ 793; 974; 1040 …	Y 85 2,7 h β^+ 1,5; 2,1 … γ 503; 914 …; m 4,9 h β^+ 2,2 … γ 2123; 768; 1404 …; g	Y 86 48 m Iγ 208 … e^- β^+ … γ … 14,74 h ε; β^+ 1,2; 3,2 … γ 1077; 628; 1153 …	Y 87 13 h Iγ 381 β^+ … g 80,3 h ε β^+ … γ 485 m	Y 88 108 d ε β^+ … γ 1836; 898 …
				38	**Sr** 87,62 σ 1,21	Sr 78 30,6 m β^+ g	Sr 79 8,1 m β^+ γ 189; 166; 560; 116 …	Sr 80 1,8 h ε γ 589; 175; 553 …	Sr 81 25 m β^+ 2,7; 3,0 … γ 154; 148; 188; 443 … g	Sr 82 25 d ε no β^+ no γ g	Sr 83 5,0 s Iγ 259 33 h ε β^+ 1,2 … γ 763; 381; 418 …	Sr 84 0,56 σ 0,55 + 0,26	Sr 85 67,7 m Iγ 232 … ε γ 151 64,9 d ε no β^+ γ 514 …	Sr 86 9,9 σ 0,84 + ?	Sr 87 2,81 h Iγ 388 ε 7,0 σ 16
			37	**Rb** 85,4678 σ 0,37	Rb 75 21 s	Rb 76 36,8 s	Rb 77 3,9 m β^+ γ 67; 179; 394; 150 …	Rb 78 6,0 m β^+ Iγ 103 γ 455; 665; 693; 1110 … 17,7 m β^+ γ 455; 693; 3438 …	Rb 79 23,0 m β^+ 1,8; 2,5 … γ 688; 183; 506; 143; 130 … g; m	Rb 80 30 s β^+ 4,1 … γ 616 …	Rb 81 32 m β^+ 1,4 Iγ 85 e^- g 4,58 h ε β^+ 1,1 … γ 446 … m	Rb 82 6,3 h ε β^+ 0,8 … γ 776; 554; 619 … 1,3 m β^+ 3,2 … γ 776 …	Rb 83 86,2 d ε; no β^+ γ 520; 530; 553 … m; g	Rb 84 21 m Iγ 248; 465; 216 34,5 d ε; β^+ 0,8; 1,7 … β^- 0,9 γ 882 …	Rb 85 72,17 σ 0,050 + 0,41	Rb 86 1,02 m Iγ 556 18,7 d β^- 1,8 … ε γ 1077
36	**Kr** 83,80 σ 25,0	Kr 72 17 s β^+ γ 415; 310; 163; 577 …	Kr 73 26 s β^+ γ 178; 241; 455 … (p 1,5–3,0)	Kr 74 18 m β^+ 3,1 γ 90; 203; 63; 297; 307 … m	Kr 75 4,5 m β^+ 3,2 … γ 133; 155 …	Kr 76 14,6 h ε γ 316; 270; 46; 407 …	Kr 77 1,24 h β^+ 1,9 … γ 130; 147 … g; m	Kr 78 0,35 σ 0,21 + 4,5	Kr 79 50 s Iγ 130 34,9 h ε β^+ 0,6 … γ 261; 398; 606 …; g	Kr 80 2,25 σ 4,55 + 9,45	Kr 81 13,3 s Iγ 191 $2,1 \cdot 10^5$ a ε no γ	Kr 82 11,6 σ 20 + 25	Kr 83 1,83 h Iγ 9 … e^- 11,5 σ 200	Kr 84 57,0 σ 0,09 + 0,042	Kr 85 4,48 h β^- 0,8 … γ 151 … Iγ 305 10,76 a β^- 0,7 … γ … σ 1,66
35	**Br** 79,904 σ 6,8		Br 72 1,3 m β^+ γ 862; 1317; 455 …	Br 73 3,3 m β^+ 3,7 γ 65; 700; 336 … m	Br 74 26 m β^+ γ 42 m β^+ 4,7 … γ 635; 728; 1289 …	Br 75 1,6 h β^+ 1,7 … γ 287; 141 …	Br 76 16,0 h β^+ 3,7 … γ 560; 1216; 1852; 657 …	Br 77 4,3 m Iγ 106 e^- 56 h ε; β^+ … γ 239; 521; 297 …; g	Br 78 6,46 m β^+ 2,6 … γ 614 …	Br 79 4,9 s Iγ 208 50,69 σ 2,6 + 8,5	Br 80 4,42 h Iγ 37 … e^- 17,6 m β^- 2,1 … ε; β^- 0,9 γ 616; 666 …	Br 81 49,31 σ 2,43 + 0,26	Br 82 6,1 m Iγ … e^- β^- 3,1 … γ … 35,34 h β^- 0,4 … γ 776; 554; 619 …	Br 83 2,40 h β^- 0,9 … γ 530; 520 … m	Br 84 6,0 m β^- 2,2 γ 424; 882; 1463 … 31,8 m β^- 4,6 … γ 882; 1898 …

34	Se 78,96 σ 11,7	Se 69 1,8 m β^+ 5,2 γ 145 — 14,0 m β^+ 5,2 γ 166; 116; 145; 282	Se 70 38,9 m β^+ γ 427 …	Se 71 4,9 m β^+ 3,4 … γ 147; 831; 1096 …	Se 72 8,5 d ε no β^+ γ 46	Se 73 39 m Iγ 26 β^+ 1,8 … γ 254; 85; 393 … — 7,1 h β^+ 1,3; 1,7 … γ 361; 67 …	Se 74 0,9 σ 51,8	Se 75 120 d ε γ 265; 136; 280; 121 …	Se 76 9,0 σ 21 + 64	Se 77 17,5 s Iγ 162 — 7,5 σ 42	Se 78 23,5 σ 0,33 + ~0,2	Se 79 3,9 m Iγ 96 e⁻ — 6,5·10⁴ a β^- 0,2 no γ g	Se 80 50 σ 0,080 + 0,530	Se 81 57,3 m Iγ 103 e⁻ β^- … γ … — 18 m β^- 1,6 … γ 276; 290; 828; 566 …	Se 82 9,0 σ 0,039 + 0,006	Se 83 69 s β^- 3,9 … γ 1031; 357; 988; 674 … — 22,4 m β^- 0,9; 2,8 … γ 357; 510; 225 …
33	As 74,9216 σ 4,3	As 68 2,7 m β^+ γ 1017; 651; 763; 1779 …	As 69 15 m β^+ γ 233; 146; 87 …	As 70 53 m β^+ 2,1; 2,8 … γ 1040; 668; 1114; 745; 1708; 2020; 595 …	As 71 64 h ε β^+ 0,8 … γ 175 …	As 72 26,0 h β^+ 2,5; 3,3 … γ 834; 630 …	As 73 80,3 d ε; no β^+ γ 53 … e⁻	As 74 17,77 d ε; β^+ 0,9; 1,5 … β^- 1,4 … γ 596; 635 …	As 75 100 σ 4,3	As 76 26,4 h β^- 3,0 … γ 560; 657; 1216 …	As 77 38,8 h β^- 0,7 … γ 239; 521; 250 …; g	As 78 1,5 h β^- 4,4 … γ 614; 695; 1309 …	As 79 8,2 m β^- 2,1 … γ 96; 365; 432; 879 … m	As 80 15,2 s β^- 6,0 … γ 666; 1645; 1207 …	As 81 34 s β^- 3,8 … γ 468; 491 … g	As 82 14,0 s β^- ~4,3; ~5,4 … γ 655; 344; 1896 … — 19,1 s β^- 7,1 … γ 655; 1731; 1080 …
Ge 65 31 s β^+ 4,6; 5,2 … γ 650; 62; 809; 191 …	Ge 66 2,3 h ε β^+ 0,7; 1,1 … γ 382; 44; 109; 273 …	Ge 67 18,7 m β^+ 3,0 … γ 167; 1473 …	Ge 68 287 d ε no β^+ no γ	Ge 69 39 h ε β^+ 1,2 … γ 1107; 574; 872; 1336 …	Ge 70 20,7 σ 3,2	Ge 71 11,2 d ε no γ	Ge 72 27,5 σ 0,98	Ge 73 7,7 σ 15	Ge 74 36,4 σ 0,143+0,24	Ge 75 48 s Iγ 140 e⁻ — 83 m β^- 1,2 … γ 265; 199 …	Ge 76 7,7 σ 0,092+0,05	Ge 77 54 s β^- 2,9 … γ 216 … Iγ 159 — 11,3 h β^- 2,2 … γ 264; 211; 216; 416 …	Ge 78 88 m β^- 0,7 … γ 277; 294	Ge 79 19 s γ — 42 s β^- 4,2 … γ 230; 543 …	Ge 80 24,5 s β^- γ 266; 110; 1564; 937 …	Ge 81 10,1 s β^- ~5,3; ~5,6 γ 336; 198 …
Ga 64 2,6 m β^+ 2,9; 6,1 … γ 991; 3366; 1387; 808; 2195 …	Ga 65 15 m β^+ 2,1; 2,2 … γ 115; 61; 153; 752 …	Ga 66 9,3 h β^+ 4,2 … γ 1039; 2752; 834; 2190; 4296 …	Ga 67 78,3 h ε; no β^+ γ 93; 185; 299 …	Ga 68 68,3 m β^+ 1,9 … γ 1077 …	Ga 69 60 σ 1,68	Ga 70 21,1 m β^- 1,7 … γ …	Ga 71 40 σ 4,7	Ga 72 14,1 h β^- 1,0; 3,2 … γ 834; 2202; 630; 2508 …	Ga 73 4,8 h β^- 1,2; 1,5 … γ 295; 328 … e⁻	Ga 74 8,3 m β^- 2,6; 4,9 … γ 596; 2354; 608 …	Ga 75 2,1 m β^- 3,3 … γ 253; 575 … g	Ga 76 27,1 s β^- 5,9 … γ 563; 546; 1108 …	Ga 77 13 s β^- m; g	Ga 78 5,1 s β^-	Ga 79 2,9 s β^-	Ga 80 1,7 s β^-
Zn 63 38,4 m β^+ 2,3 … γ 670; 962; 1412 …	Zn 64 48,9 σ 0,78	Zn 65 244 d ε β^+ 0,3 γ 1115 …	Zn 66 27,8 σ 0,85	Zn 67 4,1 σ 6,9	Zn 68 18,6 σ 0,072 + 1,0	Zn 69 13,9 h Iγ 439 β^- γ … — 56 m β^- 0,9 … γ …	Zn 70 0,62 σ 0,0087 + 0,083	Zn 71 3,9 h β^- 1,5; 2,5 … γ 386; 487; 620 … — 2,4 m β^- 2,8 … γ 512; 910; 390 …	Zn 72 46,5 h β^- 0,3 … γ 145; 192 … e⁻	Zn 73 23,5 s β^- 4,7 γ 216; 496; 911 …	Zn 74 98 s β^- 2,1 γ 57; 140; 190; 50 …	Zn 75 10,2 s β^-	Zn 76 5,7 s β^-	Zn 77 1,4 s β^-	0,020 48	0,056 0,12
Cu 62 9,76 m β^+ 2,9 … γ …	Cu 63 69,1 σ 4,5	Cu 64 12,70 h ε β^- 0,6 β^+ 0,7 γ …	Cu 65 30,9 σ 2,17	Cu 66 5,1 m β^- 2,6 … γ 1039 … σ 135	Cu 67 61,9 h β^- 0,4; 0,6 … γ 185; 93; 91 …	Cu 68 3,8 m Iγ 526; 84; 110 … β^- 1,7; 1,9 γ 1077; 1340; 1041 … — 30 s β^- 3,5; 4,6 … γ 1077; 1261 …	Cu 69 3,0 m β^- 2,5 … γ 1007; 834; 531 … g	Cu 70 42 s β^- ~3,3; ~4,5 … γ 885; 902; 1252 … — 5 s β^- ~5,4; ~6,3 γ 885 …			0,00010 44	0,00035	0,0012 46	0,0035	0,0082	
Ni 61 1,16 σ 2,5	Ni 62 3,71 σ 14,2	Ni 63 100 a β^- 0,07 no γ σ 23	Ni 64 0,95 σ 1,49	Ni 65 2,52 h β^- 2,1 … γ 1482; 1115; 366 … σ 24,3	Ni 66 54,6 h β^- 0,2 no γ	Ni 67 18 s β^- γ 1072; 1654; 709; 874 …										
Co 60 10,5 m Iγ 59 e⁻ β … γ … σ 58 — 5,272 a β^- 0,3; 1,5 … γ 1332; 1173 … σ 2,0	Co 61 1,6 h β^- 1,2 … γ 67 …	Co 62 1,5 m β^- 4,1 … γ 1173; 2302; 1129 … — 14,0 m β^- 2,9 … γ 1173; 1163; 2003 …	Co 63 27,5 s β^- 3,6 … γ 87; 982 …	Co 64 0,4 s β^- 7,0			40		42							
Fe 59 44,6 d β^- 0,5; 1,6 … γ 1099; 1292 …	Fe 60 ~10^5 a β^- 0,1 m	Fe 61 6,0 m β^- 2,6; 2,8 … γ 1025; 1204; 297 …														
Mn 58 3,0 s β^- 6,1 no γ — 65 s β^- 3,9 … γ 811; 1323; 460; 864 …			36		38											
	34															

<table>
<tr><td></td><td></td><td></td><td></td><td></td><td></td><td></td><td></td><td></td><td></td><td></td><td>56</td><td>Ba
137,34
σ 1,2</td><td>Ba 119
5,4 s
β+
(p)</td><td></td><td>Ba 121
29,7 s
β+
(p)</td><td></td></tr>
<tr><td></td><td></td><td></td><td></td><td></td><td></td><td></td><td></td><td></td><td>55</td><td>Cs
132,9054
σ 29,0</td><td>Cs 116
3,9 s
β+
(p)</td><td>Cs 117
8 s</td><td>Cs 118
16,4 s
β+
(p)</td><td>Cs 119
37,7 s</td><td>Cs 120
60 s
β+
γ 323; 473
(p)</td><td>Cs 121
2,1 m
β+</td></tr>
<tr><td></td><td></td><td></td><td></td><td></td><td></td><td></td><td>54</td><td>Xe
131,30
σ 24,5</td><td>Xe 113
2,8 s
β+
(p)</td><td></td><td>Xe 115
18 s
ε
β+
(p 2,4–5,2)</td><td>Xe 116
55 s
ε
β+ 3,4
γ 104 ...</td><td>Xe 117
65 s
ε; β+
γ 117; 225;
295 ...; (p)</td><td>Xe 118
6 m
β+
γ 53; 117 ...
g</td><td>Xe 119
6 m
β+
γ 98; 231 ...</td><td>Xe 120
40 m
ε; β+
γ 25; 73; 178;
176; 763 ...; g</td></tr>
<tr><td></td><td></td><td></td><td></td><td></td><td></td><td></td><td>53</td><td>J
126,9045
σ 6,2</td><td></td><td></td><td></td><td>J 115
1,3 m
β+</td><td>J 116
2,2 s
β+ 6,3
γ 679; 541</td><td>J 117
2,4 m
β+ 3,5 ...
γ 326; 274 ...</td><td>J 118
~8,5 m | 13 m
β+ ~4,9 ... | β+ 5,5 ...
γ 605; 600; 612 ... | γ 605; 545; 600; 1338 ...
Iγ 104 ?</td><td>J 119
19 m
β+ 2,4 ...
γ 258 ...
g</td></tr>
<tr><td></td><td></td><td></td><td>52</td><td>Te
127,60
σ 4,7</td><td>Te 107?
2,2 s
α 3,28</td><td>Te 108? ←→ 4,4 s
β+
(p 3,3; 3,7 ...)</td><td>Te 109?</td><td></td><td>Te 111
19,3 s
β+
(p 2,82; 2,66 ...)</td><td>Te 112
1,8 m</td><td>Te 113
1,4 m
β+ 4,5
γ 814; 1172;
1018; 643</td><td>Te 114
16 m
γ 327</td><td>Te 115
6,0 m
β+ 2,8 ...
γ 723; 1380;
1326; 1098 ...</td><td>Te 116
2,5 h
ε
β+ ?
γ 94; 103 ...
g</td><td>Te 117
1,1 h
ε
β+ 1,8 ...
γ 720; 1716;
2300 ...</td><td>Te 118
6,0 d
ε
no γ
g</td></tr>
<tr><td></td><td></td><td></td><td>51</td><td>Sb
121,75
σ 5,4</td><td></td><td></td><td></td><td></td><td>Sb 110
23,0 s
β+ 6,9 ...
γ 1213; 986;
1245; 828 ...</td><td>Sb 111
74,1 s
β+ 3,3 ...
γ 155; 489;
1032 ...</td><td>Sb 112
53,5 s
β+ 4,8 ...
γ 1257; 991 ...</td><td>Sb 113
6,74 m
β+ 2,4; 2,5 ...
γ 498; 331 ...
g; m</td><td>Sb 114
~8 m | 3,5 m
β+ | β+ 4,0 ...
γ 1300; 888 ...</td><td>Sb 115
32,3 m
ε
β+ 1,5
γ 497 ...</td><td>Sb 116
60 m | 16 m
ε | ε
β+ 1,1 | β+ 2,3 ...
γ 1294; 973; 543 ... | γ 1294; 932; 2225 ...</td><td>Sb 117
2,8 h
ε
β+ 0,6
γ 159 ...
g</td></tr>
<tr><td></td><td></td><td></td><td>50</td><td>Sn
118,69
σ 0,63</td><td></td><td>Sn 106
10 s
γ 134</td><td>Sn 107
1,3 m
β+
γ 679; 1128 ...
m; g</td><td>Sn 108
10,5 m
ε; β+
γ 340; 271;
168; 670 ...
g</td><td>Sn 109
1,5 m | 18,0 m
β+ ? | β+ 2,5 ...
Iγ ? | γ 1098; 1321; 1484 ...
g; m</td><td>Sn 110
4,0 h
ε
γ 283
g</td><td>Sn 111
35,3 m
ε; β+ 1,5
γ 1153; 1914;
761; 1610 ...
g</td><td>Sn 112
1,0
σ 0,35 + 0,8</td><td>Sn 113
20 m | 115,1 d
Iγ ... | ε
e− | γ 255 ...
ε
g | m</td><td>Sn 114
0,66
σ 1,2</td><td>Sn 115
0,35
σ 50</td><td>Sn 116
14,4
σ 0,006 + ?</td></tr>
<tr><td></td><td></td><td></td><td>49</td><td>In
114,82
σ 193,5</td><td>In 104
25 m | 4,5 m
β+ | β+
γ 658; 835 | γ 658; 835</td><td>In 105
55 s | 5,1 m
Iγ 674 | β+
γ 131; 260;
604 ...</td><td>In 106
6,26 m | 5,33 m
β+ | β+ 2,7; 4,9 ...
γ 633; 861; 999 ... | γ 633; 861; 1716 ...</td><td>In 107
52 s | 32,4 m
Iγ 679 | ε
β+ 2,3 ...
γ 205; 505;
321 ...</td><td>In 108
58 m | 40 m
ε
β+ 1,3 ... | β+ 3,5 ...
γ 633; 876; 245 ... | γ 633; 1987 ...</td><td>In 109
1,34 m | 4,2 h
Iγ 650 | ε
β+ 0,8 ...
γ 204;
624 ...</td><td>In 110
4,9 h | 69,1 m
ε | β+ 2,3 ...
γ 658; 885; 937 ... | γ 658 ...</td><td>In 111
7,6 m | 2,83 d
Iγ 536 | ε
γ 245;
171 ...
g</td><td>In 112
20,8 m | 14,4 m
Iγ 157 | β− 0,7; ε
e− | β+ 1,6 ...
γ 617;
606 ...</td><td>In 113
99,48 m | 4,3
Iγ 392
σ 7,5 + 3,9</td><td>In 114
49,5 d | 71,9 s
Iγ 192 | β− 2,0 ...
e−; ε | ε
γ 558; 725 | β+ ...
γ ...</td><td>In 115
95,7
4,50 h | 6·10^14 a
Iγ 336 | β− 0,5
β− 0,8 | no γ
γ | σ 92 + 65 + 45</td></tr>
<tr><td>48</td><td>Cd
112,40
σ 2450</td><td>Cd 100
1,1 m
β+
γ 97–935
m</td><td>Cd 101
1,2 m
β+
m</td><td>Cd 102
5,5 m
β+
γ 481; 1037;
505; 415 ...
m</td><td>Cd 103
7,3 m
β+
γ 1080–1463</td><td>Cd 104
57,7 m
ε
β+ ...
γ 84; 709 ...
m</td><td>Cd 105
55 m
ε; β+ 1,7 ...
γ 962; 1302; 1693;
347; 607 ...
m; g</td><td>Cd 106
1,2
σ 1,0</td><td>Cd 107
6,5 h
ε
β+ ...
γ ...
m</td><td>Cd 108
0,9
σ 1,1</td><td>Cd 109
453 d
ε
no γ
m
σ 650</td><td>Cd 110
12,4
σ 0,10 + 11</td><td>Cd 111
12,8
49 m
Iγ 245; 151
σ 24,3</td><td>Cd 112
24,0
σ 0,06 + 2,2</td><td>Cd 113
12,3
14,6 a | 9·10^15 a
β− 0,6 | β− 0,3
Iγ ... | g
g | σ 19910</td><td>Cd 114
28,8
σ 0,036 + 0,300</td></tr>
<tr><td>Ag
107,868
σ 63,6</td><td></td><td>Ag 99
1,8 m
β+ 2,4 ...
γ 1040; 3315;
1525</td><td>Ag 100
2,3 m | 8 m
β+ 3,3 | β+ 6,5
γ 665; 750; 1694 ... | γ 728</td><td>Ag 101
< 4 s | 10,8 m
Iγ 98; 176 | β+ 2,7; 3,4 ...
γ 283; 1184; 850 ...</td><td>Ag 102
8 m | 13 m
β− 1,5; 3,3 ... | β+ 2,3 ...
Iγ 9
γ 557; 1835 | γ 557; 719; 1745 ...</td><td>Ag 103
5,7 s | 1,1 h
Iγ 135 | β+ 1,7 ...
e− | γ 119; 148; 267; 1275 ...</td><td>Ag 104
33,5 m | 69,2 h
β+ 2,7 ... | ε; β+ 1,0
γ 556 | γ 556; 768; 942 ...
Iγ</td><td>Ag 105
7,2 m | 41,2 d
Iγ 26 | ε; β+ ...
e− | γ 345; 281; 443 ...
ε
γ ...</td><td>Ag 106
8,3 d | 24 m
ε; no β+ | β+ 2,0 ...
γ 512; 1046; 717; 451 ... | γ 512 ...</td><td>Ag 107
51,83
44,3 s
Iγ 93
e−
σ 30 + 34,7</td><td>Ag 108
127 a | 2,41 m
ε | β− 1,7 ...
γ 614; 723; 434 | ε; β+ ...
Iγ 79 ... | γ 633; 434 ...</td><td>Ag 109
48,17
39,6 s
Iγ 88
e−
σ 4,5 + 89</td><td>Ag 110
250,4 d | 24,6 s
β− 0,09; 1,5 ... | β− 2,9 ...
γ 658; 885; 937 ... | γ 658
Iγ 118 ... | ε
σ 82</td><td>Ag 111
1,2 m | 7,5 d
Iγ 60 | β− 1,0 ...
β− ... | γ 342; 245 ...
γ ...
g | g; σ 3</td><td>Ag 112
3,12 h
β− 3,9 ...
γ 617; 1387 ...</td><td>Ag 113
1,1 m | 5,37 h
β 1,5; 1,9 | β− 2,0 ...
γ 316; 392; 299 | γ 299; 259; 316 ...
Iγ 65; g | g</td></tr>
<tr><td>Pd
106,4
σ 6,9</td><td>Pd 97
3,3 m
β+
γ 265; 475;
792 ...
m</td><td>Pd 98
18 m
β+ 2,3 ...
γ 112; 663;
838; 107 ...
g</td><td>Pd 99
21,4 m
β+ 2,2 ...
γ 136; 263;
673 ...
g</td><td>Pd 100
3,7 d
ε
no β+
γ 84; 75; 126 ...</td><td>Pd 101
8,47 h
ε; β+ 0,8 ...
γ 296; 590;
270 ...
m</td><td>Pd 102
1,0
σ 4,8</td><td>Pd 103
17 d
ε
γ ...
e−
m</td><td>Pd 104
11,0</td><td>Pd 105
22,2
σ 14</td><td>Pd 106
27,3
σ 0,013 + 0,279</td><td>Pd 107
21,3 s | 6,5·10^6 a
Iγ 214 | β− 0,03
no γ
g</td><td>Pd 108
26,7
σ 0,20 + 12</td><td>Pd 109
4,69 m | 13,46 h
Iγ 189 | β− 1,0 ...
γ ...
m</td><td>Pd 110
11,8
σ 0,020 + 0,20</td><td>Pd 111
5,5 h | 22 m
Iγ 172 | β− 2,1 ...
β− 2,2 ... | γ 580; 70; 651 ...
γ 70; 391; 575 ...
m; g | m</td><td>Pd 112
20,1 h
β− 0,3
γ 19
e−</td></tr>
<tr><td>Rh 95
4,8 m
β+
γ 942; 1346 ...</td><td>Rh 96
1,5 m | 9,3 m
β+ | β+
γ 832; 1098 ... | γ 832; 685; 631 ...</td><td>Rh 97
30 m | 40 m
β+ | β+ 2,5 ...
γ 421 | γ 421; 840; 878; 190 ...
Iγ ?</td><td>Rh 98
3 m? | 9,05 m
Iγ | β+ 2,8; 3,5 ...
γ 652 ...</td><td>Rh 99
4,7 h | 16 d
ε; β+ 0,7 | β+ 1,0 ...
γ 341; 618; 1261 ... | γ 528; 353; 90 ...</td><td>Rh 100
20 h
ε
β+ 2,6 ...
γ 540; 2376;
1553; 823 ...</td><td>Rh 101
4,4 d | 3 a
ε | ε
γ 307; 545 ... | γ 127; 198; 325 ...
Iγ 157</td><td>Rh 102
2,9 a | 206 d
ε | ε; β+ 1,3
γ 475; 631; 697 ... | β− 1,2
γ 475; 628; 1103 ...</td><td>Rh 103
100
56,1 m
Iγ 40
e−
σ 11 + 135</td><td>Rh 104
4,4 m | 42 s
Iγ 51 | β− 2,5 ...
e−; β | ε; γ 556; 1237 ...
γ
σ 800 | σ 40</td><td>Rh 105
45 s | 35,5 h
Iγ 129 | β− 0,6 ...
e− | γ 319; 306 ...
σ 4700 + 11250</td><td>Rh 106
2,2 h | 30 s
β− 0,9; 1,7 ... | β− 3,6 ...
γ 512; 1046; 717; 451 ... | γ 512; 622 ...</td><td>Rh 107
22 m
β− 1,2 ...
γ 303; 393 ...
g</td><td>Rh 108
5,9 m | 16,8 s
β− 1,6; 3,4 ... | β− 4,5 ...
γ 434; 581; 947 ... | γ 434; 619 ...</td><td>Rh 109
50 s | 80 s
Iγ 110 | β−
γ 327; 114 ...</td><td>Rh 110
3,0 s | 27,7 s
β− 5,5 ... | β− 2,6 ...
γ 374 ... | γ 374; 547; 688; 440 ...</td><td>Rh 111 ?
62,7 s
β−
γ 291; 249
g</td></tr>
<tr><td>Ru 94
51,8 m
ε
γ 367; 891 ...</td><td>Ru 95
1,65 h
ε; β+ 1,2 ...
γ 336; 1097;
627 ...
g</td><td>Ru 96
5,5
σ 0,25</td><td>Ru 97
2,9 d
ε
γ 216; 324 ...
g</td><td>Ru 98
1,9
σ 8</td><td>Ru 99
12,7
σ 5,0</td><td>Ru 100
12,6
σ 5,8</td><td>Ru 101
17,1
σ 3,1</td><td>Ru 102
31,6
σ 1,30</td><td>Ru 103
39,35 d
β− 0,2; 0,7 ...
γ 497; 610 ...
m</td><td>Ru 104
18,6
σ 0,47</td><td>Ru 105
4,44 h
β− 1,2; 1,8 ...
γ 725; 469;
676; 317 ...
g; m; σ 0,20</td><td>Ru 106
368 d
β− 0,04
no γ
g
σ 0,146</td><td>Ru 107
3,8 m
β− 3,2 ...
γ 194; 848;
463; 375 ...</td><td>Ru 108
4,5 m
β− 1,3 ...
γ 165; 91 ...
g</td><td>Ru 109
34,5 s
β−
γ 206; 226;
359 ...
g</td><td>Ru 110
13 s
β−
γ 276</td></tr>
</table>

Tc 93 — 43,5 m: Iγ 390; ε; γ 2845; g — 2,7 h: ε; β+ 0,8 …; γ 1363; 1521; 1478 …; g	Tc 94 — 53 m: β+ 2,5 …; γ 871 … — 4,9 h: ε; β+ 0,8; γ 871; 703; 850 …	Tc 95 — 60 d: ε; β+ …; γ 204; 582; 835 …; Iγ … — 20 h: ε; no β+; γ 766; 1074 …	Tc 96 — 52 m: Iγ …; e−; ε; γ 778; 1200 … — 4,3 d: ε; no β+; γ 778; 850; 813 …	Tc 97 — 91 d: Iγ …; e− — 2,6 · 10^6 a: ε; no γ	Tc 98 — 4,2 · 10^6 a; β− 0,4; γ 745; 652; σ 0,9 + ?	Tc 99 — 6,0 h: Iγ 141 …; e−; β− …; γ … — 2,1 · 10^5 a: β− 0,3 …; γ …; σ 19	Tc 100 — 15,8 s; β− 3,4 …; γ 540; 591 …	Tc 101 — 14 m; β− 1,3 …; γ 307; 545 …
Mo 92 — 14,8; σ < 0,006 + ~ 0,045	Mo 93 — 6,9 h: Iγ 1477; 685; 263 … — 3,5 · 10^3 a: ε; m	Mo 94 — 9,1; σ 0,016	Mo 95 — 15,9; σ 14,5	Mo 96 — 16,7; σ 1,0	Mo 97 — 9,5; σ 2,2	Mo 98 — 24,4; σ 0,130	Mo 99 — 66,0 h; β− 1,2 …; γ 740; 181; 778 …; m; g	Mo 100 — 9,6; σ 0,199
Nb 91 — 62 d: Iγ 105; e−; ε; γ 1205 — long: ε	Nb 92 — 10,15 d: ε; β+ …; γ 934 … — ~10^8 a: ε; γ 561; 934	Nb 93 — 13,6 a: Iγ …; e− — 100; σ 0,15+1,0	Nb 94 — 6,26 m: Iγ …; e−; β− …; γ … — 2 · 10^4 a: β− 0,5; γ 871; 703; σ? + ~15	Nb 95 — 86,6 h: Iγ 235; e−; β; γ 204 … — 35,15 d: β− 0,2; γ 766; σ < 7	Nb 96 — 23,4 h; β− 0,7 …; γ 778; 569; 1091 …	Nb 97 — 53 s: Iγ 743 — 74 m: β− 1,3 …; γ 658 …	Nb 98 — 2,9 s: β− 5,0 …; γ 787; 1025 … — 51 m: β− 1,9; 3,1 …; γ 787; 722; 1169 …	Nb 99 — 15 s: β− 3,5; γ 138; 98 — 2,6 m: β− 3,8; γ 98; 254; 2642; 2852 …
Zr 90 — 51,4; σ 0,10	Zr 91 — 11,2; σ 1,03	Zr 92 — 17,1; σ 0,26	Zr 93 — 1,5 · 10^6 a; β− 0,06 …; m; σ ~ 2	Zr 94 — 17,5; σ 0,056	Zr 95 — 64,0 d; β− 0,4; 0,9 …; γ 757; 724; g	Zr 96 — 2,8; σ 0,017	Zr 97 — 16,8 h; β− 1,9 …; γ 508; 1148; 355 …; m	Zr 98 — 30,7 s; β− 2,1; no γ; m
Y 89 — 16,0 s: Iγ 909 — 100; σ 0,001 + 1,28	Y 90 — 3,19 h: Iγ 203; 480 …; β …; γ … — 64,1 h: β− 2,3 …; γ …	Y 91 — 49,7 m: Iγ 556 — 58,5 d: β− 1,5 …; γ …; σ 1,4	Y 92 — 3,54 h; β− 3,6 …; γ 934; 1405; 561; 449 …	Y 93 — 10,1 h; β− 2,9 …; γ 267; 947; 1918 …	Y 94 — 19 m; β− 4,9 …; γ 918; 1139; 550 …	Y 95 — 10,3 m; β− 4,3 …; γ 954; 2176; 3577; 1324; 2633 …	Y 96 — 9,3 s; β−; γ 1750; 915; 1107; 617 …	Y 97 — 1,1 s; β−; γ 1104; 162
Sr 88 — 82,6; σ 0,0058	Sr 89 — 50,5 d; β− 1,5 …; g; σ 0,42	Sr 90 — 28,5 a; β− 0,5; no γ; g; σ 0,9	Sr 91 — 9,5 h; β− 1,1; 2,7 …; γ 1024; 750; 653 …; m; g	Sr 92 — 2,71 h; β− 0,6; 1,9; γ 1384 …	Sr 93 — 7,45 m; β− 2,7; 3,7 …; γ 590; 876; 888; 710; 169 …	Sr 94 — 74 s; β− 2,1; 3,4 …; γ 1428 …	Sr 95 — 24,4 s; β− 6,1 …; γ 686; 2717; 2933; 2247 …	Sr 96 — 1,0 s; β−; γ 123; 809; 930 …
Rb 87 — 27,83; 4,7 · 10^{10} a; β− 0,3; no γ; g; σ 0,12	Rb 88 — 17,8 m; β− 5,3 …; γ 1836; 898 …; σ 1,0	Rb 89 — 15,2 m; β− 1,3; 4,5 …; γ 1032; 1248; 2196 …	Rb 90 — 4,3 m: β−; Iγ 107; γ 832; 1375 … — 2,6 m: β− 5,8; 6,6; γ 832; 1061; 4366; 4136; 3833 …	Rb 91 — 58 s; β− 5,4; 5,8 …; γ 93; 2564; 3600; 346 …	Rb 92 — 4,5 s; β− 8,2 …; γ 815; 2821; 570 …; (n)	Rb 93 — 5,8 s; β− 7,6 …; γ 433; 214 …; (n)	Rb 94 — 2,69 s; β− 9,3 …; γ 837; 1578; 1089; 1309 …; (n)	Rb 95 — 383 ms; β− 8,6; (n)
Kr 86 — 17,3; σ 0,060	Kr 87 — 76,3 m; β− 3,5; 3,9 …; γ 403; 2555; 846 …; σ < 600	Kr 88 — 2,80 h; β− 0,5; 2,9 …; γ 2392; 196; 2196; 835; 1530 …	Kr 89 — 3,18 m; β− 3,5; 4,9 …; γ 221; 586; 1473 …	Kr 90 — 32,3 s; β− 2,6; 4,2 …; γ 1119; 122; 540 …; g	Kr 91 — 8,6 s; β− 4,6; 5,0 …; γ 109; 507; 613; 1109 …	Kr 92 — 1,84 s; β− 4,6; 6,0 …; γ 142; 1219; 813; 548 …; (n)	Kr 93 — 1,29 s; β− 6,2; 8,3; γ 1120; 253; 323; 267; 182; (n)	Kr 94 — 0,20 s; β−; γ 629; 220; 359 …
Br 85 — 2,87 m; β− 2,5 …; γ 832; 802; 925; m	Br 86 — 4,5 s?: Iγ? — 54 s: β− 3,0; 7,4 …; γ 1565; 2751 …	Br 87 — 55,7 s; β− 6,1 …; γ 1419; 604; 1465; 1476 …; (n 0,3)	Br 88 — 16,2 s; β−; γ 776; 1054 …; (n)	Br 89 — 4,5 s; β−; γ 602; 243 …; (n 0,5)	Br 90 — 1,63 s; β−; γ 362 …; (n)	Br 91 — 0,64 s; β−; γ; (n)	Br 92 — 0,25 s; β−; (n)	
Se 84 — 3,1 m; β− 1,4; γ 407; g	Se 85 — 33 s; β−; γ 345 …	Se 86 — 16,1 s; β−; γ 1208; 1081; 941; 1340 …	Se 87 — 5,60 s; β−; (n)	Se 88 — 1,5 s; β−; (n)	Se 89 — 0,4 s; β−; (n)	5,89 — 56	5,93	5,97 — 58
As 83 — 13,3 s; β−; γ 735; 1113 …; m; g	As 84 — 0,65 s: β− — 5,3 s: β−; γ 1455; 667 …; (n)	As 85 — 2,05 s; β−; γ 1112; 462 …; (n 0,5; 0,9 …)	As 86 — 0,9 s; β−; γ 704; (n)	As 87 — ~0,3 s; β−	3,62	4,80		
Ge 82 — 4,6 s; β−; γ 793; 1092; g	Ge 83 — 1,9 s; β−	Ge 84 — 1,2 s; β−	1,32	1,95 — 54				
0,21 — 50	0,34	0,533 — 52	0,986					

Tc 102 — 4,3 m: β− 1,6; 3,2 …; γ 475; 831; 628 …; Iγ — 5,3 m: β− 4,2 …; γ 475 …	Tc 103 — 50 s; β− 2,2 …; γ 136; 346; 210 …	Tc 104 — 18,0 m; β− 2,4; 4,3 …; γ 358; 530; 884; 535 …	Tc 105 — 7,6 m; β− 3,4 …; γ 143; 108; 159; 321 …	Tc 106 — 36 s; β−; γ 270; 522; 793; 721 …	Tc 107 — 21 s; β−; γ 103; 106; 177 …	Tc 108 — 5,0 s; β−; γ 242; 466; 708; 732 …	Tc 109 — ~1 s; β−
Mo 101 — 14,6 m; β− 0,8; 2,6 …; γ 192; 591; 509; 1012 …	Mo 102 — 11,5 m; β− 1,2; g	Mo 103 — 62 s; β−	Mo 104 — 1,1 m; β− 2,2; 4,8 …; γ 70	Mo 105 — 42 s; β−; γ 69; 424 …	Mo 106 — 10 s; β−	Mo 107 — ~5 s; β−	Mo 108 — 1,5 s; β−; γ 259; 126 …
Nb 100 — 1,5 s: β− 5,3 …; γ 535; 528; 159 … — 3,1 s: β−; γ 535; 600; 528; 1280 …	Nb 101 — 7,1 s; β− 4,3 …; γ 276; 158 …	Nb 102 — 1,3 s: β−; γ 296; 551; 401 — 4,3 s: β−; γ 296; 447 …		Nb 104 — 0,8 s: β−; γ 192; 369 — 4,8 s: β−; γ 192			0,17 — 66 — 0,070
Zr 99 — 2,35 s; β− 3,5; 3,9; γ 468; 544; 594 …; m	Zr 100 — 7,1 s; β−; m	Zr 101 — 2,0 s; β−; γ 594	Zr 102 — 2,9 s; β−; m				0,390
Y 98 — 1,0 s; β−; γ 1224; 1591; 212; 268	Y 99 — 1,4 s; β−; γ 122; 130	Y 100 — 0,5 s; β−; γ 212; 351		4,19	3,12 — 64	1,33	0,927
Sr 97 — ≲ 200 ms; β−	Sr 98 — 0,6 s; β−; γ 120; 36; 427; 444 …			5,05			
Rb 96 — 207 ms; β− 10,8 …; γ 815; (n)	Rb 97 — 176 ms; β−; (n)	Rb 98 — 106 ms; β−; (n)	Rb 99 — 76 ms; β−	6,30			
Kr 95 — < 0,5 s; β−		6,03	5,79 — 62	6,13			
	6,50 — 60	6,28					
6,40	6,44						

66													Dy 162,50 σ 930		
65													Tb 158,9254 σ 25,5		Tb 146 23 s β^+ γ 1580; 1079; 1417 . . .
64										Gd 157,25 σ 49000		Gd 142 1,5 m β^+ γ 179 g	Gd 143 1,9 m: β^+ γ 272; 588; 668; 799 . . . / 41 s	Gd 144 4,5 m β^+ 3,3 . . . γ 333; 347; 630 . . .	Gd 145 85 s: Iγ 721 . . . β^- γ 387; 330 / 21,8 m: ε; β^+ γ 1758; 1881; 1042; 808 . . .
63									Eu 151,96 σ 4600	Eu 139 22 s β^+ γ 111	Eu 140 1,3 s β^+ γ 531	Eu 141 ~4 s: Iγ / 37 s: β^+ γ 384; 394 . . . g	Eu 142 1,2 m: ε; β^+ γ 768; 557; 1023 . . . / 2,4 s: β^+ γ 629; 768 . . .	Eu 143 2,6 m β^+ 4,0 γ 1107; 1537; 1913; 108; 1805 . . .; g	Eu 144 10,5 s β^+ 5,2 . . . γ 1660; 818
62							Sm 150,4 σ 5800		Sm 137 44 s β^+	Sm 138 3,0 m β^+ γ 54; 75	Sm 139 9,5 s: Iγ 190; 267; 155; 112 / 2,6 m: β^+ γ 274; 306; 597; g	Sm 140 14,7 m β^+ 1,9 . . . γ 226; 140 . . . g	Sm 141 22,6 m: ε; β^+ γ 197; 432; 777 . . . / 11,3 m: β^+ γ 404; 438; 1292; 1601 . . .	Sm 142 72,4 m ε β^+ 1,0	Sm 143 65 s: Iγ 754 β^+ . . . γ . . . / 8,83 m: β^+ 2,5 γ 1057; 1515 . . .
61							Pm		Pm 136 1,5 m β^+ γ 374; 603 . . .	Pm 137 2,4 m ε; β^+ γ 581; 108; 178; 269 . . . m; g	Pm 138 3,5 m β^+ γ 521; 729; 493 . . .	Pm 139 0,5 s: Iγ 189 . . . / 4,1 m: β^+ γ 403; 367; 463 g	Pm 140 5,8 m: ε; β^+ γ 419; 774; 1028 . . . / 9,2 s: β^+ γ 774; 716; 1499 . . .	Pm 141 20,9 m β^+ 2,7 . . . γ 1223; 886; 194; 1346 . . . g	Pm 142 40,5 s β^+ 3,8 . . . γ 1576 . . .
60							Nd 144,24 σ 50,5	Nd 134 8,5 m β^+ γ 163 . . . g	Nd 135 5,5 m / 12,0 m: β^+; γ 204; 42; 441; 502; 476 . . .	Nd 136 50,7 m ε β^+ 0,9 γ 109; 575; 149 . . .	Nd 137 1,6 s: Iγ 234; 178; 108; 286 / 38,0 m: β^+ 2,4 . . . γ 76; 581; 307; 782 . . .	Nd 138 5,1 h ε β^+ γ 326; 200 . . . g	Nd 139 5,5 h: ε β^+ 1,2 γ 114; 738; 982; 708 . . . Iγ 231 / 29,7 m: ε β^+~1,7 . . . γ 405; 1074 . . .	Nd 140 3,38 d ε	Nd 141 62 s: Iγ 757 β^+ . . . γ . . . / 2,5 h: ε; β^+ 0,8 γ 1127; 1293; 1147 . . .
59			Pr 140,9077 σ 11,5				Pr 132 1,6 m γ 325; 496; 533	Pr 133 6,5 m β^+ γ 134; 316; 74; 465 . . . m	Pr 134 ~11 m: β^+ γ 409; 640 . . . / 18 m: β^+ γ 409; 1904; 2136 . . .	Pr 135 25 m β^+ 2,5 . . . γ 296; 82; 213; 538 . . . g	Pr 136 13,1 m β^+ 3,0; 4,1 . . . γ 552; 540; 1092; 461 . . .	Pr 137 76,6 m ε; β^+ 1,7 . . . γ 837; 434; 514; 160 . . . g	Pr 138 2,02 h: ε; β^+ 1,7 . . . γ 789; 1038; 302 . . . / 1,44 m: β^+ 3,4 . . . γ 789; 688; 1551 . . .	Pr 139 4,5 h ε β^+ 1,1 γ 1347 . . . g	Pr 140 3,4 m β^+ 2,3 . . . γ . . .
58			Ce 140,12 σ 0,63	Ce 128 ? 5,5 m	Ce 129 3,5 m γ 68	Ce 130 25 m γ 129	Ce 131 5,0 m: β^+ γ 396; 421; 230 / 8,5 m: ε β^+ 2,8 . . . γ 169; 414; 119; 1469 . . .	Ce 132 4,2 h ε γ 182; 155; 217 . . . g	Ce 133 97 m: ε γ 97; 77 . . . / 5,4 h: ε; β^+ 1,3 γ 477; 131; 58 . . .	Ce 134 72 h ε no γ	Ce 135 20 s: Iγ 213; 150; 82; 296 / 17,0 h: ε γ 265; 300; 606; 517 . . .	Ce 136 0,19 σ 0,95 + 6,3	Ce 137 34,4 h: Iγ 255 e⁻ ε γ . . . / 9,0 h: ε γ 446 . . . e⁻	Ce 138 0,26 σ 0,015 + 1,1	Ce 139 56,5 s: Iγ 754 / 137,5 d: ε γ 166
57		La 138,9055 σ 9,0	La 126 1,0 m β^+ γ 256 . . .	La 127 3,8 m g	La 128 4,9 m β^+ 3,2 . . . γ 284; 479 . . .	La 129 10 m g	La 130 8,7 m β^+ γ 357; 551; 452 . . .	La 131 59 m ε; β^+ 1,4; 1,9 . . . γ 108; 418; 365; 285 . . . g	La 132 25 m: Iγ 135 β^+ γ 285 / 4,5 h: β^+ 3,2; 3,7 . . . γ 464; 568; 1032; 664 . . .	La 133 4,0 h ε; β^+ 1,2 γ 279; 302; 290; 632 . . . g	La 134 6,8 m β^+ 2,7 . . . γ 605 . . .	La 135 19,4 h ε; β^+ . . . γ 481; 875; 588 . . . g	La 136 9,9 m ε β^+ 1,9 . . . γ 819; 761; 1323 . . .	La 137 $6 \cdot 10^4$ a ε no γ g	La 138 0,09 $1,3 \cdot 10^{11}$ a ε; β^- 0,4 γ 1436; 789 σ 172
	Ba 123 2,7 m γ 95; 124; 116; 93 . . .; g	Ba 124 11,9 m β^+ γ 170; 1216; 189; 272 . . .	Ba 125 8,0 m: β^+ 4,5 / 3,5 m: β^+ 3,4 γ 78; 141; 85 . . .	Ba 126 97 m ε γ 234; 258; 241 . . .	Ba 127 10,0 m: β^+ 3,5 / 18 m: β^+ 3,1 γ 110; 180; 70 . . .	Ba 128 2,43 d ε no β^+ γ 273	Ba 129 2,13 h: ε γ 182; 1459; 202 . . . / 2,20 h: β^+ 1,4 . . . γ 214; 221; 129 . . .	Ba 130 0,10 σ 2,5 + 11	Ba 131 14,5 m: Iγ 108; 79 e⁻ / 11,5 d: ε γ 496; 124; 216	Ba 132 0,095 σ 0,68 + 7,8	Ba 133 38,9 h: Iγ 276; 12 e⁻ ε γ . . . / 10,5 a: ε γ 356; 81; 303 . . .	Ba 134 2,4 σ 0,158 + 1,8	Ba 135 28,7 h: Iγ 268 e⁻ / 6,5 σ 5,8	Ba 136 7,8 σ 0,010 + ~0,4	Ba 137 2,55 m: Iγ 662 / 11,2 σ 5,1
	Cs 122 4,2 m: β^+ γ 331; 497; 638; 750 / 21,0 s: β^+ γ 331	Cs 123 1,6 s: Iγ 96; 63 / 5,9 m: β^+ 2,6 . . . γ 98	Cs 124 26,5 s β^+ γ 354; 916; 493; 847	Cs 125 45 m ε; β^+ 2,1 . . . γ 526; 112; 412 . . . g	Cs 126 1,6 m β^+ 3,8 . . . γ 389; 926; 491 . . .	Cs 127 6,25 h ε; β^+ 0,7; 1,1 . . . γ 411; 125; 462 . . . g	Cs 128 3,8 m β^+ 2,9 . . . γ 443; 527 . . .	Cs 129 32,06 h ε γ 372; 411; 549 . . . g	Cs 130 29,9 m ε β^+ 2,0 . . . β^- 0,4 γ 536 . . .	Cs 131 9,70 d ε no β^+ no γ g	Cs 132 6,47 d ε; β^+ . . . β^- 0,8 γ 668; 465; 630; 506 . . .	Cs 133 100 σ 2,5 + 26,5	Cs 134 2,90 h: Iγ 127 e⁻ / 2,06 a: β^- 0,7 . . . γ 605; 796; 569 . . . β^- . . . σ 140	Cs 135 53 m: Iγ 781; 840 / $2 \cdot 10^6$ a: β^- 0,2 no γ g σ 8,7	Cs 136 13,0 d β^- 0,3; 0,7 . . . γ 819; 1048; 341 . . . σ 1,3

Xe 121 38,8 m β⁺ 2,8 γ 253; 133; 445 ...	Xe 122 20,1 h ε γ 350; 149; 417 ...	Xe 123 2,08 h ε β⁺ 1,5 ... γ 149; 178; 330 ...	Xe 124 0,10 σ 22,0 + 106	Xe 125 57 s: Iγ 112; 140 / 16,8 h: ε; β⁺ ... γ 188; 243; 55 ... σ <5600	Xe 126 0,09 σ 0,26 + 3,7	Xe 127 70 s: Iγ 125; 173 / 36,4 d: ε γ 203; 172; 375 ...	Xe 128 1,9 σ 0,36 + 5	Xe 129 8,89 d: Iγ 40; 197; e⁻ / 26,4: σ 18	Xe 130 3,9 σ 0,42 + 6,0	Xe 131 12,0 d: Iγ 164; e⁻ / 21,2: σ 90	Xe 132 27,0 σ 0,025 + 0,36	Xe 133 2,2 d: Iγ 233; e⁻ / 5,29 d: β⁻ 0,3 ... γ 81 ... e⁻ σ 190	Xe 134 10,5 σ 0,003 + 0,25	Xe 135 15,3 m: Iγ 527 / 9,17 h: β⁻ 0,9 ... γ 250; 608 ... g σ 2,65·10⁶
J 120 53 m: ε β⁺ 3,8 ... γ 560; 601; 614 ... / 1,35 h: ε β⁺ 4,6 ... γ 560; 1523; 614 ...	J 121 2,12 h ε β⁺ 1,1 ... γ 213 ... g	J 122 3,6 m β⁺ 3,1 ... γ 564 ...	J 123 13,2 h ε no β⁺ γ 159 ... g	J 124 4,15 d ε β⁺ 2,1 ... γ 603; 1691; 723 ...	J 125 60,14 d ε γ 35; e⁻ g σ 894	J 126 13,0 d ε; β⁻ 0,9; 1,3 ... β⁺ 1,1 ... γ 389; 666 ... σ 5960	J 127 100 σ 6,2	J 128 25,0 m β⁻ 2,1 ... ε β⁺ γ 443; 527 ...	J 129 1,57 · 10⁷ a β⁻ 0,2 γ 40 e⁻; g σ 18 + 9	J 130 9,0 m: Iγ ... e⁻ β⁻ 2,5 ... γ 536 ... / 12,36 h: β⁻ 1,0; 1,8 ... γ 536; 669; 739 ... σ 18	J 131 8,04 d β⁻ 0,6; 0,8 ... γ 364; 637; 284 ... g; σ ~0,7	J 132 84 m: Iγ 98 β⁻ ... γ 668; 773; 600; 175 / 2,38 h: β⁻ 2,1 ... γ 668; 773; 955; 523 ...	J 133 9 s: Iγ 913; 647; 73 / 20,8 h: β⁻ 1,2; 1,5 ... γ 530; 875 ...; g	J 134 3,5 m: Iγ 272; 44 β⁻ 2,5 γ 847; 884; 234 / 52,0 m: β⁻ 1,3; 2,4 ... γ 847; 884 ...
Te 119 4,7 d: ε γ 153; 1213; 271 ... / 16 h: ε β⁺ 0,6 ... γ 644; 700 ...	Te 120 0,09 σ 0,34 + 2,0	Te 121 154 d: Iγ 212 ... e⁻ ε; β⁺ ... γ 1102 ... / 16,8 d: ε γ 573; 508 ...	Te 122 2,4 σ 1,1 + 1,7	Te 123 0,87 119,7 d: Iγ 159 ... e⁻ / 1,24·10¹³ a: ε no γ σ 406	Te 124 4,6 σ 0,04 + 6,8	Te 125 58 d: I, 35 e⁻ / 7,0: σ 1,55	Te 126 18,7 σ 0,135 + 0,90	Te 127 109 d: Iγ ... e⁻ β⁻ 0,7 ... γ ... / 9,35 h: β⁻ 0,7 ... γ 418 ...	Te 128 31,8 σ 0,015 + 0,200	Te 129 33,6 d: Iγ ... β⁻ 1,6 ... γ 696 ... / 69,6 m: β⁻ 1,5 ... γ 28; 460; 487 ...	Te 130 34,5 σ 0,02 + 0,270	Te 131 30 h: β⁻ 0,5; 2,5 ... Iγ 182 γ 774; 150; 852 ... / 25,0 m: β⁻ 2,1 ... γ 150; 452 ...	Te 132 78 h β⁻ 0,2 γ 228; 50 ... g	Te 133 55,4 m: β⁻ 1,3; 2,4 Iγ 334 γ 913; 647; 864 ... g / 12,5 m: β⁻ 2,8; 3,2 ... γ 312; 408; 1333 ...; g
Sb 118 5,0 h: ε; β⁺ ... γ 1230; 1051; 254 ... / 3,5 m: β⁺ 2,7 ... γ 1230; 1267 ...	Sb 119 38,5 h ε γ 24 e⁻ g	Sb 120 5,76 d: ε no β⁺ γ 197; 1172; 1023; 90 / 15,9 m: ε β⁺ 1,7 ... γ 1172 ...	Sb 121 57,3 σ 0,055 + 6,2	Sb 122 4,2 m: Iγ 61; 76 e⁻ / 2,70 d: β⁻ 1,4; 2,0 ... ε; β⁺ ... γ 564; 693 ...	Sb 123 42,7 σ 0,011 + 0,035 + 4,28	Sb 124 20 m: I, e⁻ / 1,6 m: I, e⁻ ... 1,2; 1,7 ... 603; 646; 498 / 60,3 d: β⁻ 0,6; 2,3 γ 603; 1691 σ 6,5	Sb 125 2,77 a β⁻ 0,3; 0,6 ... γ 428; 601; 636; 463 ... g; m	Sb 126 19,0 m: β⁻ 1,9 ... 415; 666; 695 ... Iγ ...; e⁻ / 12,4 d: β⁻ 0,5; 1,9 ... γ 666; 695; 415 ...	Sb 127 3,85 d β⁻ 0,9; 1,5 ... γ 686; 473; 784 ... g; m	Sb 128 10,4 m: β⁻ 2,5 ... γ 743; 754; 314 ... / 9,0 h: β⁻ 2,0 γ 743; 754; 314; 527 ...	Sb 129 4,32 h β⁻ 0,6; 2,2 ... γ 813; 915; 544 ... g; m	Sb 130 6,5 m: β⁻ 3,0 ... γ 839; 793; 182 ... / 40 m: β⁻ 3,8 ... γ 839; 793; 331; 182 ...	Sb 131 23 m β⁻ 1,5; 3,2 ... γ 944; 933; 642 ... g; m	Sb 132 4,1 m: β⁻ 3,7 ... γ 974; 697; 151; 103 ... / 2,8 m: β⁻ 3,9 ... γ 974; 697; 989 ...
Sn 117 14,0 d: Iγ 159 e⁻ / 7,6: σ 2,6	Sn 118 24,1 σ 0,016	Sn 119 245 d: Iγ 24 e⁻ / 8,6: σ 2,3	Sn 120 32,8 σ ~0,001 + 0,14	Sn 121 ~50 a: β⁻ 0,35 γ 37 e⁻ / 27,0 h: β⁻ 0,38 no γ	Sn 122 4,7 σ 0,180 + 0,001	Sn 123 40,1 m: β⁻ 1,3 ... γ 160 ... / 129,2 d: β⁻ 1,4 ... γ ...	Sn 124 5,8 σ 0,13 + 0,004	Sn 125 9,5 m: β⁻ 2,0 ... γ 332 ... / 9,64 d: β⁻ 2,4 ... γ 1067; 1089; 823; 916 ...	Sn 126 ~10⁵ a β⁻ 0,3 γ 88; 64; 87 ... m	Sn 127 4,4 m: β⁻ 2,7 γ 491 / 2,1 h: β⁻ 3,2 ... γ 1114; 1096; 823 ...	Sn 128 59 m β⁻ 0,8 ... γ 482; 75; 557 ... m	Sn 129 2,5 m: β⁻ 3,3 ... γ 642; 2100 / 7,5 m: β⁻ γ 1161	Sn 130 1,7 m: β⁻ ~4 γ 145; 899; 85; 311 ...; g / 3,7 m: β⁻ 2,1; 2,5 γ 193; 780; 70 ...; m	Sn 131 59 s β⁻ 3,5 γ 1226; 783; 450; 1229; 304 ...
In 116 2,2 s: Iγ 164 e⁻ / 54 m: β⁻ 1,0 γ 1294; 1097; 417 / 14 s: β⁻ 3,3 γ 1294	In 117 1,95 h: β⁻ 1,8 ... γ 159 ... Iγ 315 e⁻; g / 38 m: β⁻ 0,7; 1,1 γ 553; 159 ...; g	In 118 8,5 s: Iγ 138 e⁻ β⁻ 1,8 γ 254 / 4,4 m: β⁻ 1,3; 2,0 γ 1230; 1051; 683 ... / 5 s: β⁻ 4,2 γ 1230; 528 ...	In 119 18 m: β⁻ 2,7 ... γ 1065; 1250 ... Iγ 311; g / 2,3 m: β⁻ 1,6 γ 763 ... g	In 120 44 s: β⁻ 2,2; 3,1 ... γ 1172; 1023; 864 ... / 3,0 s: β⁻ 5,6 ... γ 1172 ...	In 121 3,8 m: β⁻ 3,7 ... γ 60 ... Iγ 321 g / 25 s: β⁻ 2,2 ... γ 927; 658; 262 ...; g	In 122 10,0 s: β⁻ 4,4 γ 1141; 1003; 1194 ... / 1,5 s: β⁻ 6,3 ... γ 1141	In 123 47,8 s: β⁻ 4,6 ... γ 127; 3234; 1170; 3127 ...; m / 6 s: β⁻ 3,3 γ 1130 m	In 124 3,2 s β⁻ 5,3 ... γ 1132 ...	In 125 12,2 s: β⁻ / 2,3 s: β⁻ 5,3 γ 140	In 126 ? 1,53 s β⁻	In 127 3,6 s ?: β⁻ / 2,0 s ?: β⁻	In 128 ? 3,7 s β⁻	In 129 ? 0,8 s β⁻	In 130 0,53 s β⁻ 7,3 γ 775; 1217; 127; 409 g
Cd 115 44,8 d: β⁻ 1,6 ... γ 934; 1291; 484 ... g / 53,38 h: β⁻ 1,1 ... γ 528; 492 ... m	Cd 116 7,6 σ 0,027 + 0,050	Cd 117 3,31 h: β⁻ 0,7; 1,9 ... γ 1235; 1997; 1066 ...; g / 2,42 h: β⁻ 2,5 ... γ 273; 345; 1303 ... m; g	Cd 118 50,3 m β⁻ ~0,8 g	Cd 119 1,9 m: β⁻ γ 1025; 2021 ... g / 2,6 m: β⁻ γ 293; 343 ... m; g	Cd 120 50,8 s β⁻ γ g	Cd 121 4,8 s: β⁻ γ 1023; 1042; 99 ...; g / 12,8 s: β⁻ γ 325; 349; 990 ...; m	Cd 122 5,5 s β⁻ g							
Ag 114 4,5 s β⁻ 4,9 ... γ 558; 575 ...	Ag 115 19 s: β⁻ γ 230; 131 ... m / 20,0 m: β⁻ 1,1; 3,2 γ 230; 214; 473; 2157 ... g; m	Ag 116 10,4 s: β⁻ γ 514; 706; 700; 1030 ... Iγ 81 / 2,7 m: β⁻ γ 514; 700 ...	Ag 117 5,3 s: β⁻ γ 135 ... m / 73 s: β⁻ γ 135; 311; 337 ... g; m	Ag 118 2,8 s: β⁻ γ 488; 677 ... Iγ 128 / 3,7 s: β⁻ γ 488; 677 ...	Ag 119 6 s β⁻	Ag 120 0,32 s: ... 506; 698; 926 ... Iγ 203 / 1,17 s: β⁻ γ 506; 698 ...	Ag 121 0,8 s β⁻ γ 194–448	Ag 122 1,5 s β⁻ γ 570; 760						
Pd 113 1,6 m β⁻ γ 96; 643; 739; 483 ... g	Pd 114 2,4 m β⁻ γ 127; 137; 222; 232 ...	Pd 115 38 s β⁻ γ 343; 89; 255 ... m	Pd 116 13,6 s β⁻ γ 115; 101; 178	Pd 117 4,8 s β⁻	Pd 118 3,1 s β⁻									
Rh 112 4,7 s β⁻ γ 349	Rh 113 ? 0,9 s β⁻ γ 129	Rh 114 1,7 s β⁻ γ 333												
	Ru 112 ~0,7 s β⁻													
Tc 110 0,83 s β⁻ γ 241														

0,030 0,020 0,017 0,012 0,012 0,0109 0,011 0,011 0,011 0,011 0,012 0,014 0,014 0,016 0,022 0,030 0,055 0,125 0,35 0,65

68 70 72 74 76 78 80

				74	W 183,85 σ 18,5	W 162 <0,25 s α 5,53	W 163 2,5 s α 5,39	W 164 6,3 s α 5,15						W 170 4 m	W 171 9,0 m	W 172 6,7 m γ 458; 36…
				73	Ta 180,9479 σ 21							Ta 167 2,9 m	Ta 168 3,4 m γ 124; 262; 371; 750	Ta 169 5,0 m γ 69–463	Ta 170 6,3 m γ 221; 101	Ta 171 2,0 m ? γ 365 | 23,3 m γ 50; 506; 502; 166…
72	Hf 178,49 σ 102		Hf 157 0,12 s α 5,68	Hf 158 2,8 s α 5,27	Hf 159 5,6 s ε α 5,09	Hf 160 ~12 s ε α 4,77	Hf 161 17 s α 4,60					Hf 166 5,8 m	Hf 167 2,05 m ε β^+ γ 315…	Hf 168 25,95 m ε; β^+ ? γ 184; 157 m	Hf 169 3,25 m ε; β^+ 1,9 γ 493; 370… g	Hf 170 16,0 h ε γ 165; 621; 120; 573…
71	Lu 174,97 σ 77	Lu 155 0,07 s α 5,63	Lu 156 ~0,5 s α 5,43 | 0,23 s α 5,54						Lu 162 1,4 m		Lu 164 3,1 m γ 124; 262…	Lu 165 11,8 m γ 121; 132; 174; 204…	Lu 166 3,3 m ε; β^+ γ 228; 338; 102; 369…	Lu 167 55 m ε; β^+… γ 239; 1507; 1268; 278…	Lu 168 6,7 m ε β^+ 1,5; 2,7… Iγ γ 198; 979; 896; 885 | 5,5 m β^+ 1,2… γ 198; 299; 88…	Lu 169 2,7 m Iγ 29 e^- | 1,42 d ε; β^+… γ 960; 191; 1450… g; m
70	Yb 173,04 σ 36,6	Yb 154 0,39 s α 5,33	Yb 155 1,65 s α 5,19	Yb 156 24 s α 4,80	Yb 157 34 s α 4,50	Yb 158 1,5 m	Yb 159 ? 4,6 m γ 174; 216	Yb 160 4,8 m	Yb 161 ? 4,1 m γ 78; 600; 632…	Yb 162 18,9 m ε γ 163; 119…	Yb 163 11,1 m γ 860; 64; 123…	Yb 164 75,8 m ε γ 675; 445… g	Yb 165 10,5 m ε; β^+ 1,6 γ 80; 69; 1090…	Yb 166 56,7 h ε γ 82 e^-	Yb 167 17,7 m ε; β^+… γ 113; 106; 176…	Yb 168 0,14 σ 3470
69	Tm 168,9342 σ 103	Tm 153 1,6 s α 5,11	Tm 154 5 s α 4,96 | 3,0 s α 5,03	Tm 155 39 s α 4,45 m	Tm 156 19 s α 4,46 | 80 s α 4,23 g		Tm 158 4,3 m γ 193; 335; 1151	Tm 159 12 m γ 290; 220; 348; 272; 85…	Tm 160 9,2 m γ 126; 728; 264; 853; 860	Tm 161 37 m ε γ 46–354	Tm 162 21,8 m β^+ 2,3; 3,8… γ 102; 900; 799…	Tm 163 1,81 h ε; β^+… γ 104; 241; 1434; 1398…	Tm 164 5,1 m Iγ ε γ 208; 315… | 2,0 m ε β^+ 2,9… γ 91; 1155; 769…	Tm 165 30,06 h ε; β^+… γ 243; 297; 806…	Tm 166 7,70 h ε; β^+ 1,9… γ 779; 2052; 184; 1274…	Tm 167 9,25 d ε γ 208; 532… m
Er 167,26 σ 162	Er 151 23 s m	Er 152 9,8 s α 4,80 ε	Er 153 36 s α 4,67 ε	Er 154 5,8 m ε α 4,17 g	Er 155 5,3 m ε α 4,01		Er 157 ~25 m ε ? β^+ γ 117–2000	Er 158 2,3 h ε; β^+ 0,8 γ 72; 387… m_1	Er 159 36 m ε γ 624; 649… g; m	Er 160 28,6 h ε no γ m_1	Er 161 3,1 h ε; β^+… γ 827… g; m	Er 162 0,14 σ 19	Er 163 75 m ε β^+…; γ… g	Er 164 1,6 σ 13	Er 165 10,3 h ε no γ	Er 166 33,4 σ 15 + 5
Ho 164,9304 σ 66,5	Ho 150 ~30 s	Ho 151 42 s ε α 4,61 | 35,6 s ε α 4,52	Ho 152 52,3 s ε α 4,45 γ 613; 647; 683… | 2,4 m ε α 4,39	Ho 153 2,0 m α 3,91 | 9,3 m ε α 4,01	Ho 154 3,3 m β^+ α 3,72 γ 335; 413; 477… | 11,8 m β^+ α 3,93 γ 335	Ho 155 48 m β^+ 1,8 γ 240; 136…	Ho 156 ~1 h γ 266; 138… | 55 m β^+ 1,8; 2,9… γ 138; 266; 366…	Ho 157 12,6 m ε β^+ 1,2; 1,5… γ 280; 341; 193; 87…	Ho 158 19,5 m γ 218; 99; 320… | 29 m Iγ 67 β^+ γ 218; 99; 1791 | 11 m β^+ 1,3; 2,9… γ 218; 99; 949; 848…	Ho 159 8,3 s Iγ 206… | 33 m ε γ 121; 132; 310; 253…	Ho 160 ~1 h γ 87; 197; 297… | 5,0 h Iγ 60 ε β^+ γ 87; 197… | 26 m ε; β^+… γ 728; 879; 962; 966…	Ho 161 6,7 s Iγ 211 | 2,5 h ε γ 26; 78… e^-	Ho 162 68 m Iγ 59; 38… e^-; ε γ 185; 1220; 283; 937… | 15 m ε β^+ 1,1… γ 81; 1319… e^-	Ho 163 1,1 s Iγ 299 | ~33 a ε	Ho 164 37 m Iγ 37; 57… e^- | 29 m ε β^- 1,0… γ 91; 73… e^-	Ho 165 100 σ 3,5 + 63,0
Dy 148 3,1 m β^+ γ 621 g	Dy 149 4,1 m γ g	Dy 150 7,2 m β^+ α 4,23 γ 390 g	Dy 151 17 m ε ? β^+ α 4,07 γ 146	Dy 152 2,4 h ε α 3,63 γ 257 g	Dy 153 6,29 h ε α 3,46 γ 82; 100…	Dy 154 10^6 a α 2,87	Dy 155 9,59 h ε β^+ 0,9; 1,1 γ 227…	Dy 156 0,06 σ 33	Dy 157 8,1 h ε γ 326…	Dy 158 0,10 σ 43	Dy 159 144,4 d ε γ 58… e^-	Dy 160 2,3 σ 61	Dy 161 18,9 σ 585	Dy 162 25,5 σ 180	Dy 163 24,9 σ 130	Dy 164 28,2 σ 1700 + 1000
Tb 147 1,83 m β^+ γ 1397; 1798… | 1,61 h ε; β^+ γ 1153; 139; 894…	Tb 148 2,2 m β^+ γ 784; 632; 882; 395… | 70 m ε β^+ 4,6… γ 784; 489; 632; 1077…	Tb 149 4,2 m ε α 3,99 γ 796; 164… | 4,1 h ε β^+ 1,8… α 3,97… γ 352; 165; 388…	Tb 150 5,8 m β^+ γ 638; 650; 827; 438… | 3,5 h ε β^+ 3,7… α 3,49 γ 638; 496; 792…	Tb 151 17,6 h ε α 3,41… γ 252; 287; 108…	Tb 152 4,2 m Iγ 283; 160… β^+ γ 344; 411; 472… | 17,5 h ε β^+ 2,8… γ 344; 271; 587…	Tb 153 2,34 d ε γ 212; 170; 110; 102; 83…	Tb 154 22,8 h ε Iγ γ 123; 248; 1420… | 9,0 h ε Iγ γ 123; 248; 540… | 21,4 h ε β^+ 2,5 … γ 123; 1274…	Tb 155 5,32 d ε γ 87; 105; 180; 263…	Tb 156 24,4 h Iγ 50 | 5,4 h Iγ 88 e^- ε β^- ? β^-… | 5,35 d ε γ 534; 199; 1222 …	Tb 157 150 a. ε γ… e^-	Tb 158 10,5 s Iγ… e^- | 150 a ε β^- 0,8… γ 944; 962; 80…	Tb 159 100 σ 25,5	Tb 160 72,1 d β^- 0,6; 1,7… γ 879; 299; 966… σ 525	Tb 161 6,90 d β^- 0,5; 0,6… γ 49; 75… e^-	Tb 162 7,6 m β^- 1,3; 2,4… γ 260; 807; 888…	Tb 163 19,5 m β^- 0,8; 1,3… γ 351; 390; 494…
Gd 146 48,3 d ε β^+… γ 155; 116; 115…	Gd 147 38,1 h ε β^+… γ 229; 396; 929…	Gd 148 ~90 a α 3,183	Gd 149 9,5 d ε α 3,01 γ 150; 299; 347…	Gd 150 $1,8 \cdot 10^6$ a α 2,72	Gd 151 120 d ε α 2,60 γ 154; 243; 175…	Gd 152 0,20 $1,1 \cdot 10^{14}$ a α 2,14 σ 1100	Gd 153 241,6 d ε γ 97; 103…	Gd 154 2,2 σ 85	Gd 155 14,9 σ 61000	Gd 156 20,6 σ 1,5	Gd 157 15,7 σ 254000	Gd 158 24,7 σ 2,5	Gd 159 18,56 h β^- 0,9… γ 364; 58…	Gd 160 21,7 σ 0,77	Gd 161 3,6 m β^- 1,6; 1,7… γ 361; 315; 102… σ 31000	Gd 162 8,2 m β^- 1,0 γ 442; 403…
Eu 145 5,93 d ε β^+ 1,7… γ 894; 1659; 654…	Eu 146 4,65 d ε β^+ 1,5; 2,1… γ 747; 633; 634…	Eu 147 24 d ε; β^+… α 2,91 γ 197; 121; 678…	Eu 148 54 d ε; β^+… α 2,63 γ 550; 630; 611…	Eu 149 93,1 d ε γ 328; 277…	Eu 150 12,6 h β^- 1,0 ε; β^+… γ 334; 407… | ~35 a ε γ 334; 439; 584…	Eu 151 47,8 σ 4,0 + 3300 + 5900	Eu 152 96 m Iγ 40; 90… | 9,3 h β^- 1,9 … ε β^+ γ 841; 963 | 12,4 a ε; β^+… β^- 0,7; 1,5 γ 122; 344; 1408…	Eu 153 52,2 σ 390	Eu 154 8,5 a β^- 0,6; 1,8 ε γ 123; 1274; 723; 1005…	Eu 155 4,96 a β^- 0,1; 0,2… γ 87; 105… σ 4040	Eu 156 15,2 d β^- 0,5; 2,4… γ 812; 89; 1231…	Eu 157 15,15 h β^- 1,3… γ 413; 64; 373…	Eu 158 46 m β^- 2,4; 3,4… γ 944; 977; 80; 898…	Eu 159 18,7 m β^- 2,6… γ 68; 79; 96; 146; 665…	Eu 160 42 s β^- 3,9… γ 174; 412; 516; 822…	0,000088
Sm 144 3,1 σ ~0,7	Sm 145 340 d ε γ 61… e^- σ ~110	Sm 146 $7 \cdot 10^7$ a α 2,46	Sm 147 15,0 $1,06 \cdot 10^{11}$ a α 2,233 σ 64	Sm 148 11,2 $7 \cdot 10^{15}$ a α 1,96 σ 2,7	Sm 149 13,8 σ 41000	Sm 150 7,4 σ 102	Sm 151 93 a β^- 0,1 γ… e^- σ 15000	Sm 152 26,7 σ 206	Sm 153 46,75 h β^- 0,7; 0,8… γ 103; 70…	Sm 154 22,8 σ 5,5	Sm 155 22,4 m β^- 1,5… γ 104; 246; 141…	Sm 156 9,4 h β^- 0,7… γ 204; 88; 166… e^-	Sm 157 8,0 m β^- 2,4 γ 198; 196; 394…			0,00035
												0,0325	0,0133	0,0064	0,0031	0,00102
												94		96		98

Pm 143 265 d ε no β^+ γ 742	Pm 144 1,0 a ε no β^+ γ 618; 696; 477…	Pm 145 17,7 a ε α 2,24 γ 72; 67 e^-	Pm 146 5,53 a ε; β^- 0,8… γ 454; 747; 736… σ 8400	Pm 147 2,62 a β^- 0,2 γ … / σ 85 + 96	Pm 148 41,3 d: β^- 0,4; 1,0… γ 550; 630; 726… Iγ…; e^- σ 22 000 5,37 d: β^- 2,5… γ 550; 1465; 915… σ 2000	Pm 149 53,1 h β^- 1,1… γ 286… σ 1400	Pm 150 2,7 h β^- 2,3; 3,4… γ 334; 1325; 1166…	Pm 151 28 h β^- 0,8; 1,2… γ 340; 168… σ < 700	Pm 152 15,0 m: β^- γ 122; 231; 245; 340… 7,5 m: β 1,9 Iγ γ 122; 245… 4,2 m: β^- 3,5… γ 122; 841; 961; 963…	Pm 153 5,3 m β^- 1,7… γ 36; 127; 28; 120…	Pm 154 2,6 m: β^- 1,9; 2,5… γ 185; 82… 1,6 m: β^- 4,0… γ 2059; 1393; 840…
Nd 142 27,1 σ 18,7	Nd 143 12,2 σ 325	Nd 144 23,9 2,1 · 10^{15} a α 1,83 σ 3,6	Nd 145 8,3 σ 42	Nd 146 17,2 σ 1,3	Nd 147 10,98 d β^- 0,8… γ 91; 531… e^-	Nd 148 5,7 σ 2,48	Nd 149 1,73 h β^- 1,4; 1,6… γ 211; 114; 270…	Nd 150 5,6 σ 1,2	Nd 151 12,4 m β^- 1,2; 2,3… γ 117; 256; 1181…	Nd 152 11,4 m β^- ~0,7; ~1,0 γ 279; 250… g	
Pr 141 100 σ 3,9 + 7,6	Pr 142 14,6 m: Iγ… e^- 19,2 h: β^- 2,2… ε γ 1576… σ 20	Pr 143 13,57 d β^- 0,9… γ… σ 89	Pr 144 7,2 m: Iγ 59 β^-… γ… 17,3 m: β^- 3,0… γ 696…	Pr 145 5,98 h β^- 1,8… γ 676; 748…	Pr 146 24,0 m β^- 4,1… γ 454; 1525…	Pr 147 12,0 m β^- 2,1; 2,7… γ 575; 331; 640; 312…	Pr 148 1,98 m β^- 4,4 γ 302; 452; 698…	Pr 149 2,3 m: β^- 3,0… γ 110; 139; 165… 28 s: β^- γ 70; 138…	Pr 150 10 s β^- 5,7… γ 130; 258…	Pr 151 4 s β^- γ 164	
Ce 140 88,5 σ 0,57	Ce 141 32,51 d β^- 0,4; 0,6 γ 145 σ 29	Ce 142 11,1 σ 0,95	Ce 143 33,0 h β^- 1,4… γ 293; 57; 665; 722… σ 6,0	Ce 144 284,8 d β^- 0,3… γ 134; 80… g σ 1,0	Ce 145 3,0 m β^- 1,7; 2,1… γ 67; 725; 1148; 285; 440…	Ce 146 13,9 m β^- 0,8… γ 317; 218; 134; 265…	Ce 147 57 s β^- γ 99; 121; 269…	Ce 148 48 s β^- γ 196; 292; 397…	Ce 149 5 s β^- γ 58; 380 g	Ce 150 3,5 s β^-	
La 139 99,91 σ 9,0	La 140 40,2 h β^- 1,4; 2,2… γ 1596; 487; 816; 329… σ 2,7	La 141 3,93 h β^- 2,4… γ 1355…	La 142 92,5 m β^- 2,1; 4,5… γ 641; 2398; 2543…	La 143 14,3 m β^- 3,3… γ 619; 1160; 800…	La 144 39,8 s β^- γ 397; 541; 845…	La 145 29 s β^- γ 160	La 146 8,3 s β^- γ 259…	La 147 1,6 s β^-	La 148 1,3 s β^- γ 158		
Ba 138 71,9 σ 0,35	Ba 139 82,7 m β^- 2,4… γ 166… σ 6	Ba 140 12,79 d β^- 0,5; 1,0… γ 537; 163; 305… σ 1,6	Ba 141 18,3 m β^- 2,8; 3,0… γ 190; 304; 277; 344…	Ba 142 10,7 m β^- 1,0; 1,7… γ 255; 1204; 895…	Ba 143 20 s β^- γ 211; 799…	Ba 144 11,9 s β^- γ 291; 156…	Ba 145 5,6 s β^- γ 545; 571; 298	Ba 146 2,2 s β^- γ 327…			
Cs 137 30,1 a β^- 0,5; 1,7 m; g σ 0,110	CS 138 2,90 m: Iγ 80 β 3,0… γ 1436; 463; 192… 32,2 m: β^- 2,8; 3,9… γ 1436; 1010; 463; 2218…	Cs 139 9,3 m β^- 4,3… γ 1284; 627; 1421…	Cs 140 64 s β^- 5,7; 6,3… γ 602; 909; 1201…	Cs 141 24,7 s β^- 5,0… γ 562; 588; 1043… (n)	Cs 142 1,68 s β^- γ 360; 967; 1326… (n)	Cs 143 1,68 s β^- γ 211 (n)	Cs 144 1,0 s β^- γ 199 (n)	Cs 145 0,61 s β^- (n)	Cs 146 0,19 s β^- (n)		
Xe 136 8,9 σ 0,16	Xe 137 3,83 m β^- 4,1… γ 455…	Xe 138 14,1 m β^- 0,8; 2,8… γ 258; 435; 1768; 2016… g	Xe 139 39,7 s β^- 4,6; 4,8… γ 219; 297; 175…	Xe 140 13,5 s β^- 2,6… γ 806; 1414; 1315; 622…	Xe 141 1,79 s β^- 4,9… γ 119; 909; 106… (n)	Xe 142 1,24 s β^- γ 572; 657; 618; 538… (n)	Xe 143 0,83 s β^- γ 139; 194	Xe 144 <1 s β^-	Xe 145 0,9 s		
J 135 6,59 h β^- 1,5; 2,2… γ 1260; 1132; 1678; 1458; 1792… g; m	J 136 46 s: β^- 4,4; 4,8… γ 1313; 381; 197… 83 s: β^- 3,7; 5,0 γ 1313; 1321…	J 137 24,2 s β^- γ 1219; 602 (n 0,38…)	J 138 6,3 s β^- γ 589; 483 (n)	J 139 2,3 s β^- γ 271 (n)	J 140 0,87 s β^- γ 377; 458 (n)	J 141 0,45 s β^- γ (n)					
Te 134 41,8 m β^- γ 767; 211; 79; 278; 566… g	Te 135 18 s β^- γ 603; 870; 267	Te 136 20,9 s β^- γ 333–629 (n)	Te 137 3,5 s β^- (n)								
Sb 133 2,3 m β^- 1,2; 2,4… γ 1096; 632; 817; 2752; 839… g; m	Sb 134 0,85 s: β^- 8,4 no γ 10,5 s: β^- ~6,0; 6,8 γ 1279; 297; 706; 115	Sb 135 1,7 s β^- (n)									
Sn 132 40 s β^- 1,7 γ 86; 340; 247; 899; 992… g	Sn 133 1,5 s β^- 7,5 γ 963										
In 131 0,3 s β^-	In 132 0,12 s β^- ~5 γ 4041										

0,075

0,164

0,270

0,420

0,648

1,07

92

1,07

1,68

2,25

2,97

6,50 6,36 5,82 5,87 5,95 5,39 3,93

88 90 92

6,80

6,26

7,65 6,60 6,18

86

1,7 2,82 4,20 6,75

82 84

												86	Rn		Rn 200 1,0 s α 6,91 ε
										85	At	At 196 0,3 s α 7,06	At 197 0,4 s α 6,959	At 198 1,5 s α 6,849 4,9 s α 6,748	At 199 7,2 s α 6,639
							84	Po		Po 193 short α 6,98	Po 194 0,6 s α 6,85	Po 195 2,0 s α 6,699 4,5 s α 6,609	Po 196 5,5 s α 6,520 ε	Po 197 26 s α 6,38 ε 56 s α 6,281 ε	Po 198 1,76 m α 6,182 ε g
					83	Bi 208,9804 σ 0,033	Bi 189 <1,5 s α 6,67	Bi 190 5,4 s α 6,45 ε	Bi 191 20 s α 6,86; 6,63 13 s α 6,32	Bi 192 42 s α 6,06	Bi 193 3,5 s α 6,48… 64 s α 5,90	Bi 194 1,8 m α 5,61 γ 965; 575; 280…	Bi 195 90 s α 6,11 2,8 m α 5,43	Bi 196 4,5 m ε γ 1049; 688; 372…	Bi 197 ~10 m α 5,77
			82	Pb 207,2 σ 0,170	Pb 186 8 s α 6,32	Pb 187 17 s α 6,08	Pb 188 24,5 s α 5,98	Pb 189 51 s α 5,72	Pb 190 1,2 m α 5,58	Pb 191 1,3 m α 5,29	Pb 192 2,3 m α 5,06		Pb 194 11 m ε γ 204 g	Pb 195 17 m ε γ 384; 708; 394… m; g ?	Pb 196 37 m ε γ 192–503 g
			81	Tl 204,37 σ 3,4		Tl 186 25 s γ 405; 403… ~1 m γ 405; 403	Tl 187 ? 13 s γ 300…	Tl 188 1,2 m γ 412; 592; 504… 1,2 m γ 412…	Tl 189 1,4 m β^+ γ 216; 335; 229… 2,3 m γ 335	Tl 190 3,7 m γ 416; 625; 731… 2,6 m γ 416; 625…	Tl 191 5,2 m ε γ 216; 326; 265… g	Tl 192 11 m ε γ 423; 635; 786… 9,5 m ε γ 423…	Tl 193 2,1 m Iγ 365 21,0 m ε; β^+ γ 324; 1045; 676; 1579; 344…; g	Tl 194 32,8 m ε γ 428; 636; 749… 33 m ε β^+ γ 428…	Tl 195 3,6 s Iγ 384… 1,2 h ε; β^+ 1,8 γ 564; 885… g
Hg 179 1,09 s α 6,27 (p)	Hg 180 2,9 s α 6,12	Hg 181 3,6 s ε α 6,00; 5,92 (p)	Hg 182 11,2 s α 5,87; 5,70 γ 129; 217; 413	Hg 183 8,8 s α 5,91; 5,83 (p)	Hg 184 30,6 s α 5,54 γ 237; 156; 296…	Hg 185 21 s Iγ ? α 5,38 γ 211; 292 50 s α 5,65; 5,58 γ 222; 258…; m	Hg 186 1,4 m ε α 5,09 γ 252; 112… m	Hg 187 1,6 m α 5,04 γ 335; 112 2,4 m α 4,87 γ 233; 378… g	Hg 188 3,25 m ε α 5,14 γ 191; 253; 115…	Hg 189 8,7 m ε γ 566; 388… m 7,7 m ε γ 238; 248… g	Hg 190 20,0 m ε γ 143; 172; 155…	Hg 191 50,8 m ε γ 420; 579… m ~50 m ε γ 196…	Hg 192 4,9 h ε γ 275; 157; 307…	Hg 193 11,1 h ε; β^+ 1,2 γ 408; 573 Iγ …; m ~6 h ε γ 186; 258; 861…; g	Hg 194 ≥15 a ε no γ
Au 178 2,6 s α 5,92	Au 179 7,2 s α 5,85		Au 181 11,5 s α 5,62; 5,48	Au 182 22 s α 5,46 ? γ 155; 264	Au 183 42 s α 5,34	Au 184 53 s ε; β^+ α 5,17; 5,11… γ 163; 273; 362…	Au 185 6,8 m ε γ 145 g 4,2 m ε α 5,07 γ 311	Au 186 <2 m ε γ 192 10,7 m ε; β^+ γ 192; 299; 765; 416…	Au 187 6,4 m ? ε γ 181 8,5 m ε; α 4,69 γ 185; 190; 251…	Au 188 8,8 m ε γ 266; 340; 606…	Au 189 4,6 m ε β^+ γ 167; 321 28,7 m ε γ 713; 813; 348; 448…	Au 190 42,8 m ε β^+ 3,4… γ 296; 302; 598…	Au 191 1 s Iγ 241; 253… 3,18 h ε γ 24–1102	Au 192 5,0 h ε β^+ 2,5… γ 316; 296; 612…	Au 193 3,9 s Iγ 258… ε e^- m 17,65 h ε γ 186; 256; 268…; g
Pt 177 11 s α 5,52; 5,48	Pt 178 20 s α 5,44; 5,31	Pt 179 33 s α 5,16	Pt 180 50 s α 5,14	Pt 181 51 s α 5,02	Pt 182 2,6 m α 4,84 γ 136; 146; 210…	Pt 183 6,5 m α 4,73	Pt 184 17,3 m ε α 4,50 γ 155; 192; 548; 731…	Pt 185 33 m ε γ 255; 198; 641; 721… 1,2 h ε γ 461; 841; 153…	Pt 186 2,0 h ε α 4,23 γ 689; 612… m	Pt 187 2,35 h ε γ 106; 202; 110; 285; 709…	Pt 188 10,2 d ε α 3,92 γ 188; 195; 382; 424…	Pt 189 11 h ε γ 721; 608; 569; 243; 545…	Pt 190 0,013 $6,1 \cdot 10^{11}$ a α 3,17 σ 150	Pt 191 2,8 d ε γ 539; 409; 360… g	Pt 192 0,78 σ 2,2 + <14
Ir 176 8 s α 5,12	Ir 177 21 s α 5,01	Ir 178 12 s ε γ 266; 132; 363…	Ir 179 4 m	Ir 180 1,5 m ε γ 132; 276…	Ir 181 5,0 m ε γ 49; 108; 228; 319; 1640…	Ir 182 15 m ε γ 273; 127…	Ir 183 58 m ε γ 238 m; g	Ir 184 3,0 h ε β^+ 2,9… γ 264; 120; 390…	Ir 185 14 h ε γ 254; 97; 101; 158…	Ir 186 1,75 h β^+ γ 137; 767; 630 15,8 h ε; β^+ 1,9… γ 297; 137; 435…	Ir 187 11 h ε γ 913; 427; 401; 611…	Ir 188 41,5 h ε β^+… γ 155; 633; 478…	Ir 189 13,3 d ε γ 245; 70; 59… g; m	Ir 190 3,1 h ε Iγ 149 m 1,2 h I, e 12,1 d ε; γ 187; 605; 518; 558… g	Ir 191 37,4 4,88 s Iγ 129 e^- σ ~0,1 + 300 ··624
Os 175 1,4 m γ 125; 181…	Os 176 3,6 m γ 1291; 776; 1209…	Os 177 3,5 m γ	Os 178 5,0 m γ 969; 1331; 595…	Os 179 3 m γ 219; 309… 9 m γ 219; 1312; 760; 166…	Os 180 21,7 m ε γ 20; 668; 329… e^-	Os 181 2,7 m ε β^+ 1,8 γ 145; 118… 1,8 h ε γ 239; 827; 118…	Os 182 22,1 h ε γ 510; 180; 263; 56… m	Os 183 10 h ε; γ 1102; 1108; 1035… Iγ 171; e^- 14 h ε γ 382; 114; 168…	Os 184 0,02 σ 3000	Os 185 94 d ε γ 646; 875; 880; 717…	Os 186 1,6 $2,0 \cdot 10^{15}$ a α 2,76	Os 187 1,6 σ 336	Os 188 13,3 σ 4,3	Os 189 16,1 6 h Iγ… e^- σ 0,00026 + 23	Os 190 26,4 9,9 m Iγ 503; 617; 361; 187… σ 9,1 + 3,9
Re 174 2,1 m γ 112; 243; 349	Re 175 5 m γ 185	Re 176 5,7 m ε γ 241; 109…	Re 177 14 m ε γ 197; 80; 84; 96…	Re 178 13,2 m ε β^+ 3,3; 3,5… γ 237; 106; 939…	Re 179 19,7 m ε γ 430; 290; 1680… g; m	Re 180 2,5 m ε β^+ 1,8 γ 902; 104…	Re 181 20 h ε γ 366; 361; 639…	Re 182 13 h ε; β^+… γ 1121; 1221; 1189; 100… 64 h ε; γ 229; 1121; 1221; 100…	Re 183 71 d ε; γ 162; 47; 292; 209; 110; 99… g	Re 184 165 d Iγ 105 e^-, ε γ 216; 921; 111 38 d ε γ 903; 792; 111; 895	Re 185 37,40 σ 112	Re 186 $2 \cdot 10^{5}$ a Iγ 59; 40; 99 90,64 h β^- 1,1… ε γ 137	Re 187 62,60 $5 \cdot 10^{10}$ a β^- 0,0026 no γ σ 1,6 + 73	Re 188 18,6 m Iγ 64; 106… e^- 16,98 h β^- 2,1… γ 155; 633…	Re 189 24,3 h β^- 1,0… γ 217; 219; 245… g; m

W 173 16,5 m ε	W 174 29 m ε	W 175 34 m ε γ	W 176 2,5 h ε γ 34–100	W 177 2,25 h ε γ 116; 186; 427; 1036…	W 178 22 d ε no γ m	W 179 6,7 m Iγ 222… ε γ… ‖ 38 m ε γ 31… e^-	W 180 0,13 σ ~3,5	W 181 121,2 d ε γ… e^-	W 182 26,3 σ 20,7	W 183 5,3 s Iγ 108; 47; 53; 99… ‖ 14,3 σ 10,2	W 184 30,7 σ 0,002 + 1,8	W 185 1,68 m Iγ 66; 132; 174… ‖ 75,1 d β^- 0,4… γ…	W 186 28,6 σ 37,8	W 187 23,8 h β^- 0,6; 1,3… γ 686; 480; 72… σ 64	W 188 69 d β^- 0,3… γ… g
Ta 172 37,0 m ε β^+ 2,3; 2,5… γ 214; 95; 1109…	Ta 173 3,6 h ε β^+ 2,5; 2,6… γ 172; 70; 90; 160; 181…	Ta 174 1,2 h ε β^+ 2,5; 2,8… γ 206; 91; 1206…	Ta 175 10,5 h ε γ 207; 349; 267; 82; 126; 1793…	Ta 176 8,1 h ε β^+… γ 1159; 88; 1225…	Ta 177 56,6 h ε β^+… γ 113; 208… g	Ta 178 9,25 m ε; β^+ 0,9… γ 93; 1351; 1341… g ‖ 2,2 h ε γ 332 m_1	Ta 179 ~600 d ε no γ g	Ta 180 8,1 h ε β^- 0,7… γ 93; 104 g ‖ 0,012 σ 700	Ta 181 99,988 σ 0,0103 + 21,0	Ta 182 16 m Iγ 172; 147; 185… ‖ 115,0 d β^- 0,4; 1,7… γ 68; 1121; 1221… σ 8200	Ta 183 5,0 d β^- 0,6; 0,8… γ 246; 108; 354; 161… g	Ta 184 8,7 h β^- 1,2; 1,6… γ 414; 253; 921…	Ta 185 49 m β^- 1,8… γ 174; 177… g	Ta 186 10,5 m β^- 2,2; 2,6… γ 197; 214; 511; 737; 615…	
Hf 171 12,2 h ε γ 662; 122; 1071… g; m	Hf 172 1,87 a ε γ 24; 126; 67… m	Hf 173 23,6 h ε γ 124; 297; 140; 311…	Hf 174 0,18 $2{,}0 \cdot 10^{15}$ a α 2,50 σ 390	Hf 175 70,0 d ε γ 343…	Hf 176 5,2 σ_{abs} 38	Hf 177 51 m Iγ 277; 295; 327… ‖ 1,1 s Iγ 208; 229; 379… ‖ 18,5 σ 10^{-7} + 1,1 + 365	Hf 178 31 a Iγ 574; 495; 217… ‖ 4,3 s Iγ 426; 326; 213; 89… ‖ 27,2 σ ? + 53 + 33	Hf 179 25 d Iγ 454; 363; 123; 146… ‖ 18,7 s Iγ 214… ‖ 13,8 σ 0,34 + 45	Hf 180 5,5 h Iγ 332; 443; 215; 57… ‖ 35,1 σ 12,6	Hf 181 42,4 d β^- 0,4… γ 482; 133; 346 σ ~40	Hf 182 65 m β γ 172… Iγ 344; 224…; g ‖ $9 \cdot 10^6$ a β^- <0,4 γ 270; 156… g	Hf 183 64 m β^- 1,2; 1,5… γ 784; 73; 459…	Hf 184 4,12 h β^- 1,1… γ 139; 345; 181…		
Lu 170 2,0 d ε; β^+… γ 84; 1280; 2042; 985…	Lu 171 76 s Iγ… e^- ‖ 8,22 d ε; β^+… γ 740; 667; 781…	Lu 172 3,7 m Iγ 42 e^- ‖ 6,7 d ε; γ 1094; 901; 181; 810; 912…	Lu 173 1,37 a ε γ 272; 79; 101… e^-	Lu 174 142 d Iγ 67… e^- ε γ… ‖ 3,31 a ε β^+… γ 1242; 76…	Lu 175 97,4 σ 16,4 + 7	Lu 176 2,6 3,68 h β^- 1,2; 1,3… γ 88… e^- ‖ $3{,}3 \cdot 10^{10}$ a β^- 0,6… γ 307; 202; 88… σ 7 + 2100	Lu 177 161 d β 0,2 Iγ 414; 319; 122… m_1 ‖ 6,71 d β^- 0,5… γ 208; 113… g	Lu 178 22,7 m β^- 1,2… γ 332 m_1 ‖ 28,4 m β^- 2,0… γ 93; 1341; 1310; 1269…; g	Lu 179 4,6 h β^- 1,4… γ 214… g	Lu 180 5,7 m β^- 1,5… γ 408; 1199; 1107; 215… g					
Yb 169 46 s Iγ 24 e^- ‖ 30,7 d ε γ 63; 198; 177; 110…	Yb 170 3,0 σ 10	Yb 171 14,3 σ 50	Yb 172 21,9 σ 1,3	Yb 173 16,2 σ 19	Yb 174 31,8 σ 65	Yb 175 4,2 d β^- 0,5… γ 396; 283; 114…	Yb 176 12 s Iγ 293; 390; 190; 96… ‖ 12,7 σ 2,4	Yb 177 6,5 s Iγ 104; 228 e^- ‖ 1,9 h β^- 1,4… γ 150; 1080; 122; 1241…; g	Yb 178 74 m β^- 0,6… γ 391; 348… g		110		112		114
Tm 168 93,1 d ε γ 198; 816; 447…	Tm 169 100 σ 103	Tm 170 128,6 d β^- 1,0… ε γ 84… e^-; σ 92	Tm 171 1,92 a β^- 0,1… γ… e^- σ 4,5	Tm 172 63,6 h β^- 1,8; 1,9… γ 79; 1094; 1387; 1530; 1466; 1609…	Tm 173 8,2 h β^- 0,9; 1,3… γ 399; 461…	Tm 174 5,4 m β^- 1,2… γ 366; 992; 273; 177…	Tm 175 15,2 m β^- 0,9; 1,9… γ 515; 941; 364…	Tm 176 1,9 m β^- 2,0; 2,8… γ 82; 190; 1069; 382… g					80	Hg 200,59 σ 375	Hg 178 0,47 s α 6,43
Er 167 2,3 s Iγ 208 e^- ‖ 22,9 σ 670	Er 168 27,0 σ 1,95	Er 169 9,3 d β^- 0,3… γ… e^-	Er 170 15,0 σ 5,7	Er 171 7,5 h β^- 1,1; 1,5… γ 308; 296; 112; 124… σ 280	Er 172 49 h β^- 0,4; 0,9… γ 610; 407…	Er 173 1,4 m β^- γ 895; 199; 193…			108			79	Au 196,9665 σ 98,8		Au 177 1,3 s α 6,12; 6,15
Ho 166 1200 a β^- 0,07… γ 184; 810; 712… ‖ 26,7 h β^- 1,9… γ 81… e^-	Ho 167 3,1 h β^- 0,3; 1,0… γ 347; 321… g; m	Ho 168 3,0 m β^- 2,0… γ 741; 821; 816…	Ho 169 4,6 m β^- 1,2; 2,0… γ 788; 853; 761; 778…	Ho 170 2,9 m β^- 2,0… γ 932; 182… ‖ 42 s β^- 4… γ 811; 1893…			106		78	Pt 195,09 σ 10,0		Pt 173 short α 6,19	Pt 174 0,7 s α 6,03	Pt 175 2,52 s α 5,96	Pt 176 6,33 s α 5,74
Dy 165 1,3 m Iγ 108; e^- β^- 0,9; 1,0… γ 515; 362… ‖ 2,35 h β^- 1,3… γ 95; 362… σ 3900	Dy 166 81,5 h β^- 0,4; 0,5 γ 82; 426… g	Dy 167 ~4,5 m β^- γ 569; 258; 249			104				77	Ir 192,22 σ 426	Ir 171 1,0 s α 5,91	Ir 172 1,7 s α 5,81	Ir 173 3,0 s α 5,67	Ir 174 4,0 s α 5,48	Ir 175 4,5 s α 5,39
Tb 164 3,0 m β^- 1,7; 3,0… γ 169; 755; 215; 688; 611…								76	Os 190,2 σ 15,3	Os 169 3,0 s α 5,56	Os 170 7,1 s ε α 5,40	Os 171 8,2 s α 5,24	Os 172 19 s ε α 5,11	Os 173 16 s ε α 4,94	Os 174 45 s ε α 4,76 γ 118; 325…
	100		102					75	Re 186,207 σ 88			Re 170 9 s γ 306; 156; 413…		Re 172 23 s γ 254; 123; 350…	

								91	Pa 231,0359	Pa 216 0,20 s α 7,72; 7,82; 7,92	Pa 217 ? short α 8,34					Pa 222 5,7 ms α 8,21; 8,54; 8,33 . . .
						90	Th 232,0381 α 7,40	Th 213 ? 0,15 s α 7,69	Th 214 ? 0,13 s α 7,68	Th 215 1,2 s α 7,39; 7,52 . . .	Th 216 28 ms α 7,92	Th 217 252 μs α 9,25	Th 218 0,1 μs α 9,68	Th 219 1,0 μs α 9,34	Th 220 9,7 μs α 8,79	Th 221 1,68 ms α 8,15; 8,47 . . .
			89	Ac	Ac 209 0,10 s α 7,59	Ac 210 0,35 s α 7,46	Ac 211 0,25 s α 7,48	Ac 212 0,93 s α 7,38	Ac 213 0,80 s α 7,36	Ac 214 8,2 s α 7,214; 7,082 . . . ε	Ac 215 0,17 s α 7,60 ε	Ac 216 0,33 ms: α 9,03; 9,11 . . . ~0,3 ms: α 9,07; 8,99	Ac 217 0,11 μs α 9,65	Ac 218 0,27 μs α 9,21	Ac 219 7 μs α 8,67	Ac 220 26 ms α 7,85; 7,61; 7,68 . . . γ 134 . . .
Ra 226,0254			Ra 206 0,4 s α 7,272	Ra 207 1,3 s α 7,133	Ra 208 1,4 s α 7,133	Ra 209 4,6 s α 7,010	Ra 210 3,7 s α 7,020	Ra 211 13 s α 6,912	Ra 212 14 s α 6,90	Ra 213 2,74 m α 6,62; 6,73; 6,52 ε	Ra 214 2,5 s α 7,14 ε g	Ra 215 1,6 ms α 8,697 . . .	Ra 216 0,18 μs α 9,35	Ra 217 1,6 μs α 8,99	Ra 218 14 μs α 8,39	Ra 219 10 ms α 7,68; 7,98
Fr	Fr 203 0,7 s α 7,132	Fr 204 2,2 s: α 7,03 3,3 s: α 6,975	Fr 205 3,7 s α 6,919	Fr 206 15,6 s α 6,794 ε	Fr 207 14,7 s α 6,774	Fr 208 59 s α 6,648	Fr 209 54 s α 6,648	Fr 210 3,18 m α 6,57 ε	Fr 211 3,08 m α 6,534	Fr 212 19,3 m ε; α 6,261; 6,383; 6,407 . . . γ 1272; 227; 1184 . . .	Fr 213 34,7 s α 6,774 ε	Fr 214 3,4 ms: α 8,48; 8,55 . . . 5,0 ms: α 8,43; 8,36 . . .	Fr 215 0,09 μs α 9,36	Fr 216 0,70 μs α 9,01	Fr 217 22 μs α 8,32	Fr 218 0,7 ms α 7,87; 7,56 . . .
Rn 201 3,8 s: α 6,77 ε m 7,0 s: α 6,72 ε g	Rn 202 9,85 s α 6,64 ε m	Rn 203 28 s: α 6,548 m 45 s: α 6,498 ε g	Rn 204 1,24 m α 6,417 ε	Rn 205 2,83 m ε α 6,263 . . . γ 266 . . .; g	Rn 206 5,67 m α 6,260 . . . ε	Rn 207 9,3 m ε; α 6,126 . . . γ 345; 747 . . . g	Rn 208 24,4 m α 6,140 . . . ε γ . . .	Rn 209 30 m ε; β^+ α 6,039 . . . γ 408; 746; 338; 689 . . .	Rn 210 2,4 h α 6,040 . . . ε γ 458 . . .	Rn 211 14,6 h ε; α 5,783; 5,851 . . . γ 674; 1363; 678 . . .; g	Rn 212 24 m α 6,264 . . . γ . . .	Rn 213 25 ms α 8,09 . . . γ . . .	Rn 214 0,27 μs α 9,04	Rn 215 2,3 μs α 8,67 g	Rn 216 45 s α 8,05 g	Rn 217 0,54 ms α 7,742 . . .
At 200 4,3 s: α 6,536 42 s: α 6,413; 6,464	At 201 1,5 m α 6,343 ε	At 202 2,6 m: α 6,23 3,0 m: ε α 6,135 γ 675; 570; 441	At 203 7,3 m ε; α 6,088 γ 1034; 639; 1002 . . . ε→g; m; α→g	At 204 9,3 m ε α 5,951 γ 683; 515; 425 . . .	At 205 26,2 m ε; β^- α 5,902 γ 719; 669; 629 . . .; g	At 206 31,4 m ε α 5,703 γ 700; 476; 395 . . .	At 207 1,8 h ε; α 5,758 γ 814; 588; 301 . . . g	At 208 1,63 h ε α 5,641 . . . γ 685; 660; 177 . . .	At 209 5,5 h ε α 5,647 . . . γ 545; 782; 790 . . .	At 210 8,3 h ε; α 5,524; 5,442; 5,361 . . . γ 1181; 245; 1483 . . .	At 211 7,2 h ε α 5,867 . . . γ . . . g	At 212 122 ms: α 7,84; 7,90 . . . γ . . . 315 ms: α 7,68; 7,62 . . . γ . . .	At 213 0,11 μs α 9,08	At 214 ~2 μs α 8,82 . . . g	At 215 ~0,1 ms α 8,03 . . . γ . . .	At 216 0,3 ms α 7,81; 7,71 . . . γ . . .
Po 199 4,2 m: ε α 6,059 γ 473 5,2 m: ε α 5,952	Po 200 11,5 m ε α 5,863	Po 201 8,9 m: ε Iγ 418 . . . e^- α 5,786 α→m 15,3 m: ε α 5,683 γ 239 . . . ε→m α→g	Po 202 45,0 m ε α 5,588 γ 689; 316; 166; 790 . . .	Po 203 1,2 m: Iγ 641 e^- 37 m: ε; β^+ α 5,384 γ 909; 1091; 894; 215 . . .	Po 204 3,52 h ε α 5,377 γ 884; 270; 1016 . . .	Po 205 1,8 h ε; α 5,22 γ 872; 1001; 850; 837 . . . g	Po 206 8,8 d ε; α 5,223 γ 1033; 808; 511; 287 . . . g	Po 207 2,8 s: Iγ 814; 301; 268 5,84 h: ε; β^+ . . . α 5,12 γ 992; 743; 912 . . .; g	Po 208 2,898 a α 5,116 . . . ε γ . . . g	Po 209 102 a α 4,882 . . . ε γ . . .	Po 210 138,38 d α 5,3045 . . . γ . . . σ <0,0005 + <0,030	Po 211 25,5 s: α 7,27; 8,88 . . . γ 570; 1064 . . . 0,56 s: α 7,450 . . . γ . . .	Po 212 46 s: α 11,65 . . . γ 2615; 583 0,3 μs: α 8,784	Po 213 4,2 μs α 8,375 . . . γ . . .	Po 214 164 μs α 7,687 . . . γ . . .	Po 215 1,78 ms α 7,3864 . . . β^- . . . γ . . .
Bi 198 7,7 s: Iγ 248 11,8 m: ε γ 1063; 562; 198; 318 . . .	Bi 199 24,7 m: ε α 5,49 27 m: ε	Bi 200 36 m ε β^+ . . . γ 1026; 462; 420; 245 . . .	Bi 201 59 m: ε Iγ 846 α 5,24 g 1,7 h: ε γ 902; 936 . . . m	Bi 202 1,8 h ε; β^+ . . . γ 961; 422; 658 . . . g	Bi 203 11,76 h ε; β^+ 1,4 . . . γ 820; 825; 897; 1848 . . . g; m	Bi 204 11,3 h ε γ 899; 375; 984 . . . g	Bi 205 15,31 d ε β^+ . . . γ 1764; 703; 988 . . .	Bi 206 6,24 d ε; β^+ . . . γ 803; 881; 516; 1719; 537 . . .	Bi 207 38 a ε β^+ . . . γ 570; 1064; 1770 . . .	Bi 208 $3{,}68 \cdot 10^5$ a ε γ 2615	Bi 209 100 σ 0,014 + 0,019	Bi 210 $3{,}5 \cdot 10^6$ a: α 4,946; 4,908 . . . γ 266; 305 . . . σ 0,054 5,01 d: β^- 1,2 α 4,650; 4,687 γ . . .	Bi 211 2,13 m α 6,623; 6,278 γ 351 . . . β^- . . . α→g; β^-→g	Bi 212 60,55 m β^- 2,3 . . . α 6,051; 6,090 . . . γ 727; 1621 . . . g	Bi 213 45,59 m β^- 1,4 . . . α 5,87 . . . γ 440 . . .	Bi 214 19,8 m β^- 3,3 . . . α 5,448; 5,512 . . . γ 609; 1764; 1120 . . .
Pb 197 42 m: ε γ 386; 222; 773 . . . Iγ 85 . . . <42 m	Pb 198 2,40 h ε γ 290; 173 . . . g	Pb 199 12 m: Iγ 425 e^- ε γ 145–2613 1,5 h: ε; β^+ 2,8 γ 367; 353; 1135; 720	Pb 200 21,5 h ε γ 148; 257; 236; 268 . . .	Pb 201 61 s: Iγ 629 9,4 h: ε; β^+ . . . γ 331; 361; 946 . . .	Pb 202 3,62 h: Iγ 961, 422; 787 . . . ε γ 490; 460; 390 . . . ~$3 \cdot 10^5$ a: ε no γ	Pb 203 6,2 s: Iγ 825; 820 . . . 52,1 h: ε γ 279; 401 . . .	Pb 204 1,4 66,9 m: Iγ 899; 912; 375 . . . $1{,}4 \cdot 10^{17}$ a: α 2,6 σ 0,661	Pb 205 $1{,}4 \cdot 10^7$ a ε no γ	Pb 206 24,1 σ 0,0305	Pb 207 22,1 σ 0,709	Pb 208 52,4 σ 0,000487	Pb 209 3,25 h β^- 0,6 no γ	Pb 210 22,3 a β^- 0,02; 0,06 γ 47 e^-; α 3,72 g; σ 0,5	Pb 211 36,1 m β^- 1,4 . . . γ 405; 832; 427 . . .	Pb 212 10,64 h β^- 0,3; 0,6 . . . γ 239; 300 . . .	Pb 213 10,2 m β^-

Tl 196 1,4 h ε γ 426; 635; 895 … Iγ 34–275 1,8 h ε γ 426 …	Tl 197 2,84 h ε γ 426; 152 … g	Tl 198 1,87 h ε Iγ 283 … γ 637; 412; 587 … 5,3 h ε β^+ … γ 412; 676; 637 …	Tl 199 7,42 h ε γ 455; 208; 247; 158 … g	Tl 200 26,1 h ε β^+ … γ 368; 1206; 579; 828 …	Tl 201 73,5 h ε γ 167; 135 …	Tl 202 12,2 d ε γ 440 …	Tl 203 29,5 σ 11,0	Tl 204 3,78 a β^- 0,8 ε no γ g; σ 21,6	Tl 205 70,5 σ 0,10	Tl 206 4,20 m β^- 1,5 … γ …	Tl 207 1,3 s Iγ 1000; 350 4,8 m β^- 1,4 … γ …	Tl 208 3,054 m β^- 1,8; 2,4 … γ 2615; 583; 511 …	Tl 209 2,2 m β^- 1,8 γ 1566; 117; 465	Tl 210 1,3 m β^- 1,9; 2,3 … γ 795; 296 … (n)		
Hg 195 40 h ε; γ 560; 388 … Iγ … e^-; m 9,5 h ε γ 780; 61 … g	Hg 196 0,15 σ 120 + 3080	Hg 197 23,8 h Iγ 134 … e^- ε m 64,1 h ε γ 77; 192 … e^-; g	Hg 198 10,1 σ 0,018 + 1,9	Hg 199 42,6 m Iγ 158; 374 e^- 16,9 σ 2000	Hg 200 23,1 σ < 60	Hg 201 13,2 σ < 60	Hg 202 29,7 σ 4,9	Hg 203 46,6 d β^- 0,2 γ 279	Hg 204 6,8 σ 0,43	Hg 205 5,2 m β^- 1,5 … γ 204 …	Hg 206 8,2 m β^- 1,3 … γ 305; 650 …					
Au 194 39,5 h ε β^+ 1,5 … γ 329; 294; 1469 …	Au 195 30,5 s Iγ 262 … e^- 183 d ε γ 99 … e^- g	Au 196 9,7 h Iγ 148; 188 8,2 s Iγ … e^- 6,2 d ε β^- 0,3 γ 356; 333; 426	Au 197 7,8 s Iγ 279 … e^- 100 σ 98,8	Au 198 2,3 d Iγ 215; 97; 180; 204 … 2,695 d β^- 1,0; 1,4 … γ 412 … σ 25100	Au 199 3,13 d β^- 0,3; 0,5 … γ 158; 208 … g; σ 30	Au 200 18,7 h β 0,6 γ 368; 498; 579; 256 … Iγ 333 … 48,4 m β^- 2,3 … γ 368; 1226; 1263 …	Au 201 26,4 m β^- 1,3 … γ 543; 517; 613; 167 …	Au 202 28 s β^- 3,5 … γ 440; 1125; 1307; 1204 …		Au 204 40 s β^- γ 437; 1511 …						
Pt 193 4,33 d Iγ … e^- ~50 a ε no γ g	Pt 194 32,9 σ 0,090 + 1,2	Pt 195 4,02 d Iγ 99; 130 … e^- 33,8 σ 27	Pt 196 25,3 σ 0,050 + 0,7	Pt 197 81 m Iγ 346 … e^- β 0,7 m 20,0 h β^- 0,7 … γ 77; 192 … e^-; g	Pt 198 7,2 σ 0,027 + 3,7	Pt 199 14 s Iγ 393 … e^- 30,8 m β^- 1,7 … γ 543; 493; 317; 186 … σ ~15	Pt 200 11,5 h β^- g	Pt 201 2,5 m β^- 2,7 γ 1760 …								
Ir 192 241 a Iγ 161 1,4 m Iγ … e^- β … γ … 74,8 d β 0,7 … ε; β^+ … γ 316; 468 … σ 1100 + ? g	Ir 193 11,9 d Iγ … e^- σ > 0,035 + ? 62,6 σ ?+110	Ir 194 171 d β^- γ 483; 329 … 19,4 h β^- 2,2 … γ 329; 294 …	Ir 195 3,8 h β^- 0,4; 1,0 … γ 320; 433; 685; 365 … g; m 2,5 h β^- 1,0; 1,1 … γ 99; 130 … g	Ir 196 1,40 h β^- 1,2 … γ 394; 521; 356; 447; 647 … 52 s β^- 3,2 … γ 356; 779; 447; 333 …	Ir 197 7 m β^- 2,0 … γ ~500 g	Ir 198 8 s β^- γ 407; 507										
Os 191 13,03 h Iγ … e^- 15,4 d β^- 0,1 m	Os 192 5,9 s Iγ 206; 302; 569; 453 … 41,0 σ 2,0	Os 193 30,0 h β^- 1,1 … γ 139; 460; 73 … g; σ 1540	Os 194 6,0 a β^- 0,1 … γ 43 … e^-; g	Os 195 6,5 m β^- 2 g												
Re 190 3,0 h β^- ~1,8 … Iγ 119 γ 187; 558; 569 …; g 3,1 m β^- 1,8 γ 187; 558; 829 … g	Re 191 9,8 m β^- 1,8	Re 192 16 s β^- ~4 γ 206–751														
W 189 11 m β^- 2,5 … γ 258; 417; 550 …																
	116		118		120		122		124		126		128		130	

101

100												Fm	Fm 244 3,3 ms sf	Fm 245 4,2 s α 8,15
99												Es	Es 243 20 s α 7,90	Es 244 40 s ε
98										Cf	Cf 240 1,06 m α 7,59	Cf 241 3,78 m ε α 7,34	Cf 242 3,68 m α 7,39; 7,35 ε	Cf 243 10,7 m ε α 7,06; 7,17
97										Bk				Bk 242 sf
96									Cm		Cm 238 2,4 h ε α 6,52	Cm 239 3 h ε γ 187 g	Cm 240 27 d α 6,291; 6,248 ... sf; g	Cm 241 sf | 36d ε α 5,939 ... γ 475 ... g
95					Am	Am 232 ? 1,4 m sf ?		Am 234 2,6 m sf ?			Am 237 sf | 75 m ε α ~ 6,01 γ 280; 439 ... g	Am 238 sf | 1,6 h ε; α 5,94 γ 963; 919; 561; 605 ... g	Am 239 sf | 11,9 h ε α 5,776 ... γ 278; 228 ... g	Am 240 sf | 50,8 h ε α 5,378 ... γ 988; 889 ... g
94			Pu				Pu 232 34 m ε α 6,60; 6,54	Pu 233 20,9 m ε α 6,31 γ 235; 535 ...	Pu 234 8,8 h ε α 6,202; 6,151 ... γ ...; e^-	Pu 235 sf | 25 m ε α 5,85 γ 49; 756 ...	Pu 236 sf | 2,85 a α 5,768; 5,721 ... sf; γ ...; e^- σ_f 165	Pu 237 sf | 45,6 d ε α 5,35; 5,65 ... γ 60 ...; e^- σ_f 2400	Pu 238 sf | 87,75 a α 5,499; 5,457 ... sf; γ ...; e^- σ 547; σ_f 16,5	Pu 239 sf | 2,439·10^4 a α 5,155; 5,143 sf; m; γ ...; e^- σ 268,8 σ_f 742,5
93		Np 237,0482	Np 227 — Np 228 60 s sf ?		Np 229 4,0 m α 6,89	Np 230 4,6 m ε α 6,66	Np 231 48,8 m ε α 6,26 γ 371; 348; 264 ...	Np 232 14,7 m ε γ 327; 820; 867; 864; 282 ...	Np 233 36,2 m ε α 5,54 γ 312; 299; 547 ...	Np 234 4,4 d ε; β^+ ... γ 1559; 1528; 1602 ... σ_f ~ 900	Np 235 396 d ε α 5,022 ... γ ...; g σ 1600 + 184	Np 236 22,5 h ε β^- 0,5 ... γ 642 ... e^-; g | 1,29·10^6 a β^- g σ_f 2500	Np 237 sf | 2,14·10^6 a α 4,788; 4,770 ... γ 87; 29 ...; e^- σ 169; σ_f 0,019	Np 238 50,8 h β^- 1,3 ... γ 984; 1029; 1026; 924 ... g; σ_f 2070
92		U 238,029 σ_{abs} 7,59	U 226 0,5 s α 7,43	U 227 1,1 m α 6,87	U 228 9,2 m α 6,68; 6,59 ... ε γ ...; e^-	U 229 58 m ε α 6,360; 6,332; 6,297 ...	U 230 20,8 d α 5,889; 5,818 ... γ ...; e^- σ_f 25	U 231 4,2 d ε α 5,46 γ 26; 84; 220 ... e^-; σ_f 400	U 232 71,7 a α 5,320; 5,264 ... sf; γ ...; e^- σ 73,1; σ_f 75,2	U 233 1,59 · 10^5 a α 4,824; 4,783 ... γ ...; e^- σ 47,7 σ_f 531,1	U 234 0,0055 2,44 · 10^5 a α 4,774; 4,722 ... sf; γ ...; e^- σ 100,2; σ_f < 0,65	U 235 0,720 26 m Iγ e^- | 7,04·10^8 a α 4,401; 4,365 ...; sf γ 186; 144 σ 98,6 σ_f 582,2	U 236 sf | 2,342·10^7 a α 4,494; 4,445 ... sf; γ ...; e^- σ 5,2	U 237 6,75 d β^- 0,2 ... γ 60; 208 ... e^-; σ 411 σ_f < 0,35
	Pa 223 6,5 ms α 8,01; 8,20	Pa 224 0,95 s α 7,49 ...	Pa 225 1,8 s α 7,25; 7,20	Pa 226 1,8 m α 6,86; 6,82 ... ε g	Pa 227 38,3 m α 6,466; 6,416 ... ε γ 65; 110 ...	Pa 228 26 h ε α 6,078; 6,105; 5,799; 6,118 ... γ 911; 463; 966; 970 ...	Pa 229 1,4 d ε α 5,579; 5,669; 5,614 ... γ ...; e^-	Pa 230 17,4 d ε; β^- 0,4 α 5,345; 5,326 ... γ 952; 918; 455; 899; 444 ... σ_f 1500	Pa 231 3,25 · 10^4 a α 5,014; 4,952; 5,030 ... γ 27; 303; 300 ... e^- σ 210; σ_f 0,010	Pa 232 1,32 d β^- 0,3; 1,3 ... γ 969; 894; 150 ... e^- σ 760; σ_f 700	Pa 233 27,0 d β^- 0,3; 0,6 ... γ 312; 300; 341 ... e^- σ 21+20; σ_f < 0,1	Pa 234 1,18 m β^- 2,3 ... γ ... Iγ ... e^- σ_f < 500 | 6,75 h β^- 0,5; 1,2 γ 131; 881; 883; 946 ...; e^- σ_f < 5000	Pa 235 24,2 m β^- 1,4 ... γ 128 – 659 m	Pa 236 9,1 m β^- 2,0; 3,1 ... γ 642; 687; 1763 ... g
	Th 222 2,8 ms α 7,98	Th 223 0,66 s α 7,29; 7,32	Th 224 1,03 s α 7,17; 7,00 ... γ 177 ... e^-	Th 225 8 m α 6,478; 6,441; 6,501 ...; ε γ 322; 246; 362 ...	Th 226 31 m α 6,335; 6,225 ... γ 111; 242 ... e^-	Th 227 18,72 d α 6,038; 5,978; 5,757 ... γ 236; 50; 256 ... e^-; σ_f 200	Th 228 1,913 a α 5,423; 5,341 ... γ 84; 216 ... e^- σ 123; σ_f < 0,3	Th 229 7340 a α 4,845; 4,901; 4,815 ... γ 194; 31; 125 ... e^-; σ 54 σ_f 30,5	Th 230 7,7 · 10^4 a α 4,688; 4,621 ... γ ... e^-; σ 23,2 σ_f < 0,0012	Th 231 25,6 h β^- 0,3 ... γ 26; 84 ... e^-	Th 232 100 1,405 · 10^{10} a α 4,012; 3,953 sf; γ ...; e^- σ 7,40 σ_f 0,000039	Th 233 22,3 m β^- 1,2 ... γ 87; 29; 459 ... e^- σ 1500; σ_f 15	Th 234 24,10 d β^- 0,2 ... γ 93; 63 ...; e^- σ 1,8; σ_f < 0,01 m	Th 235 6,9 m β^- γ 416 – 932
	Ac 221 52 ms α 7,65; 7,44; 7,38 ...	Ac 222 66 s α 6,81; 6,75; 6,89; 7,00 Iγ ?; ε | 4,2 s α 7,00; 6,96 ε	Ac 223 2,2 m α 6,647; 6,662; 6,564 ... ε; γ ...	Ac 224 2,9 h ε; α 6,1385; 6,0566; 6,2106 ... γ 216; 132 ...	Ac 225 10,0 d α 5,830; 5,794; 5,732 ... γ 100; 150 ...	Ac 226 29 h β^- 0,9; 1,1 ε; α ? γ 230; 159; 255 ...	Ac 227 21,8 a β^- 0,04 ... α 4,954; 4,942 ... γ ...; e^-; σ 515 σ_f < 0,002	Ac 228 6,13 h β^- 1,2; 2,1 ... γ 911; 969; 338; 964 ...	Ac 229 62,7 m β^- 1,1 γ 165; 569; 262; 146; 135 ...	Ac 230 80 s β^- 2,2 γ 455; 508	Ac 231 7,5 m β^- γ 282; 307; 221; 186; 369 ...	Ac 232 35 s β^-		
	Ra 220 23 ms α 7,46 ... γ 465	Ra 221 28 s α 6,610; 6,758; 6,665 ... γ 89; 152 ...	Ra 222 38 s α 6,556 ... γ 325 ...	Ra 223 11,43 d α 5,7164; 5,6076 γ 270; 154; 144; 338 σ 130; σ_f 0,7	Ra 224 3,64 d α 5,68556; 5,4489 ... γ 241 ... σ 12	Ra 225 14,8 d β^- 0,3 ... γ 40 e^-	Ra 226 1600 a α 4,78450; 4,6019 ... γ 186 ...; σ 11,5 σ_f < 0,0001	Ra 227 41 m β^- 1,3 ... γ 27; 300; 303 ...	Ra 228 5,75 a β^- 0,05 ... γ ...; e^- σ 36; σ_f < 2	Ra 229 4,0 m β^-	Ra 230 1 h β^-		144	
	Fr 219 20 ms α 7,32 ... γ ...	Fr 220 27,5 s α 6,68; 6,64; 6,58 ... β ... γ 45; 106; 162	Fr 221 4,8 m α 6,341; 6,126 ... γ 218 ...	Fr 222 14,8 m β^-	Fr 223 21,8 m β^- 1,2 ... α 5,34 γ 50; 80; 235 ...	Fr 224 2,7 m β^-	Fr 225 3,9 m β^-	Fr 226 48 s β	Fr 227 2,4 m β^-	Fr 228 39 s β^-	Fr 229 50 s β^-			
	Rn 218 35 ms α 7,132 ... γ ...	Rn 219 3,96 s α 6,819; 6,553; 6,425 ... γ 271; 402 ...	Rn 220 55,6 s α 6,288 ... γ ... σ < 0,2	Rn 221 25 m β α ~ 6,0	Rn 222 3,824 d α 5,48966 ... γ ... σ 0,72	Rn 223 43 m β^-	Rn 224 1,9 h β^-	Rn 225 4,5 m β^-	Rn 226 6,0 m β^-					
	At 217 32,3 ms α 7,067 ... β ... γ ...	At 218 ~ 2 s α 6,694; 6,654 ... β ...; γ ...	At 219 0,9 m α 6,27 β				138		140		142			
	Po 216 0,15 s α 6,7785 ... γ ...	Po 217 < 10 s α 6,55	Po 218 3,05 m α 6,0026 ... β^- ... γ ...											
	Bi 215 7,4 m β^-				136									
	Pb 214 26,8 m β^- 0,7; 1,0 ... γ 352; 295; 242 ...													
	132		134											

Z	146	147	148	149	150	151	152	153	154	155	156	157	158
105										260 1,6 s α 9,06; 9,10; 9,14 sf ?	261 1,8 s α 8,93 sf ?	262 ~40 s α 8,45; 8,66 sf ?	
104								257 4,5 s α 9,00; 8,95 … γ 127	258 ? 11 ms sf	259 3 s α 8,77; 8,86	260 0,1 s sf	261 65 s α 8,28	
103						Lr	Lr 255 22 s α 8,37; 8,35	Lr 256 31 s α 8,43; 8,39; 8,52 …	Lr 257 0,6 s α 8,87; 8,81	Lr 258 4,2 s α 8,62; 8,59; 8,65 …	Lr 259 5,4 s α 8,45	Lr 260 3 m α 8,03	
102		No		No 251 0,8 s α 8,60; 8,68	No 252 2,3 s α 8,41 sf	No 253 1,6 m α 8,01	No 254 0,28 s; Iγ ‖ 55 s; α 8,10; sf; g	No 255 3,3 m α 8,11; 7,92; 7,76 …	No 256 3,5 s α 8,42 sf	No 257 26 s α 8,22; 8,27; 8,32	No 258 ? 1,2 ms sf	No 259 58 m α 7,60; 7,53.. sf	
	Md	Md 248 7 s ε α 8,32; 8,36	Md 249 24 s ε α 8,03	Md 250 52 s ε α 7,75; 7,82	Md 251 4,0 m ε α 7,55	Md 252 2,3 m ε		Md 254 10 m; ε ‖ ~28 m; ε	Md 255 27 m ε α 7,333 γ 430	Md 256 1,3 h ε α 7,23; 7,16 … γ 400	Md 257 5,0 h ε α 7,08	Md 258 55 d α 6,72; 6,79 g	
	Fm 246 1,2 s α 8,23 sf	Fm 247 9,2 s; α 8,18 ‖ 35 s; α 7,87; 7,93; ε ?	Fm 248 37 s α 7,88; 7,83 sf	Fm 249 2,6 m ε α 7,53	Fm 250 1,8 s; Iγ ‖ 30 m; α 7,43; sf	Fm 251 5,30 h ε α 6,833; 6,782 … γ	Fm 252 22,8 h α 7,04 … sf	Fm 253 3,0 d ε α 6,943; 6,875 … γ 272 …	Fm 254 3,24 h α 7,187; 7,145 … sf γ …; e^- σ ~76	Fm 255 20,1 h α 7,016 … sf γ 81; 58 …; e^- σ 26; σ_f 3400	Fm 256 2,63 h sf α 6,93 σ ~45	Fm 257 100,5 d α 6,519 … sf γ 242; 180 …; e^- σ_{abs} 6100; σ_f 2950	Fm 258 0,38 ms sf
	Es 245 1,3 m ε α 7,73	Es 246 7,3 m ε α 7,36	Es 247 4,7 m ε α 7,31	Es 248 28 m ε α 6,87 α → g	Es 249 1,7 h ε α 6,77 γ 379; 812; 375 α → g	Es 250 2,1 h; ε; γ 989; 1032 ‖ 8,3 h; ε; γ 829; 303; 349 …	Es 251 33 h ε α 6,49 … γ 178; 153 …	Es 252 401 d α 6,631; 6,561 … ε γ 785; 139 … g	Es 253 20,47 d α 6,633; 6,592 … sf γ …; e^- σ 155 + <3	Es 254 39,3 h; β^- 0,5; 1,1 …; α 6,382 …; ε; γ 649; 694 …; σ ~1,3; σ_f 1840 ‖ 276 d; α 6,429; γ …; e^-; σ <40; σ_f 2900	Es 255 39,8 d β^- α 6,300 … sf σ 43		
	Cf 244 19,7 m α 7,218; 7,178	Cf 245 43,6 m ε α 7,137 … g	Cf 246 35,7 h α 6,758; 6,719 … sf; γ … e^-; g	Cf 247 2,45 h ε γ 295; 417 … e^-	Cf 248 333,5 d α 6,26; 6,22 sf γ …; e^- g	Cf 249 350,6 a α 5,812 …; sf γ 388; 333 … g σ 465; σ_f 1660	Cf 250 13,08 a α 6,031; 5,989 … sf; γ …; e^- σ 2030; σ_f <350	Cf 251 898 a α 5,680; 5,846; 6,008 … γ 177 … σ 2850; σ_f 4300	Cf 252 2,62 a α 6,118; 6,076 … sf γ …; e^- σ 20,4; σ_f 32	Cf 253 17,8 d β^- 0,3 α 5,979 … σ 17,6 σ_f 1300	Cf 254 60,5 d sf α 5,834 … σ_{abs} 90		
	Bk 243 4,5 h ε α 6,574; 6,542 … γ 755; 946 … g	Bk 244 sf ‖ 4,35 h; ε; α 6,667; 6,625; γ 892; 218; 922 …; g	Bk 245 sf ‖ 4,98 d; ε; α 5,892; 6,155 …; γ 253; 381 …; e^-; g	Bk 246 1,83 d ε γ 800; 1082; 835; 1124 …	Bk 247 1380 a α 5,531; 5,710; 5,688 … γ 84; 265 … g	Bk 248 18 h; β^- 0,7; ε ‖ >9 a	Bk 249 314 d β^- 0,1 α 5,417 5,930 … sf; γ … σ_{abs} 1300	Bk 250 3,22 h β^- 0,7; 1,8 … γ 989; 1032 … σ_f 960	Bk 251 57 m β^- ~0,5; ~1,0 γ 37; 140 …				
	Cm 242 sf ‖ 163,0 d; α 6,113; 6,070; sf; γ …; e^-; g; σ 16; σ_f <5	Cm 243 sf ‖ 30 a; α 5,785; 5,742; ε; γ 278; 228; 210 …; e^-; g; σ 225; σ_f 600	Cm 244 sf ‖ 18,099 a; α 5,805; 5,763 …; sf; γ …; e^-; g; σ 13,9; σ_f 1,2	Cm 245 sf ‖ 8532 a; α 5,360; 5,306; γ 175; 133; g; σ 345; σ_f 2020	Cm 246 4820 a α 5,386; 5,343 sf γ …; e^-; g σ 1,3 σ_f 0,17	Cm 247 $1,56 \cdot 10^7$ a α 4,869; 5,266 … γ 402; 278 … g σ 60; σ_f 90	Cm 248 $3,61 \cdot 10^5$ a α 5,078; 5,034 sf γ …; e^- σ 4; σ_f 0,34	Cm 249 64,2 m β^- 0,9 γ σ 1,6	Cm 250 $1,13 \cdot 10^4$ a sf				
	Am 241 sf ‖ 433 a; α 5,486; 5,443; sf; γ 60; 26 …; e^-; g; σ 0,00010 + 83,8 + 748; σ_f 3,15	Am 242 sf ‖ 152 a; Iγ; α 5,205; sf; e^-; γ; σ 1400; σ_f 6600 ‖ 16 h; β^- 0,6; 0,7; ε; γ …; e^-; g; σ_f 2900	Am 243 sf ‖ 7400 a; α 5,275; 5,234; sf; γ 75; 44; σ ? + 75,2 + 4,1; σ_f < 0,07	Am 244 sf ‖ 26 m; β^- 1,5; ε; γ; e^-; σ_f 1600 ‖ 10,1 h; β^- 0,4; γ 744; 898; 154; e^-; σ_f 2300	Am 245 sf ‖ 2,05 h; β^- 0,9 …; γ 253 …; e^-	Am 246 sf ‖ 25 m; β^- 1,2; 2,2; γ 1079; 799; 1062; 1036 … ‖ 39 m; β^-; γ 679; 205; 154; 756	Am 247 22 m β^- γ 285; 225						
	Pu 240 sf ‖ 6537 a; α 5,168; 5,124; sf; γ …; e^-; g; σ 289,5; σ_f 0,030	Pu 241 sf ‖ 14,89 a; β^- 0,02; α 4,897 …; γ …; e^-; σ 368; σ_f 1009	Pu 242 sf ‖ $3,87 \cdot 10^5$ a; α 4,901; 4,857; sf; γ …; e^-; g; σ 18,5; σ_f <0,2	Pu 243 sf ‖ 4,96 h; β^- 0,6 …; γ 84 …; σ 60; σ_f 196	Pu 244 $8,26 \cdot 10^7$ a α 4,589; 4,546 sf γ …; e^- σ 1,7	Pu 245 10,48 h β^- 0,9; 1,2 … γ 327; 560; 308 σ 150	Pu 246 10,9 d β^- 0,2; 0,3 γ 44; 224; 180 … m						
	Np 239 2,355 d β^- 0,3; 0,7 … γ 106; 278; 228 … e^-; g σ 31+14; σ_f <1	Np 240 7,4 m; β^- 2,2 …; γ 555; 597 …; Iγ …; g ‖ 65 m; β^- 0,9; γ 567; 974; 601; 448 …; g	Np 241 16,0 m β^- 1,3 … γ 133; 174 g										
	U 238 99,28 sf ‖ $4,47 \cdot 10^9$ a; α 4,198; 4,149 …; sf; γ …; e^-; σ 2,70	U 239 23,5 m β^- 1,2; 1,3 … γ 75 … σ 22; σ_f 14	U 240 14,1 h β^- 0,4 … γ 44 m										
	Pa 237 8,7 m β^- 1,4; 2,3 … γ 854; 865; 529; 541 …	Pa 238 2,3 m β^- 1,7; 2,9 … γ 1015; 635; 449; 680 … g											
	Th 236 37,5 m β^- 1,0; 1,1 … γ 111; 113												
	146		148		150		152		154		156		158

Author Index

(Page numbers in italic figures refer to the literature references at the end of each chapter; other page numbers refer to citations in the text)

N.C.—W

Element and Nuclide Index

Subject Index

APPENDIX I. PERIODIC TABLE OF THE ELEMENTS

IA	IIA	IIIB	IVB	VB	VIB	VIIB	VIII	VIII	VIII	IB	IIB	IIIA	IVA	VA	VIA	VIIA	
○ 1 H 1.0079																1 H 1.0079	He 4.00260
3 Li 6.941	○ 4 Be 9.01218											5 B 10.81	○ 6 C 12.011	7 N 14.0067	8 O 15.9994	9 F 18.99840	10 Ne 20.179
11 Na 22.98977	12 Mg 24.305											13 Al 26.98154	14 Si 28.086	15 P 30.97376	16 S 32.06	17 Cl 35.453	18 Ar 39.948
○ 19 K 39.098	20 Ca 40.08	21 Sc 44.9559	22 Ti 47.90	○ 23 V 50.9414	24 Cr 51.996	25 Mn 54.9380	26 Fe 55.847	27 Co 58.9332	28 Ni 58.70	29 Cu 63.546	30 Zn 65.38	31 Ga 69.72	32 Ge 72.59	33 As 74.9216	34 Se 78.96	35 Br 79.904	36 Kr 83.80
○ 37 Rb 85.4678	38 Sr 87.62	39 Y 88.9059	40 Zr 91.22	41 Nb 92.9064	42 Mo 95.94	● 43 Tc 98.9062	44 Ru 101.07	45 Rh 102.9055	46 Pd 106.4	47 Ag 107.868	48 Cd 112.40	○ 49 In 114.82	50 Sn 118.69	51 Sb 121.75	○ 52 Te 127.60	53 I 126.9045	54 Xe 131.30
55 Cs 132.9054	56 Ba 137.34	○ 57 La* 138.9055	○ 72 Hf 178.49	73 Ta 180.9479	74 W 183.85	○ 75 Re 186.207	○ 76 Os 190.2	77 Ir 192.22	○ 78 Pt 195.09	79 Au 196.9665	80 Hg 200.59	81 Tl 204.37	○ 82 Pb 207.2	○ 83 Bi 208.9804	● 84 Po (210)	● 85 At (210)	● 86 Rn (222)
● 87 Fr (223)	● 88 Ra 226.0254	● 89 Ac** (227)	● 104 Ku (261)	● 105 Ha (262)	106	107	108	109	110	111	112	113	114	115	116	117	118

*Lanthanum Series

58 Ce 140.12	59 Pr 140.9077	○ 60 Nd 144.24	● 61 Pm (147)	○ 62 Sm 150.4	63 Eu 151.96	○ 64 Gd 157.25	65 Tb 158.9254	66 Dy 162.50	67 Ho 164.9304	68 Er 167.26	69 Tm 168.9342	70 Yb 173.04	○ 71 Lu 174.97

**Actinium Series

● 90 Th 232.0381	● 91 Pa 231.0359	● 92 U 238.029	● 93 Np 237.0482	● 94 Pu (244)	● 95 Am (243)	● 96 Cm (247)	● 97 Bk (247)	● 98 Cf (251)	● 99 Es (252)	● 100 Fm (257)	● 101 Md (258)	● 102 No (259)	● 103 Lr (260)

○ Stable and radioactive natural isotopes

● No stable isotopes

APPENDIX II. *Quantities and units*

Quantity	Symbol	Unit	Symbol	Dimensions (within brackets), derived units etc.
Length	l	meter	m	Basic SI unit; 1 fermi = 10^{-15} m (= 1 fm); 1 $\mu = 10^{-6}$ m; 1 Å = 10^{-10} m;
Mass	m	kilogram	kg	" ; 1 tonne (t) = 10^3 kg;
Time interval	t	second	s	" ; 1 (calendar) year (y or a) = 365 days (d) = 8760 hours (h) = 3.1536×10^7 s
Electric current	I	ampere	A	
Thermodynamic temperature	T	kelvin	K	" ; $t_c = t_k - 273.15$, t_c = temp. in degree Celcius ("centigrade"), °C.
Luminous intensity	I*	candela	cd	
Atomic (molecular) weight	M	atomic mass unit	u	$M = m_i \times 10^3 \times N_A$, m_i mass of atom (molecule); 1 mole = M *g; molarity (mol/lit)* = $m_a M^{-1} V^{-1}$, m_a mass of pure substance; see also pp. 2, 13 ff.
Density	ρ		kg m^{-3}	1 g $cm^{-3} = 10^3$ kg m^{-3}
Pressure	P	pascal	Pa	[Pa = Nm^{-2} = kg s^{-2} m^{-1}]; 1 atm = 1.013×10^5 Pa; 1 bar = 10^5 Pa; 1 torr = 133.3 Pa
Energy	E	joule	J	[J = N m = kg m^2 s^{-2}]; [J = Ws]
Frequency	f, ν	hertz	Hz	[Hz = s^{-1}]
Force	F	newton	N	[N = kg m s^{-2}]; 1 dyne = 10^{-5} N; 1 kp = 9.8067 N
Angle, flat		radian	rad	= 57.30°; full circle 360° = 2π rad
" space		steradian	sr	full space angle = 4π steradians
Energy flux (power)	P	watt	W	[W = J s^{-1} = N m s^{-1} = kg m^2 s^{-3}]
Angular frequency	ω		rad s^{-1}	
Electric potential (voltage)	U	volt	V	[V = W A^{-1} = kg m^3 s^{-3} A^{-1}]
resistance	R	ohm	Ω	[Ω = V A^{-1} = W A^{-2} = kg m^2 s^{-3} A^{-2}]
charge	q	coulomb	C	[C = A s]; 1 C = 0.1 c statcoulomb (esu)
capacitance	C	farad	F	[F = C V^{-1} = s A^2 W^{-1} = s^4 A^2 kg^{-1} m^{-2}]
Magnetic inductance	L*	henry	H	[H = Wb A^{-1} = V s A^{-1} = kg m^2 s^{-2} A^{-2}]
induction (flux density)	B	tesla	T	[T = Wb m^{-2} = V s m^{-2} = kg s^{-2} A^{-1}]; 1 gauss = 10^{-4} T
flux	ϕ_B*	weber	Wb	[Wb = V s = W s A^{-1} = kg m^2 s^{-2} A^{-1}]
Radioactivity	A	becquerel	Bq	[Bq = (radioactive events) s^{-1}]; 1 curie (Ci) = 3.7×10^{10} Bq
Radiation exposure		röntgen	R	1 R = 2.58×10^{-4} C kg^{-1}
Radiation dose absorbed	D	gray	Gy	[Gy = J kg^{-1}]; 1 Gy = 100 rad**
Volume	V	cubic meter	m^3	1 $m^3 = 10^3$ lit (l), 1 lit = 10^3 milliliters (ml), 1 ml = 1 cm^3

*Not used in this text. **Radiation unit (see 15.2).

APPENDIX III. *Fundamental constants*

Quantity	Symbol	Value	SI unit	Auxiliary value
Speed of light in vacuum	$\mathbf{c}$	$2.997\,925 \times 10^{8}$	$m\ s^{-1}$	
Elementary charge	$\mathbf{e}$	$1.602\,189 \times 10^{-19}$	C	
Planck constant	$\mathbf{h}$	$6.626\,18 \times 10^{-34}$	J s	$= 4.135\,70 \times 10^{-15}$ eV s $\hbar = \mathbf{h}/2\pi = 1.054\,589\ 10^{-34} \times$ J s
Avogadro constant	$\mathbf{N_A}$	$6.022\,04 \times 10^{23}$	mol^{-1}	
Atomic mass unit	1 u	$1.660\,566 \times 10^{-27}$	kg	$= 931.501\,6$ MeV; mass of $^{12}C = 12$ u
Electron rest mass	$\mathbf{m_e}$	$0.910\,953 \times 10^{-30}$	kg	$\mathbf{M_e} = \mathbf{N_A} \cdot \mathbf{m_e} = 0.000\,548\,580$ u $= 0.511\,003\,4$ MeV
Proton rest mass	$\mathbf{m_p}$	$1.672\,649 \times 10^{-27}$	kg	$\mathbf{M_p} = \mathbf{N_A} \cdot \mathbf{m_p} = 1.007\,276\,5$ u $= 938.279\,6$ MeV
Neutron rest mass	$\mathbf{m_n}$	$1.674\,954 \times 10^{-27}$	kg	$\mathbf{M_n} = \mathbf{N_A} \cdot \mathbf{m_n} = 1.008\,665\,0$ u $= 939.573\,1$ MeV
Faraday constant	$\mathbf{F}$	$9.648\,46 \times 10^{4}$	$C\ mol^{-1}$	$= \mathbf{N_A} \cdot \mathbf{e}$
Rydberg constant	$\mathbf{R_\infty}$	$1.097\,373 \times 10^{7}$	m^{-1}	$\mathbf{R_\infty}\,\mathbf{h} \cdot \mathbf{c} = 13.605\,8$ eV
Fine structure constant	α^{-1}	$137.036\,0$		$= \mu^0 \cdot \mathbf{c} \cdot \mathbf{e}^2/2\,\mathbf{h}$; μ^0 (permeability of vacuum) $= 4\,\pi\ 10^{-7}\ H\ m^{-1}$
Bohr radius	$\mathbf{a_0}$	$0.529\,177 \times 10^{-10}$	m	$= \alpha/4\pi \mathbf{R_\infty}$
Electron magnetic moment	μ_e	$9.284\,83 \times 10^{-24}$	$J\ T^{-1}$	
Proton magnetic moment	μ_p	$1.410\,617 \times 10^{-26}$	$J\ T^{-1}$	
Bohr magneton	μ_B	$9.274\,08 \times 10^{-24}$	$J\ T^{-1}$	$= \mathbf{e} \cdot \hbar/2 \cdot \mathbf{m_e}$ ($1\ J\ T^{-1} = 10^3$ erg gauss^{-1})
Nuclear magneton	μ_N	$5.050\,82 \times 10^{-27}$	$J\ T^{-1}$	$= \mathbf{e} \cdot \hbar/2 \cdot \mathbf{m_p}$
Molar gas constant	$\mathbf{R}$	$8.314\,41$	$J\ mol^{-1}\ K^{-1}$	$= 0.082\,057$ lit atm $mol^{-1}\ K^{-1} = 1.987\,2$ cal $mol^{-1}\ K^{-1}$
Molar volume of ideal gas (s.t.p.)	$\mathbf{V_m}$	$0.022\,413\,8$	$m^3\ mol^{-1}$	$= \mathbf{R} \cdot T_0/p_0$, $T_0 = 273.15$ K, $p_0 = 1$ atm
Boltzmann constant	$\mathbf{k}$	$1.380\,662 \times 10^{-23}$	$J\ K^{-1}$	$= \mathbf{R}/\mathbf{N_A}$; $= 8.617\,35 \times 10^{-5}$ eV K^{-1}; $1/\mathbf{k} = 11\,604.5$ K eV^{-1}
Gravitational constant	$\mathbf{G}$	$6.672\,0 \times 10^{-11}$	$N\ m^2\ kg^{-2}$	
Standard acceleration of free fall	$\mathbf{g}$	$9.806\,65$	$m\ s^{-2}$	

APPENDIX IV. *Energy conversion factors*

Joule (J) (newton meter, Nm) (watt seconds, Ws)	Kilowatthour (kWh)	Erg (dyne cm)	Kilocalorie (kcal)	Electronvolt (eV)
1	$2.777\,78 \times 10^{-7}$	$1.000\,00 \times 10^{7}$	$2.388\,46 \times 10^{-4}$	$6.241\,46 \times 10^{18}$
$3.600\,00 \times 10^{6}$	1	$3.600\,00 \times 10^{13}$	$8.598\,46 \times 10^{2}$	$2.246\,93 \times 10^{25}$
$1.000\,00 \times 10^{-7}$	$2.777\,78 \times 10^{-14}$	1	$2.388\,46 \times 10^{-11}$	$6.241\,46 \times 10^{11}$
$4.186\,80 \times 10^{3}$	$1.163\,00 \times 10^{-3}$	$4.186\,80 \times 10^{10}$	1	$2.613\,17 \times 10^{22}$
$1.602\,19 \times 10^{-19}$	$4.450\,53 \times 10^{-26}$	$1.602\,19 \times 10^{-12}$	$3.826\,77 \times 10^{-23}$	1

1 eV = $1.602\,189 \times 10^{-19}$ J; 1 eV atom^{-1} = 23.045 0 kcal mol^{-1} = 96.485 kJ mol^{-1}
Energy-wavelength product ($\Delta E\,\lambda$) = 12 398.5 eV/Å
1 Q = 10^{18} Btu; 1 Btu (British thermal unit) = 1.055 06 kJ; 1 hp (horse power) = 0.746 kJ s^{-1}
1 toe (tonne oil equivalent) = 10 Gcal = 11.63 MWh = 41.87 GJ
1 tonne hard coal (tce) = 0.65 toe = 27.2 GJ; 1000 m^3 natural gas = 0.80 toe
1 g ^{235}U fissioned at 200 MeV/fission = 82.11 GJ = 0.95 MWd (heat)

Prefixes for powers of ten

E	exa	10^{18}
P	peta	10^{15}
T	tera	10^{12}
G	giga	10^{9}
M	mega	10^{6}
k	kilo	10^{3}
c	centi	10^{-2}
m	milli	10^{-3}
μ	micro	10^{-6}
n	nano	10^{-9}
p	pico	10^{-12}
f	femto	10^{-15}
a	atto	10^{-18}

Some numerical values

e	2.718 28
log e	0.434 29
ln 2	0.693 15
ln 10	2.302 59
ln 2/ln 10	0.301 03
π	3.141 59
ln a = ln 10 × log a	

Some English measures

1 inch	0.025 4 m
1 (statute) mile	1 609.34 m
1 (int.) nautical mile	1852 m
1 (US liq.) gallon	3.785 lit
1 barrel	0.159 0 m^3
1 cubic foot	0.028 32 m^3
1 pound (mass)	0.4536 kg

lwks
ks